HOLT MIDDLE SCHOOL Math

Course 3

Jennie M. Bennett
David J. Chard
Audrey Jackson
Jim Milgram
Janet K. Scheer
Bert K. Waits

HOLT, RINEHART AND WINSTON
A Harcourt Education Company
Austin • Orlando • Chicago • New York • Toronto • London • San Diego

Staff Credits

Editorial

Lila Nissen, *Editorial Vice President*
Robin Blakely, *Associate Director*
Joseph Achacoso, *Assistant Managing Editor*
Threasa Boyar, *Editor*

Student Edition
Glenn Worthman, *Senior Editor*
Tessa Henry, *Editor*
Kristi Smith, *Associate Editor*

Teacher's Edition
Kelli Flanagan, *Senior Editor*
Thomas Hamilton, *Editor*

Ancillaries
Mary Fraser, *Executive Editor*
Higinio Dominguez, *Associate Editor*

Technology Resources
John Kerwin, *Executive Editor*
Robyn Setzen, *Senior Editor*
Patricia Platt, *Senior Technology Editor*
Manda Reid, *Technology Editor*

Copyediting
Denise Nowotny, *Copyediting Supervisor*
Patrick Ricci, *Copyeditor*

Support
Jill Lawson, *Senior Administrative Assistant*
Benny Carmona, III, *Editorial Coordinator*

Design

Book Design
Marc Cooper, *Design Director*
Tim Hovde, *Senior Designer*
Lisa Woods, *Designer*
Teresa Carrera-Paprota, *Designer*
Bruce Albrecht, *Design Associate*
Ruth Limon, *Design Associate*
Holly Whittaker, *Senior Traffic Coordinator*

Teacher's Edition
José Garza, *Designer*
Charlie Taliaferro, *Design Associate*

Cover Design
Pronk & Associates

Image Acquisition
Curtis Riker, *Director*
Tim Taylor, *Photo Research Supervisor*
David Knowles, *Photo Researcher*
Elaine Tate, *Art Buyer Supervisor*
Nicole McLeod, *Art Buyer*
Sam Dudgeon, *Senior Staff Photographer*
Victoria Smith, *Staff Photographer*
Lauren Eischen, *Photo Specialist*

New Media Design
Ed Blake, *Design Director*

Media Design
Dick Metzger, *Design Director*
Chris Smith, *Senior Designer*

Graphic Services

Kristen Darby, *Director*
Eric Rupprath, *Ancillary Designer*
Linda Wilbourn, *Image Designer*

Prepress and Manufacturing

Mimi Stockdell, *Senior Production Manager*
Susan Mussey, *Production Supervisor*
Sara Downs, *Production Coordinator*
Jevara Jackson, *Senior Manufacturing Coordinator*
Ivania Lee, *Inventory Analyst*
Wilonda Ieans, *Manufacturing Coordinator*

Printed in the United States of America

ISBN 0-03-071141-X

3 4 5 6 7 8 9 048 10 09 08 07 06 05 04

Authors

Jennie M. Bennett, Ed.D., is the Instructional Mathematics Supervisor for the Houston Independent School District and president of the Benjamin Banneker Association.

David J. Chard, Ph.D., is an Assistant Professor and Director of Graduate Studies in Special Education at the University of Oregon. He is the President of the Division for Research at the Council for Exceptional Children, is a member of the International Academy for Research on Learning Disabilities, and is the Principal Investigator on two major research projects for the U.S. Department of Education.

Audrey Jackson is a Principal in St. Louis, Missouri, and has been a curriculum leader and staff developer for many years.

Jim Milgram, Ph.D., is a Professor of Mathematics at Stanford University. He is a member of the Achieve Mathematics Advisory Panel and leads the Accountability Works Analysis of State Assessments funded by The Fordham and Smith-Richardson Foundations. Most recently, he has been named lead advisor to the Department of Education on the implementation of the Math-Science Initiative, a key component of the No Child Left Behind legislation.

Janet K. Scheer, Ph.D., Executive Director of Create A Vision™, is a motivational speaker and provides customized K-12 math staff development. She has taught internationally and domestically at all grade levels.

Bert K. Waits, Ph.D., is a Professor Emeritus of Mathematics at The Ohio State University and co-founder of T^3 (Teachers Teaching with Technology), a national professional development program.

Consulting Authors

Paul A. Kennedy is a Professor in the Mathematics Department at Colorado State University and has recently directed two National Science Foundation projects focusing on inquiry-based learning.

Mary Lynn Raith is the Mathematics Curriculum Specialist for Pittsburgh Public Schools and co-directs the National Science Foundation project PRIME, Pittsburgh Reform in Mathematics Education.

Reviewers

Thomas J. Altonjy
Assistant Principal
Robert R. Lazar Middle School
Montville, NJ

Jane Bash, M.A.
Math Education
Eisenhower Middle School
San Antonio, TX

Charlie Bialowas
District Math Coordinator
Anaheim Union High School District
Anaheim, CA

Lynn Bodet
Math Teacher
Eisenhower Middle School
San Antonio, TX

Louis D'Angelo, Jr.
Math Teacher
Archmere Academy
Claymont, DE

Troy Deckebach
Math Teacher
Tredyffrin-Easttown Middle School
Berwyn, PA

Mary Gorman
Math Teacher
Sarasota, FL

Brian Griffith
Supervisor of Mathematics, K–12
Mechanicsburg Area School District
Mechanicsburg, PA

Ruth Harbin-Miles
District Math Coordinator
Instructional Resource Center
Olathe, KS

Kim Hayden
Math Teacher
Milford Jr. High School
Milford, OH

Susan Howe
Math Teacher
Lime Kiln Middle School
Fulton, MD

Paula Jenniges
Austin, TX

Ronald J. Labrocca
District Mathematics Coordinator
Manhasset Public Schools
Manhasset, NY

Victor R. Lopez
Math Teacher
Washington School
Union City, NJ

George Maguschak
Math Teacher/Building Chairperson
Wilkes-Barre Area
Wilkes-Barre, PA

Dianne McIntire
Math Teacher
Garfield School
Kearny, NJ

Kenneth McIntire
Math Teacher
Lincoln School
Kearny, NJ

Francisco Pacheco
Math Teacher
IS 125
Bronx, NY

Vivian Perry
Edwards, IL

Vicki Perryman Petty
Math Teacher
Central Middle School
Murfreesboro, TN

Jennifer Sawyer
Math Teacher
Shawboro, NC

Russell Sayler
Math Teacher
Longfellow Middle School
Wauwatosa, WI

Raymond Scacalossi
Math Chairperson
Hauppauge Schools
Hauppauge, NY

Richard Seavey
Math Teacher–Retired
Metcalf Jr. High
Eagan, MN

Sherry Shaffer
Math Teacher
Honeoye Central School
Honeoye Falls, NY

Gail M. Sigmund
Math Teacher
Charles A. Mooney Preparatory School
Cleveland, OH

Jonathan Simmons
Math Teacher
Manor Middle School
Killeen, TX

Jeffrey L. Slagel
Math Department Chair
South Eastern Middle School
Fawn Grove, PA

Karen Smith, Ph.D.
Math Teacher
East Middle School
Braintree, MA

Bonnie Thompson
Math Teacher
Tower Heights Middle School
Dayton, OH

Mary Thoreen
Mathematics Subject Area Leader
Wilson Middle School
Tampa, FL

Paul Turney
Math Teacher
Ladue School District
St. Louis, MO

Welcome to Holt Middle School Math North Carolina Edition!

Have you ever wondered if anyone really uses the math that you learn in school? As you work through this book, you will find out that they certainly do. Not only that, but people are using math almost right in your backyard.

We have found the places where math is being used in your home state. In each chapter, you will "visit" one or two North Carolina locations, or you will learn about a specific aspect of North Carolina's culture, history, or industry to see how math is being used near you. It will be like taking a math tour of North Carolina without leaving your classroom!

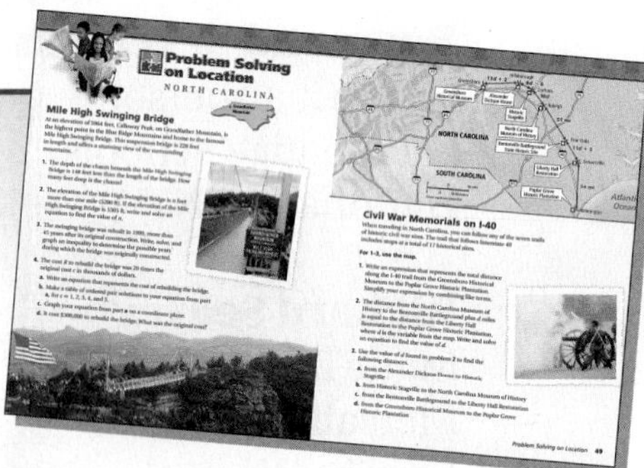

Chapter 1 Use equations to learn about the famous Mile-High Swinging Bridge, located at the highest point in the Blue Ridge Mountains.

Chapter 2 Bog turtles, the smallest turtles in the U.S., can be found in North Carolina. Use integers to describe changes in their population.

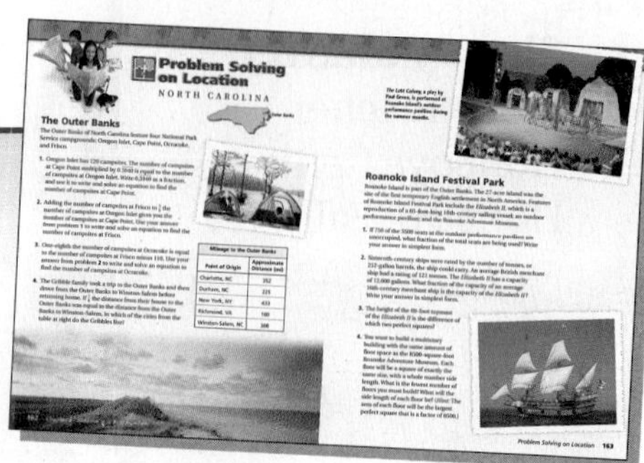

Chapter 3 Use rational numbers to learn about camping in the Outer Banks.

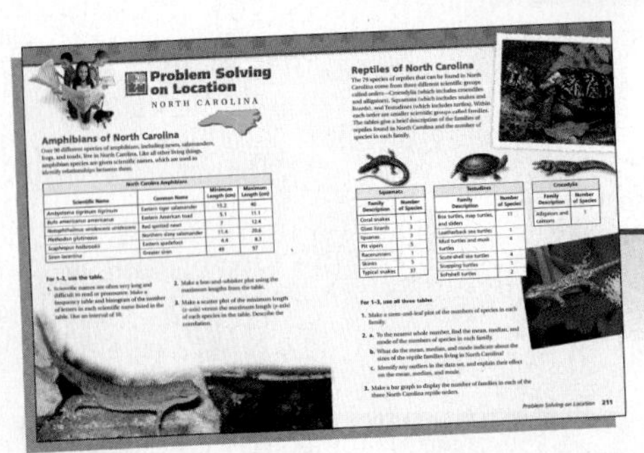

Chapter 4 Analyze data about North Carolina's amphibians and reptiles.

Chapter 5 View the six-mural art project at the Raleigh-Durham airport, which has many examples of geometry.

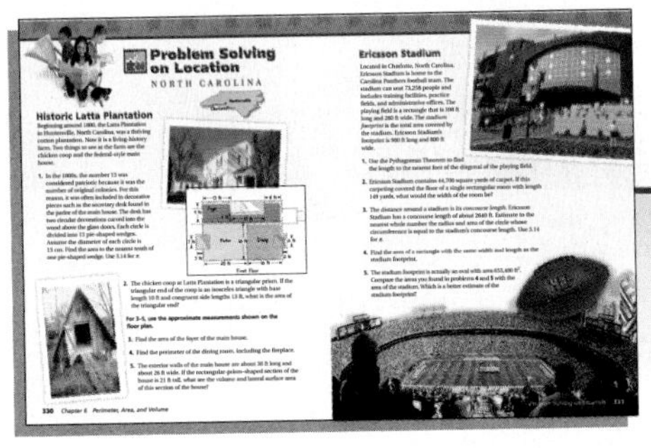

Chapter 6 Take some measurements of Ericsson Stadium, home of the Carolina Panthers.

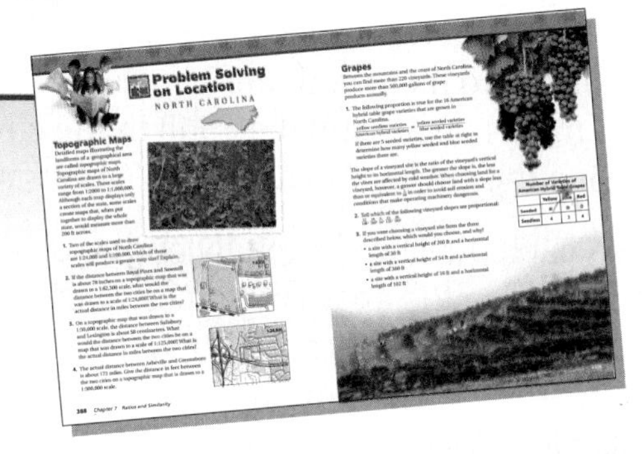

Chapter 7 Use ratios and proportions to describe the different varieties of grapes grown in North Carolina.

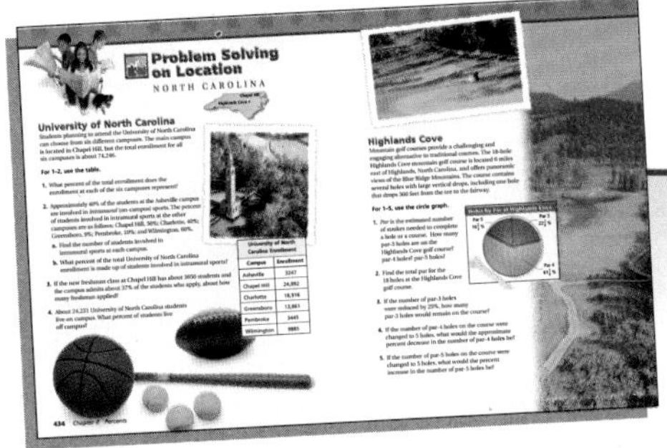

Chapter 8 Percents are used to describe the students at the University of North Carolina.

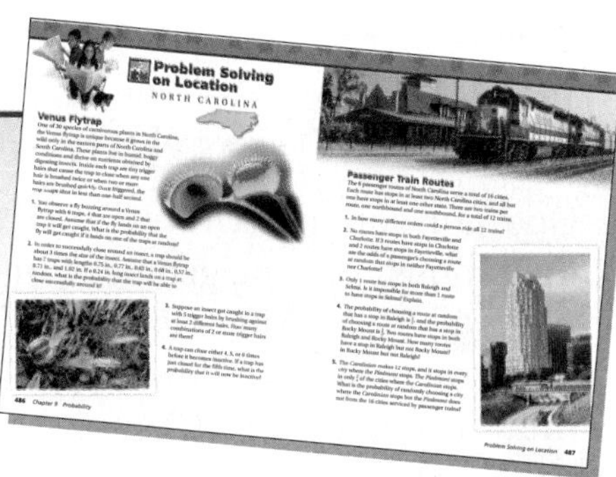

Chapter 9 North Carolina is one of only two states where you can find the carnivorous Venus flytrap growing wild. Find probabilities about the Venus flytrap.

Chapter 10 Relationships between distance, rate, and time are used to describe trips on the Blue Ridge Parkway.

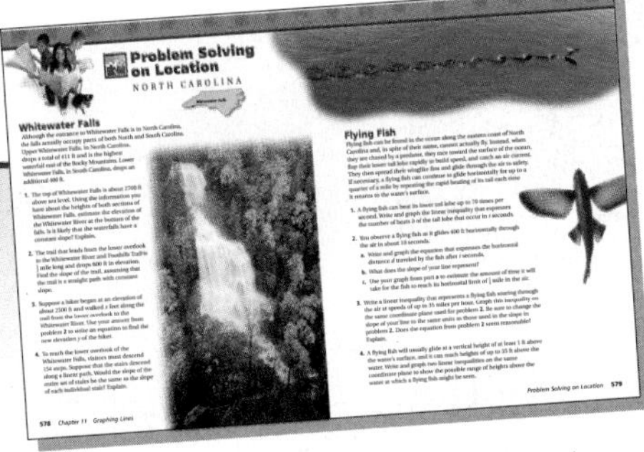

Chapter 11 Have you ever seen a fish fly? Use equations and inequalities to describe the "flight" of North Carolina's flying fish.

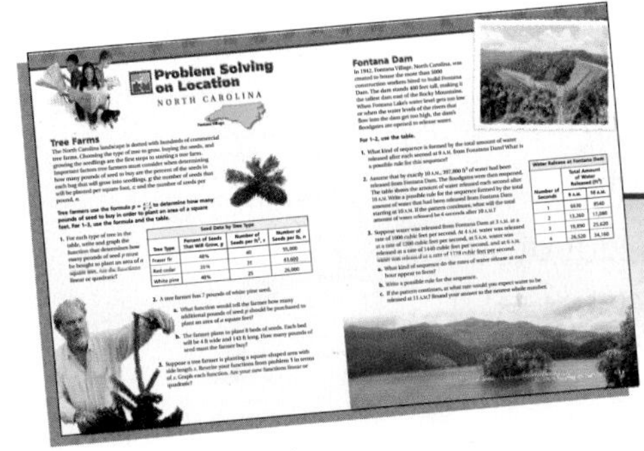

Chapter 12 Sequences are used to describe the amount of water released from Fontana Dam.

North Carolina End-of-Grade (EOG) Test for the Middle Grades

The North Carolina End-of-Grade (EOG) Test is a multiple-choice test. To answer questions on the EOG test, you will fill in an answer sheet. It is very important to fill in your answer sheet correctly. When shading in circles, make your marks heavy and dark. Fill in the circles completely, but do not shade outside the circles. Do not make any stray marks on your answer sheet.

Read each question carefully and work the problem. Choose your answer from among the answer choices given, and fill in the corresponding circle on your answer sheet. If your answer is not one of the choices, read the question again. Be sure that you understand the problem. Check your work for possible errors.

Some questions on the EOG test are calculator inactive, and others are calculator active. You may not use a calculator on the calculator inactive questions, but you may on the calculator active questions.

Sample Question

Questions on the EOG test may require an understanding of number and operations, algebra, geometry, measurement, and data analysis and probability. Drawings, grids, or charts may be included for certain types of questions. Try the following practice question to prepare for taking a mutliple-choice test. Choose the best answer from the choices given.

In a group of 30 students, 27 are middle school students and the others are high school students. If one person is selected at random from this group, what is the probability that the person selected will be a high school student?

A $\frac{1}{30}$

B $\frac{1}{10}$

C $\frac{3}{10}$

D $\frac{9}{10}$

Think About the Solution

First consider the total number of people in the group (30). If 27 out of 30 are middle school students, how many of them are high school students? (3) If one person is selected at random, there is a probability of 3 out of 30. This can be written as a ratio (3:30), a fraction $\left(\frac{3}{30}\right)$, a decimal (0.1), or a percent

(10%). None of these solutions is listed as one of the choices, so you must look for a solution that is equivalent. The fraction $\frac{3}{30}$ can be simplified to $\frac{1}{10}$. Since $\frac{1}{10}$ is given as one of your answer choices, B is the correct response.

Indicate your response by filling in the circle that contains B.

Test-Taking Tip

Sometimes you can find the best solution to a test question by understanding what is wrong with some of the choices. Read the sample question again. Why are A, C, and D incorrect?

Response A is $\frac{1}{30}$.

You might think A is correct because there are 30 people and you are selecting one. However, this answer indicates that only 1 of the 30 people is a high school student. Since that is not what the problem states, A cannot be correct.

Response C is $\frac{3}{10}$.

This answer indicates that 3 out of 10 people are high school students. The numerator is correct, since there are three high school students in the group. However, the denominator must show the relationship 3 out of 30. C is not correct.

Response D is $\frac{9}{10}$.

If you chose D, read the problem again. The problem asks you to find the probability that a high school student will be selected. Answer D would be the best choice if you wanted to find the probability that a *middle school* student will be selected, but D does not match the question that was asked.

Practice, Practice, Practice

On the following pages, you will find a practice standardized test. This test has been designed to resemble the EOG test and the types of questions it contains. Use this test to practice answering these questions, as well as to review some of the math that you will be tested on. The more comfortable and familiar you can become with the EOG test, the better your chances of success!

EOG Test Practice

Directions: Read each question and choose the correct response. You may *not* use a calculator on this part of the test.

1. What is the value of $6 \times 7 - 2^3 \div 4$?

   A 8.5   C 28

   B 10   D 40

2. Which is a solution of $y < 3x + 2$?

   A (1, 5)   C $\left(\frac{1}{3}, 1\right)$

   B (0, 3)   D $(^{-}1, 0)$

3. If $6n - 3 = 33$, what is the value of $5n + 4$?

   A 6   C 29

   B 15   D 34

4. The bar graph below shows the average amount of water, in gallons, used each day by a typical family during certain activities.

   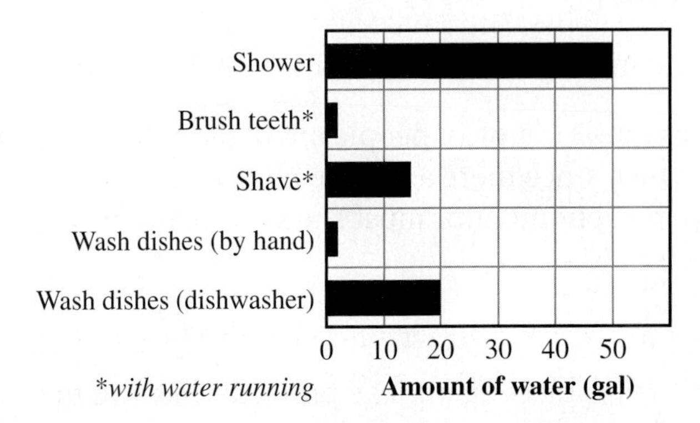

   If a family switched from using a dishwasher to washing their dishes by hand, about how much water would they save each week?

   A 14 gallons   C 140 gallons

   B 126 gallons   D 252 gallons

5. A bag of cat food holds 160 ounces. The label on the side of the bag states that one serving is $1\frac{1}{2}$ ounces. How many full servings are in one bag?

   A 80   C 120

   B 106   D 145

6. Which number is between $^-3.7$ and $^-3\frac{3}{5}$?

A $^-3\frac{4}{5}$

B $^-3.71$

C $^-3.65$

D $^-3.59$

7. Merrick is buying a board to use as a diagonal brace on a rectangular gate. The gate is 36 inches wide and 42 inches tall.

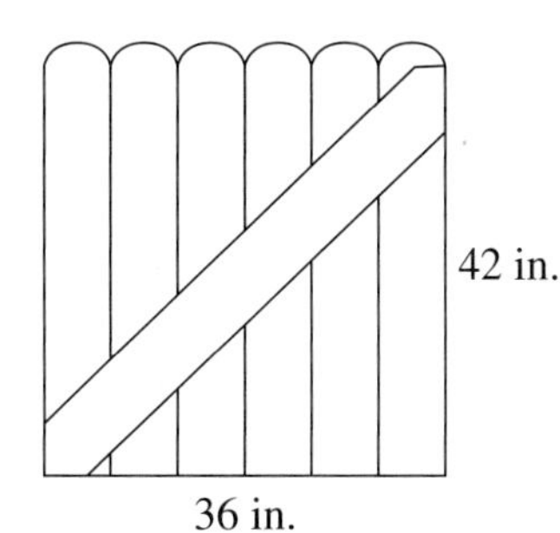

What length of board should be used for the brace?

A 5.5 feet

B 6.5 feet

C 42.7 inches

D 55.3 inches

8. A rental car company offers car rentals for \$29.99 per day plus \$0.15 per mile. Jim rented a car for one day and paid \$35.54. How many miles did Jim drive?

A 5.6 miles

B 15 miles

C 35.5 miles

D 37 miles

9. A public plaza has an area of 8,100 square meters. The plaza is in the shape of a square. How long is each side?

A 30 meters

B 45 meters

C 81 meters

D 90 meters

10. Simon bought b bags of ice that cost \$0.99 each. The total cost can be represented by the expression $0.99b$. What is the total cost if $b = 4$?

A \$3.96

B \$4.04

C \$4.95

D \$5.94

Go On

11. Carlo mows lawns in his neighborhood. He charges a fixed fee per square foot. For a lawn that is 75 feet by 75 feet, Carlo charges $20. How much should Carlo charge for a lawn that is 150 feet by 150 feet?

A Carlo should charge twice as much.

B Carlo should charge one-half as much.

C Carlo should charge three times as much.

D Carlo should charge four times as much.

12. The table shows the shipping rates for an overnight delivery company.

OVERNIGHT SHIPPING RATES

Weight	Cost
Less than 2 pounds	$12.19
2 pounds up to 5 pounds	$18.35
5 pounds–10 pounds	$24.07

Frieda sent three items for $42.42. The first item weighed 4.77 pounds, the second 0.25 pound, and the third 2.19 pounds. Which statement is true?

A She sent all three items together.

B She sent the first and third items together, and the second item separately.

C She sent the first and second items together, and the third item separately.

D She sent all three items separately.

13. A scientist wants to know the number of fish in a certain lake. She captures 15 fish, places tags on them, and re-releases them into the lake. The next day, she captures 15 more fish. Two of these fish have tags.

About how many fish are in the lake?

A 100

B 113

C 125

D 200

14. During December, 21,828 passengers flew on a certain airline. This was four times the number of passengers that flew the same airline in October. How many passengers flew on the airline in October?

A 5,457

B 5,500

C 78,866

D 87,312

15. Fran has a rectangular pen for her puppies, shown below. Fran bought more fencing to double the length of the shorter side of the pen. The length of the longer side was unchanged. What is the area of the new pen?

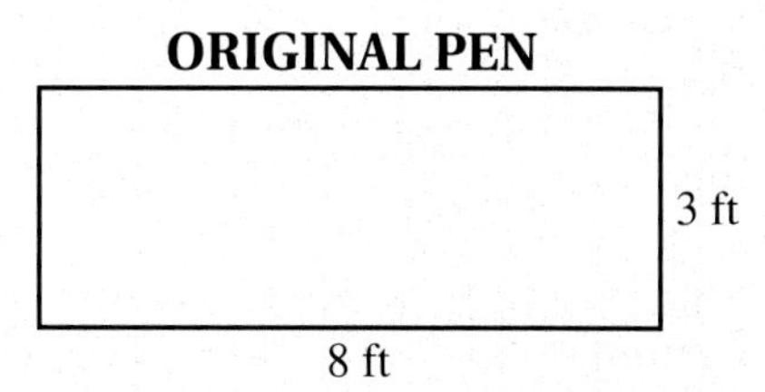

A 22 square feet

B 24 square feet

C 48 square feet

D 96 square feet

16. What is the distance across the lake?

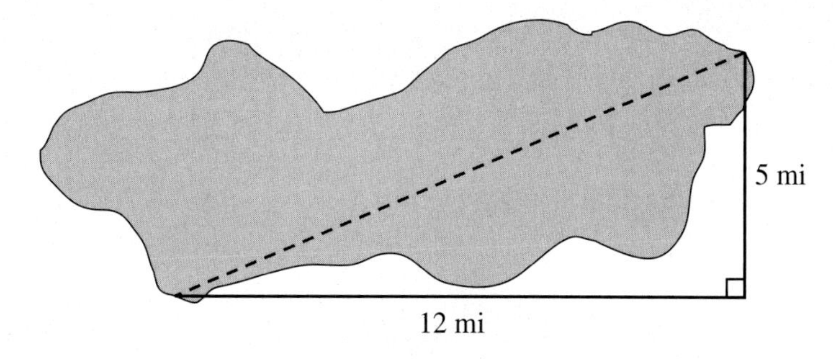

A 12.4 miles

B 13 miles

C 34 miles

D 169 miles

17. A coat costs $65 more than a dress. Together they cost $113. How much does the dress cost?

A $24

B $48

C $65

D $89

18. What is the slope of the line that passes through ($^{-}1$, 1) and (2, 5)?

A $\frac{3}{4}$

B $\frac{4}{3}$

C $\frac{2}{3}$

D $\frac{3}{2}$

19. Between what two whole numbers is $\sqrt{51}$?

A 5 and 6

B 6 and 7

C 7 and 8

D 8 and 9

Go On

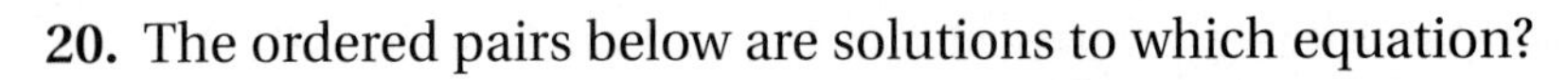

20. The ordered pairs below are solutions to which equation?

$$(^-2, ^-7), (^-1, ^-5), (0, ^-3), (1, ^-1), (4, 5)$$

A $y = 2x - 3$

B $y = \frac{1}{2}x + 3$

C $y = x - 5$

D $y = x - 3$

21. Which of the following is the graph of the solutions of $^-3x > 9$?

A x
−5 −4 −3 −2 −1 0 1

B x
−5 −4 −3 −2 −1 0 1

C x
−5 −4 −3 −2 −1 0 1

D x
−5 −4 −3 −2 −1 0 1

22. What is the volume of the shape below?

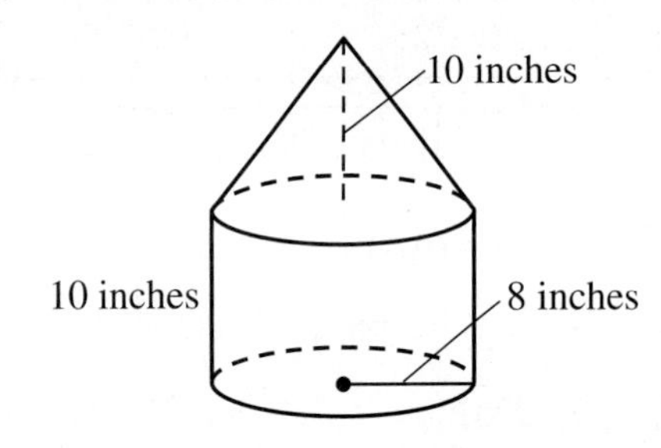

A 334.93 cubic inches

B 800 cubic inches

C 2,512 cubic inches

D 2,679.47 cubic inches

23. What is the surface area of this cylinder, to the nearest hundredth?

6.5 in.
26 in.

A 301.67 square inches

B 563.83 square inches

C 596.99 square inches

D 1,098.50 square inches

Directions: Read each question and choose the correct response. You may use a calculator on this part of the test.

24. One year, the U.S. Department of Agriculture reported that nearly 60 million poinsettia plants were purchased in the United States. During the same year, poinsettia sales totaled $22 million.

Suppose all of the poinsettia plants cost the same amount. What equation could be used to determine the price (p) of each plant?

A $22p = 60$

B $p = 60 \div 22$

C $p = 22 \times 60$

D $60p = 22$

25. A sports store orders jackets from a wholesale company. Two options for ordering the jackets are given in the table.

ORDERING OPTIONS

	Service Charge	Cost per Jacket
Option 1	$200.00	$30.00
Option 2	$0.00	$40.00

What number of jackets is the same price under both options?

A 15

B 20

C 25

D 30

26. In Friday's basketball game, Laura made 8 baskets. This brought her season total for baskets to 15.

What equation could be used to determine the number of baskets (b) that Laura made before Friday's game?

A $15 + 8 = b$

B $8 - b = 15$

C $b + 8 = 15$

D $15 + b = 8$

27. Which of the following numbers is irrational?

A $\frac{1}{3}$

B $\sqrt{2}$

C $\sqrt{4}$

D $6.\overline{78}$

Go On

28. What is the slope of the line that is graphed below?

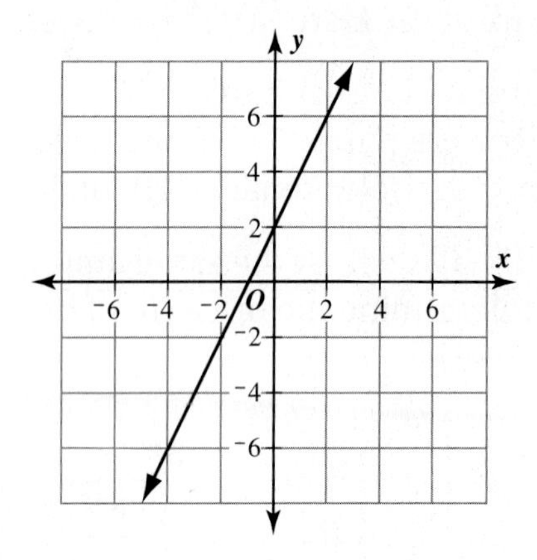

A 2

B $\frac{1}{2}$

C $-\frac{1}{2}$

D $^{-}2$

29. One reason camels can live in the desert is that their body temperatures change to adjust to changes in the desert temperature.

The table shows a camel's body temperature at six different times during one day.

CAMEL'S BODY TEMPERATURE

Time	Temperature (°C)
2 A.M.	34
6 A.M.	36
10 A.M.	39
2 P.M.	40
6 P.M.	38
10 P.M.	36

During which time period did the camel's body temperature change the *most*?

A 2 A.M. to 6 A.M.

B 2 P.M. to 6 P.M.

C 6 P.M. to 10 P.M.

D 6 A.M. to 10 A.M.

30. Maria wondered how the growth of plants was affected by the amount of sunlight they receive. She recorded the number of hours of sunlight her eight houseplants received each day, as well as their monthly growth in inches. Her results are shown in the table below.

GROWTH OF HOUSEPLANTS

Plant	Daily sunlight (h)	Monthly growth (in.)
1	1	0.25
2	1.5	0.5
3	2	0.5
4	3	1
5	3.5	1.5
6	4	1.25
7	5	2
8	5	2.5

Which statement is true?

A The correlation between sunlight and growth cannot be determined.

B The correlation between sunlight and growth is positive.

C The correlation between sunlight and growth is negative.

D There is no correlation between sunlight and growth of the plants.

31. Look at the function table. What is the value of y when $x = 2$?

x	y
$^-1$	2
0	0
1	$^-2$
2	

A $^-4$

B 0

C 2

D 4

32. Luisa and her father stand next to each other on a sunny day. Luisa is 130 centimeters tall, and her father is 200 centimeters tall. Luisa's shadow is 100 centimeters long.

How long is Luisa's father's shadow, to the nearest centimeter?

A 65 centimeters

B 154 centimeters

C 170 centimeters

D 260 centimeters

Go On

EOG Test Practice

33. If you double the height of this triangle, how will the area change?

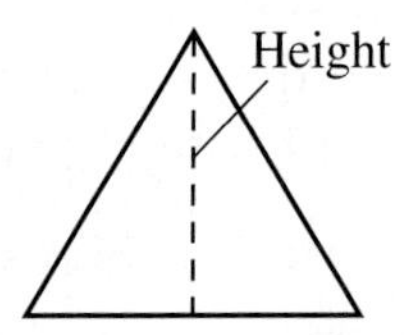

A The area will remain the same.

B The area will be twice as large.

C The area will be one-half as large.

D The area will be four times as large.

34. The area of this square is 10 square meters. How long is each side?

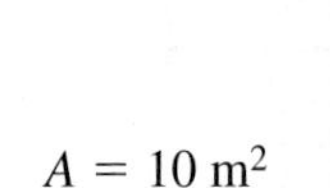

A 10 meters

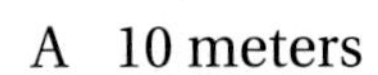

B $\sqrt{10}$ meters

C 100 meters

D $\sqrt{20}$ meters

35. A shoe company featured the bar graph below in one of its advertisements. Why is this graph misleading?

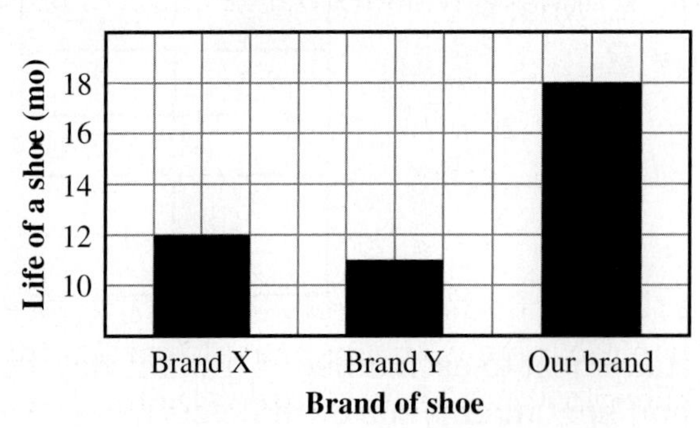

A The scale on the vertical axis does not start at 0.

B The bars are not of equal width.

C The vertical scale has an interval of 2.

D Two of the bars are close to the same height.

36. Which ordered pair is a solution of the equation that is graphed below?

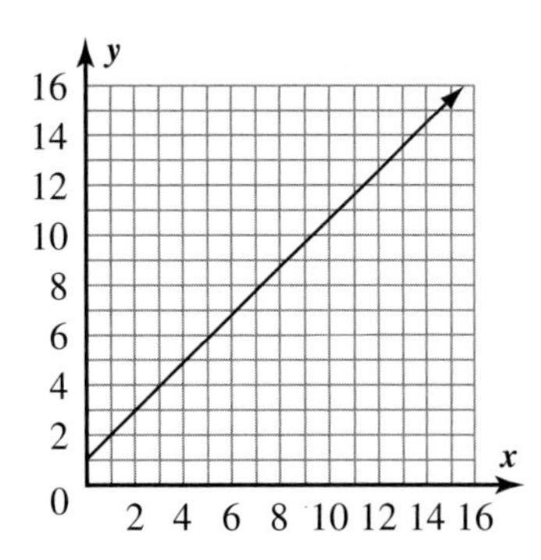

A (1, 2)

B (3, 2)

C (4, 4)

D (4, 3)

37. The public library recorded the number of books that were checked out each year. The results are given in the table below.

BOOKS CHECKED OUT

Year	Number of Books (thousands)
1	33.5
2	34.4
3	35.5
4	36.8

If this pattern continues, how many books will be checked out in Year 5?

A 37,000

B 37,800

C 38,000

D 38,300

38. An 8-piece pizza was shared by 5 people. Each person got the same number of pieces. Then the remaining amount of pizza was divided equally among the 5 people. What fraction of the original pizza does this second portion represent?

A $\frac{3}{40}$

B $\frac{3}{15}$

C $\frac{3}{8}$

D $\frac{3}{5}$

Go On

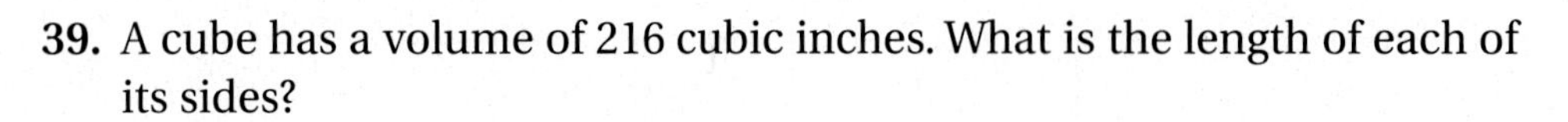

39. A cube has a volume of 216 cubic inches. What is the length of each of its sides?

A 6 inches

B 36 inches

C 72 inches

D 648 inches

40. What is the lower quartile of this data set?

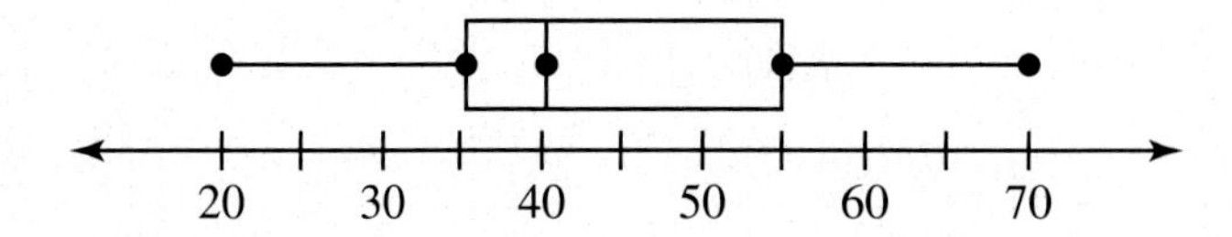

A 20

B 35

C 40

D 55

41. Which of the following equations is linear?

A $y = 4x^2$

B $y = 2x^2 + 1$

C $y = x^3$

D $y = \frac{1}{2}x$

42. The water level of a river is 34 feet. After one day, the water level has decreased to 33.5 feet. After two days, the water level is 33 feet, and after three days, the water level is 32.5 feet.

The equation $w = 34 - 0.5d$, where w is the water level and d is the number of days, can be used to describe this situation. Which part of this equation describes the amount by which the water level is changing each day?

A 34

B 0.5

C $^{-}0.5$

D $^{-}0.5d$

43. What will be the perimeter of this triangle after it has been dilated by a scale factor of 0.8?

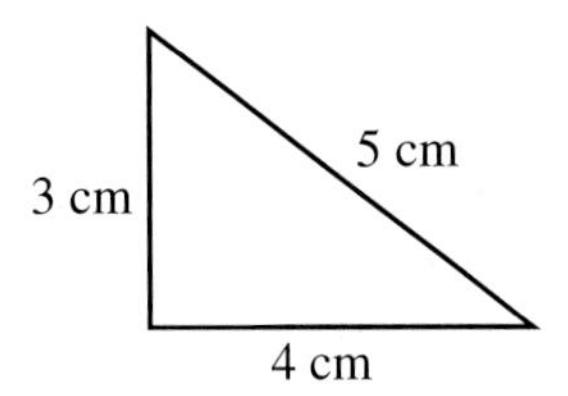

A 50 centimeters

B 40 centimeters

C 12 centimeters

D 9.6 centimeters

44. Which three lengths could be the lengths of the sides of a right triangle?

A 2 inches, 9 inches, 11 inches

B 2 inches, 4 inches, 16 inches

C 6 inches, 8 inches, 10 inches

D 6 inches, 6 inches, 36 inches

45. The student council is planning an all-school picnic. To decide what food to serve, council members will survey a sample of students. Which of the following is the *best* sampling method?

A Each council member will survey his or her friends.

B Council members will survey students whose locker numbers end in 5.

C Council members will survey students at the local pizza parlor.

D Each council member will survey the members of his or her first-period class.

46. The marketing department of a radio station wants to determine what times of day people under the age of 20 listen to the radio. They have several different ideas of ways to survey people. Which of the following survey methods is *most* appropriate?

A randomly surveying 100 listeners that are girls

B randomly surveying 100 listeners under the age of 20

C randomly surveying 100 listeners under the age of 16

D randomly surveying 100 listeners

Stop

Problem Solving Handbook

CHAPTER 1 Algebra Toolbox

Assessment

Interdisciplinary LINKS

Student Help

internet connect go.hrw.com

Homework Help Online

KEYWORD: MT4 HWHelp

Indicates algebra included in lesson development

Integers an[illegible] Exponents

Rational and Real Numbers

Assessment

Interdisciplinary LINKS

Student Help

internet connect go.hrw.com

Homework Help Online

KEYWORD: MT4 HWHelp

Algebra *Indicates algebra included in lesson development*

Collecting, Displaying, and Analyzing Data

Assessment

Interdisciplinary LINKS

Student Help

Homework Help Online

KEYWORD: MT4 HWHelp

Plane Geometry

Algebra *Indicates algebra included in lesson development*

Perimeter, Area, and Volume

Assessment

EOG Test Preparation Online . . . . . **KEYWORD: MT4 TestPrep**

Interdisciplinary LINKS

Student Help

internet connect
Homework Help Online
283, 287, 292, 296, 304, 310, 314, 318, 322, 326
KEYWORD: MT4 HWHelp

Ratios and Similarity

Assessment

go.hrw.com EOG Test Preparation Online . . . . . KEYWORD: MT4 TestPrep

Interdisciplinary LINKS

Student Help

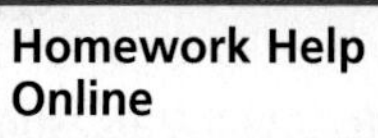

go.hrw.com

Homework Help Online

344, 348, 353, 358, 364, 370, 374, 378, 384

KEYWORD: MT4 HWHelp

Algebra *Indicates algebra included in lesson development*

Percents

Assessment

Interdisciplinary LINKS

Student Help

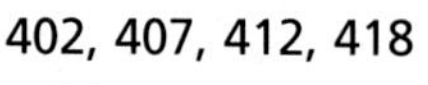

Homework Help Online

402, 407, 412, 418

KEYWORD: MT4 HWHelp

Probability

Assessment

Interdisciplinary LINKS

Student Help

internet connect go.hrw.com

Homework Help Online

KEYWORD: MT4 HWHelp

Algebra *Indicates algebra included in lesson development*

More Equations and Inequalities

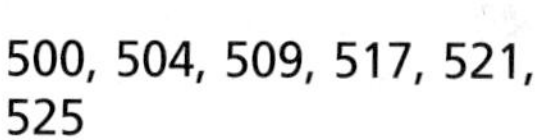

Problem Solving Handbook

The Problem Solving Plan

In order to be a good problem solver, you need to use a good problem-solving plan. The plan used in this book is detailed below. If you have another plan that you like to use, you can use it as well.

UNDERSTAND the Problem

- **What are you asked to find?** Restate the question in your own words.
- **What information is given?** Identify the important facts in the problem.
- **What information do you need?** Determine which facts are needed to answer the question.
- **Is all the information given?** Determine whether all the facts are given.
- **Is there any information given that you will not use?** Determine which facts, if any, are unnecessary to solve the problem.

Make a PLAN

- **Have you ever solved a similar problem?** Think about other problems like this that you successfully solved.
- **What strategy or strategies can you use?** Determine a strategy that you can use and how you will use it.

SOLVE

- Follow your plan. Show the steps in your solution. Write your answer as a complete sentence.

LOOK BACK

- **Have you answered the question?** Be sure that you answered the question that is being asked.
- **Is your answer reasonable?** Your answer should make sense in the context of the problem.
- **Is there another strategy you could use?** Solving the problem using another strategy is a good way to check your work.
- **Did you learn anything that could help you solve similar problems in the future?** Try to remember the problems you have solved and the strategies you used to solve them.

Using the Problem Solving Plan

Roy has a rectangular piece of land that he wants to put a fence around. He will place a post every 9 ft along the perimeter. Each post is 5 ft tall. The land is 63 ft long and 45 ft wide. How many posts does Roy need?

UNDERSTAND the Problem

Roy has a piece of land that is 63 ft by 45 ft. He wants a post every 9 ft along the perimeter. You must find out how many posts Roy needs.

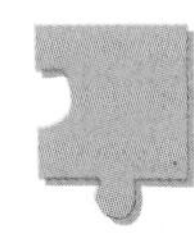

Make a PLAN

You can **draw a diagram** to show how many posts Roy needs for his fence.

SOLVE

Draw a rectangle that is similar to Roy's land. Place marks along the perimeter of the rectangle to represent the posts to be placed every 9 ft.

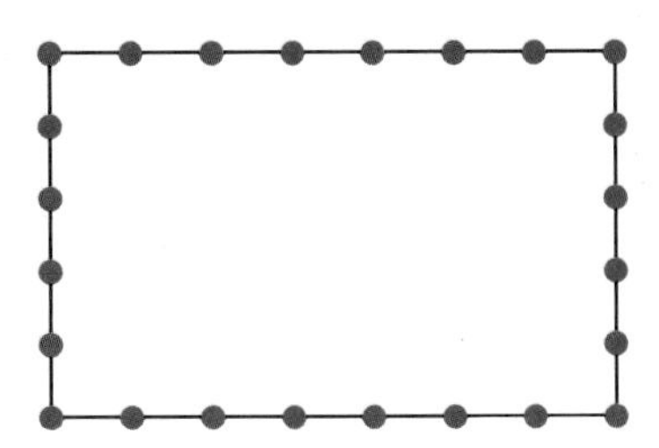

Count the number of marks you placed around the rectangle. Each corner should only have one mark.

Roy needs 24 posts for his fence.

LOOK BACK

The perimeter is $63 + 45 + 63 + 45 = 216$ ft. If a post is placed every 9 ft, there will be $216 \div 9 = 24$ posts. The answer is reasonable.

Draw a Diagram

When problems involve objects, distances, or places, drawing a diagram can make the problem clearer. You can **draw a diagram** to help understand the problem and to solve the problem.

Problem Solving Strategies

Draw a Diagram	Make a Table
Make a Model	Solve a Simpler Problem
Guess and Test	Use Logical Reasoning
Work Backward	Use a Venn Diagram
Find a Pattern	Make an Organized List

June is moving her cat, dog, and goldfish to her new apartment. She can only take 1 pet with her on each trip. She cannot leave the cat and the dog or the cat and the goldfish alone together. How can she get all of her pets safely to her new apartment?

Understand the Problem The answer will be the description of the trips to her new apartment. At no time can the cat be alone with the dog or the goldfish.

Make a Plan **Draw a diagram** to represent each trip to and from the apartment.

Solve In the beginning, the cat, dog, and goldfish are all at her old apartment.

Old Apartment		New Apartment	
June, Cat, Dog, Fish	June, Cat →	June, Cat	*Trip 1: She takes the cat and returns alone.*
June, Dog, Fish	← June	Cat	
June, Dog, Fish	June, Dog →	June, Dog, Cat	*Trip 2: She takes the dog and returns with the cat.*
June, Cat, Fish	← June, Cat	Dog	
June, Cat, Fish	June, Fish →	June, Dog, Fish	*Trip 3: She takes the fish and returns alone.*
June, Cat	← June	Dog, Fish	
June, Cat	June, Cat →	June, Cat, Dog, Fish	*Trip 4: She takes the cat.*

Look Back Check to make sure that the cat is never alone with either the fish or the dog.

PRACTICE

1. There are 8 flags evenly spaced around a circular track. It takes Ling 15 s to run from the first flag to the third flag. At this pace, how long will it take her to run around the track twice?

2. A frog is climbing a 22-foot tree. Every 5 minutes, it climbs up 3 feet, but slips back down 1 foot. How long will it take it to climb the tree?

Make a Model

A problem that involves objects may be solved by making a model out of similar items. **Make a model** to help you understand the problem and find the solution.

Problem Solving Strategies

Draw a Diagram	Make a Table
Make a Model	Solve a Simpler Problem
Guess and Test	Use Logical Reasoning
Work Backward	Use a Venn Diagram
Find a Pattern	Make an Organized List

The volume of a rectangular prism can be found by using the formula $V = \ell wh$, where ℓ is the length, w is the width, and h is the height of the prism. Find all possible rectangular prisms with a volume of 16 cubic units and dimensions that are all whole numbers.

Understand the Problem

You need to find the different possible prisms. The length, width, and height will be whole numbers whose product is 16.

Make a Plan

You can use unit cubes to make a model of every possible rectangular prism. Work in a systematic way to find all possible answers.

Solve

Begin with a $16 \times 1 \times 1$ prism.

$16 \times 1 \times 1$

Keeping the height of the prism the same, explore what happens to the length as you change the width. Then try a height of 2. Notice that an $8 \times 2 \times 1$ prism is the same as an $8 \times 1 \times 2$ prism turned on its side.

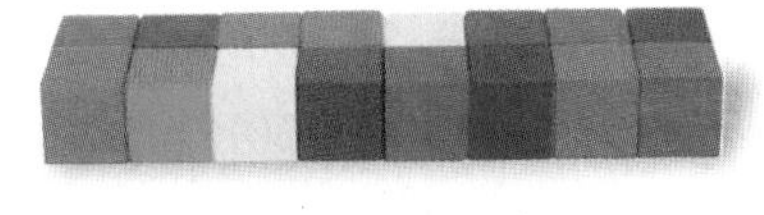

$8 \times 2 \times 1$

Not a rectangular prism

$4 \times 4 \times 1$

$4 \times 2 \times 2$

The possible dimensions are $16 \times 1 \times 1$, $8 \times 2 \times 1$, $4 \times 4 \times 1$, and $4 \times 2 \times 2$.

Look Back

The product of the length, width, and height must be 16. Look at the prime factorization of the volume: $16 = 2 \cdot 2 \cdot 2 \cdot 2$. Possible dimensions:

$1 \cdot 1 \cdot (2 \cdot 2 \cdot 2 \cdot 2) = 1 \cdot 1 \cdot 16$

$1 \cdot 2 \cdot (2 \cdot 2 \cdot 2) = 1 \cdot 2 \cdot 8$

$1 \cdot (2 \cdot 2) \cdot (2 \cdot 2) = 1 \cdot 4 \cdot 4$

$2 \cdot 2 \cdot (2 \cdot 2) = 2 \cdot 2 \cdot 4$

PRACTICE

1. Four unit squares are arranged so that each square shares a side with another square. How many different arrangements are possible?

2. Four triangles are formed by cutting a rectangle along its diagonals. What possible shapes can be formed by arranging these triangles?

Guess and Test

When you think that guessing may help you solve a problem, you can use **guess and test.** Using clues to make guesses can narrow your choices for the solution. Test whether your guess solves the problem, and continue guessing until you find the solution.

Problem Solving Strategies

Draw a Diagram	Make a Table
Make a Model	Solve a Simpler Problem
Guess and Test	Use Logical Reasoning
Work Backward	Use a Venn Diagram
Find a Pattern	Make an Organized List

North Middle School is planning to raise $1200 by sponsoring a car wash. They are going to charge $4 for each car and $8 for each minivan. How many vehicles would have to be washed to raise $1200 if they plan to wash twice as many cars as minivans?

Understand the Problem You must determine the number of cars and the number of minivans that need to be washed to make $1200. You know the charge for each vehicle.

Make a Plan You can **guess and test** to find the number of cars and minivans. Guess the number of cars, and then divide it by 2 to find the number of minivans.

Solve You can organize your guesses in a table.

	Cars	Minivans	Money Raised	
First guess	200	100	$4(200) + $8(100) = $1600	*Too high*
Second guess	100	50	$4(100) + $8(50) = $800	*Too low*
Third guess	150	75	$4(150) + $8(75) = $1200	

They should wash 150 cars and 75 minivans, or 225 vehicles.

Look Back The total raised is $4(150) + $8(75) = $1200, and the number of cars is twice the number of minivans. The answer is reasonable.

PRACTICE

1. At a baseball game, adult tickets cost $15 and children's tickets cost $8. Twice as many children attended as adults, and the total ticket sales were $2480. How many people attended the game?

2. Angie is making friendship bracelets and pins. It takes her 6 minutes to make a bracelet and 4 minutes to make a pin. If she wants to make three times as many pins as bracelets, how many pins and bracelets can she make in 3 hours?

Make an Organized List

Sometimes a problem involves many possible ways in which something can be done. To find a solution to this kind of problem, you need to **make an organized list.** This will help you to organize and count all the possible outcomes.

Problem Solving Strategies

Draw a Diagram	Make a Table
Make a Model	Solve a Simpler Problem
Guess and Test	Use Logical Reasoning
Work Backward	Use a Venn Diagram
Find a Pattern	**Make an Organized List**

What is the greatest amount of money you can have in coins (quarters, dimes, nickels, and pennies) without being able to make change for a dollar?

Understand the Problem

You are looking for an amount of money. You cannot have any combinations of coins that make a dollar, such as 4 quarters or 3 quarters, 2 dimes, and a nickel.

Make a Plan

Make an organized list, starting with the maximum possible number of each type of coin. Consider all the ways you can add other types of coins without making exactly one dollar.

Solve

List the maximum number of each kind of coin you can have.

3 quarters = 75¢   9 dimes = 90¢   19 nickels = 95¢   99 pennies = 99¢

Next, list all the possible combinations of two kinds of coins.

3 quarters and 4 dimes = 115¢
3 quarters and 4 nickels = 95¢
3 quarters and 24 pennies = 99¢
9 dimes and 1 quarter = 115¢
9 dimes and 1 nickel = 95¢
9 dimes and 9 pennies = 99¢

19 nickels and 4 pennies = 99¢

Look for any combinations from this list that you could add another kind of coin to without making exactly one dollar.

3 quarters, 4 dimes, and 4 pennies = 119¢
3 quarters, 4 nickels, and 4 pennies = 99¢
9 dimes, 1 quarter, and 4 pennies = 119¢
9 dimes, 1 nickel, and 4 pennies = 99¢

The largest amount you can have is 119¢, or $1.19.

Look Back

Try adding one of any type of coin to either combination that makes $1.19, and then see if you could make change for a dollar.

PRACTICE

1. How can you arrange the numbers 2, 6, 7, and 12 with the symbols +, ×, and ÷ to create the expression with the greatest value?

2. How many ways are there to arrange 24 desks in 3 or more equal rows if each row must have at least 2 desks?

Chapter 1

Algebra Toolbox

TABLE OF CONTENTS

Toxic Gases Released By Fires

Gas	Danger Level (ppm)	Source
Carbon monoxide (CO)	1200	Incomplete burning
Hydrogen chloride (HCl)	50	Plastics
Hydrogen cyanide (HCN)	50	Wool, nylon, polyurethane foam, rubber, paper
Phosgene ($COCl_2$)	2	Refrigerants

Career Firefighter

A firefighter approaching a fire should be aware of ventilation, space, what is burning, and what could be ignited. Oxygen, fuel, heat, and chemical reactions are at the core of a fire, but the amounts and materials differ.

The table above lists some of the toxic gases that firefighters frequently encounter.

internet connect

Chapter Opener Online
go.hrw.com
KEYWORD: MT4 Ch1

ARE YOU READY?

Choose the best term from the list to complete each sentence.

1. __?__ is the __?__ of addition.
2. The expressions $3 \cdot 4$ and $4 \cdot 3$ are equal by the __?__.
3. The expressions $1 + (2 + 3)$ and $(1 + 2) + 3$ are equal by the __?__.
4. Multiplication and __?__ are opposite operations.
5. __?__ and __?__ are commutative.

addition
Associative Property
Commutative Property
division
opposite operation
multiplication
subtraction

Complete these exercises to review skills you will need for this chapter.

✔ Whole Number Operations

Simplify each expression.

6. $8 + 116 + 43$ **7.** $2431 - 187$ **8.** $204 \cdot 38$ **9.** $6447 \div 21$

✔ Compare and Order Whole Numbers

Order each sequence of numbers from least to greatest.

10. 1050; 11,500; 105; 150 **11.** 503; 53; 5300; 5030 **12.** 44,400; 40,040; 40,400; 44,040

✔ Inverse Operations

Rewrite each expression using the inverse operation.

13. $72 + 18 = 90$ **14.** $12 \cdot 9 = 108$ **15.** $100 - 34 = 66$ **16.** $56 \div 8 = 7$

✔ Order of Operations

Simplify each expression.

17. $2 + 3 \cdot 4$ **18.** $50 - 2 \cdot 5$ **19.** $6 \cdot 3 \cdot 3 - 3$ **20.** $(5 + 2)(5 - 2)$

21. $5 - 6 \div 2$ **22.** $16 \div 4 + 2 \cdot 3$ **23.** $(8 - 3)(8 + 3)$ **24.** $12 \div 3 \div 2 + 5$

✔ Evaluate Expressions

Determine whether the given expressions are equal.

25. $(4 \cdot 7) \cdot 2$ and $4 \cdot (7 \cdot 2)$ **26.** $(2 \cdot 4) \div 2$ and $2 \cdot (4 \div 2)$ **27.** $2 \cdot (3 - 3) \cdot 2$ and $(2 \cdot 3) - 3$ **28.** $5 \cdot (50 - 44)$ and $5 \cdot 50 - 44$

29. $9 - (4 \cdot 2)$ and $(9 - 4) \cdot 2$ **30.** $2 \cdot 3 + 2 \cdot 4$ and $2 \cdot (3 + 4)$ **31.** $(16 \div 4) + 4$ and $16 \div (4 + 4)$ **32.** $5 + (2 \cdot 3)$ and $(5 + 2) \cdot 3$

1-1 Variables and Expressions

Learn to evaluate algebraic expressions.

Vocabulary
- variable
- coefficient
- algebraic expression
- constant
- evaluate
- substitute

This nautilus shell has about 34 chambers. Using this information, you can determine its approximate age.

The nautilus is a sea creature whose shell has a series of chambers. Every lunar month (about 30 days), the nautilus creates and moves into a new chamber of the shell.

Let n be the number of chambers in the shell. You can approximate the age, in days, of the nautilus using the following expression:

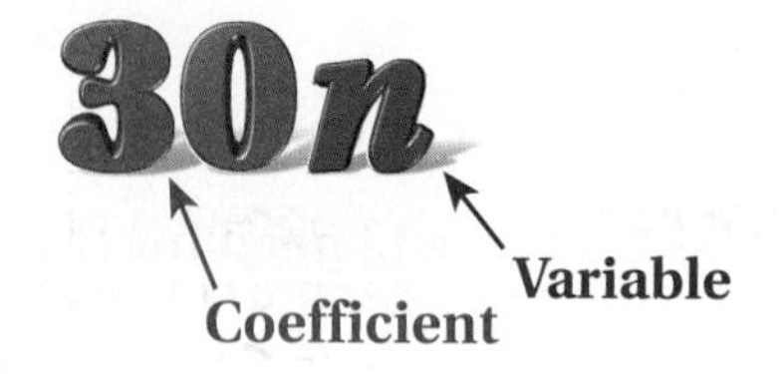

A **variable** is a letter that represents a value that can change or vary. The **coefficient** is the number multiplied by the variable. An **algebraic expression** has one or more variables.

In the algebraic expression $x + 6$, 6 is a **constant** because it does not change. To **evaluate** an algebraic expression, **substitute** a given number for the variable, and find the value of the resulting numerical expression.

EXAMPLE 1 Evaluating Algebraic Expressions with One Variable

Evaluate each expression for the given value of the variable.

A $x + 6$ for $x = 13$

$13 + 6$ *Substitute 13 for x.*

19 *Add.*

Remember!

Order of Operations PEMDAS:
1. Parentheses
2. Exponents
3. Multiply and Divide from left to right.
4. Add and Subtract from left to right.

B $2a + 3$ for $a = 4$

$2(4) + 3$ *Substitute 4 for a.*

$8 + 3$ *Multiply.*

11 *Add.*

C $3(5 + n) - 1$ for $n = 0, 1, 2$

n	Substitute	Parentheses	Multiply	Subtract
0	3(5 + 0) − 1	3(5) − 1	15 − 1	**14**
1	3(5 + 1) − 1	3(6) − 1	18 − 1	**17**
2	3(5 + 2) − 1	3(7) − 1	21 − 1	**20**

EXAMPLE 2 Evaluating Algebraic Expressions with Two Variables

Evaluate each expression for the given values of the variables.

A $2x + 3y$ for $x = 15$ and $y = 12$

$2(15) + 3(12)$	*Substitute 15 for x and 12 for y.*
$30 + 36$	*Multiply.*
66	*Add.*

B $1.5p - 2q$ for $p = 18$ and $q = 7.5$

$1.5(18) - 2(7.5)$	*Substitute 18 for p and 7.5 for q.*
$27 - 15$	*Multiply.*
12	*Subtract.*

EXAMPLE 3 *Physical Science Application*

If c is a temperature in degrees Celsius, then $1.8c + 32$ can be used to find the temperature in degrees Fahrenheit. Convert each temperature from degrees Celsius to degrees Fahrenheit.

A freezing point of water: 0°C

$1.8c + 32$	
$1.8(0) + 32$	*Substitute 0 for c.*
$0 + 32$	*Multiply.*
32	*Add.*
$0°C = 32°F$	

Water freezes at 32°F.

B world's highest recorded temperature (El Azizia, Libya): 58°C

$1.8c + 32$	
$1.8(58) + 32$	*Substitute 58 for c.*
$104.4 + 32$	*Multiply.*
136.4	*Add.*
$58°C = 136.4°F$	

The highest recorded temperature in the world is 136.4°F.

Think and Discuss

1. **Give an example** of an expression that is algebraic and of an expression that is not algebraic.
2. **Tell** the steps for evaluating an algebraic expression for a given value.
3. **Explain** why you cannot find a numerical value for the expression $4x - 5y$ for $x = 3$.

1-1 Exercises

FOR EOG PRACTICE
see page 644

internet connect
Homework Help Online
go.hrw.com Keyword: MT4 1-1

GUIDED PRACTICE

See Example 1 Evaluate each expression for the given value of the variable.

1. $x + 5$ for $x = 12$
2. $3a + 5$ for $a = 6$
3. $2(4 + n) - 5$ for $n = 0$

See Example 2 Evaluate each expression for the given values of the variables.

4. $3x + 2y$ for $x = 8$ and $y = 10$
5. $1.2p - 2q$ for $p = 3.5$ and $q = 1.2$

See Example 3 You can make cornstarch slime by mixing $\frac{1}{2}$ as many tablespoons of water as cornstarch. How many tablespoons of water do you need for each number of tablespoons of cornstarch?

6. 10 tbsp
7. 16 tbsp
8. 23 tbsp
9. 34 tbsp

INDEPENDENT PRACTICE

See Example 1 Evaluate each expression for the given value of the variable.

10. $x + 7$ for $x = 23$
11. $5t + 3$ for $t = 6$
12. $6(2 + k) - 5$ for $k = 0$

See Example 2 Evaluate each expression for the given values of the variables.

13. $5x + 4y$ for $x = 7$ and $y = 8$
14. $4m - 2n$ for $m = 25$ and $n = 2.5$

See Example 3 If q is the number of quarts, then $\frac{1}{4}q$ can be used to find the number of gallons. Find the number of gallons for each of the following.

15. 16 quarts
16. 24 quarts
17. 8 quarts
18. 32 quarts

PRACTICE AND PROBLEM SOLVING

Evaluate each expression for the given value of the variable.

19. $12d$ for $d = 0$
20. $x + 3.2$ for $x = 5$
21. $30 - n$ for $n = 8$
22. $5t + 5$ for $t = 1$
23. $2a - 5$ for $a = 7$
24. $3 + 5b$ for $b = 1.2$
25. $12 - 2m$ for $m = 3$
26. $3g + 8$ for $g = 14$
27. $x + 7.5$ for $x = 2.5$
28. $15 - 5y$ for $y = 3$
29. $4y + 2$ for $y = 3.5$
30. $2(z + 8)$ for $z = 5$

Evaluate each expression for $t = 0, x = 1.5, y = 6$, and $z = 23$.

31. $y + 5$
32. $2y + 7$
33. $z - 2x$
34. $3z - 3y$
35. $2z - 2y$
36. xy
37. $2.6y - 2x$
38. $1.2z - y$
39. $4(y - x)$
40. $3(4 + y)$
41. $4(2 + z) + 5$
42. $2(y - 6) + 3$
43. $3(6 + t) - 1$
44. $y(4 + t) - 5$
45. $x + y + z$
46. $10x + z - y$
47. $3y + 4(x + t)$
48. $3(z - 2t) + 1$
49. $7tyz$
50. $z - 2xy$

51. ***LIFE SCIENCE*** Measuring your heart rate is one way to check the intensity of exercise. Studies show that a person's maximum heart rate depends on his or her age. The expression $220 - a$ approximates a person's maximum heart rate in beats per minute, where a is the person's age. Find your maximum heart rate.

52. ***LIFE SCIENCE*** In the Karvonen Formula, a person's resting heart rate r, age a, and desired intensity I are used to find the number of beats per minute the person's heart rate should be during training.

$$\text{training heart rate (THR)} = I(220 - a - r) + r$$

What is the THR of a person who is 45 years old, and who has a resting heart rate of 85 and a desired intensity of 0.5?

53. ***ENTERTAINMENT*** There are 24 frames, or still shots, in one second of movie footage.
    a. Write an expression to determine the number of frames in a movie.
    b. Using the running time of *E.T. the Extra-Terrestrial,* determine how many frames are in the movie.

E.T. the Extra-Terrestrial (1982) has a running time of 115 minutes, or 6900 seconds.

54. ***CHOOSE A STRATEGY*** A baseball league has 192 players and 12 teams, with an equal number of players on each team. If the number of teams were reduced by four but the total number of players remained the same, there would be ________ players per team.

    A four more B eight fewer C four fewer D eight more

55. ***WRITE ABOUT IT*** A student says that for any value of x the expression $5x + 1$ will always give the same result as $1 + 5x$. Is the student correct? Explain.

56. ***CHALLENGE*** Can the expressions $2x$ and $x + 2$ ever have the same value? If so, what must the value of x be?

Spiral Review

Identify the odd number(s) in each list of numbers. (Previous course)

57. 15, 18, 22, 34, 21, 62, 71, 100

58. 101, 114, 122, 411, 117, 121

59. 4, 6, 8, 16, 18, 20, 49, 81, 32

60. 9, 15, 31, 47, 65, 93, 1, 3, 43

61. **EOG PREP** Which is *not* a multiple of 21? (Previous course)

A 7 C 42
B 21 D 105

62. **EOG PREP** Which is a factor of 12? (Previous course)

A 4 C 24
B 8 D 36

1-2 Write Algebraic Expressions

Problem Solving Skill

Learn to write algebraic expressions.

Each 30-second block of commercial time during Super Bowl XXXV cost an average of $2.2 million.

This information can be used to write an algebraic expression to determine how much a given number of 30-second blocks would have cost.

Eighty-three commercials aired during the 2002 Super Bowl.

	Word Phrases	Expression
+	• a number plus 5 • add 5 to a number • sum of a number and 5 • 5 more than a number • a number increased by 5	$n + 5$
−	• a number minus 11 • subtract 11 from a number • difference of a number and 11 • 11 less than a number • a number decreased by 11	$x - 11$
×	• 3 times a number • 3 multiplied by a number • product of 3 and a number	$3m$
÷	• a number divided by 7 • 7 divided into a number • quotient of a number and 7	$\frac{a}{7}$ or $a \div 7$

EXAMPLE 1 Translating Word Phrases into Math Expressions

Write an algebraic expression for each word phrase.

A a number *n* decreased by 11

n decreased by 11

$n \quad - \quad 11$

$n - 11$

Write an algebraic expression for each word phrase.

B **the quotient of 3 and a number h**

	quotient of	3 and h
3	$\div$	h

$\frac{3}{h}$

Helpful Hint

In Example 1C parentheses are not needed because multiplication is performed first by the order of operations.

C **1 more than the product of 12 and p**

1	more than	the	product of	12 and p
1	+	(12	$\cdot$	p)

$1 + 12p$

D **3 times the sum of q and 1**

3	times	the	sum of	q and 1
3	$\cdot$	(q	+	1)

$3(q + 1)$

To solve a word problem, you must first interpret the action you need to perform and then choose the correct operation for that action. When a word problem involves groups of equal size, use multiplication or division. Otherwise, use addition or subtraction. The table gives more information to help you decide which operation to use to solve a word problem.

Action	Operation	Possible Question Clues
Combine	Add	How many altogether?
Combine equal groups	Multiply	How many altogether?
Separate	Subtract	How many more? How many less?
Separate into equal groups	Divide	How many equal groups?

EXAMPLE 2 Interpreting Which Operation to Use in Word Problems

A **Monica got a 200-minute calling card and called her brother at college. After talking with him for t minutes, she had t less than 200 minutes remaining on her card. Write an expression to determine the number of minutes remaining on the calling card.**

$200 - t$ *Separate t minutes from the original 200.*

B **If Monica talked with her brother for 55 minutes, how many minutes does she have left on her calling card?**

$200 - 55 = 145$ *Evaluate the expression for $t = 55$.*

There are 145 minutes remaining on her calling card.

EXAMPLE 3 Writing and Evaluating Expressions in Word Problems

Write an algebraic expression to evaluate each word problem.

A **Rob and his friends buy a set of baseball season tickets. The 81 tickets are to be divided equally among p people. If he divides them among 9 people, how many tickets does each person get?**

$81 \div p$ — *Separate the tickets into p equal groups.*

$81 \div 9 = 9$ — *Evaluate for $p = 9$.*

Each person gets 9 tickets.

B **A company airs its 30-second commercial n times during Super Bowl XXXV at a cost of $2.2 million each time. What will the cost be if the commercial is aired 2, 3, 4, and 5 times?**

$2.2 million · n — *Combine n equal amounts of $2.2 million.*

$2.2n$ — *In millions of dollars*

n	$2.2n$	Cost
2	2.2(2)	$4.4 million
3	2.2(3)	$6.6 million
4	2.2(4)	$8.8 million
5	2.2(5)	$11 million

Evaluate for $n = 2, 3, 4,$ and 5.

Helpful Hint

Some word problems give more numbers than are necessary to find the answer. In Example 3B, 30 seconds describes the length of a commercial, and the number is not needed to solve the problem.

C **Before Benny took his road trip, his car odometer read 14,917 miles. After the trip, his odometer read m miles more than 14,917. If he traveled 633 miles on the trip, what did the odometer read after his trip?**

$14{,}917 + m$ — *Combine 14,917 miles and m miles.*

$14{,}917 + \mathbf{633} = 15{,}550$ — *Evaluate for $m = 633$.*

The odometer read 15,550 miles after the trip.

Think and Discuss

1. **Give** two words or phrases that can be used to express each operation: addition, subtraction, multiplication, and division.
2. **Express** $5 + 7n$ in words in at least two different ways.

1-2 Exercises

FOR EOG PRACTICE
see page 644

internet connect
Homework Help Online
go.hrw.com Keyword: MT4 1-2

5.01a

GUIDED PRACTICE

See Example 1 **Write an algebraic expression for each word phrase.**

1. the quotient of 6 and a number t

2. a number y decreased by 25

3. 7 times the sum of m and 6

4. the sum of 7 times m and 6

See Example 2

5. a. Carl walked n miles for charity at a rate of $8 per mile. Write an expression to find out how much money Carl raised.

b. How much money would Carl have raised if he had walked 23 miles?

See Example 3 **Write an algebraic expression to evaluate the word problem.**

6. Cheryl and her friends buy a pizza for $15.00 plus a delivery charge of d dollars. If the delivery charge is $2.50, what is the total cost?

INDEPENDENT PRACTICE

See Example 1 **Write an algebraic expression for each word phrase.**

7. a number k increased by 34

8. the quotient of 12 and a number h

9. 5 plus the product of 5 and z

10. 6 times the difference of x and 4

See Example 2

11. a. Mr. Gimble's class is going to a play. The 42 students will be seated equally among p rows. Write an expression to determine how many people will be seated in each row.

b. If there are 6 rows, how many students will be in each row?

See Example 3 **Write an algebraic expression and evaluate each word problem.**

12. Julie bought a card good for 35 visits to a health club and began a workout routine. After y visits, she had y fewer than 35 visits remaining on her card. After 18 visits, how many visits did she have left?

13. Myron bought n dozen eggs for $1.75 per dozen. If he bought 8 dozen eggs, how much did they cost?

PRACTICE AND PROBLEM SOLVING

Write an algebraic expression for each word phrase.

14. 7 more than a number y

15. 6 times the sum of 4 and y

16. 11 less than a number t

17. half the sum of m and 5

18. 9 more than the product of 6 and a number y

19. 6 less than the product of 13 and a number y

20. 2 less than a number m divided by 8

21. twice the quotient of a number m and 35

Translate each algebraic expression into words.

22. $4b - 3$ **23.** $t + 12$ **24.** $3(m + 4)$

25. ***ENTERTAINMENT*** Ron bought two comic books on sale. Each comic book was discounted $1 off the regular price r. Write an expression to find what Ron paid before taxes. If each comic book was regularly $2.50, what was the total cost before taxes?

26. ***SPORTS*** In basketball, players score 2 points for each field goal, 3 points for each three-point shot, and 1 point for each free throw made. Write an expression for the total score for a team that makes g field goals, t three-point shots, and f free throws. Find the total score for a team that scores 23 field goals, 6 three-pointers, and 11 free throws.

27. At age 2, a cat or dog is considered 24 "human" years old. Each year after age 2 is equivalent to 4 "human" years. Fill in the expression $[24 + \square(a - 2)]$ so that it represents the age of a cat or dog in human years. Copy the chart and use your expression to complete it.

Age	$24 + \square(a - 2)$	Age (human years)
2		
3		
4		
5		
6		

28. ***WHAT'S THE ERROR?*** A student says $3(n - 5)$ is equal to $3n - 5$. What's the error?

29. ***WRITE ABOUT IT*** Paul used addition to solve a word problem about the weekly cost of commuting by toll road for $1.50 each day. Fran solved the same problem by multiplying. They both had the correct answer. How is this possible?

30. ***CHALLENGE*** Write an expression for the sum of 1 and twice a number n. If you let n be any odd number, will the result always be an odd number?

Spiral Review

Find each sum, difference, product, or quotient. (Previous course)

31. $200 + 2$ **32.** $200 \div 2$ **33.** $200 \cdot 2$ **34.** $200 - 2$

35. $200 + 0.2$ **36.** $200 \div 0.2$ **37.** $200 \cdot 0.2$ **38.** $200 - 0.2$

39. **EOG PREP** Which is *not* a factor of 24? (Previous course)

A 8 C 24

B 12 D 48

40. **EOG PREP** Which is a multiple of 15? (Previous course)

A 1 C 5

B 3 D 15

1-3 Solving Equations by Adding or Subtracting

Learn to solve equations using addition and subtraction.

Vocabulary

- equation
- solve
- solution
- inverse operation
- isolate the variable
- Addition Property of Equality
- Subtraction Property of Equality

Mexico City is built on top of a large underground water source. Over the 100 years between 1900 and 2000, as the water was drained, the city sank as much as 30 feet in some areas.

If you know the altitude of Mexico City in 2000 was 7350 feet above sea level, you can use an *equation* to estimate the altitude in 1900.

In 1910, the Monumento a la Independencia was built at ground level. It now requires 23 steps to reach the base because the ground around the monument has sunk.

An **equation** uses an equal sign to show that two expressions are equal. All of these are equations.

$3 + 8 = 11$ $\quad r + 6 = 14$ $\quad 24 = x - 7$ $\quad 9n = 27$ $\quad \frac{100}{2} = 50$

To **solve** an equation that contains a variable, find the value of the variable that makes the equation true. This value of the variable is called the **solution** of the equation.

EXAMPLE 1 Determining Whether a Number Is a Solution of an Equation

Determine which value of x is a solution of the equation.

$x - 4 = 16; x = 12, 20, \text{ or } 21$

Substitute each value for x in the equation.

$x - 4 = 16$

$12 - 4 \stackrel{?}{=} 16$ *Substitute 12 for x.*

$8 \stackrel{?}{=} 16$ ✗

So 12 **is not** a solution.

$x - 4 = 16$

$20 - 4 \stackrel{?}{=} 16$ *Substitute 20 for x.*

$16 \stackrel{?}{=} 16$ ✓

So 20 **is** a solution.

$x - 4 = 16$

$21 - 4 \stackrel{?}{=} 16$ *Substitute 21 for x.*

$17 \stackrel{?}{=} 16$ ✗

So 21 **is not** a solution.

Helpful Hint

The phrase "subtraction 'undoes' addition" can be understood with this example:
If you start with 3 and add 4, you can get back to 3 by subtracting 4.

$$\begin{array}{r} 3+4 \\ -4 \\ \hline 3 \end{array}$$

Addition and subtraction are **inverse operations**, which means they "undo" each other. To solve an equation, use inverse operations to **isolate the variable**. In other words, get the variable alone on one side of the equal sign.

To solve a subtraction equation, like $y - 15 = 7$, you would use the **Addition Property of Equality**.

ADDITION PROPERTY OF EQUALITY

Words	Numbers	Algebra
You can add the same number to both sides of an equation, and the statement will still be true.	$\begin{array}{rr} 2+3= & 5 \\ +4 & +4 \\ \hline 2+7= & 9 \end{array}$	$x = y$ $x + z = y + z$

There is a similar property for solving addition equations, like $x + 9 = 11$. It is called the **Subtraction Property of Equality**.

SUBTRACTION PROPERTY OF EQUALITY

Words	Numbers	Algebra
You can subtract the same number from both sides of an equation, and the statement will still be true.	$\begin{array}{rr} 4+7= & 11 \\ -3 & -3 \\ \hline 4+4= & 8 \end{array}$	$x = y$ $x - z = y - z$

EXAMPLE 2 Solving Equations Using Addition and Subtraction Properties

Solve.

A $3 + t = 11$

$$\begin{array}{rcl} 3 + t &=& 11 \\ \underline{-3} & & \underline{-3} \\ 0 + t &=& 8 \\ t &=& 8 \end{array}$$

Subtract 3 from both sides.

Identity Property of Zero: 0 + t = t

Check

$$3 + t = 11$$
$$3 + \mathbf{8} \stackrel{?}{=} 11 \quad \textit{Substitute 8 for t.}$$
$$11 \stackrel{?}{=} 11 \checkmark$$

Solve.

B $m - 7 = 11$

$$\begin{aligned} m - 7 &= 11 \\ +7 \quad & \quad +7 \\ m + 0 &= 18 \\ m &= 18 \end{aligned}$$

Add 7 to both sides.

C $15 = w + 14$

$$\begin{aligned} 15 &= w + 14 \\ 15 - 14 &= w + 14 - 14 \\ 1 &= w + 0 \\ 1 &= w \\ w &= 1 \end{aligned}$$

Subtract 14 from both sides.

Definition of Equality

EXAMPLE 3 *Geography Applications*

A **The altitude of Mexico City in 2000 was about 7350 ft above sea level. What was the approximate altitude of Mexico City in 1900 if it sank 30 ft during the 100-year period?**

beginning altitude	$-$	altitude sank	$=$	altitude in 2000
Solve: x	$-$	30	$=$	7350

$$\begin{aligned} x - 30 &= 7350 \\ +30 \quad & \quad +30 \\ x + 0 &= 7380 \\ x &= 7380 \end{aligned}$$

Add 30 to both sides.

In 1900, Mexico City was at an altitude of 7380 ft.

B **From 1954 to 1999, shifting plates increased the height of Mount Everest from 29,028 ft to 29,035 ft. By how many feet did Mount Everest's altitude increase during the 45-year period?**

Solve: $29{,}028 \text{ ft} + h = 29{,}035 \text{ ft}$

$$\begin{aligned} 29{,}028 + h &= 29{,}035 \\ -29{,}028 \quad & \quad -29{,}028 \\ 0 + h &= 7 \\ h &= 7 \end{aligned}$$

Subtract 29,028 from both sides.

Mount Everest's altitude increased 7 ft between 1954 and 1999.

Geography LINK

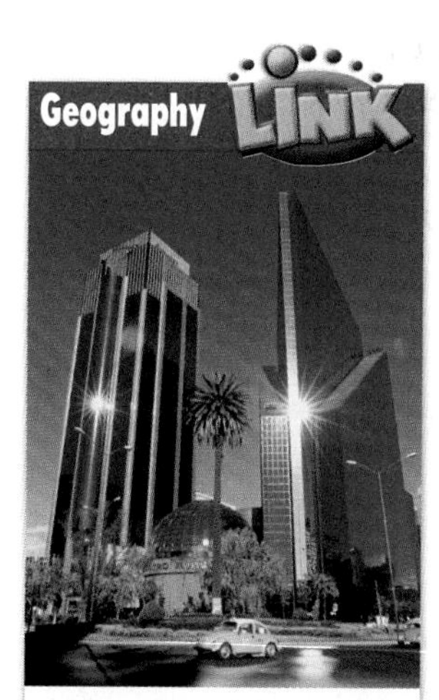

Mexico City, above, sank 19 inches in one year while Venice, Italy, possibly the most famous sinking city, has sunk only 9 inches in the last century.

go.hrw.com
KEYWORD: MT4 Sinking

CNN Student News

Think and Discuss

1. **Explain** whether you would use addition or subtraction to solve $x - 9 = 25$.
2. **Explain** what it means to isolate the variable.

1-3 Exercises

FOR EOG PRACTICE

see page 644

internet connect

Homework Help Online

go.hrw.com Keyword: MT4 1-3

go.hrw.com

5.03, 5.04

GUIDED PRACTICE

See Example 1

Determine which value of x is a solution of each equation.

1. $x + 9 = 14$; $x = 2, 5,$ or 23

2. $x - 7 = 14$; $x = 2, 7,$ or 21

See Example 2

Solve.

3. $m - 9 = 23$

4. $8 + t = 13$

5. $13 = w - 4$

See Example 3

6. At what altitude did a climbing team start if it descended 3600 feet to a camp at an altitude of 12,035 feet?

INDEPENDENT PRACTICE

See Example 1

Determine which value of x is a solution of each equation.

7. $x - 14 = 8$; $x = 6, 22,$ or 32

8. $x + 7 = 35$; $x = 5, 28,$ or 42

See Example 2

Solve.

9. $9 = w + 8$

10. $m - 11 = 33$

11. $4 + t = 16$

See Example 3

12. If a team camps at an altitude of 18,450 feet, how far must it ascend to reach the summit of Mount Everest at an altitude of 29,035 feet?

PRACTICE AND PROBLEM SOLVING

Determine which value of the variable is a solution of the equation.

13. $d + 4 = 24$; $d = 6, 20,$ or 28

14. $m - 2 = 13$; $m = 11, 15,$ or 16

15. $y - 7 = 23$; $y = 30, 26,$ or 16

16. $k + 3 = 4$; $k = 1, 7,$ or 17

17. $12 + n = 19$; $n = 7, 26,$ or 31

18. $z - 15 = 15$; $z = 0, 15,$ or 30

19. $x + 48 = 48$; $x = 0, 48,$ or 96

20. $p - 2.5 = 6$; $p = 3.1, 3.5,$ or 8.5

Solve the equation and check the solution.

21. $7 + t = 12$

22. $h - 21 = 52$

23. $15 = m - 9$

24. $m - 5 = 10$

25. $h + 8 = 11$

26. $6 + t = 14$

27. $1785 = t - 836$

28. $m + 35 = 172$

29. $x - 29 = 81$

30. $p + 8 = 23$

31. $n - 14 = 31$

32. $20 = 8 + w$

33. $0.8 + t = 1.3$

34. $5.7 = c - 2.8$

35. $9.87 = w + 7.97$

36. ***SOCIAL STUDIES*** In 1990, the population of Cheyenne, Wyoming, was 73,142. By 2000, the population had increased to 81,607. Write and solve an equation to find n, the increase in Cheyenne's population from 1990 to 2000.

37. ***SOCIAL STUDIES*** In 1804, explorers Lewis and Clark began their journey to the Pacific Ocean at the mouth of the Missouri River. Use the map to determine the following distances.

a. from Blackbird Hill, Nebraska, to Great Falls, Montana

b. from the meeting point, or confluence, of the Missouri and Yellowstone Rivers to Great Falls, Montana

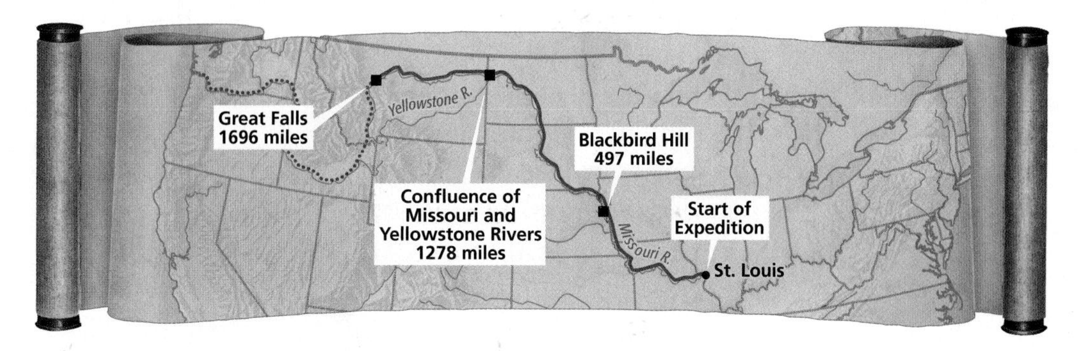

38. ***SOCIAL STUDIES*** The United States flag had 15 stars in 1795. How many stars have been added since then to make our present-day flag with 50 stars? Write and solve an equation to find s, the number of stars that have been added to the United States flag since 1795.

39. ***ENTERTAINMENT*** Use the bar graph about movie admission costs to write and solve an equation for each of the following.

a. Find c, the increase in cost of a movie ticket from 1940 to 1990.

b. The cost c of a movie ticket in 1950 was \$3.82 less than in 1995. Find the cost of a movie ticket in 1995.

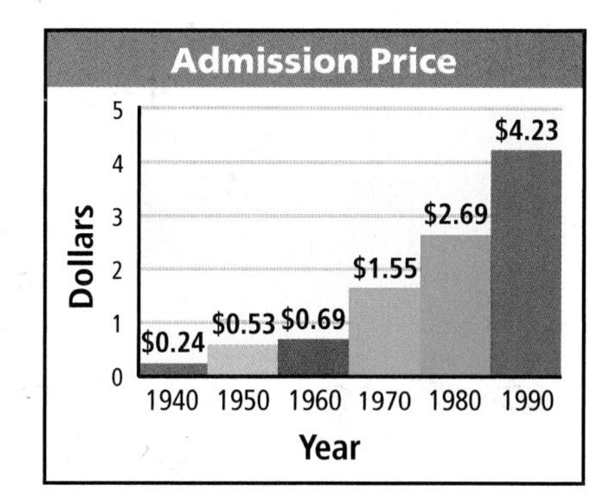

40. ***WRITE A PROBLEM*** Write a subtraction problem using the graph about admission costs. Explain your solution.

41. ***WRITE ABOUT IT*** Write a set of rules to use when solving addition and subtraction equations.

42. ***CHALLENGE*** Explain how you could solve for h in the equation $14 - h = 8$ using algebra. Then find the value of h.

Spiral Review

Evaluate each expression for the given value of the variable. (Lesson 1-1)

43. $x + 9$ for $x = 13$

44. $x - 8$ for $x = 18$

45. $14 + x$ for $x = 12$

46. **EOG PREP** Which is "3 times the difference of y and 4"? (Lesson 1-2)

A $3 \cdot y - 4$ B $3 \cdot (y + 4)$ C $3 \cdot (y - 4)$ D $3 - (y - 4)$

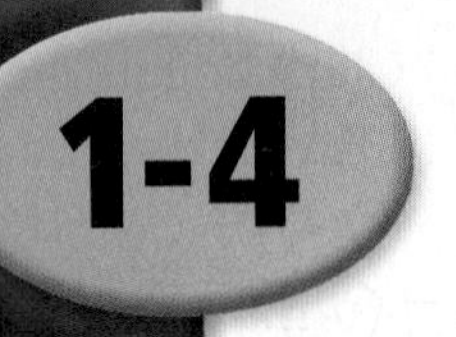

1-4 Solving Equations by Multiplying or Dividing

Learn to solve equations using multiplication and division.

Vocabulary

Division Property of Equality

Multiplication Property of Equality

In 1912, Wilbur Scoville invented a way to measure the hotness of chili peppers. The unit of measurement became known as the Scoville unit.

You can use Scoville units to write and solve multiplication equations for substituting one kind of pepper for another in a recipe.

A Thai pepper is 90,000 Scoville units. This means it takes 90,000 cups of sugar water to neutralize the hotness of one cup of Thai peppers.

You can solve a multiplication equation using the **Division Property of Equality**.

DIVISION PROPERTY OF EQUALITY

Words	Numbers	Algebra
You can divide both sides of an equation by the same nonzero number, and the statement will still be true.	$4 \cdot 3 = 12$ $\frac{4 \cdot 3}{2} = \frac{12}{2}$ $\frac{12}{2} = 6$	$x = y$ $\frac{x}{z} = \frac{y}{z}$

EXAMPLE 1 Solving Equations Using Division

Solve $7x = 35$.

$7x = 35$

$\frac{7x}{7} = \frac{35}{7}$ *Divide both sides by 7.*

$1x = 5$ *$1 \cdot x = x$*

$x = 5$

Check

$7x = 35$

$7(5) \stackrel{?}{=} 35$ *Substitute 5 for x.*

$35 \stackrel{?}{=} 35$ ✔

Remember!

Multiplication and division are inverse operations.

$\frac{8 \cdot 3}{3} = 8$

You can solve division equations using the **Multiplication Property of Equality**.

MULTIPLICATION PROPERTY OF EQUALITY		
Words	**Numbers**	**Algebra**
You can multiply both sides of an equation by the same number, and the statement will still be true.	$2 \cdot 3 = 6$ $4 \cdot 2 \cdot 3 = 4 \cdot 6$ $8 \cdot 3 = 24$	$x = y$ $zx = zy$

EXAMPLE 2 **Solving Equations Using Multiplication**

Solve $\frac{h}{3} = 6$.

$\frac{h}{3} = 6$

$3 \cdot \frac{h}{3} = 3 \cdot 6$ *Multiply both sides by 3.*

$h = 18$

EXAMPLE 3 ***Food Application***

A recipe calls for 1 tabasco pepper, but Jennifer wants to use jalapeño peppers. How many jalapeño peppers should she substitute in the dish to equal the Scoville units of 1 tabasco pepper?

Scoville Units of Selected Peppers	
Pepper	**Scoville Units**
Ancho (Poblano)	1,500
Bell	100
Cayenne	30,000
Habanero	360,000
Jalapeño	5,000
Serrano	10,000
Tabasco	30,000
Thai	90,000

Scoville units of 1 jalapeño	$\cdot$	number of jalapeños	$=$	Scoville units of 1 tabasco
5000	$\cdot$	n	$=$	30,000

$5{,}000n = 30{,}000$ *Write the equation.*

$\frac{5{,}000n}{5{,}000} = \frac{30{,}000}{5{,}000}$ *Divide both sides by 5000.*

$n = 6$

Six jalapeños are about as hot as one tabasco pepper. Jennifer should substitute 6 jalapeños for the tabasco pepper in her recipe.

EXAMPLE 4 *Money Application*

Helene's band needs money to go to a national competition. So far, band members have raised $560, which is only one-third of what they need. What is the total amount needed?

fraction of total amount raised so far	·	total amount needed	=	amount raised so far
$\frac{1}{3}$	·	x	=	$560

$\frac{1}{3}x = 560$ *Write the equation.*

$3 \cdot \frac{1}{3}x = 3 \cdot 560$ *Multiply both sides by 3.*

$x = 1680$

The band needs to raise a total of $1680.

Sometimes it is necessary to solve equations by using two inverse operations. For instance, the equation $6x - 2 = 10$ has multiplication and subtraction.

Variable term

Multiplication → $6x - 2 = 10$

Subtraction

To solve this equation, add to isolate the term with the variable in it. Then divide to solve.

EXAMPLE 5 Solving a Simple Two-Step Equation

Solve $2x + 1 = 7$.

Step 1:

$$\begin{aligned} 2x + 1 &= 7 \\ -1 &= -1 \\ \hline 2x &= 6 \end{aligned}$$

Subtract 1 from both sides to isolate the term with x in it.

Step 2:

$\frac{2x}{2} = \frac{6}{2}$ *Divide both sides by 2.*

$x = 3$

Think and Discuss

1. **Explain** what property you would use to solve $\frac{k}{2.5} = 6$.
2. **Give** the equation you would solve to figure out how many ancho peppers are as hot as one cayenne pepper.

1-4 Exercises

FOR EOG PRACTICE

see page 645

internet connect

Homework Help Online
go.hrw.com Keyword: MT4 1-4

5.03, 5.04

GUIDED PRACTICE

See Example 1

Solve.

1. $4x = 28$
2. $7t = 49$
3. $3y = 42$
4. $2w = 26$

See Example 2

5. $\frac{l}{15} = 4$
6. $\frac{k}{8} = 9$
7. $\frac{h}{19} = 3$
8. $\frac{m}{6} = 1$

See Example 3

9. One serving of milk contains 8 grams of protein, and one serving of steak contains 32 grams of protein. Write and solve an equation to find the number of servings of milk n needed to get the same amount of protein as there is in one serving of steak.

See Example 4

10. Gary needs to buy a suit to go to a formal dance. Using a coupon, he can save \$60, which is only one-fourth of the cost of the suit. Write and solve an equation to determine the cost c of the suit.

See Example 5

Solve.

11. $3x + 2 = 23$
12. $\frac{k}{5} - 1 = 7$
13. $3y - 8 = 1$
14. $\frac{m}{6} + 4 = 10$

INDEPENDENT PRACTICE

See Example 1

Solve.

15. $3d = 57$
16. $7x = 105$
17. $4g = 40$
18. $16y = 112$

See Example 2

19. $\frac{n}{9} = 63$
20. $\frac{h}{27} = 2$
21. $\frac{a}{6} = 102$
22. $\frac{j}{8} = 12$

See Example 3

23. An orange contains about 80 milligrams of vitamin C, which is 10 times as much as an apple contains. Write and solve an equation to find n, the number of milligrams of vitamin C in an apple.

See Example 4

24. Fred gathered 150 eggs on his family's farm today. This is one-third the number he usually gathers. Write and solve an equation to determine the number n that he usually gathers.

See Example 5

Solve.

25. $6x - 5 = 7$
26. $\frac{n}{3} - 4 = 1$
27. $2y + 5 = 9$
28. $\frac{h}{7} + 2 = 2$

PRACTICE AND PROBLEM SOLVING

Solve.

29. $2x = 14$
30. $4y = 80$
31. $6y = 12$
32. $9m = 9$
33. $\frac{k}{8} = 7$
34. $\frac{1}{5}x = 121$
35. $\frac{b}{6} = 12$
36. $\frac{n}{15} = 1$
37. $3x = 51$
38. $15g = 75$
39. $16y + 18 = 66$
40. $3z - 14 = 58$
41. $\frac{b}{4} = 12$
42. $\frac{m}{24} = 24$
43. $\frac{n}{5} - 3 = 4$
44. $\frac{a}{2} + 8 = 14$

Social Studies LINK

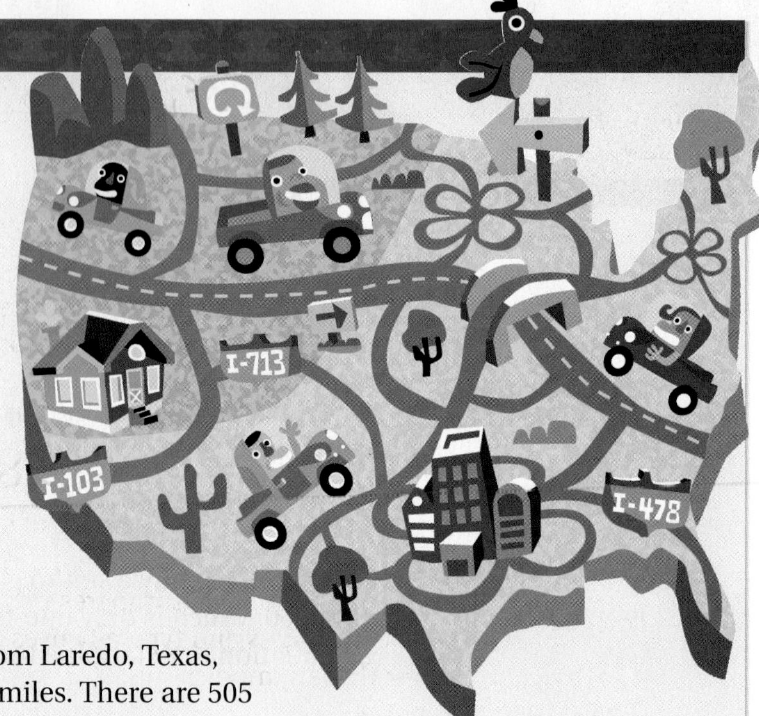

In 1956, during President Eisenhower's term, construction began on the United States interstate highway system. The original plan was for 42,000 miles of highways to be completed within 16 years. It actually took 37 years to complete. The last part, Interstate 105 in Los Angeles, was completed in 1993.

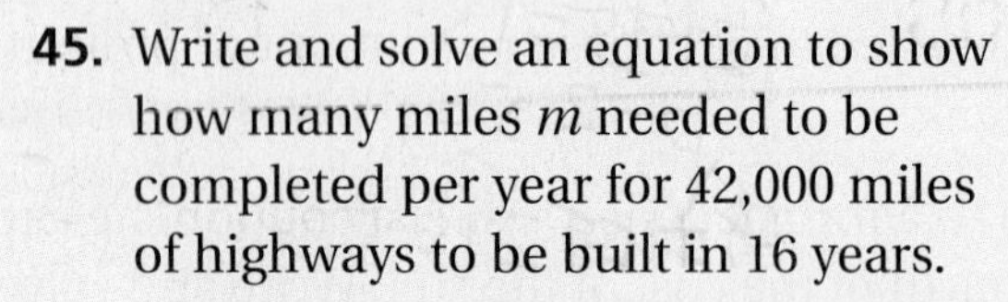

45. Write and solve an equation to show how many miles m needed to be completed per year for 42,000 miles of highways to be built in 16 years.

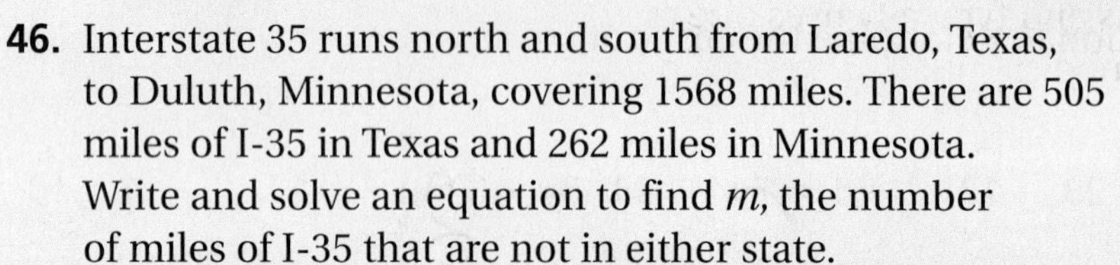

46. Interstate 35 runs north and south from Laredo, Texas, to Duluth, Minnesota, covering 1568 miles. There are 505 miles of I-35 in Texas and 262 miles in Minnesota. Write and solve an equation to find m, the number of miles of I-35 that are not in either state.

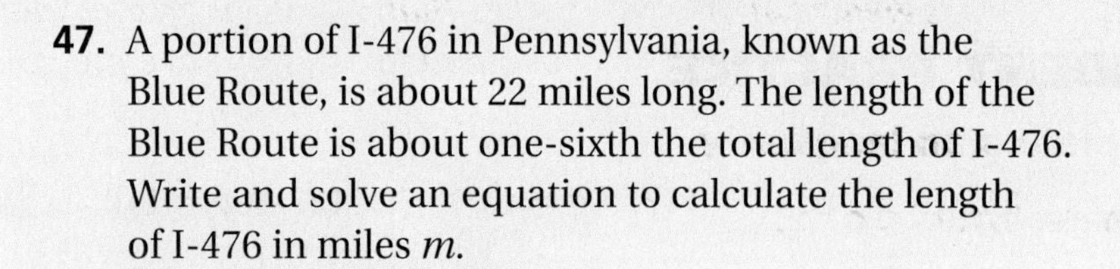

47. A portion of I-476 in Pennsylvania, known as the Blue Route, is about 22 miles long. The length of the Blue Route is about one-sixth the total length of I-476. Write and solve an equation to calculate the length of I-476 in miles m.

48. ***CHALLENGE*** Interstate 80 extends from California to New Jersey. At right are the number of miles of Interstate 80 in each state the highway passes through.

a. __?__ has 134 more miles than __?__.

b. __?__ has 174 fewer miles than __?__.

Number of I-80 Miles

State	Miles
California	195 mi
Nevada	410 mi
Utah	197 mi
Wyoming	401 mi
Nebraska	455 mi
Iowa	301 mi
Illinois	163 mi
Indiana	167 mi
Ohio	236 mi
Pennsylvania	314 mi
New Jersey	68 mi

Spiral Review

Solve. (Lesson 1-3)

49. $3 + x = 11$

50. $y - 6 = 8$

51. $13 = w + 11$

52. $5.6 = b - 4$

53. **EOG PREP** Which is the prime factorization of 72? (Previous Course)

A $3 \cdot 3 \cdot 2 \cdot 2 \cdot 2$

B $3^3 \cdot 2^2$

C $3 \cdot 2 \cdot 2 \cdot 6$

D $3^2 \cdot 4 \cdot 2$

54. **EOG PREP** What is the value of the expression $3x + 4$ for $x = 2$? (Lesson 1-1)

A 4

B 6

C 9

D 10

1-5 Solving Simple Inequalities

Learn to solve and graph inequalities.

Vocabulary

- **inequality**
- **algebraic inequality**
- **solution of an inequality**
- **solution set**

Laid end to end, the paper used by personal computer printers each year would circle the earth *more than* 800 times.

number of times around the earth	> 800

An **inequality** compares two quantities and typically uses one of these symbols:

$<$ *is less than*

is greater than

is less than or equal to

is greater than or equal to

EXAMPLE 1 Completing an Inequality

Compare. Write $<$ or $>$.

A $12 - 7 \;\blacksquare\; 6$

$5 \;\blacksquare\; 6$

$5 < 6$

B $3(8) \;\blacksquare\; 16$

$24 \;\blacksquare\; 16$

$24 > 16$

Helpful Hint

An open circle means that the corresponding value is not a solution. A solid circle means that the value is part of the solution set.

An inequality that contains a variable is an **algebraic inequality**. A number that makes an inequality true is a **solution of the inequality**.

The set of all solutions is called the **solution set**. The solution set can be shown by graphing it on a number line.

Word Phrase	Inequality	Sample Solutions	Solution Set
x is less than 5	$x < 5$	$x = 4$ $4 < 5$ $x = 2.1$ $2.1 < 5$	0 1 2 3 4 5 6 7
a is greater than 0 a is more than 0	$a > 0$	$a = 7$ $7 > 0$ $a = 25$ $25 > 0$	−3 −2 −1 0 1 2 3
y is less than or equal to 2 y is at most 2	$y \leq 2$	$y = 0$ $0 \leq 2$ $y = 1.5$ $1.5 \leq 2$	−3 −2 −1 0 1 2 3 4 5
m is greater than or equal to 3 m is at least 3	$m \geq 3$	$m = 17$ $17 \geq 3$ $m = 3$ $3 \geq 3$	−1 0 1 2 3 4 5 6

Most inequalities can be solved the same way equations are solved. Use inverse operations on both sides of the inequality to isolate the variable. (There are special rules when multiplying or dividing by a negative number, which you will learn in the next chapter.)

EXAMPLE 2 Solving and Graphing Inequalities

Solve and graph each inequality.

> **Remember!**
> The inequality symbol opens to the side with the greater number.
> $2 < 10$

A $x + 7.5 < 10$

$\underline{-7.5 \quad -7.5}$ *Subtract 7.5 from both sides.*

$x < 2.5$

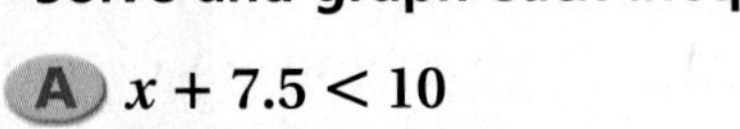

According to the graph, 2.4 should be a solution, since $2.4 < 2.5$, and 3 should not be a solution because $3 > 2.5$.

Check $x + 7.5 < 10$

$2.4 + 7.5 \overset{?}{<} 10$ *Substitute 2.4 for x.*

$9.9 \overset{?}{<} 10$ ✔

So 2.4 is a solution.

Check $x + 7.5 < 10$

$3 + 7.5 \overset{?}{<} 10$ *Substitute 3 for x.*

$10.5 \overset{?}{<} 10$ ✘

And 3 is not a solution.

B $6n \geq 18$

$\frac{6n}{6} \geq \frac{18}{6}$ *Divide both sides by 6.*

$n \geq 3$

C $t - 3 \leq 22$

$\underline{+3 \quad +3}$ *Add 3 to both sides.*

$t \leq 25$

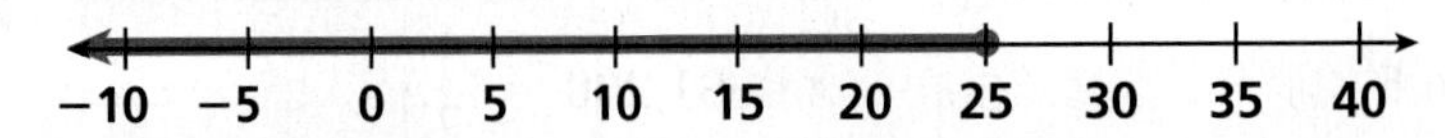

D $5 > \frac{w}{2}$

$2 \cdot 5 > 2 \cdot \frac{w}{2}$ *Multiply both sides by 2.*

$10 > w$ *$10 > w$ is the same as $w < 10$.*

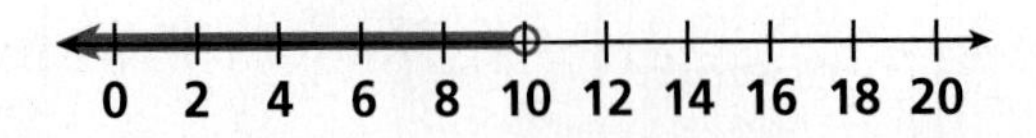

EXAMPLE 3 PROBLEM SOLVING APPLICATION

If all of the sheets of paper used by personal computer printers each year were laid end to end, they would circle the earth more than 800 times. The earth's circumference is about 25,120 mi (1,591,603,200 in.), and one sheet of paper is 11 in. long. How many sheets of paper are used each year?

1 Understand the Problem

The **answer** is the number of sheets of paper used by personal computer printers in one year. **List the important information:**

- The amount of paper would circle the earth *more than* 800 times.
- Once around the earth is 1,591,603,200 in.
- One sheet of paper is 11 in. long.

Show the relationship of the information:

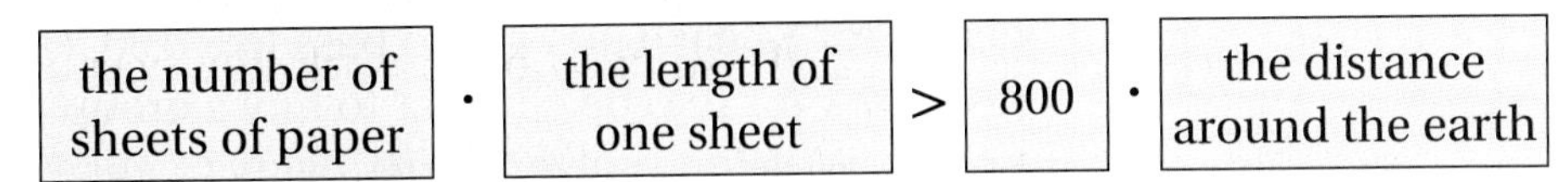

the number of sheets of paper
·
the length of one sheet
>
800
·
the distance around the earth

2 Make a Plan

Use the relationship to *write an inequality.* Let x represent the number of sheets of paper.

$$\boxed{x} \cdot \boxed{11 \text{ in.}} > \boxed{800} \cdot \boxed{1{,}591{,}603{,}200 \text{ in.}}$$

3 Solve

$11x > 800 \cdot 1{,}591{,}603{,}200$

$11x > 1{,}273{,}282{,}560{,}000$ *Multiply.*

$\frac{11x}{11} > \frac{1{,}273{,}282{,}560{,}000}{11}$ *Divide both sides by 11.*

$x > 115{,}752{,}960{,}000$

More than 115,752,960,000 sheets of paper are used by personal computer printers in one year.

4 Look Back

To circle the earth once takes $\frac{1{,}591{,}603{,}200}{11} = 144{,}691{,}200$ sheets of paper; to circle it 800 times would take $800 \cdot 144{,}691{,}200 = 115{,}752{,}960{,}000$ sheets.

Think and Discuss

1. **Give** all the symbols that make $5 + 8$ ▢ 13 true. Explain.
2. **Explain** which symbols make $3x$ ▢ 9 false if $x = 3$.

1-5 Exercises

FOR EOG PRACTICE see page 645

internet connect
Homework Help Online
go.hrw.com Keyword: MT4 1-5

1.01b, 5.03, 5.04

GUIDED PRACTICE

See Example 1 **Compare. Write < or >.**

1. $4 + 8 \blacksquare 13$ **2.** $4(2) \blacksquare 7$ **3.** $27 - 13 \blacksquare 11$ **4.** $5(9) \blacksquare 42$

5. $9 + 2 \blacksquare 10$ **6.** $3(8) \blacksquare 27$ **7.** $52 - 37 \blacksquare 14$ **8.** $8(7) \blacksquare 54$

See Example 2 **Solve and graph each inequality.**

9. $x + 3 < 4$ **10.** $4b \geq 20$ **11.** $m - 4 \leq 28$ **12.** $5 > \frac{x}{3}$

13. $y + 8 \geq 25$ **14.** $6f < 30$ **15.** $z - 8 > 13$ **16.** $7 \leq \frac{x}{2}$

See Example 3 **17.** For a field trip to the museum, the science club can purchase individual tickets for $4 each or a group pass for $160. How many club members are necessary for it to be cheaper to buy a group pass than to buy individual tickets? Write and solve an inequality to answer the question.

INDEPENDENT PRACTICE

See Example 1 **Compare. Write < or >.**

18. $4 + 7 \blacksquare 12$ **19.** $6(4) \blacksquare 25$ **20.** $15 - 9 \blacksquare 4$ **21.** $7(6) \blacksquare 40$

22. $13 + 5 \blacksquare 17$ **23.** $5(2.3) \blacksquare 12$ **24.** $7 \blacksquare 19 - 13$ **25.** $12 \blacksquare 3(4.2)$

See Example 2 **Solve and graph each inequality.**

26. $b + 4 < 8$ **27.** $7x \geq 49$ **28.** $h - 2 \geq 3$ **29.** $1 < \frac{t}{4}$

30. $6 + a > 9$ **31.** $3x \geq 12$ **32.** $f - 9 \leq 2$ **33.** $2 < \frac{a}{3}$

See Example 3 **34.** There are 88 keys on a new piano. If there are 12 pianos in a room of broken pianos, some of which may have missing keys, how many piano keys could be on the pianos? Write and solve an inequality to answer the question.

PRACTICE AND PROBLEM SOLVING

Write the inequality shown by each graph.

35. −4 −2 0 2 4 6 8

36. 0 2 4 6 8 10 12

37. −4 −2 0 2 4 6 8

38. −4 −2 0 2 4 6 8

39. −6 −4 −2 0 2 4 6

40.
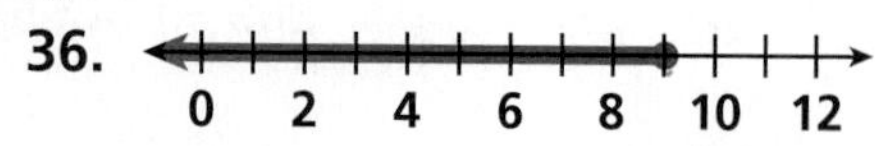

41. −4 −2 0 2 4 6 8

42.
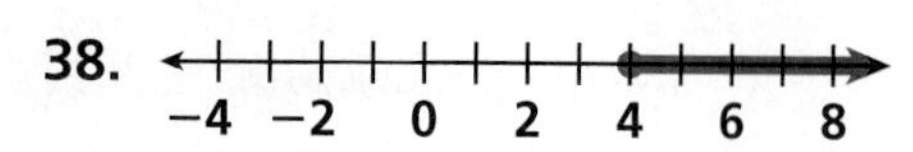

Sports LINK

The BT Global Challenge 2000 began on September 10, 2000, and ended June 30, 2001. Of that almost 10-month period, the crews spent about 177 days, 19 hours, and 20 minutes at sea.

43. Reginald's cement truck can carry up to 2200 pounds of cargo. He needs to haul 50 bags of cement that weigh 50 pounds each. Write and solve an inequality to determine whether Reginald will be able to carry all of the cement in one trip.

44. ***SPORTS*** There were 7 legs of the BT Global Challenge 2000 yacht race. If the crew of the winning boat, the *LG Flatron*, sailed at a rate of at least 6 knots (6 nautical miles per hour) continuously, how many hours would it have taken them to sail the leg from Cape Town, South Africa, to La Rochelle, France?

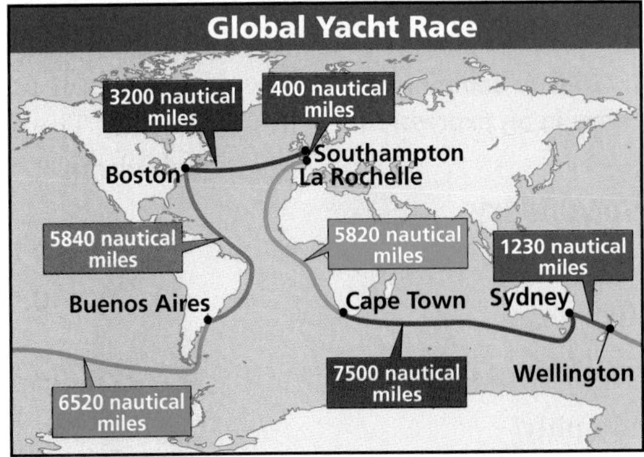

45. Suly earned an 87 on her first test. She needs a total of 140 points on her first two tests to pass the class. What score must Suly make on her second test to ensure that she passes the class?

46. ***BUSINESS*** A rule of thumb for electronic signs is that a sign with letters n inches tall is readable from up to $50n$ feet away. How tall should the letters be to be readable from 900 feet away?

47. ***WRITE A PROBLEM*** The weight limit for an elevator is 2500 pounds. Write, solve, and graph a problem about the elevator and the number of 185-pound passengers it can safely carry.

48. ***WRITE ABOUT IT*** In mathematics, the conventional way to write an inequality is with the variable on the left, such as $x > 5$. Explain how to rewrite the inequality $4 \leq x$ in the conventional manner.

49. ***CHALLENGE*** $3 \leq x < 5$ means both $3 \leq x$ and $x < 5$ are true at the same time. Solve and graph $6 < x \leq 12$.

Spiral Review

Evaluate each expression for the given values of the variable. (Lesson 1-1)

50. $2(4 + x) - 3$ for $x = 0, 1, 2, 3$

51. $3(8 - x) - 2$ for $x = 0, 1, 2, 3$

52. $5(x - 1) - 1$ for $x = 5, 6, 7, 8$

53. $4(x + 2) - 3$ for $x = 2, 4, 6, 8$

54. $3(7 + x) + 4$ for $x = 2, 4, 6, 8$

55. $2(9 - x) + 3$ for $x = 3, 4, 5, 6$

56. **EOG PREP** A company prints n books at a cost of \$9 per book. What is the total cost of the books? (Lesson 1-2)

A $\$9 - n$

B $n + \$9$

C $\frac{n}{\$9}$

D $\$9n$

57. **EOG PREP** Which value of x is the solution of the equation $x - 5 = 8$? (Lesson 1-3)

A 3

B 11

C 13

D 15

1-6 Combining Like Terms

Learn to combine like terms in an expression.

Vocabulary

term

like term

equivalent expression

simplify

The district choir festival combines choirs from all three high schools in the district. The festival director has received the following rosters from each choir.

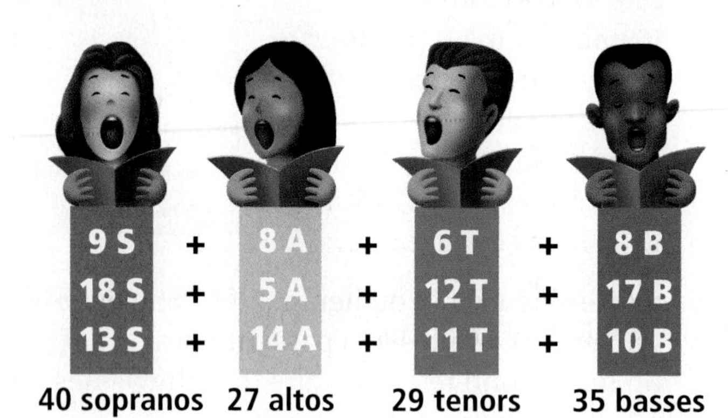

To find the total number in each section, the director groups together like parts from each school. Students from different schools who sing in the same section are similar to *like terms* in an expression.

Helpful Hint

Constants such as 4, 0.75, and 11 are like terms because none of them have a variable.

Terms in an expression are separated by plus or minus signs.

$$7x + 5 - 3y + 2x$$

Like terms can be grouped together because they have the same variable raised to the same power. Often, like terms have different coefficients. When you combine like terms, you change the way an expression looks, but not the value of the expression. **Equivalent expressions** have the same value for all values of the variables.

EXAMPLE 1 Combining Like Terms to Simplify

Combine like terms.

A $5x + 3x$ — *Identify like terms.*

$8x$ — *Combine coefficients: $5 + 3 = 8$*

B $5m - 2m + 8 - 3m + 6$ — *Identify like terms.*

$0m + 14$ — *Combine coefficients: $5 - 2 - 3 = 0$ and $8 + 6 = 14$*

14

EXAMPLE 2 Combining Like Terms in Two-Variable Expressions

Combine like terms.

A $6a + 8a + 4b + 7$

$\boxed{6a} + \boxed{8a} + (4b) + 7$ — *Identify like terms.*

$14a + 4b + 7$ — *Combine coefficients: 6 + 8 = 14*

B $k + 3n - 2n + 4k$

$\boxed{1k} + (3n) - (2n) + \boxed{4k}$ — *Identify like terms; the coefficient of k is 1, because 1k = k.*

$5k + n$ — *Combine coefficients.*

C $4f - 12g + 16$

$\boxed{4f} - (12g) + 16$ — *No like terms*

To **simplify** an expression, perform all possible operations, including combining like terms.

EXAMPLE 3 Simplifying Algebraic Expressions by Combining Like Terms

Remember!

The Distributive Property states that $a(b + c) = ab + ac$ for all a, b, and c. For instance, $2(3 + 5) = 2(3) + 2(5)$.

Simplify $4(y + 9) - 3y$.

$4(y + 9) - 3y$

$4(y) + 4(9) - 3y$ — *Distributive Property*

$4y + 36 - 3y$ — *4y and 3y are like terms.*

$1y + 36$ — *Combine coefficients: 4 − 3 = 1*

$y + 36$

EXAMPLE 4 Solving Algebraic Equations by Combining Like Terms

Solve $8x - x = 112$.

$8x - x = 112$ — *Identify like terms. The coefficient of x is 1.*

$7x = 112$ — *Combine coefficients: 8 − 1 = 7*

$\frac{7x}{7} = \frac{112}{7}$ — *Divide both sides by 7.*

$x = 16$

Think and Discuss

1. **Describe** the first step in simplifying the expression $2 + 8(3y + 5) - y$.
2. **Tell** how many sets of like terms are in the expression in Example 1B. What are they?
3. **Explain** why $8x + 8y + 8$ is already simplified.

1-6 Exercises

FOR EOG PRACTICE

see page 645

internet connect

Homework Help Online
go.hrw.com Keyword: MT4 1-6

5.0?, 5.04

GUIDED PRACTICE

See Example 1 **Combine like terms.**

1. $7x - 3x$
2. $2z + 5 + 3z$
3. $4f + 2 - 2f + 6 + 6f$
4. $9g + 8g$
5. $5p - 8 - p$
6. $2x + 7 - x + 5 + 3x$

See Example 2

7. $4x + 3y - x + 2y$
8. $5x + 2y - y + 4x$
9. $3x + 4y + 2x - 3y$
10. $7p + 2p + 5z - 2z$
11. $7g + 5h - 12$
12. $2h + 3m + 8h - 3m$

See Example 3 **Simplify.**

13. $3(r + 2) - 2r$
14. $5(2 + x) + 3x$
15. $7(t + 8) - 5t$

See Example 4 **Solve.**

16. $4n - 2n = 84$
17. $y + 3y = 96$
18. $5p - 2p = 51$

INDEPENDENT PRACTICE

See Example 1 **Combine like terms.**

19. $8y + 5y$
20. $5z - 6 - 3z$
21. $2a + 4 - a + 7 + 6a$
22. $4z - z$
23. $8x + 2 - 5x$
24. $9b + 6 - 3b - 3 - b$
25. $12p - 7p$
26. $7a + 8 - 3a$
27. $2x + 8 + 2x - 5 + 5x$

See Example 2

28. $2z + 5z + b - 7$
29. $4a + a + 3z - 2z$
30. $9x + 8y + 2x - 8 - 4y$
31. $5x + 3 + 2x + 5q$
32. $7d - d + 3e + 12$
33. $15a + 6c + 3 - 6a + c$

See Example 3 **Simplify.**

34. $5(y + 2) - y$
35. $3(4y - 6) + 8y$
36. $4(x + 8) + 9x$
37. $2(3y + 4) + 9$
38. $6(2x + 8) - 9x$
39. $3(3x - 3) + 2x$

See Example 4 **Solve.**

40. $5x - x = 48$
41. $8p - 3p = 25$
42. $p + 2p = 18$
43. $3y + 5y = 64$
44. $a + 5a = 72$
45. $9x - 5x = 56$

PRACTICE AND PROBLEM SOLVING

Simplify.

46. $7(3l + 5k) - 14l + 12$
47. $6d + 8 + 5d - 3d - 7$

Solve.

48. $13(g + 2) = 78$
49. $7x - 12 = x + 2 + 2x - 3x$

Write and simplify an expression for each situation.

50. ***BUSINESS*** A museum charges $5 for each adult ticket, plus an additional $1 per ticket for tax. What is the total cost of x tickets?

51. ***SPORTS*** Use the information below to find how many medals of each kind were won by the four countries in the 2000 Summer Olympics.

United States	Great Britain	Brazil	Lithuania
39 Gold	11 Gold	0 Gold	2 Gold
25 Silver	10 Silver	6 Silver	0 Silver
33 Bronze	7 Bronze	6 Bronze	3 Bronze

Write and solve an equation for each situation.

52. ***BUSINESS*** The accounting department ordered 12 cases of paper, and the marketing department ordered 20 cases of paper. If the total cost of the combined order was $896 before taxes, what is the price of each case of paper?

53. ***WHAT'S THE ERROR?*** A student said that $2x + 3y$ can be simplified to $5xy$ by combining like terms. What error did the student make?

54. ***WRITE ABOUT IT*** Write an expression that can be simplified by combining like terms. Then write an expression that cannot be simplified, and explain why it is already in simplest form.

55. ***CHALLENGE*** Simplify and solve $2(7x + 5 - 3x) + 4(2x - 2) = 50$.

Spiral Review

Solve each equation. (Lesson 1-3)

56. $4 + x = 13$

57. $x - 4 = 9$

58. $17 = x + 9$

59. $19 = x + 11$

60. $5 + x = 22$

61. $x - 24 = 8$

62. $x - 7 = 31$

63. $41 = x + 25$

64. $x + 8 = 15$

65. **EOG PREP** Determine which value of x is a solution of the equation $3x + 2 = 11$. (Lesson 1-3)

A $x = 2.2$

B $x = 3$

C $x = 3.6$

D $x = 4.3$

66. **EOG PREP** Determine which value of x is a solution of the equation $4x - 3 = 13$. (Lesson 1-3)

A $x = 2.5$

B $x = 3$

C $x = 3.5$

D $x = 4$

Chapter 1

Mid-Chapter Quiz

Mid-Chapter Quiz

LESSON 1-1 (pp. 4–7)

Evaluate each expression for the given values of the variables.

1. $4x + 7y$ for $x = 7$ and $y = 5$
2. $5(r - 8t)$ for $r = 100$ and $t = 4$
3. $2(3m + 7n)$ for $m = 13$ and $n = 8$

LESSON 1-2 (pp. 8–12)

Write an algebraic expression for each word phrase.

4. 12 more than twice a number n
5. 5 less than 3 times a number b
6. 6 times the sum of p and 3
7. 10 plus the product of 16 and m

Write an algebraic expression to represent the problem situation.

8. Sami has a calendar with 365 pages of cartoons. After she tears off p pages, how many pages of cartoons remain?

LESSON 1-3 (pp. 13–17)

Solve.

9. $5 + x = 26$
10. $p - 8 = 16$
11. $32 = h + 21$
12. $60 = k - 33$

Write and solve an algebraic equation for the word problem.

13. The deepest location in Lake Superior is 1333 feet, which is 1123 feet deeper than the deepest location in Lake Erie. What is the deepest location in Lake Erie?

LESSON 1-4 (pp. 18–22)

Solve.

14. $4m = 88$
15. $\frac{w}{50} = 50$
16. $100y = 50$
17. $\frac{1}{2}x = 16$
18. $3x + 4 = 10$
19. $4z - 1 = 11$
20. $\frac{1}{3}y - 2 = 7$
21. $16 = 10 + 2m$

LESSON 1-5 (pp. 23–27)

Solve and graph each inequality.

22. $x + 2.3 < 12$
23. $3n > 15$
24. $y - 4.1 \geq 3$
25. $6 \leq \frac{z}{2}$

LESSON 1-6 (pp. 28–31)

Solve.

26. $7y - 4y = 6$
27. $\frac{5x + 3x}{2} = 20$
28. $2(t + 5t) = 48$

Focus on Problem Solving

Solve

- **Choose an operation: Addition or Subtraction**

To decide whether to add or subtract, you need to determine what action is taking place in the problem. If you are combining numbers or putting numbers together, you need to add. If you are taking away or finding out how far apart two numbers are, you need to subtract.

Action	Operation	Illustration
Combining or putting together	Add	
Removing or taking away	Subtract	
Finding the difference	Subtract	

Jan has 10 red marbles. Joe gives her 3 more. How many marbles does Jan have now? The action is combining marbles. Add 10 and 3.

Determine the action in each problem. Use the actions to restate the problem. Then give the operation that must be used to solve the problem.

1. The state of Michigan is made up of two parts, the Lower Peninsula and the Upper Peninsula. The Upper Peninsula has an area of about 16,400 mi^2, and the Lower Peninsula has an area of about 40,400 mi^2. Estimate the area of the state.

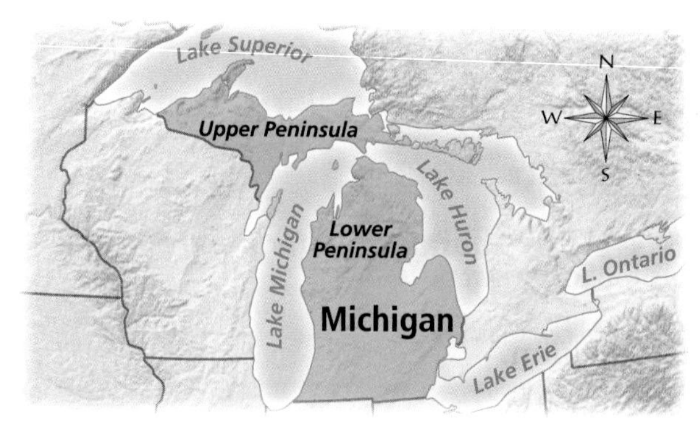

2. The average temperature in Homer, Alaska, is 53.4°F in July and 24.3°F in December. Find the difference between the average temperature in Homer in July and in December.

3. Einar has $18 to spend on his friend's birthday presents. He buys one present that costs $12.35. How much does he have left to spend?

4. Dinah got 87 points on her first test and 93 points on her second test. What is her combined point total for the first two tests?

Create a Table of Solutions

Use with Lesson 1-8

The *Table* feature on a graphing calculator can help you make a table of values quickly.

Activity

1 Make a table of solutions of the equation $y = 2x - 3$. Then find the value of y when $x = 29$.

To enter the equation, press the Y= key. Then press 2 X,T,θ,n − 3.

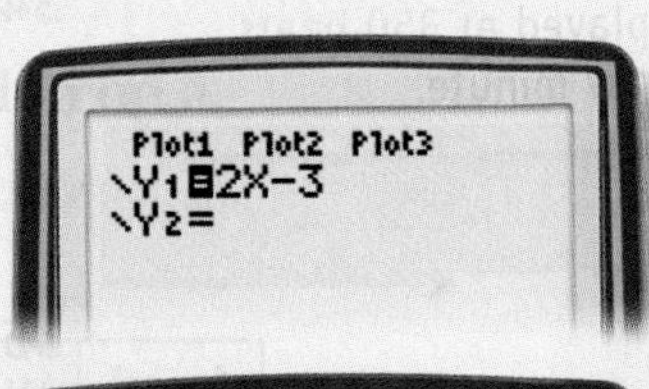

Press 2nd WINDOW (TBLSET) to go to the Table Setup menu. In this menu, **TblStart** shows the starting x-value, and **ΔTbl** shows how the x-values increase. If you need to change these values, use the arrow keys to highlight the number you want to change and then type a new number.

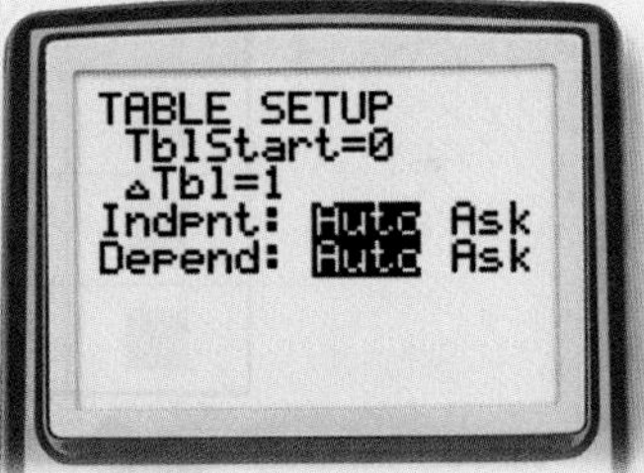

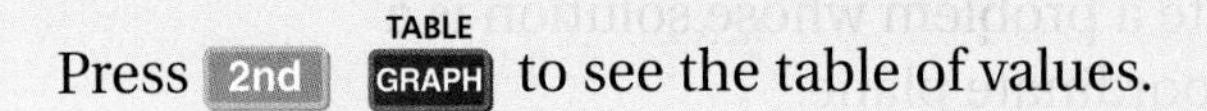

Press 2nd GRAPH (TABLE) to see the table of values.

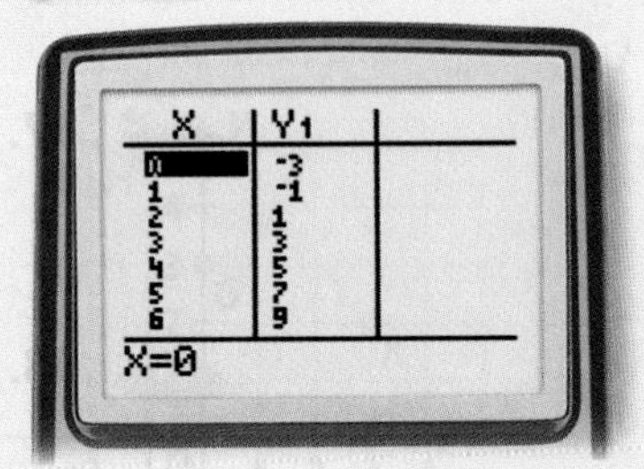

On this screen, you can see that $y = 7$ when $x = 5$.

Use the arrow keys to scroll down the list. You can see that $y = 55$ when $x = 29$.

To check, substitute 29 into $y = 2x - 3$.

$y = 2x - 3$
$= 2(29) - 3 = 58 - 3 = 55$

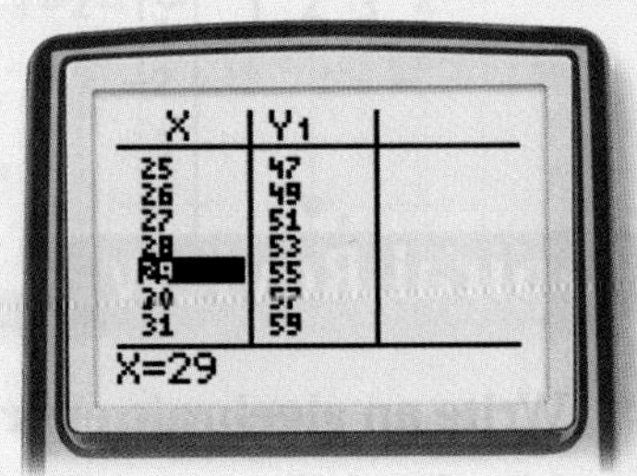

Think and Discuss

1. On an Internet site, pencils can be purchased for 17¢ each, but they only come in boxes of 12. You decide to make a table to compare x, the number of pencils, to y, the total cost of the pencils. What **TblStart** and **ΔTbl** values will you use? Explain.

Try This

For each equation, use a table to find the y-values for the given x-values. Give the **TblStart** and **ΔTbl** values you used.

1. $y = 3x + 6$ for $x = 1, 3,$ and 7 **2.** $y = \frac{x}{4}$ for $x = 5, 10, 15,$ and 20

1-9 Interpreting Graphs and Tables

Learn **to interpret information given in a graph or table and to make a graph to solve problems.**

The table below shows how quickly the temperature can increase in a car that is left parked during an afternoon of errands when the outside temperature is 93°F.

Location	Temperature on Arrival	Temperature on Departure
Home	—	140° at 1:05
Cleaners	75° at 1:15	95° at 1:25
Mall	72° at 1:45	165° at 3:45
Market	80° at 4:00	125° at 4:20

EXAMPLE 1

Matching Situations to Tables

The table gives the speeds of three dogs in mi/h at given times. Tell which dog corresponds to each situation described below.

Time	12:00	12:01	12:02	12:03	12:04
Dog 1	8	8	20	3	0
Dog 2	0	10	0	7	0
Dog 3	0	4	4	0	12

A **David's dog chews on a toy, then runs to the backyard, then sits and barks, and then runs back to the toy and sits.**

Dog 2—The dog's speed is 0 to start, while he sits and barks, and when he gets back to the toy. It is positive while he is running.

B **Kareem's dog runs with him and then chases a cat until Kareem calls for him to come back. The dog returns to his side and sits.**

Dog 1—The dog is running at the start, so his speed is positive. His speed increases while he chases the cat and then decreases to 0 when he sits.

C **Janelle's dog sits on top of a pool slide, slides into the swimming pool, and swims to the ladder. He gets out of the pool and shakes and then runs around the pool.**

Dog 3—The dog's speed is 0 at the top of the slide, 4 while swimming, and 12 while he runs around the pool.

EXAMPLE 2 Matching Situations to Graphs

Tell which graph corresponds to each situation described in Example 1.

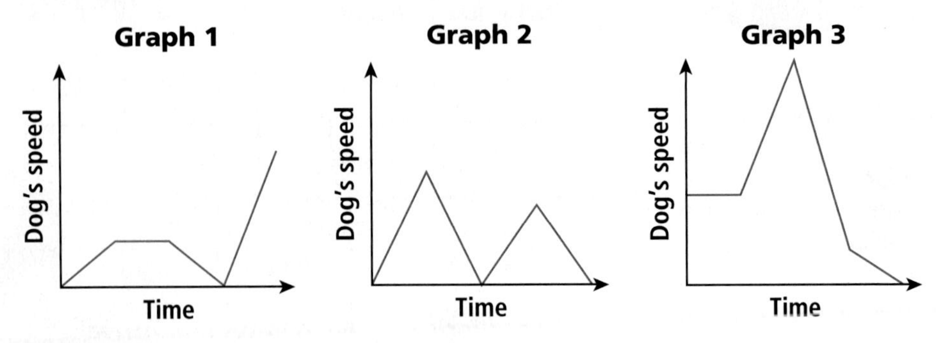

A **David's dog**

Graph 2—The dog's speed is 0 when the graph is on the x-axis.

B **Kareem's dog**

Graph 3—The dog's speed is not 0 when the graph starts.

C **Janelle's dog**

Graph 1—The dog is running at the end, so his speed is not 0.

EXAMPLE 3 Creating a Graph of a Situation

The temperature inside a car can get dangerously high. Create a graph that illustrates the temperature inside a car.

Location	Temperature (°F)	
	On Arrival	On Departure
Home	—	140° at 1:00
Cleaners	75° at 1:10	95° at 1:20
Mall	72° at 1:40	165° at 3:40
Market	80° at 3:55	125° at 4:15

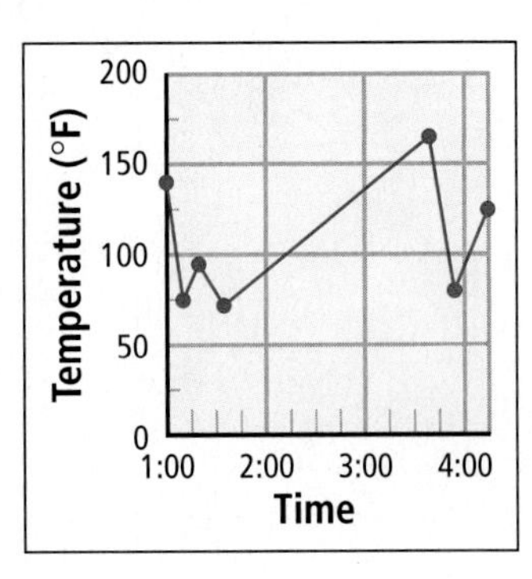

Think and Discuss

1. **Describe** what it means when a graph of speed starts at (0, 0).
2. **Give** a situation that, when graphed, would include a horizontal segment.

1-9 Exercises

4.01

FOR EOG PRACTICE see page 647

internet connect
Homework Help Online
go.hrw.com Keyword: MT4 1-9

GUIDED PRACTICE

See Example 1

1. Tell which table corresponds to the situation described below.

Jerry rides his bike to the end of the street and then rides quickly down a steep hill. At the bottom of the hill, Jerry stops to talk to Ryan. After a few minutes, Jerry rides over to Reggie's house and stops.

Table 1

Time	Speed (mi/h)
3:00	0
3:05	8
3:10	0
3:15	5
3:20	3

Table 2

Time	Speed (mi/h)
3:00	5
3:05	12
3:10	0
3:15	5
3:20	0

Table 3

Time	Speed (mi/h)
3:00	6
3:05	3
3:10	2
3:15	0
3:20	5

See Example 

2. Tell which table from Exercise 1, if any, corresponds to each graph.

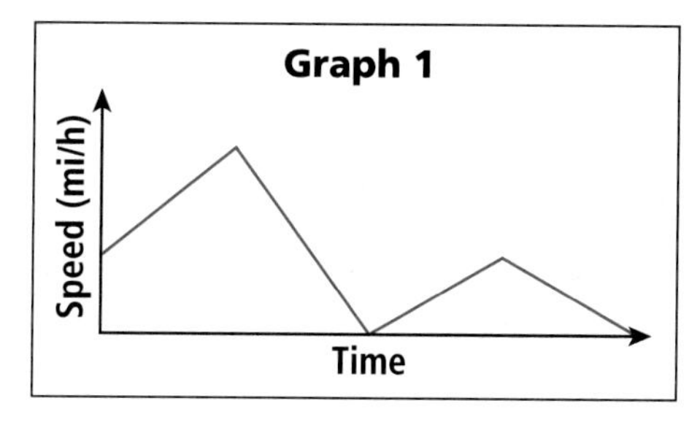

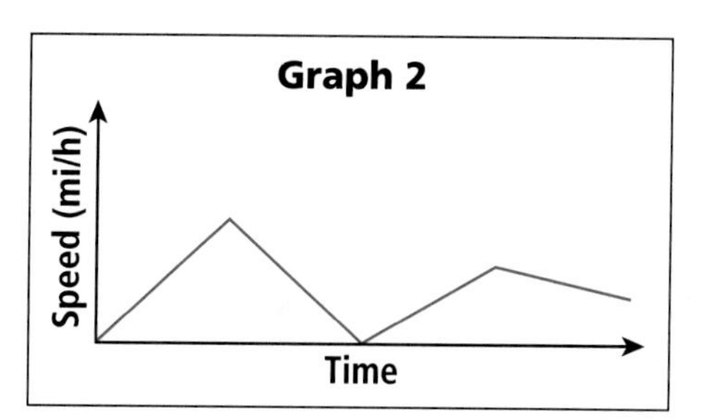

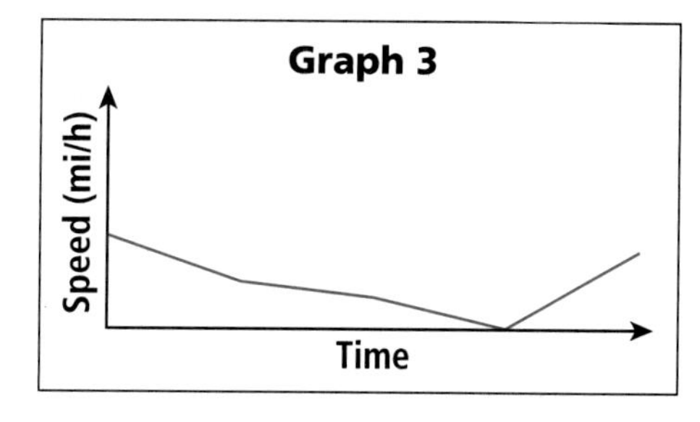

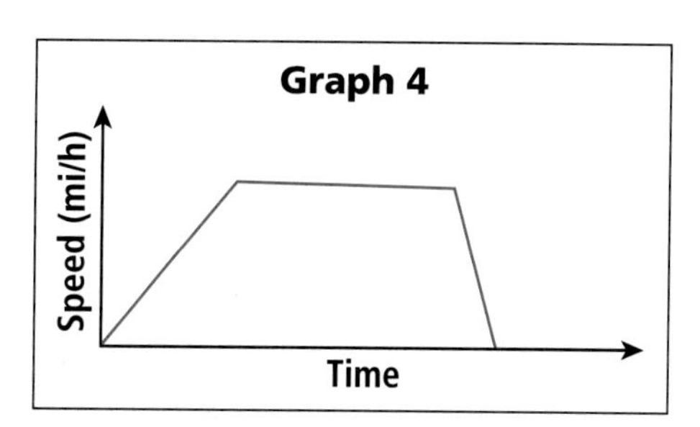

See Example 

3. Create a graph that illustrates the information in the table about a ride at an amusement park.

Time	3:20	3:21	3:22	3:23	3:24	3:25
Speed (mi/h)	0	14	41	62	8	0

INDEPENDENT PRACTICE

See Example 1

4. **Tell which table corresponds to the situation.**

An airplane sits at the gate while the passengers get on. Then the airplane taxis away from the gate and out to the runway. The plane waits at the end of the runway for clearance to take off. Then the plane takes off and continues to accelerate as it ascends.

Table 1

Time	Speed (mi/h)
6:00	0
6:10	20
6:20	40
6:30	0
6:40	80

Table 2

Time	Speed (mi/h)
6:00	20
6:10	0
6:20	10
6:30	80
6:40	300

Table 3

Time	Speed (mi/h)
6:00	0
6:10	10
6:20	0
6:30	80
6:40	350

See Example 2

5. **Tell which graph corresponds to each table described in Exercise 4.**

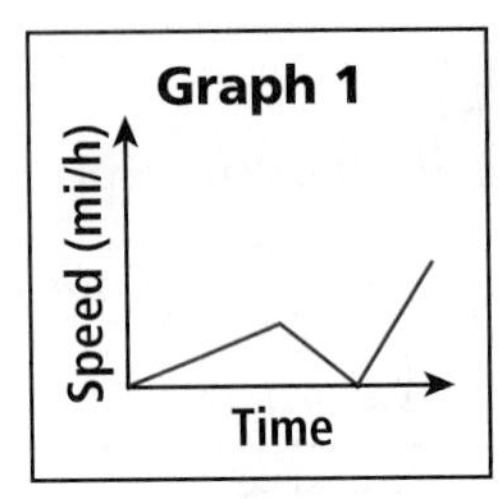

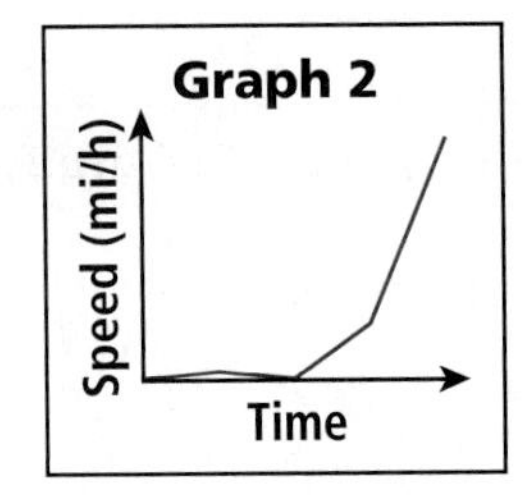

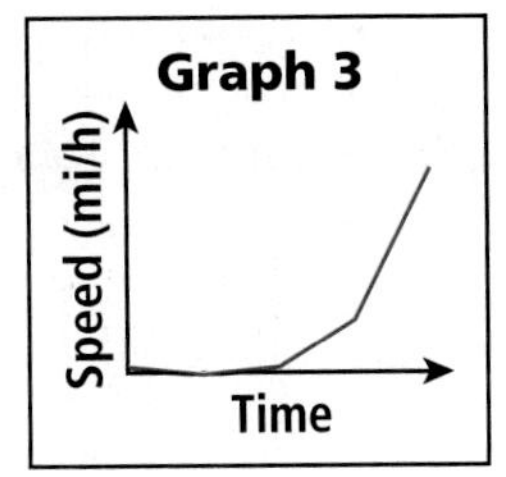

See Example 3

6. **Create a graph that illustrates the information in the table about Mr. Schwartz's commute from work to home.**

Time	Speed (mi/h)	Time	Speed (mi/h)
5:12	7	5:15	46
5:13	35	5:16	12
5:14	8	5:17	0

PRACTICE AND PROBLEM SOLVING

7. **Use the table to graph the movement of an electronic security gate.**

Time (s)	0	5	10	15	20	25	30	35
Gate Opening (ft)	0	3	6	9	12	12	12	9

Time (s)	40	45	50	55	60	65	70	75
Gate Opening (ft)	8	12	12	12	9	6	3	0

Earth Science LINK

Geyser is an Icelandic word meaning "to gush or rush forth." Geysers erupt because underground water begins to boil. Pressure builds as the temperature rises until the geyser erupts as a fountain of steam and water.

Old Faithful is the most famous geyser at Yellowstone National Park.

go.hrw.com
KEYWORD: MT4 Geyser
CNN Student News

8. Explain what the data tells about Beehive geyser. Make a graph.

Average Water Height of Beehive Geyser								
Time	1:00	1:01	1:02	1:03	1:04	1:05	1:06	1:07
Average Height (ft)	0	147	153	155	152	148	0	0

9. Use the chart to choose the correct geyser name to label each graph.

Yellowstone National Park Geysers				
Geyser Name	Old Faithful	Grand	Riverside	Pink Cone
Duration (min)	1.5 to 5	10	20	80

a.

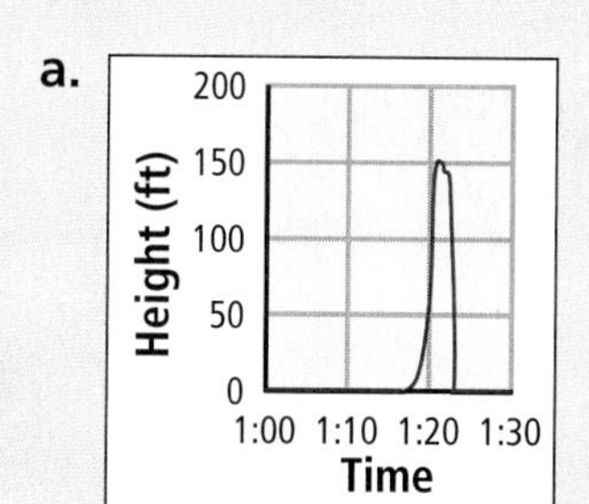

b.

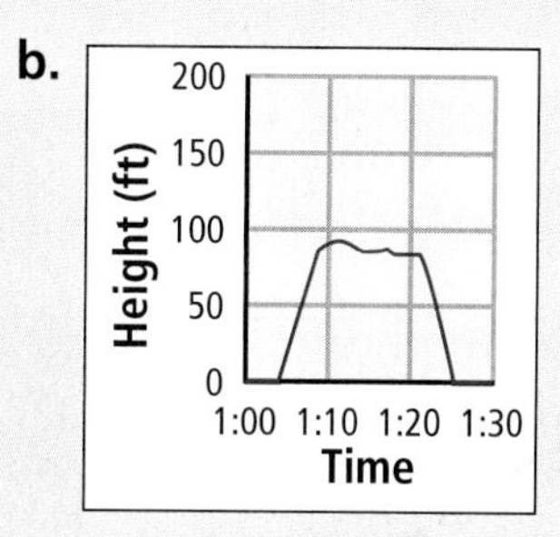

10. ***CHALLENGE*** Old Faithful erupts to heights between 105 ft and 184 ft. It erupted at 7:34 A.M. for 4.5 minutes. Later it erupted for 2.5 minutes. It then erupted a third time for 3 minutes. Use the table to determine how many minutes followed each of the three eruptions. Sketch a possible graph.

Old Faithful Eruption Information	
Duration	Time Until Next Eruption
1.5 min	48 min
2 min	55 min
2.5 min	70 min
3 min	72 min
3.5 min	74 min
4 min	82 min
4.5 min	93 min
5 min	100 min

Spiral Review

Solve. (Lesson 1-3)

11. $4 + x = 13$
12. $13 = 9 + x$
13. $x - 9 = 2$
14. $x - 2 = 5$

15. **EOG PREP** What is the solution of $x + 7 < 15$? (Lesson 1-5)

A $x > 22$ C $x < 22$
B $x > 8$ D $x < 8$

16. **EOG PREP** What is the solution of $4 \leq \frac{x}{2}$? (Lesson 1-5)

A $x \geq 8$ C $x \leq 8$
B $x \geq 2$ D $x \leq 2$

Problem Solving on Location

NORTH CAROLINA

Mile High Swinging Bridge

At an elevation of 5964 feet, Calloway Peak, on Grandfather Mountain, is the highest point in the Blue Ridge Mountains and home to the famous Mile High Swinging Bridge. This suspension bridge is 228 feet in length and offers a stunning view of the surrounding mountains.

1. The depth of the chasm beneath the Mile High Swinging Bridge is 148 feet less than the length of the bridge. How many feet deep is the chasm?

2. The elevation of the Mile High Swinging Bridge is n feet more than one mile (5280 ft). If the elevation of the Mile High Swinging Bridge is 5305 ft, write and solve an equation to find the value of n.

3. The swinging bridge was rebuilt in 1999, more than 45 years after its original construction. Write, solve, and graph an inequality to determine the possible years during which the bridge was originally constructed.

4. The cost R to rebuild the bridge was 20 times the original cost c in thousands of dollars.
   - **a.** Write an equation that represents the cost of rebuilding the bridge.
   - **b.** Make a table of ordered pair solutions to your equation from part **a**, for $c = 1, 2, 3, 4$, and 5.
   - **c.** Graph your equation from part **a** on a coordinate plane.
   - **d.** It cost \$300,000 to rebuild the bridge. What was the original cost?

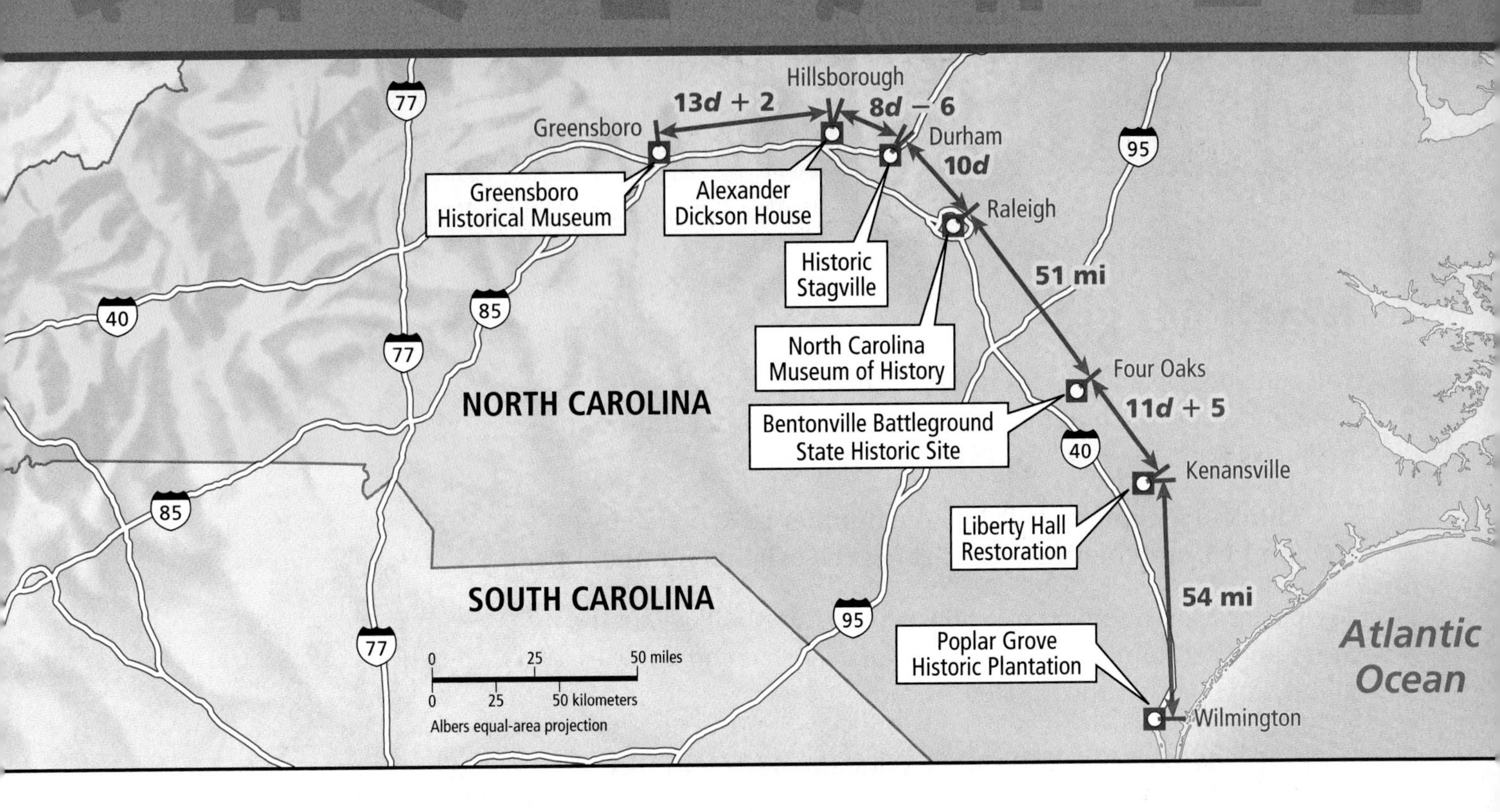

Civil War Memorials on I-40

When traveling in North Carolina, you can follow any of the seven trails of historic civil war sites. The trail that follows Interstate 40 includes stops at a total of 17 historical sites.

For 1–3, use the map.

1. Write an expression that represents the total distance along the I-40 trail from the Greensboro Historical Museum to the Poplar Grove Historic Plantation. Simplify your expression by combining like terms.

2. The distance from the North Carolina Museum of History to the Bentonville Battleground plus d miles is equal to the distance from the Liberty Hall Restoration to the Poplar Grove Historic Plantation, where d is the variable from the map. Write and solve an equation to find the value of d.

3. Use the value of d found in problem **2** to find the following distances.
   a. from the Alexander Dickson House to Historic Stagville
   b. from Historic Stagville to the North Carolina Museum of History
   c. from the Bentonville Battleground to the Liberty Hall Restoration
   d. from the Greensboro Historical Museum to the Poplar Grove Historic Plantation

MATH-ABLES

Math Magic

You can guess what your friends are thinking by learning to "operate" your way into their minds! For example, try this math magic trick.

> **Think of a number. Multiply the number by 8, divide by 2, add 5, and then subtract 4 times the original number.**

No matter what number you choose, the answer will always be 5. Try another number and see. You can use what you know about variables to prove it. Here's how:

	What you say:	What the person thinks:	What the math is:
Step 1:	Pick any number.	6 (for example)	n
Step 2:	Multiply by 8.	$8(6) = 48$	$8n$
Step 3:	Divide by 2.	$48 \div 2 = 24$	$8n \div 2 = 4n$
Step 4:	Add 5.	$24 + 5 = 29$	$4n + 5$
Step 5:	Subtract 4 times the original number.	$29 - 4(6) = 29 - 24 = 5$	$4n + 5 - 4n = 5$

Invent your own math magic trick that has at least five steps. Show an example using numbers and variables. Try it on a friend!

Crazy Cubes

This game, called The Great Tantalizer around 1900, was reintroduced in the 1960s as "Instant Insanity™." The goal is to line up four cubes so that each row of faces has four different sides showing. Make four cubes with paper and tape, numbering each side as shown.

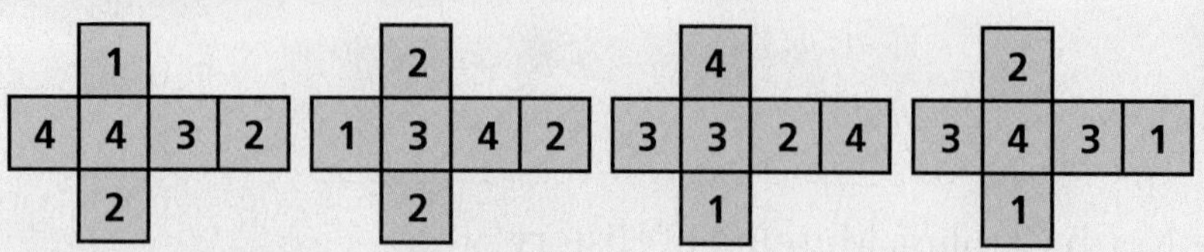

Line up the cubes so that 1, 2, 3, and 4 can be seen along the top, bottom, front, and back of the row of cubes. They can be in any order, and the numbers do not have to be right-side up.

Graph Points

On a graphing calculator, the WINDOW menu settings determine which points you see and the spacing between those points. In the standard viewing window, the x- and y-values each go from -10 to 10, and the tick marks are one unit apart. The boundaries are set by **Xmin, Xmax, Ymin**, and **Ymax. Xscl** and **Yscl** give the distance between the tick marks.

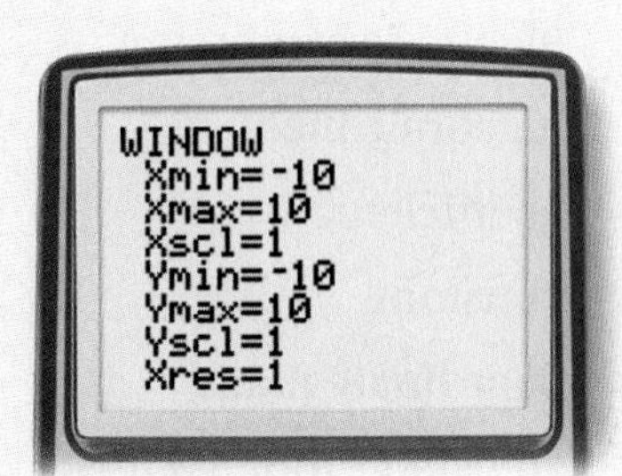

Activity

1 Plot the points (2, 5), $(-2, 3)$, $(-\frac{3}{2}, 4)$, and $(1.75, -2)$ in the standard window. Then change the minimum and maximum x- and y-values of the window to -5 and 5.

Press WINDOW to check that you have the standard window settings.

To plot (2, 5), press 2nd PRGM (DRAW) **POINTS** ENTER.

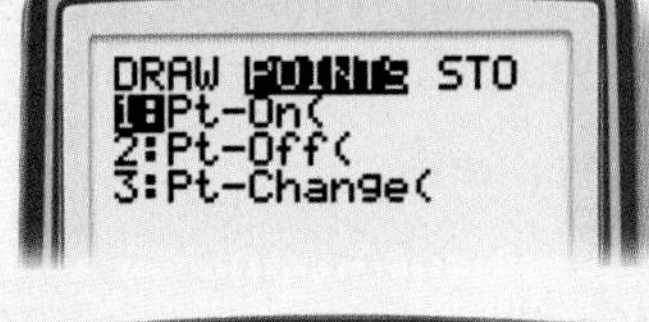

Then press 2 , 5 ENTER. After you see the grid with a point at (2, 5), press 2nd MODE (QUIT) to quit. Repeat the steps above to graph $(-2, 3)$, $(-\frac{3}{2}, 4)$, and $(1.75, -2)$.

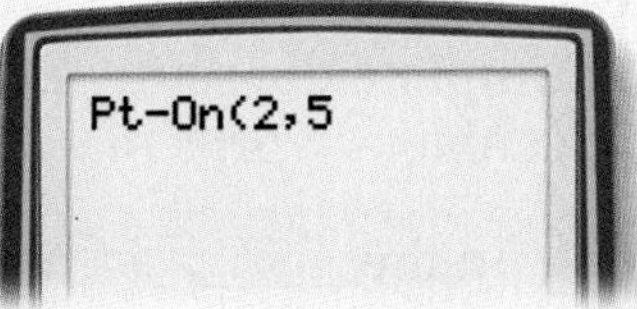

This is the graph in the standard window.

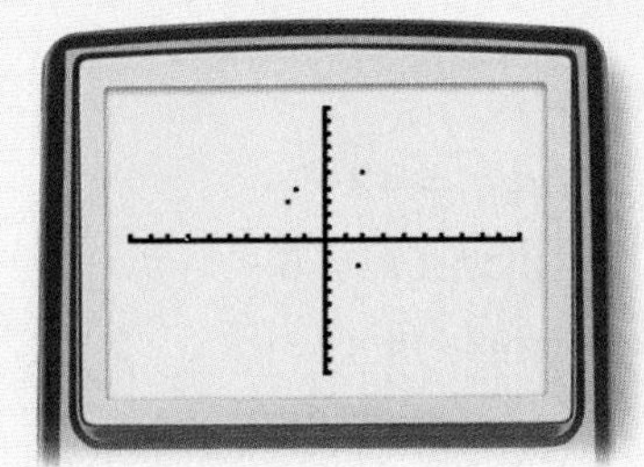

Press WINDOW. Change the **Xmin, Xmax, Ymin,** and **Ymax** values as shown.

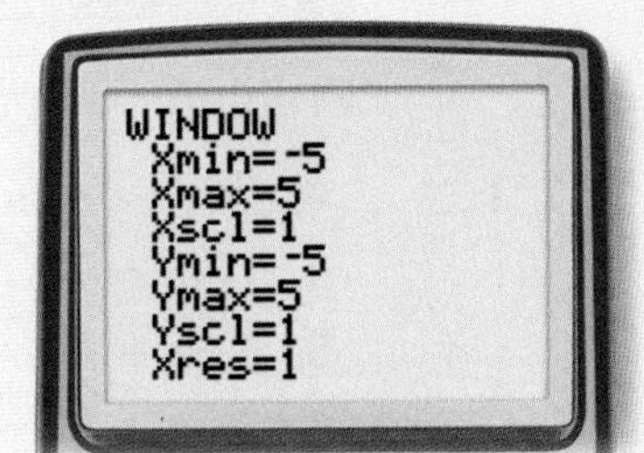

Repeat the steps above to graph the points in the new window.

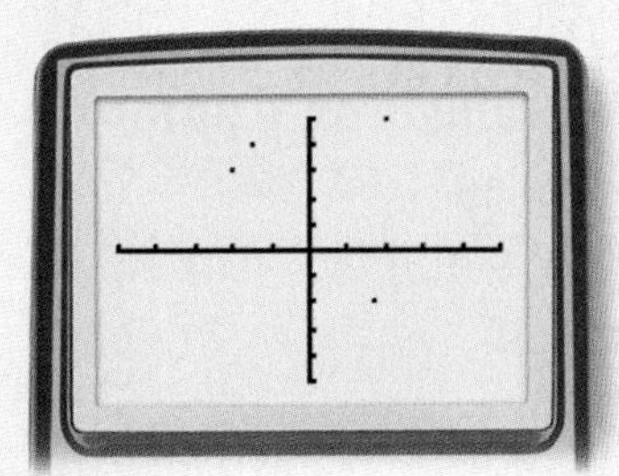

Think and Discuss

1. Compare the two graphs above. Describe and explain any differences you see.

Try This

Graph the points $(-4, -8)$, (1, 2), (2.5, 7), (3, 8), and $(-4.5, 12)$ in each window.

1. standard window

2. **Xmin** $= -5$; **Xmax** $= 5$; **Ymin** $= -20$; **Ymax** $= 20$; **Yscl** $= 5$

Chapter 1

Study Guide and Review

Vocabulary

Addition Property of Equality 14	inequality 23	solution set 23
algebraic expression 4	inverse operation 14	solve 13
algebraic inequality 23	isolate the variable 14	substitute 4
coefficient 4	like term 28	Subtraction Property of Equality ... 14
constant 4	Multiplication Property of Equality 19	term 28
coordinate plane 38	ordered pair 34	variable 4
Division Property of Equality 18	origin 38	x-axis 38
equation 13	simplify 29	x-coordinate 38
equivalent expression ... 28	solution of an equation 13	y-axis 38
evaluate 4	solution of an inequality 23	y-coordinate 38
graph of an equation 39		

Complete the sentences below with vocabulary words from the list above. Words may be used more than once.

1. In the __?__ (4, 9), 4 is the __?__ and 9 is the __?__.

2. $x < 3$ is the __?__ to the __?__ $x + 5 < 8$.

1-1 Variables and Expressions (pp. 4–7)

EXAMPLE

■ Evaluate $4x + 9y$ for $x = 2$ and $y = 5$.

$4x + 9y$

$4(2) + 9(5)$ *Substitute 2 for x and 5 for y.*

$8 + 45$ *Multiply.*

53 *Add.*

EXERCISES

Evaluate each expression.

3. $9a + 7b$ for $a = 7$ and $b = 12$

4. $17m - 3n$ for $m = 10$ and $n = 6$

5. $1.5r + 19s$ for $r = 8$ and $s = 14$

1-2 Writing Algebraic Expressions (pp. 8–12)

EXAMPLE

■ Write an algebraic expression for the word phrase "2 less than a number n."

$n - 2$ *Write as a subtraction.*

EXERCISES

Write an algebraic expression for each phrase.

6. twice the sum of k and 4

7. 5 more than the product of 4 and t

1-3 Solving Equations by Adding or Subtracting (pp. 13–17)

EXAMPLE

Solve.

■ $x + 7 = 12$

$\underline{-7} \quad \underline{-7}$ *Subtract 7 from each side.*

$x + 0 = 5$

$x = 5$ *Identity Property of Zero*

■ $y - 3 = 1.5$

$\underline{+3} \quad \underline{+3}$ *Add 3 to each side.*

$y + 0 = 4.5$

$y = 4.5$ *Identity Property of Zero*

EXERCISES

Solve and check.

8. $z - 9 = 14$ **9.** $t + 3 = 11$

10. $6 + k = 21$ **11.** $x + 2 = 13$

Write an equation and solve.

12. A polar bear weighs 715 lb, which is 585 lb less than a sea cow. How much does the sea cow weigh?

13. The Mojave Desert, at 15,000 mi^2, is 11,700 mi^2 larger than Death Valley. What is the area of Death Valley?

1-4 Solving Equations by Multiplying or Dividing (pp. 18–22)

EXAMPLE

Solve.

■ $4h = 24$

$\frac{4h}{4} = \frac{24}{4}$ *Divide each side by 4.*

$1h = 6$ *$4 \div 4 = 1$*

$h = 6$ *$1 \cdot h = h$*

■ $\frac{t}{4} = 16$

$4 \cdot \frac{t}{4} = 4 \cdot 16$ *Multiply each side by 4.*

$1t = 64$ *$4 \div 4 = 1$*

$t = 64$ *$1 \cdot t = t$*

EXERCISES

Solve and check.

14. $7g = 56$ **15.** $108 = 12k$ **16.** $0.1p = 8$

17. $\frac{w}{4} = 12$ **18.** $20 = \frac{y}{2}$ **19.** $\frac{z}{2.4} = 8$

Write an equation to solve.

20. The Lewis family drove 235 mi toward their destination. This was $\frac{2}{3}$ of the total distance. What was the total distance?

21. Luz will pay a total of $9360 on her car loan. Her monthly payment is $390. For how many months is the loan?

1-5 Solving Simple Inequalities (pp. 23–27)

EXAMPLE

Solve and graph.

■ $x + 5 \leq 8$

$\underline{-5} \quad \underline{-5}$

$x \leq 3$

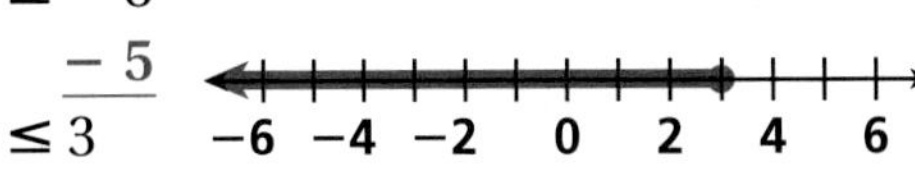

■ $3w > 18$

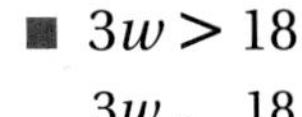

$\frac{3w}{3} > \frac{18}{3}$

$w > 6$

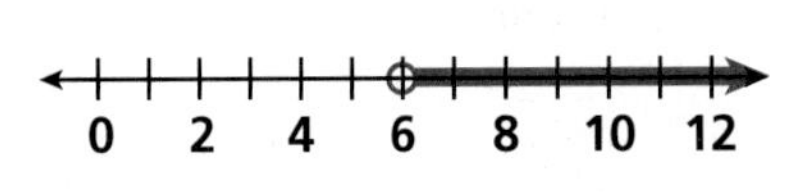

EXERCISES

Solve and graph.

22. $h + 3 < 7$ **23.** $y - 2 > 5$ **24.** $2x \geq 8$

25. $4p < 2$ **26.** $2m > 4.6$ **27.** $3q \leq 0$

28. $\frac{w}{2} \geq 4$ **29.** $\frac{x}{3} \leq 1$ **30.** $\frac{y}{4} > 4$

31. $4 < x + 1$ **32.** $2 < y - 4$ **33.** $8 \geq 4x$

Study Guide and Review

1-6 Combining Like Terms (pp. 28–31)

EXAMPLE

- **Simplify.**

$3(z - 6) + 2z$
$3z - 3(6) + 2z$ — *Distributive Property*
$3z - 18 + 2z$ — *3z and 2z are like terms.*
$5z - 18$ — *Combine coefficients.*

EXERCISES

Simplify.

34. $4(2m - 1) + 3m$ **35.** $12w + 2(w + 3)$

Solve.

36. $6y + y = 35$ **37.** $9z - 3z = 48$

1-7 Ordered Pairs (pp. 34–37)

EXAMPLE

- **Determine whether (8, 3) is a solution of the equation $y = x - 6$.**

$y = x - 6$
$3 \stackrel{?}{=} 8 - 6$
$3 \stackrel{?}{=} 2$ ✗
(8, 3) is not a solution.

EXERCISES

Determine whether the ordered pair is a solution of the given equation.

38. $(27, 0)$; $y = 81 - 3x$ **39.** $(4, 5)$; $y = 5x$

Use the values to make a table of solutions.

40. $y = 3x + 2$ for $x = 0, 1, 2, 3, 4$

1-8 Graphing on a Coordinate Plane (pp. 38–41)

EXAMPLE

- **Graph $A(3, -1)$, $B(0, 4)$, $C(-2, -3)$, and $D(1, 0)$ on a coordinate plane.**

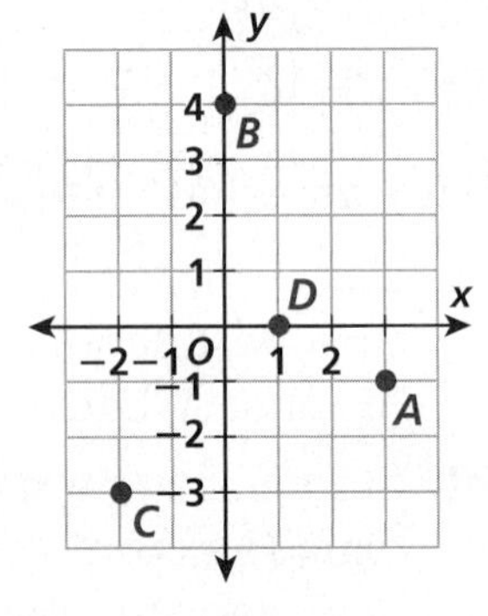

EXERCISES

Graph each point on a coordinate plane.

41. $A(3, 2)$ **42.** $B(-1, 0)$ **43.** $C(0, -5)$
44. $D(1, -3)$ **45.** $E(0, 4)$ **46.** $F(-3, -5)$

Give the missing coordinate for the solutions of $y = 3x + 5$.

47. $(0, y)$ **48.** $(1, y)$ **49.** $(5, y)$

1-9 Interpreting Graphs and Tables (pp. 43–47)

EXAMPLE

- **Which car has the faster acceleration?**

Acceleration	Car A (s)	Car B (s)
0 to 30 mi/h	1.8	3.2
0 to 40 mi/h	2.8	4.7
0 to 50 mi/h	3.9	6.4
0 to 60 mi/h	5.1	8.8

Car A accelerates from 0 to each measured speed in fewer seconds than car B.

EXERCISES

50. Which oven had not been preheated?

Time (min)	Oven D (°F)	Oven E (°F)
0	450°	70°
1	435°	220°
2	445°	450°
3	455°	440°
4	450°	450°

Chapter Test

Evaluate each expression for the given values of the variables.

1. $4x + 5y$ for $x = 9$ and $y = 7$

2. $5k(6 - 6m)$ for $k = 2$ and $m = \frac{1}{2}$

Write an algebraic expression for each word phrase.

3. 3 more than twice p

4. 4 times the sum of t and 5

5. 6 less than half of n

Solve.

6. $m + 15 = 25$

7. $4d = 144$

8. $50 = h - 3$

9. $\frac{x}{3} = 18$

10. $y - 4 \geq 1.1$

11. $\frac{x}{3} < 6$

12. $w + 1 < 4.5$

13. $2p > 15$

Graph each inequality.

14. $x > 4$

15. $y \leq 8$

Write and solve an equation for each problem.

16. Acme Sporting Products manufactures 3216 tennis balls a day. Each container holds 3 balls. How many tennis ball containers are needed daily?

17. In the 1996 presidential election, Bill Clinton received 2,459,683 votes in Texas. This was 177,868 more votes than he had received in 1992 in Texas. How many votes did Bill Clinton get in Texas in 1992?

Solve.

18. $4x + 3 = 19$

19. $\frac{y}{2} - 5 = 1$

20. $10z + 2z = 108$

21. $26 = 3f + 10f$

Determine whether the ordered pair is a solution of the given equation.

22. $(6, 5)$; $y = 5x - 25$

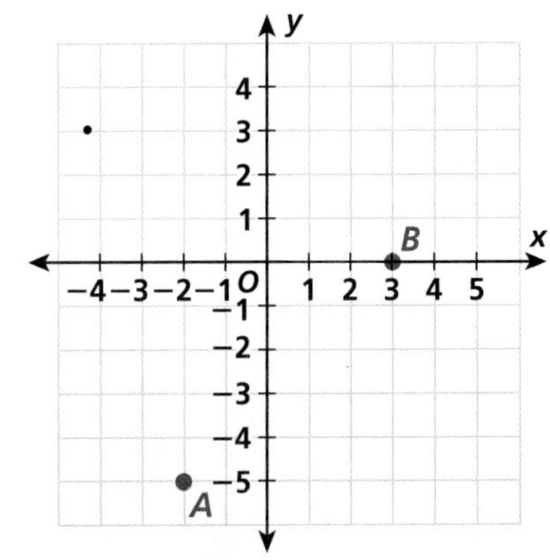

Give the coordinates of each point shown on the coordinate plane.

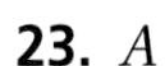

23. A

24. B

25. Use the table to graph the speed of the car over time.

Time (s)	0	5	10	15
Speed (mi/h)	0	20	30	35

Chapter Test

Chapter 1

Performance Assessment

Show What You Know

Create a portfolio of your work from this chapter. Complete this page and include it with your four best pieces of work from Chapter 1. Choose from your homework or lab assignments, mid-chapter quiz, or any journal entries you have done. Put them together using any design you want. Make your portfolio represent what you consider your best work.

Short Response

1. Find the solution set for the equation $x + 6 = 10$. Find the solution set for the inequality $2x \geq 8$. Explain what the solution sets have in common and then explain why they are different.

2. The average rise and fall of the tides in Eastport, Maine, is 5 ft 10 in. more than twice the average in Philadelphia, Pennsylvania. Write an algebraic expression you can use to find the measurement for Eastport, Maine. Then find that measurement.

Average Rise and Fall of Tides

Place	ft	in.
Boston, MA	10	4
Charleston, SC	5	10
Eastport, ME	■	■
Fort Pulaski, GA	7	6
Key West, FL	1	10
Philadelphia, PA	6	9

Extended Problem Solving

Choose any strategy to solve each problem.

3. A soft-drink company is running a contest. A whole number less than 100 is printed on each bottle cap. If you collect a set of caps with a sum of exactly 100, you win a prize. Below are some typical bottle caps.
   a. Write the prime factorization of each of the numbers on the bottle caps.
   b. What do all of the numbers on the bottle caps have in common?
   c. Do the bottle caps contain a winning set? Explain.

Performance Assessment

internet connect

EOG Test Practice Online
go.hrw.com Keyword: MT4 TestPrep

Getting Ready for EOG

Chapter 1

Cumulative Assessment, Chapter 1

1. Which algebraic equation represents the word sentence "15 less than the number of computers c is 32"?

A $\frac{c}{15} = 32$ C $15 - c = 32$

B $15c = 32$ D $c - 15 = 32$

2. Which inequality is represented by this graph?

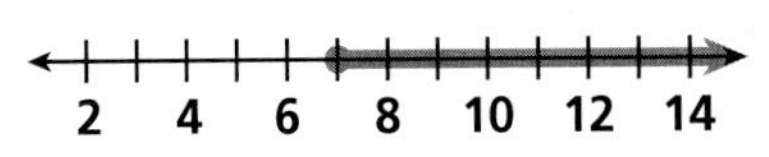

A $x < 7$ C $7 < x$

B $x \leq 7$ D $7 \leq x$

3. Bill is 3 years older than his cat. The sum of their ages is 25. If c represents the cat's age, which equation could be used to find c?

A $c + 25 = c + 3$ C $c + 3c = 25$

B $c + 25 = 3c$ D $c + (c + 3) = 25$

4. What is the solution of $k + 3(k - 2) = 34$?

A $k = 10$ C $k = 8$

B $k = 9$ D $k = 7$

5. Jamal brings $20 to a pizza restaurant where a plain slice costs $2.25, including tax. Which inequality can he use to find the number of plain slices he can buy?

A $2.25 + s \leq 20$ C $2.25s \leq 20$

B $2.25 + s \geq 20$ D $2.25s \geq 20$

6. When twice a number is decreased by 4, the result is 236. What is the number?

A 29.5 C 116

B 59 D 120

7. A number n is increased by 5 and the result is multiplied by 5. This result is decreased by 5. What is the final result?

A $5n$ C $5n + 10$

B $5n + 5$ D $5n + 20$

8. Which has the greatest value?

A $(2 + 3)(2 + 3)$ C $(2 \cdot 3)(2 \cdot 3)$

B $2 + 3 \cdot 3$ D $2 \cdot 2 + 3 \cdot 3$

TIP!

TEST TAKING TIP!

To convert from a larger unit of measure to a smaller unit, multiply by the conversion factor. To convert from a smaller unit of measure to a larger unit, divide by the conversion factor.

9. ***SHORT RESPONSE*** Jo has 197 fund-raising posters. She decides to use four 5-inch strips of tape to hang each poster. Each roll of tape is 250 feet long. Estimate the number of whole rolls Jo will need to hang all of the posters. Explain in words how you determined your estimate. (*Hint:* 12 in. = 1 ft)

10. ***SHORT RESPONSE*** Mrs. Morton recorded the lengths of the telephone calls she made this week.

Length of call (min)	2	5	7	12	15
Number of calls	7	x	2	2	3

The number of calls shorter than 6 minutes is equal to the number of calls longer than 6 minutes. Write an equation that could be used to determine the number of 5-minute calls Mrs. Morton made. Solve your equation.

Chapter 2

Integers and Exponents

TABLE OF CONTENTS

Atomic Particle	Independent Life Span (s)
Electron	Indefinite
Proton	Indefinite
Neutron	920
Muon	2.2×10^{-6}

internet connect

Chapter Opener Online
go.hrw.com
KEYWORD: MT4 Ch2

Career *Nuclear Physicist*

The atom was defined by the ancient Greeks as the smallest particle of matter. We now know that atoms are made up of many smaller particles.

Nuclear physicists study these particles using large machines—such as linear accelerators, synchrotrons, and cyclotrons—that can smash atoms to uncover their component parts.

Nuclear physicists use mathematics along with the data they discover to create models of the atom and the structure of matter.

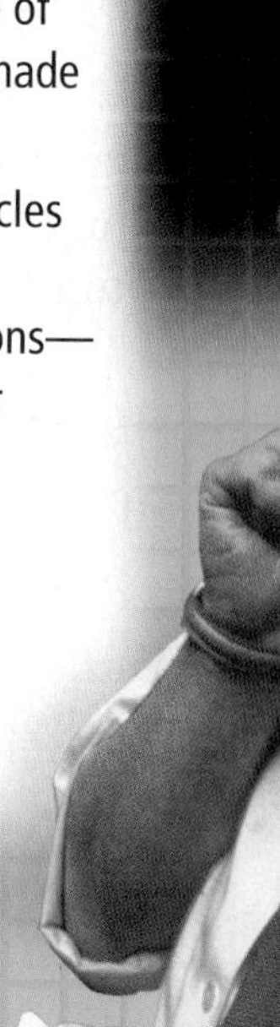

ARE YOU READY?

Choose the best term from the list to complete each sentence.

1. According to the __?__, you must multiply or divide before you add or subtract when simplifying a numerical __?__.
2. An algebraic expression is a mathematical sentence that has at least one __?__.
3. In a(n) __?__, an equal sign is used to show that two quantites are the same.
4. You use a(n) __?__ to show that one quantity is greater than another quantity.

expression
inequality
order of operations
variable
equation

Complete these exercises to review skills you will need for this chapter.

✔ Order of Operations

Simplify by using the order of operations.

5. $(12) + 4(2)$ **6.** $12 + 8 \div 4$ **7.** $15(14 - 4)$

8. $(23 - 5) - 36 \div 2$ **9.** $12 \div 2 + 10 \div 5$ **10.** $40 \div 2 \cdot 4$

✔ Equations

Solve.

11. $x + 9 = 21$ **12.** $3z = 42$ **13.** $\frac{w}{4} = 16$

14. $24 + t = 24$ **15.** $p - 7 = 23$ **16.** $12m = 0$

✔ Match a Number Line to an Inequality

Write an inequality that describes the set of points shown on each number line.

17. −6 −4 −2 0 2 4 6

18. −6 −4 −2 0 2 4 6

19. −4 −2 0 2 4 6 8

20. −4 −2 0 2 4 6 8

✔ Multiply and Divide by Powers of Ten

Multiply or divide.

21. $358(10)$ **22.** $358(1000)$ **23.** $358(100{,}000)$

24. $\frac{358}{10}$ **25.** $\frac{358}{1000}$ **26.** $\frac{358}{100{,}000}$

Model Solving Equations

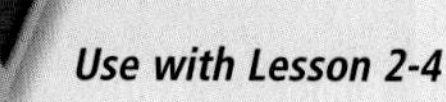

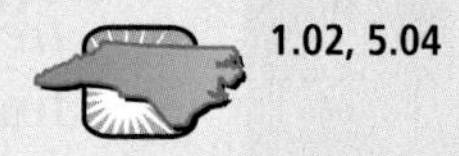

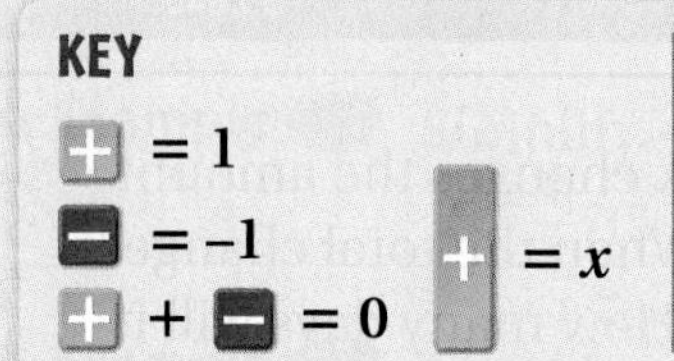

REMEMBER
It will not change the value of an expression if you add or remove zero.

You can use algebra tiles to help you solve equations.

Activity

To solve the equation $x + 3 = 5$, you need to get x alone on one side of the equal sign. You can add or remove tiles as long as you add the same amount or remove the same amount on both sides.

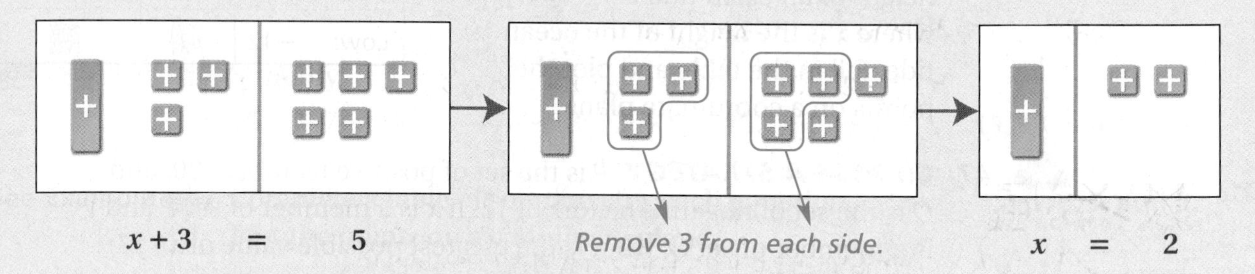

1 Use algebra tiles to model and solve each equation.

a. $x + 1 = 2$ **b.** $x + 2 = 7$ **c.** $x + (-6) = -9$ **d.** $x + 4 = 4$

The equation $x + 4 = 2$ is more difficult to solve because there are not enough yellow tiles on the right side. You can use the fact that $1 + (-1) = 0$ to help you solve the equation.

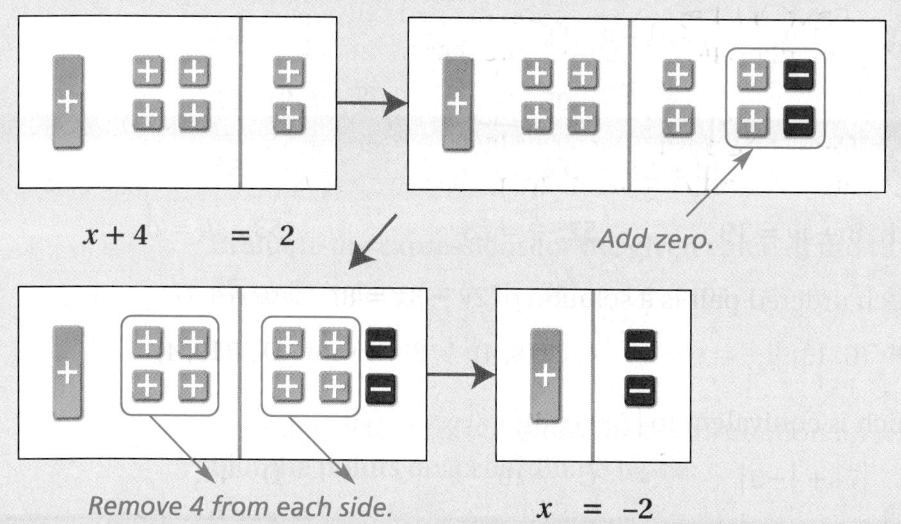

Focus on Problem Solving

Look Back

- **Is your answer reasonable?**

After you solve a word problem, ask yourself if your answer makes sense. You can round the numbers in the problem and estimate to find a reasonable answer. It may also help to write your answer in sentence form.

Read the problems below and tell which answer is most reasonable.

1. Tonia makes $1836 per month. Her total expenses are $1005 per month. How much money does she have left each month?
   A. about −$800 per month
   B. about $1000 per month
   C. about $800 per month
   D. about −$1000 per month

2. The Qin Dynasty in China began about 2170 years before the People's Republic of China was formed in 1949. When did the Qin Dynasty begin?
   A. before 200 B.C.
   B. between 200 B.C. and A.D. 200
   C. between A.D. 200 and A.D. 1949
   D. after A.D 1949

3. On Mercury, the coldest temperature is about 600°C below the hottest temperature of 430°C. What is the coldest temperature on the planet?
   A. about 1030°C
   B. about −1030°C
   C. about −170°C
   D. about 170°C

4. Julie is balancing her checkbook. Her beginning balance is $325.46, her deposits add up to $285.38, and her withdrawals add up to $683.27. What is her ending balance?
   A. about −$70
   B. about −$600
   C. about $700
   D. about $1300

2-6 Exponents

Learn to evaluate expressions with exponents.

Vocabulary

power

exponential form

exponent

base

Fold a piece of $8\frac{1}{2}$-by-11-inch paper in half. If you fold it in half again, the paper is 4 sheets thick. After the third fold in half, the paper is 8 sheets thick. How many sheets thick is the paper after 7 folds?

With each fold the number of sheets doubles.

$$2 \cdot 2 \cdot 2 \cdot 2 \cdot 2 \cdot 2 \cdot 2 = 128 \text{ sheets thick after 7 folds.}$$

This multiplication problem can also be written in *exponential form.*

$$2 \cdot 2 \cdot 2 \cdot 2 \cdot 2 \cdot 2 \cdot 2 = 2^7$$

The number 2 is a factor 7 times.

The term 2^7 is called a **power**. If a number is in **exponential form**, the **exponent** represents how many times the **base** is to be used as a factor.

EXAMPLE 1 Writing Exponents

Write in exponential form.

A $3 \cdot 3 \cdot 3 \cdot 3 \cdot 3 \cdot 3$

$3 \cdot 3 \cdot 3 \cdot 3 \cdot 3 \cdot 3 = 3^6$ — *Identify how many times 3 is a factor.*

Reading Math

Read 3^6 as "3 to the 6th power."

B $(-2) \cdot (-2) \cdot (-2) \cdot (-2)$

$(-2) \cdot (-2) \cdot (-2) \cdot (-2) = (-2)^4$ — *Identify how many times −2 is a factor.*

C $n \cdot n \cdot n \cdot n \cdot n$

$n \cdot n \cdot n \cdot n \cdot n = n^5$ — *Identify how many times n is a factor.*

D 12

$12 = 12^1$ — *12 is used as a factor 1 time, so $12 = 12^1$.*

EXAMPLE 2 Evaluating Powers

Helpful Hint

Always use parentheses to raise a negative number to a power.

$(-8)^2 = (-8) \cdot (-8)$
$= 64$

$-8^2 = -(8 \cdot 8)$
$= -64$

Evaluate.

A 2^6

$2^6 = 2 \cdot 2 \cdot 2 \cdot 2 \cdot 2 \cdot 2$ — *Find the product of six 2's.*

$= 64$

B $(-8)^2$

$(-8)^2 = (-8) \cdot (-8)$ — *Find the product of two −8's.*

$= 64$

Evaluate.

C $(-5)^3$

$(-5)^3 = (-5) \cdot (-5) \cdot (-5)$ *Find the product of three −5's.*

$= -125$

EXAMPLE 3 Simplifying Expressions Containing Powers

Simplify $50 - 2(3 \cdot 2^3)$.

$50 - 2(3 \cdot 2^3)$

$= 50 - 2(3 \cdot 8)$ *Evaluate the exponent.*

$= 50 - 2(24)$ *Multiply inside the parentheses.*

$= 50 - 48$ *Multiply from left to right.*

$= 2$ *Subtract from left to right.*

EXAMPLE 4 *Geometry Application*

The number of diagonals of an n-sided figure is $\frac{1}{2}(n^2 - 3n)$. Use the formula to find the number of diagonals for a 5-sided figure.

$\frac{1}{2}(n^2 - 3n)$

$\frac{1}{2}(5^2 - 3 \cdot 5)$ *Substitute the number of sides for n.*

$\frac{1}{2}(25 - 3 \cdot 5)$ *Evaluate the exponent.*

$\frac{1}{2}(25 - 15)$ *Multiply inside the parentheses.*

$\frac{1}{2}(10)$ *Subtract inside the parentheses.*

5 diagonals *Multiply.*

Verify your answer by sketching the diagonals.

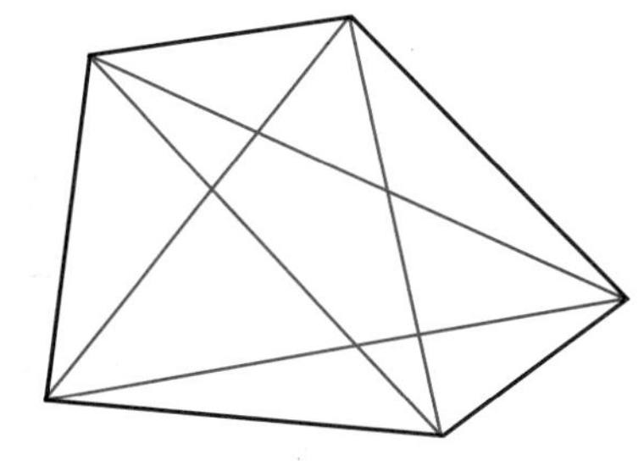

Think and Discuss

1. Describe a rule for finding the sign of a negative number raised to a whole number power.

2. Compare $3 \cdot 2$, 3^2, and 2^3.

3. Show that $(4 - 11)^2$ is not equal to $4^2 - 11^2$.

2-6 Exercises

FOR EOG PRACTICE

see page 650

internet connect

Homework Help Online
go.hrw.com Keyword: MT4 2-6

1.01, 1.02

GUIDED PRACTICE

See Example 1 **Write in exponential form.**

1. 14
2. $15 \cdot 15$
3. $b \cdot b \cdot b \cdot b$
4. $(-1) \cdot (-1) \cdot (-1)$

See Example 2 **Evaluate.**

5. 3^4
6. $(-5)^2$
7. $(-3)^5$
8. 7^4

See Example 3 **Simplify.**

9. $(3 - 6^2)$
10. $42 + (3 \cdot 4^2)$
11. $(8 - 5^3)$
12. $61 - (4 \cdot 3^3)$

See Example 4

13. The sum of the first n positive integers is $\frac{1}{2}(n^2 + n)$. Check the formula for the first four positive integers. Then use the formula to find the sum of the first 12 positive integers.

INDEPENDENT PRACTICE

See Example 1 **Write in exponential form.**

14. $6 \cdot 6 \cdot 6 \cdot 6 \cdot 6 \cdot 6 \cdot 6$
15. $(-7) \cdot (-7) \cdot (-7)$
16. -6
17. $c \cdot c \cdot c \cdot c \cdot c$

See Example 2 **Evaluate.**

18. 6^6
19. $(-4)^4$
20. 8^4
21. $(-2)^9$

See Example 3 **Simplify.**

22. $(1 - 7^2)$
23. $27 + (2 \cdot 5^2)$
24. $(8 - 10^3)$
25. $45 - (5 \cdot 3^4)$

See Example 4

26. A circle can be divided by n lines into a maximum of $\frac{1}{2}(n^2 + n) + 1$ regions. Use the formula to find the maximum number of regions for 7 lines.

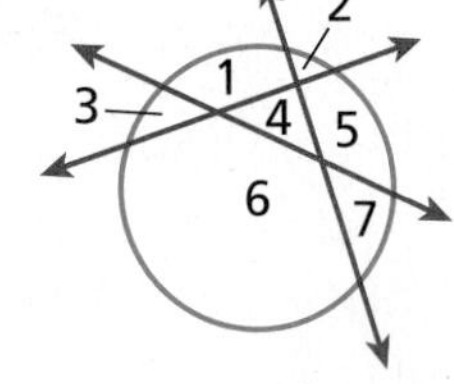

3 lines → 7 regions

PRACTICE AND PROBLEM SOLVING

Write in exponential form.

27. $(-2) \cdot (-2) \cdot (-2)$
28. $h \cdot h \cdot h \cdot h$
29. $4 \cdot 4 \cdot 4 \cdot 4$
30. $(5)(5)(5)(5)(5)$

Evaluate.

31. 7^3
32. 8^2
33. $(-12)^3$
34. $(-6)^5$
35. $(-3)^6$
36. $(-9)^3$
37. 4^1
38. 2^9

Simplify.

39. $(9 - 5^3)$ **40.** $(18 - 7^3)$ **41.** $42 + (8 - 6^3)$ **42.** $16 + (2 + 8^3)$

43. $32 - (4 \cdot 3^2)$ **44.** $(5 + 5^5)$ **45.** $(5 - 6^1)$ **46.** $86 - [6 - (-2)^5]$

Evaluate each expression for the given value of the variable.

47. a^3 for $a = 6$ **48.** x^7 for $x = -1$ **49.** $n^4 + 1$ for $n = 4$ **50.** $1 - y^5$ for $y = 2$

Life Science LINK

Most bacteria reproduce by a type of simple cell division known as binary fission. Each species reproduces best at a specific temperature and moisture level.

51. ***LIFE SCIENCE*** Bacteria can divide every 20 minutes, so one bacterium can multiply to 2 in 20 minutes, 4 in 40 minutes, 8 in 1 hour, and so on. How many bacteria will there be in 6 hours? Write your answer using exponents, and then evaluate.

52. Make a table with the column headings n, n^2, and $2n$. Complete the table for $n = -5, -4, -3, -2, -1, 0, 1, 2, 3, 4,$ and 5.

53. For any whole number n, $5^n - 1$ is divisible by 4. Verify this for $n = 3$ and $n = 5$.

54. The chart shows Han's genealogy. Each generation consists of twice as many people as the generation after it.

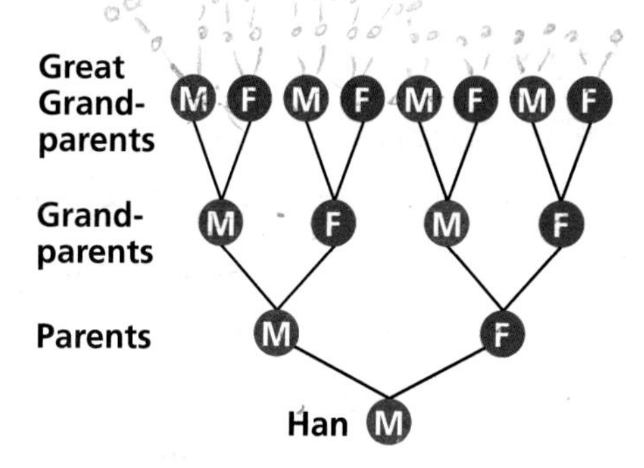

a. Write the number of Han's great-grandparents using an exponent.

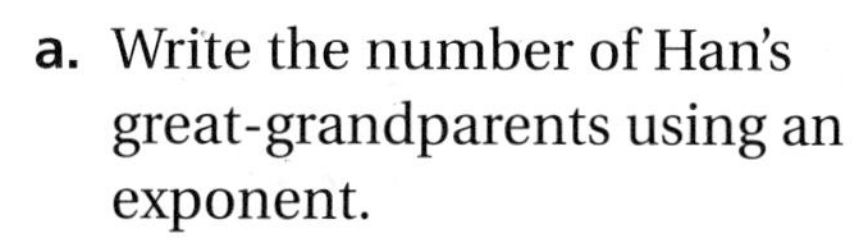

b. How many ancestors were in the fifth generation back from Han?

55. ***CHOOSE A STRATEGY*** Place the numbers 1, 2, 3, 4, and 5 in the boxes to make a true statement: $\square \cdot \square^3 = \square^2 - \square^{\square}$

56. ***WRITE ABOUT IT*** Compare 10^3 and 3^{10}. For any two numbers, which usually gives the greater number, using the larger number as the base or as the exponent? Give at least one exception.

57. ***CHALLENGE*** Write $(3^2)^3$ using a single exponent.

Spiral Review

Multiply or divide. (Lesson 2-3)

58. $7(-8)$ **59.** $\frac{-63}{-7}$ **60.** $\frac{38}{-19}$ **61.** $-8(-13)$ **62.** $-6(15)$

63. **EOG PREP** Which represents the phrase *the difference of a number and 32*? (Lesson 1-2)

A $n + 32$ C $n - 32$

B $n \times 32$ D $32 \div n$

64. **EOG PREP** Which of the values $-2, -1,$ and 0 are solutions of $x - 2 > -3$? (Lesson 2-5)

A -1 and 0 C -2 and -1

B only 0 D $-2, -1,$ and 0

2-7 Properties of Exponents

Learn to apply the properties of exponents and to evaluate the zero exponent.

The factors of a power, such as 7^4, can be grouped in different ways. Notice the relationship of the exponents in each product.

$$7 \cdot 7 \cdot 7 \cdot 7 = 7^4$$
$$(7 \cdot 7 \cdot 7) \cdot 7 = 7^3 \cdot 7^1 = 7^4$$
$$(7 \cdot 7) \cdot (7 \cdot 7) = 7^2 \cdot 7^2 = 7^4$$

MULTIPLYING POWERS WITH THE SAME BASE

Words	Numbers	Algebra
To multiply powers with the same base, keep the base and add the exponents.	$3^5 \cdot 3^8 = 3^{5+8} = 3^{13}$	$b^m \cdot b^n = b^{m+n}$

EXAMPLE 1 Multiplying Powers with the Same Base

Multiply. Write the product as one power.

A $3^5 \cdot 3^2$

$3^5 \cdot 3^2$

3^{5+2} *Add exponents.*

3^7

B $a^{10} \cdot a^{10}$

$a^{10} \cdot a^{10}$

a^{10+10} *Add exponents.*

a^{20}

C $16 \cdot 16^7$

$16 \cdot 16^7$

$16^1 \cdot 16^7$ *Think: $16 = 16^1$*

16^{1+7} *Add exponents.*

16^8

D $6^4 \cdot 4^4$

$6^4 \cdot 4^4$ *Cannot combine; the bases are not the same.*

Notice what occurs when you divide powers with the same base.

$$\frac{5^5}{5^3} = \frac{5 \cdot 5 \cdot 5 \cdot 5 \cdot 5}{5 \cdot 5 \cdot 5} = \frac{\cancel{5} \cdot \cancel{5} \cdot \cancel{5} \cdot 5 \cdot 5}{\cancel{5} \cdot \cancel{5} \cdot \cancel{5}} = 5 \cdot 5 = 5^2$$

DIVIDING POWERS WITH THE SAME BASE

Words	Numbers	Algebra
To divide powers with the same base, keep the base and subtract the exponents.	$\frac{6^9}{6^4} = 6^{9-4} = 6^5$	$\frac{b^m}{b^n} = b^{m-n}$

EXAMPLE 2 Dividing Powers with the Same Base

Divide. Write the quotient as one power.

A $\frac{100^9}{100^3}$

$\frac{100^9}{100^3}$

100^{9-3} *Subtract exponents.*

100^6

B $\frac{x^8}{y^5}$

$\frac{x^8}{y^5}$ *Cannot combine; the bases are not the same.*

Helpful Hint

0^0 does not exist because 0^0 represents a quotient of the form $\frac{0^n}{0^n}$.

But the denominator of this quotient is 0, which is impossible, since you cannot divide by 0.

When the numerator and denominator of a fraction have the same base and exponent, subtracting the exponents results in a 0 exponent.

$$1 = \frac{4^2}{4^2} = 4^{2-2} = 4^0 = 1$$

This result can be confirmed by writing out the factors.

$$\frac{4^2}{4^2} = \frac{(4 \cdot 4)}{(4 \cdot 4)} = \frac{(\not{4} \cdot \not{4})}{(\not{4} \cdot \not{4})} = \frac{1}{1} = 1$$

THE ZERO POWER		
Words	**Numbers**	**Algebra**
The zero power of any number except 0 equals 1.	$100^0 = 1$ $(-7)^0 = 1$	$a^0 = 1$, if $a \neq 0$

EXAMPLE 3 *Physical Science Application*

There are about 10^{25} molecules in a cubic meter of air at sea level, but only 10^{23} molecules at a high altitude (33 km). How many times more molecules are there at sea level than at 33 km?

You want to find the number that you must multiply by 10^{23} to get 10^{25}. Set up and solve an equation. Use x as your variable.

$(10^{23})x = 10^{25}$ *"10^{23} times some number x equals 10^{25}."*

$\frac{(10^{23})x}{10^{23}} = \frac{10^{25}}{10^{23}}$ *Divide both sides by 10^{23}.*

$x = 10^{25-23}$ *Subtract the exponents.*

$x = 10^2$

There are 10^2 times more molecules per cubic meter of air at sea level than at 33 km.

Think and Discuss

1. **Explain** why the exponents cannot be added in the product $14^3 \cdot 18^3$.
2. **List** two ways to express 4^5 as a product of powers.

2-7 Exercises

FOR EOG PRACTICE

see page 650

internet connect
Homework Help Online
go.hrw.com Keyword: MT4 2-7

1.01, 1.02

GUIDED PRACTICE

See Example 1

Multiply. Write the product as one power.

1. $3^4 \cdot 3^7$
2. $12^3 \cdot 12^2$
3. $m \cdot m^5$
4. $14^5 \cdot 8^5$

See Example 2

Divide. Write the quotient as one power.

5. $\frac{8^7}{8^5}$
6. $\frac{a^9}{a^1}$
7. $\frac{12^5}{12^5}$
8. $\frac{7^{18}}{7^6}$

See Example 3

9. A scientist estimates that a sweet corn plant produces 10^8 grains of pollen. If there are 10^{10} grains of pollen, how many plants are there?

INDEPENDENT PRACTICE

See Example 1

Multiply. Write the product as one power.

10. $10^{10} \cdot 10^7$
11. $2^3 \cdot 2^3$
12. $r^5 \cdot r^4$
13. $16 \cdot 16^3$

See Example 2

Divide. Write the quotient as one power.

14. $\frac{7^{12}}{7^8}$
15. $\frac{m^{10}}{d^3}$
16. $\frac{t^8}{t^5}$
17. $\frac{10^8}{10^8}$

See Example 3

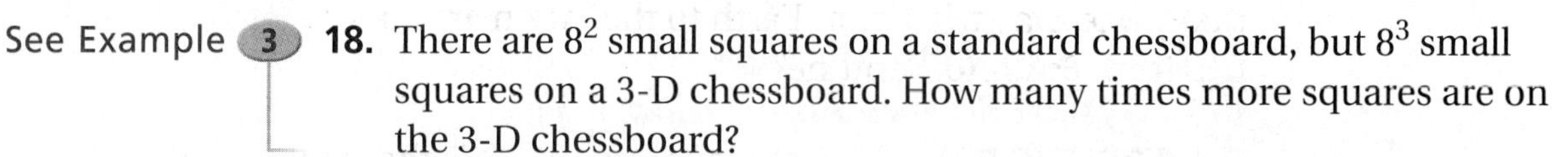

18. There are 8^2 small squares on a standard chessboard, but 8^3 small squares on a 3-D chessboard. How many times more squares are on the 3-D chessboard?

PRACTICE AND PROBLEM SOLVING

Multiply or divide. Write the product or quotient as one power.

19. $\frac{6^8}{6^5}$
20. $7^9 \cdot 7^1$
21. $\frac{a^3}{a^2}$
22. $\frac{10^{18}}{10^9}$
23. $x^3 \cdot x^7$
24. $a^7 \cdot b^8$
25. $6^4 \cdot 6^2$
26. $4 \cdot 4^2$
27. $\frac{12^5}{6^3}$
28. $\frac{11^7}{11^6}$
29. $\frac{y^9}{y^9}$
30. $\frac{2^9}{2^3}$
31. $x^5 \cdot x^3$
32. $c^9 \cdot d^3$
33. $4^4 \cdot 4^2$
34. $9^2 \cdot 9^2$
35. $10^5 \cdot 10^9$
36. $\frac{k^6}{p^2}$
37. $n^8 \cdot n^8$
38. $\frac{9^{11}}{9^6}$
39. $4^9 \div 4^5$
40. $2^{12} \div 2^6$
41. $6^2 \cdot 6^3 \cdot 6^4$
42. $5^3 \cdot 5^6 \cdot 5^0$

43. There are 26^3 ways to make a 3-letter "word" (from *aaa* to *zzz*) and 26^5 ways to make a 5-letter word. How many times more ways are there to make a 5-letter word than a 3-letter word?

44. ***ASTRONOMY*** The mass of the known universe is about 10^{23} solar masses, which is 10^{50} metric tons. How many metric tons is one solar mass?

45. ***BUSINESS*** Using the manufacturing terms below, tell how many dozen are in a great gross. How many gross are in a great gross?

1 dozen	$= 12^1$ items
1 gross	$= 12^2$ items
1 great gross	$= 12^3$ items

46. A googol is the number 1 followed by 100 zeros.
   a. What is a googol written as a power?
   b. What is a googol times a googol written as a power?

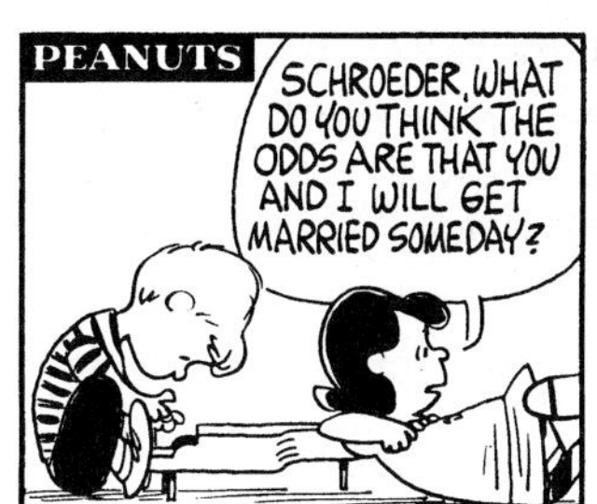

47. ***ASTRONOMY*** The distance from Earth to the moon is about 22^4 miles. The distance from Earth to Neptune is about 22^7 miles. How many one-way trips from Earth to the moon are about equal to one trip from Earth to Neptune?

48. ***WHAT'S THE ERROR?*** A student said that $\frac{4^7}{8^7}$ is the same as $\frac{1}{2}$. What mistake has the student made?

49. ***WRITE ABOUT IT*** Why do you add exponents when multiplying powers with the same base?

50. ***CHALLENGE*** A number to the 10th power divided by the same number to the 7th power equals 125. What is the number?

Spiral Review

Evaluate each expression for $m = -3$. (Lesson 2-1)

51. $m + 6$ **52.** $m + -5$ **53.** $-9 + m$ **54.** $m + 3$

Subtract. (Lesson 2-2)

55. $-8 - 8$ **56.** $-3 - (-7)$ **57.** $-10 - 2$ **58.** $11 - (-9)$

59. **EOG PREP** Which is *not* a solution to $-3x > 15$? (Lesson 2-5)

A -20 B -100 C -6 D -5

2-8 Look for a Pattern in Integer Exponents

 Problem Solving Skill

Learn to evaluate expressions with negative exponents.

The nanoguitar is the smallest guitar in the world. It is no larger than a single cell, at about 10^{-5} meters long. Can you imagine 10^{-5} meters?

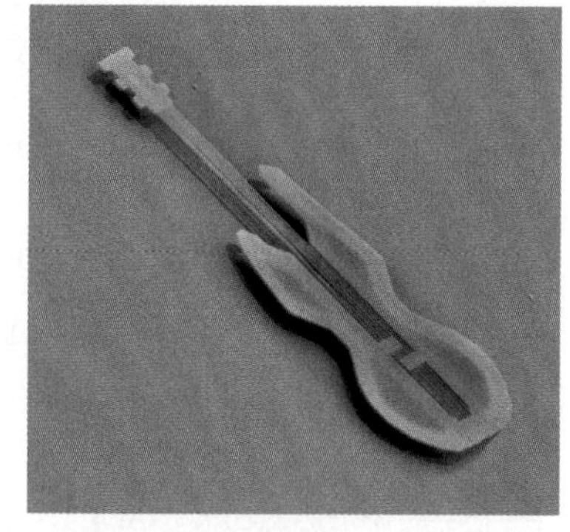

The nanoguitar is carved from crystalline silicon. It has 6 strings that are each about 100 atoms wide.

Look for a pattern in the table to extend what you know about exponents to include negative exponents. Start with what you know about positive and zero exponents.

10^2	10^1	10^0	10^{-1}	10^{-2}
$10 \cdot 10$	10	1	$\frac{1}{10}$	$\frac{1}{10 \cdot 10}$
100	10	1	$\frac{1}{10} = 0.1$	$\frac{1}{100} = 0.01$

÷ 10 ÷ 10 ÷ 10 ÷ 10

EXAMPLE 1 Using a Pattern to Evaluate Negative Exponents

Evaluate the powers of 10.

A 10^{-3}

$$10^{-3} = \frac{1}{10 \cdot 10 \cdot 10}$$ *Extend the pattern from the table.*

$$10^{-3} = \frac{1}{1000} = 0.001$$

B 10^{-4}

$$10^{-4} = \frac{1}{10 \cdot 10 \cdot 10 \cdot 10}$$ *Extend the pattern from Example 1A.*

$$10^{-4} = \frac{1}{10{,}000} = 0.0001$$

C 10^{-5}

$$10^{-5} = \frac{1}{10 \cdot 10 \cdot 10 \cdot 10 \cdot 10}$$ *Extend the pattern from Example 1B.*

$$10^{-5} = \frac{1}{100{,}000} = 0.00001$$

So how long is 10^{-5} meters?

$10^{-5}\text{ m} = \frac{1}{100{,}000}\text{ m}$ ⟶ "one hundred-thousandth of a meter"

NEGATIVE EXPONENTS		
Words	**Numbers**	**Algebra**
A power with a negative exponent equals 1 divided by that power with its opposite exponent.	$5^{-3} = \frac{1}{5^3} = \frac{1}{125}$	$b^{-n} = \frac{1}{b^n}$

EXAMPLE 2 Evaluating Negative Exponents

Remember!

The reciprocal of a number is 1 divided by that number.

Evaluate $(-2)^{-3}$.

$(-2)^{-3}$

$\frac{1}{(-2)^3}$ *Write the reciprocal; change the sign of the exponent.*

$\frac{1}{(-2)(-2)(-2)}$

$-\frac{1}{8}$

EXAMPLE 3 Evaluating Products and Quotients of Negative Exponents

Evaluate.

A $10^3 \cdot 10^{-3}$

$10^3 \cdot 10^{-3}$

$10^{3 + (-3)}$ *Bases are the same, so add the exponents.*

$10^0 = 1$ *Check* $10^3 \cdot 10^{-3} = 10^3 \cdot \frac{1}{10^3} = \frac{10^3}{10^3} = \frac{\cancel{10} \cdot \cancel{10} \cdot \cancel{10}}{\cancel{10} \cdot \cancel{10} \cdot \cancel{10}} = 1$

B $\frac{2^4}{2^7}$

$\frac{2^4}{2^7}$

2^{4-7} *Bases are the same, so subtract the exponents.*

2^{-3}

$\frac{1}{2^3}$ *Write the reciprocal; change the sign of the exponent.*

$\frac{1}{8}$ *Check* $\frac{2^4}{2^7} = \frac{\cancel{2} \cdot \cancel{2} \cdot \cancel{2} \cdot \cancel{2}}{\cancel{2} \cdot \cancel{2} \cdot \cancel{2} \cdot \cancel{2} \cdot 2 \cdot 2 \cdot 2} = \frac{1}{8}$

Think and Discuss

1. Express $\frac{1}{2}$ using an exponent.

2. Tell whether the statement is true or false: If a power has a negative exponent, then the power is negative. Justify your answer.

3. Tell whether an integer raised to a negative exponent can ever be greater than 1.

2-8 Exercises

1.01, 1.02

FOR EOG PRACTICE

see page 651

internet connect

Homework Help Online

go.hrw.com Keyword: MT4 2-8

GUIDED PRACTICE

See Example 1 Evaluate the powers of 10.

1. 10^{-7}
2. 10^{-3}
3. 10^{-6}
4. 10^{-1}

See Example 2 Evaluate.

5. $(-2)^{-4}$
6. $(-3)^{-2}$
7. 2^{-3}
8. $(-2)^{-5}$

See Example 3

9. $10^{7} \cdot 10^{-4}$
10. $3^{5} \cdot 3^{-7}$
11. $\frac{6^8}{6^5}$
12. $\frac{3^6}{3^9}$

INDEPENDENT PRACTICE

See Example 1 Evaluate the powers of 10.

13. 10^{-2}
14. 10^{-9}
15. 10^{-5}
16. 10^{-11}

See Example 2 Evaluate.

17. $(-4)^{-3}$
18. 3^{-2}
19. $(-10)^{-4}$
20. $(-2)^{-1}$

See Example 3

21. $10^{5} \cdot 10^{-1}$
22. $\frac{2^3}{2^5}$
23. $\frac{5^2}{5^2}$
24. $\frac{3^7}{3^2}$
25. $\frac{2^1}{2^4}$
26. $4^{2} \cdot 4^{-3}$
27. $10^{3} \cdot 10^{-6}$
28. $6^{4} \cdot 6^{-2}$

PRACTICE AND PROBLEM SOLVING

Evaluate.

29. 2^7
30. $\frac{5^7}{5^5}$
31. $\frac{m^9}{m^2}$
32. $x^{-5} \cdot x^{7}$
33. $\frac{(-3)^2}{(-3)^4}$
34. $8^{4} \cdot 8^{-4}$
35. $4^{9} \cdot 4^{-4}$
36. $\frac{7^2}{8^6}$
37. $2^{-2} \cdot 2^{-2} \cdot 2^{3}$
38. $\frac{(7-3)^3}{(5-1)^6}$
39. $(5-3)^{-7} \cdot (7-5)^{5}$
40. $\frac{(4-11)^5}{(1-8)^2}$
41. $(2 \cdot 6)^{-5} \cdot (4 \cdot 3)^{3}$
42. $\frac{(3+2)^4}{5(7-2)^3}$
43. $(2+2)^{-5} \cdot (1+3)^{6}$
44. ***COMPUTER SCIENCE*** Computer files are measured in bytes. One byte contains approximately 1 character of text.

	Byte	Kilobyte (KB)	Megabyte (MB)	Gigabyte (GB)
Value (bytes)	$2^0 = 1$	2^{10}	2^{20}	2^{30}

a. If a hard drive on a computer holds 2^{35} bytes of data, how many gigabytes does the hard drive hold?

b. A Zip® disk holds about 2^8 MB of data. How many bytes is that?

Science LINK

Prefixes for the International System of Units										
Factor	10^3	10^2	10^1	10^{-1}	10^{-2}	10^{-3}	10^{-6}	10^{-9}	10^{-12}	10^{-15}
Prefix	kilo-	hecto-	deca-	deci-	centi-	milli-	micro-	nano-	pico-	femto-
Symbol	k	h	da	d	c	m	μ	n	p	f

45. The sperm whale is the deepest diving whale. It can dive to depths greater than 10^{12} nanometers. How many kilometers is that?

46. The greatest known depth of the Arctic ocean is about 10^6 millimeters. How many hectometers is that?

47. The primary food in the blue whale's diet is a crustacean called a krill. One krill weighs approximately 10^{-5} kg.

a. How many grams does one krill weigh?

b. If a blue whale eats 10^7 krill, how many grams of krill is that?

c. How many decagrams do 10^7 krill weigh?

48. The tropical brittlestar is a sea creature that lives in the coral reef. It is covered with 20,000 crystal eyes that are each about 100 micrometers wide.

a. How many meters wide is one crystal eye?

b. How long would a row of 10^5 crystal eyes be in meters?

49. ***CHALLENGE*** A cubic centimeter is the same as 1 mL. If a humpback whale has more than 1 kL of blood, how many cubic centimeters of blood does the humpback whale have?

Krill may be up to 2 in. long, nearly $\frac{1}{288}$ of the length of a humpback whale, pictured above.

Spiral Review

Evaluate each expression for the given values of the variables. (Lesson 1-1)

50. $2x - 3y$ for $x = 8$ and $y = 4$

51. $6s - t$ for $s = 7$ and $t = 12$

52. $7w + 2z$ for $w = 3$ and $z = 0$

53. $5x + 4y$ for $x = 9$ and $y = 10$

54. **EOG PREP** Which of the following numbers is greater than 1? (previous course)

A -235 B 1.000008 C 0.99999 D -5.88

2-9 Scientific Notation

Learn to express large and small numbers in scientific notation.

Vocabulary

scientific notation

An ordinary penny contains about 20,000,000,000,000,000,000,000 atoms. The average size of an atom is about 0.00000003 centimeter across.

The length of these numbers in standard notation makes them awkward to work with.

Scientific notation is a shorthand way of writing such numbers.

$$1.8 \times 10^4$$

In scientific notation the number of atoms in a penny is 2.0×10^{22}, and the size of each atom is 3.0×10^{-8} centimeters across.

EXAMPLE 1 **Translating Scientific Notation to Standard Notation**

Write each number in standard notation.

A 2.64×10^7

2.64×10^7

$2.64 \times 10{,}000{,}000$ *$10^7 = 10{,}000{,}000$*

$26{,}400{,}000$ *Think: Move the decimal right 7 places.*

B 1.35×10^{-4}

1.35×10^{-4}

$1.35 \times \frac{1}{10{,}000}$ *$10^{-4} = \frac{1}{10{,}000}$*

$1.35 \div 10{,}000$ *Divide by the reciprocal.*

0.000135 *Think: Move the decimal left 4 places.*

C -5.8×10^6

-5.8×10^6

$-5.8 \times 1{,}000{,}000$ *$10^6 = 1{,}000{,}000$*

$-5{,}800{,}000$ *Think: Move the decimal right 6 places.*

Helpful Hint

The sign of the exponent tells which direction to move the decimal. A positive exponent means move the decimal to the right, and a negative exponent means move the decimal to the left.

EXAMPLE 2 Translating Standard Notation to Scientific Notation

Write 0.000002 in scientific notation.

0.000002

2 — *Move the decimal to get a number between 1 and 10.*

$2 \times 10^{\blacksquare}$ — *Set up scientific notation.*

Think: The decimal needs to move left to change 2 to 0.000002, so the exponent will be negative.

Think: The decimal needs to move 6 places.

So 0.000002 written in scientific notation is 2×10^{-6}.

Check $2 \times 10^{-6} = 2 \times 0.000001 = 0.000002$

EXAMPLE 3 *Money Application*

Suppose you have a million dollars in pennies. A penny is 1.55 mm thick. How tall would a stack of all your pennies be? Write the answer in scientific notation.

$1.00 = 100 pennies, so $1,000,000 = 100,000,000 pennies.

$1.55 \text{ mm} \times 100{,}000{,}000$ — *Find the total height.*

155,000,000 mm

$1.55 \times 10^{\blacksquare}$ — *Set up scientific notation.*

Think: The decimal needs to move right to change 1.55 to 155,000,000, so the exponent will be positive.

Think: The decimal needs to move 8 places.

In scientific notation the total height of one million dollars in stacked pennies is 1.55×10^8 mm. This is about 96 miles tall.

Think and Discuss

1. **Explain** the benefit of writing numbers in scientific notation.
2. **Describe** how to write 2.977×10^6 in standard notation.
3. **Determine** which measurement would be least helpful in scientific notation: size of bacteria, speed of a car, or number of stars.

2-9 Exercises

FOR EOG PRACTICE

see page 651

internet connect
Homework Help Online
go.hrw.com Keyword: MT4 2-9

1.01, 1.02

GUIDED PRACTICE

See Example 1 **Write each number in standard notation.**

1. 3.15×10^3 **2.** 1.25×10^{-7} **3.** 4.1×10^5 **4.** 3.9×10^{-4}

See Example 2 **Write each number in scientific notation.**

5. 0.000057 **6.** 0.0003 **7.** 4,890,000 **8.** 0.00000014

See Example 3 **9.** The temperature on the Sun's surface is about 5500°C. Scientists believe that the temperature at the center of the Sun is 270 times hotter. What is the temperature at the center of the Sun? Write the answer in scientific notation.

INDEPENDENT PRACTICE

See Example 1 **Write each number in standard notation.**

10. 8.3×10^5 **11.** 6.7×10^{-4} **12.** 2.1×10^{-3} **13.** 6.37×10^7

See Example 2 **Write each number in scientific notation.**

14. 0.000009 **15.** 7,800,000 **16.** 1,000,000,000 **17.** 0.00000003

See Example 3 **18.** Protons and neutrons make up the nucleus of an atom and are the most massive particles in the atom. In fact, if a nucleus were the size of an average grape, it would have a mass greater than 9 million metric tons. A metric ton is 1000 kg. What would the mass of a grape-size nucleus be in kilograms? Write your answer in scientific notation.

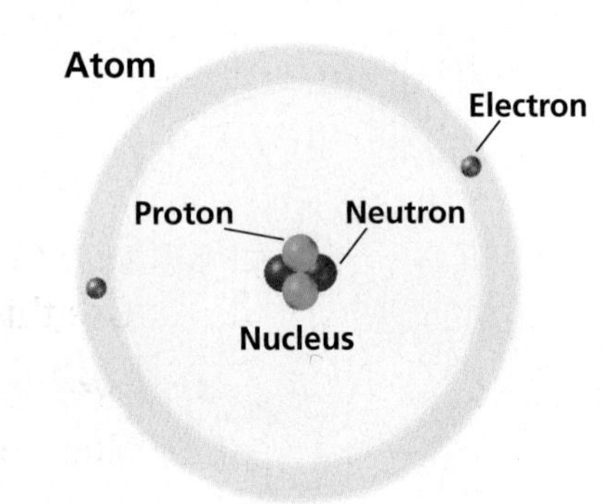

PRACTICE AND PROBLEM SOLVING

Write each number in standard notation.

19. 1.3×10^4 **20.** 4.45×10^{-2} **21.** 5.6×10^1 **22.** 1.3×10^{-7}

23. 5.3×10^{-8} **24.** 9.567×10^{-5} **25.** 8.58×10^6 **26.** 7.1×10^3

27. 9.112×10^6 **28.** 3.4×10^{-1} **29.** 2.9×10^{-4} **30.** 6.8×10^2

Write each number in scientific notation.

31. 0.00467 **32.** 0.00000059 **33.** 56,000,000 **34.** 8,079,000,000

35. 0.0076 **36.** 0.0000000002 **37.** 3500 **38.** 0.0000000091

39. 900 **40.** 0.000005 **41.** 6,000,000 **42.** 0.0095678

43. ***SOCIAL STUDIES***

a. Express the population and area of Taiwan in scientific notation.

b. Divide the number of square miles by the population to find the number of square miles per person in Taiwan. Express your answer in scientific notation.

Taiwan

Population:	22,113,250
Area:	14,032 mi^2
Capital:	Taipei
Number of televisions:	10,800,000
Languages:	Taiwanese (Min), Mandarin, Hakka dialects

Life Science LINK

This frog is covered with duckweed plants. Duckweed plants can grow both in sunlight and in shade. They produce tiny white flowers that are nearly invisible to the human eye.

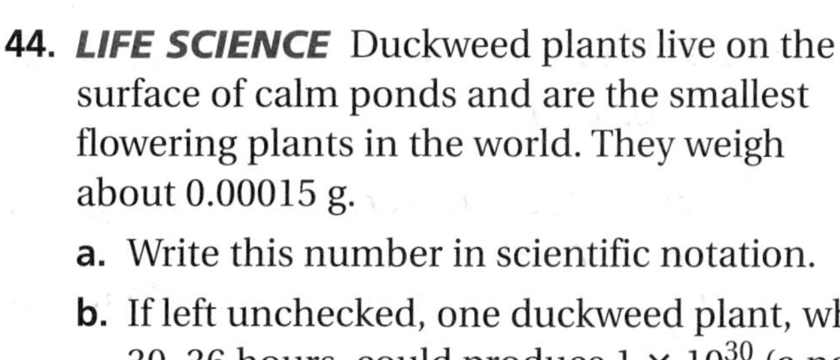

44. ***LIFE SCIENCE*** Duckweed plants live on the surface of calm ponds and are the smallest flowering plants in the world. They weigh about 0.00015 g.

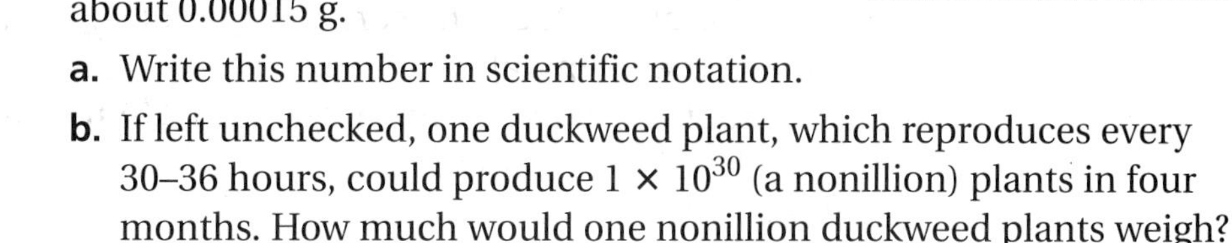

a. Write this number in scientific notation.

b. If left unchecked, one duckweed plant, which reproduces every 30–36 hours, could produce 1×10^{30} (a nonillion) plants in four months. How much would one nonillion duckweed plants weigh?

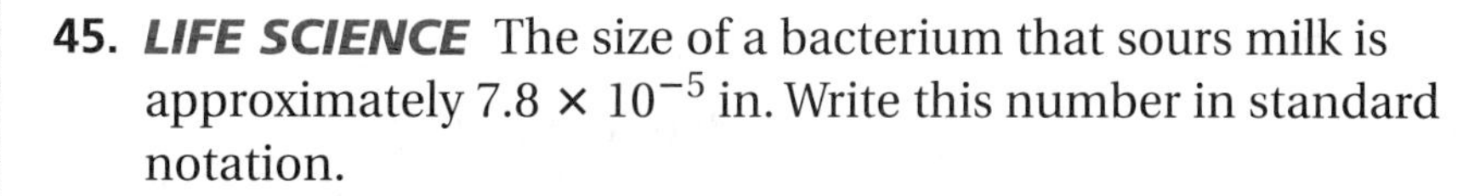

45. ***LIFE SCIENCE*** The size of a bacterium that sours milk is approximately 7.8×10^{-5} in. Write this number in standard notation.

46. ***PHYSICAL SCIENCE*** The *atomic mass* of an element is the mass, in grams, of one *mole* (mol), or 6.02×10^{23} atoms.

a. How many atoms are there in 3.5 mol of carbon?

b. If you know that 3.5 mol of carbon weighs 42 grams, what is the atomic mass of carbon?

c. Using your answer from part **b,** find the approximate mass of one atom of carbon.

47. ***WRITE A PROBLEM*** A proton has a mass of about 1.7×10^{-24} g. Use this information to write a problem.

48. ***WRITE ABOUT IT*** Two numbers are written in scientific notation. How can you tell which number is greater?

49. ***CHALLENGE*** Where on a number line does the value of a positive number in scientific notation with a negative exponent lie?

Spiral Review

Simplify. (Lesson 2-3)

50. $-3(6 - 8)$ **51.** $4(-3 - 2)$ **52.** $-5(3 + 2)$ **53.** $-3(1 - 8)$

Solve. (Lesson 2-4)

54. $m - 2 = 7$ **55.** $8 + t = -1$ **56.** $y - 24 = -19$ **57.** $b + 4 = -23$

58. **EOG PREP** Which number is equivalent to -64? (Lesson 2-6)

A $(-4)^3$ B $(-4)^{-3}$ C 4^3 D 4^{-3}

NORTH CAROLINA

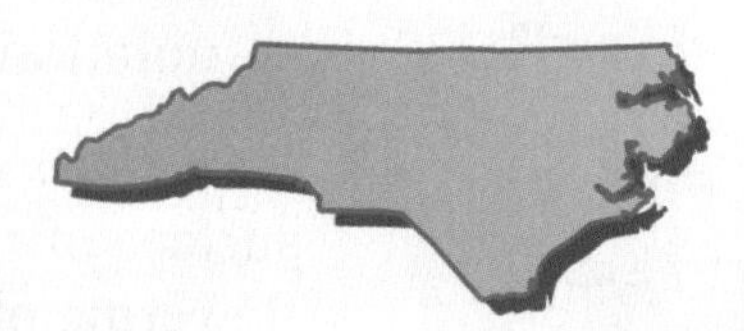

Artificial Reefs

A number of artificial reefs have been built off the coast of North Carolina. Referred to as *hard-bottom area,* each reef is begun when environmentally safe objects made from steel or concrete are dropped into the ocean and allowed to settle to the bottom. Over time, the structures develop reefs comparable to the natural *live-bottom reefs* of the Florida Keys.

1. A diver dove to reef AR-225, which is made from train cars, and then swam 17 ft up toward the surface of the water before stopping to rest. At what depth did the diver stop to rest?

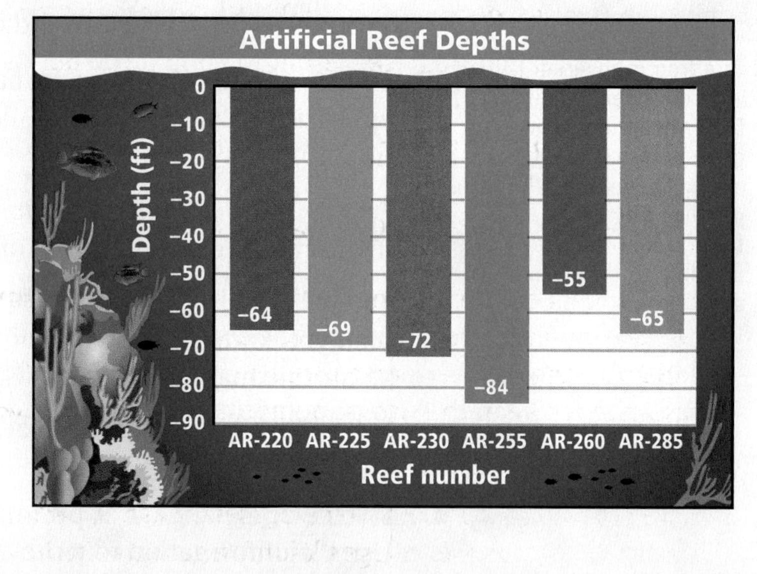

For 2–3, write and solve equations to find the answers.

2. The 130 ft yard freighter at AR-230 sits 8 ft deeper than which reef?

3. The boxcars of AR-220 rest at a depth 20 ft shallower than which reef?

4. George went on a diving expedition to the artificial reefs. First, he dove to the sunken crewboat *Miss Clara* at reef AR-260. Then he dove to a different reef on his second dive. The difference in depth between his first dive and his second dive was less than 17 ft. Write and solve an inequality to determine which of the reefs on the graph George could have chosen for his second dive.

Bog Turtles

Ranging from 3 to 3.5 inches in length, bog turtles are the smallest species of turtles in the United States and the second smallest in the world. A majority of the bog turtle population in the South can be found at altitudes of up to 4000 ft, in the Blue Ridge Mountains.

1. Bog turtles were placed on the federal list of threatened species in 1997, due to a rapid decrease in population over a 20-year period. Suppose the North Carolina population of bog turtles in 1977 was 10,000, and the population decreased by 250 turtles each year for 20 years.
   a. Write an equation to represent the bog turtle population y after x years.
   b. Make a table of ordered pairs for your equation from part **a** for $x = 5, 10, 15,$ and 20.
   c. Graph your equation from part **a** on a coordinate plane.

2. Researchers divide the bog turtle population into a northern population and a southern population. Geographically, 250 miles separates the northern and southern populations. What is the distance between the two populations in feet? Write your answer in both standard and scientific notation.

3. A bog turtle can lay from 1 to 6 eggs per year. Suppose a scientist recorded the number of eggs laid by a group of turtles every eighth year. The original bog turtle laid 6 eggs in the first recorded year, and then each of those 6 turtles laid 6 eggs during the second recorded year. Then, in the third recorded year, each of the 36, or 6^2, turtles from the second recorded year laid 6 eggs, and so on. How many new turtles were there in the third, fourth, and fifth recorded years? Write your answers in exponential form. What was the total number of turtles after 5 recorded years, written as a sum of powers?

MATH-ABLES

Magic Squares

A *magic square* is a square with numbers arranged so that the sums of the numbers in each row, column, and diagonal are the same.

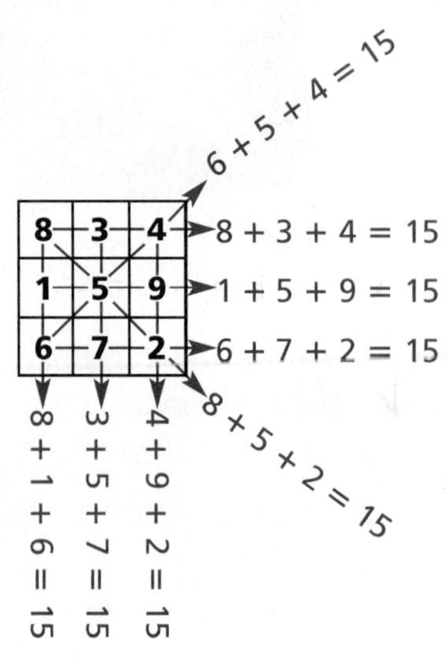

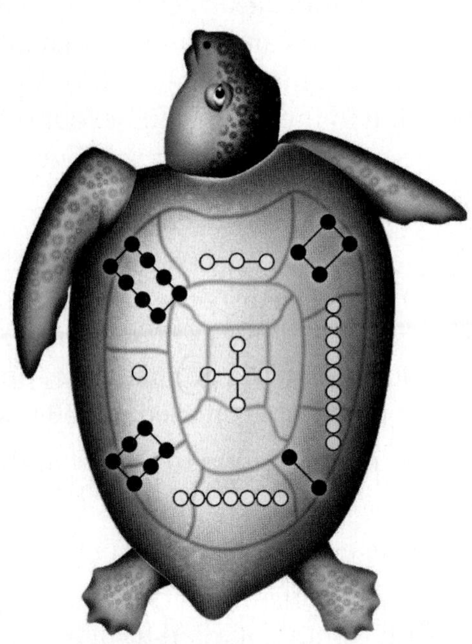

According to an ancient Chinese legend, a tortoise from the Lo river had the pattern of this magic square on its shell.

1. Complete each magic square below.

6		4
1	3	
	7	

	−6	−1
−4		0
−3	2	

−7		6	−4
4	−2		1
	2	3	−3
5	−5	−6	

2. Use the numbers −4, −3, −2, −1, 0, 1, 2, 3, and 4 to make a magic square with row, column, and diagonal sums of 0.

Equation Bingo

Each bingo card has numbers on it. The caller has a collection of equations. The caller reads an equation, and then the players solve the equation for the variable. If players have the solution on their cards, they place a chip on it. The winner is the first player with a row of chips either down, across, or diagonally.

internet connect

For a complete set of rules and cards, visit ***go.hrw.com***
KEYWORD: MT4 Game2

Evaluate Expressions

Use with Lesson 2-8

A graphing calculator can be used to evaluate expressions that have negative exponents.

Activity

1 Use the STO▶ button to evaluate x^{-3} for $x = 2$. View the answer as a decimal and as a fraction.

2 STO▶ X,T,θ,n ENTER   X,T,θ,n ^ (−) 3 ENTER   MATH ENTER ENTER

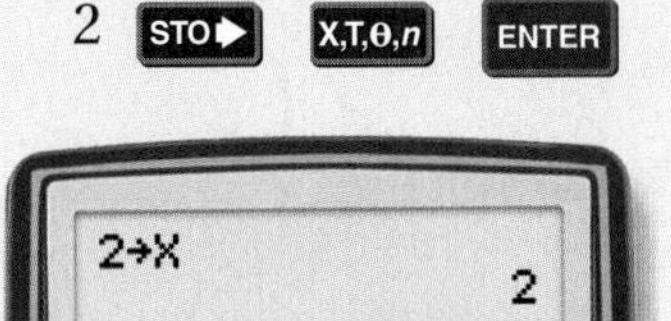

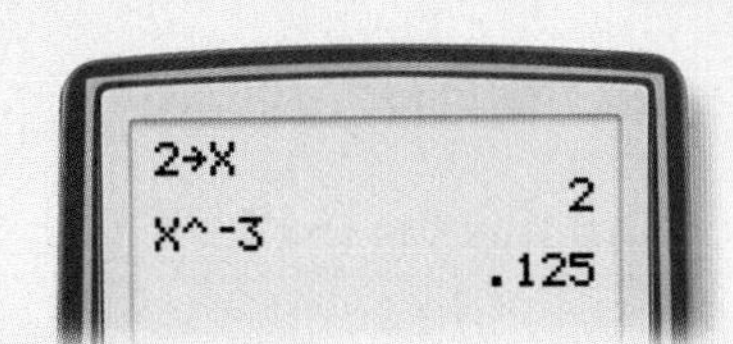

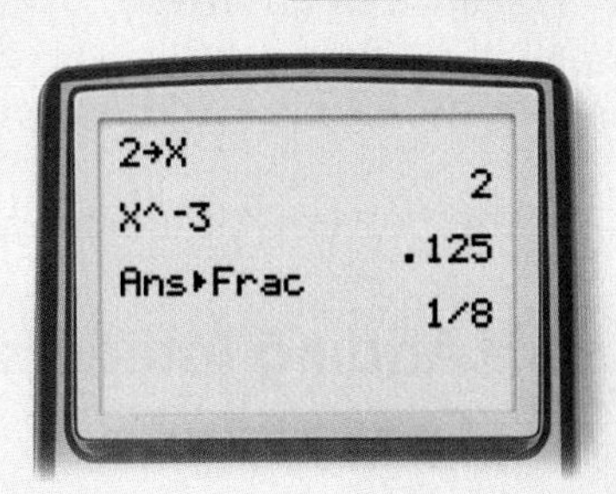

Notice that $2^{-3} = 0.125$, which is equivalent to $\frac{1}{2^3}$, or $\frac{1}{8}$.

2 Use the **TABLE** feature to evaluate 2^{-x} for several x-values. Match the settings shown.

Y= 2 ^ (−) X,T,θ,n ENTER   2nd WINDOW (TBLSET)   2nd GRAPH (TABLE)

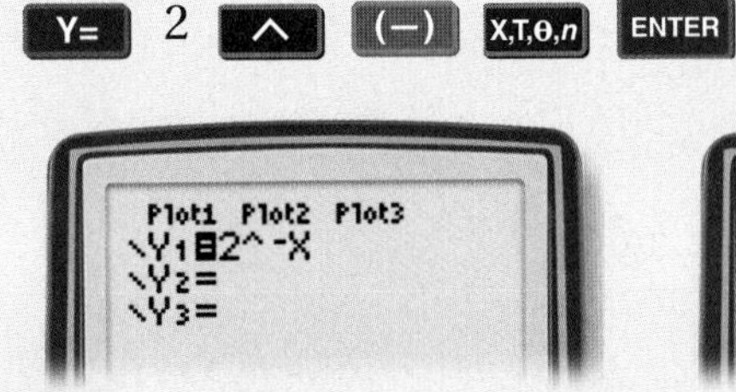

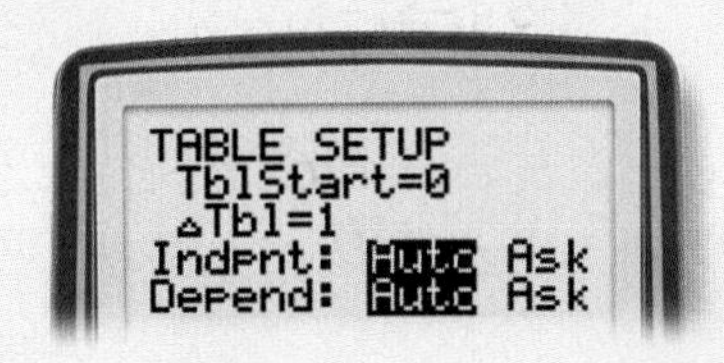

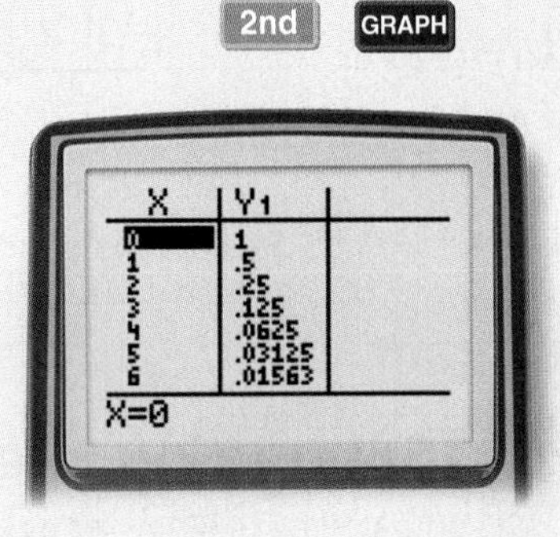

The **Y1** list shows the value of 2^{-x} for several x-values.

Think and Discuss

1. When you evaluated 2^{-3} in Activity 1, the result was not a negative number. Is this surprising? Why or why not?

Try This

Evaluate each expression for the given x-value(s). Give your answers as fractions and as decimals rounded to the nearest hundredth.

1. 4^{-x}; $x = 2$
2. 3^{-x}; $x = 1, 2$
3. x^{-2}; $x = 1, 2, 5$

Chapter 2

Study Guide and Review

Vocabulary

Complete the sentences below with vocabulary words from the list above. Words may be used more than once.

1. The sum of an integer and its ___?___ is 0.

2. A number in ___?___ is a number from 1 to 10 times a(n) ___?___ of 10.

3. In the power 3^5, the 5 is the ___?___ and the 3 is the ___?___ .

2-1 Adding Integers (pp. 60–63)

EXAMPLE

■ **Add.**

$-8 + 2$ — *Find the difference of 8 and 2.*
-6 — *8 > 2; use the sign of the 8.*

■ **Evaluate.**

$-4 + a$ for $a = -7$
$-4 + (-7)$ — *Substitute.*
-11 — *Same sign*

EXERCISES

Add.

4. $-6 + 4$
5. $-3 + (-9)$
6. $4 + (-7)$
7. $4 + (-3)$
8. $-11 + (-5) + (-8)$

Evaluate.

9. $k + 11$ for $k = -3$
10. $-6 + m$ for $m = -2$

2-2 Subtracting Integers (pp. 64–67)

EXAMPLE

■ **Subtract.**

$-3 - (-5)$
$-3 + 5$ — *Add the opposite of −5.*
2 — *5 > 3; use the sign of the 5.*

■ **Evaluate.**

$-9 - d$ for $d = 2$
$-9 - 2$ — *Substitute.*
$-9 + (-2)$ — *Add the opposite of 2.*
-11 — *Same sign*

EXERCISES

Subtract.

11. $-7 - 9$
12. $8 - (-9)$
13. $-2 - (-5)$
14. $13 - (-2)$
15. $-5 - 17$
16. $16 - 20$

Evaluate.

17. $9 - h$ for $h = -7$
18. $12 - z$ for $z = 17$

2-3 Multiplying and Dividing Integers (pp. 68–71)

EXAMPLE

Multiply or divide.

- $4(-9)$ — *The signs are **different**.*
  -36 — *The answer is **negative**.*
- $\frac{-33}{-11}$ — *The signs are the **same**.*
  3 — *The answer is **positive**.*

EXERCISES

Multiply or divide.

19. $7(-5)$ **20.** $\frac{72}{-4}$ **21.** $-4(-13)$

22. $\frac{-100}{-4}$ **23.** $8(-3)(-5)$ **24.** $\frac{10(-5)}{-25}$

2-4 Solving Equations with Integers (pp. 74–77)

EXAMPLE

Solve.

- $x - 9 = -12$
  $\underline{+9 = +9}$
  $x = -3$
- $y + 4 = -11$
  $\underline{-4 = -4}$
  $y = -15$
- $4m = 20$
  $\frac{4m}{4} = \frac{20}{4}$
  $m = 5$
- $\frac{t}{-2} = 10$
  $(-2) \cdot \frac{t}{-2} = (-2) \cdot 10$
  $t = -20$

EXERCISES

Solve.

25. $p - 8 = 1$ **26.** $t + 4 = 7$

27. $6 + k = 9$ **28.** $-7g = 42$

29. $\frac{w}{-4} = 20$ **30.** $10 = \frac{b}{-2}$

31. $8 = -2a$ **32.** $-13 = \frac{h}{7}$

33. $-15 + s = 23$

Study Guide and Review

2-5 Solving Inequalities with Integers (pp. 78–81)

EXAMPLE

Solve and graph.

- $x + 5 \le -1$
  $\underline{-5 \quad -5}$
  $x \le -6$

−10 −8 −6 −4 −2

- $-3q > 21$
  $\frac{-3q}{-3} > \frac{21}{-3}$
  $q < -7$

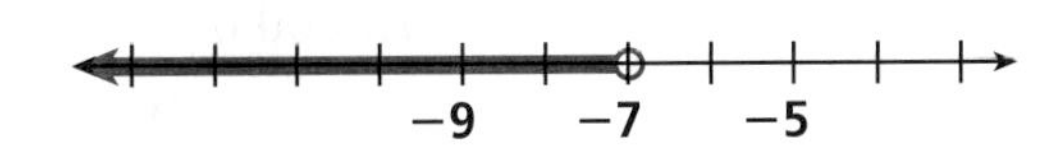

EXERCISES

Solve and graph.

34. $b + 3 < 1$ **35.** $r - 2 > 4$

36. $2m \ge 6$ **37.** $4p < -8$

38. $-2z > 10$ **39.** $-3q \le -9$

40. $\frac{m}{2} \ge 2$ **41.** $\frac{x}{-3} < 1$

42. $\frac{y}{-1} > -4$ **43.** $4 + x > 1$

44. $-3b \ge 0$ **45.** $-2 + y < 4$

2-6 Exponents (pp. 84–87)

EXAMPLE

■ Write in exponential form.

$4 \cdot 4 \cdot 4$

4^3

■ Evaluate the power.

$(-2)^3$

$(-2) \cdot (-2) \cdot (-2)$

-8

EXERCISES

Write in exponential form.

46. $7 \cdot 7 \cdot 7$

47. $(-3) \cdot (-3)$

48. $k \cdot k \cdot k \cdot k$

Evaluate each power.

49. 5^4

50. $(-2)^5$

51. $(-1)^9$

2-7 Properties of Exponents (pp. 88–91)

EXAMPLE

Write the product or quotient as one power.

■ $2^5 \cdot 2^3$

2^{5+3}

2^8

■ $\frac{10^9}{10^2}$

10^{9-2}

10^7

EXERCISES

Write the product or quotient as one power.

52. $4^2 \cdot 4^5$

53. $9^2 \cdot 9^4$

54. $p \cdot p^3$

55. $\frac{8^5}{8^2}$

56. $\frac{9^3}{9}$

57. $\frac{m^7}{m^2}$

58. $5^0 \cdot 5^3$

59. $y^6 \div y$

60. $k^4 \div k^4$

2-8 Looking for a Pattern in Integer Exponents (pp. 92–95)

EXAMPLE

Evaluate.

■ $(-3)^{-2}$

$\frac{1}{(-3)^2}$

$\frac{1}{9}$

■ $\frac{2^5}{2^5}$

2^{5-5}

2^0

1

EXERCISES

Evaluate.

61. 5^{-3}

62. $(-4)^{-3}$

63. 11^{-1}

64. $\frac{7^4}{7^4}$

65. $\frac{5^7}{5^7}$

66. $\frac{x^3}{x^3}$

67. $(9 - 7)^{-3}$

68. $(6 - 9)^{-3}$

2-9 Scientific Notation (pp. 96–99)

EXAMPLE

Write in standard notation.

■ 3.58×10^4

$3.58 \times 10,000$

$35,800$

■ 3.58×10^{-4}

$3.58 \times \frac{1}{10,000}$

$3.58 \div 10,000$

0.000358

Write in scientific notation.

■ $0.000007 = 7 \times 10^{-6}$

■ $62,500 = 6.25 \times 10^4$

EXERCISES

Write in standard notation.

69. 1.62×10^3

70. 1.62×10^{-3}

71. 9.1×10^5

72. 9.1×10^{-5}

Write in scientific notation.

73. 0.000000008

74. 73,000,000

75. 0.0000096

76. 56,400,000,000

Chapter 2

Chapter Test

Perform the given operations.

1. $-9 + (-12)$ **2.** $11 - 17$ **3.** $6(-22)$ **4.** $(-20) \div (-4)$

5. $42 - (-5)$ **6.** $-18 \div 3$ **7.** $-9 - (-13)$ **8.** $12 - (-6) + (-5)$

9. $-2(-21 - 17)$ **10.** $(-15 + 3) \div (-4)$ **11.** $(54 \div 6) - (-1)$ **12.** $-(16 + 4) - 20$

13. The temperature on a winter day increased 37°F. If the beginning temperature was −9°F, what was the temperature after the increase?

Evaluate each expression for the given value of the variable.

14. $16 - p$ for $p = -12$ **15.** $t - 7$ for $t = -14$

16. $13 - x + (-2)$ for $x = 4$ **17.** $-8y + 27$ for $y = -9$

Solve.

18. $y + 19 = 9$ **19.** $4z = -32$ **20.** $52 = p - 3$ **21.** $\frac{w}{3} = 9$

22. $t + 1 < 7$ **23.** $z - 4 \geq 7$ **24.** $\frac{m}{-2} \leq 6$ **25.** $-3q > 15$

Graph each inequality.

26. $x > -4$ **27.** $n \leq 3$

Evaluate each power.

28. 4^3 **29.** $(-5)^4$ **30.** $(-3)^5$

Multiply or divide. Write the product or quotient as one power.

31. $7^4 \cdot 7^5$ **32.** $\frac{12^5}{12^2}$ **33.** $x \cdot x^3$

Evaluate.

34. $(12 - 3)^2$ **35.** $40 + 5^3$ **36.** $\frac{3^4}{3^7}$ **37.** $10^4 \cdot 10^{-4}$

Write each number in standard notation.

38. 3×10^6 **39.** 3.1×10^{-6} **40.** 4.52×10^5

Write each number in scientific notation.

41. 3000 **42.** 42,000,000 **43.** 0.00000092

44. A sack of cocoa beans weighs about 132 lb. How much would one thousand sacks of cocoa beans weigh? Write the answer in scientific notation.

Chapter Test

Chapter 2

Performance Assessment

Show What You Know

Create a portfolio of your work from this chapter. Complete this page and include it with your four best pieces of work from Chapter 2. Choose from your homework or lab assignments, mid-chapter quiz, or any journal entries you have done. Put them together using any design you want. Make your portfolio represent what you consider your best work.

Short Response

1. a. Complete the following rules for operations involving odd and even numbers:

even + even = __?__ odd + odd = __?__ odd + even = __?__

even · even = __?__ odd · odd = __?__ even · odd = __?__

b. Compare the rules from part **a** with the rules for finding the sign when multiplying two integers.

2. Write the subtraction equation $4 - 6 = -2$ as an addition equation. Draw a number-line diagram to illustrate the addition equation.

3. Consider the statement "Half of a number is less than or equal to -2." Write an inequality for this word sentence and solve it. Show your work.

Extended Problem Solving

4. The formula for converting degrees Celsius (°C) to degrees Fahrenheit (°F) is $F = \frac{9}{5}C + 32$. A way to estimate the temperature in degrees Fahrenheit is to double the temperature in degrees Celsius and add 30.

a. Write the way of estimating as a formula.

b. Compare the results for the exact formula and the estimate formula for -10°C, 0°C, 30°C, and 100°C.

c. For which of the values was the estimate closest to the exact answer? Find a temperature in degrees Celsius for which the estimate and the exact answer are the same. Show your work.

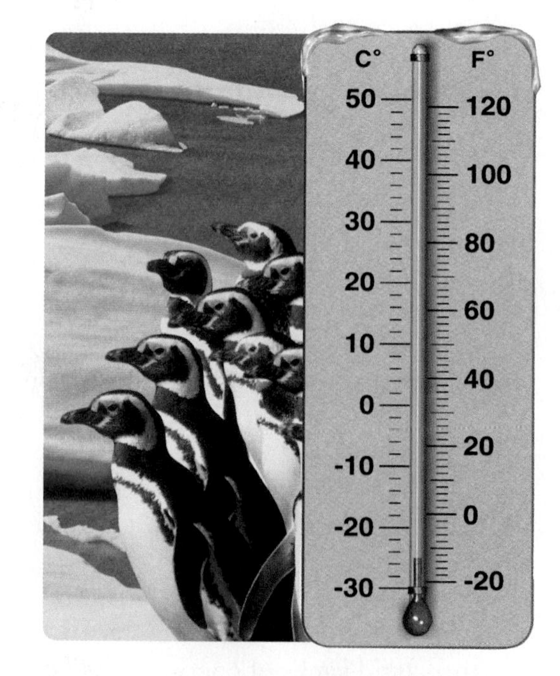

Performance Assessment

internet connect
EOG Test Practice Online
go.hrw.com Keyword: MT4 TestPrep

Getting Ready for EOG

Chapter 2

Cumulative Assessment, Chapters 1–2

1. If $(n + 3)(9 - 5) = 16$, then what does n equal?

   A 1 C 7
   B 4 D 9

2. If $x = -\frac{1}{4}$, which is least?

   A $1 - x$ C x
   B $x - 1$ D $1 \div x$

TEST TAKING TIP!

Make comparisons: Express quantities in a common number base.

3. Which ratio compares the value of a hundred \$1000 bills with the value of a thousand \$100 bills?

   A 1 to 10 C 5 to 1
   B 1 to 1 D 10 to 1

4. Which is $3 \times 3 \times 3 \times 3 \times 11 \times 11 \times 11$ expressed in exponential form?

   A $4^3 \times 3^{11}$ C 33^7
   B $3^4 \times 11^3$ D 33^3

5. Which number is equivalent to 2^{-5}?

   A $\frac{1}{10}$ C $-\frac{1}{32}$
   B $\frac{1}{32}$ D $-\frac{1}{10}$

6. Which is 8.1×10^{-5}?

   A 8,100,000 C 0.000081
   B 810,000 D 0.0000081

7. Which power is equivalent to $5^{12} \div 5^4$?

   A 1^3 C 5^3
   B 1^8 D 5^8

8. The bar graph shows the average daily temperatures in Sturges, Michigan, for five months. Between which two months did the average temperature change by the greatest amount?

   A January and February
   B February and March
   C March and April
   D April and May

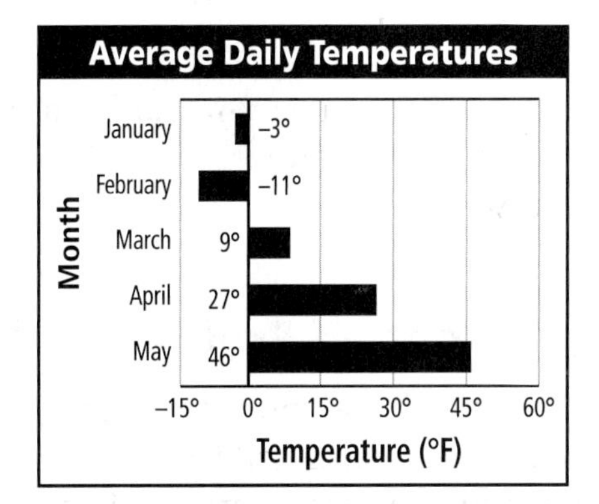

9. ***SHORT RESPONSE*** Linda takes her grandson Colin to the ice cream parlor every Wednesday and spends \$6.50. During a 30-day month that began on a Monday, how much money did Linda spend at the ice cream parlor? Explain how you found your answer.

10. ***SHORT RESPONSE*** An elevator begins 7 floors above ground level and descends to a floor that is 2 floors below ground level. Each floor is 12 feet high. Draw a diagram to determine the number of feet the elevator traveled.

Chapter 3

Rational and Real Numbers

TABLE OF CONTENTS

internet connect

Chapter Opener Online
go.hrw.com
KEYWORD: MT4 Ch3

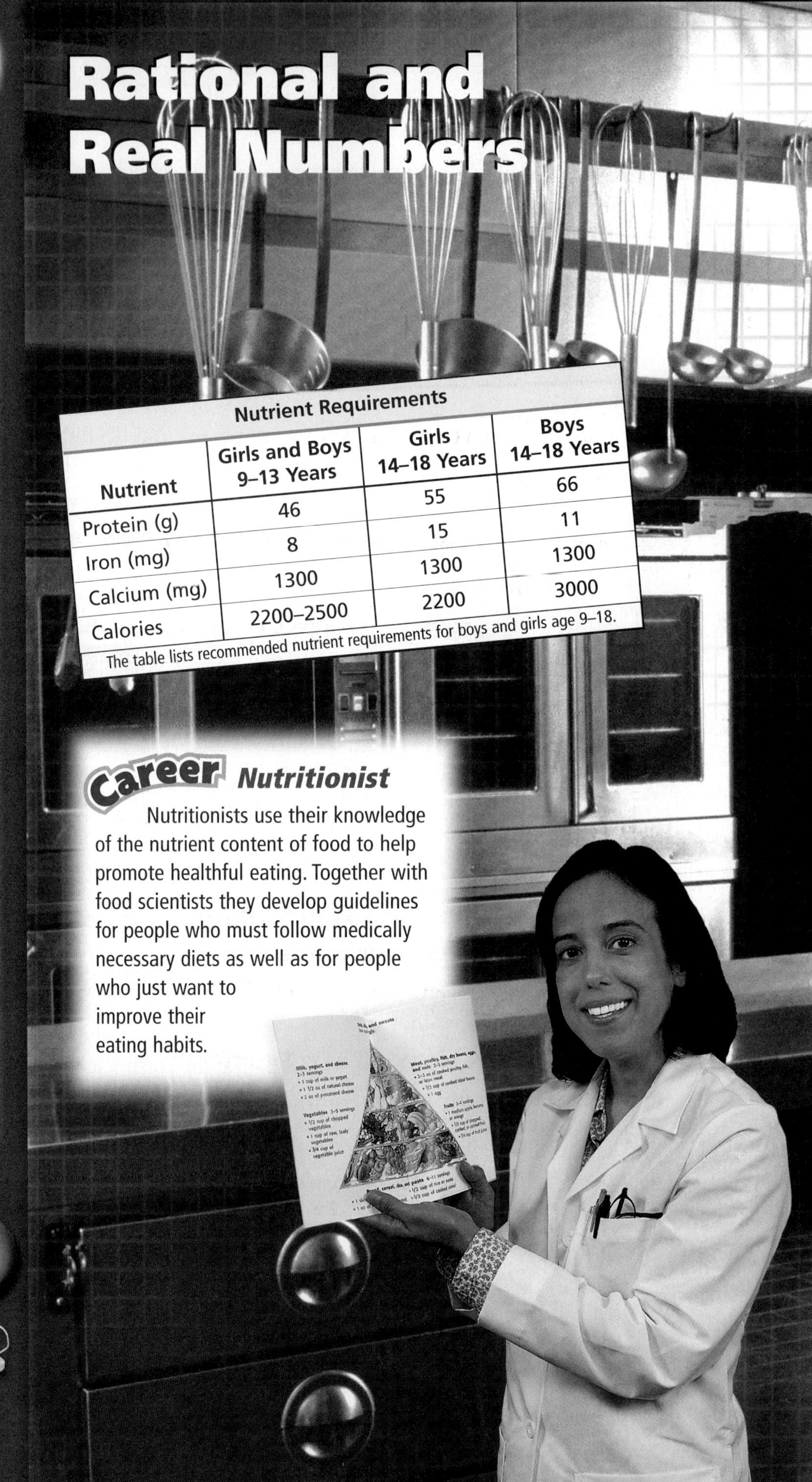

Nutrient Requirements

Nutrient	Girls and Boys 9–13 Years	Girls 14–18 Years	Boys 14–18 Years
Protein (g)	46	55	66
Iron (mg)	8	15	11
Calcium (mg)	1300	1300	1300
Calories	2200–2500	2200	3000

The table lists recommended nutrient requirements for boys and girls age 9–18.

Career ***Nutritionist***

Nutritionists use their knowledge of the nutrient content of food to help promote healthful eating. Together with food scientists they develop guidelines for people who must follow medically necessary diets as well as for people who just want to improve their eating habits.

ARE YOU READY?

Choose the best term from the list to complete each sentence.

equivalent fraction
fraction
improper fraction
mixed number
proper fraction

1. A number that consists of a whole number and a fraction is called a(n) __?__.
2. A(n) __?__ is a number that represents a part of a whole.
3. A fraction whose absolute value is greater than 1 is called a(n) __?__, and a fraction whose absolute value is between 0 and 1 is called a(n) __?__.
4. A(n) __?__ names the same value.

Complete these exercises to review skills you will need for this chapter.

✔ Model Fractions

Write a fraction to represent the shaded portion of each diagram.

5.

6.

7.

8.

✔ Write a Fraction as a Mixed Number

Write each improper fraction as a mixed number.

9. $\frac{22}{7}$
10. $\frac{18}{5}$
11. $\frac{104}{25}$
12. $\frac{65}{9}$

✔ Write a Mixed Number as a Fraction

Write each mixed number as an improper fraction.

13. $7\frac{1}{4}$
14. $10\frac{3}{7}$
15. $5\frac{3}{8}$
16. $11\frac{1}{11}$

✔ Write Equivalent Fractions

Supply the missing information.

17. $\frac{3}{8} = \frac{■}{24}$
18. $\frac{5}{13} = \frac{■}{52}$
19. $\frac{7}{12} = \frac{■}{36}$
20. $\frac{8}{15} = \frac{■}{45}$

3-1 Rational Numbers

Learn to write rational numbers in equivalent forms.

Vocabulary

rational number

relatively prime

In 2001, the Wimbledon tennis tournament increased the number of "seeds" from 16 to 32. Since Wimbledon has 128 total players, $\frac{32}{128}$ of the players are seeded.

A **rational number** is any number that can be written as a fraction $\frac{n}{d}$, where n and d are integers and $d \neq 0$.

Decimals that terminate or repeat are rational numbers.

Some ranked players, like Venus and Serena Williams, are "seeded" so that they will not meet until late in a tournament.

Rational Number	Description	Written as a Fraction
−1.5	Terminating decimal	$\frac{-15}{10}$
$0.8\overline{3}$	Repeating decimal	$\frac{5}{6}$

The goal of simplifying fractions is to make the numerator and the denominator *relatively prime*. **Relatively prime** numbers have no common factors other than 1.

You can often simplify fractions by dividing both the numerator and denominator by the same nonzero integer. You can simplify the fraction $\frac{12}{15}$ to $\frac{4}{5}$ by dividing both the numerator and denominator by 3.

12 of the 15 boxes are shaded.

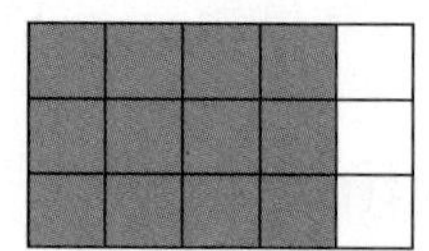

$$\frac{12}{15} = \frac{4}{5}$$

4 of the 5 boxes are shaded.

The same total area is shaded.

EXAMPLE 1 Simplifying Fractions

Simplify.

A $\frac{6}{9}$

$\frac{6}{9} = \frac{6 \div 3}{9 \div 3}$ *$6 = 2 \cdot 3$, $9 = 3 \cdot 3$; 3 is a common factor.*

$= \frac{2}{3}$ *Divide the numerator and denominator by 3.*

Simplify.

B $\frac{21}{25}$

$21 = 3 \cdot 7$, $25 = 5 \cdot 5$; *there are no common factors.*

$\frac{21}{25} = \frac{21}{25}$ *21 and 25 are relatively prime.*

C $\frac{-24}{32}$

$\frac{-24}{32} = \frac{-24 \div 8}{32 \div 8}$ $24 = (2 \cdot 2 \cdot 2) \cdot 3$, $32 = (2 \cdot 2 \cdot 2) \cdot 2 \cdot 2$ *8 is a common factor.*

$= \frac{-3}{4}$, or $-\frac{3}{4}$ *Divide the numerator and denominator by 8.*

Remember!

$\frac{0}{a} = 0$ for $a \neq 0$

$\frac{a}{a} = 1$ for $a \neq 0$

$\frac{-3}{4} = \frac{3}{-4} = -\frac{3}{4}$

To write a finite decimal as a fraction, identify the place value of the digit farthest to the right. Then write all of the digits after the decimal point as the numerator with the place value as the denominator.

Place Value

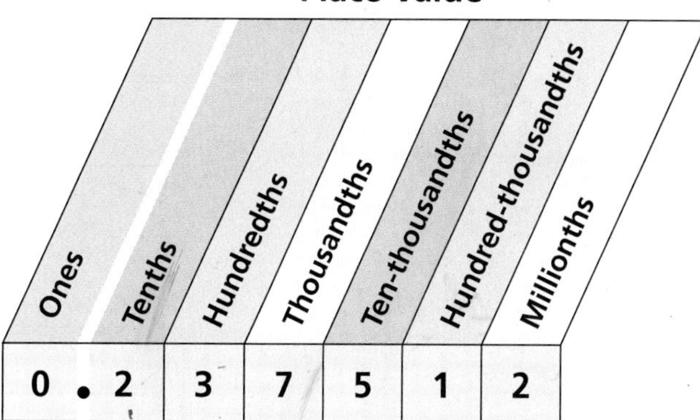

Ones
Tenths
Hundredths
Thousandths
Ten-thousandths
Hundred-thousandths
Millionths
0 • 2 3 7 5 1 2

EXAMPLE 2 Writing Decimals as Fractions

Write each decimal as a fraction in simplest form.

A **0.5**

0.5

$= \frac{5}{10}$ *5 is in the tenths place.*

$= \frac{1}{2}$ *Simplify by dividing by the common factor 5.*

B **−2.37**

-2.37

$= -2\frac{37}{100}$ *7 is in the hundredths place.*

C **0.8716**

0.8716

$= \frac{8716}{10{,}000}$ *6 is in the ten-thousandths place.*

$= \frac{2179}{2500}$ *Simplify by dividing by the common factor 4.*

To write a fraction as a decimal, divide the numerator by the denominator. You can use long division.

When writing a long division problem from a fraction, put the numerator inside the "box," or division symbol. It may help to write the numerator first and then say "divided by" to yourself as you write the division symbol.

$$\frac{\text{numerator}}{\text{denominator}} \rightarrow \text{denominator}\overline{)\text{numerator}}$$

EXAMPLE 3 Writing Fractions as Decimals

Write each fraction as a decimal.

A $\frac{5}{4}$

$$\begin{array}{r} 1.25 \\ 4\overline{)5.00} \\ -4 \\ \hline 10 \\ -8 \\ \hline 20 \\ -20 \\ \hline 0 \end{array}$$

The remainder is 0. This is a terminating decimal.

The fraction $\frac{5}{4}$ is equivalent to the decimal 1.25.

B $\frac{1}{6}$

$$\begin{array}{r} 0.1\overline{6} \\ 6\overline{)1.000} \\ -6 \\ \hline 40 \\ -36 \\ \hline 40 \end{array}$$

The pattern repeats, so draw a bar over the 6 to indicate that this is a repeating decimal.

The fraction $\frac{1}{6}$ is equivalent to the decimal $0.1\overline{6}$.

Think and Discuss

1. **Explain** how you can be sure that a fraction is simplified.
2. **Give** the sign of a fraction in which the numerator is negative and the denominator is negative.

3-1 Exercises

1.01

FOR EOG PRACTICE

see page 652

internet connect

Homework Help Online
go.hrw.com Keyword: MT4 3-1

GUIDED PRACTICE

See Example 1 Simplify.

1. $\frac{12}{15}$ 2. $\frac{6}{10}$ 3. $-\frac{16}{24}$ 4. $\frac{11}{27}$

5. $\frac{57}{69}$ 6. $-\frac{20}{24}$ 7. $-\frac{7}{27}$ 8. $\frac{49}{112}$

See Example 2 Write each decimal as a fraction in simplest form.

9. 0.75 10. 1.125 11. 0.431 12. 0.8

13. −2.2 14. 0.625 15. 3.21 16. −0.3878

See Example 3 Write each fraction as a decimal.

17. $\frac{7}{8}$ 18. $\frac{3}{5}$ 19. $\frac{5}{12}$ 20. $\frac{3}{4}$

21. $\frac{16}{4}$ 22. $\frac{1}{8}$ 23. $\frac{12}{5}$ 24. $\frac{9}{4}$

INDEPENDENT PRACTICE

See Example 1 Simplify.

25. $\frac{21}{28}$ 26. $\frac{25}{60}$ 27. $-\frac{17}{34}$ 28. $-\frac{18}{21}$

29. $\frac{13}{17}$ 30. $\frac{22}{35}$ 31. $\frac{64}{76}$ 32. $-\frac{78}{126}$

See Example 2 Write each decimal as a fraction in simplest form.

33. 0.4 34. 3.5 35. 0.71 36. −0.183

37. 1.377 38. 1.450 39. −1.4 40. −2.9

See Example 3 Write each fraction as a decimal.

41. $\frac{3}{8}$ 42. $\frac{11}{12}$ 43. $\frac{7}{5}$ 44. $\frac{9}{20}$

45. $\frac{34}{50}$ 46. $\frac{23}{5}$ 47. $\frac{29}{25}$ 48. $\frac{7}{3}$

PRACTICE AND PROBLEM SOLVING

49. Make up a fraction that cannot be simplified that has 36 as its denominator.

50. Make up a fraction that cannot be simplified that has 27 as its denominator.

51. **a.** Simplify each fraction below.

$\frac{9}{12}$ $\frac{5}{30}$ $\frac{15}{27}$ $\frac{68}{80}$

$\frac{39}{96}$ $\frac{22}{50}$ $\frac{57}{72}$ $\frac{32}{60}$

b. Write the denominator of each simplified fraction as the product of prime factors.

c. Write each simplified fraction as a decimal. Label each as a terminating or repeating decimal.

52. The ruler is marked at every $\frac{1}{16}$ in. Do the labeled measurements convert to terminating or repeating decimals?

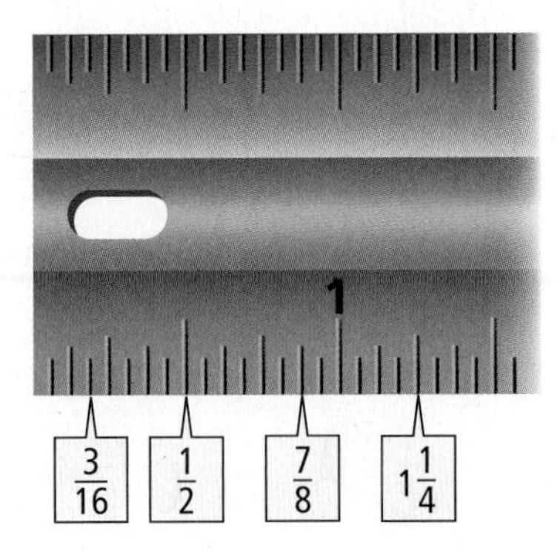

53. Remember that the greatest common factor, GCF, is the largest common factor of two or more given numbers. Find and remove the GCF of 48 and 76 from the fraction $\frac{48}{76}$. Can the resulting fraction be further simplified? Explain.

54. Prices on one stock market are shown using decimal equivalents for fractions or mixed numbers. Write the stock price 13.625 as a mixed number.

55. ***WHAT'S THE ERROR?*** A student simplified a fraction in this manner: $\frac{-12}{-18} = -\frac{2}{3}$. What error did the student make?

56. ***WRITE ABOUT IT*** Using your answers to Exercise 51, examine the prime factors in the denominators of the simplified fractions that are equivalent to terminating decimals. Then examine the prime factors in the denominators of the simplified fractions that are equivalent to repeating decimals. What pattern do you see?

57. ***CHALLENGE*** A student simplified a fraction to $-\frac{3}{7}$ by removing the common factors, which were 3 and 7. What was the original fraction?

Spiral Review

Evaluate each expression for the given values of the variable. (Lesson 1-1)

58. $3x + 5$ for $x = 2$ and $x = 3$

59. $4(x + 1)$ for $x = 6$ and $x = 11$

60. $2x - 4$ for $x = 5$ and $x = 7$

61. $7(3x + 2)$ for $x = 1$ and $x = 0$

62. **EOG PREP** What is the solution to the inequality $7 > \frac{x}{3}$? (Lesson 1-5)

A $21 < x$ B $x < 21$ C $2.333 > x$ D $\frac{7}{3} > x$

63. **EOG PREP** What is the solution to the inequality $8x \leq 24$? (Lesson 1-5)

A $x \leq 32$ B $x \leq 16$ C $x \leq 3$ D $x < 3$

3-2 Adding and Subtracting Rational Numbers

Learn to add and subtract decimals and rational numbers with like denominators.

The 100-meter dash is measured in thousandths of a second, so runners must react quickly to the starter pistol.

If you subtract a runner's reaction time from the total race time, you can find the amount of time the runner took to run the actual 100-meter distance.

Pressurized pads in the starting blocks ensure that a runner does not "jump the gun."

EXAMPLE 1 *Sports Application*

In the 2001 World Championships 100-meter dash, it took Maurice Green 0.132 seconds to react to the starter pistol. His total race time, including this reaction time, was 9.82 seconds. How long did it take him to run the actual 100 meters?

$$\begin{array}{r} 9.820 \\ -\ 0.132 \\ \hline 9.688 \end{array}$$

9.820 ← *Add a zero so the decimals align.*

The time he spent running the actual distance was 9.688 seconds.

EXAMPLE 2 Using a Number Line to Add Rational Numbers

Use a number line to find each sum.

A $-0.4 + 1.3$

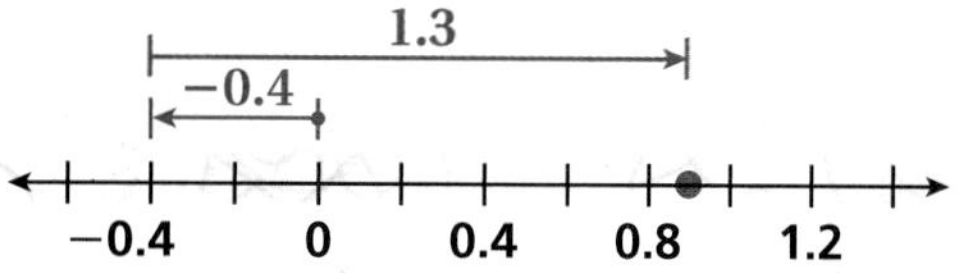

Move left 0.4 units. From −0.4, move right 1.3 units.

You finish at 0.9, so $-0.4 + 1.3 = 0.9$.

B $-\frac{5}{8} + \left(-\frac{7}{8}\right)$

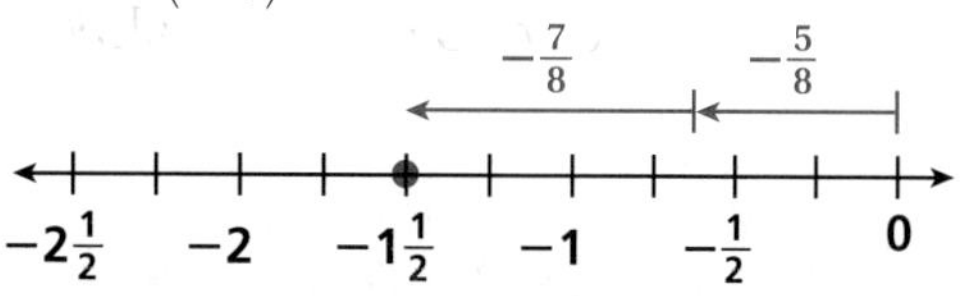

Move left $\frac{5}{8}$ units. From $-\frac{5}{8}$, move left $\frac{7}{8}$ units.

You finish at $-\frac{12}{8}$, which simplifies to $-\frac{3}{2} = -1\frac{1}{2}$.

ADDING AND SUBTRACTING WITH LIKE DENOMINATORS		
Words	**Numbers**	**Algebra**
To add or subtract rational numbers with the same denominator, add or subtract the numerators and keep the denominator.	$\frac{2}{7} + -\frac{4}{7} = \frac{2 + (-4)}{7}$ $= \frac{-2}{7}$, or $-\frac{2}{7}$	$\frac{a}{d} + \frac{b}{d} = \frac{a + b}{d}$

EXAMPLE 3 Adding and Subtracting Fractions with Like Denominators

Add or subtract.

A $\frac{6}{11} + \frac{9}{11}$

$\frac{6}{11} + \frac{9}{11} = \frac{6 + 9}{11}$ *Add numerators. Keep the denominator.*

$= \frac{15}{11}$, or $1\frac{4}{11}$

B $-\frac{3}{8} - \frac{5}{8}$

$-\frac{3}{8} - \frac{5}{8} = \frac{-3}{8} + \frac{-5}{8}$ *$-\frac{5}{8}$ can be written as $\frac{-5}{8}$.*

$= \frac{-3 + (-5)}{8} = \frac{-8}{8} = -1$

EXAMPLE 4 Evaluating Expressions with Rational Numbers

Evaluate each expression for the given value of the variable.

A $23.8 + x$ **for** $x = -41.3$

$23.8 + (-41.3)$ *Substitute −41.3 for x.*

-17.5 *Think: 41.3 − 23.8. 41.3 > 23.8. Use sign of 41.3.*

B $-\frac{1}{8} + t$ **for** $t = 2\frac{5}{8}$

$-\frac{1}{8} + 2\frac{5}{8}$ *Substitute $2\frac{5}{8}$ for t.*

$= \frac{-1}{8} + \frac{21}{8}$ *$2\frac{5}{8} = \frac{2(8) + 5}{8} = \frac{21}{8}$*

$= \frac{-1 + 21}{8} = \frac{20}{8}$ *Add numerators. Keep the denominator.*

$= \frac{5}{2}$, or $2\frac{1}{2}$

Think and Discuss

1. **Give an example** of an addition problem that involves simplifying an improper fraction in the final step.
2. **Explain** why $\frac{7}{9} + \frac{7}{9}$ does not equal $\frac{14}{18}$.

3-2 Exercises

FOR EOG PRACTICE

see page 652

internet connect

Homework Help Online

go.hrw.com Keyword: MT4 3-2

1.01

GUIDED PRACTICE

See Example 1

1. In the World Championships for the 100-meter dash in Edmonton, Alberta, Canada, on August 5, 2001, Tim Montgomery had a reaction time of 0.157 seconds. His total race time was 9.85 seconds. How long did it take him to run the actual distance?

See Example 2

Use a number line to find each sum.

2. $-0.7 + 2.1$
3. $-\frac{3}{4} + (-\frac{5}{4})$
4. $-1.3 + 0.9$
5. $-\frac{1}{2} + \left(-\frac{4}{2}\right)$
6. $-1.8 + 0.3$
7. $-\frac{1}{9} + (-\frac{4}{9})$
8. $-3.6 + 1.7$
9. $-\frac{2}{3} + \left(-\frac{7}{3}\right)$

See Example 3

Add or subtract.

10. $\frac{4}{9} - \frac{7}{9}$
11. $-\frac{5}{12} - \frac{11}{12}$
12. $\frac{1}{10} + \frac{7}{10}$
13. $-\frac{3}{20} + \frac{11}{20}$
14. $\frac{5}{8} - \frac{1}{8}$
15. $-\frac{4}{17} + \frac{9}{17}$
16. $\frac{13}{5} + \frac{8}{5}$
17. $-\frac{17}{18} - \frac{29}{18}$

See Example 4

Evaluate each expression for the given value of the variable.

18. $17.3 + x$ for $x = -13.1$
19. $-\frac{1}{5} + x$ for $x = \frac{3}{5}$
20. $35.3 + x$ for $x = -13.9$
21. $-\frac{3}{5} + x$ for $x = 1$

INDEPENDENT PRACTICE

See Example 1

22. In the men's 5000 m short-track speed-skating relay in the 2002 Olympics, the Canadian team won the gold medal with a time of 411.579 seconds, defeating the second place Italian team by 4.748 seconds. How long did it take the Italian team to finish the race?

See Example 2

Use a number line to find each sum.

23. $-3.4 + 1.8$
24. $-\frac{3}{4} + \left(-\frac{3}{4}\right)$
25. $-0.9 + 2.5$
26. $-\frac{1}{12} + \left(-\frac{7}{12}\right)$
27. $-1.7 + 3.6$
28. $-\frac{7}{10} + \left(-\frac{3}{10}\right)$
29. $-4 + 1.3$
30. $-\frac{15}{16} + \left(-\frac{9}{16}\right)$

See Example 3

Add or subtract.

31. $\frac{8}{11} - \frac{3}{11}$
32. $-\frac{4}{13} - \frac{8}{13}$
33. $\frac{9}{17} + \frac{16}{17}$
34. $-\frac{19}{25} + \frac{13}{25}$
35. $\frac{11}{32} - \frac{27}{32}$
36. $-\frac{1}{15} + \frac{13}{15}$
37. $\frac{8}{21} + \frac{15}{21}$
38. $-\frac{31}{57} - \frac{49}{57}$

See Example 4

Evaluate each expression for the given value of the variable.

39. $47.3 + x$ for $x = -18.6$
40. $-\frac{9}{10} + x$ for $x = \frac{3}{10}$
41. $13.95 + x$ for $x = -30.29$
42. $-\frac{16}{23} + x$ for $x = \frac{11}{23}$

PRACTICE AND PROBLEM SOLVING

43. ***DESIGN*** In a mechanical drawing, a hidden line is represented by dashes $\frac{4}{32}$ inch long with $\frac{1}{32}$-inch spaces between them. Without measuring, how long is each set of dashes?

    a. - - - - - -   **b.** - - - - - - - - -   **c.** - - - -

44. ***SPORTS*** A college football must be between $10\frac{14}{16}$ inches and $11\frac{7}{16}$ inches long. What is the greatest possible difference in length between two college footballs that meet these standards?

45. ***ENERGY*** The circle graph shows the sources of renewable energy and their use in the United States in British thermal units (Btu).

    a. How many quadrillion Btu's created by hydroelectric, solar, and wind methods combined were used?

    b. How many more Btu's created by wood and waste were used than those created by geothermal, solar, and wind sources combined?

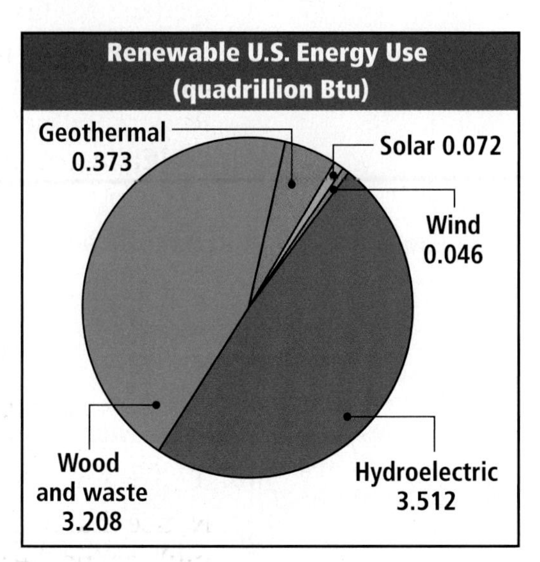

46. ***WRITE A PROBLEM*** Write a problem that requires a decimal to be converted to a fraction and that also involves addition or subtraction of fractions.

47. ***WRITE ABOUT IT*** When a student was adding fractions, the denominators were not added. Explain why.

48. ***CHALLENGE*** The gutter of a bowling lane measures $9\frac{5}{16}$ inches wide. This is $\frac{3}{16}$ inch less than the widest gutter permitted and $\frac{5}{16}$ inch greater than the narrowest gutter permitted. What is the greatest possible difference in the width of two gutters?

Spiral Review

Combine like terms. (Lesson 1-6)

49. $7x - 5y + 18$

50. $3x + y + 5y - 2x$

51. $34x + 17y + 3 - 18x + 5y + 8$

52. $48x + 23y + 5x + 6 - 3y + 15$

53. **EOG PREP** What is the value of $-4 - (-12)$? (Lesson 2-2)

    A -16 B -8 C 8 D 16

54. **EOG PREP** What is the value of $-15 - (-8)$? (Lesson 2-2)

    A -23 B -7 C 7 D 23

3-3 Multiplying Rational Numbers

Learn to multiply fractions, mixed numbers and decimals.

Kendall invited 36 people to a party. She needs to triple the recipe for a dip, or multiply the amount of each ingredient by 3. Remember that multiplication by a whole number can be written as repeated addition.

Favorite Vegetable Dip
1 c sour cream
1/2 c mayonnaise
1 envelope dry Italian dressing mix
1/2 tsp thyme
1/4 tsp curry powder
Mix and chill 24 hours. Serves 12.

Repeated addition	Multiplication
$\frac{1}{4} + \frac{1}{4} + \frac{1}{4} = \frac{3}{4}$	$3\left(\frac{1}{4}\right) = \frac{3 \cdot 1}{4} = \frac{3}{4}$

Notice that multiplying a fraction by a whole number is the same as multiplying the whole number by just the numerator of the fraction and keeping the same denominator.

RULES FOR MULTIPLYING TWO RATIONAL NUMBERS

If the signs of the factors are the same, the product is positive.

$(+) \cdot (+) = (+)$ **or** $(-) \cdot (-) = (+)$

If the signs of the factors are different, the product is negative.

$(+) \cdot (-) = (-) \cdot (+) = (-)$

EXAMPLE 1 Multiplying a Fraction and an Integer

Helpful Hint

To write $\frac{12}{5}$ as a mixed number, divide:

$\frac{12}{5} = 2\text{ R}2$

$= 2\frac{2}{5}$

Multiply. Write each answer in simplest form.

A $6\left(\frac{2}{3}\right)$

$6\left(\frac{2}{3}\right)$

$= \frac{6 \cdot 2}{3}$

$= \frac{12}{3}$

$= 4$

B $-4\left(2\frac{3}{5}\right)$

$-4\left(2\frac{3}{5}\right)$

$= -4\left(\frac{13}{5}\right)$ $\quad 2\frac{3}{5} = \frac{2(5) + 3}{5} = \frac{13}{5}$

$= -\frac{52}{5}$ *Multiply.*

$= -10\frac{2}{5}$ *Simplify.*

A model of $\frac{3}{5} \cdot \frac{2}{3}$ is shown. Notice that to multiply fractions, you multiply the numerators and multiply the denominators.

If you place the first rectangle on top of the second, the number of green squares represents the numerator, and the number of total squares represents the denominator.

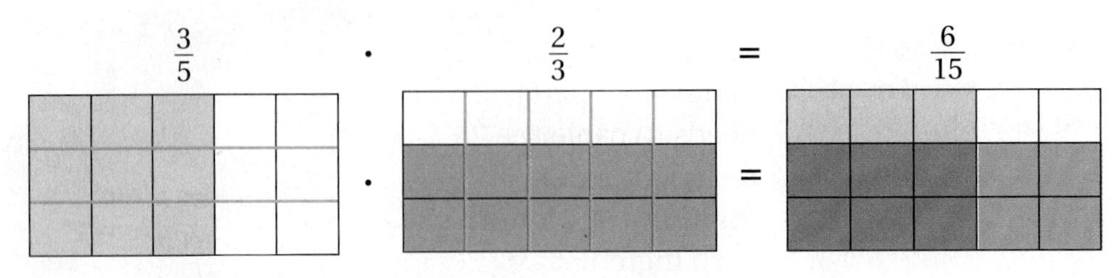

To simplify the product, rearrange the six green squares into the first two columns. You can see that this is $\frac{2}{5}$.

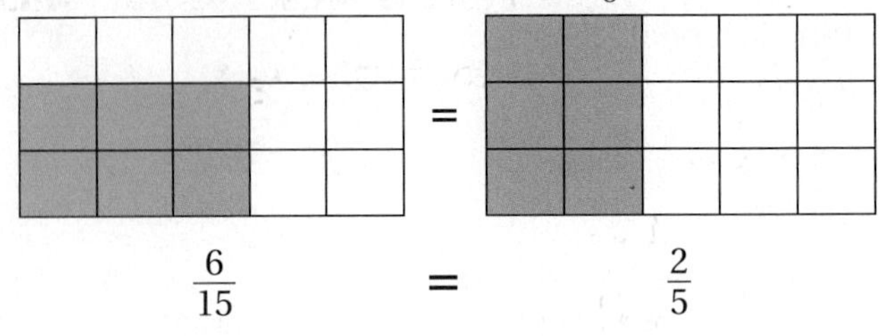

EXAMPLE 2 Multiplying Fractions

Multiply. Write each answer in simplest form.

A $-\frac{1}{2}\left(-\frac{3}{5}\right)$

$$-\frac{1}{2}\left(-\frac{3}{5}\right) = \frac{-1}{2}\left(\frac{-3}{5}\right)$$

$$= \frac{(-1)(-3)}{2(5)}$$ *Multiply numerators. Multiply denominators.*

$$= \frac{3}{10}$$ *Simplest form*

B $\frac{5}{12}\left(-\frac{12}{5}\right)$

$$\frac{5}{12}\left(-\frac{12}{5}\right) = \frac{5}{12}\left(\frac{-12}{5}\right)$$

$$= \frac{\overset{1}{\cancel{5}}(\overset{-1}{\cancel{-12}})}{\underset{1}{\cancel{12}}(\underset{1}{\cancel{5}})}$$ *Look for common factors: 12, 5.*

$$= \frac{-1}{1} = -1$$ *Simplest form*

C $6\frac{2}{3}\left(\frac{7}{20}\right)$

$$6\frac{2}{3}\left(\frac{7}{20}\right) = \frac{20}{3}\left(\frac{7}{20}\right)$$ *Write as an improper fraction.*

$$= \frac{\overset{1}{\cancel{20}}(7)}{3(\underset{1}{\cancel{20}})}$$ *Look for common factors: 20.*

$$= \frac{7}{3}, \text{ or } 2\frac{1}{3}$$ *7 ÷ 3 = 2 R1*

Helpful Hint

A fraction is in lowest terms, or simplest form, when the numerator and denominator have no common factors.

EXAMPLE 3 Multiplying Decimals

Multiply.

A $-2.5(-8)$

$-2.5 \cdot (-8) = 20.0$ *Product is positive with 1 decimal place.*

You can drop the zero after the decimal point.

$= 20$

B $-0.07(4.6)$

$-0.07 \cdot 4.6 = -0.322$ *Product is negative with 3 decimal places.*

EXAMPLE 4 Evaluating Expressions with Rational Numbers

Evaluate $-5\frac{1}{2}t$ for each value of t.

A $t = -\frac{2}{3}$

$-5\frac{1}{2}t$

$= -5\frac{1}{2}\left(-\frac{2}{3}\right)$ *Substitute $-\frac{2}{3}$ for t.*

$= -\frac{11}{2}\left(-\frac{2}{3}\right)$ *Write as an improper fraction.*

$= \frac{11 \cdot \cancel{2}^{1}}{{}_{1}\cancel{2} \cdot 3}$ *The product of 2 negative numbers is positive.*

$= \frac{11}{3}$, or $3\frac{2}{3}$ *11 ÷ 3 = 3 R2*

B $t = 8$

$-5\frac{1}{2}t$

$= -\frac{11}{2}(8)$ *Substitute 8 for t.*

$= -\frac{88}{2}$

$= -44$

Think and Discuss

1. **Name** the number of decimal places in the product of 5.625 and 2.75.
2. **Explain** why products of fractions are like products of integers.
3. **Give an example** of two fractions whose product is an integer due to common factors.

3-3 Exercises

FOR EOG PRACTICE

see page 652

internet connect

Homework Help Online
go.hrw.com Keyword: MT4 3-3

1.01

GUIDED PRACTICE

See Example 1 Multiply. Write each answer in simplest form.

1. $4\left(\frac{1}{3}\right)$
2. $-6\left(2\frac{2}{5}\right)$
3. $3\left(\frac{5}{8}\right)$
4. $-2\left(1\frac{9}{10}\right)$
5. $7\left(\frac{4}{9}\right)$
6. $-5\left(1\frac{8}{11}\right)$
7. $9\left(\frac{3}{4}\right)$
8. $3\left(2\frac{1}{8}\right)$

See Example 2 Multiply. Write each answer in simplest form.

9. $-\frac{1}{3}\left(-\frac{4}{7}\right)$
10. $\frac{3}{8}\left(-\frac{7}{10}\right)$
11. $6\frac{2}{5}\left(\frac{5}{9}\right)$
12. $-\frac{2}{3}\left(-\frac{3}{8}\right)$
13. $\frac{5}{13}\left(-\frac{5}{6}\right)$
14. $4\frac{7}{8}\left(\frac{5}{12}\right)$
15. $-\frac{7}{8}\left(-\frac{2}{3}\right)$
16. $\frac{5}{12}\left(-\frac{11}{16}\right)$

See Example 3 Multiply.

17. $-3.1(-4)$
18. $0.04(3.6)$
19. $-7.3(-5)$
20. $-0.15(2.8)$
21. $-5.9(-7)$
22. $0.5(7.3)$
23. $-4.7(-3)$
24. $-0.08(5.2)$

See Example 4 Evaluate $3\frac{2}{7}x$ for each value of x.

25. $x = 4$
26. $x = 1\frac{3}{4}$
27. $x = -2$
28. $x = -\frac{3}{7}$
29. $x = 7$
30. $x = 2\frac{1}{3}$
31. $x = -3$
32. $x = -\frac{3}{10}$

INDEPENDENT PRACTICE

See Example 1 Multiply. Write each answer in simplest form.

33. $3\left(\frac{1}{5}\right)$
34. $-4\left(1\frac{5}{8}\right)$
35. $2\left(\frac{9}{16}\right)$
36. $-5\left(1\frac{3}{4}\right)$
37. $9\left(\frac{14}{15}\right)$
38. $-2\left(4\frac{7}{8}\right)$
39. $6\left(\frac{2}{3}\right)$
40. $-7\left(3\frac{1}{5}\right)$

See Example 2 Multiply. Write each answer in simplest form.

41. $-\frac{2}{3}\left(-\frac{5}{6}\right)$
42. $\frac{2}{5}\left(-\frac{9}{10}\right)$
43. $2\frac{5}{7}\left(\frac{2}{9}\right)$
44. $-\frac{1}{2}\left(-\frac{11}{12}\right)$
45. $\frac{4}{5}\left(-\frac{3}{8}\right)$
46. $5\frac{1}{3}\left(\frac{13}{16}\right)$
47. $-\frac{3}{4}\left(-\frac{1}{8}\right)$
48. $\frac{7}{8}\left(\frac{3}{5}\right)$

See Example 3 Multiply.

49. $-2.9(-3)$
50. $-0.02(5.9)$
51. $-6.2(-7)$
52. $-0.25(3.5)$
53. $-4.8(-7)$
54. $-0.07(4.8)$
55. $-3.6(-8)$
56. $-0.04(9.2)$

See Example 4 Evaluate $2\frac{3}{4}x$ for each value of x.

57. $x = 6$
58. $x = 2\frac{1}{3}$
59. $x = -4$
60. $x = -\frac{3}{8}$
61. $x = 3$
62. $x = 4\frac{7}{8}$
63. $x = -7$
64. $x = -\frac{7}{9}$

PRACTICE AND PROBLEM SOLVING

Becoming a veterinarian requires at least two years at an undergraduate college and four years at a veterinary college. There are fewer than 30 veterinary colleges in the United States.

65. ***HEALTH*** As a rule of thumb, people should drink $\frac{1}{2}$ ounce of water for each pound of body weight per day. How much water should a 145-pound person drink per day?

66. People who are physically active should increase the daily amount of water they drink to $\frac{2}{3}$ ounce per pound of body weight. How much water should a 245-pound football player drink per day?

67. ***ANIMALS*** The label on a bottle of pet vitamins lists dosage guidelines. What dosage would you give to each of these animals?
   - **a.** a 50 lb adult dog
   - **b.** a 12 lb cat
   - **c.** a 40 lb pregnant dog

Do-Good Pet Vitamins

- **Adult dogs:** $\frac{1}{2}$ tsp per 20 lb body weight
- **Puppies, pregnant dogs, or nursing dogs:** $\frac{1}{2}$ tsp per 10 lb body weight
- **Cats:** $\frac{1}{4}$ tsp per 2 lb body weight

68. ***CONSUMER ECONOMICS*** At a clothing store, the ticketed price of a sweater is $\frac{1}{2}$ the original price. You have a discount coupon for $\frac{1}{2}$ off the ticketed price. What fraction of the original price is the additional discount?

69. ***WHAT'S THE ERROR?*** A student multiplied two mixed numbers in the following fashion: $3\frac{3}{8} \cdot 4\frac{1}{3} = 12\frac{1}{8}$. What's the error?

70. ***WRITE ABOUT IT*** In the pattern $\frac{1}{3} + \frac{1}{4} + \frac{1}{5} + \ldots$, which fraction makes the sum greater than 1? Explain.

71. ***CHALLENGE*** On January 20, 2001, George W. Bush was inaugurated as the forty-third president of the United States. Of the 42 presidents before him, $\frac{1}{3}$ had served as vice-president. Of those previous vice-presidents, $\frac{3}{7}$ served as president for more than four years. What fraction of the first 42 presidents were former vice-presidents who also served more than four years as president?

Spiral Review

Solve. (Lesson 1-3)

72. $7 + x = 13$ **73.** $x - 5 = 7$ **74.** $x + 8 = 19$

75. $12 + x = 46$ **76.** $x - 27 = 54$ **77.** $x + 31 = 75$

78. **EOG PREP** What is the solution to the inequality $-3a \geq 24$? (Lesson 2-5)

A $a \geq -8$ B $a \leq -8$ C $a > 8$ D $a < 8$

79. **EOG PREP** What is the solution to the inequality $\frac{a}{2} < -22$? (Lesson 2-5)

A $a < -44$ B $a > -44$ C $a > -11$ D $a < 11$

3-4 Dividing Rational Numbers

Learn to divide fractions and decimals.

Vocabulary
reciprocal

A number and its **reciprocal** have a product of 1. To find the reciprocal of a fraction, exchange the numerator and the denominator. Remember that an integer can be written as a fraction with a denominator of 1.

Number	Reciprocal	Product
$\frac{3}{4}$	$\frac{4}{3}$	$\frac{3}{4}\left(\frac{4}{3}\right) = 1$
$-\frac{5}{12}$	$-\frac{12}{5}$	$-\frac{5}{12}\left(-\frac{12}{5}\right) = 1$
6	$\frac{1}{6}$	$6\left(\frac{1}{6}\right) = 1$

Multiplication and division are inverse operations. They undo each other.

$$\frac{1}{3}\left(\frac{2}{5}\right) = \frac{2}{15} \longrightarrow \frac{2}{15} \div \frac{2}{5} = \frac{1}{3}$$

Notice that multiplying by the reciprocal gives the same result as dividing.

$$\left(\frac{2}{15}\right)\left(\frac{5}{2}\right) = \frac{2 \cdot 5}{15 \cdot 2} = \frac{1}{3}$$

DIVIDING RATIONAL NUMBERS IN FRACTION FORM		
Words	**Numbers**	**Algebra**
To divide by a fraction, multiply by the reciprocal.	$\frac{1}{5} \div \frac{2}{3} = \frac{1}{5} \cdot \frac{3}{2} = \frac{3}{10}$	$\frac{a}{b} \div \frac{c}{d} = \frac{a}{b} \cdot \frac{d}{c} = \frac{ad}{bc}$

EXAMPLE  1 Dividing Fractions

Divide. Write each answer in simplest form.

A $\frac{7}{12} \div \frac{2}{3}$

$\frac{7}{12} \div \frac{2}{3} = \frac{7}{12} \cdot \frac{3}{2}$ *Multiply by the reciprocal.*

$= \frac{7 \cdot \cancel{3}^{1}}{{}_{4}\cancel{12} \cdot 2}$ *Reduce common factors.*

$= \frac{7}{8}$ *Simplest form*

Divide. Write each answer in simplest form.

B $3\frac{1}{4} \div 4$

$3\frac{1}{4} \div 4 = \frac{13}{4} \div \frac{4}{1}$ *Write as improper fractions.*

$= \frac{13}{4}\left(\frac{1}{4}\right)$ *Multiply by the reciprocal.*

$= \frac{13 \cdot 1}{4 \cdot 4}$ *No common factors.*

$= \frac{13}{16}$ *Simplest form*

When dividing a decimal by a decimal, multiply both numbers by a power of 10 so you can divide by a whole number. To decide which power of 10 to multiply by, look at the denominator. The number of decimal places is the number of zeros to write after the 1.

$$\frac{1.32}{0.4} = \frac{1.32}{0.4}\left(\frac{10}{10}\right) = \frac{13.2}{4}$$

1 decimal place *1 zero*

EXAMPLE 2 Dividing Decimals

Divide.

2.92 ÷ 0.4

$2.92 \div 0.4 = \frac{2.92}{0.4}\left(\frac{10}{10}\right) = \frac{29.2}{4}$

$= 7.3$ *Divide.*

EXAMPLE 3 Evaluating Expressions with Fractions and Decimals

Evaluate each expression for the given value of the variable.

A $\frac{7.2}{n}$ **for** $n = 0.24$

$\frac{7.2}{0.24} = \frac{7.2}{0.24}\left(\frac{100}{100}\right)$ *0.24 has 2 decimal places, so use $\frac{100}{100}$.*

$= \frac{720}{24}$ *Divide.*

$= 30$

When $n = 0.24$, $\frac{7.2}{n} = 30$.

B $m \div \frac{3}{8}$ **for** $m = 7\frac{1}{2}$

$7\frac{1}{2} \div \frac{3}{8} = \frac{15}{2} \cdot \frac{8}{3}$

$= \frac{\overset{5}{\cancel{15}} \cdot \overset{4}{\cancel{8}}}{\underset{1}{\cancel{2}} \cdot \underset{1}{\cancel{3}}} = \frac{20}{1} = 20$

When $m = 7\frac{1}{2}$, $m \div \frac{3}{8} = 20$.

EXAMPLE  PROBLEM SOLVING

You pour $\frac{2}{3}$ cup of a sports drink into a glass. The serving size is 6 ounces, or $\frac{3}{4}$ cup. How many servings will you consume? How many calories will you consume?

Nutrition Facts
Serving size 6 fl oz (240 mL)
Servings per container: 2

Amount per serving	
Calories	50
	% daily value
Total fat 0 g	0%
Sodium 110 mg	5%
Potassium 30 mg	1%
Total carbohydrates 0 g	5%
Sugar 14 g	5%
Protein 0 g	0%

1 Understand the Problem

The number of calories you consume is the number of calories in the fraction of a serving.

List the **important information:**

- The amount you plan to drink is $\frac{2}{3}$ cup.
- The amount of a full serving is $\frac{3}{4}$ cup.
- The number of calories in one serving is 50.

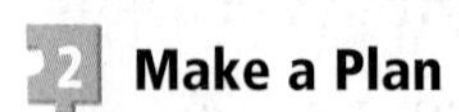

2 Make a Plan

Set up an equation to find the number of servings you will drink.

amount you drink	÷	serving size	=	number of servings

Using the number of servings, you can find the calories consumed.

number of servings	·	calories per serving	=	total calories

3 Solve

Let n = number of servings. Let c = total calories.

Servings: $\frac{2}{3} \div \frac{3}{4} = n$

$\frac{2}{3} \cdot \frac{4}{3} = n$

$\frac{8}{9} = n$

Calories: $\frac{8}{9} \cdot 50 = c$

$\frac{8 \cdot 50}{9} = c$

$\frac{400}{9} = c \approx 44.4$

You will drink $\frac{8}{9}$ of a serving, which is about 44.4 calories.

4 Look Back

You did not pour a full serving, so $\frac{8}{9}$ is a reasonable answer. It is less than 1, and 44.4 calories is less than the calories in a full serving, 50.

Think and Discuss

1. **Tell** what happens when you divide a fraction by itself. Show that you are correct using multiplication by the reciprocal.
2. **Model** the product of $\frac{2}{3}$ and $\frac{1}{4}$.

3-4 Exercises

FOR EOG PRACTICE
see page 653

internet connect
Homework Help Online
go.hrw.com Keyword: MT4 3-4

1.01

GUIDED PRACTICE

See Example 1 **Divide. Write each answer in simplest form.**

1. $\frac{2}{3} \div \frac{5}{6}$
2. $2\frac{1}{4} \div 3\frac{2}{5}$
3. $-\frac{6}{7} \div 3$
4. $\frac{7}{8} \div \frac{3}{10}$
5. $3\frac{3}{16} \div 2\frac{5}{8}$
6. $-\frac{5}{9} \div 6$
7. $\frac{9}{10} \div \frac{3}{5}$
8. $2\frac{5}{12} \div \frac{5}{6}$

See Example 2 **Divide.**

9. $3.72 \div 0.3$
10. $3.4 \div 0.05$
11. $10.71 \div 0.7$
12. $3.44 \div 0.4$
13. $3.46 \div 0.9$
14. $14.08 \div 0.8$
15. $7.86 \div 0.006$
16. $2.76 \div 0.3$

See Example 3 **Evaluate each expression for the given value of the variable.**

17. $\frac{4.5}{x}$ for $x = 0.2$
18. $\frac{8.4}{x}$ for $x = 0.4$
19. $\frac{40.5}{x}$ for $x = 0.9$
20. $\frac{9.2}{x}$ for $x = 2.3$
21. $\frac{20.8}{x}$ for $x = 1.6$
22. $\frac{21.6}{x}$ for $x = 0.08$

See Example 4

23. You drink $\frac{3}{4}$ pint of spring water. One serving of the water is $\frac{7}{8}$ pint. How much of a serving did you drink?

INDEPENDENT PRACTICE

See Example 1 **Divide. Write each answer in simplest form.**

24. $\frac{1}{8} \div \frac{2}{5}$
25. $3\frac{1}{2} \div 1\frac{7}{8}$
26. $-\frac{5}{12} \div \frac{2}{3}$
27. $\frac{9}{10} \div \frac{1}{4}$
28. $1\frac{3}{4} \div 4\frac{1}{8}$
29. $-\frac{2}{9} \div \frac{7}{12}$
30. $\frac{2}{5} \div \frac{5}{16}$
31. $2\frac{3}{8} \div 1\frac{1}{6}$
32. $-\frac{3}{11} \div \frac{4}{7}$
33. $\frac{3}{16} \div \frac{3}{4}$
34. $3\frac{11}{12} \div 2\frac{1}{4}$
35. $-\frac{3}{4} \div \frac{1}{6}$

See Example 2 **Divide.**

36. $10.86 \div 0.6$
37. $1.94 \div 0.02$
38. $9.76 \div 0.8$
39. $8.55 \div 0.5$
40. $6.52 \div 0.004$
41. $24.66 \div 0.9$
42. $9.36 \div 0.03$
43. $17.78 \div 0.7$
44. $11.128 \div 0.52$
45. $24 \div 0.75$
46. $13.608 \div 0.81$
47. $3.6864 \div 0.64$

See Example 3 **Evaluate each expression for the given value of the variable.**

48. $\frac{6.3}{x}$ for $x = 0.3$
49. $\frac{9.1}{x}$ for $x = 0.7$
50. $\frac{12}{x}$ for $x = 0.02$
51. $\frac{15.4}{x}$ for $x = 1.4$
52. $\frac{3.69}{x}$ for $x = 0.9$
53. $\frac{22.2}{x}$ for $x = 0.06$
54. $\frac{1.6}{x}$ for $x = 3.2$
55. $\frac{0.56}{x}$ for $x = 0.8$
56. $\frac{94.05}{x}$ for $x = 28.5$

See Example 4

57. The platform on the school stage is $8\frac{3}{4}$ feet wide. Each chair is $1\frac{5}{12}$ feet wide. How many chairs will fit across the platform?

PRACTICE AND PROBLEM SOLVING

58. Reba is eating her favorite cereal. There are $3\frac{2}{3}$ servings remaining in the box. Reba pours only $\frac{1}{3}$ of a serving into her bowl at a time. How many more bowls can Reba have before the box is empty?

59. The thickest vinyl floor tiles available are $\frac{1}{8}$ inch thick. The thinnest tiles are $\frac{1}{20}$ inch thick. How many thin tiles would equal the thickness of one of the thick tiles?

60. Nesting dolls called *matrushkas* are a well-known type of Russian folk art. Use the information in the picture to find the height of the largest doll.

$\frac{6}{25}x = \frac{7}{8}$ in.

61. Cal has 41 DVDs in cases that are each $\frac{5}{8}$ inch thick. Can he put all the DVDs on a shelf that is 29 inches long?

62. ***WHAT'S THE ERROR?*** A student had a recipe that called for $\frac{7}{8}$ cup of rice. Since he wanted to make only $\frac{1}{3}$ of the whole recipe, he calculated the amount of rice he would need: $\frac{7}{8} \div \frac{1}{3} = \frac{7}{8} \cdot \frac{3}{1} = 2\frac{5}{8}$ cups of rice. What was his error?

63. ***WRITE ABOUT IT*** A proper fraction with denominator 6 is divided by a proper fraction with denominator 3. Will the denominator of the quotient be odd or even? Explain.

64. ***CHALLENGE*** According to the 2000 U.S. census, about $\frac{1}{30}$ of the U.S. population resides in Los Angeles County. About $\frac{1}{8}$ of thc U.S. population resides in California. What fraction of the California population resides in Los Angeles County?

Spiral Review

Solve each equation. (Lesson 1-4)

65. $7x = 45.5$ **66.** $\frac{x}{6} = 11.2$ **67.** $1032 = 129x$

68. $\frac{x}{5} = 16.25$ **69.** $13x = 58.5$ **70.** $\frac{x}{2} = 1.38$

71. **EOG PREP** What is the value of $3^4 \cdot 3^{-2}$? (Lesson 2-8)

A $\frac{1}{9}$ B 72 C 9 D 6

72. **EOG PREP** What is the value of $\frac{2^5}{2^9}$? (Lesson 2-8)

A $\frac{1}{16}$ B 16 C 8 D -16

3-5 Adding and Subtracting with Unlike Denominators

Learn to add and subtract fractions with unlike denominators.

Vocabulary

least common denominator (LCD)

A pattern for a double-circle skirt requires $9\frac{1}{3}$ yards of 45-inch-wide material. To add a ruffle takes another $2\frac{2}{5}$ yards. If the total amount of material for the skirt and ruffle are cut from a bolt of fabric $15\frac{1}{2}$ yards long, how much fabric is left?

To solve this problem, add and subtract rational numbers with unlike denominators. First find a common denominator using one of the following methods:

Method 1 Find a common denominator by multiplying one denominator by the other denominator.

Method 2 Find the **least common denominator (LCD)**; the least common multiple of the denominators.

EXAMPLE 1 Adding and Subtracting Fractions with Unlike Denominators

Remember!

The least common multiple of two numbers is the smallest number other than zero that is a multiple of the two numbers.

Add or subtract.

A $\frac{2}{3} + \frac{1}{5}$

Method 1: $\frac{2}{3} + \frac{1}{5}$ — *Find a common denominator: 3(5) = 15.*

$= \frac{2}{3}\left(\frac{5}{5}\right) + \frac{1}{5}\left(\frac{3}{3}\right)$ — *Multiply by fractions equal to 1.*

$= \frac{10}{15} + \frac{3}{15}$ — *Rewrite with a common denominator.*

$= \frac{13}{15}$ — *Simplify.*

B $3\frac{2}{5} + \left(-3\frac{1}{2}\right)$

Method 2: $3\frac{2}{5} + \left(-3\frac{1}{2}\right)$

$= \frac{17}{5} + \left(-\frac{7}{2}\right)$ — *Write as improper fractions.*

Multiples of 5: 5, (10), 15, 20, . . .
Multiples of 2: 2, 4, 6, 8, (10), . . . — *List the multiples of each denominator and find the LCD.*

$= \frac{17}{5}\left(\frac{2}{2}\right) + \left(-\frac{7}{2}\right)\left(\frac{5}{5}\right)$ — *Multiply by fractions equal to 1.*

$= \frac{34}{10} + \left(-\frac{35}{10}\right)$ — *Rewrite with a common denominator.*

$= -\frac{1}{10}$ — *Simplify.*

EXAMPLE 2 Evaluating Expressions with Rational Numbers

Evaluate $n - \frac{11}{16}$ for $n = -\frac{1}{3}$.

$n - \frac{11}{16}$

$= \left(-\frac{1}{3}\right) - \frac{11}{16}$ *Substitute $-\frac{1}{3}$ for n.*

$= \left(-\frac{1}{3}\right)\left(\frac{16}{16}\right) - \frac{11}{16}\left(\frac{3}{3}\right)$ *Multiply by fractions equal to 1.*

$= -\frac{16}{48} - \frac{33}{48}$ *Rewrite with a common denominator: 3(16) = 48.*

$= -\frac{49}{48}$, or $-1\frac{1}{48}$ *Simplify.*

EXAMPLE 3 Consumer Application

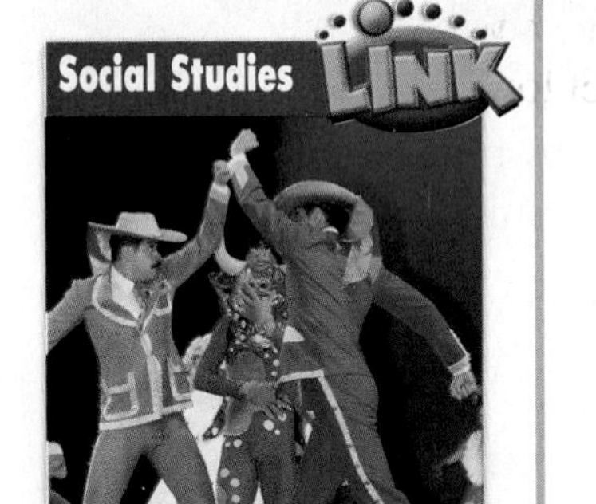

Social Studies LINK

There are three categories of folkloric dance in Mexico: *danza, mestizo,* and *bailes regionales.* These dances are performed by traveling groups such as the Ballet Folklorico.

A folkloric dance skirt pattern calls for $2\frac{2}{5}$ yards of 45-inch-wide material to make the ruffle and $9\frac{1}{3}$ yards to make the skirt. The material for the skirt and ruffle will be cut from a bolt that is $15\frac{1}{2}$ yards long. How many yards will be left on the bolt?

$2\frac{2}{5} + 9\frac{1}{3}$ *Add to find length needed for the skirt and ruffle.*

$= \frac{12}{5} + \frac{28}{3}$ *Write as improper fractions. The LCD is 15.*

$= \frac{36}{15} + \frac{140}{15}$ *Rewrite with a common denominator.*

$= \frac{176}{15}$

The amount needed for the skirt and ruffle is $\frac{176}{15}$, or $11\frac{11}{15}$ yards. Now find the number of yards remaining.

$15\frac{1}{2} - 11\frac{11}{15}$ *Subtract amount needed from bolt length.*

$= \frac{31}{2} - \frac{176}{15}$ *Write as improper fractions. The LCD is 30.*

$= \frac{465}{30} - \frac{352}{30}$ *Rewrite with a common denominator.*

$= \frac{113}{30}$, or $3\frac{23}{30}$ *Simplify.*

There will be $3\frac{23}{30}$ yards left on the bolt.

Think and Discuss

1. **Give an example** of two denominators with no common factors.
2. **Tell** if $-2\frac{1}{5} - \left(-2\frac{3}{16}\right)$ is positive or negative. Explain.
3. **Explain** how to add $2\frac{2}{5} + 9\frac{1}{3}$ without first writing them as improper fractions.

3-5 Exercises

FOR EOG PRACTICE
see page 653

internet connect
Homework Help Online
go.hrw.com Keyword: MT4 3-5

1.01

GUIDED PRACTICE

See Example 1
Add or subtract.

1. $\frac{5}{8} + \frac{1}{6}$
2. $\frac{5}{16} + \frac{2}{7}$
3. $\frac{1}{3} - \frac{7}{9}$
4. $\frac{3}{4} - \frac{5}{16}$
5. $2\frac{1}{5} + \left(-5\frac{2}{3}\right)$
6. $4\frac{11}{12} + \left(-7\frac{3}{8}\right)$
7. $3\frac{7}{12} + \left(-2\frac{4}{5}\right)$
8. $5\frac{3}{5} - 3\frac{7}{8}$

See Example 2
Evaluate each expression for the given value of the variable.

9. $2\frac{3}{5} + x$ for $x = -1\frac{1}{8}$
10. $n - \frac{4}{7}$ for $n = -\frac{5}{9}$
11. $3\frac{1}{2} + x$ for $x = -2\frac{7}{8}$
12. $n - \frac{7}{16}$ for $n = -\frac{1}{3}$

See Example 3
13. A $2\frac{1}{4}$-foot-long piece of wood is needed to replace a window sill. If this amount is cut from a piece of wood $8\frac{7}{8}$ feet long, how much remains?

INDEPENDENT PRACTICE

See Example 1
Add or subtract.

14. $\frac{5}{12} + \frac{3}{7}$
15. $\frac{1}{5} + \frac{7}{9}$
16. $\frac{15}{16} - \frac{9}{10}$
17. $\frac{1}{3} + \frac{11}{12}$
18. $5\frac{4}{5} + \left(-3\frac{2}{7}\right)$
19. $\frac{5}{7} - \frac{13}{16}$
20. $1\frac{2}{3} - 4\frac{5}{8}$
21. $\frac{1}{5} + \frac{8}{9}$

See Example 2
Evaluate each expression for the given value of the variable.

22. $1\frac{7}{8} + x$ for $x = -2\frac{5}{6}$
23. $n - \frac{2}{3}$ for $n = \frac{9}{16}$
24. $2\frac{5}{8} + x$ for $x = -1\frac{9}{10}$
25. $n - \frac{13}{15}$ for $n = \frac{3}{4}$

See Example 3
26. A DVD contains a movie that takes up $4\frac{1}{3}$ gigabytes of space. If the DVD can hold $9\frac{2}{5}$ gigabytes, how much space on the disk is unused?

PRACTICE AND PROBLEM SOLVING

27. ***MEASUREMENT*** Bernard I. Pietsch measured the sides of the base of the Washington Monument. The north side measured $661\frac{3}{8}$ inches, the west side measured 661 inches, the south side measured $660\frac{13}{25}$ inches, and the east side measured 661 inches. Find the average side length.

28. ***CONSTRUCTION*** A water pipe has an outside diameter of $1\frac{1}{4}$ inches and a wall thickness of $\frac{5}{16}$ inch. What is the inside diameter of the pipe?

$1\frac{1}{4}$ in.

$\frac{5}{16}$ in.

Earth Science LINK

Niagara Falls, on the border of Canada and the United States, has two major falls, Horseshoe Falls on the Canadian side and American Falls on the U.S. side. Surveys of the erosion of the falls began in 1842. From 1842 to 1905, Horseshoe Falls eroded $239\frac{2}{5}$ feet.

29. In 1986, Thomas Martin noted that American Falls eroded $7\frac{1}{2}$ inches and Horseshoe Falls eroded $2\frac{4}{25}$ feet. What is the difference between the two measurements?

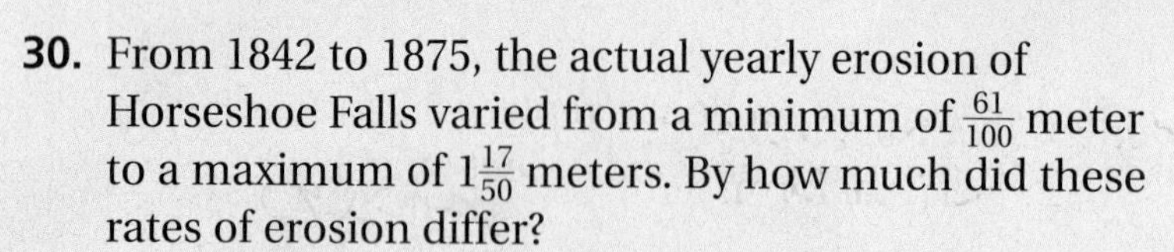

30. From 1842 to 1875, the actual yearly erosion of Horseshoe Falls varied from a minimum of $\frac{61}{100}$ meter to a maximum of $1\frac{17}{50}$ meters. By how much did these rates of erosion differ?

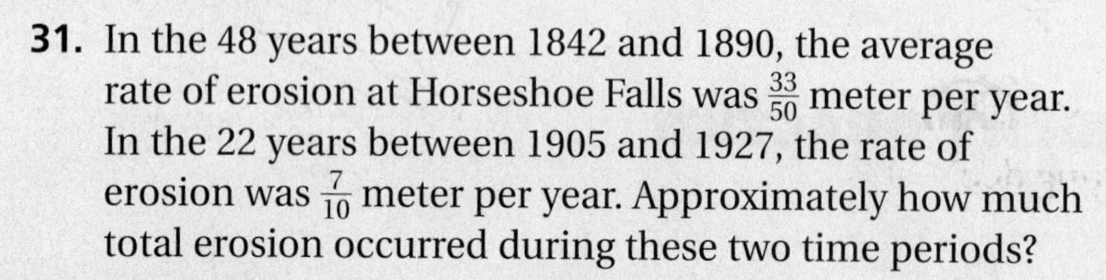

31. In the 48 years between 1842 and 1890, the average rate of erosion at Horseshoe Falls was $\frac{33}{50}$ meter per year. In the 22 years between 1905 and 1927, the rate of erosion was $\frac{7}{10}$ meter per year. Approximately how much total erosion occurred during these two time periods?

32. Lake Erie, which feeds Niagara Falls, has a six-month average precipitation rate of $48\frac{1}{2}$ centimeters. From September 1999 to February 2000, the precipitation was $40\frac{1}{5}$ centimeters. How far below the average was precipitation during this period?

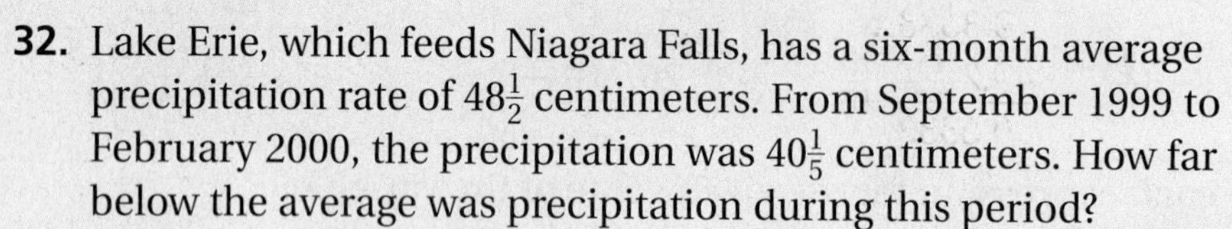

33. ***CHALLENGE*** Rates of erosion of American Falls have been recorded as $\frac{23}{100}$ meter per year for 33 years, $\frac{9}{40}$ meter per year for 48 years, and $\frac{1}{5}$ meter per year for 4 years. What is the total amount of erosion during these three time spans?

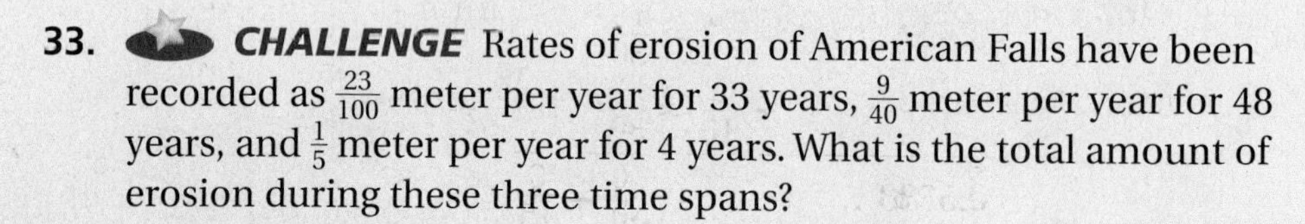

Spiral Review

Simplify. (Lesson 2-3)

34. $-4(6-8)$ **35.** $3(-5-4)$ **36.** $-2(4-9)$

37. $-8(-5-6)$ **38.** $7(2-5)$ **39.** $-3(-3-3)$

40. **EOG PREP** What is the value of $100 - 2(4 \cdot 3^2)$? (Lesson 2-6)

A 104 B 28 C −188 D −535

41. **EOG PREP** What is the value of $41 + 3(8 - 2^3)$? (Lesson 2-6)

A 0 B 41 C 89 D 689

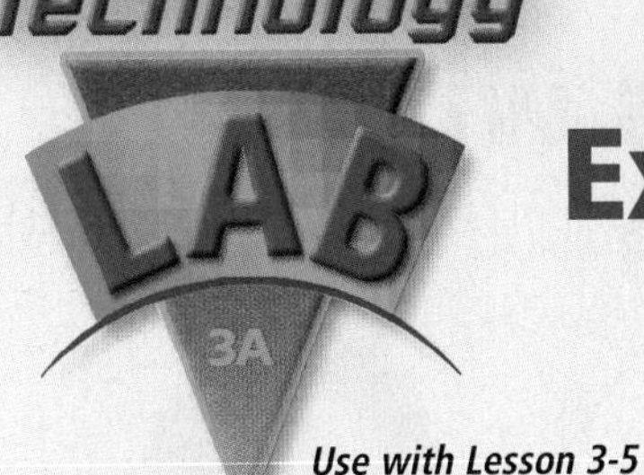

Explore Repeating Decimals

Use with Lesson 3-5

You can divide to display decimal equivalents of fractions using your graphing calculator. To display decimals as fractions, use the MATH key.

You can also use the MATH key to find fractions equivalent to repeating decimals.

Activity

1 Use a graphing calculator to find the decimal equivalent of each fraction. Look for patterns in the fraction and decimal forms.

$\frac{1}{9}$ $\frac{4}{9}$ $\frac{23}{99}$ $\frac{47}{99}$ $\frac{461}{999}$ $\frac{703}{999}$

For example, type 1 ÷ 9, and press ENTER.

1/9
.1111111111

The decimal equivalent is a repeating decimal, $0.\overline{1}$.

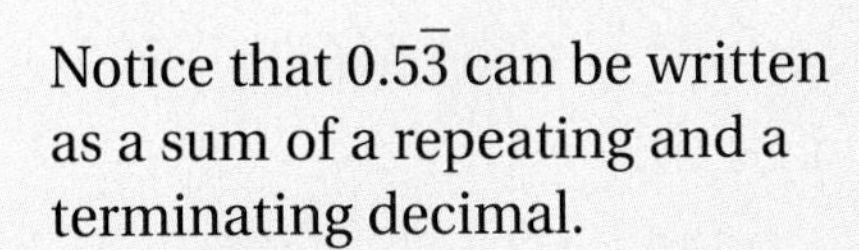

Notice that $0.5\overline{3}$ can be written as a sum of a repeating and a terminating decimal.

$$\begin{array}{r} 0.3333\ldots \\ +\ 0.2 \\ \hline 0.5333\ldots \end{array}$$

2 Find the fraction for $0.5\overline{3}$.

To find the fraction for $0.5\overline{3}$, write the decimals as fractions and add.

$$\begin{array}{rr} 0.3333\ldots & \frac{1}{3} = \frac{10}{30} \\ +\ 0.2 & +\ \frac{2}{10} = \frac{6}{30} \\ \hline 0.5333\ldots & = \frac{16}{30} = \frac{8}{15} \end{array}$$

Think and Discuss

1. Based on the pattern you found in **1**, how would you write the repeating decimal $0.\overline{3726}$ as a fraction? Divide to check your answer.

Try This

Write each decimal as a sum or difference of a repeating and a terminating decimal. Then write the repeating decimals as a fraction. Check by dividing.

1. $0.\overline{15}$ **2.** $0.1\overline{3}$ **3.** $0.6\overline{51}$ **4.** $0.9\overline{15}$ **5.** $0.4\overline{532}$

3-6 Solving Equations with Rational Numbers

Learn to solve equations with rational numbers.

One of the world's most famous jewels is the Hope diamond. The roughly cut Hope diamond was sold to King Louis XIV of France in 1668. When the king's jeweler cut it, the diamond was reduced by $45\frac{1}{16}$ carats to a steely blue $67\frac{1}{8}$ carats. You can write an equation using these fractions and solve for the weight of the roughly cut diamond.

EXAMPLE 1 Solving Equations with Decimals

Solve.

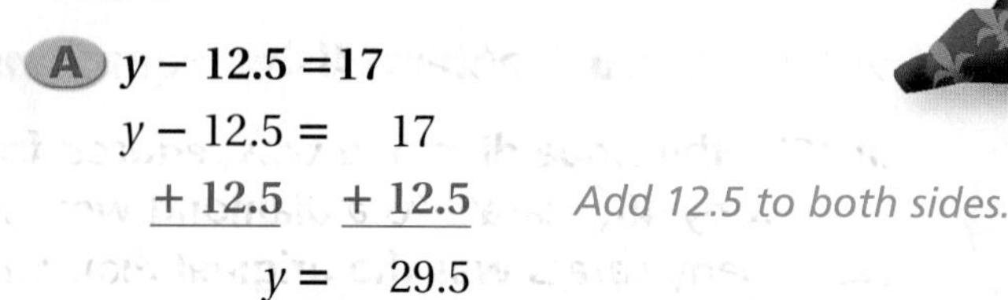

A $y - 12.5 = 17$

$$\begin{aligned} y - 12.5 &= 17 \\ \underline{+12.5} \quad & \underline{+12.5} \qquad \textit{Add 12.5 to both sides.} \\ y &= 29.5 \end{aligned}$$

Remember!

Once you have solved an equation, it is a good idea to check your answer. To check your answer, substitute your answer for the variable in the original equation.

B $-2.7p = 10.8$

$$-2.7p = 10.8$$

$$\frac{-2.7}{-2.7}p = \frac{10.8}{-2.7} \qquad \textit{Divide both sides by } -2.7.$$

$$p = -4$$

C $\frac{t}{7.5} = 4$

$$\frac{t}{7.5} = 4$$

$$7.5 \cdot \frac{t}{7.5} = 7.5 \cdot 4 \qquad \textit{Multiply both sides by 7.5.}$$

$$t = 30$$

EXAMPLE 2 Solving Equations with Fractions

Solve.

A $x + \frac{1}{5} = -\frac{2}{5}$

$$x + \frac{1}{5} = -\frac{2}{5}$$

$$x + \frac{1}{5} - \frac{1}{5} = -\frac{2}{5} - \frac{1}{5} \qquad \textit{Subtract } \tfrac{1}{5} \textit{ from both sides.}$$

$$x = -\frac{3}{5}$$

Solve.

B

$$x - \frac{1}{4} = \frac{3}{8}$$

$$x - \frac{1}{4} = \frac{3}{8}$$

$$x - \frac{1}{4} + \frac{1}{4} = \frac{3}{8} + \frac{1}{4}$$ *Add $\frac{1}{4}$ to both sides of the equation.*

$$x = \frac{3}{8} + \frac{2}{8}$$ *Find a common denominator, 8.*

$$x = \frac{5}{8}$$

C

$$\frac{3}{5}w = \frac{3}{16}$$

$$\frac{3}{5}w = \frac{3}{16}$$

$$\frac{\cancel{5}}{\cancel{3}} \cdot \frac{\cancel{3}}{\cancel{5}}w = \frac{5}{\cancel{3}} \cdot \frac{\cancel{3}}{16}$$ *Multiply both sides by $\frac{5}{3}$. Simplify.*

$$w = \frac{5}{16}$$

EXAMPLE 3 Solving Word Problems Using Equations

In 1668 the Hope diamond was reduced from its original weight by $45\frac{1}{16}$ carats to a diamond weighing $67\frac{1}{8}$ carats. How many carats was the original diamond?

Convert fractions:

$$45\frac{1}{16} = \frac{45(16) + 1}{16} = \frac{721}{16} \qquad 67\frac{1}{8} = \frac{67(8) + 1}{8} = \frac{537}{8}$$

Write an equation:

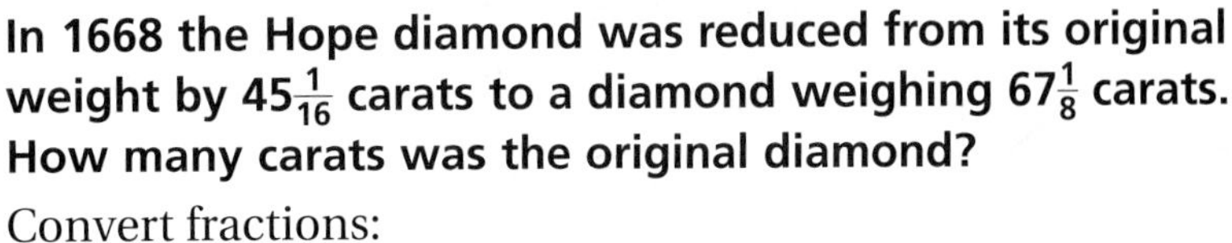

$$w \quad - \quad \frac{721}{16} \quad = \quad \frac{537}{8}$$

$$w - \frac{721}{16} = \frac{537}{8}$$

$$w - \frac{721}{16} + \frac{721}{16} = \frac{537}{8} + \frac{721}{16}$$ *Add $\frac{721}{16}$ to both sides.*

$$w = \frac{1074}{16} + \frac{721}{16}$$ *Find a common denominator, 16.*

$$w = \frac{1795}{16}, \text{ or } 112\frac{3}{16}$$ *Simplify.*

The original Hope diamond was $112\frac{3}{16}$ carats.

Think and Discuss

1. **Explain** the first step in solving an addition equation with fractions having *like* denominators.
2. **Explain** the first step in solving an addition equation with fractions having *unlike* denominators.

3-6 Exercises

FOR EOG PRACTICE

see page 653

internet connect

Homework Help Online

go.hrw.com Keyword: MT4 3-6

5.03, 5.04

GUIDED PRACTICE

See Example 1 Solve.

1. $y + 23.4 = -52$
2. $-6.3f = 44.1$
3. $\frac{m}{3.2} = -6$
4. $r - 17.9 = 36.8$
5. $\frac{s}{13.21} = 5.2$
6. $0.04g = 0.252$

See Example 2 Solve.

7. $x + \frac{1}{7} = -\frac{3}{7}$
8. $-\frac{2}{9} + k = -\frac{5}{9}$
9. $\frac{3}{5}w = -\frac{7}{15}$
10. $m - \frac{4}{3} = -\frac{4}{3}$
11. $\frac{7}{19}y = -\frac{63}{19}$
12. $t + \frac{4}{13} = \frac{12}{39}$

See Example 3

13. The Hope diamond has a width of $21\frac{39}{50}$ millimeters. Its width is equal to its length plus $3\frac{41}{50}$ millimeters. How many millimeters long is the Hope diamond?

INDEPENDENT PRACTICE

See Example 1 Solve.

14. $y + 16.7 = -49$
15. $5.8m = -52.2$
16. $-\frac{h}{6.7} = 3$
17. $k - 2.1 = -4.5$
18. $\frac{z}{10.7} = 4$
19. $c + 2.94 = 8.1$

See Example 2 Solve.

20. $j + \frac{1}{3} = \frac{3}{4}$
21. $\frac{5}{8}d = \frac{6}{18}$
22. $6h = \frac{12}{37}$
23. $x - \frac{1}{12} = \frac{5}{12}$
24. $r + \frac{5}{9} = -\frac{1}{9}$
25. $\frac{5}{6}c = \frac{7}{24}$

See Example 3

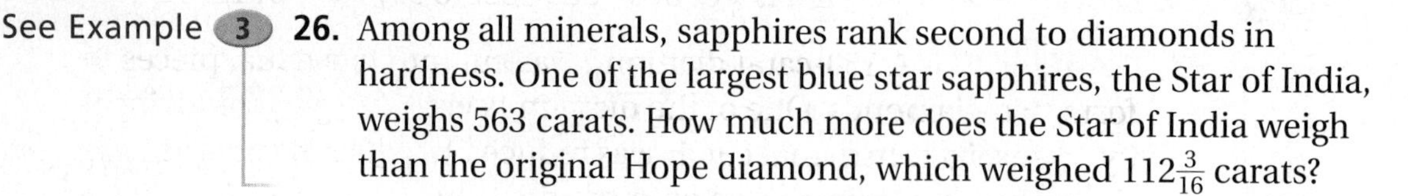

26. Among all minerals, sapphires rank second to diamonds in hardness. One of the largest blue star sapphires, the Star of India, weighs 563 carats. How much more does the Star of India weigh than the original Hope diamond, which weighed $112\frac{3}{16}$ carats?

PRACTICE AND PROBLEM SOLVING

Solve.

27. $z - \frac{5}{9} = \frac{1}{9}$
28. $-5f = -1.5$
29. $\frac{j}{8.1} = -4$
30. $t - \frac{3}{4} = 6\frac{1}{4}$
31. $-2.9g = -26.1$
32. $\frac{4}{9}d = -\frac{2}{9}$
33. $\frac{v}{5.5} = -5.5$
34. $r + \frac{5}{8} = -2\frac{3}{8}$
35. $y + 3.8 = -1.6$
36. $-\frac{1}{12} + r = \frac{3}{4}$
37. $-5c = \frac{5}{24}$
38. $m - 2.34 = 8.2$
39. $y - 68 = -3.9$
40. $-14 = -7.3 + f$
41. $\frac{2m}{0.7} = -8$

Earth Science LINK

Diamonds are found in several forms: the well-known gemstone, bort, ballas, and carbonado. Carbonado, ballas, and bort are used to cut stone and for the cutting edges of drills and other cutting tools.

42. ***EARTH SCIENCE*** The largest of all known diamonds, the Cullinan diamond, weighed 3106 carats before it was cut into 105 gems. The largest cut, Cullinan I, or the Great Star of Africa, weighs $530\frac{1}{5}$ carats. Another cut, Cullinan II, weighs $317\frac{2}{5}$ carats. Cullinan III weighs $94\frac{2}{5}$ carats, and Cullinan IV weighs $63\frac{3}{5}$ carats.

a. How many carats of the original Cullinan diamond were left after the Great Star of Africa and Cullinan II were cut?

b. How much more does Cullinan II weigh than Cullinan IV?

c. Which diamond weighs 223 carats less than Cullinan II?

43. Jack is tiling along the walls of the rectangular kitchen with the tile shown. The kitchen has a length of $243\frac{3}{4}$ inches and a width of $146\frac{1}{4}$ inches.

a. How many tiles will fit along the length of the room?

b. How many tiles will fit along its width?

c. If Jack needs 48 tiles to tile around all four walls of the kitchen, how many boxes of ten tiles must he buy? (*Hint:* He must buy whole boxes of tile.)

44. ***LIFE SCIENCE*** Each tablet in a box of allergy medicine weighs 0.3 gram. The total weight of all the tablets is 15 grams. How many tablets are in the box?

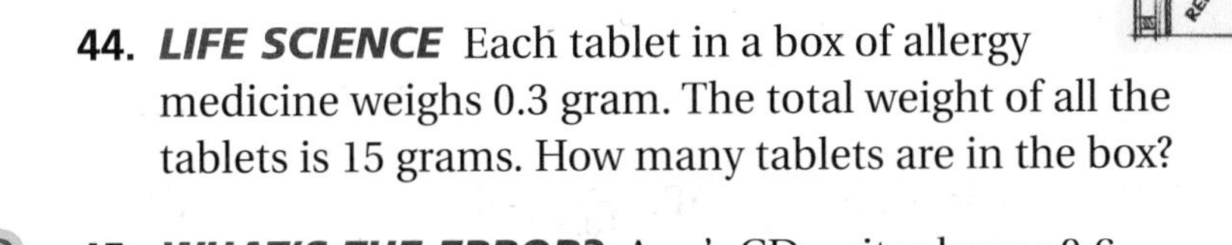

45. ***WHAT'S THE ERROR?*** Ann's CD writer burns 0.6 megabytes of information per second. A computer salesperson said that if Ann had 28.8 megabytes of information to burn, she could burn it in a little more than 15 seconds with this writer. What was the error?

46. ***WRITE ABOUT IT*** If a is $\frac{1}{3}$ of b, is it correct to say $\frac{1}{3}a = b$? Explain.

47. ***CHALLENGE*** A 150-carat diamond was cut into two equal pieces to form two diamonds. One of the diamonds was cut again, reducing it by $\frac{1}{3}$ its weight. In a final cut, it was reduced by $\frac{1}{4}$ its new weight. How many carats remained after the final cut?

Spiral Review

Evaluate each expression for the given values of the variables. (Lesson 1-1)

48. $4x + 5y$ for $x = 3$ and $y = 9$

49. $7m - 2n$ for $m = 5$ and $n = 7$

Write each number in scientific notation. (Lesson 2-9)

50. -0.000348

51. 0.00000524

52. $-4{,}870{,}000{,}000$

53. 64,000,000,000

54. **EOG PREP** If $x + y = 6$, then what is the value of $x + y - 2$? (Lesson 1-5)

A 8

B 4

C 3

D -4

3-7 Solving Inequalities with Rational Numbers

Learn to solve inequalities with rational numbers.

The minimum size for a piece of first-class mail is 5 inches long, $3\frac{1}{2}$ inches wide, and 0.007 inch thick. For a piece of mail, the combined length of the longest side and the distance around the thickest part may not exceed 108 inches. Many inequalities are used in determining postal rates.

EXAMPLE 1 Solving Inequalities with Decimals

Solve.

A $0.5x \geq 0.5$

$0.5x \geq 0.5$

$\frac{0.5}{0.5}x \geq \frac{0.5}{0.5}$ — *Divide both sides by 0.5.*

$x \geq 1$

B $t - 7.5 > 30$

$t - 7.5 > 30$

$t - 7.5 + 7.5 > 30 + 7.5$ — *Add 7.5 to both sides of the equation.*

$t > 37.5$

EXAMPLE 2 Solving Inequalities with Fractions

Solve.

A $x + \frac{1}{2} < 1$

$x + \frac{1}{2} < 1$

$x + \frac{1}{2} - \frac{1}{2} < 1 - \frac{1}{2}$ — *Subtract $\frac{1}{2}$ from both sides.*

$x < \frac{1}{2}$

Remember!

When multiplying or dividing an inequality by a *negative* number, reverse the inequality symbol.

B $-3\frac{1}{3}y \geq 10$

$-3\frac{1}{3}y \geq 10$

$-\frac{10}{3}y \geq 10$ — *Rewrite $-3\frac{1}{3}$ as the improper fraction $-\frac{10}{3}$.*

$\left(-\frac{3}{10}\right)\left(-\frac{10}{3}\right)y \leq \left(-\frac{3}{10}\right)10$ — *Multiply both sides by $-\frac{3}{10}$. Change $\geq$ to $\leq$.*

$y \leq -3$

EXAMPLE 3 PROBLEM SOLVING APPLICATION

With first-class mail, there is an extra charge in any of these cases:

- **The length is greater than $11\frac{1}{2}$ in.**
- **The height is greater than $6\frac{1}{8}$ in.**
- **The thickness is greater than $\frac{1}{4}$ in.**
- **The length divided by the height is less than 1.3 or greater than 2.5.**

The height of an envelope is 4.5 inches. What are the minimum and maximum lengths to avoid an extra charge?

1 Understand the Problem

The **answer** is the minimum and maximum lengths for an envelope to avoid an extra charge. List the **important information:**

- The height of the piece of mail is 4.5 inches.
- If the length divided by the height is between 1.3 and 2.5, there *will not be* an extra charge.

Show the **relationship** of the information:

$$1.3 \leq \frac{\textit{length}}{\textit{height}} \leq 2.5$$

2 Make a Plan

You can use the model above to write an inequality where ℓ is the length and 4.5 is the height.

$$1.3 \leq \frac{\ell}{4.5} \leq 2.5$$

3 Solve

$1.3 \leq \frac{\ell}{4.5}$	and	$\frac{\ell}{4.5} \leq 2.5$	
$4.5 \cdot 1.3 \leq \ell$	and	$\ell \leq 4.5 \cdot 2.5$	*Multiply both sides of each inequality by 4.5.*
$\ell \geq 5.85$	and	$\ell \leq 11.25$	*Simplify.*

4 Look Back

The length of the envelope must be between 5.85 in. and 11.25 in.

Think and Discuss

1. **Explain** the first steps in solving $0.5x > 7$ and solving $\frac{3}{5}x > 3$.
2. **Give** an example of an inequality with a fraction in which the sign changes during solving.

3-7 Exercises

FOR EOG PRACTICE
see page 653

internet connect
Homework Help Online
go.hrw.com Keyword: MT4 3-7

1.01, 5.03, 5.04

GUIDED PRACTICE

See Example 1 Solve.

1. $0.3x \geq 0.6$
2. $k - 7.2 > 2.1$
3. $\frac{g}{-0.5} \geq -\frac{7}{0.5}$
4. $h + 0.79 < 1.58$
5. $6.07w \leq 1.4568$
6. $z - 0.75 > -0.75$

See Example 2 Solve.

7. $k - \frac{2}{5} > \frac{3}{15}$
8. $y + \frac{7}{9} \geq \frac{56}{72}$
9. $13q \leq -\frac{1}{13}$
10. $x + \frac{1}{3} < 2$
11. $-3f < -\frac{4}{5}$
12. $3\frac{1}{4}m \geq 13$

See Example 3

13. Timothy is driving from Sampson to Williamsbery, a distance of 366.5 miles. If he averages between 45 mi/h and 55 mi/h, how long will it take him to get to Williamsbery to the nearest tenth of an hour, assuming he does not stop?

INDEPENDENT PRACTICE

See Example 1 Solve.

14. $0.6 + y \geq -0.72$
15. $m - 5.8 \leq -5.87$
16. $-0.8x \geq -0.56$
17. $\frac{g}{-2.7} \geq 9$
18. $c + 11.7 < 6$
19. $\frac{w}{-0.4} \geq \frac{3}{0.8}$

See Example 2 Solve.

20. $\frac{5}{9} + n \leq \frac{9}{5}$
21. $2\frac{2}{5}k \geq 1\frac{2}{3}$
22. $-\frac{2}{7} + x < 3$
23. $x + \frac{2}{5} \geq 5$
24. $7t < -\frac{14}{15}$
25. $-6\frac{1}{8}m \geq 7$

See Example 3

26. It takes an elevator 2 seconds to go from floor to floor. Each passenger takes 1.5 to 2.0 seconds to board or exit. Twelve passengers board on the first floor and exit on the fourth floor. How long will you wait if you are on the seventh floor and ring for the elevator just as it leaves the first floor?

PRACTICE AND PROBLEM SOLVING

Solve.

27. $-0.5d \geq 1.5$
28. $-3\frac{3}{4}m \geq 7\frac{1}{2}$
29. $\frac{2g}{0.5} \geq -\frac{4}{0.5}$
30. $x + \frac{2}{5} \geq 3$
31. $-4t < -\frac{12}{13}$
32. $r + 9.3 > 4.2$
33. $-1.6y \leq 12.8$
34. $c - 15.3 < 61.7$
35. $\frac{w}{-1.6} \geq \frac{1}{4.8}$
36. $6f > -\frac{4}{9}$
37. $5 < c + 1.9$
38. $\frac{2}{-0.4} \geq -\frac{r}{0.8}$
39. $2 > c - 1\frac{1}{3}$
40. $-f < \frac{6}{7}$
41. $3\frac{1}{4}t \leq 19.5$

Science LINK

Tsunamis, sometimes called tidal waves, move across deep oceans at high speeds with barely a ripple on the water surface. It is only when tsunamis hit shallow water that their energy moves them upward into a mammoth destructive force.

Tsunamis can be caused by earthquakes, volcanoes, landslides, or meteorites.

39. The rate of speed of a tsunami, in feet per second, can be found by the formula $r = \sqrt{32d}$, where d is the water depth in feet. Suppose the water depth is 20,000 ft. How fast is the tsunami moving?

40. The speed of a tsunami in miles per hour can be found using $r = \sqrt{14.88d}$, where d is the water depth in feet. Suppose the water depth is 25,000 ft.

a. How fast is the tsunami moving in miles per hour?

b. How long would it take a tsunami to travel 3000 miles if the water depth were a consistent 10,000 ft?

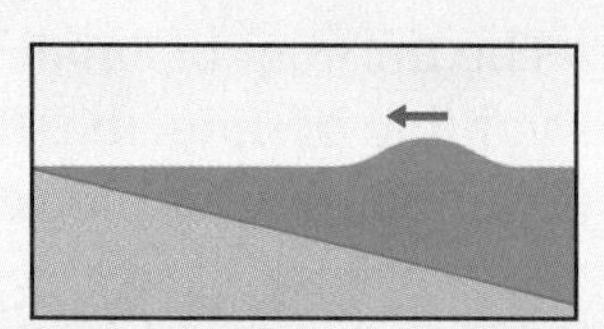

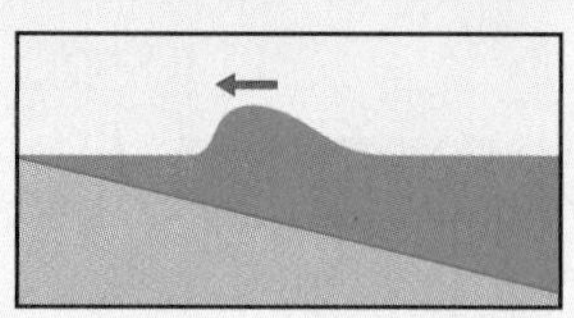

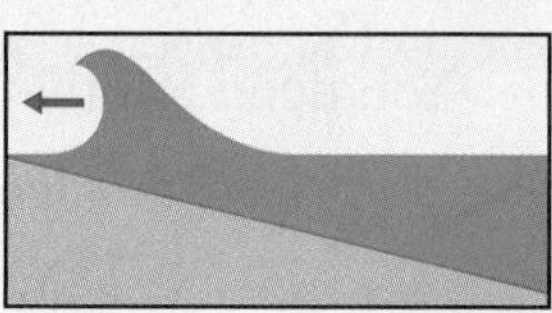

As the wave approaches the beach, it slows, builds in height, and crashes on shore.

41.  ***WHAT'S THE ERROR?*** Ashley found the speed of a tsunami, in feet per second, by taking the square root of 32 and multiplying by the depth, in feet. What was her error?

42. ***CHALLENGE*** Find the depth of the water if a tsunami's speed is 400 miles per hour.

go.hrw.com
KEYWORD: MT4 Wave
CNN Student News

Spiral Review

Solve. (Lesson 3-6)

169 1-35 odd

43. $y - 27.6 = -32$ **44.** $-5.3f = 74.2$ **45.** $\frac{m}{3.2} = -8$ **46.** $x + \frac{1}{8} = -\frac{5}{8}$

Evaluate. (Lesson 3-7)

47. $x + \frac{1}{3} < 6$ **48.** $-7f < -\frac{4}{5}$ **49.** $3\frac{1}{4}m \geq 26$ **50.** $0.7x \geq -1.4$

Find the square roots of each number. (Lesson 3-8)

51. 16 **52.** 81 **53.** 100 **54.** 1

55. **EOG PREP** It took Tina 6 minutes to saw a board into 3 equal pieces. How long would it have taken her to saw it into 9 equal pieces? (*Hint:* Think about the number of cuts she must make.) (Lesson 1-4)

A 2 min B 18 min C 21 min D 24 min

Explore Cubes and Cube Roots

Use with Lesson 3-9

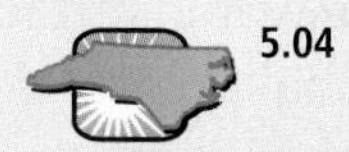

internet connect
Lab Resources Online
go.hrw.com
KEYWORD: MT4 Lab3B

WHAT YOU NEED:

Smallest base-10 blocks (Rainbow cubes or centimeter cubes will also work.)

REMEMBER

- **All edges of a cube are the same length.**
- **Volume is the number of cubic units needed to fill the space of a solid.**

The number of small unit blocks it takes to construct a cube is equal to the volume of the cube. By building a cube with edge length x and counting the number of unit blocks needed to build the cube, you can find x^3 (x-cubed), the volume.

Activity 1

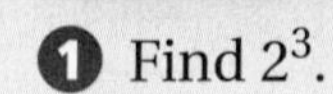

1 Find 2^3.

You need to build a cube with an edge length of 2.

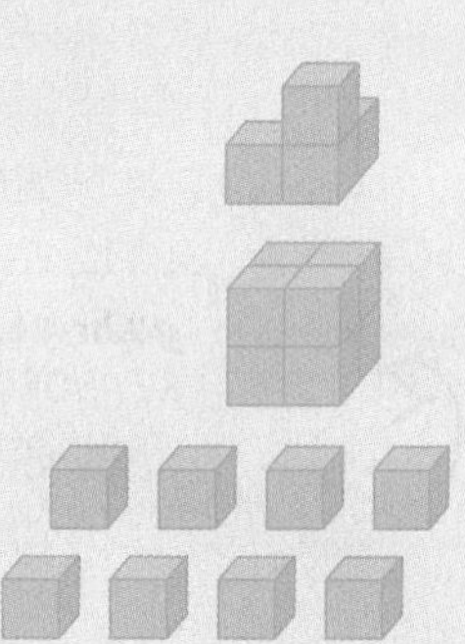

Build 3 edges of length 2.

Fill in the rest of the cube.

Count the number of unit cubes you needed to build a cube with an edge length of 2.

To make a cube with edge length 2, you need 8 unit blocks. So $2^3 = 8$.

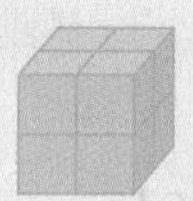

Think and Discuss

1. Why would it be difficult to model 2^4?

2. How can you find the value of a number squared from the model of that number cubed?

Try This

Model the following. How many blocks do you need to model each?

1. 1^3 **2.** 3^3 **3.** 4^3

You can determine whether any number x is a perfect cube by trying to build a cube out of x unit blocks. If you can build a cube with the given number of blocks, then that number is a perfect cube. Its *cube root* will be the length of one edge of the cube that is formed.

Activity 2

1 Try to build a cube using 27 unit blocks. Is 27 a perfect cube? If so what is its cube root?

Start by building a cube with an edge length of 2, since $1^3 = 1$ and $27 > 1$.

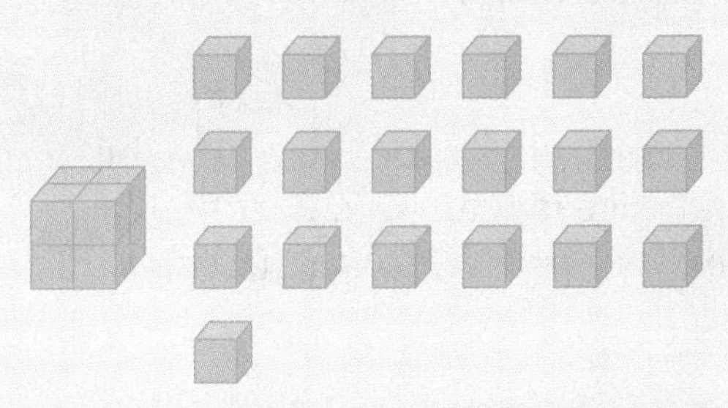

You still have 19 unit blocks left over. So try building a cube with an edge length of 3. Remember that when you add 1 unit cube to any edge you must do the same to all three edges to keep the cube shape.

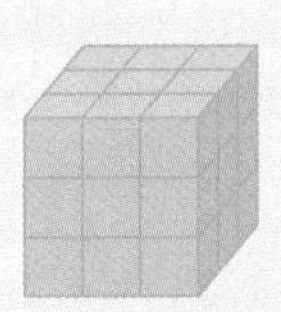

A cube with edges of length 3 can be made with 27 blocks.
length = 3
width = 3
height = 3

You can make a cube with edges of length 3 by using 27 small blocks. So 27 is a perfect cube. Its cube root is 3. We write $\sqrt[3]{27} = 3$.

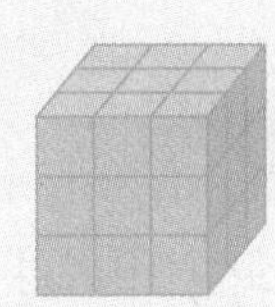

Think and Discuss

1. Is 100 a perfect cube? Why or why not?
2. How would you estimate the cube root of 100?
3. $\sqrt[3]{125} = 5$. Does $\sqrt[3]{2(125)} = \sqrt[3]{250} = 2(\sqrt[3]{125}) = 10$? Why or why not?
4. Use blocks to model a solid with a length of 3, a height of 2, and a width of 2. How many blocks did you use? Is this a perfect cube?

Try This

Model to find whether each number is a perfect cube. If the number is a perfect cube, find its cube root. If not, find the whole numbers that the cube root is between.

1. 64
2. 75
3. 125
4. 200

3-10 The Real Numbers

Learn to determine if a number is rational or irrational.

Vocabulary

irrational number

real number

Density Property

Biologists classify animals based on shared characteristics. The gray lesser mouse lemur is an animal, a mammal, a primate, and a lemur.

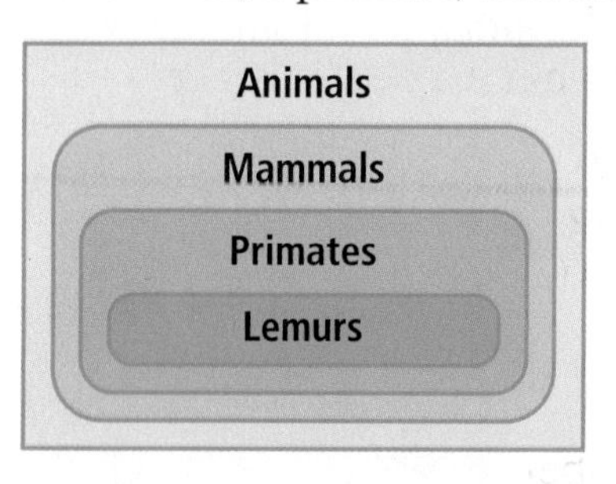

The lesser mouse lemur weighs only 2–3 oz and lives 10–15 years.

You already know that some numbers can also be classified as whole numbers, integers, or rational numbers. The number 2 is a whole number, an integer, and a rational number. It is also a *real* number.

Recall that rational numbers can be written as fractions. Rational numbers can also be written as decimals that either terminate or repeat.

$$3\frac{4}{5} = 3.8 \qquad \frac{2}{3} = 0.\overline{6} \qquad \sqrt{1.44} = 1.2$$

Helpful Hint

A repeating decimal may not appear to repeat on a calculator, because calculators show a finite number of digits.

Irrational numbers can only be written as decimals that do *not* terminate or repeat. If a whole number is not a perfect square, then its square root is an irrational number.

$\sqrt{2} \approx 1.4142135623730950488016\ldots$

The set of **real numbers** consists of the set of rational numbers and the set of irrational numbers.

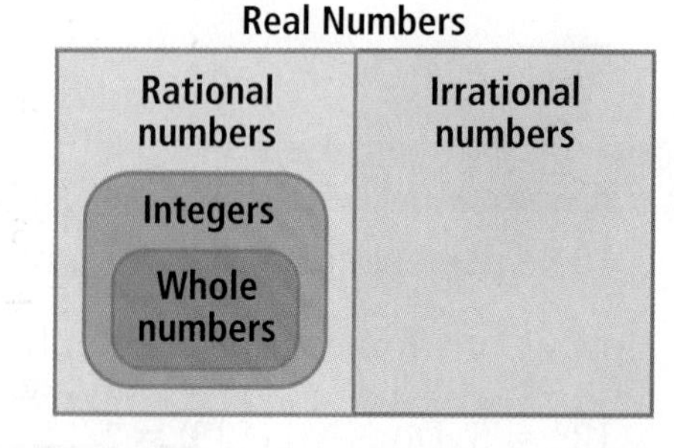

EXAMPLE 1 Classifying Real Numbers

Write all names that apply to each number.

A $\sqrt{3}$ — *3 is a whole number that is not a perfect square.*

irrational, real

B -56.85 — *−56.85 is a terminating decimal.*

rational, real

C $\frac{\sqrt{9}}{3}$ — $\frac{\sqrt{9}}{3} = \frac{3}{3} = 1$

whole, integer, rational, real

The square root of a negative number is not a real number. A fraction with a denominator of 0 is undefined. So it is not a number at all.

EXAMPLE 2 Determining the Classification of All Numbers

State if the number is rational, irrational, or not a real number.

A $\sqrt{10}$ *10 is a whole number that is not a perfect square.*

irrational

B $\frac{3}{0}$

not a number, so not a real number

C $\sqrt{\frac{1}{4}}$ $\left(\frac{1}{2}\right)\left(\frac{1}{2}\right) = \frac{1}{4}$

rational

D $\sqrt{-17}$

not a real number

The **Density Property** of real numbers states that between any two real numbers is another real number. This property is also true for rational numbers, but not for whole numbers or integers. For instance, there is no integer between −2 and −3.

EXAMPLE 3 Applying the Density Property of Real Numbers

Find a real number between $2\frac{1}{3}$ and $2\frac{2}{3}$.

There are many solutions. One solution is halfway between the two numbers. To find it, add the numbers and divide by 2.

$\left(2\frac{1}{3} + 2\frac{2}{3}\right) \div 2$

$= \left(4\frac{3}{3}\right) \div 2$

$= 5 \div 2 = 2\frac{1}{2}$

2 $2\frac{1}{3}$ $2\frac{1}{2}$ $2\frac{2}{3}$ 3

A real number between $2\frac{1}{3}$ and $2\frac{2}{3}$ is $2\frac{1}{2}$.

Think and Discuss

1. **Explain** how rational numbers are related to integers.
2. **Tell** if a number can be irrational and whole. Explain.
3. **Use** the Density Property to explain why there are infinitely many real numbers between 0 and 1.

3-10 Exercises

FOR EOG PRACTICE

see page 655

internet connect
Homework Help Online
go.hrw.com Keyword: MT4 3-10

1.01

GUIDED PRACTICE

See Example 1 **Write all names that apply to each number.**

1. $\sqrt{12}$ 2. $\sqrt{49}$ 3. 0.15 4. $-\frac{\sqrt{25}}{2}$

See Example 2 **State if the number is rational, irrational, or not a real number.**

5. $\sqrt{4}$ 6. $\sqrt{\frac{4}{25}}$ 7. $\sqrt{72}$ 8. $-\sqrt{-2}$

9. $-\sqrt{36}$ 10. $\sqrt{-4}$ 11. $\sqrt{\frac{16}{-25}}$ 12. $\frac{0}{0}$

See Example 3 **Find a real number between each pair of numbers.**

13. $5\frac{1}{6}$ and $5\frac{2}{6}$ 14. 3.14 and $\frac{22}{7}$ 15. $\frac{1}{8}$ and $\frac{1}{4}$

INDEPENDENT PRACTICE

See Example 1 **Write all names that apply to each number.**

16. $\sqrt{35}$ 17. $\frac{7}{9}$ 18. 2 19. $\frac{\sqrt{100}}{-5}$

See Example 2 **State if the number is rational, irrational, or not a real number.**

20. $\frac{-\sqrt{25}}{-5}$ 21. $-\sqrt{\frac{0}{9}}$ 22. $\sqrt{-12(-3)}$ 23. $-\sqrt{3}$

24. $\frac{\sqrt{16}}{5}$ 25. $\sqrt{18}$ 26. $\sqrt{-\frac{1}{4}}$ 27. $-\sqrt{\frac{9}{0}}$

See Example 3 **Find a real number between each pair of numbers.**

28. $3\frac{2}{5}$ and $3\frac{3}{5}$ 29. $-\frac{1}{100}$ and 0 30. 3 and $\sqrt{4}$

PRACTICE AND PROBLEM SOLVING

Write all names that apply to each number.

31. 8 32. $-\sqrt{36}$ 33. $\sqrt{20}$ 34. $\frac{2}{3}$

35. $\sqrt{3.24}$ 36. $\sqrt{25} + 5$ 37. $0.\overline{15}$ 38. $\frac{\sqrt{100}}{20}$

39. -6.5356 40. $\sqrt{4.5}$ 41. -122 42. $\frac{0}{5}$

Give an example of each type of number.

43. an irrational number that is less than -5

44. a rational number that is less than 0.5

45. a real number between $\frac{5}{9}$ and $\frac{6}{9}$

46. a real number between $-5\frac{4}{7}$ and $-5\frac{5}{7}$

47. Find a rational number between $\sqrt{\frac{1}{4}}$ and $\sqrt{1}$.

48. Find a real number between $\sqrt{2}$ and $\sqrt{3}$.

49. Find a real number between $\sqrt{5}$ and $\sqrt{11}$.

50. Find a real number between $\sqrt{70}$ and $\sqrt{75}$.

51. Find a real number between $-\sqrt{20}$ and $-\sqrt{17}$.

52. a. Find a real number between 1 and $\sqrt{2}$.
    b. Find a real number between 1 and your answer to part **a.**
    c. Find a real number between 1 and your answer to part **b.**

For what values of x is the value of each expression a real number?

53. $\sqrt{x}$

54. $5 - \sqrt{x}$

55. $\sqrt{x + 3}$

56. $\sqrt{2x - 4}$

57. $\sqrt{5x + 2}$

58. $\sqrt{1 - \frac{x}{3}}$

59. ***WHAT'S THE ERROR?*** A student said that the Density Property is true for integers because between the integers 2 and 4 is another integer, 3. Explain why the student's argument does not show that the Density Property is true for integers.

60. ***WRITE ABOUT IT*** Can you ever use a calculator to determine if a number is rational or irrational? Explain.

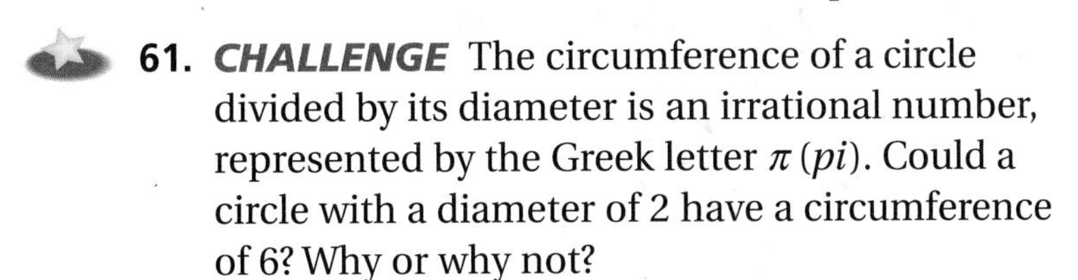

61. ***CHALLENGE*** The circumference of a circle divided by its diameter is an irrational number, represented by the Greek letter π (*pi*). Could a circle with a diameter of 2 have a circumference of 6? Why or why not?

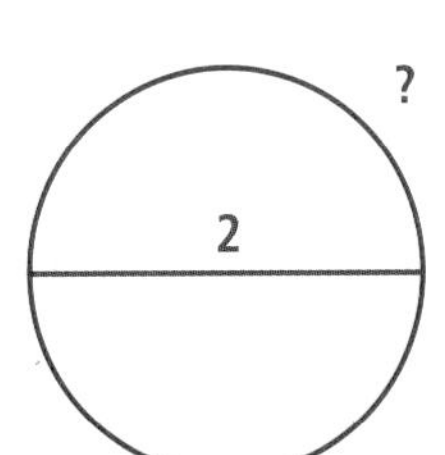

Spiral Review

Estimate each square root to two decimal places. (Lesson 3-9)

62. $\sqrt{30}$

63. $\sqrt{40}$

64. $\sqrt{50}$

65. $\sqrt{60}$

66. $\sqrt{1.8}$

67. $-\sqrt{17}$

68. $\sqrt{12}$

69. $2 \cdot \sqrt{3}$

Write each number in scientific notation. (Lesson 2-9)

70. 1,970,000,000

71. 2,500,000

72. 31,400

73. 5,680,000,000,000,000

74. **EOG PREP** If $20 \cdot 4000 = 8 \cdot 10^x$, then what is the value of x? (Lesson 2-7)

A 3 B 4 C 10 D 1000

75. **EOG PREP** If $\frac{12}{36} = 2w$, what is the value of w? (Lesson 3-6)

A $\frac{24}{72}$ B $\frac{24}{36}$ C $\frac{1}{6}$ D $\frac{1}{3}$

EXTENSION Other Number Systems

Learn to convert between bases.

Vocabulary
octal
binary

We use the base 10, or *decimal* number system, because we have ten fingers, or digits. Most cartoon characters have only eight fingers because cartoonists need to reduce detail. Cartoon characters could use the base 8, or **octal**, system.

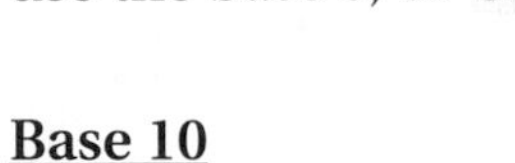

Base 10

- Place values are powers of 10.
- Digits are 0, 1, 2, 3, 4, 5, 6, 7, 8, 9.

$4316_{decimal}$ = 4 thousands · 3 hundreds · 1 ten · 6 ones

10^3	10^2	10^1	$10^0 = 1$
4	3	1	6

$$4 \times 10^3 + 3 \times 10^2 + 1 \times 10 + 6 \times 1$$

Base 8

- Place values are powers of 8.
- Digits are 0, 1, 2, 3, 4, 5, 6, 7.

4316_{octal} = 4 five hundred twelves · 3 sixty-fours · 1 eight · 6 ones

8^3	8^2	8^1	$8^0 = 1$
4	3	1	6

$$4 \times 8^3 + 3 \times 8^2 + 1 \times 8 + 6 \times 1$$

$$= 2048 + 192 + 8 + 6$$

$$= 2254_{decimal}$$

EXAMPLE 1 Changing from Base 8 to Base 10

Change 271_{octal} to base 10.

8^2	8^1	$8^0 = 1$
2	7	1

$$2 \times 8^2 + 7 \times 8^1 + 1 \times 8^0$$

$$= 128 + 56 + 1$$

$$= 185$$

$$271_{octal} = 185_{decimal}$$

EXAMPLE 2 Changing from Base 10 to Base 8

Helpful Hint

There is at least one multiple of 8^2 in 185, but no multiples of $8^3, 8^4, 8^5, \ldots$, since these are all greater than 185.

Change 185_{decimal} to base 8.

185 is between $8^2 = 64$ and $8^3 = 512$.

Do repeated divisions, by 8^2, 8^1, and finally 8^0.

$185 \div 8^2 = 2$ remainder 57
$57 \div 8^1 = 7$ remainder 1
$1 \div 8^0 = 1$ remainder 0

$185_{\text{decimal}} = 271_{\text{octal}}$

Check

$2 \times 8^2 + 7 \times 8^1 + 1 \times 8^0 = 185$

EXTENSION Exercises

Change each number in base 8 to base 10.

1. 63_{octal} **2.** 357_{octal} **3.** 1042_{octal}

Change each number in base 10 to base 8.

4. 74_{decimal} **5.** 229_{decimal} **6.** 3339_{decimal}

Base 2, or the **binary** system, is the number system used by computers. The binary system works in the same way as base 10 and base 8, except the place values are powers of 2 and the only digits are 0 and 1.

Change each number in base 2 to base 10.

7. 11_{binary} **8.** 1010_{binary} **9.** 111010_{binary}

Change each number in base 10 to base 2.

10. 13_{decimal} **11.** 222_{decimal} **12.** 1024_{decimal}

The binary system can be used in a code to represent symbols such as letters, numbers, and punctuation. There are four possible two-digit codes.

Possible Two-Digit Codes
00, 01, 10, 11

13. a. Write the possible binary three-digit codes.

b. Write the possible binary four-digit codes.

14. ***WHAT'S THE ERROR?*** The binary number 1010110_{binary} is supposed to equal 78_{decimal}. Correct the mistake in the binary number.

15. ***CHALLENGE*** What would be the digits for base 5? for base n?

Problem Solving on Location

NORTH CAROLINA

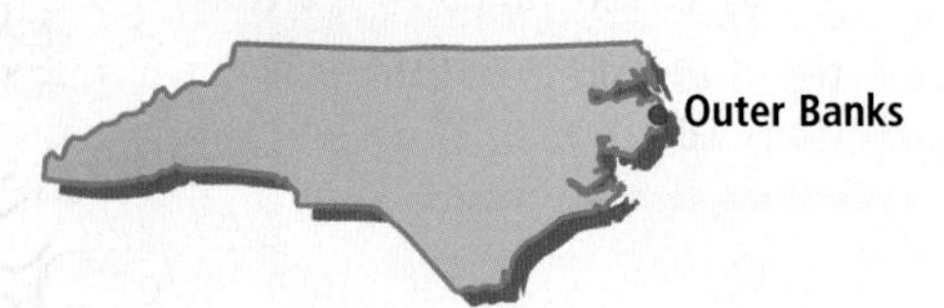

The Outer Banks

The Outer Banks of North Carolina feature four National Park Service campgrounds: Oregon Inlet, Cape Point, Ocracoke, and Frisco.

1. Oregon Inlet has 120 campsites. The number of campsites at Cape Point multiplied by $0.\overline{5940}$ is equal to the number of campsites at Oregon Inlet. Write $0.\overline{5940}$ as a fraction, and use it to write and solve an equation to find the number of campsites at Cape Point.

2. Adding the number of campsites at Frisco to $\frac{5}{8}$ the number of campsites at Oregon Inlet gives you the number of campsites at Cape Point. Use your answer from problem **1** to write and solve an equation to find the number of campsites at Frisco.

3. One-eighth the number of campsites at Ocracoke is equal to the number of campsites at Frisco minus 110. Use your answer from problem **2** to write and solve an equation to find the number of campsites at Ocracoke.

4. The Gribble family took a trip to the Outer Banks and then drove from the Outer Banks to Winston-Salem before returning home. If $\frac{7}{8}$ the distance from their house to the Outer Banks was equal to the distance from the Outer Banks to Winston-Salem, in which of the cities from the table at right do the Gribbles live?

Mileage to the Outer Banks

Point of Origin	Approximate Distance (mi)
Charlotte, NC	352
Durham, NC	225
New York, NY	433
Richmond, VA	160
Winston-Salem, NC	308

***The Lost Colony,* a play by Paul Green, is performed at Roanoke Island's outdoor performance pavilion during the summer months.**

Roanoke Island Festival Park

Roanoke Island is part of the Outer Banks. The 27-acre island was the site of the first temporary English settlement in North America. Features of Roanoke Island Festival Park include the *Elizabeth II,* which is a reproduction of a 65-foot-long 16th-century sailing vessel; an outdoor performance pavilion; and the Roanoke Adventure Museum.

1. If 750 of the 3500 seats at the outdoor performance pavilion are unoccupied, what fraction of the total seats are being used? Write your answer in simplest form.

2. Sixteenth-century ships were rated by the number of tonnes, or 252-gallon barrels, the ship could carry. An average British merchant ship had a rating of 121 tonnes. The *Elizabeth II* has a capacity of 12,600 gallons. What fraction of the capacity of an average 16th-century merchant ship is the capacity of the *Elizabeth II*? Write your answer in simplest form.

3. The height of the 69-foot topmast of the *Elizabeth II* is the difference of which two perfect squares?

4. You want to build a multistory building with the same amount of floor space as the 8500-square-foot Roanoke Adventure Museum. Each floor will be a square of exactly the same size, with a whole number side length. What is the fewest number of floors you must build? What will the side length of each floor be? (*Hint:* The area of each floor will be the largest perfect square that is a factor of 8500.)

MATH-ABLES

Egyptian Fractions

If you were to divide 9 loaves of bread among 10 people, you would give each person $\frac{9}{10}$ of a loaf. The answer was different on the ancient Egyptian Ahmes papyrus, because ancient Egyptians used only *unit fractions*, which have a numerator of 1. All other fractions were written as sums of different unit fractions. So $\frac{5}{6}$ could be written as $\frac{1}{2} + \frac{1}{3}$, but not as $\frac{1}{6} + \frac{1}{6} + \frac{1}{6} + \frac{1}{6} + \frac{1}{6}$.

Method	Example
Suppose you want to write a fraction as a sum of different unit fractions.	$\frac{9}{10}$ 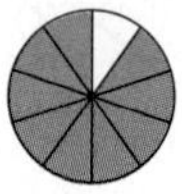
Step 1. Choose the largest fraction of the form $\frac{1}{n}$ that is less than the fraction you want. 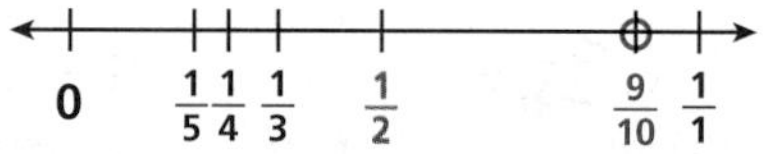 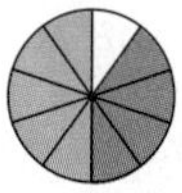	Number line: 0, $\frac{1}{5}$, $\frac{1}{4}$, $\frac{1}{3}$, $\frac{1}{2}$, $\frac{9}{10}$, $\frac{1}{1}$
Step 2. Subtract $\frac{1}{n}$ from the fraction you want.	$\frac{9}{10} - \frac{1}{2} = \frac{2}{5}$ remaining 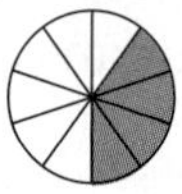
Step 3. Repeat steps 1 and 2 using the difference of the fractions until the result is a unit fraction.	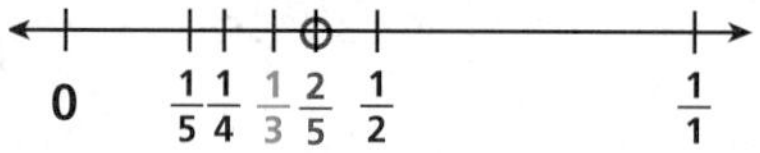 Number line: 0, $\frac{1}{5}$, $\frac{1}{4}$, $\frac{1}{3}$, $\frac{2}{5}$, $\frac{1}{2}$, $\frac{1}{1}$ $\frac{2}{5} - \frac{1}{3} = \frac{1}{15}$ remaining 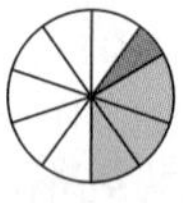
Step 4. Write the fraction you want as the sum of the unit fractions.	$\frac{9}{10} = \frac{1}{2} + \frac{1}{3} + \frac{1}{15}$ 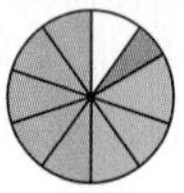

Write each fraction as a sum of different unit fractions.

1. $\frac{3}{4}$ **2.** $\frac{5}{8}$ **3.** $\frac{11}{12}$ **4.** $\frac{3}{7}$ **5.** $\frac{7}{5}$

Egg Fractions

This game is played with an empty egg carton. Each compartment represents a fraction with a denominator of 12. The goal is to place tokens in compartments with a given sum.

internet connect

Go to ***go.hrw.com*** for a complete set of rules and instructions.
KEYWORD: MT4 Game3

Add and Subtract Fractions

Use with Lesson 3-5

internet connect
Lab Resources Online
go.hrw.com
KEYWORD: MT4 TechLab3

You can add and subtract fractions using your graphing calculator. To display decimals as fractions, use the MATH key.

Activity

1 Use a graphing calculator to add $\frac{7}{12} + \frac{3}{8}$. Write the sum as a fraction.

Type 7 ÷ 12 and press ENTER.

You can see that the decimal equivalent is a repeating decimal, $0.58\overline{3}$.

Type + 3 ÷ 8 ENTER. The decimal form of the sum is displayed.

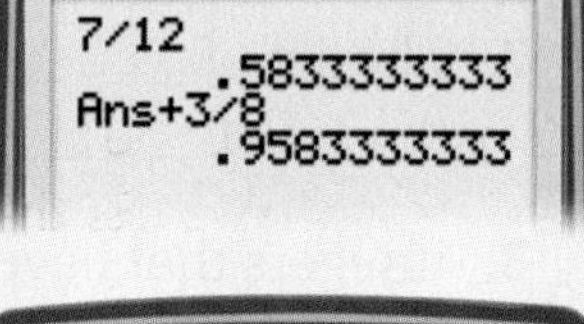

Press MATH ENTER ENTER.

The fraction form of the sum, $\frac{23}{24}$, is displayed as 23/24.

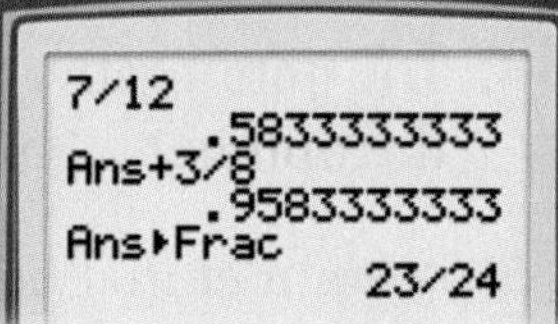

2 Use a graphing calculator to subtract $\frac{3}{5} - \frac{2}{3}$. Write the difference as a fraction.

Type 3 ÷ 5 − 2 ÷ 3 MATH ENTER ENTER.

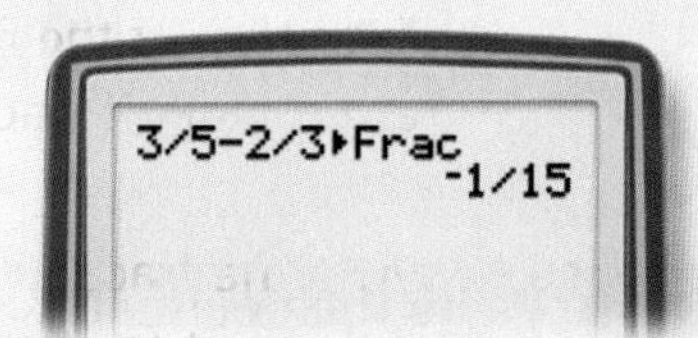

The answer is $-\frac{1}{15}$.

Think and Discuss

1. Why is the difference in **2** negative?

2. Type 0.33333... (pressing 3 at least twelve times). Press MATH ENTER ENTER to write $0.\overline{3}$ as a fraction. Now do the same for $0.\overline{9}$. What happens to $0.\overline{9}$? How does the fraction for $0.\overline{3}$ help to explain this result?

Try This

Use a calculator to add or subtract. Write each result as a fraction.

1. $\frac{1}{2} + \frac{2}{5}$ **2.** $\frac{7}{8} - \frac{2}{3}$ **3.** $\frac{7}{17} + \frac{1}{10}$ **4.** $\frac{1}{3} - \frac{5}{7}$

5. $\frac{5}{32} + \frac{2}{11}$ **6.** $\frac{33}{101} - \frac{3}{7}$ **7.** $\frac{4}{15} + \frac{7}{16}$ **8.** $\frac{1}{35} - \frac{1}{37}$

Chapter 3

Study Guide and Review

Vocabulary

Complete the sentences below with vocabulary words from the list above. Words may be used more than once.

1. Any number that can be written as a fraction $\frac{n}{d}$ (where n and d are integers and $d \neq 0$) is called a __?__.
2. The set of __?__ is made up of the set of rational numbers and the set of __?__.
3. Integers that have no common factors other than 1 are __?__.
4. The nonnegative square root of a number is called the __?__ of the number.
5. A number that has rational numbers as its square roots is a __?__.

3-1 Rational Numbers (pp. 112–116)

EXAMPLE

■ **Write the decimal as a fraction.**

$0.8 = \frac{8}{10}$ *8 is in the tenths place.*

$= \frac{8 \div 2}{10 \div 2}$ *Divide numerator and denominator by 2.*

$= \frac{4}{5}$

EXERCISES

Write each decimal as a fraction.

6. 0.6
7. 0.25
8. 0.525

Simplify.

9. $\frac{14}{21}$
10. $\frac{22}{33}$
11. $\frac{75}{100}$

3-2 Adding and Subtracting Rational Numbers (pp. 117–120)

EXAMPLE

■ **Add or subtract.**

$\frac{3}{7} + \frac{4}{7} = \frac{3+4}{7} = \frac{7}{7} = 1$

$\frac{8}{11} - \left(\frac{-2}{11}\right) = \frac{8-(-2)}{11} = \frac{8+2}{11} = \frac{10}{11}$

EXERCISES

Add or subtract.

12. $\frac{-8}{13} + \frac{2}{13}$
13. $\frac{3}{5} - \left(\frac{-4}{5}\right)$
14. $\frac{-2}{9} + \frac{7}{9}$
15. $\frac{-5}{12} - \left(\frac{-7}{12}\right)$

3-3 Multiplying Rational Numbers (pp. 121–125)

EXAMPLE

- **Multiply. Write the answer in simplest form.**

$5\left(3\frac{1}{4}\right) = \left(\frac{5}{1}\right)\left(\frac{3(4)+1}{4}\right)$

$= \left(\frac{5}{1}\right)\left(\frac{13}{4}\right)$ *Write as improper fractions.*

$= \frac{65}{4}$ *Multiply.*

$= 16\frac{1}{4}$ *Write in simplest form.*

EXERCISES

Multiply. Write each answer in simplest form.

16. $3\left(-\frac{2}{5}\right)$

17. $2\left(3\frac{4}{5}\right)$

18. $\frac{-2}{3}\left(\frac{-4}{5}\right)$

19. $\frac{8}{11}\left(\frac{-22}{4}\right)$

20. $5\frac{1}{4}\left(\frac{3}{7}\right)$

21. $2\frac{1}{2}\left(1\frac{3}{10}\right)$

3-4 Dividing Rational Numbers (pp. 126–130)

EXAMPLE

- **Divide. Write the answer in simplest form.**

$\frac{7}{8} \div \frac{3}{4} = \frac{7}{8} \cdot \frac{4}{3}$ *Multiply by the reciprocal.*

$= \frac{7 \cdot 4}{8 \cdot 3}$ *Write as one fraction.*

$\frac{7 \cdot \cancel{4}^{1}}{{}_{2}\cancel{8} \cdot 3} = \frac{7 \cdot 1}{2 \cdot 3}$ *Divide by common factor, 4.*

$\frac{7}{6} = 1\frac{1}{6}$

EXERCISES

Divide. Write each answer in simplest form.

22. $\frac{3}{4} \div \frac{1}{8}$

23. $\frac{3}{10} \div \frac{4}{5}$

24. $\frac{2}{3} \div 3$

25. $4 \div \frac{-1}{4}$

26. $3\frac{3}{4} \div 3$

27. $1\frac{1}{3} \div \frac{2}{3}$

3-5 Adding and Subtracting with Unlike Denominators (pp. 131–134)

EXAMPLE

- **Add.**

$\frac{3}{4} + \frac{2}{5}$ *Multiply denominators, $4 \cdot 5 = 20$.*

$\frac{3 \cdot 5}{4 \cdot 5} = \frac{15}{20}$ $\frac{2 \cdot 4}{5 \cdot 4} = \frac{8}{20}$ *Rename fractions with the LCD 20.*

$\frac{15}{20} + \frac{8}{20} = \frac{15 + 8}{20} = \frac{23}{20} = 1\frac{3}{20}$ *Add and simplify.*

EXERCISES

Add or subtract.

28. $\frac{5}{6} + \frac{1}{3}$

29. $\frac{5}{6} - \frac{5}{9}$

30. $3\frac{1}{2} + 7\frac{4}{5}$

31. $7\frac{1}{10} - 2\frac{3}{4}$

3-6 Solving Equations with Rational Numbers (pp. 136–139)

EXAMPLE

- **Solve.**

$x - 13.7 = -22$

$\underline{+13.7} = \underline{+13.7}$ *Add 13.7 to each side.*

$x = -8.3$

EXERCISES

Solve.

32. $y + 7.8 = -14$

33. $2.9z = -52.2$

34. $w + \frac{3}{4} = \frac{1}{8}$

35. $\frac{3}{8}p = \frac{3}{4}$

3-7 Solving Inequalities with Rational Numbers (pp. 140–143)

EXAMPLE

■ **Solve.**

$-3x > \frac{6}{7}$

$-\frac{1}{3}(-3x) > -\frac{1}{3}\left(\frac{6}{7}\right)$ *Multiply each side by $-\frac{1}{3}$.*

$x < -\frac{2}{7}$ *Change > to <, since you multiplied by a negative.*

EXERCISES

Solve.

36. $4m > -\frac{1}{3}$

37. $-2.7t \leq 32.4$

38. $7\frac{1}{2} - y \geq 10\frac{3}{4}$

39. $x + \frac{4}{5} > \frac{3}{10}$

3-8 Squares and Square Roots (pp. 146–149)

EXAMPLE

■ **Find the two square roots of 400.**

$20 \cdot 20 = 400$

$(-20) \cdot (-20) = 400$

The square roots are 20 and −20.

EXERCISES

Find the two square roots of each number.

40. 16 **41.** 900 **42.** 676

Evaluate each expression.

43. $\sqrt{4 + 21}$ **44.** $\frac{\sqrt{100}}{20}$ **45.** $\sqrt{3^4}$

3-9 Finding Square Roots (pp. 150–153)

EXAMPLE

■ **Find the side length of a square with area 359 ft^2 to one decimal place. Then find the distance around the square.**

$18^2 = 324,\ 19^2 = 361$

Side $= \sqrt{359} \approx 18.9$

Distance around $\approx 4(18.9) \approx 75.6$ feet

EXERCISES

Find the distance around each square with the area given. Answer to the nearest tenth.

46. Area of square *ABCD* is 500 in^2.

47. Area of square *MNOP* is 1750 cm^2.

3-10 The Real Numbers (pp. 156–159)

EXAMPLE

■ **State if the number is rational, irrational, or not a real number.**

$-\sqrt{2}$ real, irrational *The decimal equivalent does not repeat or end.*

$\sqrt{-4}$ not real *Square roots of negative numbers are not real.*

EXERCISES

State if the number is rational, irrational, or not a real number.

48. $\sqrt{81}$ **49.** $\sqrt{122}$ **50.** $\sqrt{-16}$

51. $-\sqrt{5}$ **52.** $\frac{0}{-4}$ **53.** $\frac{7}{0}$

Study Guide and Review

Chapter 3

Chapter Test

Simplify.

1. $\frac{36}{72}$
2. $\frac{21}{35}$
3. $\frac{16}{88}$
4. $\frac{18}{25}$

Write each decimal as a fraction in simplest form.

5. 0.225
6. 0.04
7. 0.101
8. 0.875

Write each fraction as a decimal.

9. $\frac{7}{8}$
10. $\frac{13}{25}$
11. $\frac{5}{12}$
12. $\frac{4}{33}$

Add or subtract. Write each answer in simplest form.

13. $\frac{-3}{11} - \left(\frac{-4}{11}\right)$
14. $7\frac{1}{4} - 2\frac{3}{4}$
15. $\frac{5}{6} + \frac{7}{18}$
16. $\frac{5}{6} - \frac{8}{9}$
17. $4\frac{1}{2} + 5\frac{7}{8}$
18. $8\frac{1}{5} - 1\frac{2}{3}$

Multiply or divide. Write each answer in simplest form.

19. $9\left(\frac{-2}{27}\right)$
20. $\frac{7}{8} \div \frac{5}{24}$
21. $\frac{2}{3}\left(\frac{-9}{20}\right)$
22. $3\frac{3}{7}\left(1\frac{5}{16}\right)$
23. $34 \div 3\frac{2}{5}$
24. $-4\frac{2}{3} \div 1\frac{1}{6}$

Solve.

25. $x - \frac{1}{4} = -\frac{3}{8}$
26. $-3.14y = 53.38$
27. $-2k < \frac{1}{4}$
28. $h - 3.24 \leq -1.1$

Find the two square roots of each number.

29. 196
30. 1
31. 0.25
32. 6.25

Each square root is between two integers. Name the integers.

33. $\sqrt{230}$
34. $\sqrt{125}$
35. $\sqrt{89}$
36. $-\sqrt{60}$

State whether the number is rational, irrational, or not real.

37. $-\sqrt{121}$
38. $-1.\overline{7}$
39. $\sqrt{-9}$

Solve.

40. Michelle wants to put a fence along one side of her square-shaped vegetable garden. The area of the garden is 1250 ft^2. How much fencing should she buy, to the nearest foot?

Chapter 3

Performance Assessment

Show What You Know

Create a portfolio of your work from this chapter. Complete this page and include it with your four best pieces of work from Chapter 3. Choose from your homework or lab assignments, mid-chapter quiz, or any journal entries you have done. Put them together using any design you want. Make your portfolio represent what you consider your best work.

Short Response

1. A square chessboard is made up of 64 squares. If you placed a knight in each of the squares around the edge of the board, how many knight pieces would you need? Show or explain how you determined your answer.

2. In a mechanical drawing, a hidden line is usually represented by dashes $\frac{1}{8}$ in. long with $\frac{1}{32}$ in. spaces between them. How long is a line represented by 26 dashes? Show or explain how you determined your answer.

3. Write the multiplication equation $\frac{3}{4} \cdot \frac{5}{7} = \frac{15}{28}$ as a division equation. Use your result to explain why dividing by a fraction is the same as multiplying by the fraction's reciprocal.

Extended Problem Solving

4. Use a diagram to model multiplication of fractions.
   a. Draw a diagram to model the fraction $\frac{5}{6}$.
   b. Shade $\frac{2}{5}$ of the part of your diagram that represents $\frac{5}{6}$. What product does this shaded area represent?
   c. Use your diagram to write the product in simplest form.

Performance Assessment

internet connect
EOG Test Practice Online
go.hrw.com Keyword: MT4 TestPrep

Getting Ready for EOG

Chapter 3

Cumulative Assessment, Chapters 1–3

1. Which ordered pair lies on the negative portion of the y-axis?

   A $(-4, -4)$ C $(0, -4)$
   B $(-4, 0)$ D $(4, -4)$

2. The sum of two numbers that differ by 1 is x. In terms of x, what is the value of the greater of the two numbers?

   A $\frac{x-1}{2}$ C $\frac{x+1}{2}$
   B $\frac{x}{2}$ D $\frac{x}{2}+1$

3. If the sum of the consecutive integers from -22 through x is 72, what is the value of x?

   A 23 C 50
   B 25 D 75

4. If $xy + y = x + 2z$, what is the value of y when $x = 2$ and $z = 3$?

   A $\sqrt{8}$ C $\sqrt[3]{8}$
   B $\frac{8}{3}$ D 24

5. A local library association has posted the results of community contributions to the building fund.

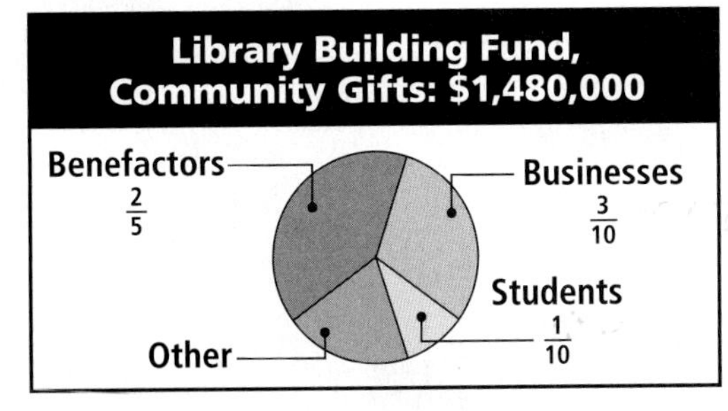

   How much money does "Other" represent?

   A $148,000 C $444,000
   B $296,000 D $592,000

6. Which number is equivalent to 3^{-3}?

   A $\frac{1}{9}$ C $-\frac{1}{27}$
   B $\frac{1}{27}$ D $-\frac{1}{9}$

7. What is the value of $10 - 2 \cdot 3^2$?

   A -26 C 72
   B -8 D 576

TEST TAKING TIP!

Making comparisons: When assigning test values, try different kinds of numbers, such as negatives and fractions.

8. If x is any real number, then which statement *must* be true?

   A $x^2 > x$
   B $x^3 > x$
   C $x^3 > x^2$
   D No relationship can be determined.

9. ***SHORT RESPONSE*** The total weight of Sam and his son Dan is 250 pounds. Sam's weight is 10 pounds more than 3 times Dan's weight. Write an equation that could be used to determine Dan's weight. Solve your equation.

10. ***SHORT RESPONSE*** There was $1000 in the bank teller's drawer when the bank opened. After the first customer's withdrawal, the drawer still had greater than $900 in it, and it had an equal number of $1, $5, $10, $20, $50, and $100 bills in it. How much money did the first customer withdraw? Show or explain how you found your answer.

Chapter 4

Collecting, Displaying, and Analyzing Data

TABLE OF CONTENTS

Chapter Opener Online
go.hrw.com
KEYWORD: MT4 CH4

Errors in Samples

Company Type	Sample Size	Errors
Software	25	2
Stoneworks	100	7
Tools	50	4
Pizza	75	3

Quality Assurance Specialist

How do manufacturers know that their products are well made? It is the job of the quality assurance specialist. QA specialists design tests and procedures that allow the companies to determine how good their products are. Because checking every product or procedure may not be possible, QA specialists use sampling to predict the margin of error.

ARE YOU READY?

Choose the best term from the list to complete each sentence.

1. A __?__ is a uniform measure where equal distances are marked to represent equal amounts.
2. __?__ is the process of approximating to a given __?__.
3. Ordered pairs of numbers are graphed on a __?__.

coordinate grid
place value
rounding
scale

Complete these exercises to review skills you will need for this chapter.

✔ Round Decimals

Round each number to the indicated place value.

4. 34.7826; nearest tenth
5. 137.5842; nearest whole number
6. 287.2872; nearest thousandth
7. 362.6238; nearest hundred

✔ Compare and Order Decimals

Order each sequence of numbers from greatest to least.

8. 3.005, 3.05, 0.35, 3.5
9. 0.048, 0.408, 0.0408, 0.48
10. 5.01, 5.1, 5.011, 5.11
11. 1.007, 0.017, 1.7, 0.107

✔ Place Value of Whole Numbers

Write each number in standard form.

12. 1.3 million
13. 7.59 million
14. 4.6 billion
15. 2.83 billion

✔ Read a Table

Use the table for problems 16–18.

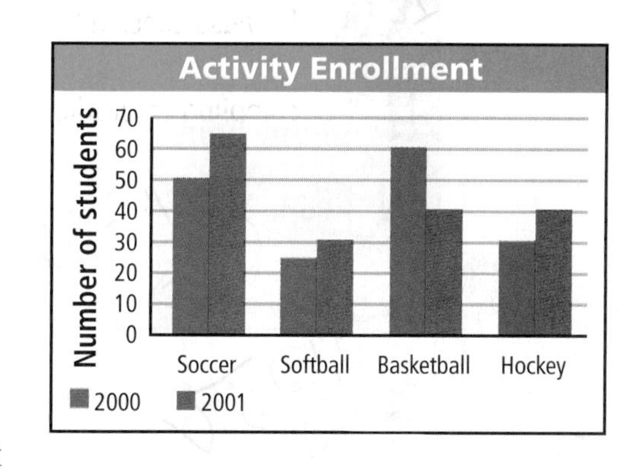

16. Which activity experienced the greatest change in participation from 2000 to 2001?
17. Which activity experienced the greatest positive change in participation from 2000 to 2001?
18. Which activity experienced the least change in participation from 2000 to 2001?

4-1 Samples and Surveys

Learn to recognize biased samples and to identify sampling methods.

Vocabulary
population
sample
biased sample
random sample
systematic sample
stratified sample

A fitness magazine printed a readers' survey. Statements 1, 2, and 3 are interpretations of the results. Which do you think the magazine would use?

1. **The average American exercises 3 times a week.**
2. **The average reader of this magazine exercises 3 times a week.**
3. **The average reader who responded to the survey exercises 3 times a week.**

The **population** is the entire group being studied. The **sample** is the part of the population being surveyed.

For statement 1, the population is all Americans and the sample is readers of the fitness magazine who chose to respond. This is a **biased sample** because it is not a good representation of the population.

People who read fitness magazines are likely to be interested in exercise. This could make the sample biased in favor of people who exercise more times per week.

EXAMPLE 1 Identifying Biased Samples

Identify the population and sample. Give a reason why the sample could be biased.

A **A radio station manager chooses 1500 people from the local phone book to survey about their listening habits.**

Population	Sample	Possible Bias
People in the local area	Up to 1500 people who take the survey	Not all people are in the phone book.

B **An advice columnist asks her readers to write in with their opinions about how to hang the toilet paper on the roll.**

Population	Sample	Possible Bias
Readers of the column	Readers who write in	Only readers with strong opinions write in.

Identify the population and sample. Give a reason why the sample could be biased.

C **Surveyors in a mall choose shoppers to ask about product preferences.**

Population	Sample	Possible Bias
All shoppers in the mall	The people who are polled	Surveyors are more likely to approach shoppers who look agreeable.

To get accurate information, it is important to use a good sampling method. In a **random sample**, every member of the population has an equal chance of being chosen. A random sample is best, but other methods are often used for convenience.

Sampling Method	How Members Are Chosen
Random	By chance
Systematic	According to a rule or formula
Stratified	At random from randomly chosen subgroups

EXAMPLE 2 Identifying Sampling Methods

Identify the sampling method used.

A **An exit poll is taken of every tenth voter.**

systematic *The rule is to question every tenth voter.*

B **In a statewide survey, five counties are randomly chosen and 100 people are randomly chosen from each county.**

stratified *The five counties are the random subgroups. People are chosen randomly from within the counties.*

C **Students in a class write their names on strips of paper and put them in a hat. The teacher draws five names.**

random *Names are chosen by chance.*

Think and Discuss

1. **Describe** ways to eliminate the possible bias in Example 1C.
2. **Decide** which sampling method would be best to find the number of times a week the average student in your school exercises.

4-1 Exercises

FOR EOG PRACTICE

see page 656

internet connect

Homework Help Online

go.hrw.com Keyword: MT4 4-1

4.01, 4.03

GUIDED PRACTICE

See Example 1

Identify the population and sample. Give a reason why the sample could be biased.

1. A pet store owner surveys 100 customers to find out what brand of dog food is purchased most frequently.

See Example 2

Identify the sampling method used.

2. People with a house number ending with 1 are polled.
3. A surveyor flips through the phone book and selects 30 names.

INDEPENDENT PRACTICE

See Example 1

Identify the population and sample. Give a reason why the sample could be biased.

4. A deli owner asks Sunday's customers to choose a favorite mustard.

See Example 2

Identify the sampling method used.

5. People at the theater seated in an even-numbered row and an odd-numbered seat are surveyed.
6. Ten study groups at the library are chosen, and one person is selected at random from each group.

PRACTICE AND PROBLEM SOLVING

Identify the population and sample. Give a reason why the sample could be biased.

7. A cafeteria worker asks students who buy the entrée if they like the food in the cafeteria.
8. A theater manager asks the last ten people to leave a movie, "Did you like the movie?"
9. A chef asks the first four customers who order the new cheese sauce if they like it.
10. A biologist studying trees samples blossoms of trees along the river.

Identify the sampling method used.

11. Every fifth name is called from a list of voters.
12. Students each write a question on a slip of paper and put it in a box. The teacher draws one question to discuss.
13. One hundred shoppers are chosen by chance from four randomly chosen computer stores.

Identify the sampling method used.

14. A manufacturer tests every sixtieth item from an assembly line.

15. Every third student signing up for an astronomy class is asked about telescope preferences.

16. Fifteen classes are randomly chosen. Ten students are randomly chosen from each class.

17. ***BUSINESS*** For an advertising campaign, you need to survey people to find out why they like to visit the San Diego Zoo.

a. How can you select an unbiased sample for this survey?

b. How can you make your sampling method systematic?

c. Why would surveying only families with children be biased?

18. ***MONEY*** Martin sorted coins he had been collecting in a jar for 15 years, and decided that most coins in circulation are dated 1980.

a. What is the population of this survey?

b. Give a reason why the sample could be biased.

19. ***WRITE ABOUT IT*** To help plan your annual class picnic, you survey students in your class about where they want to have the picnic. Choose a sampling method and explain your choice.

20. ***WHAT'S THE ERROR?*** A distributor planned to take a stratified sample of restaurants to find out what the most commonly ordered food product was. Five restaurants were chosen at random, and at each restaurant every tenth customer was surveyed. Why isn't this a stratified sample?

21. ***CHALLENGE*** The diagrams show the locations that have been chosen where soil samples will be taken to test for pollution. Identify the type of sample each diagram represents.

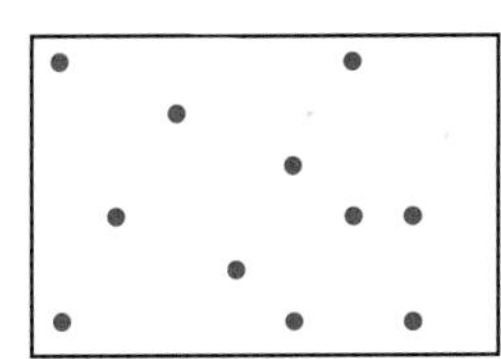

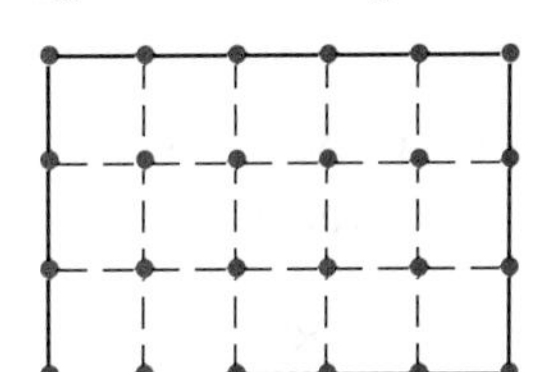

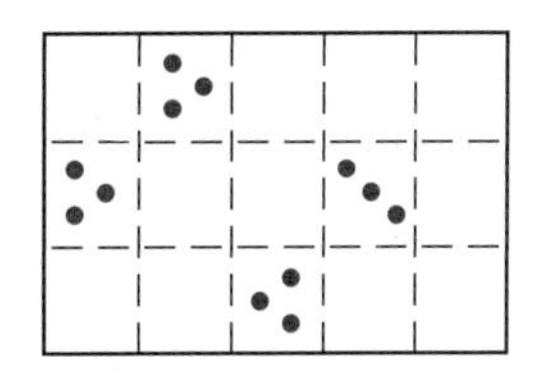

Spiral Review

Solve. (Lesson 3-6)

22. $x + \frac{1}{6} = -\frac{5}{6}$ **23.** $\frac{y}{2.4} = -3$ **24.** $y - 11.6 = -21$ **25.** $23\frac{5}{7} - 24 = c$

Solve. (Lesson 3-7)

26. $w + (-5.7) > -18.9$ **27.** $-14.9x < -381.44$

28. **EOG PREP** What is the next term in the sequence 5, 12, 26, 54, ...? (Previous course)

A 82 B 110 C 120 D 159

Explore Sampling

Use with Lesson 4-1

4.01

REMEMBER
- **Be organized before starting.**
- **Be sure that your sample reflects your population.**

You can predict data about a population by collecting data from a representative sample.

Activity

Your school district has been discussing the possibility of school uniforms. Each school will get to choose its uniform and colors. Your class has been chosen to make the selection for your school. To be fair, you want to be sure that the other students in the school have input. Therefore, you take a survey to see what the majority of the students in your school want.

1 Follow the steps below to model conducting the survey.

a. Choose your population.

- every student in the school
- all girls
- only your class
- all boys
- all 8th grade students
- teachers

b. Decide what kind of sample you will use. Discuss pros and cons of each.

- random
- systematic
- stratified

c. Decide what colors and what uniform choices to present to your sample.

- pants
- shorts
- skirts
- sweaters
- jackets
- vests
- school colors
- navy blue
- forest green

Think and Discuss

1. Explain why choosing the teachers as your population might not be the best choice.

2. How did you decide which colors to present to your sample?

Try This

1. Create forms for your survey showing the options from which you want your sample to choose. Then survey your sample. Make a table of your results. Explain what your table tells you about the population.

4-2 Organizing Data

Learn **to organize data in tables and stem-and-leaf plots.**

Vocabulary

stem-and-leaf plot

back-to-back stem-and-leaf plot

When you graduate and start looking for a job, you may have to keep track of a lot of information.

A table is one way to organize and display data so that you can understand the meaning and recognize any relationships.

Mathematics, physics, computer science, chemistry, and English are good courses to take if you want to be an airline mechanic.

EXAMPLE 1 Organizing Data in Tables

Use the given data to make a table.

Greg has received job offers as a mechanic at three airlines. The first has a salary range of \$20,000–\$34,000, benefits worth \$12,000, and 10 days' vacation. The second has 15 days' vacation, benefits worth \$10,500, and a salary range of \$18,000–\$50,000. The third has benefits worth \$11,400, a salary range of \$14,000–\$40,000, and 12 days' vacation.

	Job 1	Job 2	Job 3
Salary Range	\$20,000–\$34,000	\$18,000–\$50,000	\$14,000–\$40,000
Benefits	\$12,000	\$10,500	\$11,400
Vacation Days	10	15	12

A **stem-and-leaf plot** is another way to display data. The values are grouped so that all but the last digit is the same in each category.

Stem = first digit(s)

2 | 5 = 25

Leaf = last digit

EXAMPLE 2 Reading Stem-and-Leaf Plots

List the data values in the stem-and-leaf plot.

```
0 | 2 5
1 | 3 3 7 8
2 | 0 2 6
3 | 1 7          Key: 3|1 means 31
```

The data values are 2, 5, 13, 13, 17, 18, 20, 22, 26, 31, and 37.

EXAMPLE 3 Organizing Data in Stem-and-Leaf Plots

Use the given data to make a stem-and-leaf plot.

Heights of Tallest Trees in U.S. (m)					
Ash	47	Elm	38	Red maple	55
Beech	40	Grand fir	77	Sequoia	84
Black maple	40	Hemlock	74	Spruce	63
Cedar	67	Hickory	58	Sycamore	40
Cherry	42	Oak	61	Western pine	48
Douglas fir	91	Pecan	44	Willow	35

Heights range from 35 to 91, so stems are 3 to 9.

Stem	Leaves
3	5 8
4	0 0 0 2 4 7 8
5	5 8
6	1 3 7
7	4 7
8	4
9	1

Key: 9 | 1 means 91 m

A **back-to-back stem-and-leaf plot** is used to compare two sets of data. The stems are in the center, and the left leaves are read in reverse.

EXAMPLE 4 Organizing Data in Back-to-Back Stem-and-Leaf Plots

Use the given data to make a back-to-back stem-and-leaf plot.

Super Bowl Scores, 1990–2000											
	1990	**1991**	**1992**	**1993**	**1994**	**1995**	**1996**	**1997**	**1998**	**1999**	**2000**
Winning	55	20	37	52	30	49	27	35	31	34	23
Losing	10	19	24	17	13	26	17	21	24	19	16

Losing		Winning
9 9 7 7 6 3 0	1	
6 4 4 1	2	0 3 7
	3	0 1 4 5 7
	4	9
	5	2 5

Key: |5|2 means 52 points
1|2| means 21 points

Think and Discuss

1. **Tell** which is always the same as the number of data values in a stem-and-leaf plot: the number of stems or the number of leaves.

4-2 Exercises

FOR EOG PRACTICE
see page 656

internet connect
Homework Help Online
go.hrw.com Keyword: MT4 4-2

4.01

GUIDED PRACTICE

See Example 1

1. Use the given data to make a table.

A 100 g serving of baked potato has 2.4 g fiber, 10 mg calcium (Ca), and 27 mg magnesium (Mg).

A 100 g serving of french fries has 3.2 g fiber, 10 mg Ca, and 22 mg Mg.

A 100 g serving of potato chips has 4.5 g fiber, 24 mg Ca, and 67 mg Mg. (*Source:* USDA)

See Example 2

List the data values in the stem-and-leaf plot.

2.

Stem	Leaves
0	2 3 3 7
1	1 3 7 7 8
2	0 0 7
3	4 4 5 5

Key: 3|5 means 35

3.

Stem	Leaves
6	3 6 8
7	3 3 5 7
8	0 0 1 1
9	0 4 5 9

Key: 9|9 means 99

See Example 3

4. Use the given data to make a stem-and-leaf plot.

Atomic Numbers of Some Elements							
Hydrogen	1	Silver	47	Carbon	6	Titanium	22
Nitrogen	7	Barium	56	Argon	18	Bromine	35
Calcium	20	Iron	26	Krypton	36	Iodine	53

See Example 4

5. Use the given data to make a back-to-back stem-and-leaf plot.

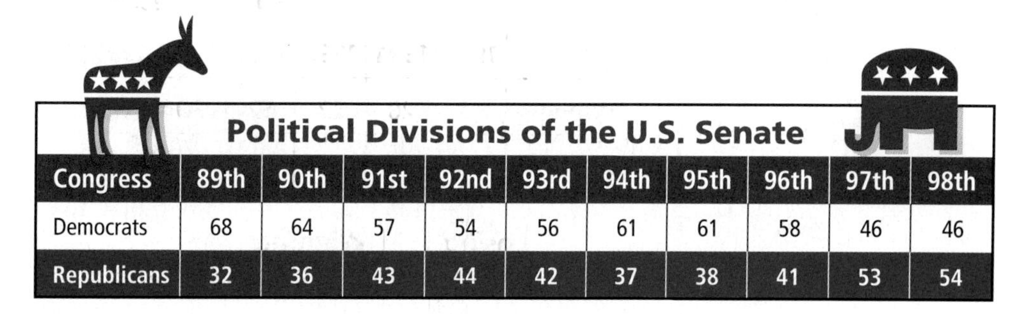

Political Divisions of the U.S. Senate										
Congress	89th	90th	91st	92nd	93rd	94th	95th	96th	97th	98th
Democrats	68	64	57	54	56	61	61	58	46	46
Republicans	32	36	43	44	42	37	38	41	53	54

INDEPENDENT PRACTICE

See Example 1

6. Use the given data to make a table.

New passenger car sales in 1970: 7,110,000 domestic, 313,000 Japanese imports, 750,000 German imports

New passenger car sales in 1980: 6,581,000 domestic, 1,906,000 Japanese imports, 305,000 German imports

New passenger car sales in 1990: 6,897,000 domestic, 1,719,000 Japanese imports, 265,000 German imports

See Example 2 **List the data values in the stem-and-leaf plot.**

7.

5	0 1 4 8
6	2 6 7
7	1 4 5 6 6
8	2

Key: 6|2 means 62

8.

0	1 5 7
1	2 4 6 8
2	0 1 7 9
3	3 3 4 6

Key: 2|1 means 21

See Example 3 **9. Use the given data to make a stem-and-leaf plot.**

Average Price per Gallon of Unleaded Regular Gasoline by Year							
1981	$1.38	1986	$0.93	1991	$1.14	1996	$1.23
1982	$1.30	1987	$0.95	1992	$1.13	1997	$1.23
1983	$1.24	1988	$0.95	1993	$1.11	1998	$1.06
1984	$1.21	1989	$1.02	1994	$1.11	1999	$1.17

See Example 4 **10.** Use the data given in the map to make a back-to-back stem-and-leaf plot.

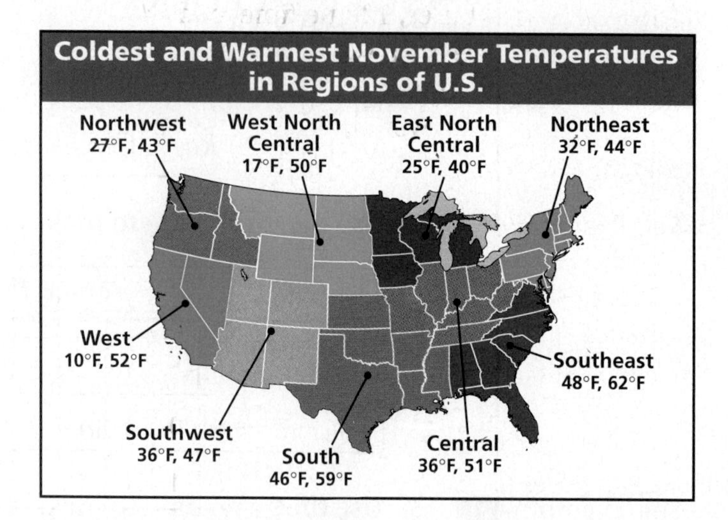

PRACTICE AND PROBLEM SOLVING

Create a stem-and-leaf plot of the data values.

11. 72, 43, 75, 57, 81, 65, 68, 72, 73, 84, 91, 76, 82, 88

12. 5.3, 6.8, 3.2, 6.4, 2.7, 4.9, 6.3, 5.5, 4.1, 3.8, 6.0, 4.1, 4.5, 5.9

13. Use the given data to make a table.
In 1980, 89% of energy used in the United States was from fossil fuels, 3% from nuclear power, and 7% from renewable sources. In 1990, 86% was from fossil fuels, 7% from nuclear power, and 7% from renewable sources. In 2000, 85% was from fossil fuels, 8% from nuclear power, and 7% from renewable sources.

14. Use the given data to make a back-to-back stem-and-leaf plot.

Miles per Gallon Ratings of a Car Company's Models										
Model	A	B	C	D	E	F	G	H	I	J
City Miles	11	17	28	19	18	15	18	22	14	20
Highway Miles	15	24	36	28	26	20	23	25	17	29

Language Arts 

An author's writing style is as unique as a fingerprint. Punctuation, spelling, and word usage can be used to determine authorship.

Don Foster used this fact to analyze the 350-year-old poem "A Funeral Elegy." The analysis confirmed that the poem of previously unknown authorship was actually written by William Shakespeare.

Don Foster's methods have also been used to analyze ransom notes and evidence in court cases.

15. Act 5 of Shakespeare's *A Midsummer Night's Dream* has the following references to numbers: 1 nine times, 2 three times, 3 six times, 10 two times, 12 one time, and 14 one time. There are also references to time: night 12 times, day 4 times, supper-time 1 time, bed-time 1 time, and evening 1 time. Use the data to make a table.

16. The table shows the punctuation in Henry Wadsworth Longfellow's poem "Paul Revere's Ride." Make a back-to-back stem-and-leaf plot of the number of commas and periods in each verse.

	Verse													
	1	2	3	4	5	6	7	8	9	10	11	12	13	14
,	4	8	6	8	10	12	15	10	7	3	5	5	5	11
—	1	1	3	0	1	2	2	0	0	0	1	1	2	2
!	0	0	1	0	0	1	3	1	0	0	0	0	0	1
.	1	1	1	1	1	1	2	1	1	2	2	3	2	1

17. ***CHALLENGE*** Select two paragraphs from a work by your favorite author and a third paragraph by a different author. Compare word choices or punctuation use in the three paragraphs, and explain the similarities and differences. Use a table or stem-and-leaf plot to support your argument.

Spiral Review

Multiply or divide. Write the product or quotient as one power. (Lesson 2-7)

18. $\frac{7^4}{7^2}$ **19.** $5^3 \cdot 5^8$ **20.** $\frac{t^8}{t^5}$ **21.** $\frac{10^9}{9^3}$

Identify the population and sample. (Lesson 4-1)

22. A cable company surveys customers whose last names begin with an *S*.

23. The principal asks every other busload of students if their ride was comfortable.

24. **EOG PREP** Which number is less than 10^3? (Lesson 2-6)

A 2^{10} B 7^5 C 8^4 D 25^2

4-3 Measures of Central Tendency

Learn to find appropriate measures of central tendency.

Vocabulary

mean

median

mode

outlier

A measure of central tendency is an attempt to describe a data set using only one number. This number represents the "middle" of the set.

Measures of Central Tendency

	Definition	Use to Answer
Mean	The sum of the values, divided by the number of values	"What is the average?" "What single number best represents the data?"
Median	If an odd number of values: the middle value If an even number of values: the average of the two middle values	"What is the halfway point of the data?"
Mode	The value or values that occur most often	"What is the most common value?"

EXAMPLE 1 Finding Measures of Central Tendency

Find the mean, median, and mode of each data set.

A 4, 8, 8, 3, 6, 8, 3

mean: $4 + 8 + 8 + 3 + 6 + 8 + 3 = 40$ *Add the values.*

$\frac{40}{7} \approx 5.7$ *Divide by 7, the number of values.*

median: 3 3 4 (6) 8 8 8 *Order the values.*

3 values *3 values*

The median is 6.

mode: 8 *The value 8 occurs three times.*

B 9, 6, 91, 5, 7, 6, 8, 8, 7, 9

mean: $9 + 6 + 91 + 5 + 7 + 6 + 8 + 8 + 7 + 9 = 156$

$\frac{156}{10} = 15.6$ *Divide by 10.*

median: 5 6 6 7 (7 8) 8 9 9 91 *Order the values.*

5 values *5 values*

$\frac{7 + 8}{2} = 7.5$ *Average the two middle values.*

mode: 6, 7, 8, 9 *Four values occur twice each.*

Find the mean, median, and mode.

C 28, 12, 101, 53

mean: $28 + 12 + 101 + 53 = 194$

$\frac{194}{4} = 48.5$

median: 12 (28 ⋮ 53) 101

$\frac{28 + 53}{2} = 40.5$

mode: No mode — *No value occurs more than any other.*

Notice that the mean in Example 1B is much greater than most of the data values. This is because 91 is so far from the other data values. An extreme value such as this is called an **outlier**. An outlier can have a strong effect on the mean of a data set.

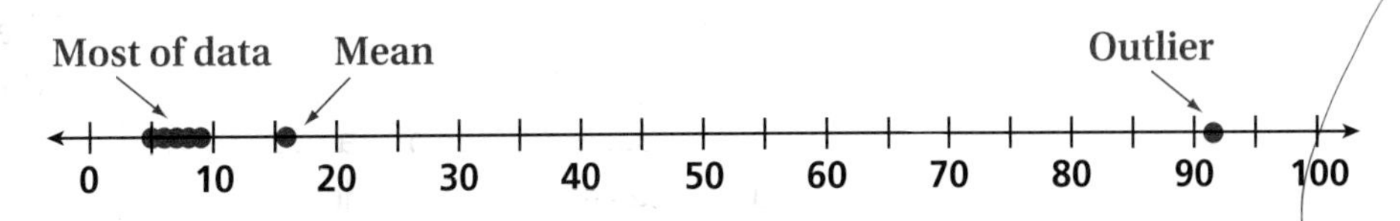

EXAMPLE 2 *Astronomy Application*

Use the data to find each answer.

Planet	Known Moons
Mercury	0
Venus	0
Earth	1
Mars	2
Jupiter	39
Saturn	30
Uranus	21
Neptune	8
Pluto	1

Source: NASA, 2002

A **Find the average number of moons for the *terrestrial planets:* Mercury, Venus, Earth, and Mars.**

Use the mean to answer, "What's the average?"

$\frac{0 + 0 + 1 + 2}{4} = \frac{3}{4} = 0.75$

B **Find the average number of moons for the *gas giants:* Jupiter, Saturn, Uranus, and Neptune.**

$\frac{39 + 30 + 21 + 8}{4} = \frac{98}{4} = 24.5$

C **Find the average number of moons per planet.**

$\frac{0 + 0 + 1 + 2 + 39 + 30 + 21 + 8 + 1}{9} = \frac{102}{9} \approx 11.33$

Astronomy LINK

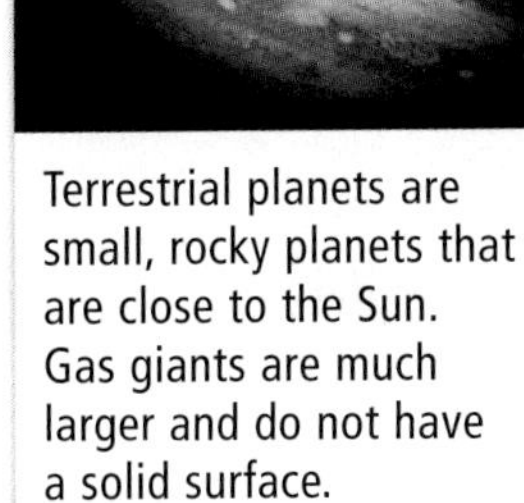

Terrestrial planets are small, rocky planets that are close to the Sun. Gas giants are much larger and do not have a solid surface.

go.hrw.com
KEYWORD: MT4 Moons

CNN Student News

Think and Discuss

1. **Compare** the mean and median of the set 1, 2, 3, and 4 to the mean and median of the set 1, 2, 3, and 40. Explain the difference.

2. **Give** a data set with the same mean, median, and mode.

4-3 Exercises

FOR EOG PRACTICE
see page 657

internet connect
Homework Help Online
go.hrw.com Keyword: MT4 4-3

4.01

GUIDED PRACTICE

See Example 1 **Find the mean, median, and mode of each data set.**

1. 35, 21, 34, 44, 36, 42, 29
2. 2.0, 4.4, 6.2, 3.2, 4.4, 6.2
3. 7, 5, 4, 6, 8, 3, 5, 2, 5
4. 23, 13, 45, 56, 72, 44, 89, 92, 67

See Example 2 **Use the data to find each answer.**

Source: 2000 U.S. Census

5. Find the average number of people per city.
6. Find the average number of people in the cities in Texas: Dallas, Houston, and San Antonio.
7. Find the average number of people in the northeastern cities: New York, Philadelphia, Chicago, and Detroit.

INDEPENDENT PRACTICE

See Example 1 **Find the mean, median, and mode of each data set.**

8. 5, 2, 12, 7, 13, 9, 8
9. 92, 88, 84, 86, 88
10. 6, 8, 6, 7, 9, 2, 4, 22
11. 4.3, 1.3, 4.5, 8.6, 9, 3, 2.1, 14

See Example 2 **Use the data to find each answer.**

12. Find the average farm acreage per state.
13. Find the average farm acreage in the southernmost states: Texas, New Mexico, and Kansas.
14. Find the average farm acreage in the northernmost states: Montana, North Dakota, South Dakota, and Nebraska.

Acres of Farmland	
State	**Acres (million)**
Texas	131
Montana	59
Kansas	46
Nebraska	46
New Mexico	46
South Dakota	44
North Dakota	39

PRACTICE AND PROBLEM SOLVING

Find the mean, median, and mode of each data set. Name any outliers.

15. 20, 17, 42, 26, 27, 12, 31

16. 4.0, 3.3, 5.6, 4.6, 3.3, 5.6

17. 15, 10, 12, 10, 13, 13, 13, 10, 3

18. 8, 5, 3, 75, 7, 3, 4, 7, 9, 2, 8, 5, 7

19. 2, 6, 29, 6, 2, 2, 1, 1, 2, 1, 2, 2, 1, 0, 0, 4, 7

20. 22, 34, 36, 18, 36, 40, 25, 23, 32, 43, 43

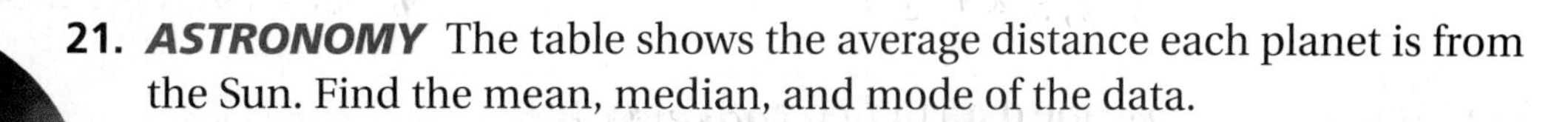

21. ***ASTRONOMY*** The table shows the average distance each planet is from the Sun. Find the mean, median, and mode of the data.

Distance from the Sun

Planet	Mercury	Venus	Earth	Mars	Jupiter	Saturn	Uranus	Neptune	Pluto
Miles (million)	36	67	93	141	484	887	1784	2796	3661

22. Teresa has taken three tests worth 100 points each. Her scores are 85, 93, and 88. She has one test left to take.

a. To get an average of 90, what must the sum of all her test scores be?

b. What score does she need on the fourth test to get an average of 90?

23. When would you use the median to describe the central tendency for these salaries? \$1350, \$1250, \$1425, \$1250, \$10,750

24. When is the median a number in the data set? When is the mode?

25. ***WRITE A PROBLEM*** Use your test scores from one course to write a problem about central tendency.

26. ***WRITE ABOUT IT*** If six friends went to dinner and split the check equally, what measure of central tendency would describe the amount each person paid? Explain.

27. ***CHALLENGE*** If $4\left(\frac{x + y + z}{3}\right) = 8$, what is the mean of x, y, and z?

Spiral Review

Simplify. (Lesson 1-6)

28. $3(p + 7) - 5p$

29. $4x + 5(2x - 9)$

30. $8 + 7(y + 5) - 3$

Solve. (Lesson 1-6)

31. $15x - 8x = 91$

32. $3j - 5j = -14$

33. $4m + 6m = 1000$

34. **EOG PREP** Which value of x is a solution of $x - 9 = 8$? (Lesson 1-3)

A -17

B -1

C 1

D 17

35. **EOG PREP** Which ordered pair is *not* a solution of $y = 3x - 2$? (Lesson 1-7)

A $(-2, -8)$

B $(0, -2)$

C $(2, 0)$

D $(2, 4)$

4-4 Variability

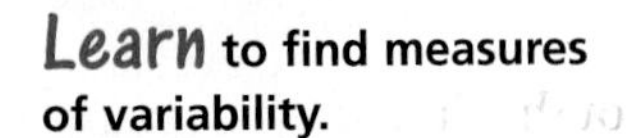

Learn to find measures of variability.

Vocabulary
variability
range
quartile
box-and-whisker plot

The table below summarizes a veterinarian's records for kitten litters born in a given year.

Litter Size	2	3	4	5	6
Number of Litters	1	6	8	11	1

The term *box-and-whisker plot* may remind you of a box of kittens. But it is a way to display data.

While central tendency describes the middle of a data set, **variability** describes how spread out the data is.

The **range** of a data set is the largest value minus the smallest value. For the kitten data, the range is $6 - 2 = 4$.

The range is affected by outliers, so another measure is often used. **Quartiles** divide a data set into four equal parts. The third quartile minus the first quartile is the range for the middle half of the data.

Kitten Data

Lower half | *Upper half*

2 3 3 3 3 3 3 (3) 4 4 4 4 4 4 (4) 4 5 5 5 5 5 (5) 5 5 5 5 5 6

First quartile: 3 median of lower half | *Median: 4 (second quartile)* | *Third quartile: 5 median of upper half*

EXAMPLE 1 Finding Measures of Variability

Find the range and the first and third quartiles for each data set.

A **85, 92, 78, 88, 90, 88, 89**

78 (85) 88 | 88 | 89 (90) 92 — *Order the values.*

range: $92 - 78 = 14$

first quartile: 85 **third quartile:** 90

B **14, 12, 15, 17, 15, 16, 17, 18, 15, 19, 20, 17**

12 14 (15 15) 15 16 | 17 17 (17 18) 19 20 — *Order the values.*

range: $20 - 12 = 8$

first quartile: $\frac{15 + 15}{2} = 15$ **third quartile:** $\frac{17 + 18}{2} = 17.5$

A **box-and-whisker plot** shows the distribution of data. The middle half of the data is represented by a "box" with a vertical line at the median. The lower fourth and upper fourth are represented by "whiskers" that extend to the smallest and largest values.

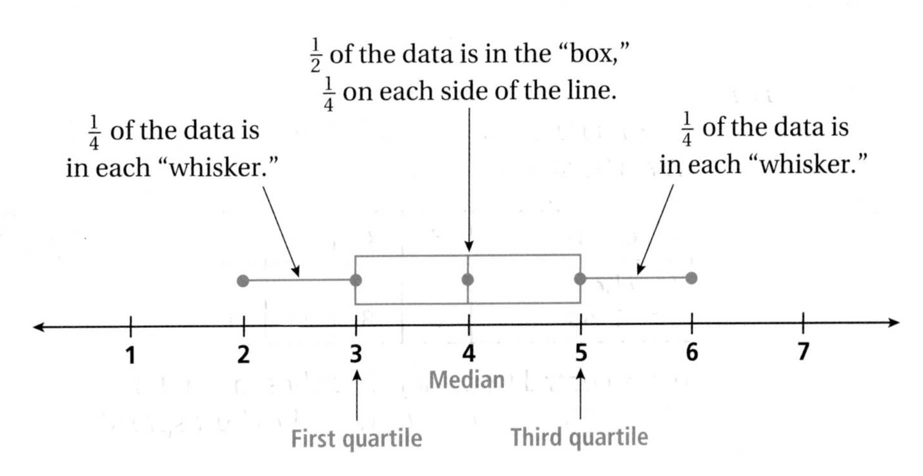

EXAMPLE 2 Making a Box-and-Whisker Plot

Use the given data to make a box-and-whisker plot.

22 17 22 49 55 21 49 62 21 16 18 44 42 48 40 33 45

Step 1: Order the data and find the smallest value, first quartile, median, third quartile, and largest value.

16 17 18 (21 21) 22 22 33 (40) 42 44 45 (48 49) 49 55 62

smallest value: 16

first quartile: $\frac{21 + 21}{2} = 21$

median: 40

third quartile: $\frac{48 + 49}{2} = 48.5$

largest value: 62

Step 2: Draw a number line and plot a point above each value from Step 1.

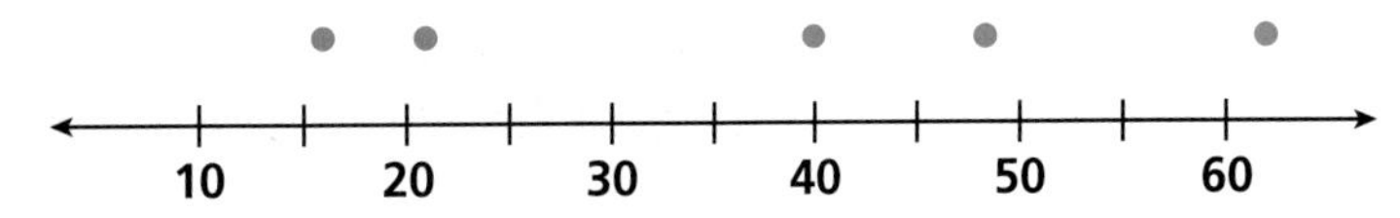

Step 3: Draw the box and whiskers.

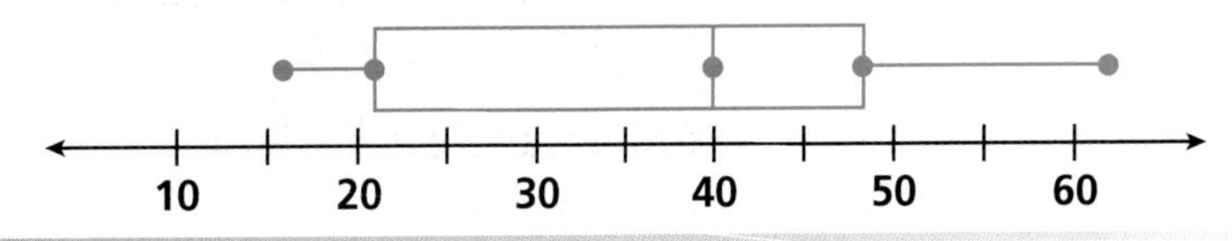

EXAMPLE 3 Comparing Data Sets Using Box-and-Whisker Plots

These box-and-whisker plots compare the number of home runs Babe Ruth hit during his 15-year career from 1920 to 1934 with the number Mark McGwire hit during the 15 years from 1986 to 2000.

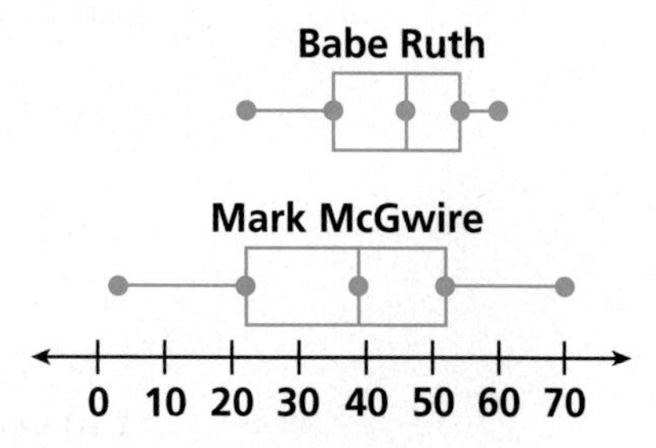

A **Compare the medians and ranges.**

Babe Ruth's median is greater than Mark McGwire's.

Mark McGwire's range is greater than Babe Ruth's.

B **Compare the ranges of the middle half of the data for each.**

The range of the middle half of the data is the length of the box, which is greater for Mark McGwire.

Think and Discuss

1. **Explain** how the range is affected by outliers.
2. **Compare** the number of data values in the box with the number of data values in the whiskers.

4.01

FOR EOG PRACTICE see page 657

internet connect
Homework Help Online
go.hrw.com Keyword: MT4 4-4

GUIDED PRACTICE

See Example 1 **Find the range and the first and third quartiles for each data set.**

1. 65, 42, 45, 20, 66, 60, 76
2. 3, 0, 4, 1, 5, 2, 6, 3, 4, 1, 5, 3

See Example 2 **Use the given data to make a box-and-whisker plot.**

3. 43, 36, 25, 22, 34, 40, 18, 32, 43
4. 21, 51, 36, 38, 45, 52, 28, 16, 41

See Example 3 **Use the box-and-whisker plots to compare the data sets.**

5. Compare the medians and ranges.
6. Compare the ranges of the middle half of the data for each set.

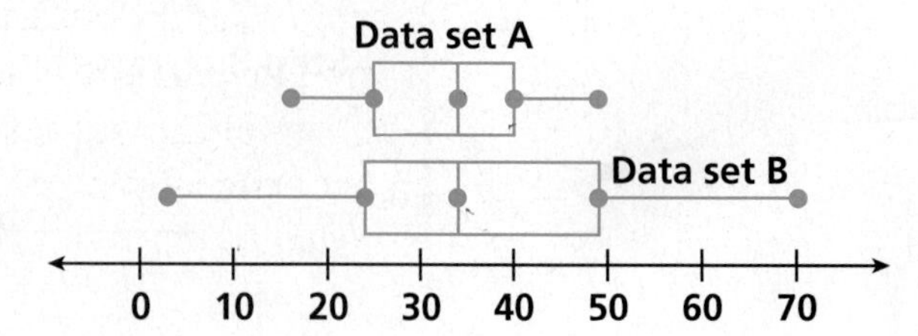

INDEPENDENT PRACTICE

See Example 1

Find the range and the first and third quartiles for each data set.

7. 37, 61, 32, 41, 37, 45, 39, 48, 31

8. 10, 15, 17, 9, 4, 20, 50, 4, 5

See Example 2

Use the given data to make a box-and-whisker plot.

9. 60, 58, 75, 64, 90, 85, 60

10. 1.2, 5.8, 5.4, 10, 8.5, 4.2, 6.7, 5, 8

See Example 3

Use the box-and-whisker plots to compare the data sets.

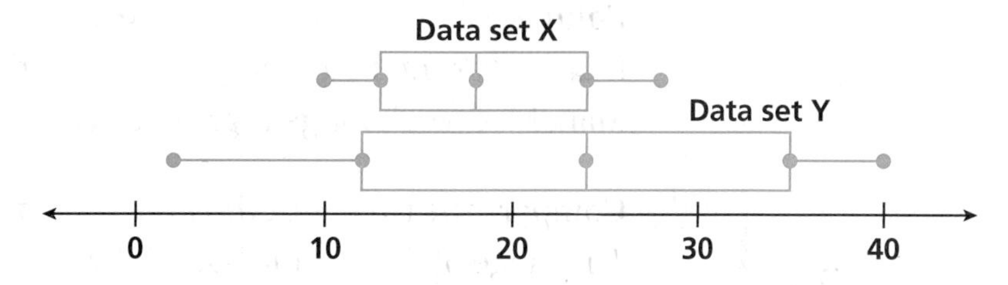

11. Compare the medians and ranges.

12. Compare the ranges of the middle half of the data for each set.

PRACTICE AND PROBLEM SOLVING

Find the range and the first and third quartiles for each data set.

13. 84, 95, 76, 88, 92, 78, 98

14. 2, 7, 9, 12, 2, 6, 8, 1

15. 46, 53, 67, 29, 35, 54, 49, 61, 35

16. 2.3, 2.4, 2.3, 2.2, 2.2, 2.2, 2.2, 2.1

17. 11, 8, 25, 27, 10, 25, 31, 8, 11, 8, 9, 22, 21, 24, 20, 16, 23

18. 13, 11, 14, 16, 14, 15, 16, 17, 14, 18, 19, 16, 25

Use the given data to make a box-and-whisker plot.

19. 56, 88, 60, 84, 72, 68, 80, 76

20. 11.5, 11.2, 14, 14, 7, 4.3, 2.3, 10, 9

21. 0, 2, 5, 2, 1, 3, 5, 2, 4, 3, 5, 4

22. 3.5, 2.2, 4.5, 2.0, 5.6, 7.0, 4.6

23. ***EARTH SCIENCE*** Hurricanes and tropical storms form in all seven ocean basins. Use a box-and-whisker plot to compare the number of tropical storms in every ocean basin per year with the number of hurricanes in every ocean basin per year.

Number of Storms Per Year

Ocean Basin	Tropical Storms	Hurricanes
NW Pacific	26	16
NE Pacific	17	9
SW Pacific	9	4
Atlantic	10	5
N Indian	5	3
SW Indian	10	4
SE Indian	7	3

24. ***GEOGRAPHY*** Find the range and quartiles of the areas of Earth's continents, in square miles: Africa, 11,700,000; Antarctica, 5,400,000; Asia, 17,400,000; Europe, 3,800,000; North America, 9,400,000; Oceania, 3,300,000; South America, 6,900,000.

25. Match each set of data with a box-and-whisker graph.

a. range: 16
first quartile: 22
third quartile: 34

b. range: 48
first quartile: 5
third quartile: 40

c. range: 35
first quartile: 10
third quartile: 35

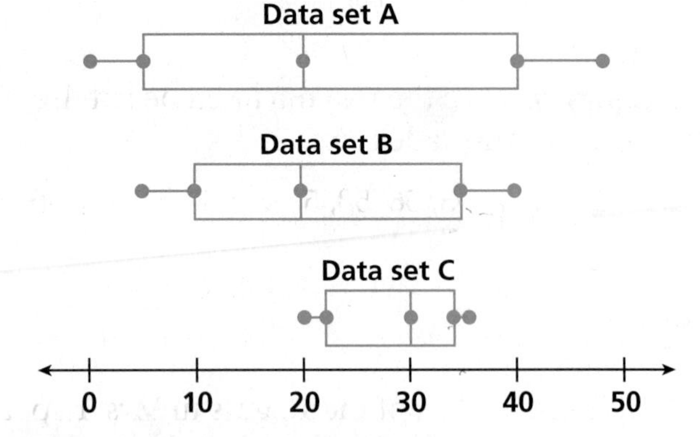

26. ***WHAT'S THE ERROR?*** A student wrote that the data set 22, 16, 45, 17, 18, 29, 22, 14, 32, 54 has a range of 32. What's the error?

27. ***WRITE ABOUT IT*** What do box-and-whisker plots tell you about data that measures of central tendency do not?

28. ***CHALLENGE*** What would an exceptionally short box with extremely long whiskers tell you about a data set?

Spiral Review

Give the missing *y*-coordinates that are solutions to $y = 4x - 2$. (Lesson 1-8)

29. $(0, y)$ **30.** $(1, y)$ **31.** $(3, y)$ **32.** $(7, y)$

Match each graph to one of the given situations. (Lesson 1-9)

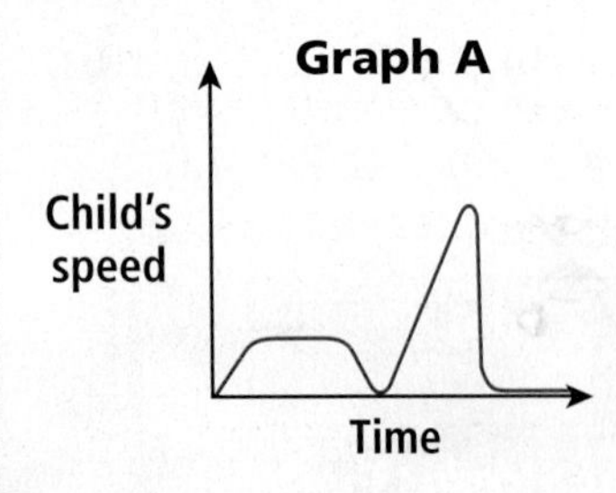

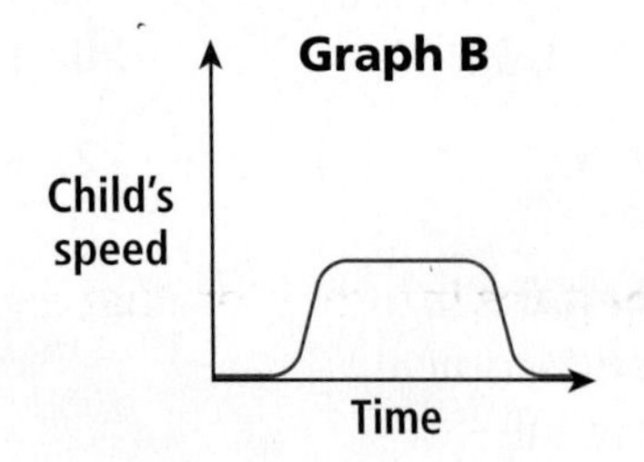

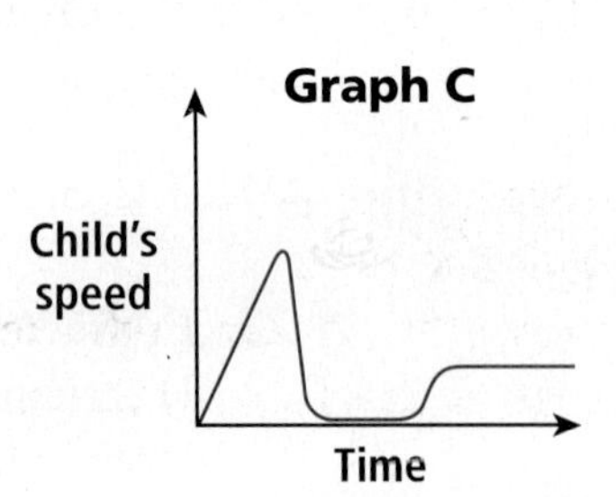

33. Emily sits on bench. Emily runs to buy treat. Emily sits to eat the treat.

34. Zen climbs ladder. Zen slides down slide. Zen sits and laughs.

35. Josh runs to catch bus. Josh sits on bus. Josh walks into school.

36. **EOG PREP** Claire visits her grandmother every 4 weeks, washes her car every 3 weeks, and gets paid every 2 weeks. How often will all three things happen in the same week? (Previous course)

A every 8 weeks B every 9 weeks C every 12 weeks D every 24 weeks

Create Box-and-Whisker Plots

1.02, 4.01

Use with Lesson 4-4

internet connect
Lab Resources Online
go.hrw.com
KEYWORD: MT4 Lab4B

The data below are the heights in inches of the 15 girls in Mrs. Lopez's 8th-grade class.

57, 62, 68, 52, 53, 56, 58, 56, 57, 50, 56, 59, 50, 63, 52

Activity

1 Graph the heights of the 15 girls in Mrs. Lopez's class on a box-and-whisker plot.

Press STAT **Edit** to enter the values into List 1 (**L1**). If necessary, press the up arrow and then CLEAR ENTER to clear old data. Enter the data from the class into **L1**. Press ENTER after each value.

Use the **STAT PLOT** editor to obtain the plot setup menu.

Press 2nd Y= (STAT PLOT) ENTER . Use the arrow keys and ENTER to select **On** and then the fifth type. **Xlist** should be **L1** and **Freq** should be 1, as shown. Press ZOOM **9:ZoomStat.**

Use the TRACE key and the ◀ and ▶ keys to see all five summary statistical values (minimum: **MinX**, first quartile: **Q1**, median: **MED**, third quartile: **Q3**, and maximum: **MaxX**). The minimum value in the data set is 50 in., the first quartile is 52 in., the median is 56 in., the third quartile is 59 in., and the maximum is 68 in.

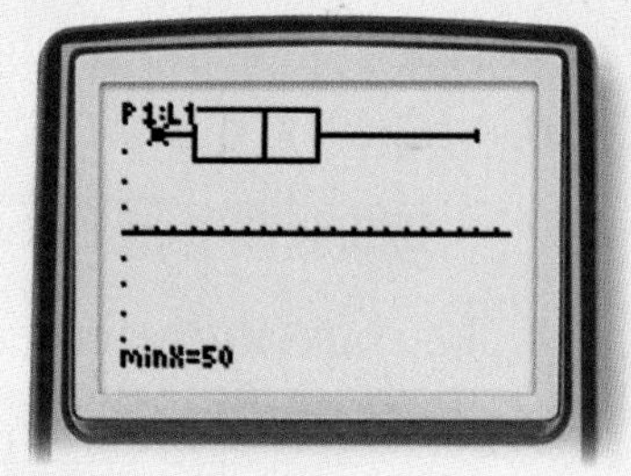

Think and Discuss

1. Explain how the box-and-whisker plot gives information that is hard to see by just looking at the numbers.

Try This

1. The shoe sizes of the 15 girls from Mrs. Lopez's 8th grade class are the following:
   5.5, 6, 7, 5, 5, 5.5, 6, 6, 6.5, 4, 6, 7, 5, 8, and 5

   Make a box-and-whisker plot of this data. What are the minimum, first quartile, median, third quartile, and maximum values of the data set?

Chapter 4

Mid-Chapter Quiz

LESSON 4-1 (pp. 174–177)

Identify the population, sample, and sampling method.

1. Every thirtieth VCR out of 500 in an assembly line is tested.
2. Names are chosen randomly from a voter registration list.

Identify the sampling method used.

3. Postcards of contest entrants are put in a revolving drum. A celebrity draws a postcard.
4. Ten schools are randomly chosen and ten students are randomly chosen from each school.
5. Every thirteenth person who enters a local video store is polled.

LESSON 4-2 (pp. 179–183)

6. Use the given data to make a stem-and-leaf plot.

Tall Buildings in Charlotte, NC			
Building Name	**Stories**	**Building Name**	**Stories**
Bank of America Center	60	One Wachovia Center	42
IJL Financial Center	30	Two Wachovia Center	32
Interstate Tower	32	Wachovia Center	32

LESSON 4-3 (pp. 184–187)

Find the mean, median, and mode of each data set.

7. 60, 70, 70, 80, 75
8. 5, 2, 1, 7, 4, 6, 9
9. 9.1, 8.7, 9.2, 9.0, 8.7, 8.9

LESSON 4-4 (pp. 188–192)

Find the range and the first and third quartiles for each data set.

10. 8, 5, 12, 9, 6, 2, 14, 7, 10, 17, 11
11. 67, 70, 72, 77, 78, 78, 80, 84, 86
12. 0, 0, 3, 3, 3, 1, 3, 1, 3, 7, 9, 9
13. 3.6, 5.0, 4.0, 4.9, 4.2, 4.5, 4.3, 4.8
14. Use box-and-whisker plots to compare the speeds of 1911–1914 with those of 1991–1994.

Indianapolis 500 Winners					
Year	**Winner**	**Speed (mi/h)**	**Year**	**Winner**	**Speed (mi/h)**
1911	Ray Harroun	75	1991	Rick Mears	176
1912	Joe Dawson	79	1992	Al Unser, Jr.	134
1913	Jules Goux	76	1993	Emerson Fittipaldi	157
1914	Rene Thomas	82	1994	Jacques Villeneuve	154

Focus on Problem Solving

Make a Plan

• Identify too much/too little information

When you read a problem, you must decide if the problem has too much or too little information. If the problem has too much information, you must decide what information to use to solve the problem. If the problem has too little information, then you should determine what additional information you need to solve the problem.

Read the problems below and decide if there is too much or too little information in each problem. If there is too much information, tell what information you would use to solve the problem. If there is too little information, tell what additional information you would need to solve the problem.

1. Mrs. Robinson has 35 students in her class. On the last test, there were 7 A's, 15 B's, 10 C's, and 2 D's. What was the average test score?

2. The average elevation in the United States is about 2500 ft above sea level. The highest point, Mt. McKinley, Alaska, has an elevation of 20,320 ft above sea level. The lowest point, in Death Valley, California, has an elevation of 282 ft below sea level. What is the range of elevations in the United States?

3. Use the table to find the median number of marriages per year in the United States for the years between 1940 and 1990.

4. George spent 1.5 hours doing homework on Tuesday, 1 hour doing homework on Wednesday, and 2.7 hours doing homework over the weekend. On Monday, Thursday, and Friday, he did not have homework and spent 1 hour each day reading or watching TV. What was the average amount of time per day George spent on homework last week?

Number of Marriages in the United States

Year	1940	1950	1960	1970	1980	1990
Number (thousands)	1596	1667	1523	2159	2390	2443

Source: National Center for Health Statistics

4-5 Displaying Data

Learn to display data in bar graphs, histograms, and line graphs.

Vocabulary

bar graph

frequency table

histogram

line graph

Usually, teenagers can hardly wait to get a driver's license. But driving can be very dangerous for teens. Many states are implementing graduated licenses that restrict driving at high-risk times, such as from midnight to 5:00 A.M.

A **bar graph** is a good way to display data that can be grouped in categories. If the data is given in the form of a list, it may help to organize the data in a **frequency table** first.

EXAMPLE 1 **Displaying Data in a Bar Graph**

Organize the data into a frequency table and make a bar graph.

The following are the ages when a randomly chosen group of 20 teenagers received their driver's licenses:

18 17 16 16 17 16 16 16 19 16 16 17 16 17 18 16 18 16 19 16

First, organize the data into a frequency table.

Age License Received	16	17	18	19
Frequency	11	4	3	2

The frequency is the number of times each value occurs.

The frequencies are the heights of the bars in the bar graph.

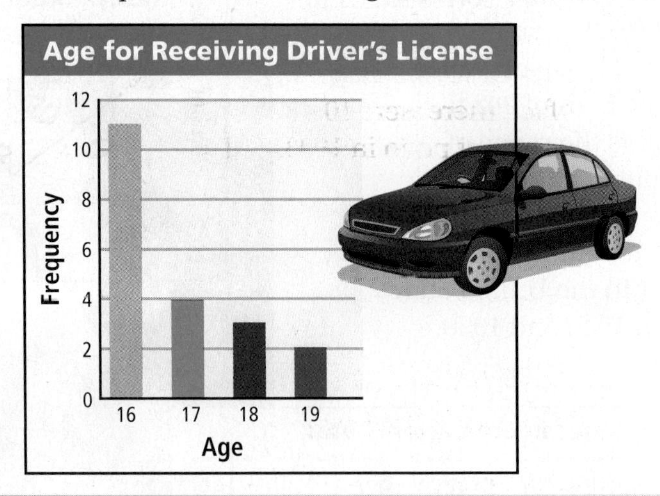

A **histogram** is a type of bar graph. The bars of a histogram represent intervals in which the data are grouped.

EXAMPLE 2 Displaying Data in a Histogram

Helpful Hint

Histograms do not have spaces between the bars.

John surveyed 15 people to find out how many pages were in the last book they read. Use the data to make a histogram.

368 153 27 187 240 636 98 114 64 212 302 144 76 195 200

First, make a frequency table with intervals of 100 pages. Then make a histogram.

Pages	Frequency
0–99	4
100–199	5
200–299	3
300–399	2
600–699	1

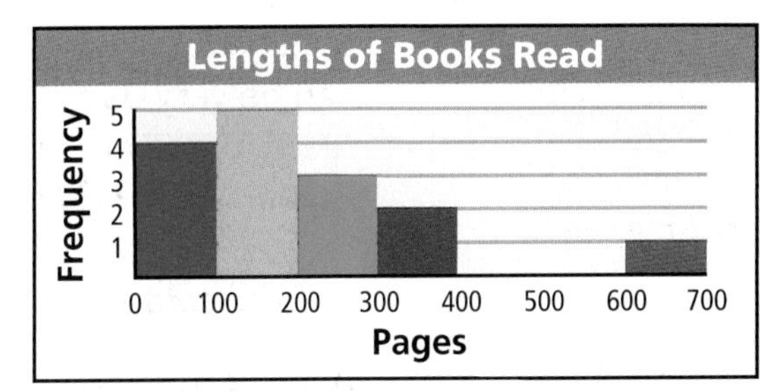

A **line graph** is often used to show trends or to make estimates for values between data points.

EXAMPLE 3 Displaying Data in a Line Graph

Make a line graph of the given data. Use the graph to estimate the number of polio cases in 1993.

Create ordered pairs from the data in the table and plot them on a grid. Connect the points with lines.

Year	Number of Polio Cases Worldwide
1975	49,293
1980	52,552
1985	38,637
1990	23,484
1995	7,035
2000	2,880

Source: World Health Organization

You can estimate the number of polio cases in 1993 by finding the point on the graph that corresponds to 1993. The graph shows about 12,000 cases. In fact, there were 10,487 cases of polio in 1993.

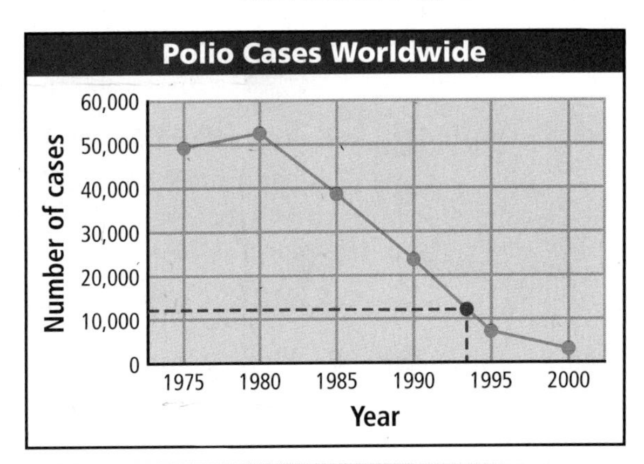

Think and Discuss

1. **Compare** a bar graph to a line graph.
2. **Explain** how changing the intervals to 200 pages in Example 2 would affect the histogram.

4-5 Exercises

FOR EOG PRACTICE
see page 658

internet connect
Homework Help Online
go.hrw.com Keyword: MT4 4-5

4.01

GUIDED PRACTICE

See Example 1 

1. Organize the data into a frequency table and make a bar graph.
13 9 10 9 11 13 12 9 11 13 9 13 13 10 9 10 9 12 10 9

See Example 2

2. Use the data to make a histogram with intervals of 50.

National Merit Scholars (1999)

School	Number of Students	School	Number of Students
Vanderbilt	98	Rice University	183
Princeton	111	Cal Tech	52
Duke	76	University of Chicago	139
Stanford	229	M.I.T.	133
Yale	170	University of Texas–Austin	244
Northwestern	128	Washington University	131

See Example 3 

3. Make a line graph of the given data. Use the graph to estimate the life expectancy of someone born in 1982.

Life Expectancy by Birth Year (U.S.)

Year	1970	1975	1980	1985	1990
Age	70.8	72.6	73.7	74.7	75.4

INDEPENDENT PRACTICE

See Example 1 

4. Organize the data into a frequency table and make a bar graph.
−34, −46, −34, −32, −25, −34, −46, −17, −32, −34, −20, −17, −2

See Example 2

5. Restaurants often break down their menus by the number of items in each price range. Use the entrée prices to make a histogram with intervals of $10.
$9 $11 $22 $22 $30 $24 $13 $16 $17 $21 $18 $25 $17 $25
$17 $21 $19 $21 $14 $19 $15 $15 $10 $16 $12 $21 $19 $17

See Example 3 

6. Make a line graph of the given data. Use the graph to estimate the number of tornados that occurred in 1995.

Number of Tornados in Illinois by Year

Year	Tornados	Year	Tornados
1988	20	1994	20
1990	50	1996	61
1992	23	1998	110

PRACTICE AND PROBLEM SOLVING

7. Organize the data into a frequency table and make a bar graph.
   1 3 6 3 1 6 1 2 1 4 1 1 5 1 2 4 4 1 2 1

8. Make a histogram of honey yield per colony with intervals of 2.

Honey-Producing Colonies						
Year	1992	1993	1994	1995	1996	1997
Yield per Colony (pounds)	72.8	80.2	78.4	79.5	77.3	74.6

Source: USDA

9. **a.** Make a line graph of the average weekly hours of a production worker by year, and estimate the average weekly hours in 1985.
   b. Make a line graph of the average hourly pay of production workers by year, and estimate the average hourly pay in 1995.

Production Worker Averages						
Year	1950	1960	1970	1980	1990	2000
Weekly Hours	39.8	38.6	37.1	35.3	34.5	34.5
Hourly Pay	\$1.34	\$2.09	\$3.23	\$6.66	\$10.01	\$13.75

10. ***WRITE A PROBLEM*** You have a class set of test scores and you want to know how many people got A's, B's, and C's. Write a problem using a histogram that would help you find this information.

11. ***WRITE ABOUT IT*** You have been asked to create a graph of the total salaries of a professional hockey team from 1980 to 2000. Which kind of graph would you choose and why?

12. ***CHALLENGE*** Using the data and the histogram, determine the size of the interval used.
    Time needed to heat a frozen dinner in the oven (minutes)
    15 25 20 17 35 28 10
    12 15 45 33 35 8 14

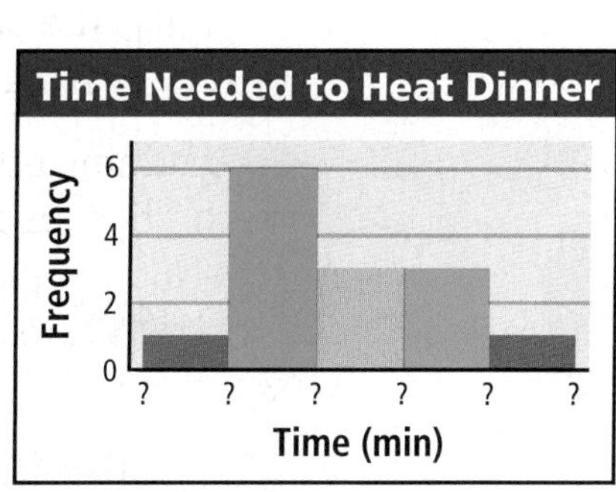

Spiral Review

Solve and graph each inequality. (Lesson 1-5)

13. $3x < 15$
14. $x + 2 \geq 4$
15. $x + 1 \leq 3$
16. $x - 4 < 4$
17. $5x > 30$
18. $x - 5 > 1$
19. $3 \geq \frac{x}{2}$
20. $8 < \frac{x}{4}$

21. **EOG PREP** What is $5^7 \cdot 5^3$ written as one power? (Lesson 2-7)

A 5^4 B 5^{10} C 5^{21} D $5^7 \cdot 5^7 \cdot 5^7$

4-6 Misleading Graphs and Statistics

Learn to recognize misleading graphs and statistics.

Graphs and statistics are often used to persuade. Advertisers and others may accidentally or intentionally present information in a misleading way.

For example, art is often used to make a graph more interesting, but it can distort the relationships in the data.

EXAMPLE 1 Identifying Misleading Graphs

Explain why each graph is misleading.

A

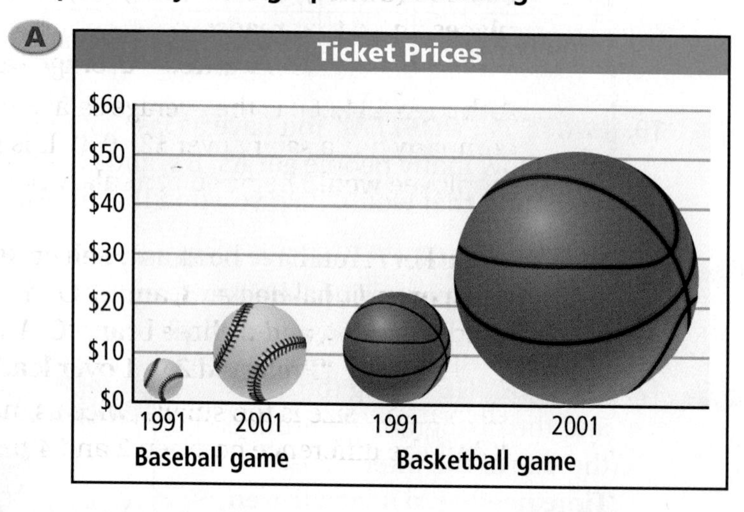

The heights of the balls are used to represent the ticket prices. However, the areas of the circles and volumes of the balls distort the comparison. The basketball prices are only about $2\frac{1}{2}$ times greater than the baseball prices, but they look like much more.

B

Because the scale does not start at 0, the bar for brand 2 is three times as tall as the bar for brand 1. In fact, the average life of brand 2 is only about 22% longer than that of brand 1.

Average Life of Running Shoes

Brand 3
Brand 2
Brand 1
400 450 500 550 600
Miles

Explain why the graph is misleading.

C

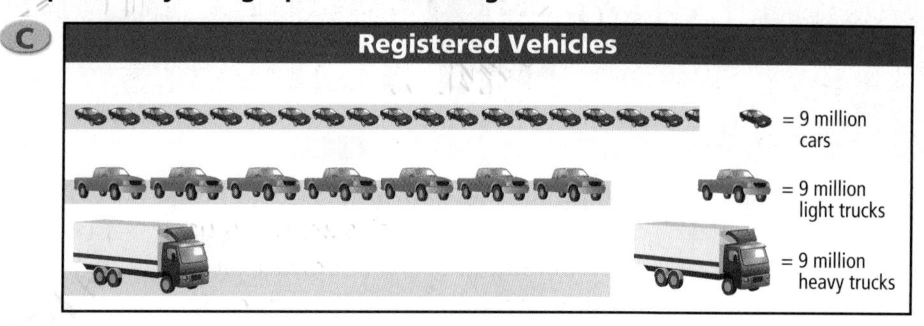

Different-sized icons represent the same number of vehicles. The number of light trucks looks like it is close to the number of cars, but it is really less than half. The number of heavy trucks is less than 5% of the total, but it appears much greater.

EXAMPLE 2 Identifying Misleading Statistics

Explain why each statistic is misleading.

A **A small business has 5 employees with the following salaries: $90,000 (owner), $18,000, $22,000, $20,000, $23,000. The owner places an ad that reads:**

"Help wanted—average salary $34,600"

Although $34,600 is the average salary, only one person in the company has a salary over $23,000. It is not likely that a new employee would be hired at a salary near $34,600.

B **A market researcher randomly selects 8 people to focus-test three brands, labeled A, B, and C. Of these, 4 chose brand A, 2 chose brand B, and 2 chose brand C. An ad for brand A states:**

"Preferred 2 to 1 over leading brands!"

The sample size is too small. Twice as many people chose brand A, but the difference between 2 and 4 people is not meaningful.

C **The total revenue at Worthman's for the three-month period from June 1 to September 1 was $72,000. The total revenue at Meilleure for the three-month period from October 1 to January 1 was $108,000.**

The revenues are measured at different times of the year, even though they are the same length of time. During a busy shopping season the revenue is greater than it would be in the summer.

Think and Discuss

1. **Give an example** of a graph that starts at zero but is still misleading.
2. **Explain** how a statistic can be accurate but still misleading.

4-6 Exercises

FOR EOG PRACTICE

see page 659

internet connect

Homework Help Online
go.hrw.com Keyword: MT4 4-6

4.01, 4.03

GUIDED PRACTICE

See Example 1 **Explain why each graph is misleading.**

1. 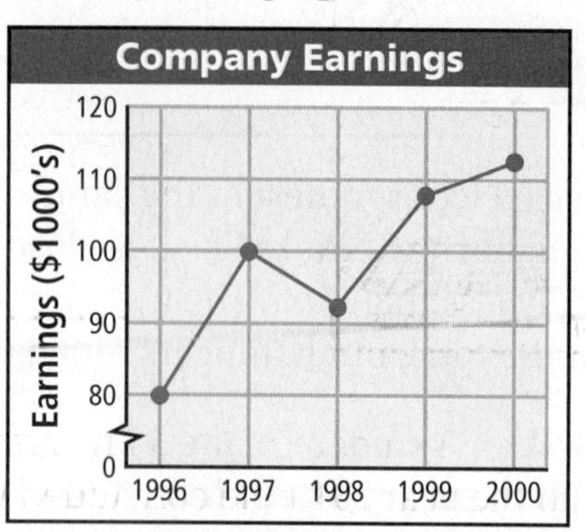

2. 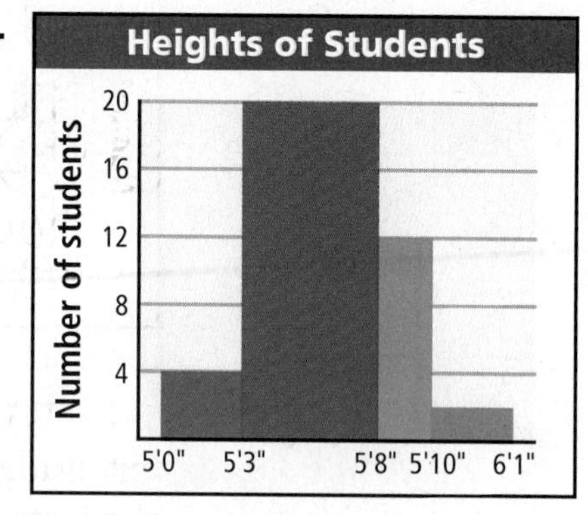

See Example 2 **Explain why each statistic is misleading.**

3. A lemon has 31 mg vitamin C. An orange has 51 mg vitamin C. A grapefruit has 114 mg vitamin C.

4. The total number of pools sold by Pool Kingdom from May 1 to August 1 was 4623. The total number of pools sold by SplashDown from June 1 to August 1 was 612.

INDEPENDENT PRACTICE

See Example 1 **Explain why each graph is misleading.**

5. 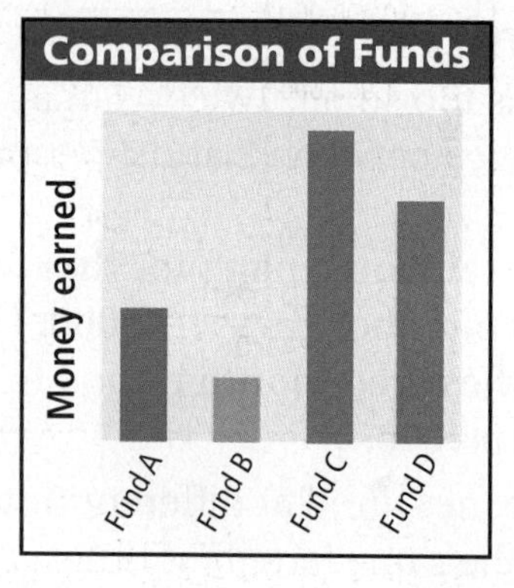

6. 

= 100 cans = 50 boxes = 20 cases

See Example 2 **Explain why each statistic is misleading.**

7. A survey of vehicle owners reported that 759 out of 1000 were satisfied with their cars and 756 out of 1000 were satisfied with their SUVs. The manufacturer claimed that car owners are happier with their vehicles.

8. A reporter asked 100 people if they went on vacation this year. Of the 45 who responded yes, 20 went to a U.S. beach, 15 visited foreign countries, and 10 vacationed other places in the United States. The reporter wrote, "Half of all people vacation at the beach."

PRACTICE AND PROBLEM SOLVING

Explain why each graph is misleading. Then redraw the graphs so that they are not misleading.

9.

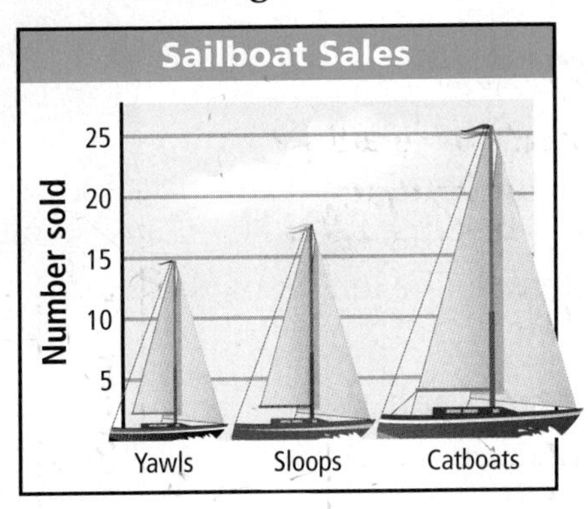

10.

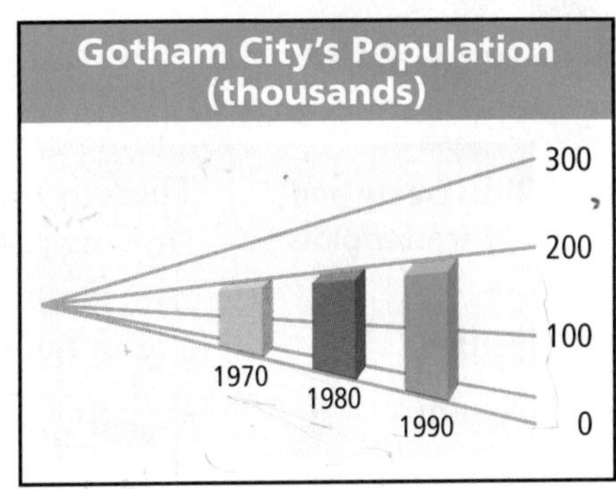

11. ***WHAT'S THE ERROR?*** A student made a line graph of the data and extended the line to the year 2010. He concluded that the record for the fastest mile in 2010 will be 3 minutes 25 seconds. Explain his error.

Year	1923	1943	1954	1975	1985
Fastest Mile	4 min 10.4 sec	4 min 2.6 sec	3 min 58.0 sec	3 min 49.4 sec	3 min 46.3 sec

12. ***WRITE ABOUT IT*** When might you want to use a scale on a graph that does not start at 0?

13. ***CHALLENGE*** This graph demonstrates Moore's Law, which states that the number of transistors on a silicon chip will double every 18 months. Why is this line graph misleading? Why do you think this type of graph was used?

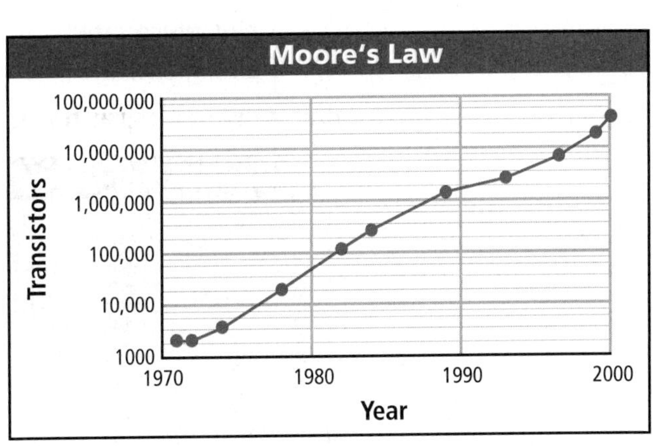

Spiral Review

Solve. (Lesson 1-4)

14. $3x = 15$

15. $\frac{b}{2} = 3$

16. $18 = 9y$

17. $\frac{a}{3} = 7$

18. $\frac{k}{4} = 4.8$

19. $7.5 = 5h$

20. $\frac{m}{2.5} = 8$

21. $4f = 6$

22. **EOG PREP** What is the median of the data set 62, 58, 47, 35, 61, 72, 58, 64? (Lesson 4-3)

A 59.5

B 58

C 57.125

D 47

4-7 Scatter Plots

Learn to create and interpret scatter plots.

Vocabulary

scatter plot

correlation

line of best fit

There is no cure for the common cold. However, studies suggest that some zinc lozenges can reduce the length of a cold by up to 7 days.

A **scatter plot** shows relationships between two sets of data.

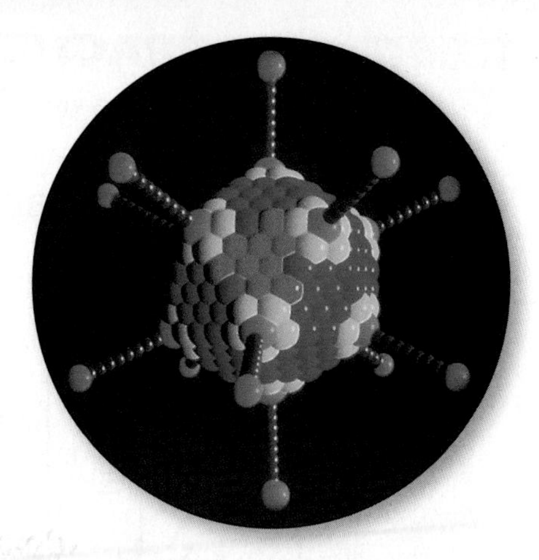

This is a model of a *rhinovirus.* Rhinoviruses make up approximately half of the more than 2000 types of cold viruses.

EXAMPLE 1 Making a Scatter Plot of a Data Set

A scientist studying the effects of zinc lozenges on colds has gathered the following data. Zinc ion availability (ZIA) is a measure of the strength of the lozenge. Use the data to make a scatter plot.

Compound	ZIA	Average Effects
Zinc gluconate	100	Reduced cold 7 days
Zinc gluconate	44	Reduced cold 4.8 days
Zinc orotate	0	None
Zinc gluconate	25	Reduced cold 1.6 days
Zinc gluconate	13.4	None
Zinc aspartate	0	None
Zinc acetate-tartarate-glycine	−55	Increased cold 4.4 days
Zinc gluconate	−11	Increased cold 1 day

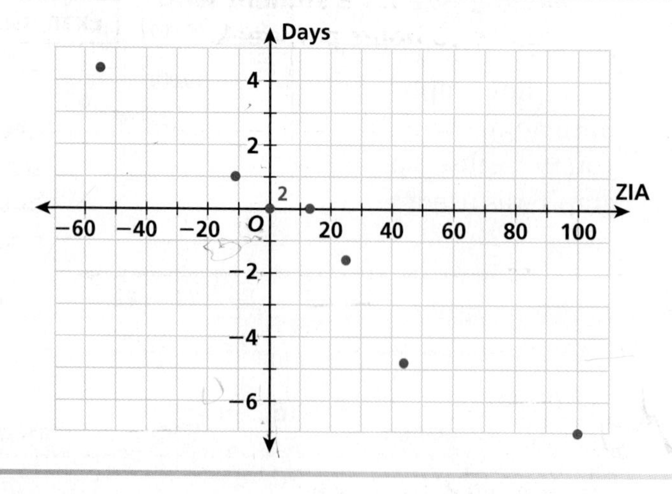

The points on the scatter plot are (100, −7), (44, −4.8), (0, 0), (25, −1.6), (13.4, 0), (0, 0), (−55, 4.4), and (−11, 1). The ***2*** *at (0,0) indicates that the point occurs twice.*

Correlation describes the type of relationship between two data sets. The **line of best fit** is the line that comes closest to all the points on a scatter plot. One way to estimate the line of best fit is to lay a ruler's edge over the graph and adjust it until it looks closest to all the points.

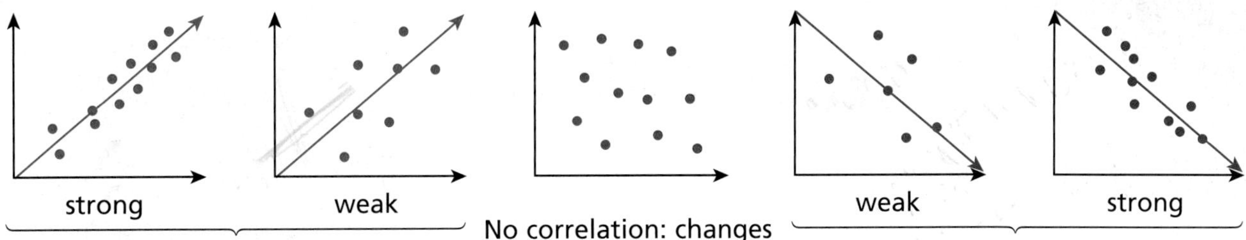

Positive correlation: both data sets increase together.

No correlation: changes in one data set do not affect the other data set.

Negative correlation: as one data set increases, the other decreases.

EXAMPLE 2 Identifying the Correlation of Data

Do the data sets have a positive, a negative, or no correlation?

A The population of a state and the number of representatives

Positive correlation: States with greater population have more representatives.

B The population of a state and the number of senators

No correlation: All states have exactly two senators.

C The population of a state and the number of senators per person

Negative correlation: The number of senators stays the same, so the ratio of senators to people decreases as population increases.

Helpful Hint

A strong correlation does not mean there is a cause-and-effect relationship. For example, your age and the price of a regular movie ticket are both increasing, so they are positively correlated.

EXAMPLE 3 Using a Scatter Plot to Make Predictions

Use the data to predict the exam grade for a student who studies 10 hours per week.

Hours Studied	5	9	3	12	1
Exam Grade	80	95	75	98	70

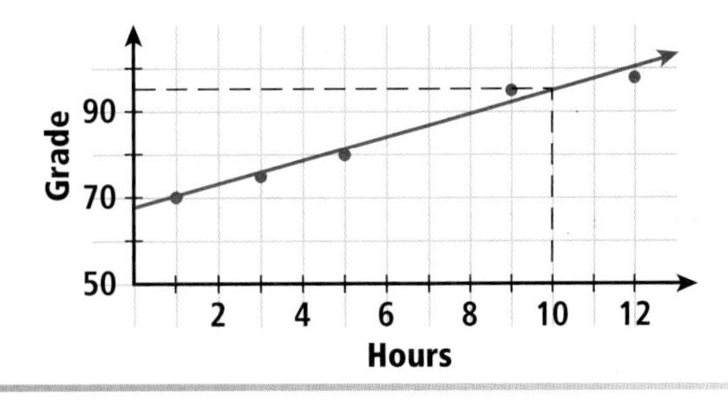

According to the graph, a student who studies 10 hours per week should earn a score of about 95.

Think and Discuss

1. **Compare** a scatter plot to a line graph.
2. **Give an example** of each type of correlation.

4-7 Exercises

FOR EOG PRACTICE

see page 659

internet connect

Homework Help Online

go.hrw.com Keyword: MT4 4-7

4.01, 4.02

GUIDED PRACTICE

See Example 1

1. Use the given data to make a scatter plot.

Country	Area (mi^2)	Population
Guatemala	42,467	12,335,580
Honduras	43,715	5,997,327
El Salvador	8,206	5,839,079
Nicaragua	50,503	4,717,132
Costa Rica	19,929	3,674,490
Panama	30,498	2,778,526

See Example 2 **Do the data sets have a positive, a negative, or no correlation?**

2. The diameter of a pizza and the price of the pizza
3. A person's age and the number of siblings

See Example 3

4. Use the data to predict the apparent temperature at 50% humidity.

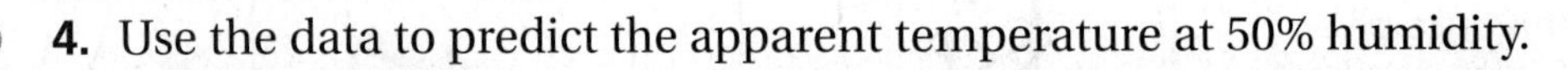

Temperature Due to Humidity at a Room Temperature of 68°F						
Humidity (%)	0	20	40	60	80	100
Apparent Temperature (°F)	61	63	65	67	69	71

INDEPENDENT PRACTICE

See Example 1

5. Use the given data to make a scatter plot.

Type of Transplant	Patients Waiting	Number Performed
Kidney	50,006	13,372
Liver	18,419	4,954
Heart	4,176	2,198
Lung	3,786	956
Pancreas	1,158	435

See Example 2 **Do the data sets have a positive, a negative, or no correlation?**

6. The number of weeks a film has been out and weekly attendance
7. The number of weeks a film has been out and total attendance

See Example 3

8. Use the data to predict the apparent temperature at 70% humidity.

Temperature Due to Humidity at a Room Temperature of 72°F						
Humidity (%)	0	20	40	60	80	100
Apparent Temperature (°F)	64	67	70	72	74	76

Life Science LINK

About 40 to 50 million Americans suffer from allergies. Airborne pollen generated by trees, grasses, plants, and weeds is a major cause of illness and disability. Because pollen grains are small and light, they can travel through the air for hundreds of miles. Pollen levels are measured in grains per cubic meter.

Some common substances that cause allergies include pollens, dust mites, and mold spores.

9. Use the given data to make a scatter plot, and describe the correlation.

Pollen Levels

Day	Weed Pollen	Grass Pollen
1	350	16
2	51	1
3	49	9
4	309	3
5	488	29
6	30	3
7	65	12

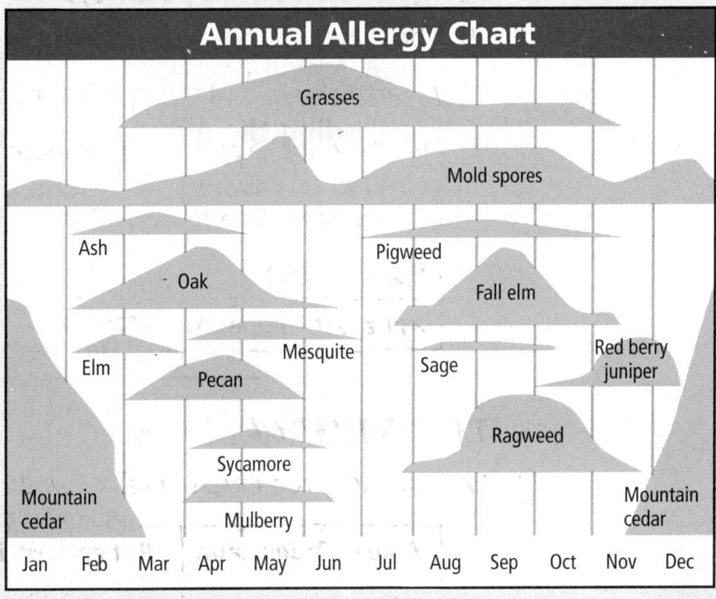

Source: Central Texas Allergy and Asthma Center

10. Explain how the pollens are compared in the chart at right.

Use the chart at right to determine if the pollens have a positive, a negative, or no correlation.

11. mountain cedar, grass
12. fall elm, ragweed

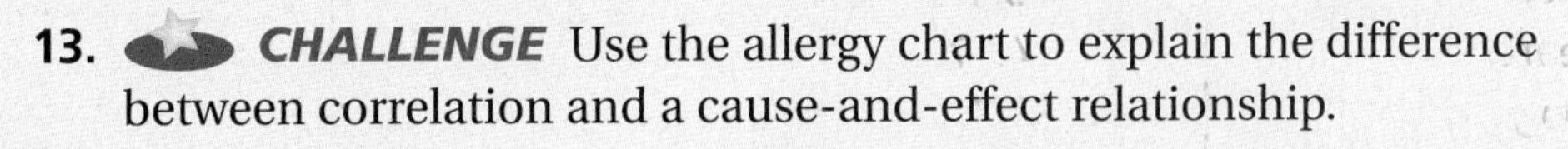

13. ***CHALLENGE*** Use the allergy chart to explain the difference between correlation and a cause-and-effect relationship.

Spiral Review

Solve. (Lesson 1-6)

14. $4x - x = 18$
15. $3(2 + x) = 21$
16. $12x - 6x = 42$
17. $4(1 + x) = 28$
18. $7x + 2x = 108$
19. $7(2x - 4) = 224$
20. **EOG PREP** What is 29,600,000,000,000 expressed in scientific notation? (Lesson 2-9)

A 2.96×10^{-10} B 29.6×10^{10} C 2.96×10^{12} D 2.96×10^{13}

EXTENSION Average Deviation

Learn to find the average deviation of a data set.

Vocabulary

average deviation

Another measure of variation is the **average deviation**, which is the average distance a data value is from the mean.

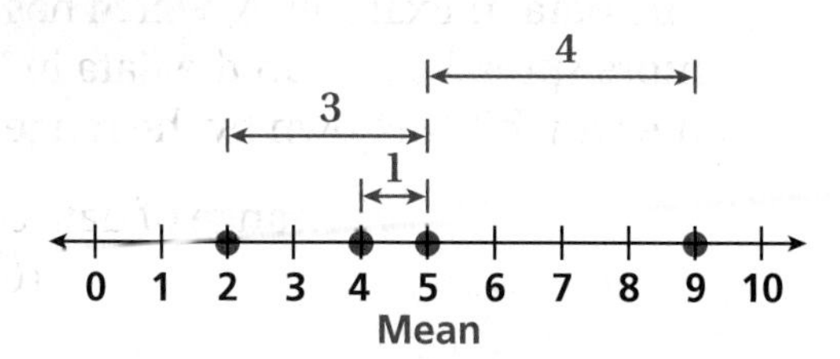

The average deviation of this data set is $\frac{3 + 1 + 4}{3} = 2.67$.

EXAMPLE 1 Finding the Average Deviation of a Data Set

Find the average deviation of each data set.

A 0, 8, 8, 16

The mean of the data set is $\frac{0 + 8 + 8 + 16}{4} = \frac{32}{4} = 8$.

Mean	Value	Difference
8	0	$\lvert 8 - 0\rvert = 8$
8	8	$\lvert 8 - 8\rvert = 0$
8	8	$\lvert 8 - 8\rvert = 0$
8	16	$\lvert 8 - 16\rvert = 8$

Subtract each data value from the mean. Distance cannot be negative, so use absolute value.

average deviation $= \frac{8 + 0 + 0 + 8}{4} = \frac{16}{4} = 4$

B 7, 8, 8, 9

The mean of the data set is $\frac{7 + 8 + 8 + 9}{4} = \frac{32}{4} = 8$.

Mean	Value	Difference
8	7	$\lvert 8 - 7\rvert = 1$
8	8	$\lvert 8 - 8\rvert = 0$
8	8	$\lvert 8 - 8\rvert = 0$
8	9	$\lvert 8 - 8\rvert = 1$

Subtract each data value from the mean. Distance cannot be negative, so use absolute value.

average deviation $= \frac{1 + 0 + 0 + 1}{4} = \frac{2}{4} = 0.5$

The two data sets have the same mean, but the average deviations show that one data set is much more spread out than the other.

A back-to-back stem-and-leaf plot compares the sets visually.

A		B
8 8 0	0	7 8 8 9
6	1	

Key: |0|9 means 9
6|1| means 16

The data in Example A, which has the greater average deviation, is more spread out than the data in Example B. This is consistent with the variability shown by the range of each set.

$$\text{range of data set A} > \text{range of data set B}$$
$$16 - 0 > 9 - 7$$
$$16 > 2$$

4.01

Find the average deviation of each data set. Round to the nearest tenth.

1. 27, 26, 25, 22, 20
2. 50, 50, 52, 52, 60, 68, 68, 70, 70
3. 10, 12, 16, 24, 30, 36, 44, 48, 50
4. 4, 4, 4, 6, 12, 20
5. 2, 4, 5, 7, 8, 10
6. even integers from 2 through 10
7. 5, 5, 5, 5, 5, 5
8. 0, 0, 0, 1, 2, 3, 4

Two sets of data are shown in each back-to-back stem-and-leaf plot. Which data set has the smaller average deviation?

9.

Set A		Set B
1 2 3 3	0	2 3 4 4 5 5
4 7	1	

Key: |0|5 means 5
7|1| means 17

10.

Set X		Set Y
1 4 5	1	
7 8 8	2	1 2 2 2 3 9

Key: |2|9 means 29
7|1| means 17

11. The data sets show the highest daily temperatures, recorded to the nearest °C, for two summer weeks.

Week 1:	37, 35, 34, 30, 32, 36, 34
Week 2:	37, 36, 40, 33, 31, 30, 31

a. For each week's data, find the average deviation to the nearest tenth.

b. Which week had readings that were more spread out?

12. Why is absolute value used when you are finding average deviation?

13. What would the average of the deviations be if the absolute value were not used?

14. What can you conclude about the average deviation of a data set in which all the values are the same?

NORTH CAROLINA

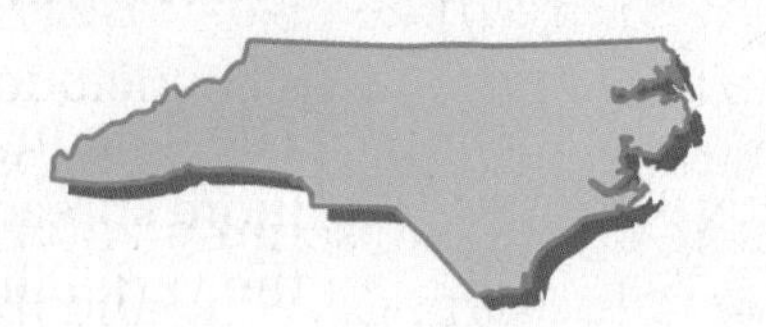

Amphibians of North Carolina

Over 90 different species of amphibians, including newts, salamanders, frogs, and toads, live in North Carolina. Like all other living things, amphibian species are given scientific names, which are used to identify relationships between them.

North Carolina Amphibians			
Scientific Name	Common Name	Minimum Length (cm)	Maximum Length (cm)
Ambystoma tigrinum tigrinum	Eastern tiger salamander	15.2	40
Bufo americanus americanus	Eastern American toad	5.1	11.1
Notophthalmus viridescens viridescens	Red spotted newt	7	12.4
Plethodon glutinosus	Northern slimy salamander	11.4	20.6
Scaphiopus holbrookii	Eastern spadefoot	4.4	8.3
Siren lacertina	Greater siren	49	97

For 1–3, use the table.

1. Scientific names are often very long and difficult to read or pronounce. Make a frequency table and histogram of the number of letters in each scientific name listed in the table. Use an interval of 10.
2. Make a box-and-whisker plot using the maximum lengths from the table.
3. Make a scatter plot of the minimum length (x-axis) versus the maximum length (y-axis) of each species in the table. Describe the correlation.

Reptiles of North Carolina

The 79 species of reptiles that can be found in North Carolina come from three different scientific groups called *orders*—Crocodylia (which includes crocodiles and alligators), Squamata (which includes snakes and lizards), and Testudines (which includes turtles). Within each order are smaller scientific groups called families. The tables give a brief description of the families of reptiles found in North Carolina and the number of species in each family.

Squamata	
Family Description	**Number of Species**
Coral snakes	1
Glass lizards	3
Iguanas	3
Pit vipers	5
Racerunners	1
Skinks	5
Typical snakes	37

Testudines	
Family Description	**Number of Species**
Box turtles, map turtles, and sliders	11
Leatherback sea turtles	1
Mud turtles and musk turtles	4
Scute-shell sea turtles	4
Snapping turtles	1
Softshell turtles	2

Crocodylia	
Family Description	**Number of Species**
Alligators and caimans	1

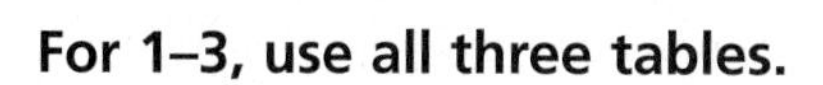

For 1–3, use all three tables.

1. Make a stem-and-leaf plot of the numbers of species in each family.

2. a. To the nearest whole number, find the mean, median, and mode of the numbers of species in each family.

b. What do the mean, median, and mode indicate about the sizes of the reptile families living in North Carolina?

c. Identify any outliers in the data set, and explain their effect on the mean, median, and mode.

3. Make a bar graph to display the number of families in each of the three North Carolina reptile orders.

MATH-ABLES

Distribution of Primes

Remember that a prime number is only divisible by 1 and itself. There are infinitely many prime numbers, but there is no algebraic formula to find them. The largest known prime number, discovered on November 14, 2001, is $2^{13,466,917} - 1$. In standard form, this number would have 4,053,946 digits.

Sieve of Eratosthenes

One way to find prime numbers is called the sieve of Eratosthenes. Use a list of whole numbers in order. Cross off 1. The next number, 2, is prime. Circle it, and then cross off all multiples of 2, because they are not prime. Circle the next number on the list, and cross off all of its multiples. Repeat this step until all of the numbers are circled or crossed off. The circled numbers will all be primes.

~~1~~	(2)	3	~~4~~	5	~~6~~	7	~~8~~	9	~~10~~
11	~~12~~	13	~~14~~	15	~~16~~	17	~~18~~	19	~~20~~
21	~~22~~	23	~~24~~	25	~~26~~	27	~~28~~	29	~~30~~
31	~~32~~	33	~~34~~	35	~~36~~	37	~~38~~	39	~~40~~
41	~~42~~	43	~~44~~	45	~~46~~	47	~~48~~	49	~~50~~

1. Use the sieve of Eratosthenes to find all prime numbers less than 50.

2. Create a scatter plot of the first 15 prime numbers. Use the prime numbers as the x-coordinates and their positions in the sequence as the y-coordinates; 2 is the 1st prime, 3 is the 2nd prime, and so on.

Prime Number	2	3	5	7											
Position in Sequence	1	2	3	4	5	6	7	8	9	10	11	12	13	14	15

3. Estimate the line of best fit and use it to guess the number of primes under 100. Use the sieve of Eratosthenes to check your guess.

Math in the Middle

This game can be played by two or more players. On your turn, roll 5 number cubes. The number of spaces you move is your choice of the mean, rounded to the nearest whole number; the median; or the mode, if it exists. The winner is the first player to land on the *Finish* square by exact count.

internet connect

go.hrw.com

Go to ***go.hrw.com*** for a complete set of rules and the game board.
KEYWORD: MT4 Game4

Mean, Median, and Mode

The National Collegiate Athletic Association (NCAA) tournaments determine the champions of women's and men's college basketball. The victory margins for the championship games from 1995 through 2001 are shown below.

Margin of Victory, NCAA Championship Games							
Year	1995	1996	1997	1998	1999	2000	2001
Men's Game (points)	11	9	5	9	3	13	10
Women's Game (points)	6	18	9	18	17	19	2

Activity

1 Use a spreadsheet to find the mean, median, and mode of the men's championship-game victory margins from the table. Fill in rows 1 and 2 with the data and labels shown in the spreadsheet below.

The **AVERAGE, MEDIAN,** and **MODE** functions find the mean, median, and mode of the data in a given range of spreadsheet cells.

- Enter **=AVERAGE(B2:H2)** into cell H3 to find the mean of the data in cells B2 through H2.
- Enter **=MEDIAN(B2:H2)** into cell H4 to find the median of the data.
- Enter **=MODE(B2:H2)** into cell H5 to find the mode of the data.

	A	B	C	D	E	F	G	H
1	Year	1995	1996	1997	1998	1999	2000	2001
2	Margin (points)	11	9	5	9	3	13	10
3							Mean	8.571429
4							Median	9
5							Mode	9

Think and Discuss

1. If an eighth game with a victory margin of 30 points were added, what would happen to these three calculated values?

Try This

1. Use a spreadsheet to find the mean, median, and mode for the women's championship games (shown in the table above).

Chapter 4

Study Guide and Review

Vocabulary

back-to-back stem-and-leaf plot 179	line graph 197	random sample 175
bar graph 196	line of best fit 204	range 188
biased sample 174	mean 184	sample 174
box-and-whisker plot ..189	median 184	scatter plot 204
correlation 204	mode 184	stem-and-leaf plot 179
frequency table 196	outlier 185	stratified sample 175
histogram 196	population 174	systematic sample 175
	quartile 188	variability 188

Complete the sentences below with vocabulary words from the list above. Words may be used more than once.

1. The ___?___ of a data set is the middle value, while the ___?___ is the value that occurs most often.

2. ___?___ describes how spread out a data set is. One measure of ___?___ is the ___?___.

3. The ___?___ is the line that comes closest to all the points on a(n) ___?___. ___?___ describes the type of relationship between two data sets.

4-1 Samples and Surveys (pp. 174–177)

EXAMPLE

- **Identify the population and sample. Give a reason why the sample could be biased.**

In a community of 1250 people, a pollster asks 250 people living near a railroad track if they want the tracks moved.

Population	Sample	Possible bias
1250 people who live in a community	250 residents living near tracks	People living near tracks are annoyed by the noise and want tracks moved.

EXERCISES

Identify the population and sample. Give a reason why the sample could be biased.

4. Of the 125 people in line for a *Star Wars* movie, 25 are asked to name their favorite type of movie.

5. Fifty parents of children attending Park Middle School are asked if the community should build a new Little League field.

6. This week, a U.S. senator asked 75 of the constituents who visited her office if she should run for reelection.

4-2 Organizing Data (pp. 179–183)

EXAMPLE

- **Make a back-to-back stem-and-leaf plot.**

American League East Final Standings 2000

Team	Wins	Losses
New York	87	74
Boston	85	77
Toronto	83	79
Baltimore	74	88
Tampa Bay	69	92

Wins		Losses
9	6	
4	7	4 7 9
7 5 3	8	8
	9	2

Key: |9|2 means 92; 9|6| means 69

EXERCISES

Make a back-to-back stem-and-leaf plot.

7.

President	Inaugural Age	Age at Death
George Washington	57	67
Thomas Jefferson	57	83
Abraham Lincoln	52	56
Franklin D. Roosevelt	51	63
John F. Kennedy	43	46

4-3 Central Tendency (pp. 184–187)

EXAMPLE

- **Find the mean, median, and mode.**

30, 41, 46, 39, 46

mean: $\frac{30 + 41 + 46 + 39 + 46}{5} = \frac{202}{5} = 40.4$

median: 30 39 (41) 46 46

mode: 46

EXERCISES

Find the mean, median, and mode.

8. 450, 500, 500, 570, 650, 700, 1950

9. 8, 8, 8.5, 10, 10, 9, 9, 11.5

10. 2, 6, 6, 10, 2, 6, 6, 10

11. 1.1, 3.1, 3.1, 3.1, 7.1, 1.1, 3.1, 3.1

4-4 Variability (pp. 188–192)

EXAMPLE

- **Find the range and quartiles.**

7, 10, 14, 16, 17, 17, 18, 20, 20

range = 20 − 7 = 13 *largest − smallest*

lower half — *upper half*

(7 (10 14) 16) (17) (17 (18 20) 20)

1st quartile — *3rd quartile*

$\frac{10 + 14}{2} = 12$ $\frac{18 + 20}{2} = 19$

EXERCISES

Find the range and quartiles.

12. 80, 80, 80, 82, 85, 87, 87, 90, 90, 90

13. 67, 68, 68, 80, 92, 99, 80, 99, 99, 99

4-5 Displaying Data (pp. 196–199)

EXAMPLE

■ **Make a histogram of the data set.**

Heights of 20 people, in inches:

72, 64, 56, 60, 66, 72, 48, 66, 58, 60, 60, 50, 68, 72, 68, 62, 72, 58, 60, 68

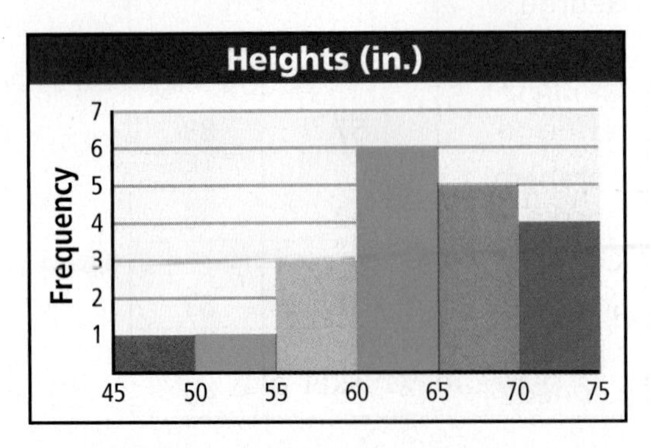

EXERCISES

Make a histogram of each data set.

14.

Test Scores	Frequency
91–100	6
81–90	8
71–80	11
61–70	4
51–60	0
41–50	3

15. TV viewing (hr/week): 19, 17, 11, 17, 3, 12, 27, 12, 20, 17, 25, 18, 23, 15, 16, 25, 23, 1, 14, 23, 17, 13, 19, 10, 21

4-6 Misleading Graphs and Statistics (pp. 200–203)

EXAMPLE

■ **Explain why the graph is misleading.**

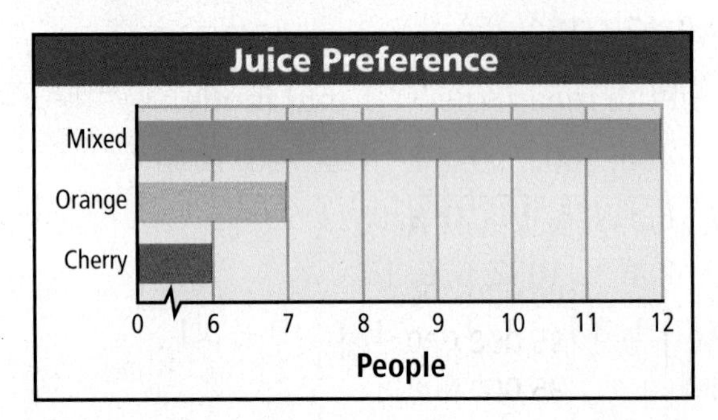

The bar for mixed juice is 7 times longer than the bar for cherry juice, but it is only preferred by 2 times as many people.

EXERCISES

Explain why the graph is misleading.

16.

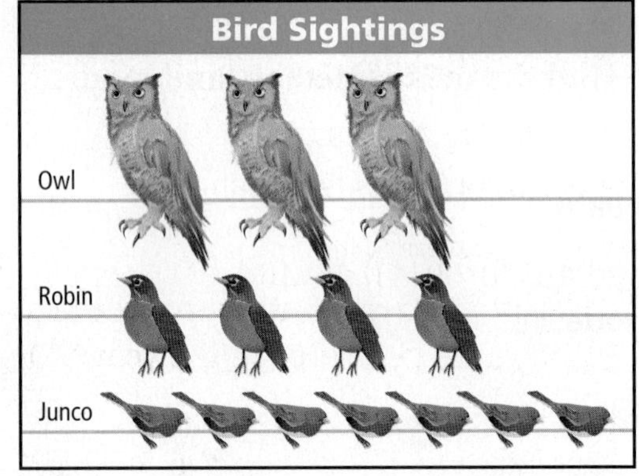

Each bird = 100 sightings

4-7 Scatter Plots (pp. 204–207)

EXAMPLE

■ **Does the data set have a positive, a negative, or no correlation? Explain.**

The age of a battery in a flashlight and the intensity of the flashlight beam.

Negative: The older the battery is, the less intense the flashlight beam will be.

EXERCISES

Do the data sets have a positive, a negative, or no correlation? Explain.

17. The price of an item and the dollar amount paid in sales tax.

18. Your height and the last digit of your phone number.

Chapter Test

Identify the sampling method used.

1. Twenty U.S. cities are randomly chosen and 100 people are randomly chosen from each city.

Use the data: **59, 21, 32, 33, 40, 51, 23, 23, 28, 26, 35, 49, 48, 41, 37, 39, 44, 54, 53, 29, 28, 29, 57, 58, 46**

2. Find the mean.
3. Find the median.
4. Find the mode.
5. Make a stem-and-leaf plot.
6. Find the range.
7. Find the first quartile.
8. Find the third quartile.
9. Make a box-and-whisker plot.

Use the data: **7, 7, 7, 7, 8, 8, 8, 5, 5, 8, 6, 6, 7, 7, 8, 8, 8, 5, 7, 5, 6, 7, 7, 6, 6, 6, 7, 7, 7, 7, 8**

10. Make a frequency table.
11. Make a bar graph.

Use the data: **155, 162, 168, 147, 152, 153, 178, 151, 180, 158, 163, 177, 171, 168, 183, 154, 180, 158, 157, 160, 171, 164, 171**

12. Make a frequency table.
13. Make a histogram.

Use the data in the table.

14. Make a line graph.
15. Use the line graph to estimate the population of Africa in the year 1800.
16. Use the line graph to estimate the population of Africa in the year 1900.

Year	Population of Africa
1650	100,000,000
1750	95,000,000
1850	95,000,000
1950	229,000,000
2000	805,000,000

17. **Give a reason why the statistic could be misleading.**

    A sign reads "Work at home—earn up to $1000 per week!"

Use the data in the table.

18. Make a scatter plot.
19. Draw the line of best fit.
20. Do the data sets have a positive, a negative, or no correlation? Explain.

Animal	Gestation Period (d)	Average Life (yr)
Baboon	187	20
Chipmunk	31	6
Elephant	645	40
Fox	52	7
Horse	330	20
Lion	100	15
Mouse	19	3

Chapter 4

Performance Assessment

Show What You Know

Create a portfolio of your work from this chapter. Complete this page and include it with your four best pieces of work from Chapter 4. Choose from your homework or lab assignments, mid-chapter quiz, or any journal entries you have done. Put them together using any design you want. Make your portfolio represent what you consider your best work.

Short Response

1. Determine the mean, median, and mode for the data set 2, 1, 8, 3, 500, 3, 1. Show your work.

2. Write a numeric expression that could be used to find the mean of the data in the frequency table. What is the mean of the data?

Number	1	2	3	4	5
Frequency	4	7	1	6	2

3. Name two ordered pairs (x, y) that satisfy these conditions: The mean of 0, x, and y is twice the median; $0 < x < y$; and $y = nx$ (y is a multiple of x). What is the value of n? Show your work or explain in words how you determined your answer.

Extended Problem Solving

4. Twenty students in a gym class kept a record of their jogging. The results are shown in the scatter plot.
   a. Describe the correlation of the data in the scatter plot.
   b. Find the average speeds of joggers who run 1, 2, 3, 4, 5, and 6 miles.
   c. Explain the relationship between your answer from part **a** and your answers from part **b**.

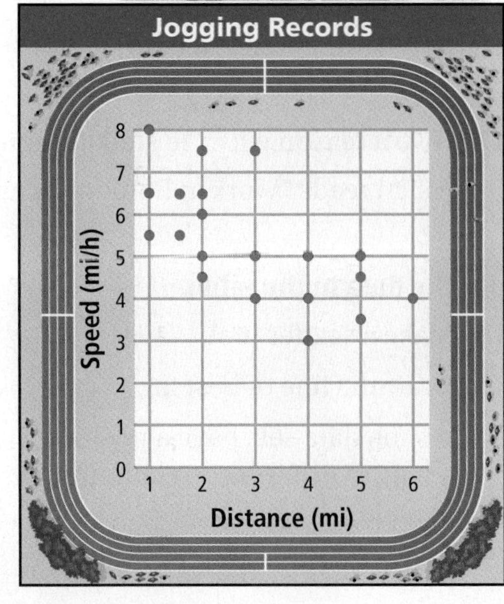

Performance Assessment

internet connect
EOG Test Practice Online
go.hrw.com Keyword: MT4 TestPrep

Getting Ready for EOG

Cumulative Assessment, Chapters 1–4

1. Dana bought 9 comic books for a total of $30.50. Which equation is equivalent to the equation $9c = 30.5$?

   A $c = 30.5 - 9$
   B $c = \frac{30.5}{9}$
   C $c = 9 - 30.5$
   D $c = \frac{9}{30.5}$

2. On the number line, what number is the coordinate of point R?

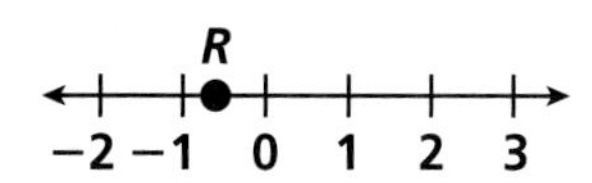

   A $-1\frac{3}{4}$
   B $-1\frac{1}{4}$
   C $-\frac{3}{4}$
   D $-\frac{1}{4}$

3. If the product of five integers is negative, then, at most, how many of the five integers could be negative?

   A five
   B four
   C three
   D two

4. Which is equivalent to $3^8 \cdot 3^4$?

   A 9^{32}
   B 9^{12}
   C 3^{32}
   D 3^{12}

5. What is the value of $32 - 2 \cdot 4^2$?

   A 14,400
   B 480
   C 0
   D −32

TEST TAKING TIP!
To calculate the median, the data must be in order.

6. For which set of data are the mean, median, and mode all the same?

   A 3, 1, 3, 3, 5
   B 1, 1, 2, 5, 6
   C 2, 1, 1, 1, 5
   D 10, 1, 3, 5, 1

7. Which is *true* for the data 6, 6, 6.5, 8, 8.5?

   A median < mode
   B median = mean
   C median < mean
   D median = mode

8. The stem-and-leaf plot shows test scores for a teacher's first and second periods. What can you conclude?

1st period		2nd period
7	6	5 8
6 4 2	7	5 6 9
9 8 6 4 2 0	8	1 3 5 7 7 8 8
9 7 7 2 1	9	0 6 7 8 9

Key: |9|0 means 90
7|6| means 67

   A More first period students scored in the 90's.
   B Fewer first period students scored 80 or below.
   C More second period students scored in the 70's.
   D More second period students scored in the 80's.

9. ***SHORT RESPONSE*** Julie wants to make homemade bows for her presents. She buys $\frac{1}{2}$ yard of red ribbon and $\frac{3}{4}$ yard of green. If each bow takes $\frac{1}{8}$ yard to make, how many total bows can she create? Justify your answer.

10. ***SHORT RESPONSE*** Max scored 75, 73, 71, 70, and 71 on his last 5 tests. Max wants to bring up his test average to a 75. What would Max need to make on his next test to bring his average up to a 75? Show your work.

Chapter 5

TABLE OF CONTENTS

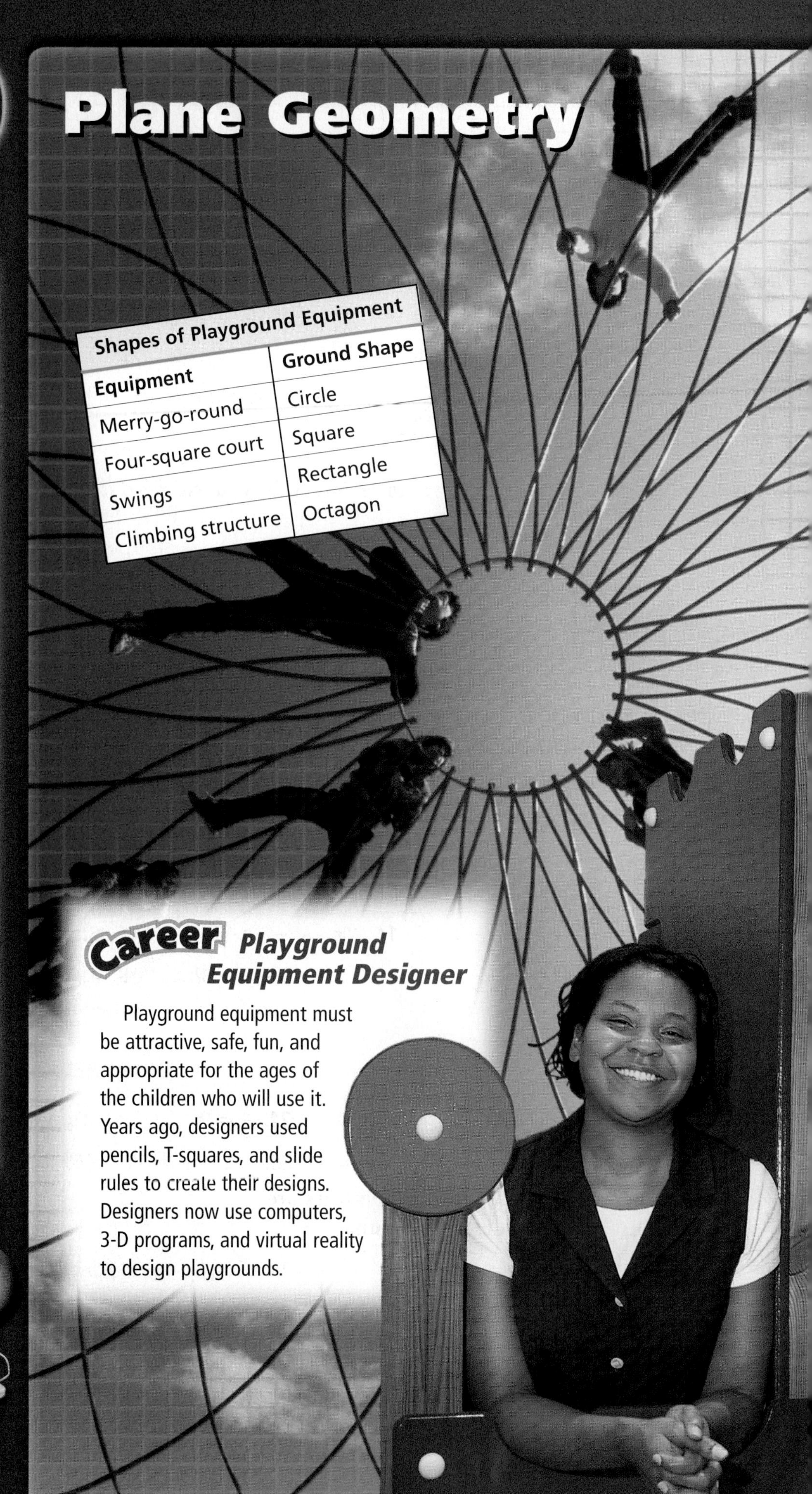

Plane Geometry

Shapes of Playground Equipment

Equipment	Ground Shape
Merry-go-round	Circle
Four-square court	Square
Swings	Rectangle
Climbing structure	Octagon

Career *Playground Equipment Designer*

Playground equipment must be attractive, safe, fun, and appropriate for the ages of the children who will use it. Years ago, designers used pencils, T-squares, and slide rules to create their designs. Designers now use computers, 3-D programs, and virtual reality to design playgrounds.

internet connect

Chapter Opener Online
go.hrw.com
KEYWORD: MT4 CH5

ARE YOU READY?

Choose the best term from the list to complete each sentence.

1. In the __?__ (4, −3), 4 is the __?__, and −3 is the __?__.
2. The __?__ divide the __?__ into four sections.
3. The point (0, 0) is called the __?__.
4. The point (0, −3) lies on the __?__, while the point (−2, 0) lies on the __?__.

coordinate axes
coordinate plane
origin
ordered pair
***x*-axis**
***y*-axis**
***x*-coordinate**
***y*-coordinate**

Complete these exercises to review skills you will need for this chapter.

✔ Ordered Pairs

Write the coordinates of the indicated points.

5. point A
6. point B
7. point C
8. point D
9. point E
10. point F
11. point G
12. point H

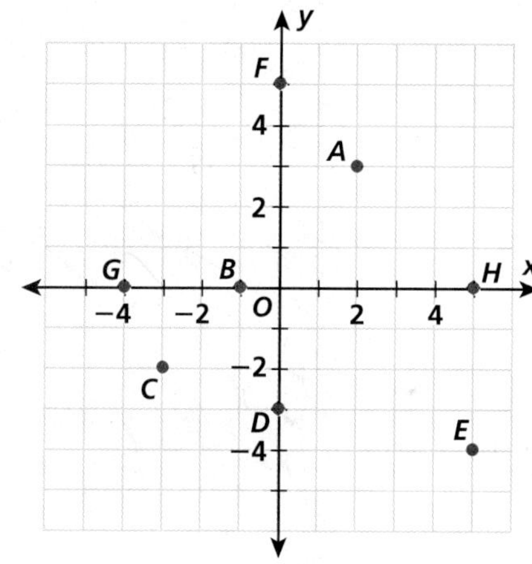

✔ Combine Like Terms

Simplify each expression by combining the like terms.

13. $5m + 7 - 2m - 1$
14. $2x - 4 - 6x + 1$
15. $6w + z - 5w - z$
16. $3r + 11s$
17. $12h - 9 + 2 - 3h$
18. $4y + 1 - 2y - x$

✔ Equations

Solve each equation.

19. $2p = 18$
20. $7 + h = 21$
21. $\frac{x}{3} = 9$
22. $y - 6 = 16$
23. $4d + 1 = 13$
24. $-2q - 3 = 3$
25. $4(z - 1) = 16$
26. $x + 3 + 4x = 23$

Determine whether the given values are solutions of the given equations.

27. $\frac{2}{3}x + 1 = 7$ $\quad x = 9$
28. $2x - 4 = 6$ $\quad x = -1$
29. $8 - 2x = -4$ $\quad x = 5$
30. $\frac{1}{2}x + 5 = -2$ $\quad x = -14$

Points, Lines, Planes, and Angles

Learn to classify and name figures.

Vocabulary
point, line, plane, segment, ray, angle, right angle, acute angle, obtuse angle, complementary angles, supplementary angles, congruent, vertical angles

Points, lines, and planes are the building blocks of geometry. Segments, rays, and angles are defined in terms of these basic figures.

Definition	Figure	Name
A **point** names a location.	• A	point A
A **line** is perfectly straight and extends forever in both directions.	ℓ, B, C	line ℓ, or $\overleftrightarrow{BC}$
A **plane** is a perfectly flat surface that extends forever in all directions.	$\mathcal{P}$, D, E, F	plane $\mathcal{P}$, or plane DEF
A **segment**, or line segment, is the part of a line between two points.	G, H	$\overline{GH}$
A **ray** is part of a line that starts at one point and extends forever in one direction.	J, K	$\overrightarrow{KJ}$

$\overleftrightarrow{BC}$ is read "line BC." $\overline{GH}$ is read "segment GH." $\overrightarrow{KJ}$ is read "ray KJ." To name a ray, always write the endpoint first.

EXAMPLE 1 Naming Points, Lines, Planes, Segments, and Rays

A **Name four points in the figure.**
point Q, point R, point S, point T

B **Name a line in the figure.**
$\overleftrightarrow{QS}$ or $\overleftrightarrow{QR}$ or $\overleftrightarrow{RS}$
Any 2 points on the line can be used.

C **Name a plane in the figure.**
plane $\mathcal{Z}$ or plane QRT *Any 3 points in the plane that form a triangle can be used.*

D **Name four segments in the figure.**
$\overline{QR}$, $\overline{RS}$, $\overline{RT}$, $\overline{QS}$

E **Name five rays in the figure.**
$\overrightarrow{RQ}$ $\overrightarrow{RS}$, $\overrightarrow{RT}$, $\overrightarrow{SQ}$, $\overrightarrow{QS}$

An **angle** ($\angle$) is formed by two rays with a common endpoint called the *vertex* (plural, *vertices*). Angles can be measured in degrees. One degree, or 1°, is $\frac{1}{360}$ of a circle. m$\angle 1$ means the measure of $\angle 1$. The angle can be named $\angle XYZ$, $\angle ZYX$, $\angle 1$, or $\angle Y$. The vertex must be the middle letter.

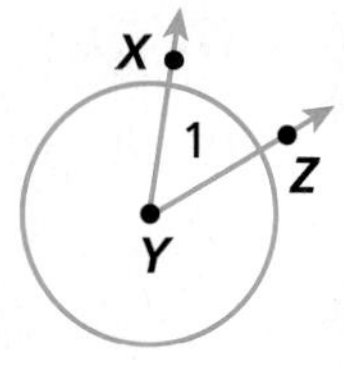

m$\angle 1 = 50°$

The measures of angles that fit together to form a straight line, such as $\angle FKG$, $\angle GKH$, and $\angle HKJ$, add to 180°.

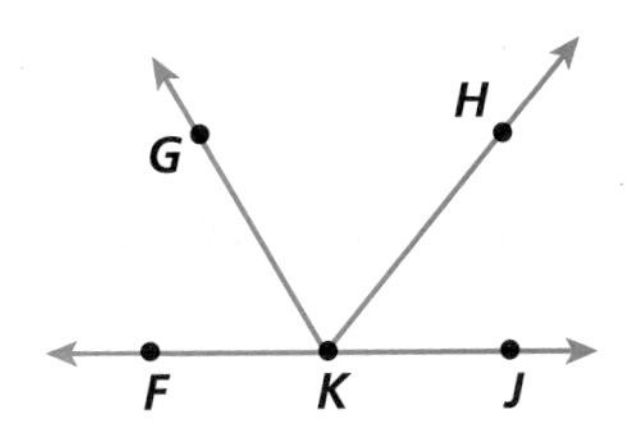

The measures of angles that fit together to form a complete circle, such as $\angle MRN$, $\angle NRP$, $\angle PRQ$, and $\angle QRM$, add to 360°.

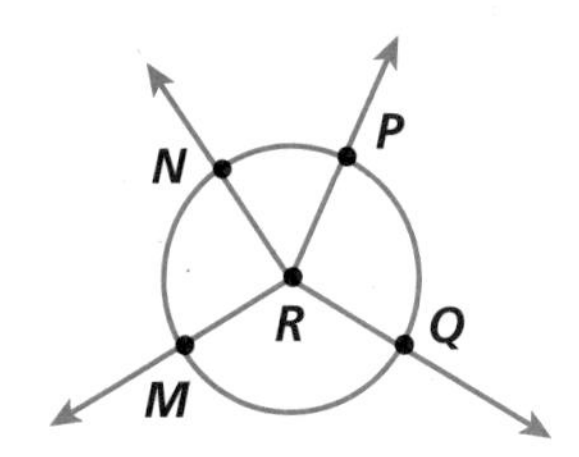

A **right angle** measures 90°. An **acute angle** measures less than 90°. An **obtuse angle** measures greater than 90° and less than 180°. **Complementary angles** have measures that add to 90°. **Supplementary angles** have measures that add to 180°.

EXAMPLE 2 Classifying Angles

Reading Math

A right angle can be labeled with a small box at the vertex.

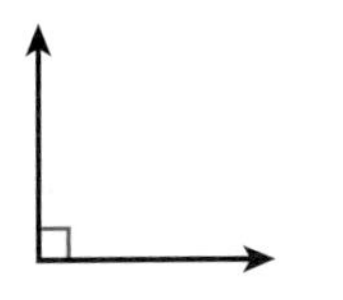

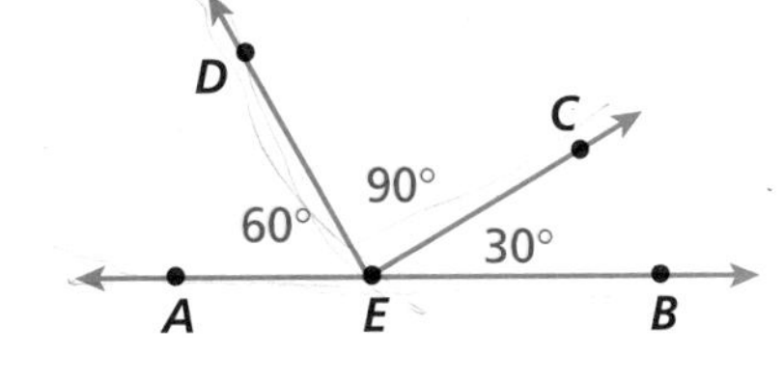

A **Name a right angle in the figure.**

$\angle DEC$

B **Name two acute angles in the figure.**

$\angle AED$, $\angle CEB$

C **Name two obtuse angles in the figure.**

$\angle AEC$, $\angle DEB$

D **Name a pair of complementary angles in the figure.**

$\angle AED$, $\angle CEB$ $\quad m\angle AED + m\angle CEB = 60° + 30° = 90°$

E **Name two pairs of supplementary angles in the figure.**

$\angle AED$, $\angle DEB$ $\quad m\angle AED + m\angle DEB = 60° + 120° = 180°$

$\angle AEC$, $\angle CEB$ $\quad m\angle AEC + m\angle CEB = 150° + 30° = 180°$

Congruent figures have the same size and shape.

- Segments that have the same length are congruent.
- Angles that have the same measure are congruent.
- The symbol for congruence is $\cong$, which is read "is congruent to."

Intersecting lines form two pairs of **vertical angles**. Vertical angles are always congruent, as shown in the next example.

EXAMPLE 3 Finding the Measures of Vertical Angles

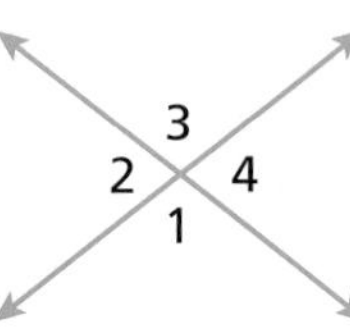

In the figure, $\angle 1$ and $\angle 3$ are vertical angles, and $\angle 2$ and $\angle 4$ are vertical angles.

A **If $m\angle 2 = 75°$, find $m\angle 4$.**

The measures of $\angle 2$ and $\angle 3$ add to 180° because they are supplementary, so $m\angle 3 = 180° - 75° = 105°$.

The measures of $\angle 3$ and $\angle 4$ add to 180° because they are supplementary, so $m\angle 4 = 180° - 105° = 75°$.

B **If $m\angle 3 = x°$, find $m\angle 1$.**

$m\angle 4 = 180° - x°$

$m\angle 1 = 180° - (180° - x°)$

$= 180° - 180° + x°$ *Distributive Property*

$= x°$ *$m\angle 1 = m\angle 3$*

Think and Discuss

1. **Tell** which statements are correct if $\angle X$ and $\angle Y$ are congruent.

   a. $\angle X = \angle Y$ b. $m\angle X = m\angle Y$ c. $\angle X \cong \angle Y$ d. $m\angle X \cong m\angle Y$

2. **Explain** why vertical angles must always be congruent.

5-1 Exercises

FOR EOG PRACTICE see page 660

internet connect
Homework Help Online
go.hrw.com Keyword: MT4 5-1

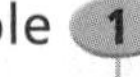
3.02

GUIDED PRACTICE

See Example 1

1. Name three points in the figure.
2. Name a line in the figure.
3. Name a plane in the figure.
4. Name three segments in the figure.
5. Name three rays in the figure.

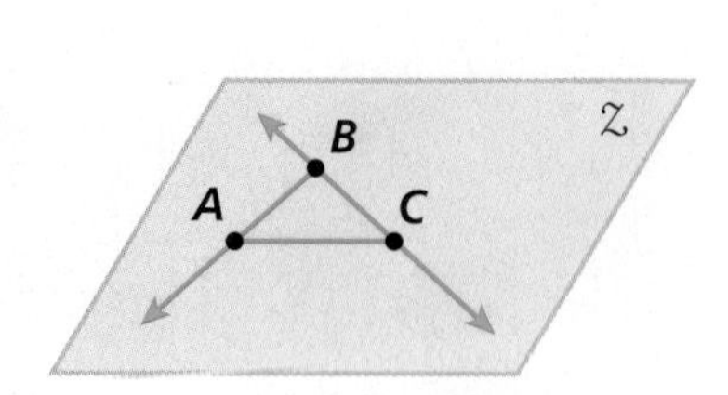

See Example 2

6. Name a right angle in the figure.
7. Name two acute angles in the figure.
8. Name an obtuse angle in the figure.
9. Name a pair of complementary angles in the figure.
10. Name two pairs of supplementary angles in the figure.

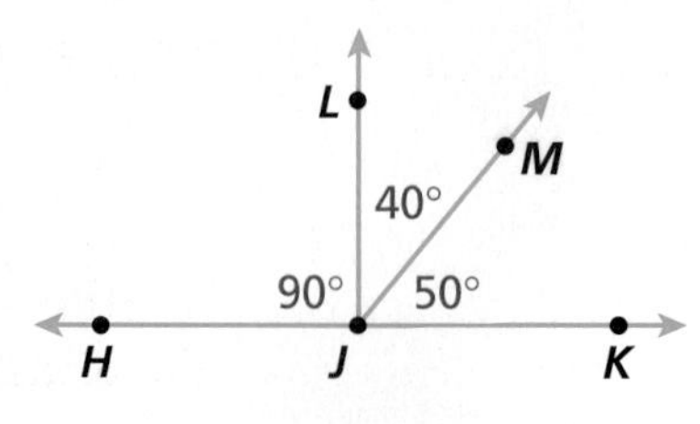

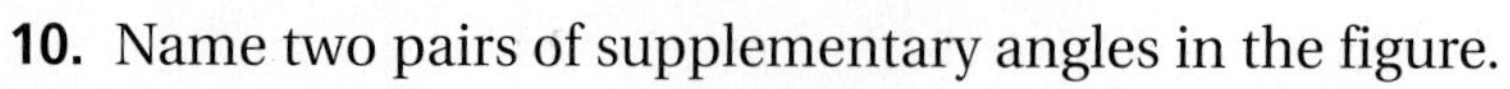

See Example 3

In the figure, $\angle 1$ and $\angle 3$ are vertical angles, and $\angle 2$ and $\angle 4$ are vertical angles.

11. If $m\angle 3 = 115°$, find $m\angle 1$.

12. If $m\angle 2 = a°$, find $m\angle 4$.

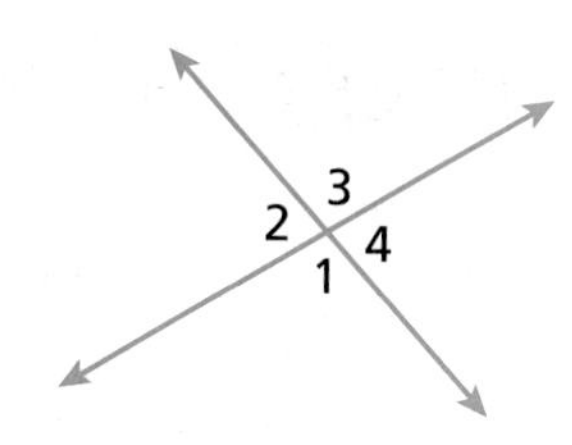

INDEPENDENT PRACTICE

See Example 1

13. Name four points in the figure.

14. Name two lines in the figure.

15. Name a plane in the figure.

16. Name three segments in the figure.

17. Name five rays in the figure.

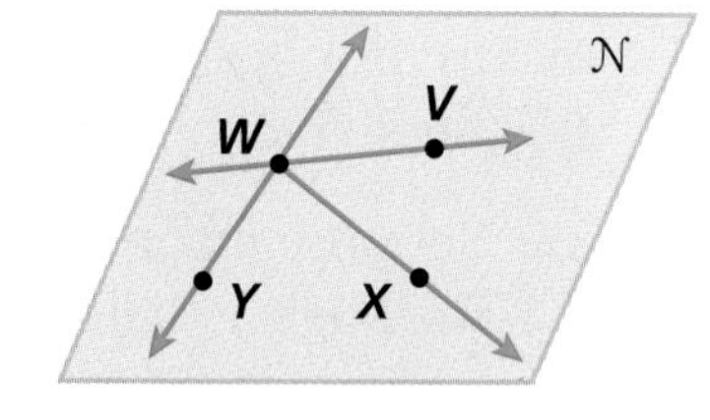

See Example 2

18. Name a right angle in the figure.

19. Name two acute angles in the figure.

20. Name two obtuse angles in the figure.

21. Name a pair of complementary angles in the figure.

22. Name two pairs of supplementary angles in the figure.

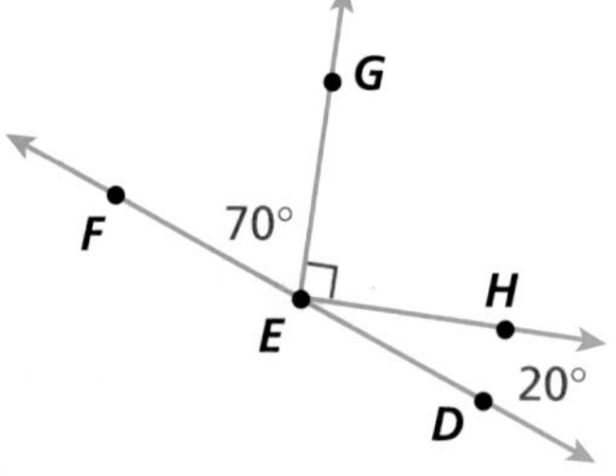

See Example 3

In the figure, $\angle 1$ and $\angle 3$ are vertical angles, and $\angle 2$ and $\angle 4$ are vertical angles.

23. If $m\angle 2 = 117°$, find $m\angle 4$.

24. If $m\angle 1 = n°$, find $m\angle 3$.

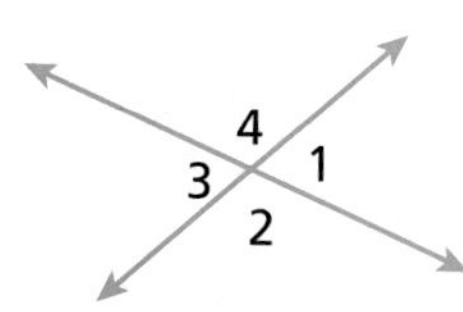

PRACTICE AND PROBLEM SOLVING

Use the figure for Exercises 25–34. Write *true* or *false*. If a statement is false, rewrite it so it is true.

25. $\overleftrightarrow{AE}$ is a line in the figure.

26. Rays $\overrightarrow{GB}$ and $\overrightarrow{GE}$ make up line $\overleftrightarrow{EB}$.

27. $\angle EGD$ is an obtuse angle.

28. $\angle 4$ and $\angle 2$ are supplementary.

29. $\angle 3$ and $\angle 5$ are supplementary.

30. $\angle 6$ and $\angle 5$ are complementary.

31. If $m\angle 1 = 30°$, then $m\angle 6 = 45°$.

32. If $m\angle FGD = 130°$, then $m\angle DGC = 130°$.

33. If $m\angle 3 = x°$, then $m\angle FGE = 180° - x°$.

34. $m\angle 1 + m\angle 3 + m\angle 5 + m\angle 6 = 180°$.

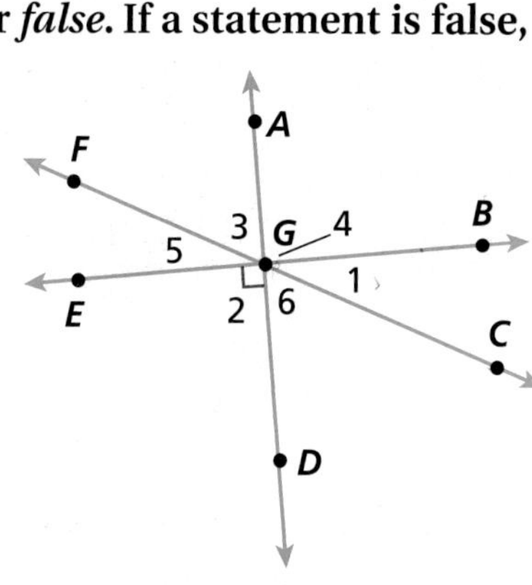

Physical Science LINK

The archerfish can spit a stream of water up to 3 meters in the air to knock its prey into the water. This job is made more difficult by *refraction,* the bending of light waves as they pass from one substance to another. When you look at an object through water, the light between you and the object is refracted. Refraction makes the object appear to be in a different location. Despite refraction, the archerfish still catches its prey.

35. Suppose that the measure of the angle between the bug's actual location and the bug's apparent location is 35°.

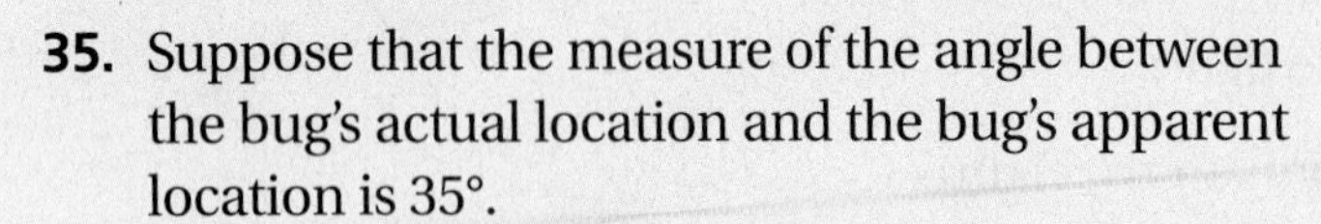

a. Refer to the diagram. Along the fish's line of vision, what is the measure of the angle between the fish and the bug's apparent location?

b. What is the relationship of the angles in the diagram?

36. In the photograph, the underwater part of the net appears to be 40° to the right of where it actually is. What is the measure of the angle formed by the image of the underwater part of the net and the part of the net above the water?

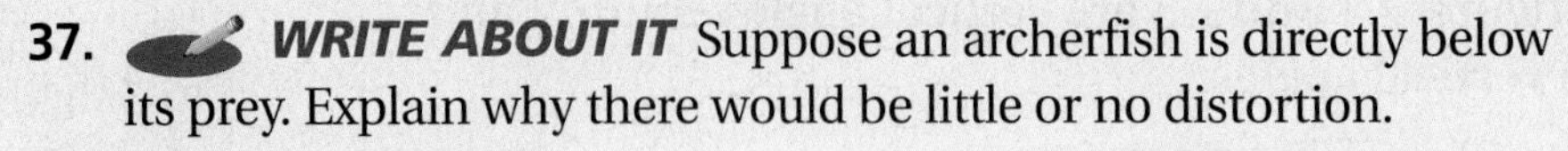

37. ***WRITE ABOUT IT*** Suppose an archerfish is directly below its prey. Explain why there would be little or no distortion.

38. ***CHALLENGE*** A person on the shore is looking at a fish in the water. At the same time, the fish is looking at the person from below the surface. Describe what each observer sees, and where the person and the fish actually are in relation to where they appear to be.

Spiral Review

Find the mean, median, and mode of each data set. Round to the nearest tenth. (Lesson 4-3)

39. 16, 16, 14, 13, 20, 29, 14, 13, 16

40. 2.1, 2.3, 3.2, 2.2, 1.9, 2.3, 2.2

Find the range and the first and third quartiles of each data set. (Lesson 4-4)

41. 32, 26, 24, 14, 20, 32, 16, 25, 26

42. 221, 223, 352, 202, 139, 243, 232

43. **EOG PREP** Which fraction is *greater* than $\frac{1}{4}$? (Previous course)

A $\frac{6}{23}$ B $\frac{12}{49}$ C $\frac{15}{68}$ D $\frac{17}{99}$

Basic Constructions

Use with Lesson 5-1

internet connect
Lab Resources Online
go.hrw.com
KEYWORD: MT4 Lab5A

When you *bisect* a figure, you divide it into two congruent parts.

Activity

1 Follow the steps below to bisect a segment.

a. Draw $\overline{JK}$ on your paper. Place your compass point on J and draw an arc. Without changing your compass opening, place your compass point on K and draw an arc.

b. Connect the intersections of the arcs with a line. Measure $\overline{JM}$ and $\overline{KM}$. What do you notice?

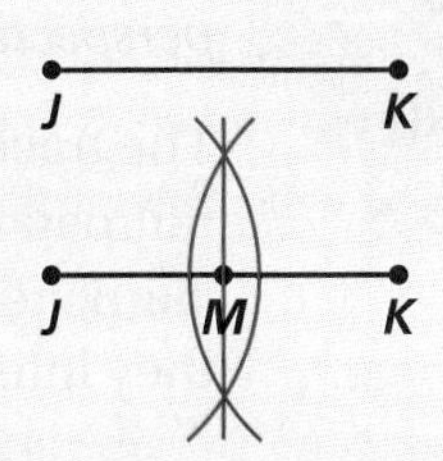

2 Follow the steps below to bisect an angle.

a. Draw acute $\angle H$ on your paper.

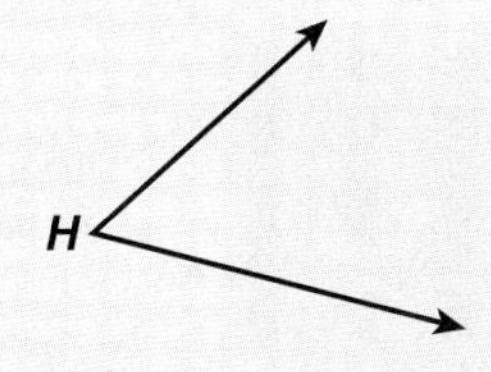

b. Place your compass point on H and draw an arc through both sides of the angle.

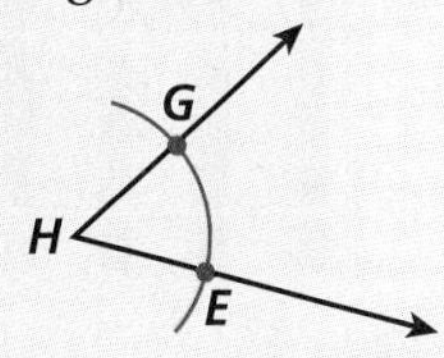

c. Without changing your compass opening, draw intersecting arcs from G and E. Label the intersection D.

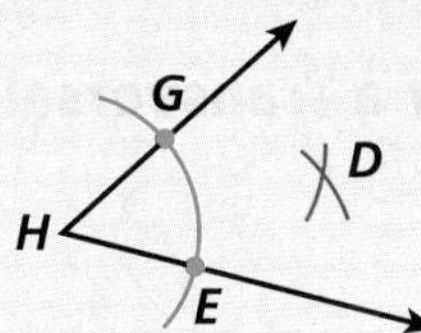

d. Draw $\overrightarrow{HD}$. Use a protractor to measure $\angle GHD$ and $\angle DHE$. What do you notice?

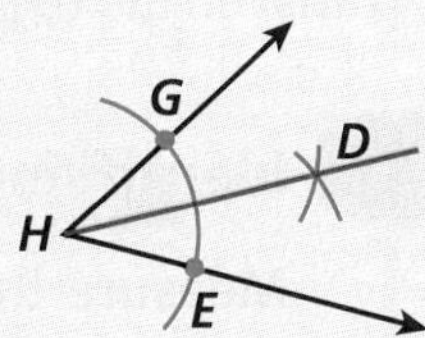

Think and Discuss

1. Explain how to use a compass and a straightedge to divide a segment into four congruent segments. Prove that the segments are congruent.

Try This

Draw each figure, and then use a compass and a straightedge to bisect it. Verify by measuring.

1. a 2-inch segment
2. a 1-inch segment
3. a 4-inch segment
4. a 64° angle
5. a 90° angle
6. a 120° angle

Advanced Constructions

Use with Lesson 5-2

internet connect
Lab Resources Online
go.hrw.com
KEYWORD: MT4 Lab5B

Copying an angle is an important step in the construction of parallel lines.

Activity

1 Follow the steps below to copy an angle.

a. Draw acute $\angle ABC$ on your paper. Draw $\overrightarrow{DE}$.

b. With your compass point on B, draw an arc through $\angle ABC$. With the same compass opening, place your compass point on D and draw an arc through $\overrightarrow{DE}$.

c. Adjust your compass to the width of the arc intersecting $\angle ABC$. Place your compass point on F and draw an arc that intersects the arc through $\overrightarrow{DE}$ at G. Draw $\overrightarrow{DG}$. Use your protractor to measure $\angle ABC$ and $\angle GDF$.

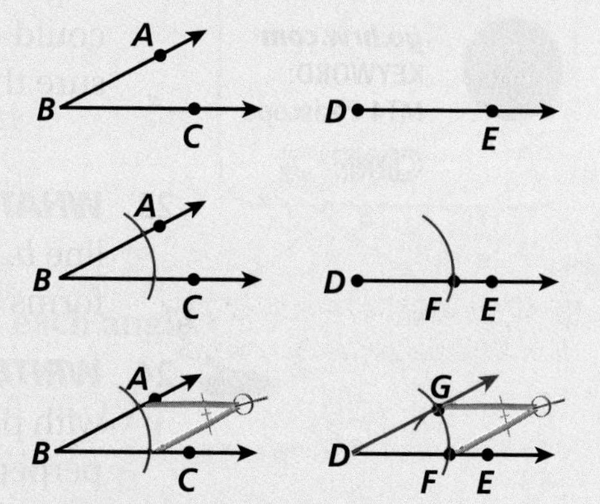

2 Follow the steps below to construct parallel lines.

1. Draw $\overleftrightarrow{QR}$ on your paper. Draw point S above or below $\overleftrightarrow{QR}$. Draw a line through point S that intersects $\overleftrightarrow{QR}$. Label the intersection T.

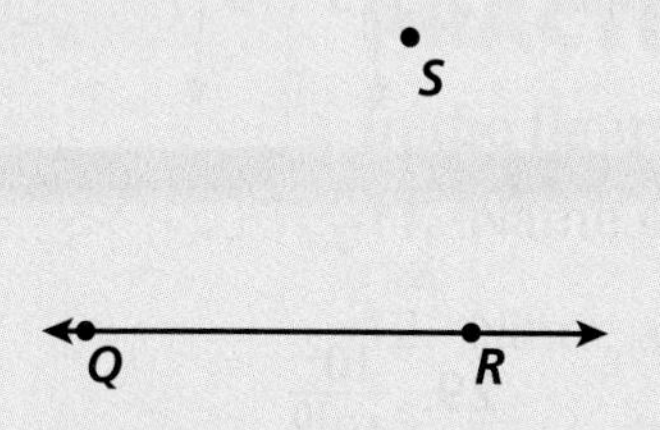

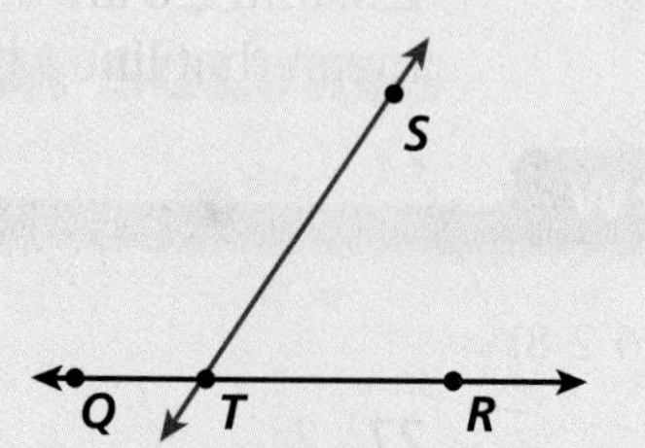

2. Make a copy of $\angle STR$ with its vertex at S using the method described in the first Activity. How do you know the lines are parallel?

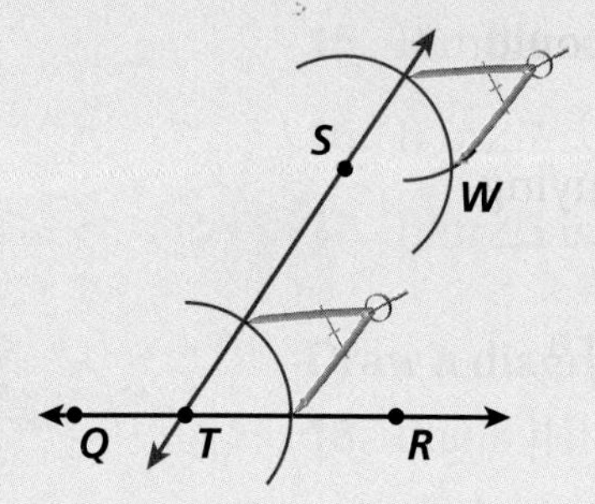

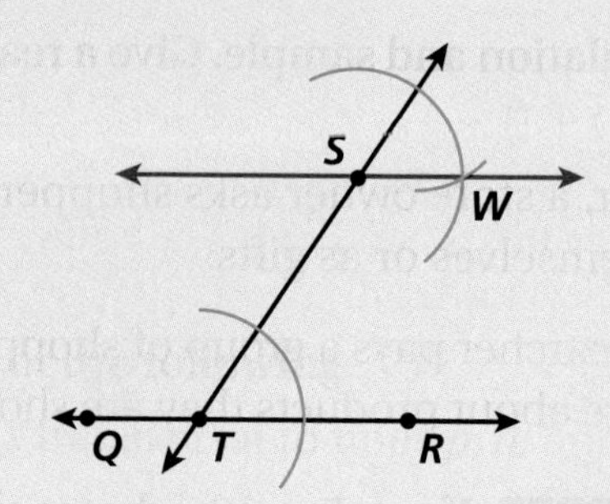

See Example 3

7. The second angle in a triangle is half as large as the first. The third angle is three times as large as the second. Find the angle measures and draw a possible picture.

INDEPENDENT PRACTICE

See Example 1

8. Find r in the acute triangle.

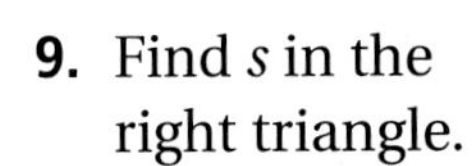

9. Find s in the right triangle.

10. Find t in the obtuse triangle.

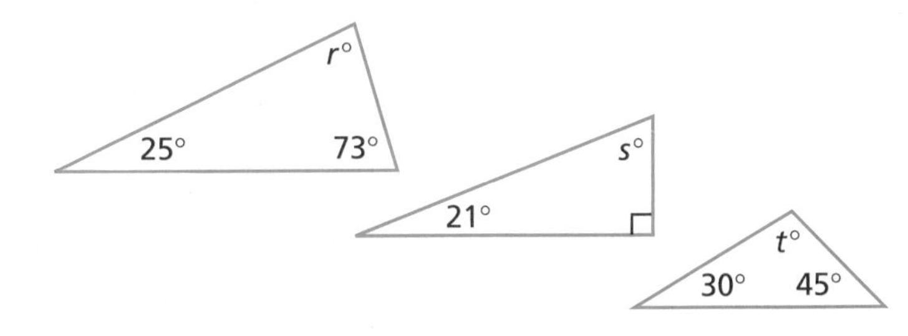

See Example 2

11. Find the angle measures in the equilateral triangle.

12. Find the angle measures in the isosceles triangle.

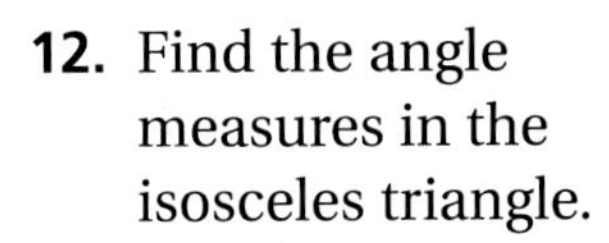

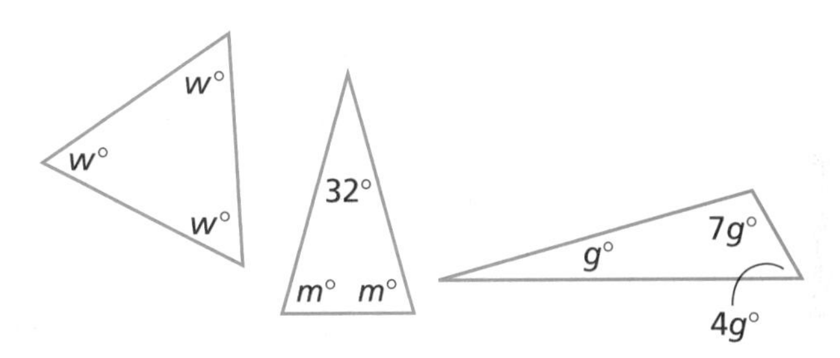

13. Find the angle measures in the scalene triangle.

See Example 3

14. The second angle in a triangle is twice as large as the first. The third angle is three-fourths as large as the first. Find the angle measures and draw a possible picture.

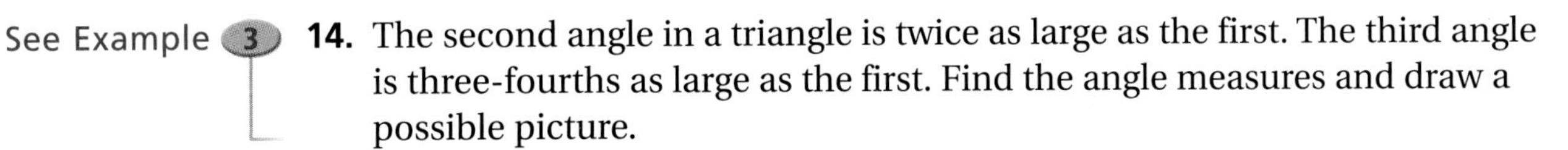

PRACTICE AND PROBLEM SOLVING

Find the value of each variable.

15.

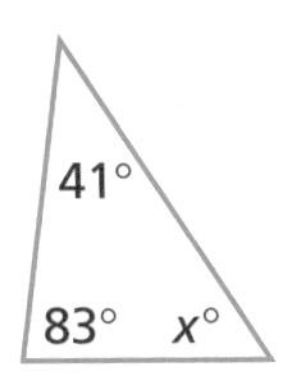

16.

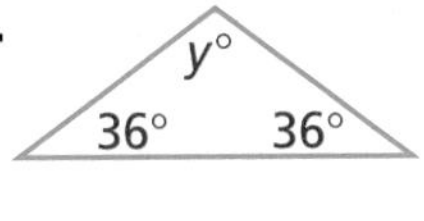

17.

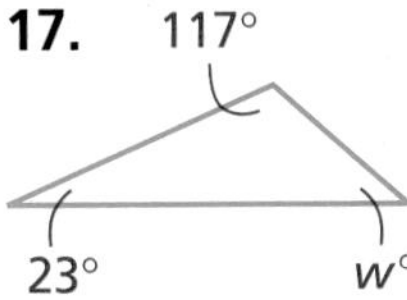

18.

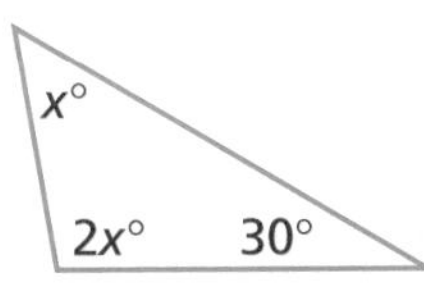

19.

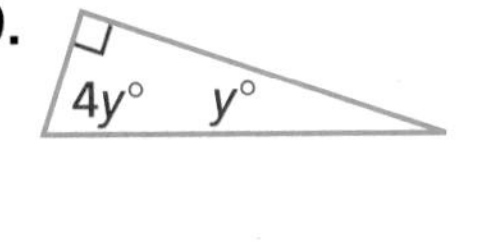

20.

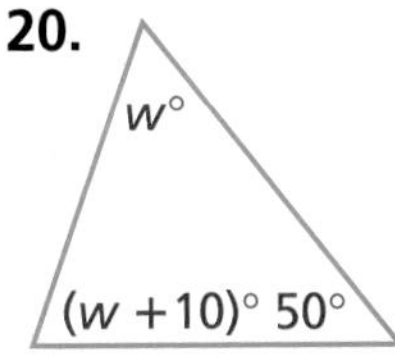

Sketch a triangle to fit each description. If no triangle can be drawn, write *not possible.*

21. acute scalene
22. obtuse equilateral
23. right scalene
24. right equilateral
25. obtuse scalene
26. acute isosceles

Describe each statement as always, sometimes, or never true.

27. An equilateral triangle is an acute triangle.

28. An equilateral triangle is an isosceles triangle.

29. An acute triangle is an equilateral triangle.

30. An isosceles triangle is an equilateral triangle.

31. A scalene triangle is an equilateral triangle.

32. An obtuse triangle is an isosceles triangle.

33. A right triangle is an obtuse triangle.

34. An obtuse triangle has two acute angles.

35. ***SOCIAL STUDIES*** American Samoa is a territory of the United States made up of a group of islands in the South Pacific Ocean, about halfway between Hawaii and New Zealand. The flag of American Samoa is shown.

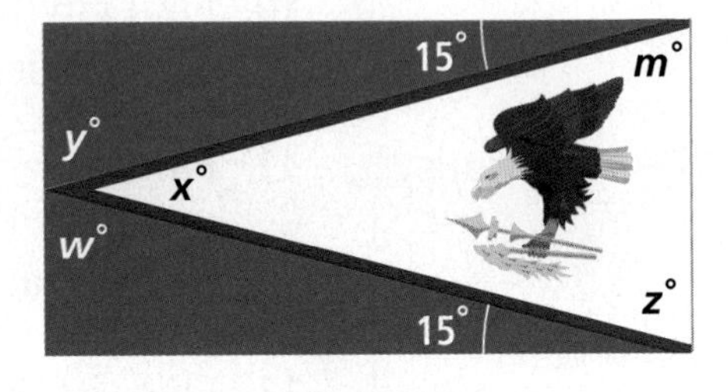

a. Find the measure of each angle in the blue triangles.

b. Use your answers to part **a** to find the angle measures in the white triangle.

c. Classify the triangles in the flag by their sides and angles.

36. ***WHAT'S THE ERROR?*** An isosceles triangle has one angle that measures 50° and another that measures 70°. Why can't this triangle be drawn?

37. ***WRITE ABOUT IT*** Explain how to cut a square or an equilateral triangle in half to form two identical triangles. What are the angle measures in the resulting triangles in each case?

38. ***CHALLENGE*** Find x, y, and z.

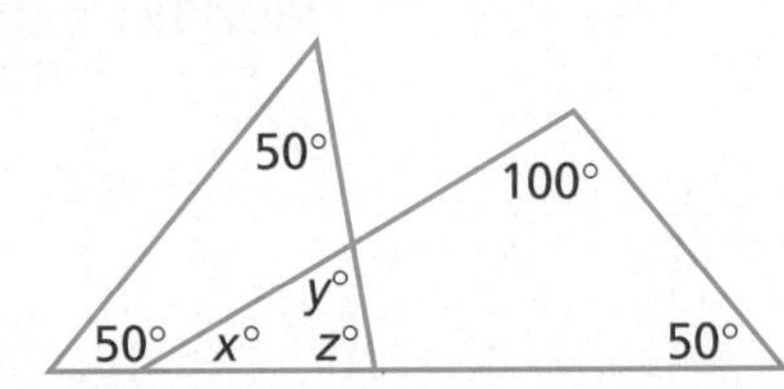

Spiral Review

Evaluate each expression for the given values of the variables. (Lesson 1-1)

39. $7x - 4y$ for $x = 5$ and $y = 6$

40. $6.5p - 9.1q$ for $p = 2.5$ and $q = 0$

41. **EOG PREP** The rectangle shown is cut by a diagonal. What two figures are formed? (Lesson 5-3)

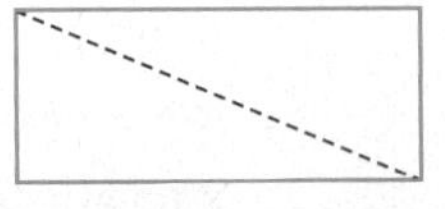

A Two acute triangles

B Two equilateral triangles

C Two right triangles

D Two isosceles triangles

5-4 Polygons

Learn to classify and find angles in polygons.

Vocabulary

polygon
regular polygon
trapezoid
parallelogram
rectangle
rhombus
square

The cross section of a brilliant-cut diamond is a *pentagon.* The most beautiful and valuable diamonds have precisely cut angles that maximize the amount of light they reflect.

Too shallow Ideal Too deep

A **polygon** is a closed plane figure formed by three or more segments. A polygon is named by the number of its sides.

Polygon	Number of Sides
Triangle	3
Quadrilateral	4
Pentagon	5
Hexagon	6
Heptagon	7
Octagon	8
n-gon	n

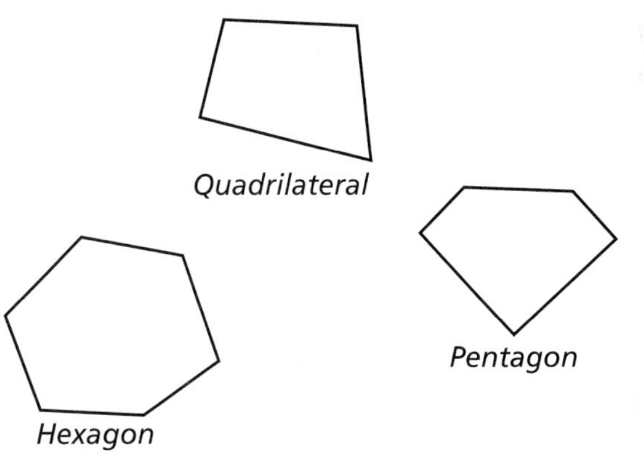

EXAMPLE 1 Finding Sums of the Angle Measures in Polygons

Find the sum of the angle measures in each figure.

A **Find the sum of the angle measures in a quadrilateral.**

Divide the figure into triangles.

$2 \cdot 180° = 360°$ *2 triangles*

B **Find the sum of the angle measures in a pentagon.**

Divide the figure into triangles.

$3 \cdot 180° = 540°$ *3 triangles*

Look for a pattern between the number of sides and the number of triangles.

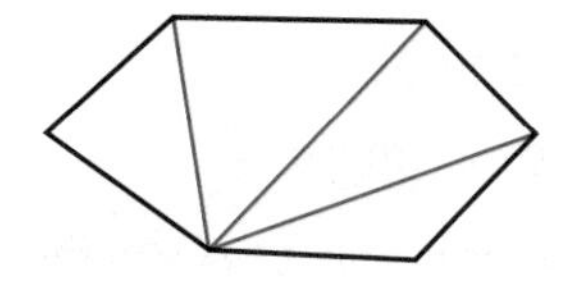

Hexagon:
6 sides
4 triangles

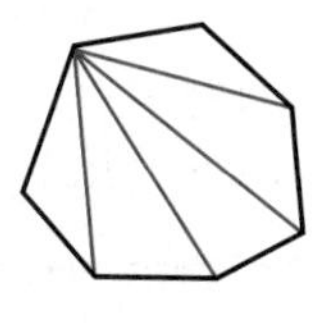

Heptagon:
7 sides
5 triangles

The pattern is that the number of triangles is always 2 less than the number of sides. So an n-gon can be divided into $n - 2$ triangles. The sum of the angle measures of any n-gon is $180°(n - 2)$.

All the sides and angles of a **regular polygon** have equal measures.

EXAMPLE 2 Finding the Measure of Each Angle in a Regular Polygon

Find the angle measures in each regular polygon.

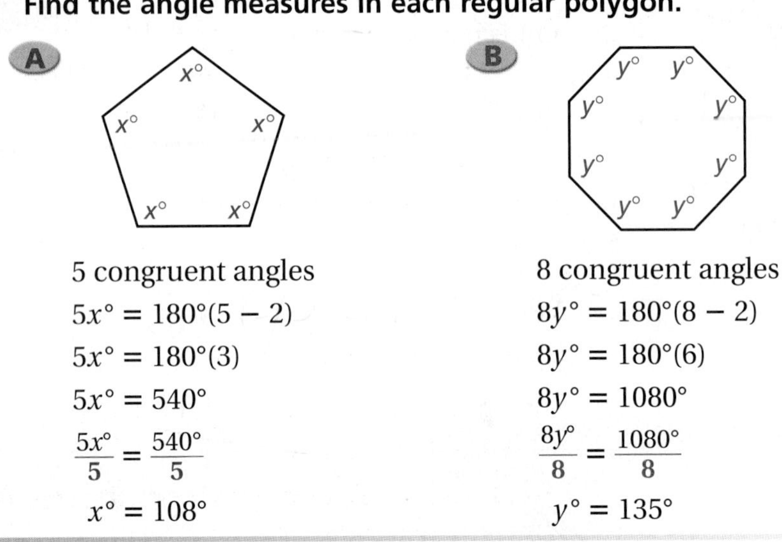

A

5 congruent angles

$5x° = 180°(5 - 2)$

$5x° = 180°(3)$

$5x° = 540°$

$\frac{5x°}{5} = \frac{540°}{5}$

$x° = 108°$

B

8 congruent angles

$8y° = 180°(8 - 2)$

$8y° = 180°(6)$

$8y° = 1080°$

$\frac{8y°}{8} = \frac{1080°}{8}$

$y° = 135°$

Quadrilaterals with certain properties are given additional names. A **trapezoid** has exactly 1 pair of parallel sides. A **parallelogram** has 2 pairs of parallel sides. A **rectangle** has 4 right angles. A **rhombus** has 4 congruent sides. A **square** has 4 congruent sides and 4 right angles.

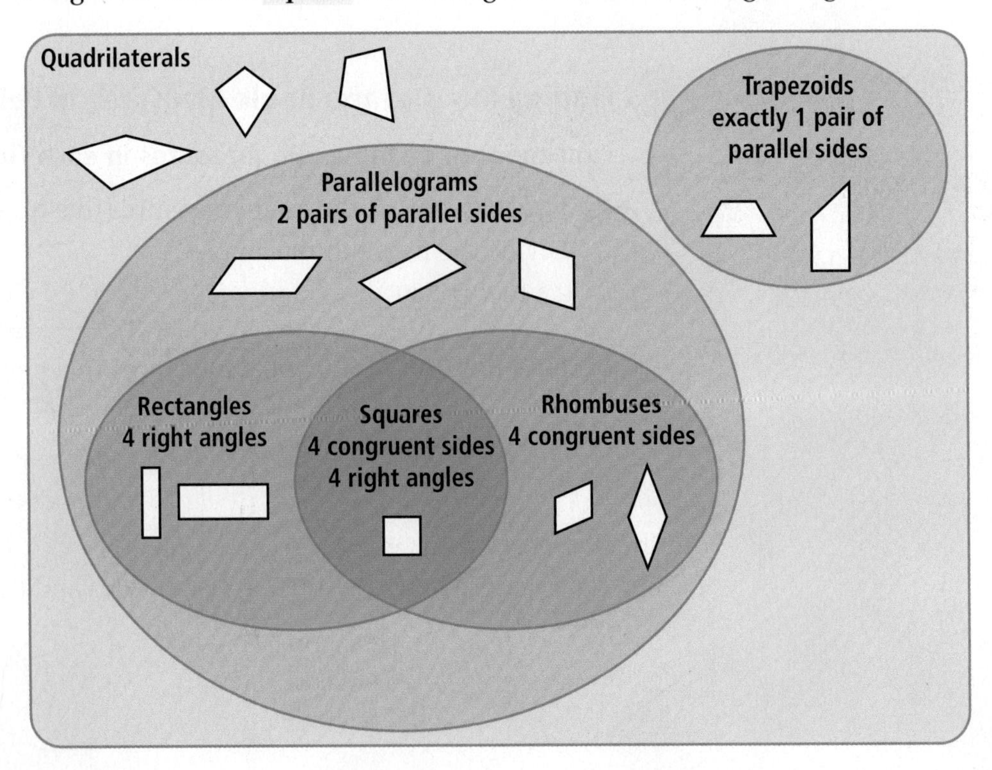

EXAMPLE 3 Classifying Quadrilaterals

Helpful Hint

Marks on the sides of a figure can be used to show congruence.

$\overline{AB} \cong \overline{CD}$ (2 marks)
$\overline{AD} \cong \overline{BC}$ (1 mark)

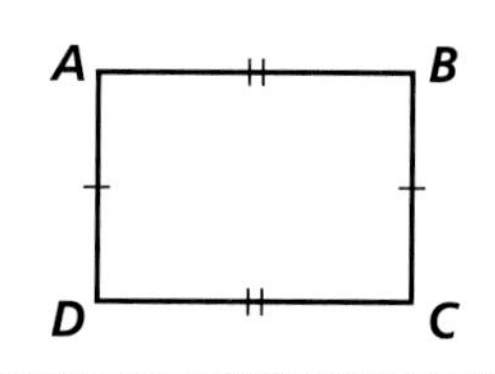

Give all of the names that apply to each figure.

A

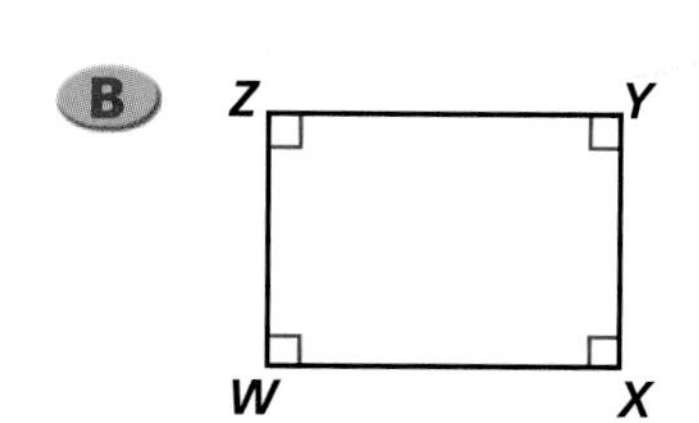

$\overline{EF} \parallel \overline{GH}$

quadrilateral — *Four-sided polygon*
trapezoid — *1 pair of parallel sides*

B

Z Y W X

quadrilateral — *Four-sided polygon*
parallelogram — *2 pairs of parallel sides*
rectangle — *4 right angles*

Think and Discuss

1. **Choose** which is larger, an angle in a regular heptagon or an angle in a regular octagon.
2. **Explain** why all rectangles are parallelograms and why all squares are rectangles.
3. **Give** another name for a regular triangle and for a regular quadrilateral.

5-4 Exercises

3.02

FOR EOG PRACTICE see page 661

GUIDED PRACTICE

See Example 1 **Find the sum of the angle measures in each figure.**

1. 

2. 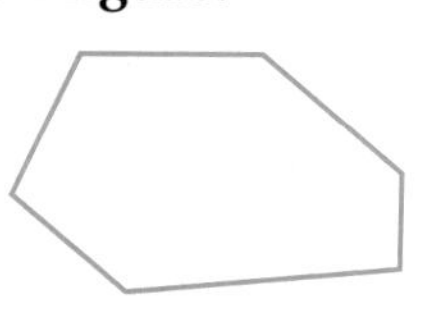

See Example 2 **Find the angle measures in each regular polygon.**

3. 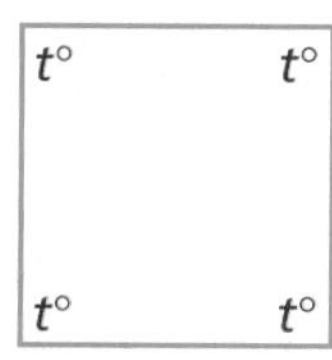

4. 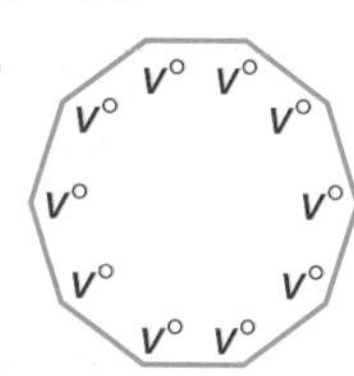

See Example 3 Give all of the names that apply to each figure.

5. 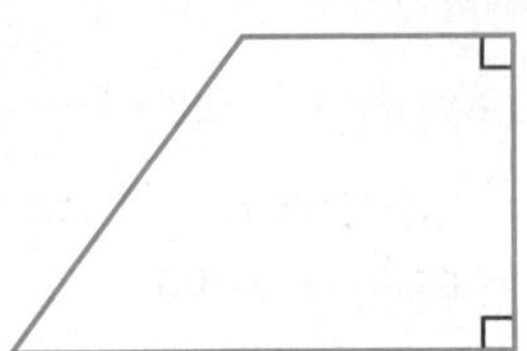

6. 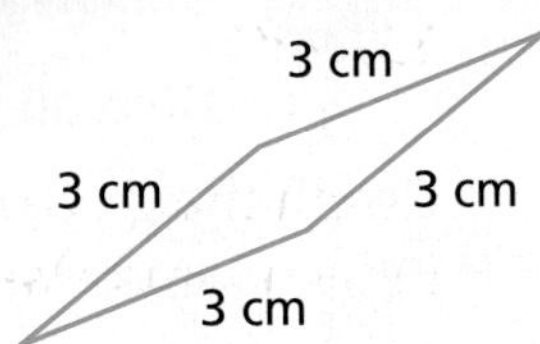

INDEPENDENT PRACTICE

See Example 1 Find the sum of the angle measures in each figure.

7. 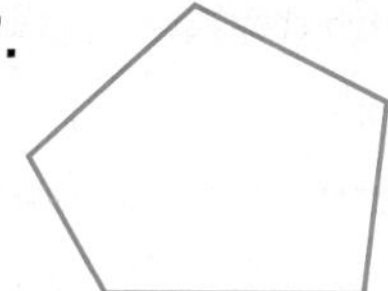

8. 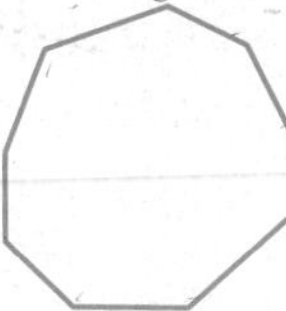

See Example 2 Find the angle measures in each regular polygon.

9. 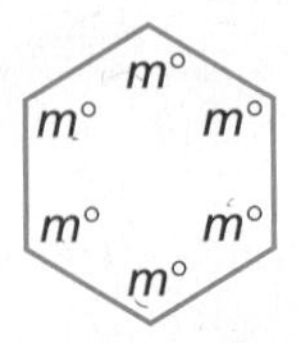

10. 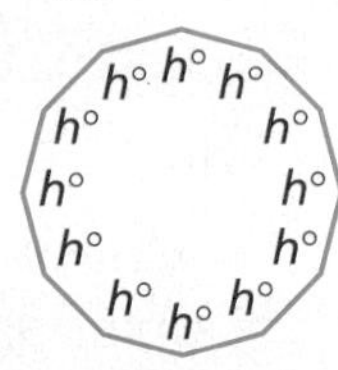

See Example 3 Give all of the names that apply to each figure.

11. 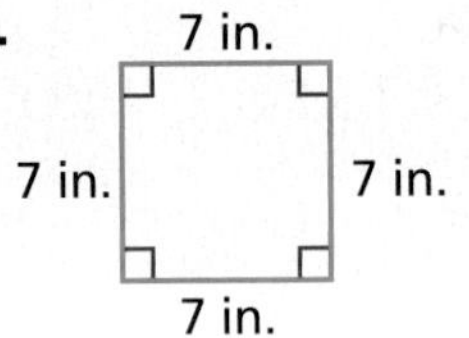

12. 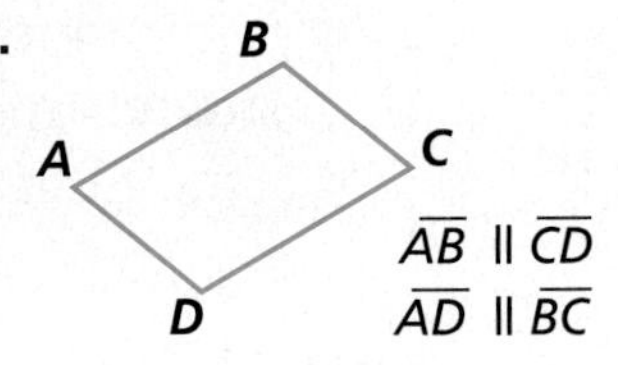

PRACTICE AND PROBLEM SOLVING

Find the sum of the angle measures in each polygon. Then, if the polygon is regular, find the measure of each angle.

13. 20-gon **14.** 11-gon **15.** 72-gon

16. pentagon **17.** 18-gon **18.** n-gon

Find the value of each variable.

19. 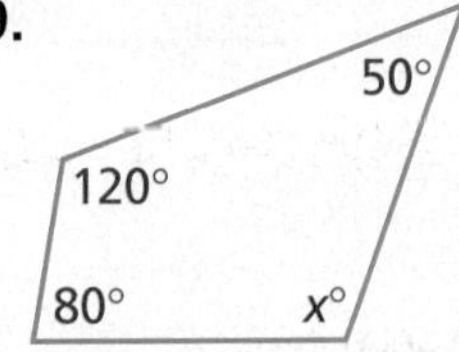

20. 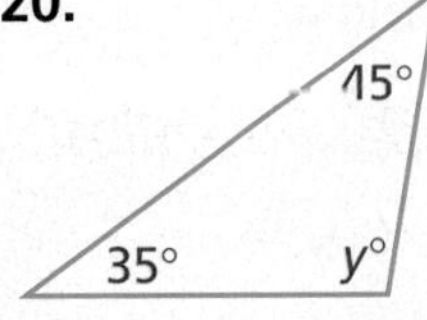

21. 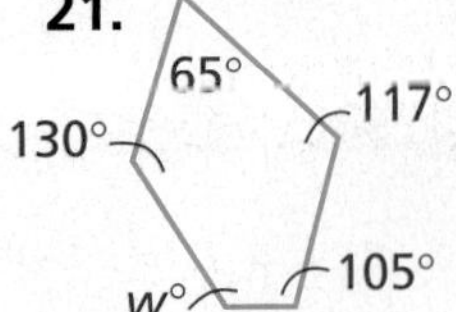

22. 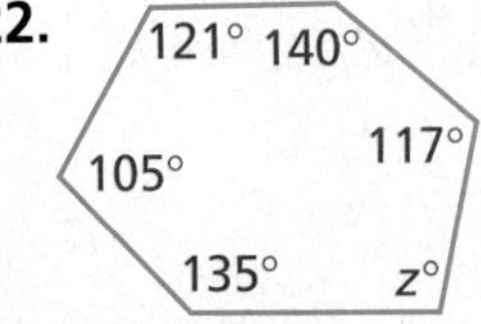

23. 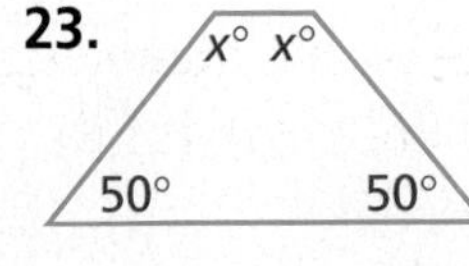

24. 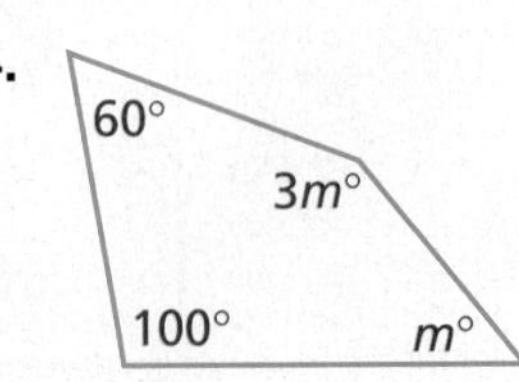

The sum of the angle measures of a polygon is given. Name the polygon.

25. 720° **26.** 360° **27.** 1980°

Graph the given vertices on a coordinate plane. Connect the points to draw a polygon and classify it by the number of its sides.

28. $A(1, 4)$, $B(2, 3)$, $C(4, 3)$, $D(5, 4)$, $E(4, 5)$, $F(2, 5)$

29. $A(-2, 1)$, $B(-2, -1)$, $C(1, -2)$, $D(3, 0)$, $E(1, 2)$

30. $A(3, 3)$, $B(5, 2)$, $C(5, 1)$, $D(3, -1)$, $E(-2, -1)$, $F(-3, 1)$, $G(-3, 2)$, $H(2, 3)$

Earth Science LINK

The master jeweler of Great Britain's crown jewels has thousands of diamonds in his care, including the world's two largest cut diamonds. The Imperial State Crown contains over 3000 precious stones, including 2800 diamonds.

Sketch a quadrilateral to fit each description. If no quadrilateral can be drawn, write *not possible.*

31. a parallelogram that is not a rectangle

32. a square that is not a rhombus

33. a quadrilateral that is not a trapezoid or a parallelogram

34. a rectangle that is not a square

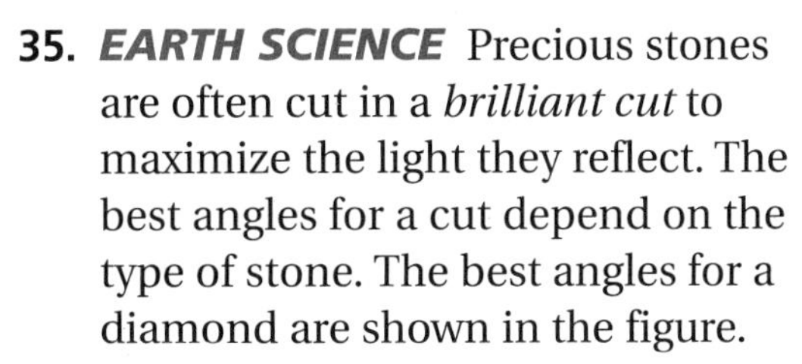

35. ***EARTH SCIENCE*** Precious stones are often cut in a *brilliant cut* to maximize the light they reflect. The best angles for a cut depend on the type of stone. The best angles for a diamond are shown in the figure.

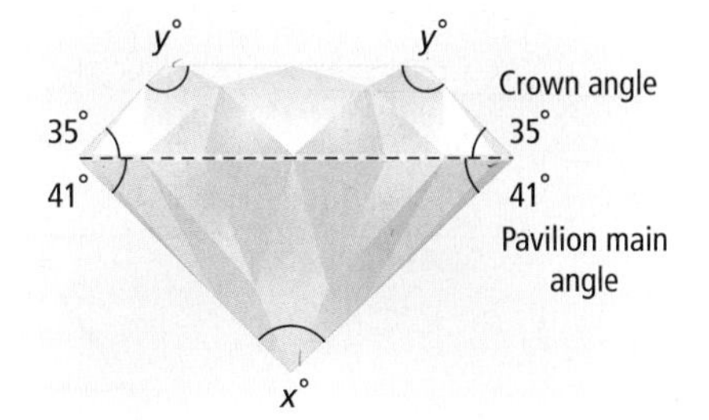

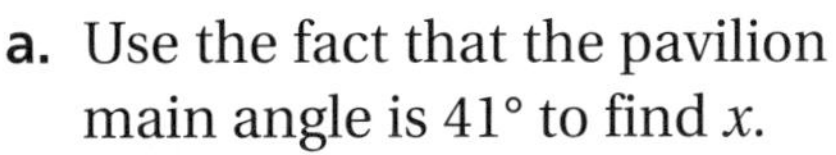

a. Use the fact that the pavilion main angle is 41° to find x.

b. Use the fact that the crown angle is 35° to find y.

36. ***WHAT'S THE ERROR?*** A student said that all squares are rectangles, but not all squares are rhombuses. What was the error?

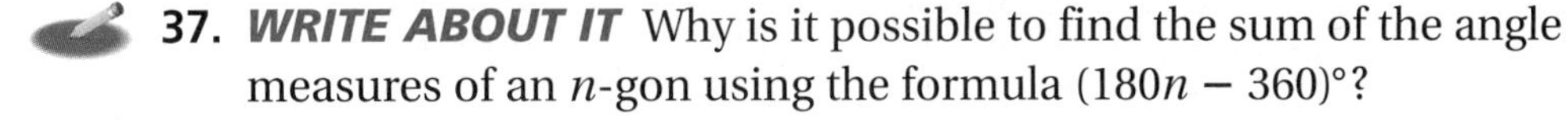

37. ***WRITE ABOUT IT*** Why is it possible to find the sum of the angle measures of an n-gon using the formula $(180n - 360)°$?

38. ***CHALLENGE*** Use properties of parallel lines to explain which angles in a parallelogram must be congruent.

Spiral Review

Write each number in scientific notation. (Lesson 2-9)

39. 0.00000064 **40.** 7,390,000,000 **41.** −0.0000016 **42.** −4,100,000

43. **EOG PREP** If the measure of one acute angle of a right triangle is 32°, then the measure of the other acute angle is ____. (Lesson 5-3)

A 32° B 48° C 58° D 148°

5-5 Coordinate Geometry

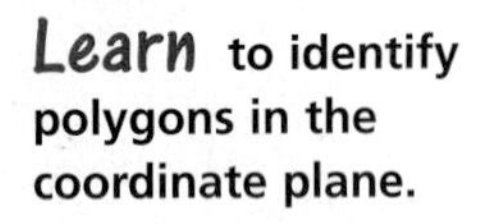

Learn to identify polygons in the coordinate plane.

Vocabulary

slope

rise

run

In computer graphics, a coordinate system is used to create images, from simple geometric figures to realistic figures used in movies.

Properties of the coordinate plane can be used to find information about figures in the plane, such as whether lines in the plane are parallel.

Slope is a number that describes how steep a line is.

$$\text{slope} = \frac{\text{vertical change}}{\text{horizontal change}} = \frac{\textbf{rise}}{\textbf{run}}$$

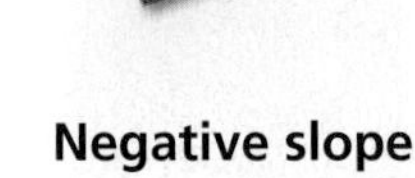

Positive slope

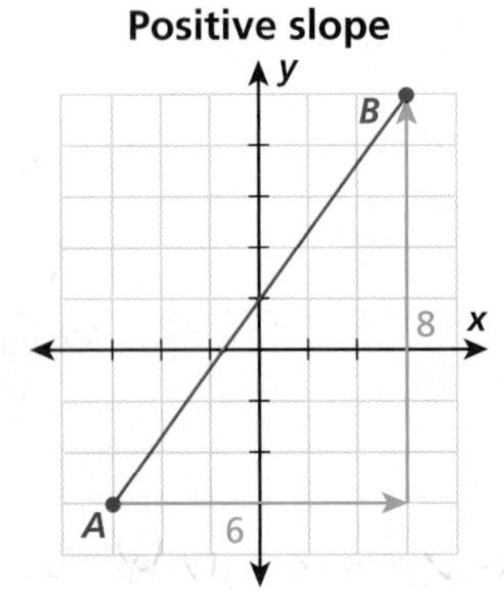

$$\text{slope of } \overline{AB} = \frac{8}{6} = \frac{4}{3}$$

$$\text{slope of } \overline{CD} = \frac{-3}{6} = \frac{-1}{2}$$

Negative slope

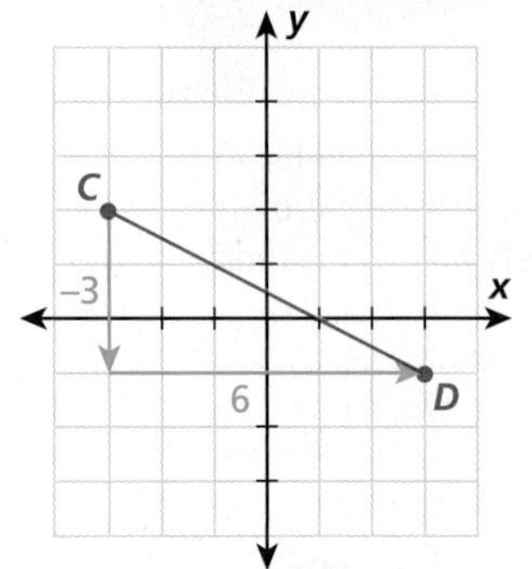

The slope of a horizontal line is 0. The slope of a vertical line is undefined.

EXAMPLE 1 Finding the Slope of a Line

Determine if the slope of each line is positive, negative, 0, or undefined. Then find the slope of each line.

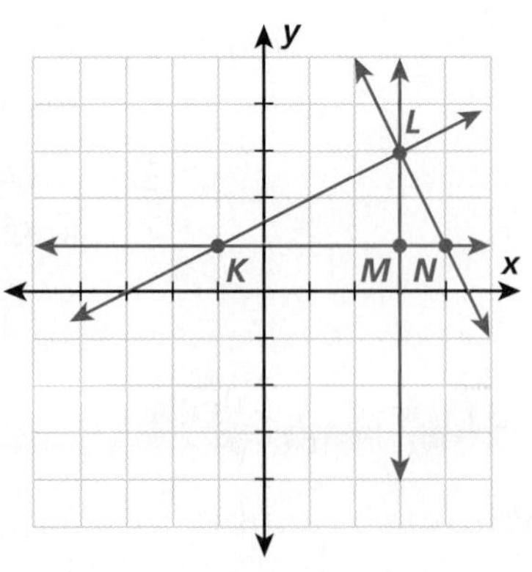

A $\overleftrightarrow{KL}$

positive slope; slope of $\overleftrightarrow{KL} = \frac{2}{4} = \frac{1}{2}$

B $\overleftrightarrow{LM}$

slope of $\overleftrightarrow{LM}$ is undefined

C $\overleftrightarrow{LN}$

negative slope; slope of $\overleftrightarrow{LN} = \frac{-2}{1} = -2$

D $\overleftrightarrow{KM}$

slope of $\overleftrightarrow{KM} = 0$

Slopes of Parallel and Perpendicular Lines
Two lines with equal slopes are parallel.
Two lines whose slopes have a product of -1 are perpendicular.

EXAMPLE 2 Finding Perpendicular and Parallel Lines

Which lines are parallel?
Which lines are perpendicular?

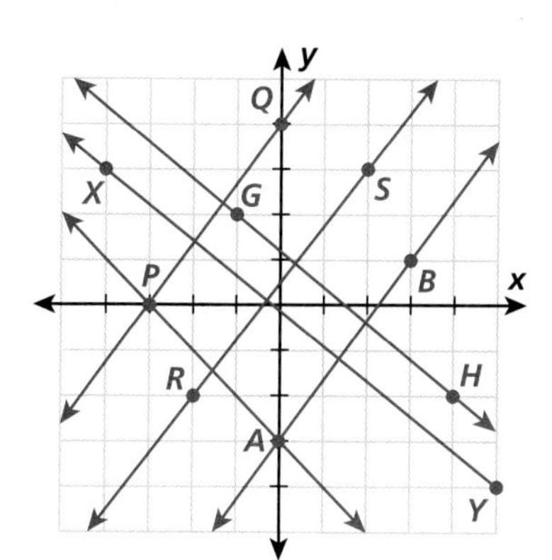

slope of $\overleftrightarrow{PQ} = \frac{4}{3}$

slope of $\overleftrightarrow{RS} = \frac{5}{4}$

slope of $\overleftrightarrow{AB} = \frac{4}{3}$

slope of $\overleftrightarrow{PA} = \frac{-3}{3}$ or -1

slope of $\overleftrightarrow{GH} = \frac{-4}{5}$

slope of $\overleftrightarrow{XY} = \frac{-7}{9}$

$\overleftrightarrow{PQ} \parallel \overleftrightarrow{AB}$ *The slopes are equal:* $\frac{4}{3} = \frac{4}{3}$

$\overleftrightarrow{RS} \perp \overleftrightarrow{GH}$ *The slopes have a product of* -1: $\frac{5}{4} \cdot \frac{-4}{5} = -1$

Helpful Hint

If a line has slope $\frac{a}{b}$, then a line perpendicular to it has slope $-\frac{b}{a}$.

EXAMPLE 3 Using Coordinates to Classify Quadrilaterals

Graph the quadrilaterals with the given vertices. Give all of the names that apply to each quadrilateral.

A $J(-6, 3), K(-2, 3), L(-2, -1), M(-6, -1)$

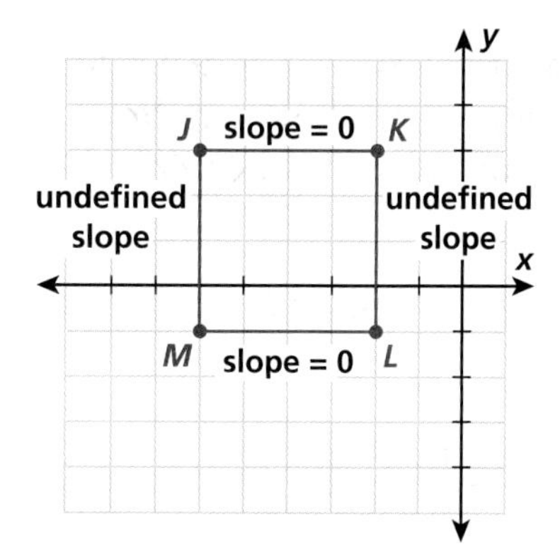

$\overleftrightarrow{JK} \parallel \overleftrightarrow{ML}$ and $\overleftrightarrow{MJ} \parallel \overleftrightarrow{LK}$
$\overleftrightarrow{JK} \perp \overleftrightarrow{LK}, \overleftrightarrow{JK} \perp \overleftrightarrow{MJ}$,
$\overleftrightarrow{ML} \perp \overleftrightarrow{LK}$ and $\overleftrightarrow{ML} \perp \overleftrightarrow{MJ}$
parallelogram, rectangle, square, rhombus

B $W(-1, 0), X(5, -4), Y(3, -7), Z(-3, -3)$

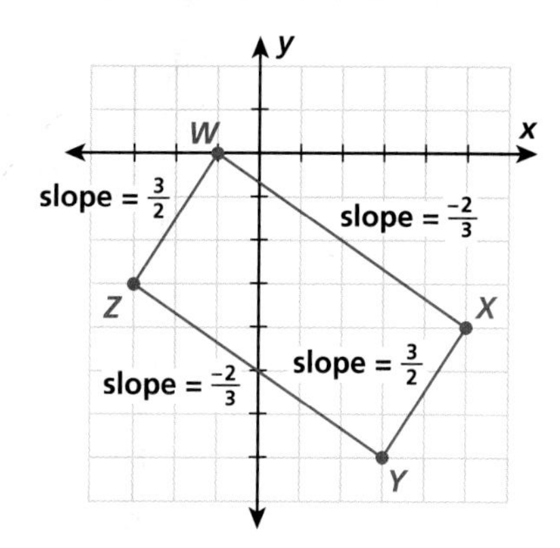

$\overleftrightarrow{WX} \parallel \overleftrightarrow{ZY}$ and $\overleftrightarrow{ZW} \parallel \overleftrightarrow{YX}$
$\overleftrightarrow{ZW} \perp \overleftrightarrow{WX}, \overleftrightarrow{ZW} \perp \overleftrightarrow{ZY}$,
$\overleftrightarrow{YX} \perp \overleftrightarrow{WX}$ and $\overleftrightarrow{YX} \perp \overleftrightarrow{ZY}$
parallelogram, rectangle

Graph the quadrilaterals with the given vertices. Give all of the names that apply to each quadrilateral.

C $E(-1, 6)$, $F(5, 6)$, $G(3, 4)$, $H(-3, 4)$

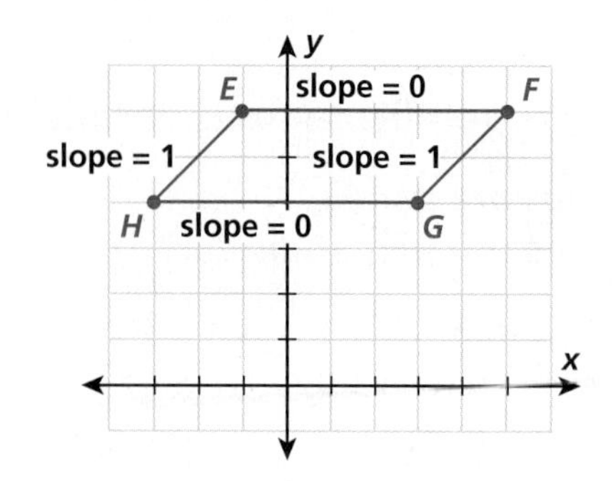

$\overleftrightarrow{EF} \parallel \overleftrightarrow{HG}$ and $\overleftrightarrow{HE} \parallel \overleftrightarrow{GF}$

parallelogram

D $P(4, 3)$, $Q(9, 2)$, $R(4, -3)$, $S(1, 0)$

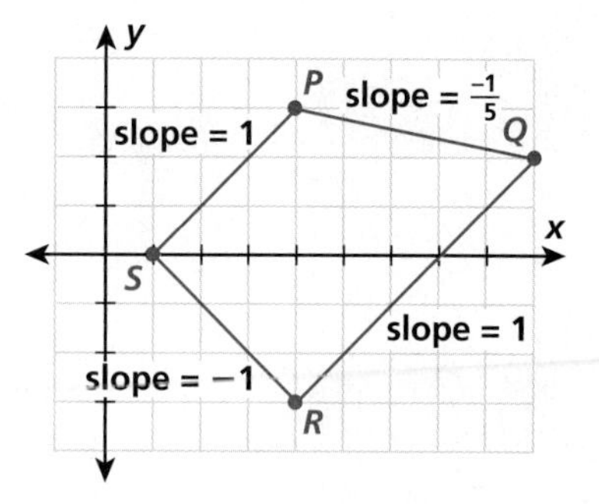

$\overleftrightarrow{SP} \parallel \overleftrightarrow{RQ}$

trapezoid

Think and Discuss

1. **Explain** why the slope of a horizontal line is 0.
2. **Explain** why the slope of a vertical line is undefined.

5-5 Exercises

FOR EOG PRACTICE

see page 661

internet connect

Homework Help Online

go.hrw.com Keyword: MT4 5-5

3.02, 5.01c

GUIDED PRACTICE

See Example 1

Determine if the slope of each line is positive, negative, 0, or undefined. Then find the slope of each line.

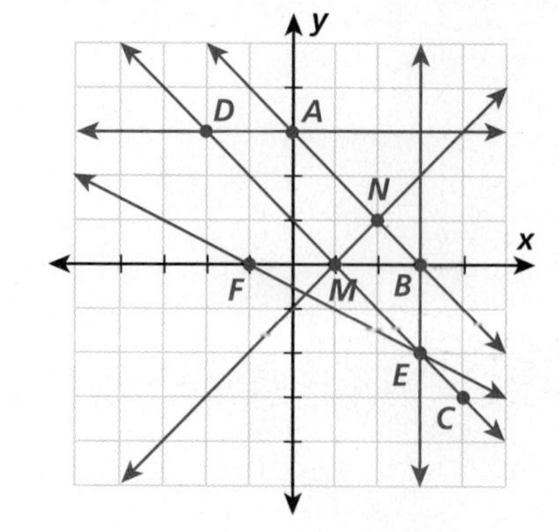

1. $\overleftrightarrow{AD}$
2. $\overleftrightarrow{BE}$
3. $\overleftrightarrow{MN}$
4. $\overleftrightarrow{EF}$

See Example 2

5. Which lines are parallel?
6. Which lines are perpendicular?

See Example 3

Graph the quadrilaterals with the given vertices. Give all of the names that apply to each quadrilateral.

7. $D(-3, -2)$, $E(-3, 3)$, $F(2, 3)$, $G(2, -2)$
8. $R(3, -2)$, $S(3, 1)$, $T(-3, 5)$, $V(-3, -2)$

INDEPENDENT PRACTICE

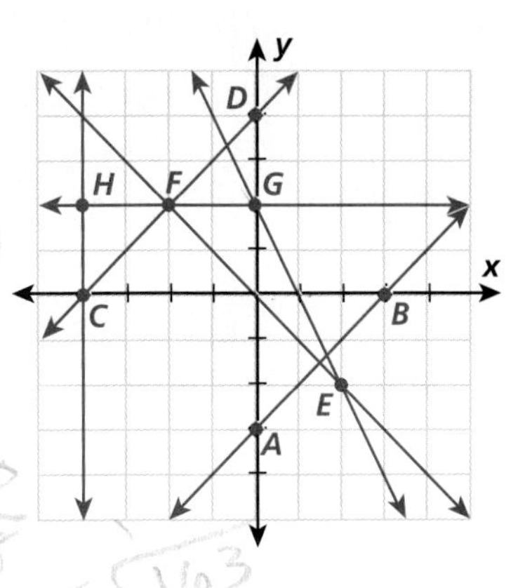

See Example 1

Determine if the slope of each line is positive, negative, 0, or undefined. Then find the slope of each line.

9. $\overleftrightarrow{AB}$
10. $\overleftrightarrow{EG}$
11. $\overleftrightarrow{HG}$
12. $\overleftrightarrow{CH}$

See Example 2

13. Which lines are parallel?
14. Which lines are perpendicular?

See Example 3

Graph the quadrilaterals with the given vertices. Give all of the names that apply to each quadrilateral.

15. $D(-3, 5)$, $E(3, 5)$, $F(3, -1)$, $G(-3, -1)$
16. $W(-2, 1)$, $X(-2, -2)$, $Y(4, 1)$, $Z(0, 2)$

PRACTICE AND PROBLEM SOLVING

Draw the line through the given points and find its slope.

17. $A(2, 1)$, $B(4, 7)$
18. $C(-2, 0)$, $D(-2, -5)$
19. $G(5, -4)$, $H(-2, -4)$
20. $E(-3, 1)$, $F(4, -2)$
21. On a coordinate grid draw a line s with slope 0 and a line t with slope 1. Then draw three lines through the intersection of lines s and t that have slopes between 0 and 1.
22. On a coordinate grid draw a line m with slope 0 and a line n with slope -1. Then draw three lines through the intersection of lines m and n that have slopes between 0 and -1.

23. ***WHAT'S THE ERROR?*** Points $P(3, 7)$, $Q(5, 2)$, $R(3, -3)$, and $S(1, 2)$ are vertices of a square. What is the error?

24. ***WRITE ABOUT IT*** Explain how using different points on a line to find the slope affects the answer.

25. ***CHALLENGE*** Use a square in a coordinate plane to explain why a line with slope 1 makes a 45° angle with the x-axis.

Spiral Review

The measures of two angles of a triangle are given. Find the measure of the third angle. (Lesson 5-3)

26. 45°, 45°
27. 30°, 60°
28. 21°, 82°
29. 105°, 42°

30. **EOG PREP** What is the value of $[(4 \cdot 5) - 5] \div 2$? (Previous course)

A 2 B 5 C 7.5 D 8

Chapter 5

Mid-Chapter Quiz

LESSON 5-1 (pp. 222–226)

Refer to the figure.

1. Name two pairs of complementary angles.
2. Name three pairs of supplementary angles.
3. Name two right angles.

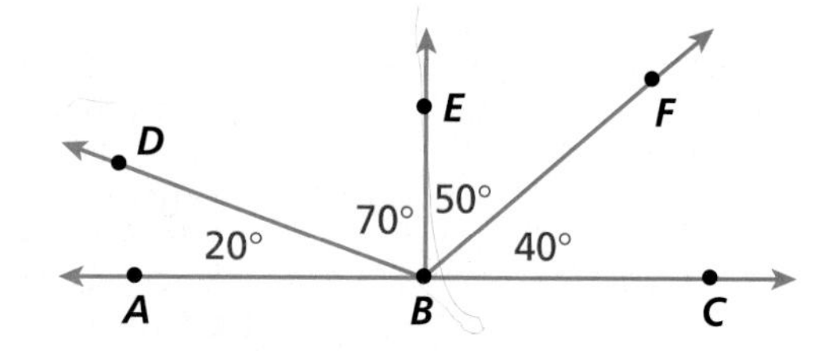

LESSON 5-2 (pp. 228–231)

In the figure, line $m \parallel$ line n. Find the measure of each angle.

4. $\angle 1$
5. $\angle 2$
6. $\angle 3$
7. $\angle 4$

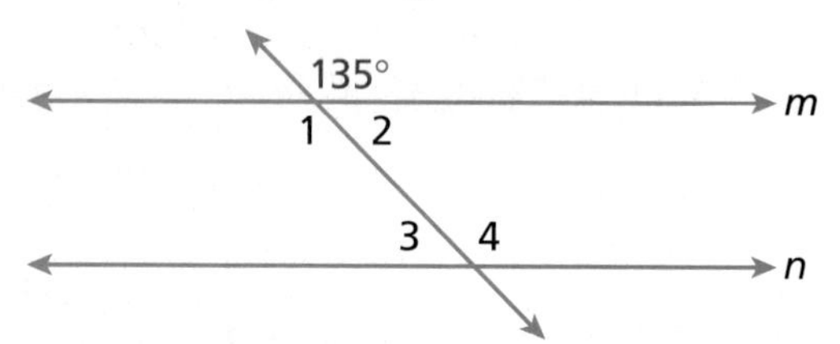

LESSON 5-3 (pp. 234–238)

Find x in each triangle.

8.

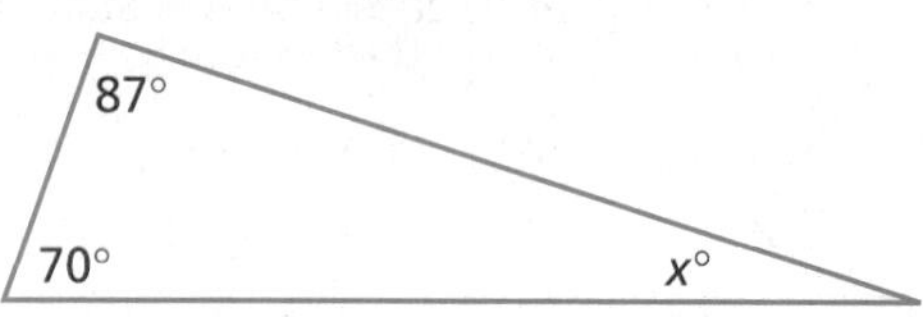

9.

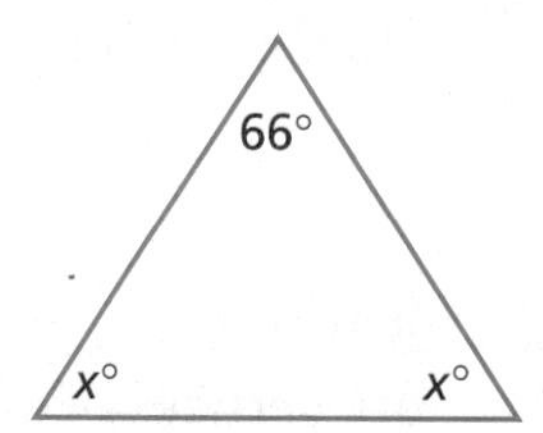

LESSON 5-4 (pp. 239–243)

Give all of the names that apply to each figure.

10.

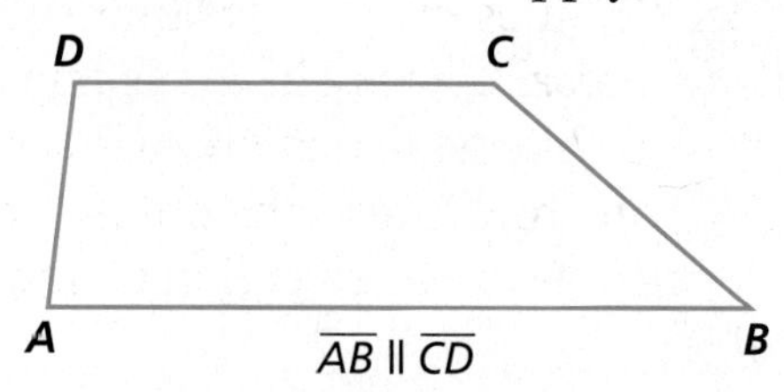

11.

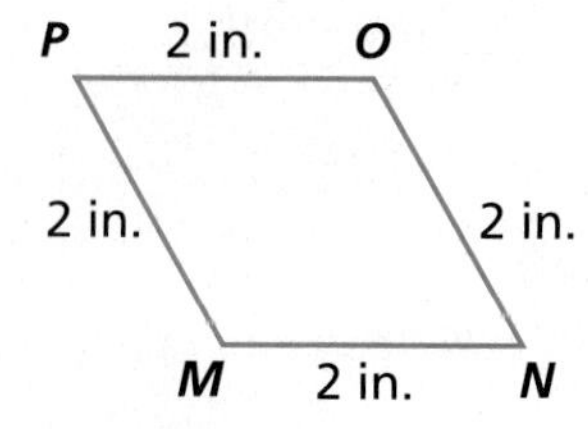

LESSON 5-5 (pp. 244–247)

Graph the quadrilaterals with the given vertices. Give all of the names that apply to each quadrilateral.

12. $A(-2, 1), B(3, 2), C(2, 0), D(-3, -1)$
13. $P(-4, 5), Q(3, 5), R(3, -2), S(-4, -2)$
14. $J(0, 2), K(4, 4), L(2, 1), M(0, 0)$
15. $U(4, 2), V(-2, 4), W(-3, 1), X(3, -1)$

Focus on Problem Solving

Understand the Problem

- **Restate the problem in your own words**

If you write a problem in your own words, you may understand it better. Before writing a problem in your own words, you may need to read it over several times—perhaps aloud, so you can hear yourself say the words.

Once you have written the problem in your own words, you may want to make sure you included all of the necessary information to solve the problem.

Write each problem in your own words. Check to make sure you have included all of the information needed to solve the problem.

1. In the figure, $\angle 1$ and $\angle 2$ are complementary, and $\angle 2$ and $\angle 3$ are supplementary. If $m\angle 2 = 50°$, find $m\angle 4 + m\angle 5$.

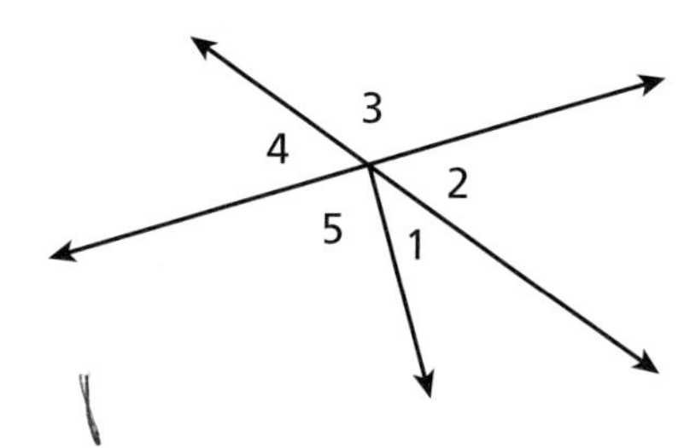

2. In triangle ABC, $m\angle A = 25°$ and $m\angle B = 65°$. Use the Triangle Sum Theorem to determine whether triangle ABC is a right triangle.

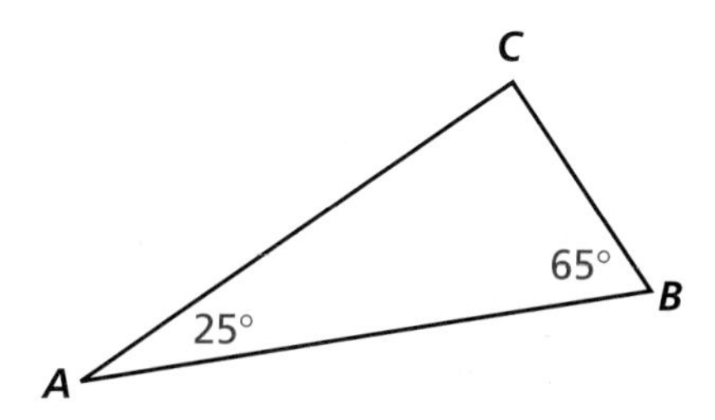

3. The second angle in a quadrilateral is six times as large as the first angle. The third angle is half as large as the second. The fourth angle is as large as the first angle and the third angle combined. Find the angle measures in the quadrilateral.

4. Parallel lines m and n are intersected by a transversal, line p. The acute angles formed by line m and line p measure 45°. Find the measure of the obtuse angles formed by the intersection of line n and line p.

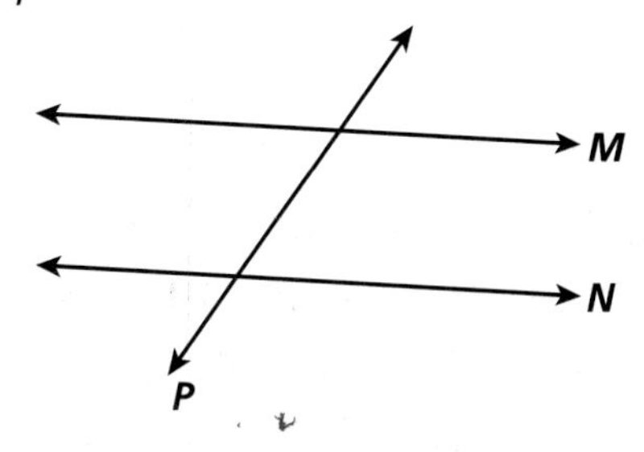

5-6 Congruence

Learn to use properties of congruent figures to solve problems.

Vocabulary

correspondence

Below are the DNA profiles of two pairs of twins. Twins A and B are identical twins. Twins C and D are fraternal twins.

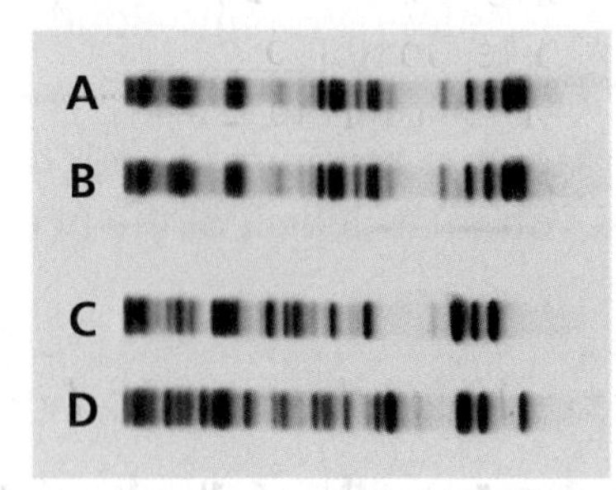

A **correspondence** is a way of matching up two sets of objects. The bands of DNA that are next to each other in each pair match up, or *correspond*. In the DNA of the identical twins, the corresponding bands are the same.

If two polygons are congruent, all of their corresponding sides and angles are congruent. In a congruence statement, the vertices in the second polygon are written in order of correspondence with the first polygon.

EXAMPLE  1 Writing Congruence Statements

Write a congruence statement for each pair of polygons.

 A

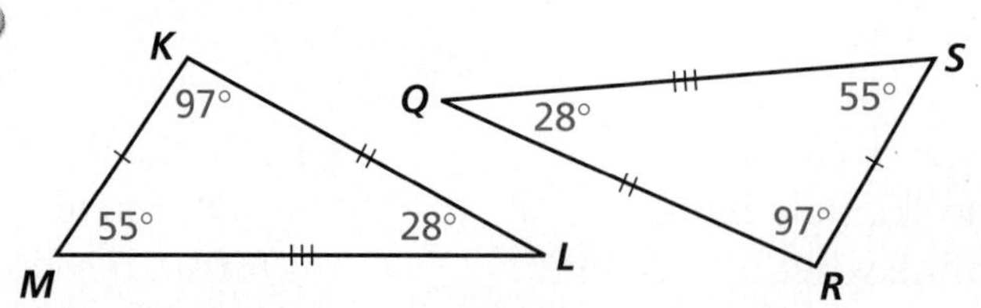

The first triangle can be named triangle *KLM*. To complete the congruence statement, the vertices in the second triangle have to be written in order of the correspondence.

$\angle K \cong \angle R$, so $\angle K$ corresponds to $\angle R$.
$\angle L \cong \angle Q$, so $\angle L$ corresponds to $\angle Q$.
$\angle M \cong \angle S$, so $\angle M$ corresponds to $\angle S$.

The congruence statement is triangle $KLM \cong$ triangle RQS.

Write a congruence statement for each pair of polygons.

B

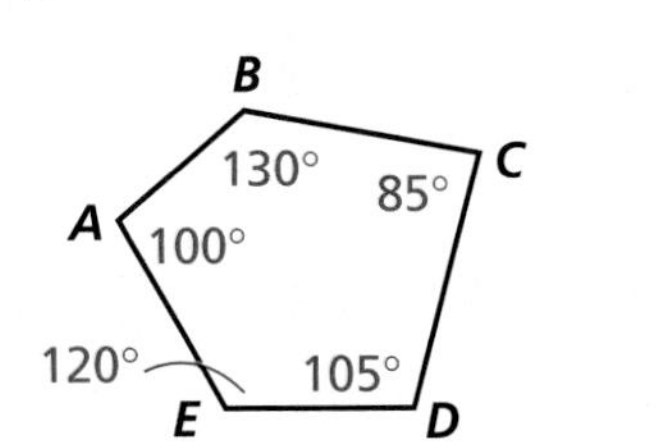

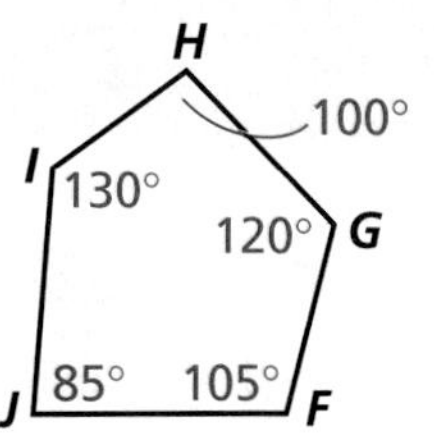

The vertices in the first pentagon are written in order around the pentagon starting at any vertex.

$\angle A \cong \angle H$, so $\angle A$ corresponds to $\angle H$.

$\angle B \cong \angle I$, so $\angle B$ corresponds to $\angle I$.

$\angle C \cong \angle J$, so $\angle C$ corresponds to $\angle J$.

$\angle D \cong \angle F$, so $\angle D$ corresponds to $\angle F$.

$\angle E \cong \angle G$, so $\angle E$ corresponds to $\angle G$.

The congruence statement is pentagon $ABCDE \cong$ pentagon $HIJFG$.

EXAMPLE 2 Using Congruence Relationships to Find Unknown Values

In the figure, quadrilateral *PQSR* ≅ quadrilateral *WTUV*.

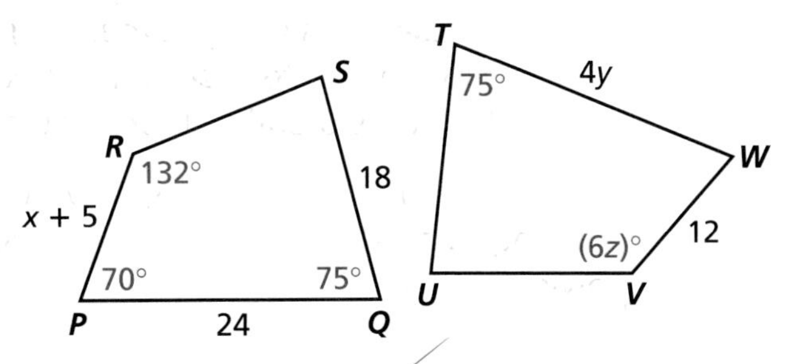

A Find *x*.

$$x + 5 = 12 \qquad \overline{PR} \cong \overline{WV}$$

$$\underline{-5 = -5} \qquad \textit{Subtract 5 from both sides.}$$

$$x = 7$$

B Find *y*.

$$4y = 24 \qquad \overline{PQ} \cong \overline{WT}$$

$$\frac{4y}{4} = \frac{24}{4} \qquad \textit{Divide both sides by 4.}$$

$$y = 6$$

C Find *z*.

$$6z = 132 \qquad \angle R \cong \angle V$$

$$\frac{6z}{6} = \frac{132}{6} \qquad \textit{Divide both sides by 6.}$$

$$z = 22$$

Think and Discuss

1. Explain what it means for two polygons to be congruent.

2. Tell how to write a congruence statement for two polygons.

5-6 Exercises

3.02

FOR EOG PRACTICE
see page 662

internet connect
Homework Help Online
go.hrw.com Keyword: MT4 5-6

GUIDED PRACTICE

See Example 1

Write a congruence statement for each pair of polygons.

1. 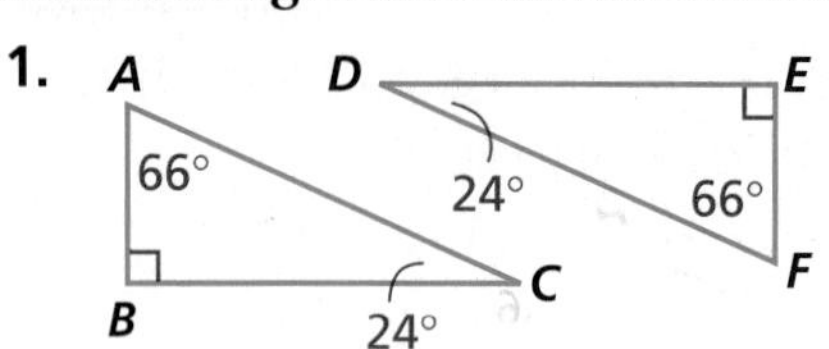

2. 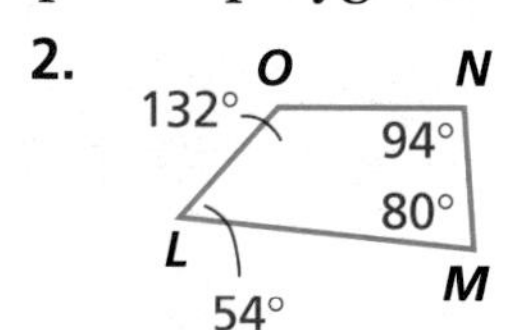

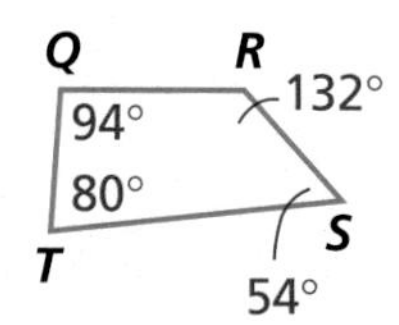

See Example 2

In the figure, triangle $ABC \cong$ triangle LMN.

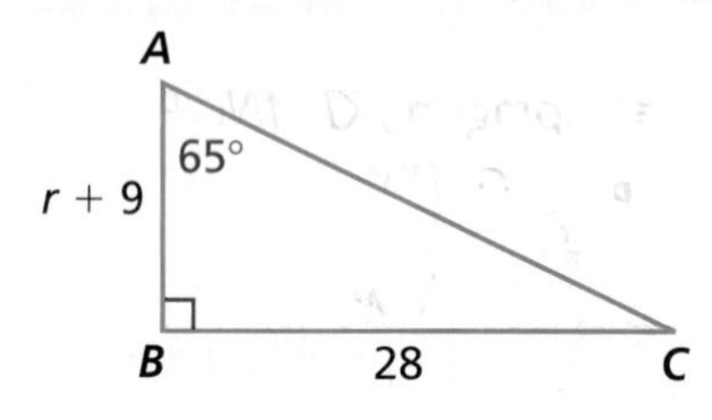

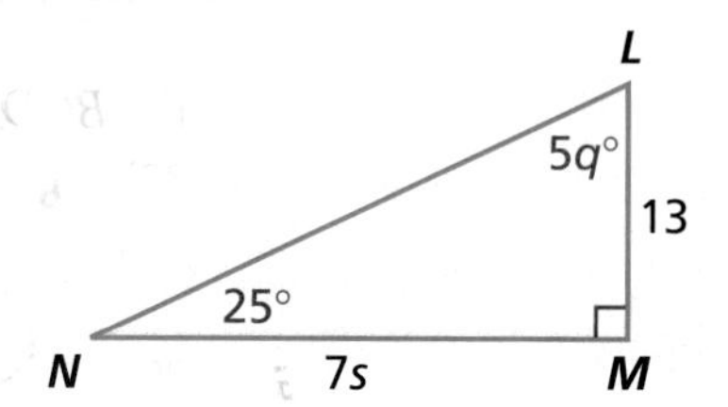

3. Find q.
4. Find r.
5. Find s.

INDEPENDENT PRACTICE

See Example 1

Write a congruence statement for each pair of poygons.

6. 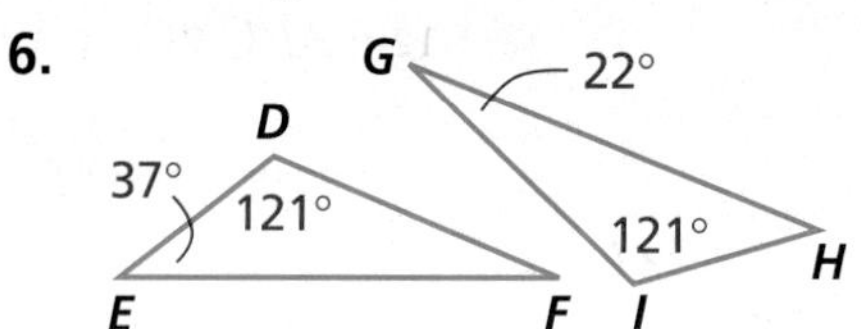

7. 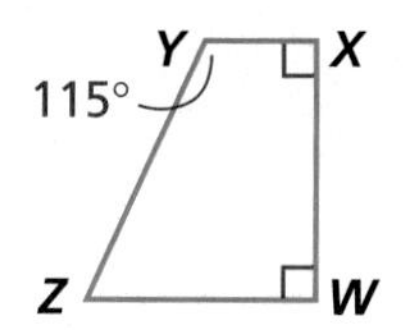

P, Q, R, S: 65°

See Example 2

In the figure, quadrilateral $ABCD \cong$ quadrilateral $LMNO$.

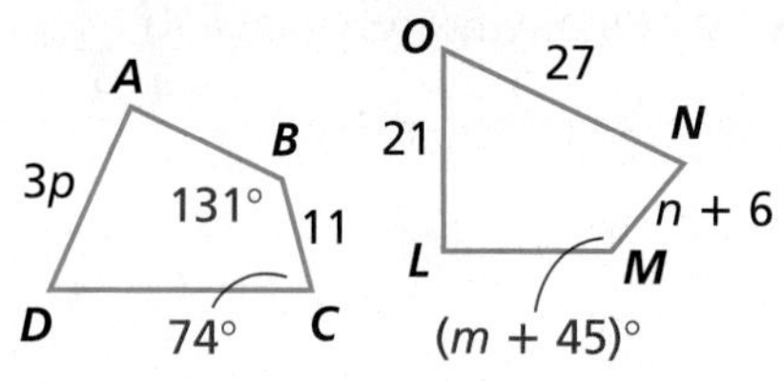

8. Find m.
9. Find n.
10. Find p.

PRACTICE AND PROBLEM SOLVING

Find the value of each variable.

11. pentagon $ABCDE \cong$ pentagon $PQRST$

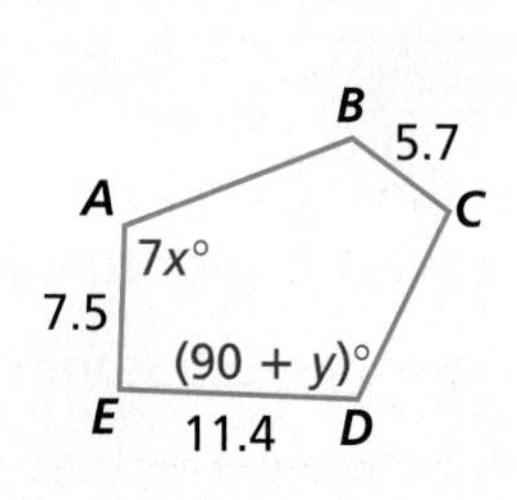

R, Q, S, 115°, $z - 2.8$, 112°, P, T

12. hexagon $ABCDEF \cong$ hexagon $LMNOPQ$

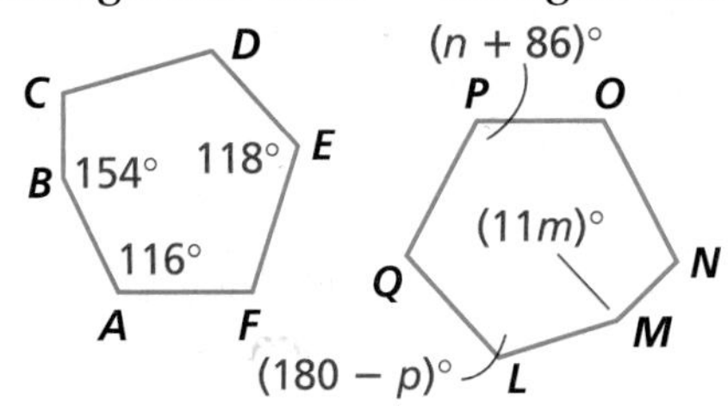

13. quadrilateral $ABCD \cong$ quadrilateral $EFGH$

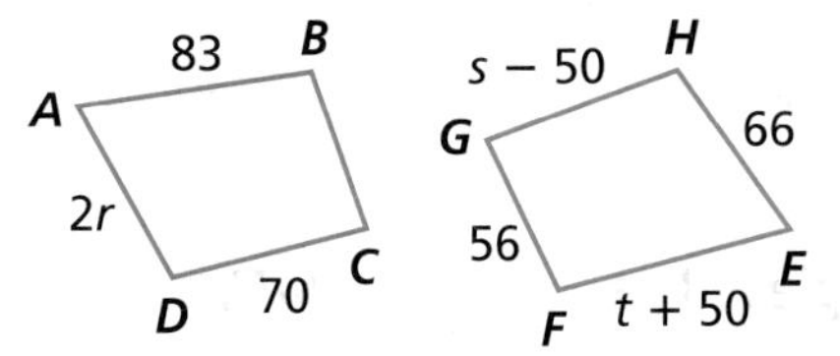

14. heptagon $ABCDEFG \cong$ heptagon $JKLMNOP$

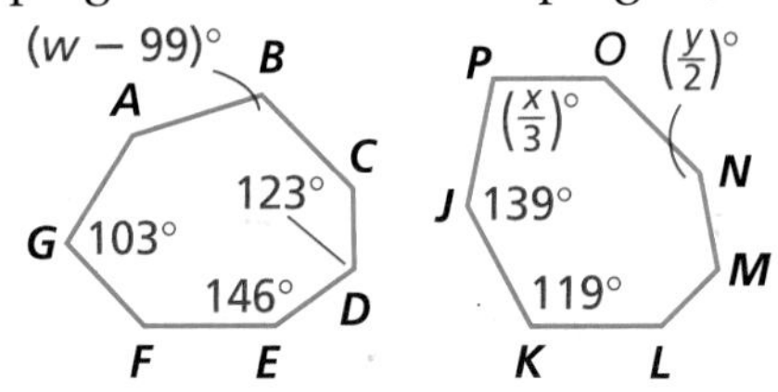

15. ***WHAT'S THE ERROR?*** Explain the error in this congruence statement and write a correct congruence statement.

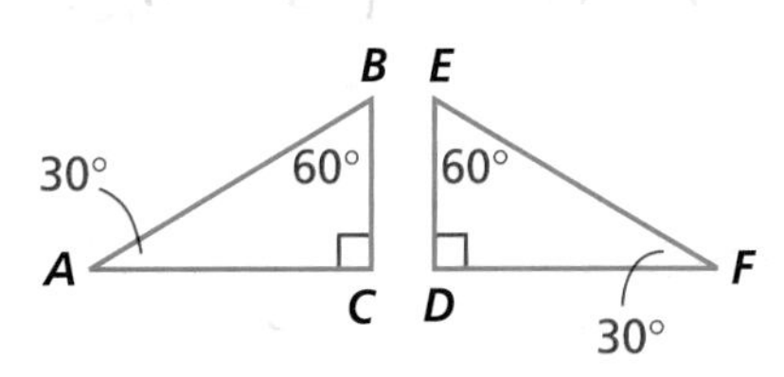

triangle $ABC \cong$ triangle DEF

16. ***WRITE ABOUT IT*** How can knowing two polygons are congruent help you find angle measures of the polygons?

17. ***CHALLENGE*** Triangle $ABC \cong$ triangle LMN and $\overline{AE} \parallel \overline{BD}$. Find $m\angle ACD$.

Spiral Review

Solve. (Lesson 2-4)

18. $\frac{m}{-3} = 4$ **19.** $64 = 4x$ **20.** $\frac{x}{-6} = -2$ **21.** $-60 = 4m$

22. $21 = 6p$ **23.** $\frac{b}{3} = -2$ **24.** $-95 = 19y$ **25.** $\frac{a}{4} = -8$

26. **EOG PREP** What are the angle measures of a triangle whose first angle is less than 90°, second angle is $\frac{3}{4}$ as large as the first angle, and third angle is $\frac{2}{3}$ as large as the second angle? (Lesson 5-3)

A 60°, 45°, 75° B 75°, 60°, 45° C 75°, 50°, 35° D 80°, 60°, 40°

5-7 Transformations

Learn to transform plane figures using translations, rotations, and reflections.

Vocabulary

transformation

translation

rotation

center of rotation

reflection

image

When you are on an amusement park ride, you are undergoing a *transformation.* Ferris wheels and merry-go-rounds are *rotations.* Free-fall rides and water slides are *translations.* Translations, rotations, and reflections are types of **transformations** .

Translation	Rotation	Reflection
A **translation** slides a figure along a line without turning.	A **rotation** turns the figure around a point, called the **center of rotation** .	A **reflection** flips the figure across a line to create a mirror image.

The resulting figure, or **image** , of a translation, rotation, or reflection is congruent to the original figure.

EXAMPLE 1 Identifying Transformations

Identify each as a translation, rotation, reflection, or none of these.

Reading Math

A' is read "A prime." The point A' is the image of point A.

A

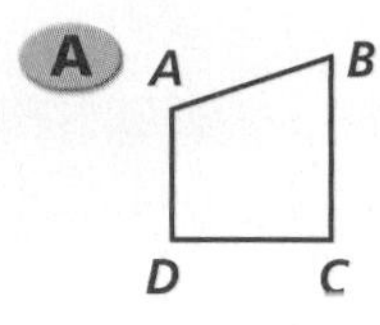

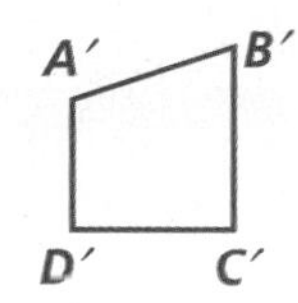

translation

B

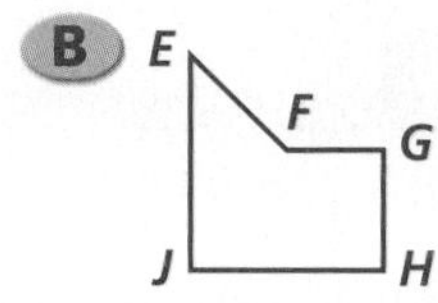

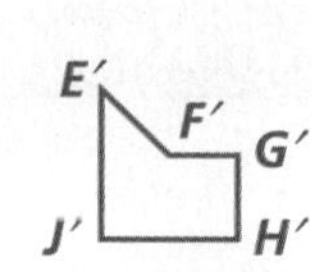

none of these

C

rotation

D

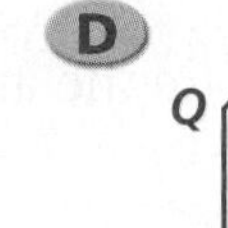

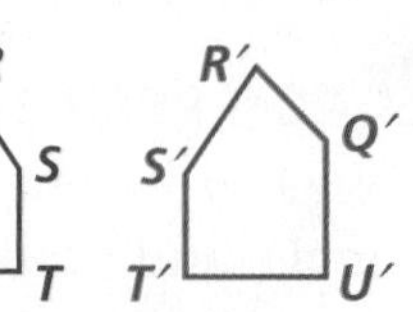

reflection

Art 

M. C. Escher created works of art by repeating interlocking shapes. He used both regular and nonregular tessellations. He often used what he called *metamorphoses*, in which shapes change into other shapes. Escher used his reptile pattern in many hexagonal tessellations. One of the most famous is entitled simply *Reptiles*.

14. The steps below show the method Escher used to make a bird out of a triangle. Use the bird to create a tessellation.

Step 1

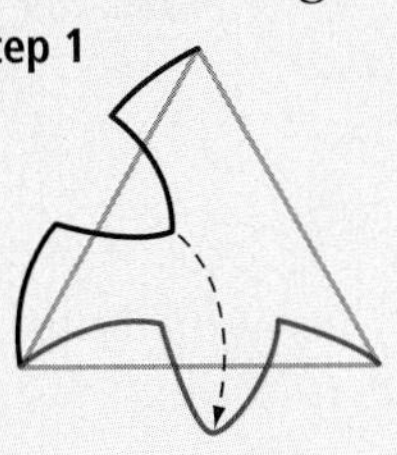

Step 2

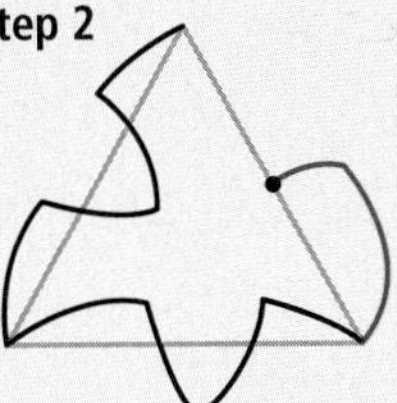

Step 3

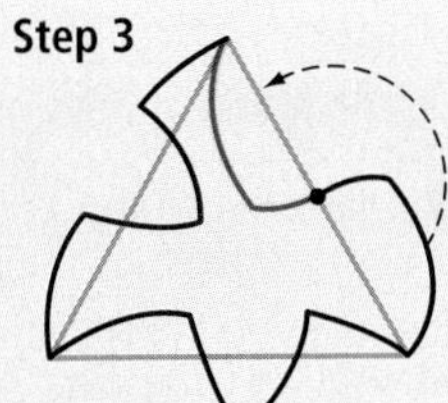

Step 4

go.hrw.com
KEYWORD: MT4 ESCHER
CNN Student News

Refer to the sketch for *Reptiles* for Exercises 15–16.

15. What regular polygon do you think Escher used to begin the sketch?

16. Describe the process he used to create each figure from the basic shape.

17. ***CHALLENGE*** Create an Escher-like tessellation of your own design.

Spiral Review

Solve and graph each inequality. (Lesson 1-5)

18. $y + 4 > 1$ **19.** $4p \leq 12$ **20.** $f - 3 \geq 2$ **21.** $4 < \frac{w}{3}$

22. $p - 1 \geq 4$ **23.** $m + 3 \leq 3$ **24.** $3 > \frac{n}{2}$ **25.** $3z < 6$

26. **EOG PREP** Which word phrase represents the expression $8 - 6p$? (Lesson 1-2)

A Eight less than six times a number

B Eight minus six, times a number

C Six times a number minus eight

D Six times a number, subtracted from eight

Problem Solving on Location

NORTH CAROLINA

Charlotte Trolley

As early as 1891, Charlotte, North Carolina, had an electric streetcar system. By 1938, gasoline-powered buses came into use, and the streetcars were retired. Then in 1998, the Charlotte City Council voted to use city funds to build a 2 mile track and put the vintage trolleys back on-line.

For 1–5, use the diagram of the trolley. Triangles *ABC* and *GJH* are congruent.

1. Write congruence statements for the angles of $\triangle ABC$ and $\triangle GJH$.

2. Assume $\triangle ABC$ is an isosceles triangle and that $\angle ACB \cong \angle BAC$. Write a congruence statement for two of the sides of $\triangle ABC$.

3. Suppose that $\angle ACB$ and $\angle BAC$ are not congruent, $m\angle BAC$ is 73°, and $m\angle ACB$ is 49°. Find $m\angle DBE$, $m\angle CBE$, and $m\angle ABD$.

4. Explain how increasing the measure of $\angle ABC$ by 15° will affect the measures of $\angle DBE$, $\angle CBE$, and $\angle ABD$.

5. Describe the transformation of $\triangle ABC$ that would result in $\triangle GJH$.

Raleigh-Durham Airport

The Raleigh-Durham Airport is the largest international airport in North Carolina. In September 2002, installation began for a six-mural art project at the airport. The central panel of each mural shows a sample landscape scene from one of the six regions of North Carolina. The side panels depict the plant and animal life that can be found in that region.

This tile mural was installed for Raleigh-Durham Airport's mural-art project.

For 1–3, use the photo of the tile mural.

1. How many sides are there on the central panel of the tile mural? What type of polygon is the central panel?

2. Is the central panel of the tile mural a regular polygon? Explain.

3. From left to right, identify each polygon in the top row of the tile mural's central panel.

The runways at Raleigh-Durham Airport are parallel to each other. The diagram at right shows two parallel runways and the road that connects them. For 4–7, use the diagram.

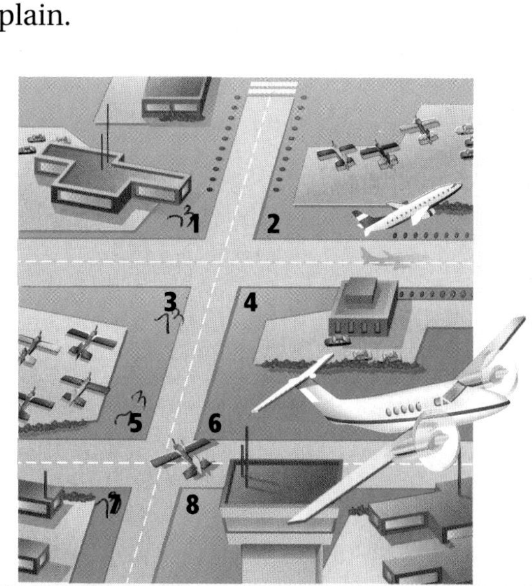

4. Which vocabulary term from this chapter describes the connecting road on the diagram?

5. Identify congruent angles in the figure.

6. If the measure of $\angle 1$ is 107°, give the measures of $\angle 2$, $\angle 3$, $\angle 4$, $\angle 5$, $\angle 6$, $\angle 7$, and $\angle 8$.

7. What is the relationship between $\angle 4$ and $\angle 7$?

MATH-ABLES

Coloring Tessellations

Two of the three regular tessellations—triangles and squares—can be colored with two colors so that no two polygons that share an edge are the same color. The third—hexagons—requires three colors.

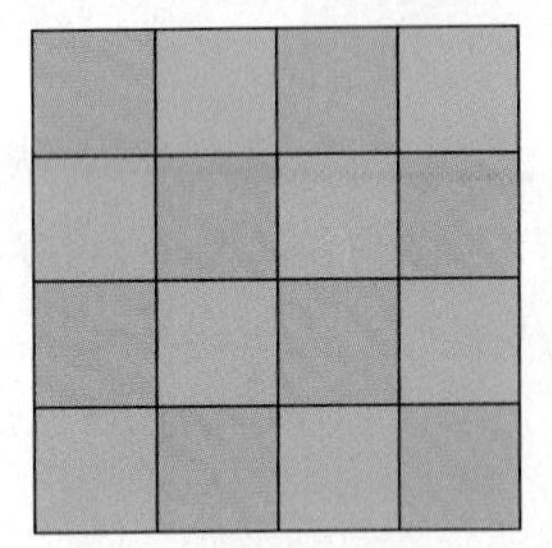

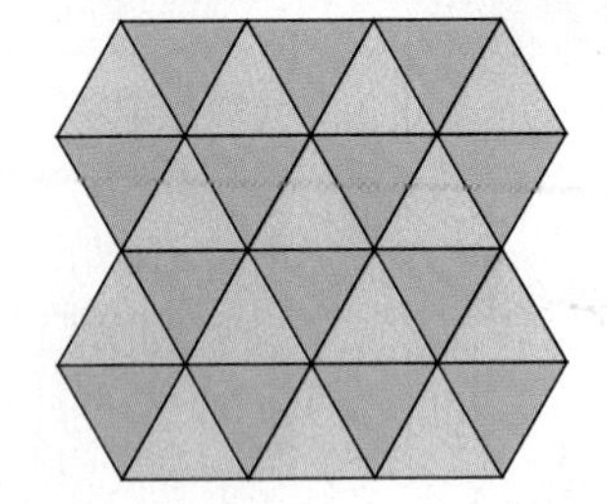

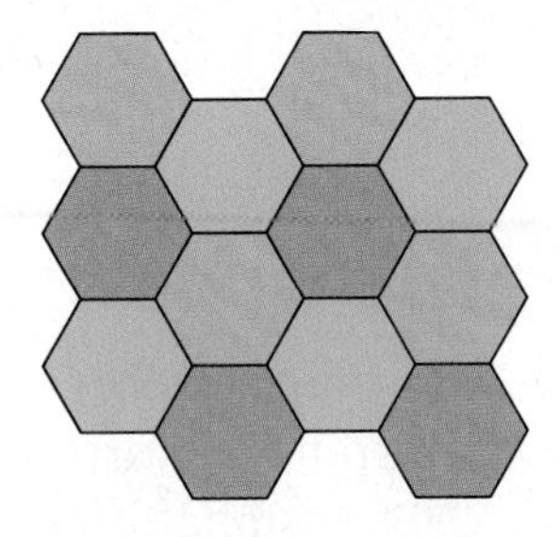

1. Determine if each semiregular tessellation can be colored with two colors. If not, tell the minimum number of colors needed.

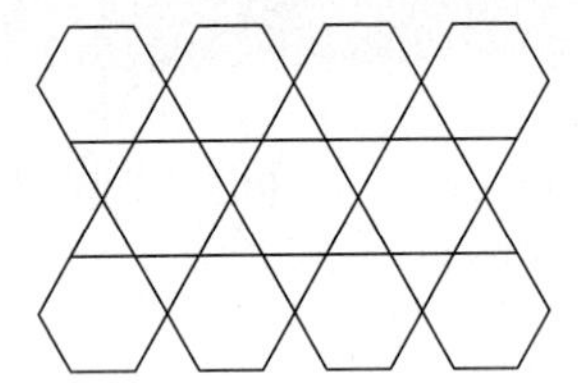

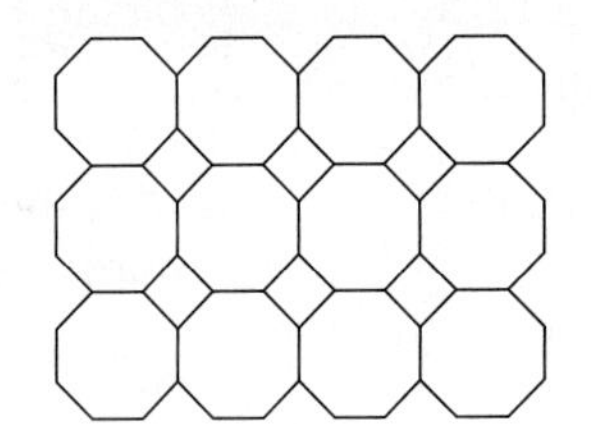

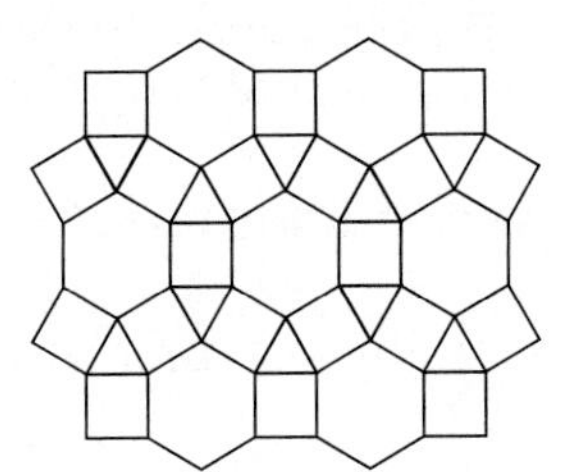

2. Try to write a rule about which tessellations can be colored with two colors.

Polygon Rummy

The object of this game is to create geometric figures. Each card in the deck shows a property of a geometric figure. To create a figure, you must draw a polygon that matches at least three cards in your hand. For example, if you have the cards "quadrilateral," "a pair of parallel sides," and "a right angle," you could draw a rectangle.

internet connect

Go to ***go.hrw.com*** for a complete set of rules and game cards.
KEYWORD: MT4 Game5

Exterior Angles of a Polygon

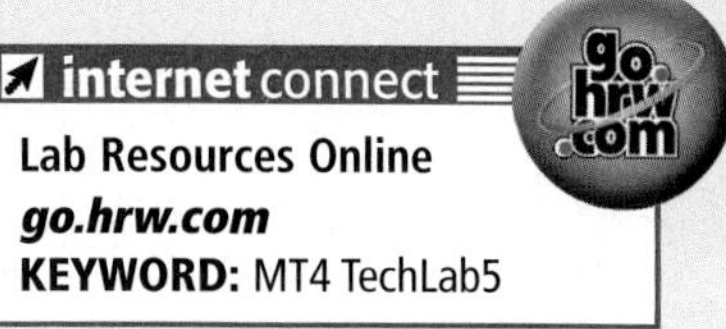

The **exterior angles** of a polygon are formed by extending the polygon's sides. Every exterior angle is supplementary to the angle next to it inside the polygon.

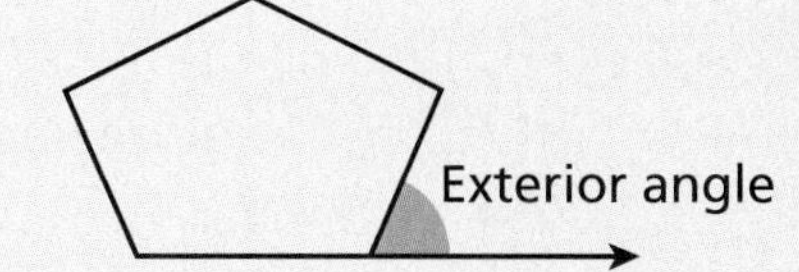

Activity

1 Follow the steps to find the sum of the exterior angle measures for a polygon.

a. Use geometry software to make a pentagon. Label the vertices A through E.

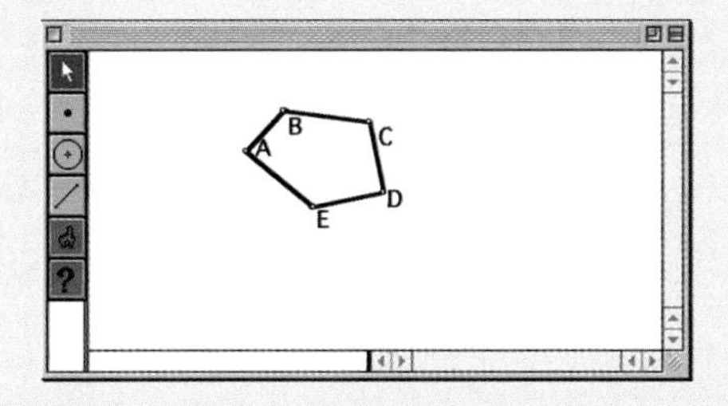

b. Use the **LINE-RAY** tool to extend the sides of the pentagon. Add points F through J as shown.

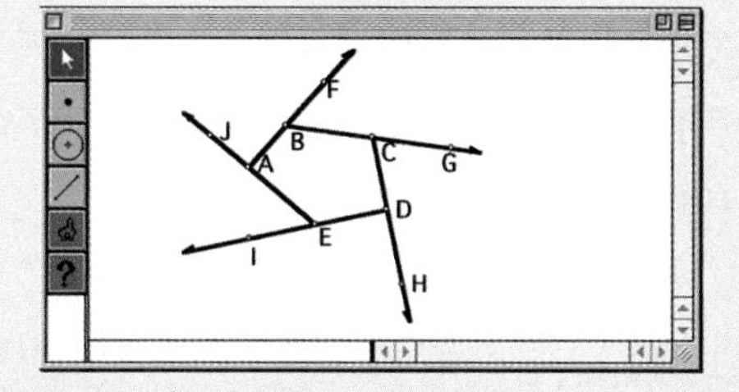

c. Use the **ANGLE MEASURE** tool to measure each exterior angle and the **CALCULATOR** tool to add the measures. Notice the sum.

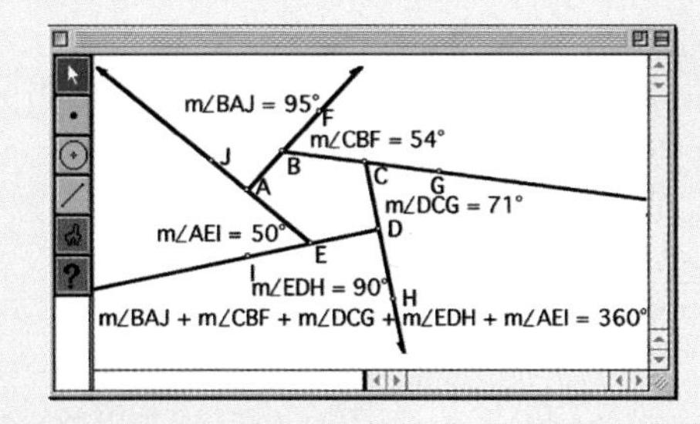

d. Drag vertices A through E and watch the sum. Notice that the sum of the angle measures is *always* 360°.

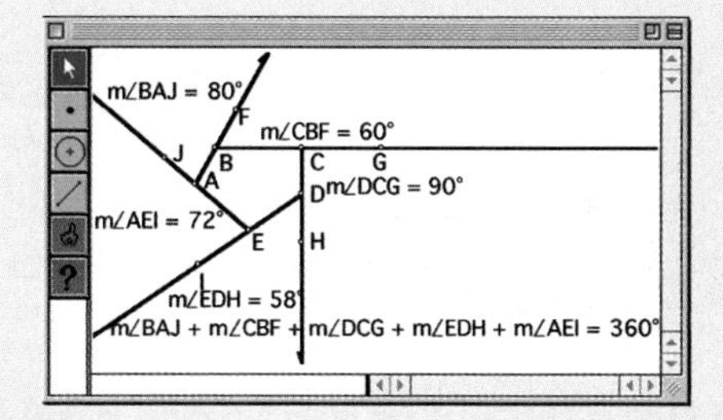

Think and Discuss

1. Suppose you were to drag the vertices of a polygon so that the polygon almost vanishes. How would this show that the sum of the exterior angle measures is 360°?

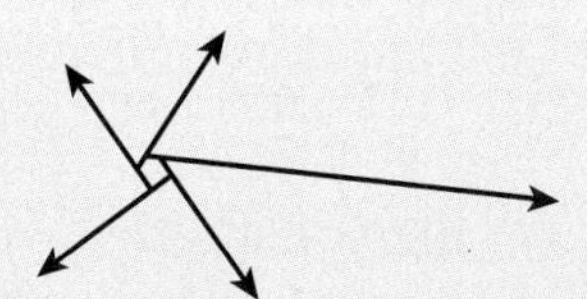

Try This

1. Use geometry software to draw a quadrilateral. Find the sum of its exterior angle measures. Drag its vertices to check that the sum is always the same.

Chapter 5

Study Guide and Review

Vocabulary

acute angle	223	parallelogram	240	run	244
acute triangle	234	perpendicular lines	228	scalene triangle	235
angle	222	plane	222	segment	222
center of rotation	254	point	222	semiregular tessellation	263
complementary angles	223	polygon	239	slope	244
congruent	223	ray	222	square	240
correspondence	250	rectangle	240	supplementary angles	223
equilateral triangle	235	reflection	254	tessellation	263
image	254	regular polygon	240	transformation	254
isosceles triangle	235	regular tessellation	263	translation	254
line	222	rhombus	240	transversal	228
line of symmetry	259	right angle	223	trapezoid	240
line symmetry	259	right triangle	234	Triangle Sum Theorem	234
obtuse angle	223	rise	244	vertical angles	223
obtuse triangle	234	rotation	254		
parallel lines	228	rotational symmetry	260		

Complete the sentences below with vocabulary words from the list above. Words may be used more than once.

1. Lines in the same plane that never meet are called ___?___. Lines that intersect at 90° angles are called ___?___.

2. A quadrilateral with 4 congruent angles is called a ___?___. A quadrilateral with 4 congruent sides is called a ___?___.

5-1 Points, Lines, Planes, and Angles (pp. 222–226)

EXAMPLE

■ **Find the angle measure.**

$m\angle 1$

$$\begin{aligned} m\angle 1 + 122^\circ &= 180^\circ \\ -122^\circ &\quad -122^\circ \\ m\angle 1 &= 58^\circ \end{aligned}$$

EXERCISES

Find each angle measure.

3. $m\angle 1$
4. $m\angle 2$
5. $m\angle 3$

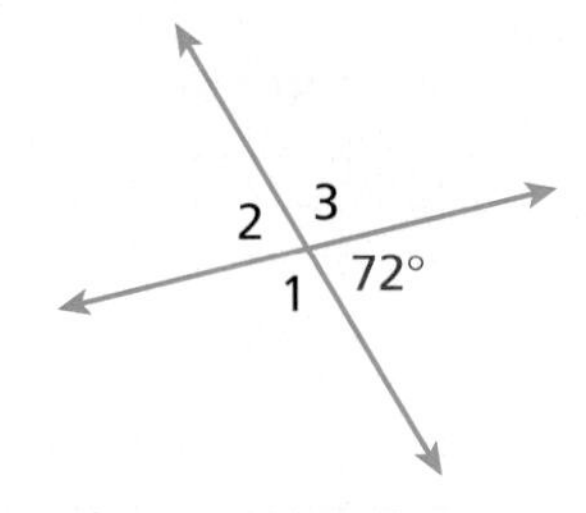

5-2 Parallel and Perpendicular Lines (pp. 228–231)

EXAMPLE

Line $j \parallel$ line k. Find each angle measure.

- m∠1

  m∠1 = 143°

- m∠2

$$\begin{aligned} m\angle 2 + 143^\circ &= 180^\circ \\ -143^\circ &\quad -143^\circ \\ m\angle 2 &= 37^\circ \end{aligned}$$

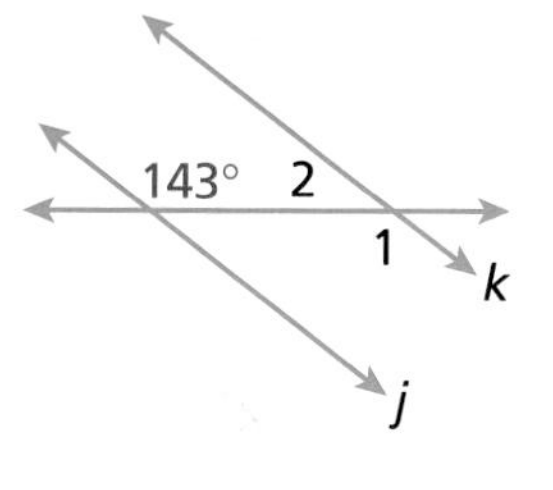

EXERCISES

Line $p \parallel$ line q. Find each angle measure.

6. m∠1

7. m∠2

8. m∠3

9. m∠4

10. m∠5

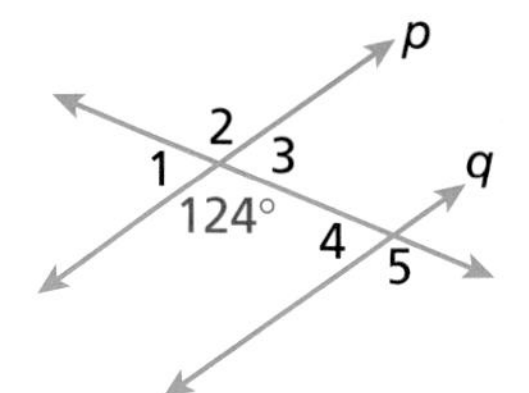

5-3 Triangles (pp. 234–238)

EXAMPLE

- **Find n.**

$$\begin{aligned} n^\circ + 50^\circ + 90^\circ &= 180^\circ \\ n^\circ + 140^\circ &= 180^\circ \\ -140^\circ &\quad -140^\circ \\ n^\circ &= 40^\circ \end{aligned}$$

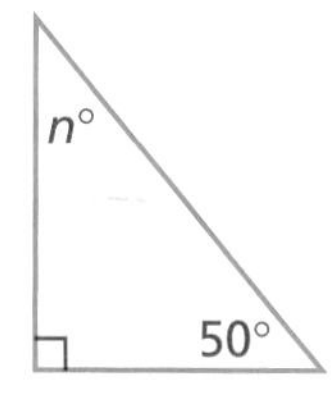

EXERCISES

11. Find m°.

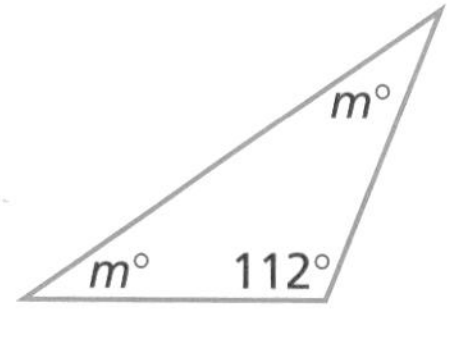

5-4 Polygons (pp. 239–243)

EXAMPLE

- **Find the sum of the angle measures in a regular 12-gon.**

  sum of angle measures = 180°(n − 2)
  = 180°(12 − 2)
  = 180°(10) = 1800°

EXERCISES

Find the angle measures in each regular polygon.

12. a regular hexagon

13. a regular 10-gon

5-5 Coordinate Geometry (pp. 244–247)

EXAMPLE

- **Graph the quadrilateral with the given vertices. Give all the names that apply.**

  $D(-2, 1)$, $E(2, 3)$, $F(3, 1)$, $G(-1, -1)$

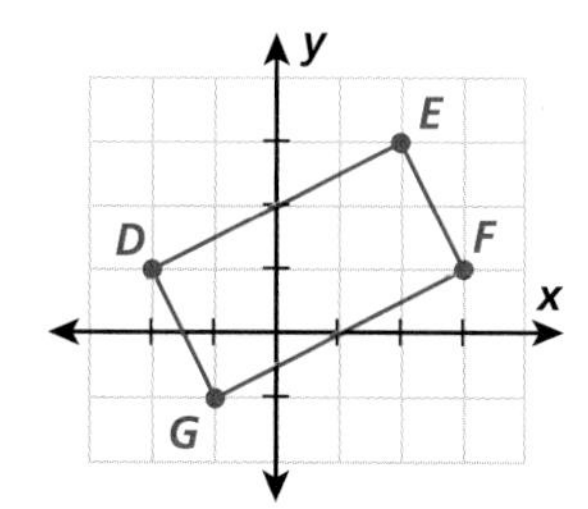

$\overline{DE} \parallel \overline{FG}$
$\overline{EF} \parallel \overline{GD}$
$\overline{DE} \perp \overline{EF}$

quadrilateral, parallelogram, rectangle

EXERCISES

Graph the quadrilaterals with the given vertices. Give all the names that apply.

14. $Q(2, 0)$, $R(-1, 1)$, $S(3, 3)$, $T(8, 3)$

15. $K(0, 3)$, $L(1, 0)$, $M(0, -3)$, $N(-1, 0)$

16. $W(2, 3)$, $X(2, -2)$, $Y(-1, -3)$, $Z(-1, 2)$

5-6 Congruence (pp. 250–253)

EXAMPLE

- **Triangle $ABC \cong$ triangle FDE. Find x.**

$\overline{AC} \cong \overline{FE}$

$$\begin{aligned} x - 4 &= 4 \\ +4 \quad & \quad +4 \\ x &= 8 \end{aligned}$$

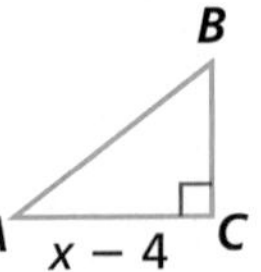

EXERCISES

Triangle $JQZ \cong$ triangle VTZ.

17. Find x.

18. Find t.

19. Find q.

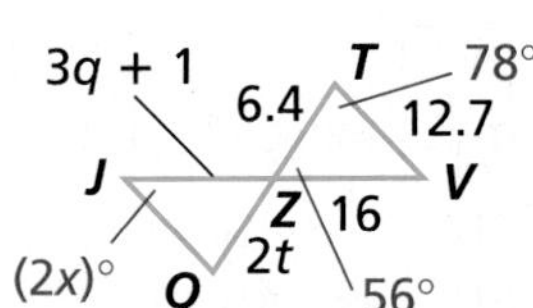

5-7 Transformations (pp. 254–257)

EXAMPLE

- **Draw the image of a triangle with vertices $(-2, 2)$, $(1, 1)$, $(-3, -2)$ after a $180°$ rotation around $(0, 0)$.**

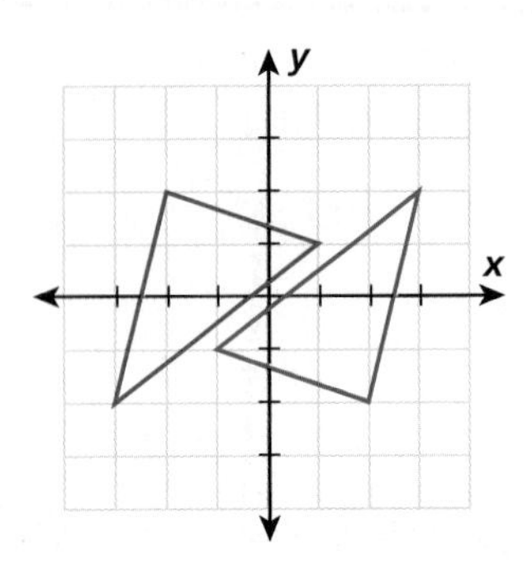

EXERCISES

Draw the image of a triangle with vertices $(1, 3)$, $(5, 1)$, $(1, 1)$ after each transformation.

20. a reflection across the x-axis

21. a reflection across the y-axis

22. a $180°$ rotation around $(0, 0)$

5-8 Symmetry (pp. 259–262)

EXAMPLE

Describe the symmetry in each letter.

- M
  line symmetry; vertical line of symmetry
- N
  2-fold rotational symmetry

EXERCISES

Describe the symmetry in each letter.

23. D

24. S

25. H

5-9 Tessellations (pp. 263–267)

EXAMPLE

- **Create a tessellation with the figure.**

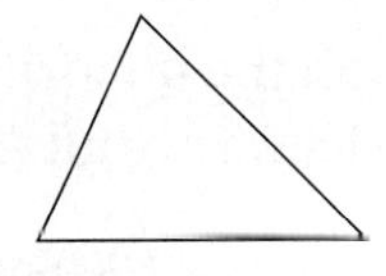

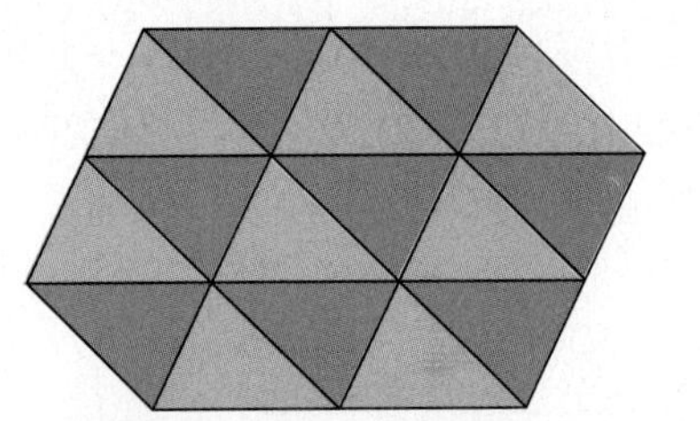

EXERCISES

Create a tessellation with each figure.

26.

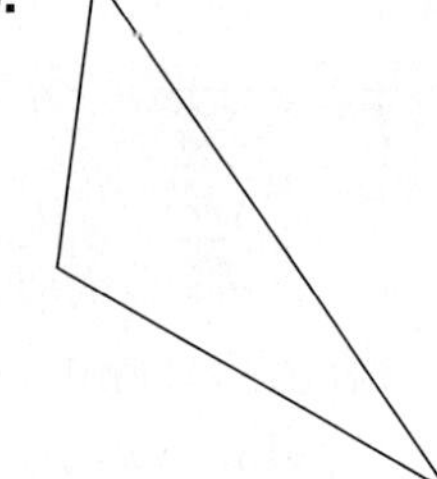

27.

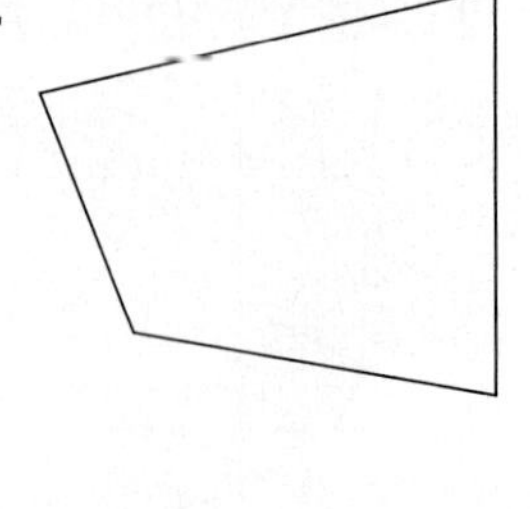

Chapter Test

Refer to the figure.

1. Name a pair of complementary angles.
2. Name a pair of supplementary angles.

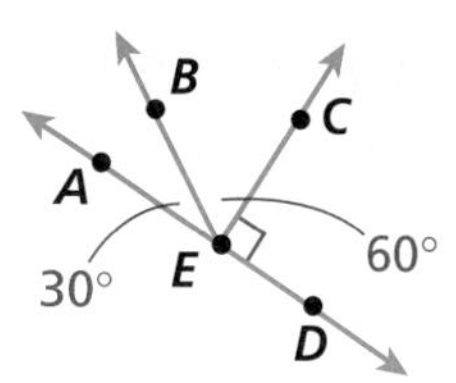

Line $w \parallel$ line v. Find each angle measure.

3. $\angle 1$ 4. $\angle 2$ 5. $\angle 3$

6. The second angle in a triangle is three times as large as the first. The measure of the third angle is 60° less than twice the measure of the first. Find the angle measures.

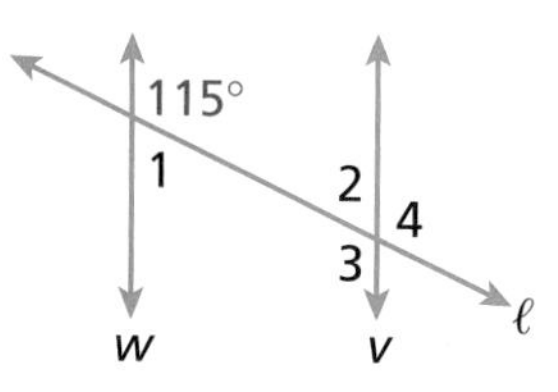

Find the angle measures in each regular polygon.

7.

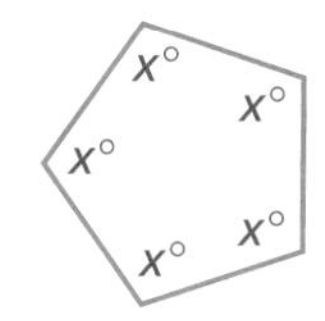

8.

Graph the quadrilaterals with the given vertices. Give all of the names that apply to each.

9. (0, 1), (−2, 2), (−1, 0), (3, −2)
10. (4, 0), (0, 4), (−4, 0), (0, −4)

Write a congruence statement for each pair of polygons.

11.

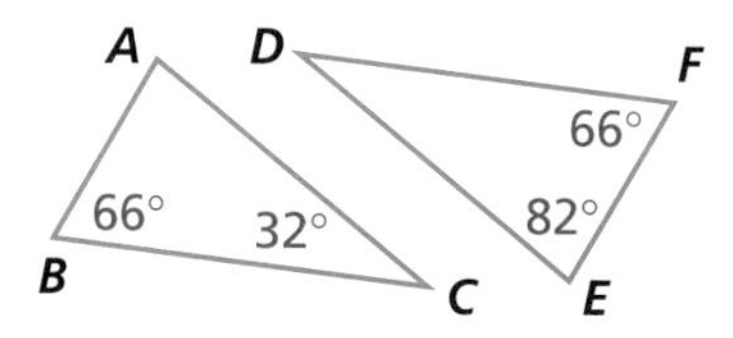

12.

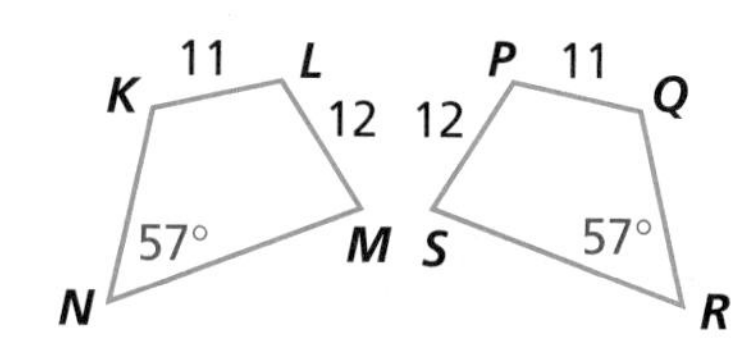

Draw the image of a triangle with vertices (0, 0), (3, 0), and (3, 4) after each transformation.

13. translation 3 units left
14. reflection across the y-axis
15. 180° rotation around (3, 0)
16. translation 2 units down

17. Complete the figure. The dashed line is the line of symmetry.

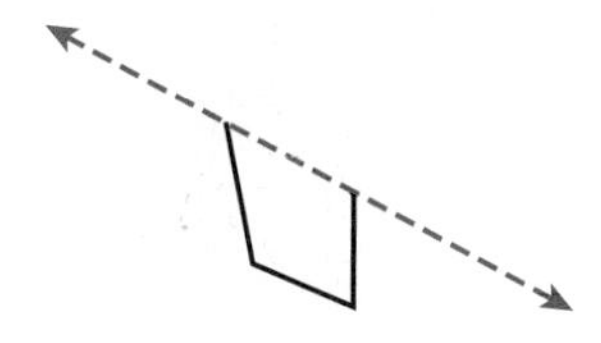

18. Create a tessellation with the given figure.

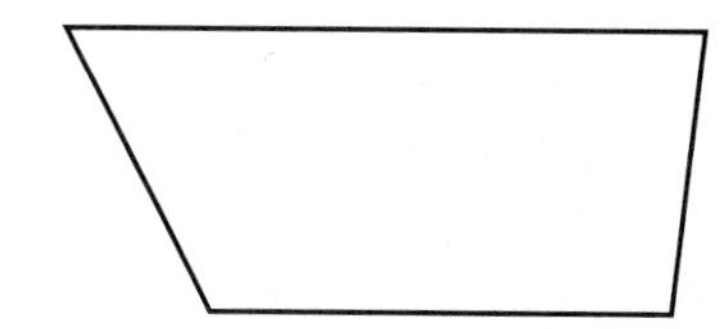

Chapter 5

Performance Assessment

Show What You Know

Create a portfolio of your work from this chapter. Complete this page and include it with your four best pieces of work from Chapter 5. Choose from your homework or lab assignments, mid-chapter quiz, or any journal entries you have done. Put them together using any design you want. Make your portfolio represent what you consider your best work.

Short Response

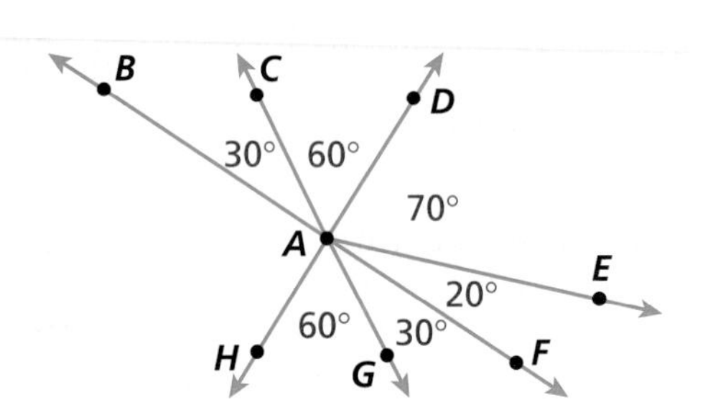

For 1–2, refer to the figure.

1. What is the measure of $\angle BAH$? Explain in words how you determined your answer.
2. Name all the pairs of supplementary angles. Explain how you know that you have named all the pairs.
3. Complete the table to show the number of diagonals for the polygons with the numbers of sides listed.

Number of Sides	3	4	5	6	7	n
Number of Diagonals	0	■	■	■	■	■

Extended Problem Solving

Choose any strategy to solve each problem.

4. Four people are introduced to each other at a party, and they all shake hands.
   a. Explain in words how the diagram can be used to determine the number of handshakes exchanged at the party.
   b. How many handshakes are exchanged?
   c. Suppose that 6 people were introduced to each other at a party. Draw a diagram similar to the one shown that could be used to determine the number of handshakes exchanged.

internet connect

EOG Test Practice Online
go.hrw.com Keyword: MT4 TestPrep

Getting Ready for EOG

Chapter 5

Cumulative Assessment, Chapters 1–5

1. Which of the following is $3.1415 \cdot 10^3$ written in standard notation?

   A 31,415,000 C 3141.5

   B 31,415 D 314.5

2. Which number is equivalent to 5^{-2}?

   A $\frac{1}{10}$ C $\frac{1}{-10}$

   B $\frac{1}{25}$ D $\frac{1}{-25}$

3. The cost of 3 sweatshirts is d dollars. At this rate, what is the cost in dollars of 30 sweatshirts?

   A $30d$ C $10d$

   B $\frac{10d}{3}$ D $\frac{30}{d}$

TEST TAKING TIP!

When a letter is used more than once in a statement, it always has the same value.

4. If $a \cdot k = a$ for all values of a, what is the value of k?

   A $-a$ C 0

   B -1 D 1

5. If $m^x \cdot m^7 = m^{28}$ and $\frac{m^y}{m^5} = m^3$ for all values of m, what is the value of $x + y$?

   A 31 C 19

   B 29 D 12

6. Laura wants to tile her kitchen floor. Which of the following shapes would *not* cover her floor with a tessellation?

   A (square) C (parallelogram)

   B (octagon) D (chevron)

7. What is the solution of $9x = -72$?

   A $x = -648$ C $x = 8$

   B $x = -8$ D $x = 648$

8. In the histogram below, which interval contains the median score?

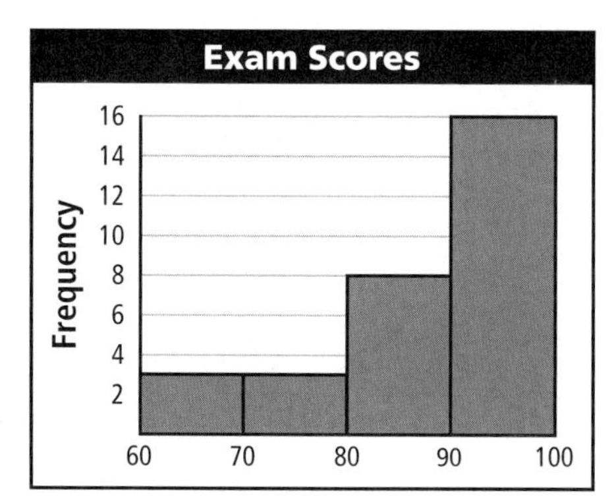

   A 60–69 C 80–89

   B 70–79 D 90–99

9. ***SHORT RESPONSE*** Triangle ABC, with vertices $A(2, 3)$, $B(4, -5)$, $C(6, 8)$, is reflected across the x-axis to trangle $A'B'C'$. On a coordinate grid, draw and label triangle ABC and triangle $A'B'C'$. Give the new coordinates for triangle $A'B'C'$.

10. ***SHORT RESPONSE*** Stephen bought 3 fish for his pond at a total cost of d dollars. At this rate what is the cost in dollars if he purchased 12 more fish? Show your work.

Getting Ready for EOG

Chapter 6

Perimeter, Area, and Volume

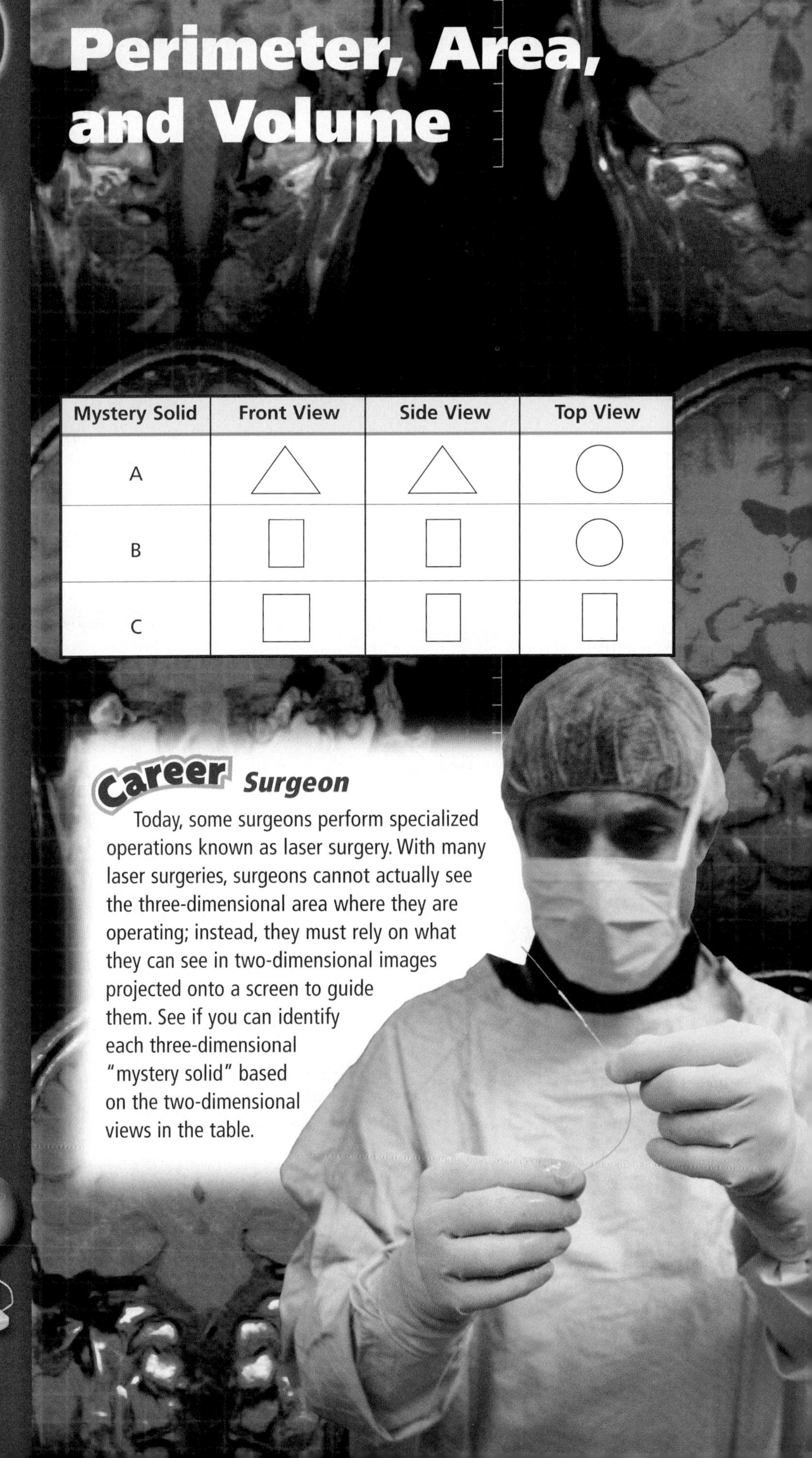

TABLE OF CONTENTS

Mystery Solid	Front View	Side View	Top View
A	△	△	○
B	▯	▯	○
C	□	▯	▯

Career *Surgeon*

Today, some surgeons perform specialized operations known as laser surgery. With many laser surgeries, surgeons cannot actually see the three-dimensional area where they are operating; instead, they must rely on what they can see in two-dimensional images projected onto a screen to guide them. See if you can identify each three-dimensional "mystery solid" based on the two-dimensional views in the table.

internet connect

Chapter Opener Online
go.hrw.com
KEYWORD: MT4 Ch6

ARE YOU READY?

Choose the best term from the list to complete each sentence.

1. A(n) ___?___ is a number that represents a part of a whole.

2. A(n) ___?___ is another way of writing a fraction.

3. To multiply 7 by the fraction $\frac{2}{3}$, multiply 7 by the ___?___ of the fraction and then divide the result by the ___?___ of the fraction.

4. To round 7.836 to the nearest tenth, look at the digit in the ___?___ place.

decimal
denominator
fraction
numerator
tenths
hundredths

Complete these exercises to review skills you will need for this chapter.

✔ Square and Cube Numbers

Evaluate.

5. 16^2 **6.** 9^3 **7.** $(4.1)^2$ **8.** $(0.5)^3$

9. $\left(\frac{1}{4}\right)^2$ **10.** $\left(\frac{2}{5}\right)^2$ **11.** $\left(\frac{1}{2}\right)^3$ **12.** $\left(\frac{2}{3}\right)^3$

✔ Multiply with Fractions

Multiply.

13. $\frac{1}{2}(8)(10)$ **14.** $\frac{1}{2}(3)(5)$ **15.** $\frac{1}{3}(9)(12)$ **16.** $\frac{1}{3}(4)(11)$

17. $\frac{1}{2}(8^2)16$ **18.** $\frac{1}{2}(5^2)24$ **19.** $\frac{1}{2}(6)(3 + 9)$ **20.** $\frac{1}{2}(5)(7 + 4)$

✔ Multiply with Decimals

Multiply. Write each answer to the nearest tenth.

21. $2(3.14)(12)$ **22.** $3.14(5^2)$ **23.** $3.14(4^2)(7)$ **24.** $3.14(2.3)^2(5)$

✔ Multiply with Fractions and Decimals

Multiply. Write each answer to the nearest tenth.

25. $\frac{1}{3}(3.14)(5^2)(7)$ **26.** $\frac{1}{3}\left(3.14\right)(5^3)$

27. $\frac{1}{3}(3.14)(3.2)^2(2)$ **28.** $\frac{4}{3}(3.14)(2.7)^3$

29. $\frac{1}{5}\left(\frac{22}{7}\right)(4^2)(5)$ **30.** $\frac{4}{11}\left(\frac{22}{7}\right)(3.2^3)$

31. $\frac{1}{2}\left(\frac{22}{7}\right)(1.7)^2(4)$ **32.** $\frac{7}{11}\left(\frac{22}{7}\right)(9.5)^3$

Perimeter & Area of Rectangles & Parallelograms

Learn **to find the perimeter and area of rectangles and parallelograms.**

Vocabulary

perimeter

area

In inlaid woodworking, artists use geometry to create a variety of beautiful patterns. One design can have thousands of pieces made from many different kinds of wood. In a design made entirely of parallelograms, the total area of the design is the sum of the areas of the parallelograms in the design.

Any side of a rectangle or parallelogram can be chosen as the base. The height is measured along a line perpendicular to the base.

Rectangle

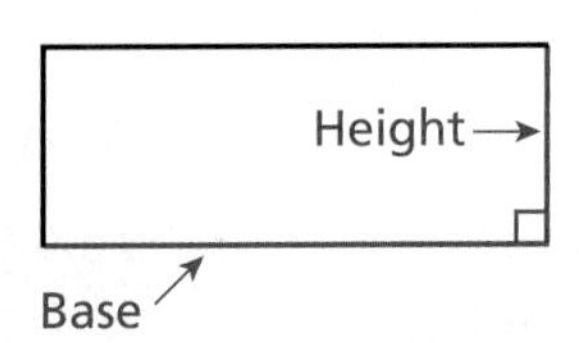

Parallelogram

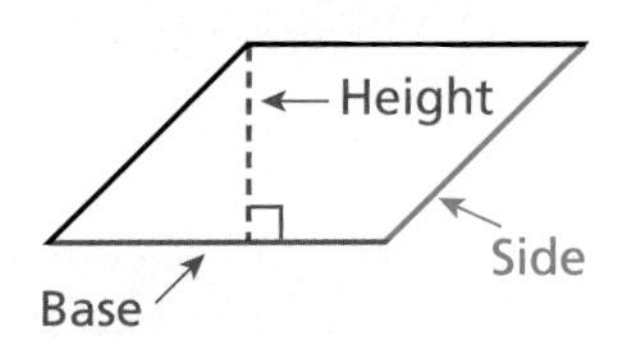

Perimeter is the distance around the outside of a figure. To find the perimeter of a figure, add the lengths of all its sides.

EXAMPLE 1 Finding the Perimeter of Rectangles and Parallelograms

Find the perimeter of each figure.

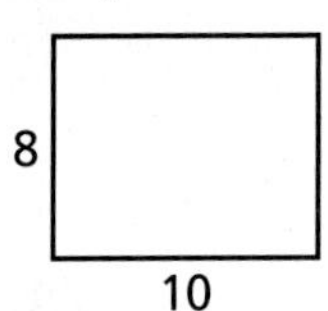

$P = 10 + 10 + 8 + 8$ *Add all side lengths.*

$= 36$ units

or $P = 2b + 2h$ *Perimeter of rectangle*

$= 2(10) + 2(8)$ *Substitute 10 for b and 8 for h.*

$= 20 + 16 = 36$ units

Helpful Hint

The formula for the perimeter of a rectangle can be written as $P = 2b + 2h$, where b is the length of the base and h is the height.

B

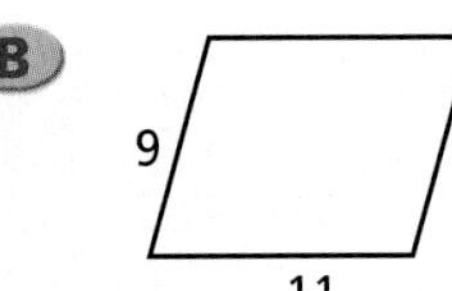

$P = 9 + 9 + 11 + 11$ *Add all side lengths.*

$= 40$ units

Area is the number of square units in a figure. A parallelogram can be cut and the cut piece shifted to form a rectangle with the same base length and height as the original parallelogram. So a parallelogram has the same area as a rectangle with the same base length and height.

AREA OF RECTANGLES AND PARALLELOGRAMS

Words	Numbers		Formula
The area A of a rectangle or parallelogram is the base length b times the height h.	5 by 3 grid; $5 \cdot 3 = 15$ units^2 **Rectangle**	5 by 3 grid; $5 \cdot 3 = 15$ units^2 **Parallelogram**	$A = bh$

EXAMPLE 2 **Using a Graph to Find Area**

Graph each figure with the given vertices. Then find the area of each figure.

A $(-2, -1), (2, -1), (2, 2), (-2, 2)$

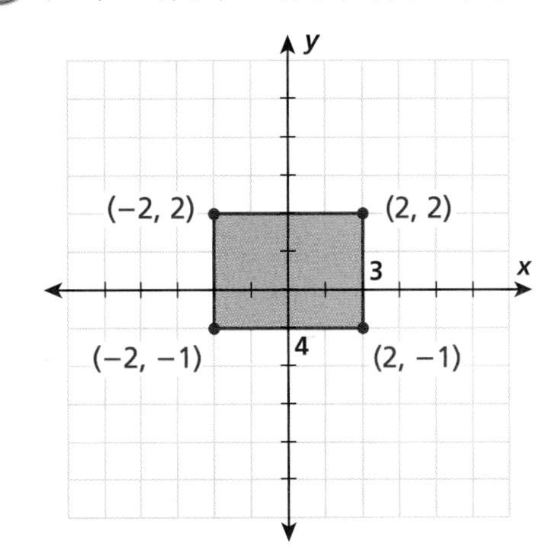

$A = bh$ *Area of rectangle*

$= 4 \cdot 3$ *Substitute 4 for b and 3 for h.*

$= 12$ units^2

Graph each figure with the given vertices. Then find the area of the figure.

B $(-4, 0), (2, 0), (4, 3), (-2, 3)$

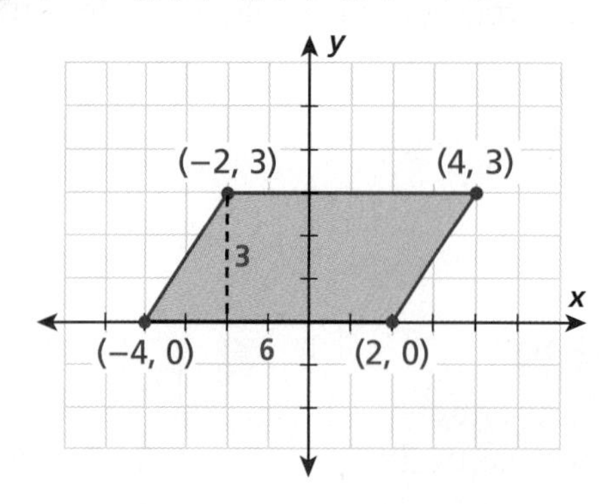

$A = bh$ *Area of parallelogram*

$= 6 \cdot 3$ *Substitute 6 for b and 3 for h.*

$= 18 \text{ units}^2$

Helpful Hint

The height of a parallelogram is not the length of its slanted side. The height of a figure is always perpendicular to the base.

EXAMPLE 3 Finding Area and Perimeter of a Composite Figure

Find the perimeter and area of the figure.

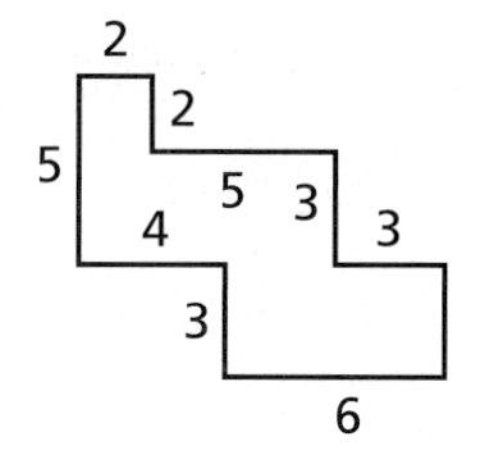

The length of the side that is not labeled is the same as the length of the opposite side, 3 units.

$P = 5 + 2 + 2 + 5 + 3 + 3 + 3 + 6 + 3 + 4$

$= 36 \text{ units}$

= 2, 5 + 5, 3 + 6, 3

$A = 5 \cdot 2 + 5 \cdot 3 + 6 \cdot 3$ *Add the areas together.*

$= 10 + 15 + 18$

$= 43 \text{ units}^2$

Think and Discuss

1. **Compare** the area of a rectangle with base b and height h with the area of a rectangle with base $2b$ and height $2h$.
2. **Express** the formulas for the area and perimeter of a square using s for the length of a side.

6-1 Exercises

FOR EOG PRACTICE

see page 664

internet connect

Homework Help Online
go.hrw.com Keyword: MT4 6-1

2.01, 3.01, 3.02

GUIDED PRACTICE

See Example 1 Find the perimeter of each figure.

1. 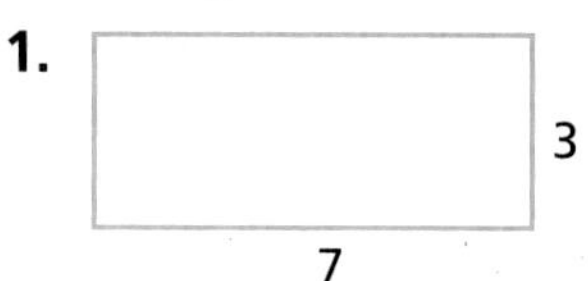

2. 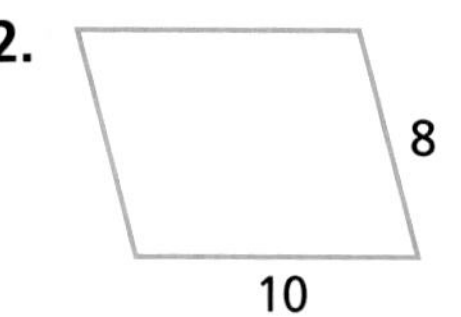

3. 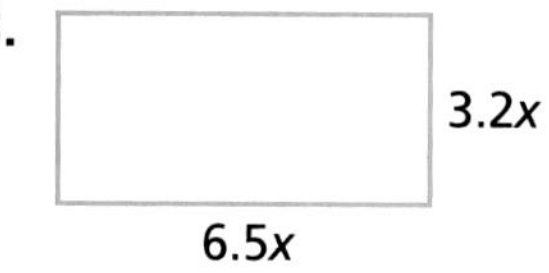

See Example 2 Graph each figure with the given vertices. Then find the area of each figure.

4. $(-3, 2), (0, 2), (3, -3), (0, -3)$
5. $(-4, 0), (-4, 4), (3, 4), (3, 0)$
6. $(-4, 1), (4, 1), (3, -3), (-5, -3)$
7. $(-2, 3), (0, 3), (0, -4), (-2, -4)$

See Example 3 8. Find the perimeter and area of the figure.

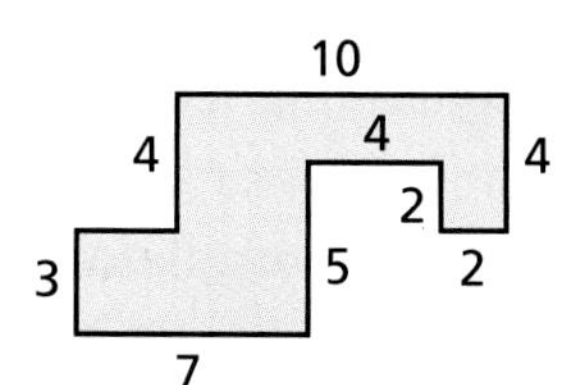

INDEPENDENT PRACTICE

See Example 1 Find the perimeter of each figure.

9. 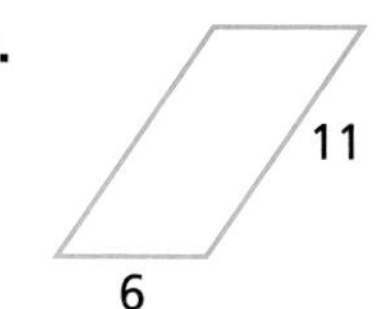

10. 

11. 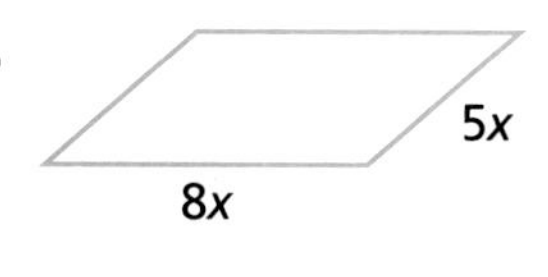

See Example 2 Graph each figure with the given vertices. Then find the area of each figure.

12. $(-5, -1), (2, -1), (2, -5), (-5, -5)$
13. $(0, 3), (6, 3), (3, -1), (-3, -1)$
14. $(3, 5), (5, 3), (-3, 3), (-5, 5)$
15. $(2, 5), (5, 5), (5, -1), (2, -1)$

See Example 3 16. Find the perimeter and area of the figure.

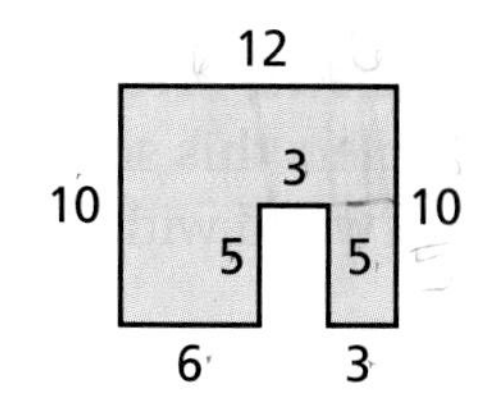

PRACTICE AND PROBLEM SOLVING

Find the perimeter of each figure.

17. 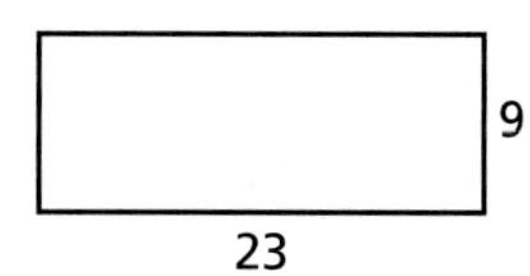

18. 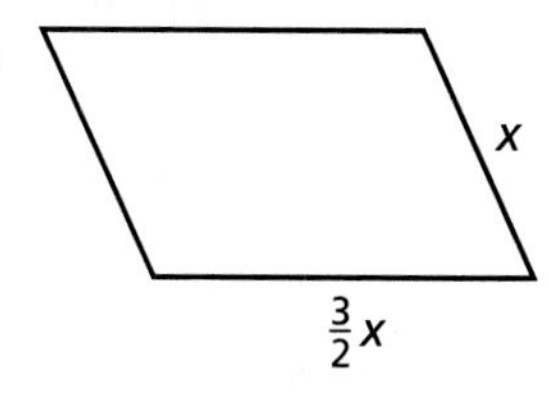

AREA OF TRIANGLES AND TRAPEZOIDS

Words	Numbers	Formula
Triangle: The area A of a triangle is one-half of the base length b times the height h.	4 8 $A = \frac{1}{2}(8)(4)$ $= 16 \text{ units}^2$	$A = \frac{1}{2}bh$
Trapezoid: The area of a trapezoid is one-half the height h times the sum of the base lengths b_1 and b_2.	3 2 7 $A = \frac{1}{2}(2)(3 + 7)$ $= 10 \text{ units}^2$	$A = \frac{1}{2}h(b_1 + b_2)$

Reading Math

In the term b_1, the number 1 is called a *subscript*.
It is read as "*b* one" or "*b* sub-one."

EXAMPLE 2 Finding the Area of Triangles and Trapezoids

Graph and find the area of each figure with the given vertices.

A (−1, 1), (3, 1), (1, 5)

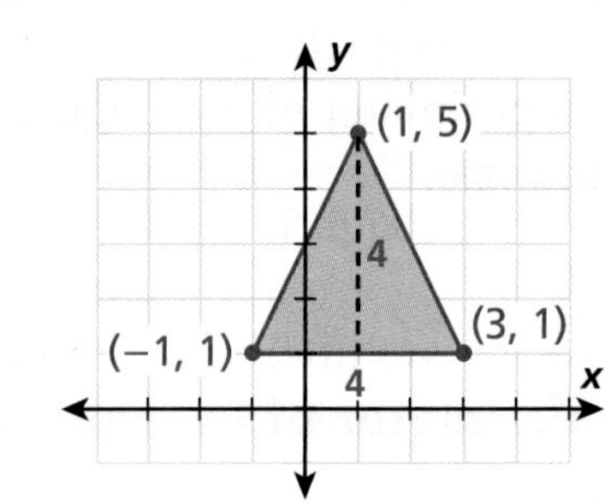

$$A = \frac{1}{2}bh$$ *Area of a triangle*

$$= \frac{1}{2} \cdot 4 \cdot 4$$ *Substitute for b and h.*

$$= 8 \text{ units}^2$$

B (−3, −2), (−3, 1), (0, 1), (2, −2)

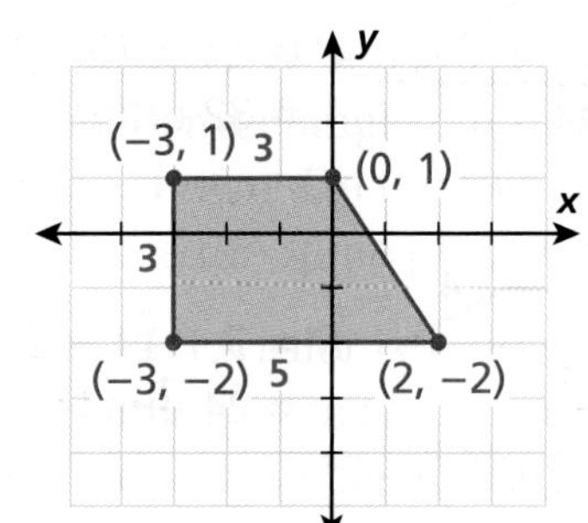

$$A = \frac{1}{2}h(b_1 + b_2)$$ *Area of a trapezoid*

$$= \frac{1}{2} \cdot 3(3 + 5)$$ *Substitute for h, b_1, and b_2.*

$$= 12 \text{ units}^2$$

Think and Discuss

1. **Describe** what happens to the area of a triangle when the base is doubled and the height remains the same.
2. **Describe** what happens to the area of a trapezoid when the length of both bases are doubled but the height remains the same.

6-2 Exercises

FOR EOG PRACTICE
see page 664

internet connect
Homework Help Online
go.hrw.com Keyword: MT4 6-2

2.01, 3.01, 3.02

GUIDED PRACTICE

See Example 1 **Find the perimeter of each figure.**

1.
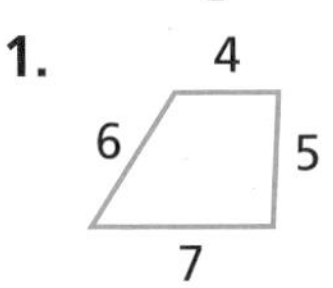

2.
$3\frac{3}{4}$, $3\frac{1}{2}$, 4

3.
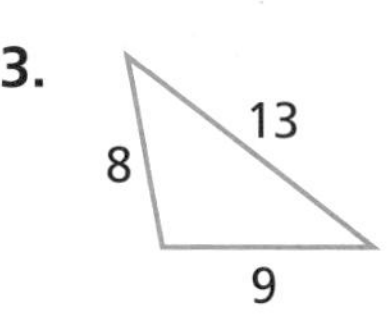

4.
9, 7.7, 6.3, 11.5

5.
27, 19, 17, 21

6.
x, $2x - 3$, $x + 4$

See Example 2 **Graph and find the area of each figure with the given vertices.**

7. (−2, 3), (2, −3), (−3, −3)

8. (5, 2), (2, −2), (−3, −2), (−4, 2)

9. (4, 2), (5, −6), (2, −6)

10. (0, −1), (−7, −1), (−5, 4), (−2, 4)

INDEPENDENT PRACTICE

See Example 1 **Find the perimeter of each figure.**

11.
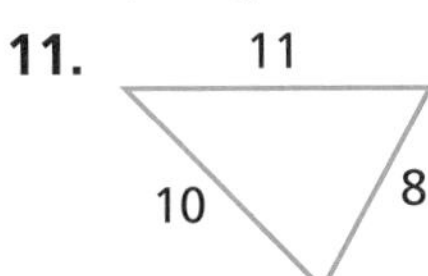

12.
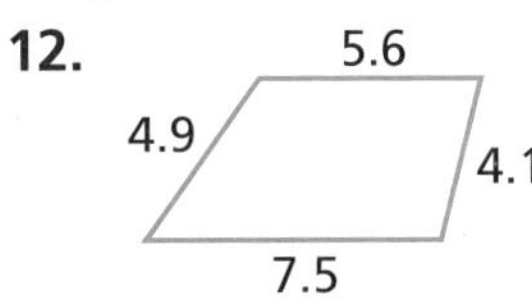

13.
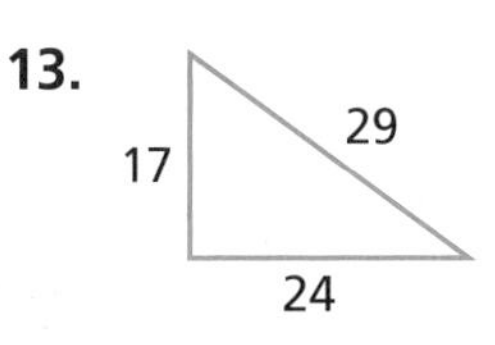

14.
4, $5\frac{1}{3}$, $2\frac{3}{4}$, $3\frac{1}{3}$

15.
$6a$, $6a$, $7a + 3$, $11a + 5$

16.
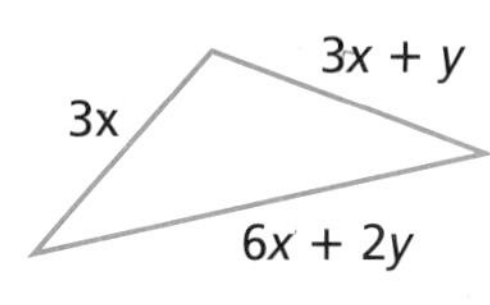

See Example 2 **Graph and find the area of each figure with the given vertices.**

17. (1, 5), (1, 1), (−3, 1), (−5, 5)

18. (−5, 2), (1, −3), (−3, −3)

19. (2, −3), (−1, −6), (−6, −3)

20. (1, 4), (4, −5), (−5, −5), (−3, 4)

PRACTICE AND PROBLEM SOLVING

Find the area of each figure with the given dimensions.

21. triangle: $b = 9$, $h = 11$

22. trapezoid: $b_1 = 6$, $b_2 = 10$, $h = 5$

23. triangle: $b = 7x$, $h = 6$

24. trapezoid: $b_1 = 4.5$, $b_2 = 8$, $h = 6.7$

25. The perimeter of a triangle is 37.4 ft. Two of its sides measure 16.4 ft and 11.9 ft, respectively. What is the length of its third side?

26. The area of a triangle is 63 mm^2. If its height is 14 mm, what is the length of its base?

Physical Science LINK

To fly, a plane must overcome gravity and achieve *lift*, the force that allows a flying object to have upward motion. The shape and size of a plane's wings affect the amount of lift that is created. The wings of high-speed airplanes are thin and usually angled back to give the plane more lift.

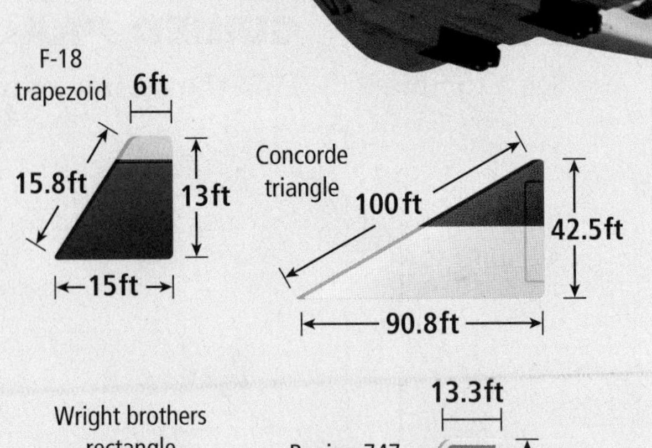

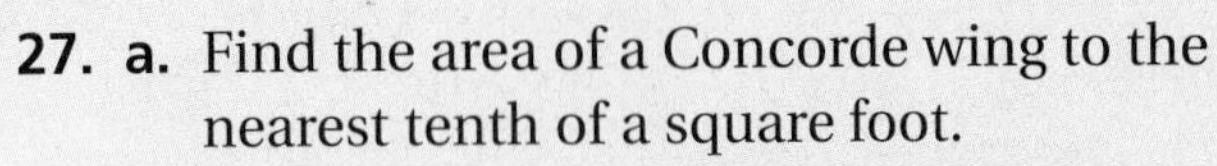

27. **a.** Find the area of a Concorde wing to the nearest tenth of a square foot.

b. Find the total perimeter of the two wings of a Concorde to the nearest tenth of a foot.

28. What is the area of a Boeing 747 wing to the nearest tenth of a square foot?

29. What is the perimeter of an F-18 wing to the nearest tenth of a foot?

30. What is the total area of the two wings of an F-18?

31. Find the area and perimeter of the wing of a space shuttle rounded to the nearest tenth.

32. ***CHALLENGE*** The wing of the Wright brothers' plane is about half the length of a Boeing 747 wing. Compare the area of the Wright brothers' wing with the area of a Boeing 747 wing. Is the area of the Wright brothers' wing half the area of the 747 wing? Explain.

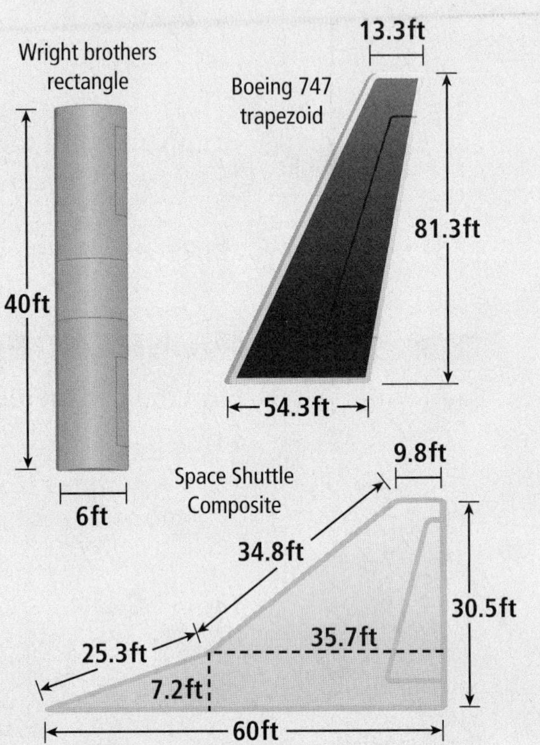

go.hrw.com
KEYWORD: MT4 Lift
CNN Student News

Spiral Review

Write each fraction as a decimal. (Lesson 3-1)

33. $\frac{3}{4}$ **34.** $\frac{1}{8}$ **35.** $\frac{10}{4}$ **36.** $\frac{9}{15}$

Do the data sets have a positive, a negative, or no correlation? (Lesson 4-7)

37. the number of shoes purchased and the amount of money left over

38. the length of a sub sandwich and the price of the sandwich

39. **EOG PREP** What name best describes a quadrilateral with vertices at (2, 4), (4, 1), (−3, 1), and (−5, 4)? (Lesson 5-5)

A Parallelogram **B** Rectangle **C** Rhombus **D** Trapezoid

Explore Right Triangles

Use with Lesson 6-3

WHAT YOU NEED
- scissors
- paper

REMEMBER
Right triangles have 1 right angle and 2 acute angles.

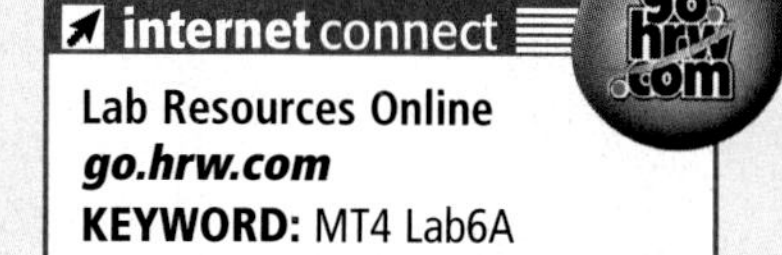

Activity

1 The Pythagorean Theorem states that if a and b are the lengths of the legs of a right triangle, then c is the length of the hypotenuse, where $a^2 + b^2 = c^2$. Prove the Pythagorean Theorem using the following steps.

a. Draw two squares side by side. Label one with side a and one with side b.

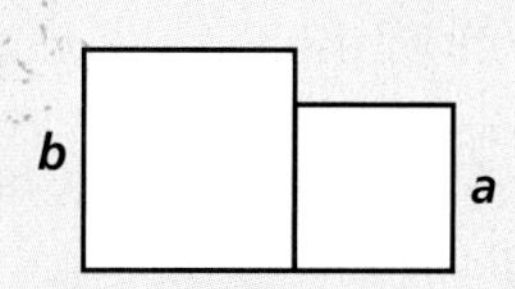

Notice that the area of this composite figure is $a^2 + b^2$.

b. Draw hypotenuses of length c, so that we have right triangles with sides a, b, and c.

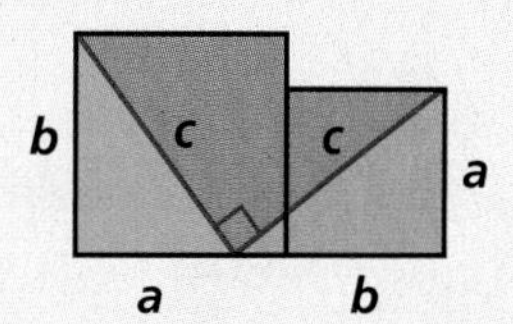

c. Cut out the triangles and the remaining piece.

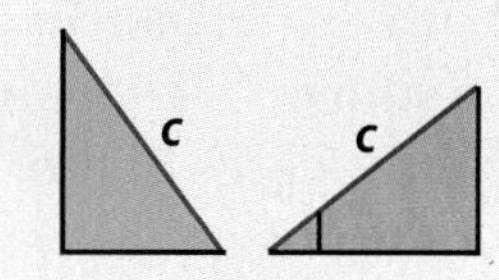

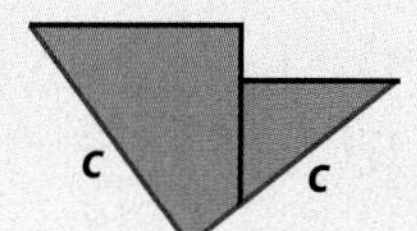

d. Fit the pieces together to make a square with sides c and area c^2. You have shown that the area $a^2 + b^2$ can be cut up and rearranged to form the area c^2, so $a^2 + b^2 = c^2$.

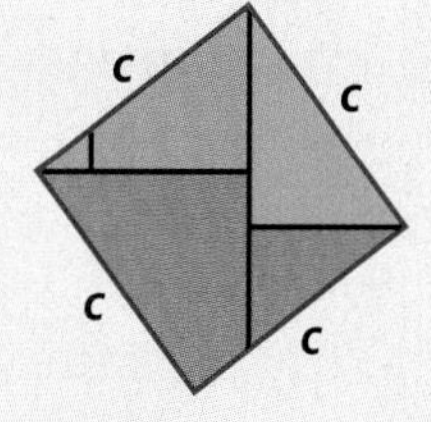

Think and Discuss

1. Does the Pythagorean Theorem work for triangles that are not right triangles?

Try This

1. If you know that the lengths of two legs of a right triangle are 9 and 12, can you find the length of the hypotenuse? Show your work.

2. Take a piece of paper and fold the right corner down so that the top edge of the paper matches the side edge. Crease the paper. Without measuring, find the diagonal's length.

6-3 The Pythagorean Theorem

Learn to use the Pythagorean Theorem and its converse to solve problems.

Vocabulary

Pythagorean Theorem

leg

hypotenuse

Pythagoras was born on the Aegean island of Samos sometime between 580 B.C. and 569 B.C. He is best known for the *Pythagorean Theorem,* which relates the side lengths of a right triangle.

A Babylonian tablet known as Plimpton 322 provides evidence that the relationship between the side lengths of right triangles was known as early as 1900 B.C. Many people, including U.S. president James Garfield, have written proofs of the Pythagorean Theorem. In 1940, E. S. Loomis presented 370 proofs of the theorem in *The Pythagorean Proposition.*

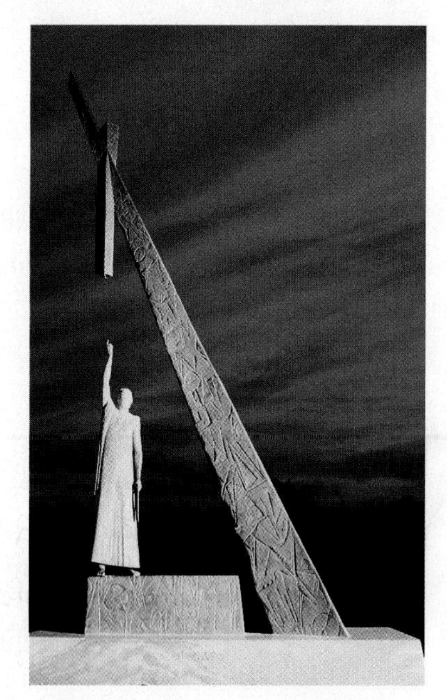

This statue of Pythagoras is located in the Pythagorion Harbor on the island of Samos.

THE PYTHAGOREAN THEOREM

Words	Numbers	Algebra
In any right triangle, the sum of the squares of the lengths of the two **legs** is equal to the square of the length of the **hypotenuse**.	4, 5, 3 $3^2 + 4^2 = 5^2$ $9 + 16 = 25$	Hypotenuse, a, c, b, Legs $a^2 + b^2 = c^2$

EXAMPLE 1 Finding the Length of a Hypotenuse

Helpful Hint

The triangle in the figure is an isosceles right triangle. It is also called a 45°-45°-90° triangle.

Find the length of the hypotenuse.

A

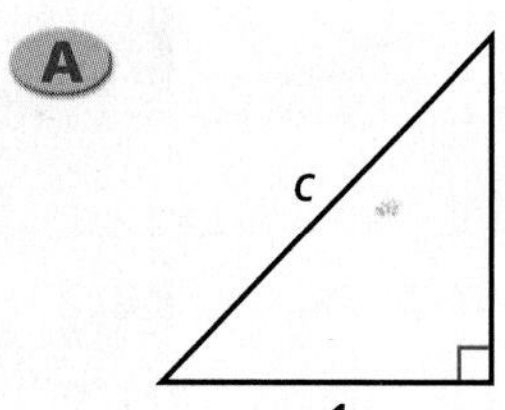

$a^2 + b^2 = c^2$ — *Pythagorean Theorem*

$1^2 + 1^2 = c^2$ — *Substitute for a and b.*

$1 + 1 = c^2$ — *Simplify powers.*

$2 = c^2$

$\sqrt{2} = c$ — *Solve for c; $c = \sqrt{c^2}$.*

$1.41 \approx c$

Find the length of the hypotenuse.

B triangle with coordinates (6, 1), (0, 9), and (0, 1)

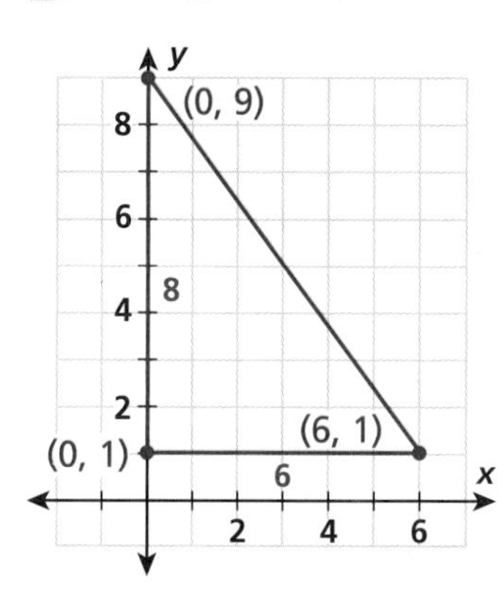

The points form a right triangle with $a = 8$ and $b = 6$.

$a^2 + b^2 = c^2$ *Pythagorean Theorem*

$8^2 + 6^2 = c^2$ *Substitute for a and b.*

$64 + 36 = c^2$ *Simplify powers.*

$100 = c^2$

$10 = c$ *$\sqrt{100} = 10$*

EXAMPLE 2 Finding the Length of a Leg in a Right Triangle

Solve for the unknown side in the right triangle.

$$a^2 + b^2 = c^2$$
$$5^2 + b^2 = 13^2$$
$$25 + b^2 = 169$$
$$\underline{-25 \qquad\quad -25}$$
$$b^2 = 144$$
$$b = 12 \qquad \sqrt{144} = 12$$

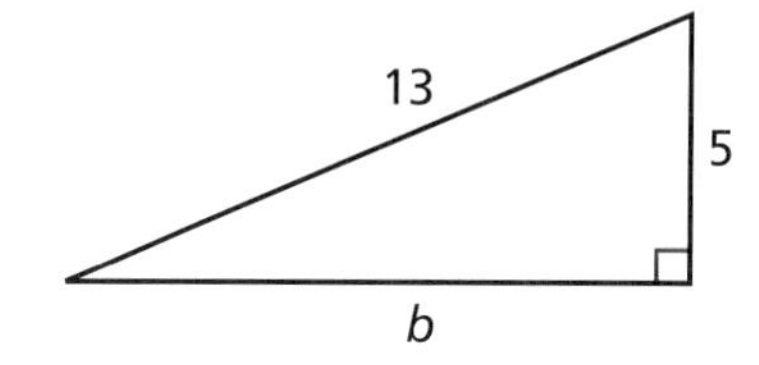

EXAMPLE 3 Using the Pythagorean Theorem to Find Area

Use the Pythagorean Theorem to find the height of the triangle. Then use the height to find the area of the triangle.

$a^2 + b^2 = c^2$

$a^2 + 1^2 = 2^2$ *Substitute 1 for b and 2 for c.*

$a^2 + 1 = 4$

$a^2 = 3$

$a = \sqrt{3}$ units ≈ 1.73 units *Find the square root of both sides.*

$A = \frac{1}{2}bh = \frac{1}{2}(2)(\sqrt{3}) = \sqrt{3}$ units$^2 \approx 1.73$ units^2

Think and Discuss

1. **Tell** how to use the Pythagorean Theorem to find the height of any isosceles triangle when the side lengths are given.
2. **Explain** if 2, 3, and 4 cm could be side lengths of a right triangle.

6-3 Exercises

FOR EOG PRACTICE
see page 665

internet connect
Homework Help Online
go.hrw.com Keyword: MT4 6-3

3.01, 3.02, 5.04

GUIDED PRACTICE

See Example 1 **Find the length of the hypotenuse in each triangle to the nearest tenth.**

1.
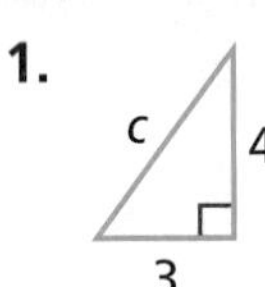

2.
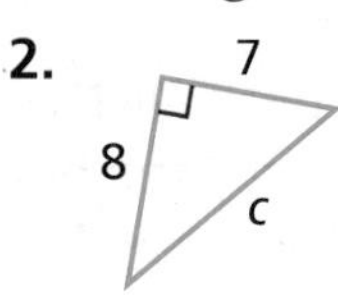

3. triangle with coordinates (−5, 0), (−5, 6), and (0, 6)

See Example 2 **Solve for the unknown side in each right triangle to the nearest tenth.**

4.
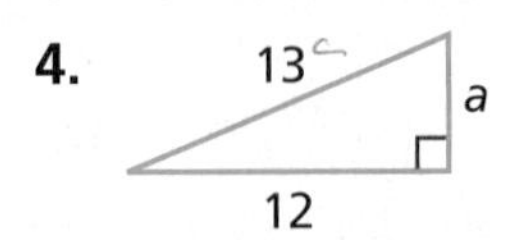

5. 6, *b*, 8

6.
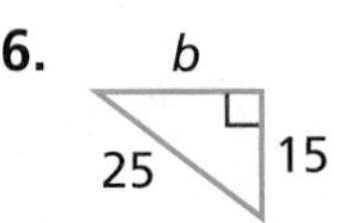

See Example 3 **7.** Use the Pythagorean Theorem to find the height of the triangle. Then use the height to find the area of the triangle.

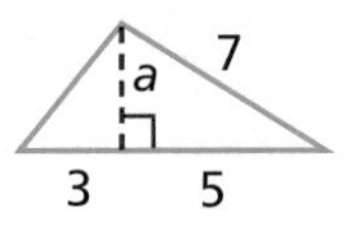

INDEPENDENT PRACTICE

See Example 1 **Find the length of the hypotenuse in each triangle to the nearest tenth.**

8.
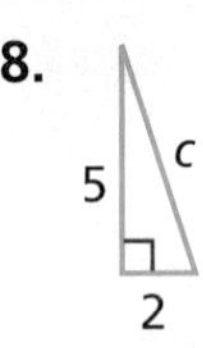

9. 15, 8, *c*

10.
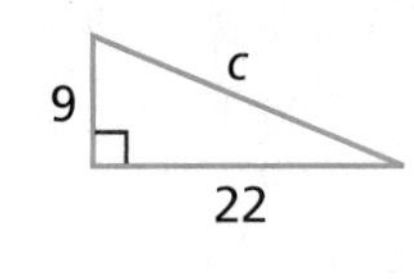

11. triangle with coordinates (−4, 2), (4, −2), and (−4, −2)

See Example 2 **Solve for the unknown side in each right triangle to the nearest tenth.**

12.
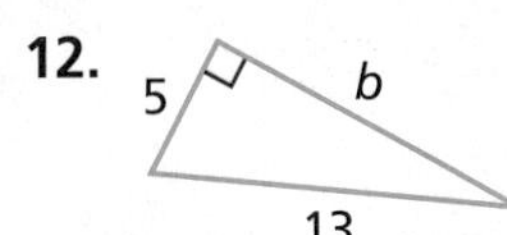

13.
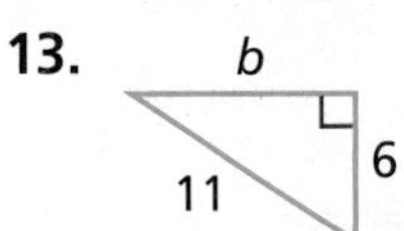

14.
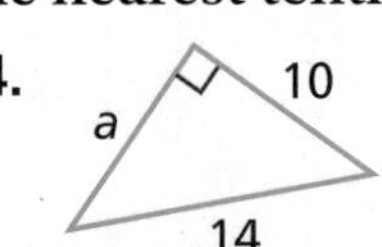

See Example 3 **15.** Use the Pythagorean Theorem to find the height of the triangle. Then use the height to find the area of the triangle.

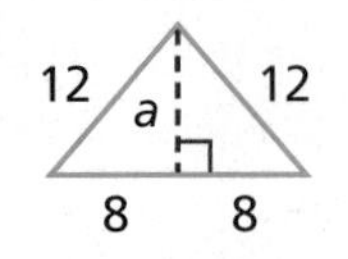

PRACTICE AND PROBLEM SOLVING

Find the missing length for each right triangle.

16. $a = 3, b = 6, c =$ ▢

17. $a =$ ▢ $, b = 24, c = 25$

18. $a = 30, b = 72, c =$ ▢

19. $a = 20, b =$ ▢ $, c = 46$

20. $a =$ ▢ $, b = 53, c = 70$

21. $a = 65, b =$ ▢ $, c = 97$

The *converse* of the Pythagorean Theorem states that any three positive numbers that make the equation $a^2 + b^2 = c^2$ true are the side lengths of a right triangle. If the side lengths are all whole numbers, they are called *Pythagorean triples.* Determine whether each set is a Pythagorean triple.

22. 2, 6, 8 **23.** 3, 4, 5 **24.** 8, 15, 17 **25.** 12, 16, 20

26. 10, 24, 26 **27.** 9, 13, 16 **28.** 11, 17, 23 **29.** 24, 32, 40

30. Use the Pythagorean Theorem to find the height of the figure. Then find the area, to the nearest whole number.

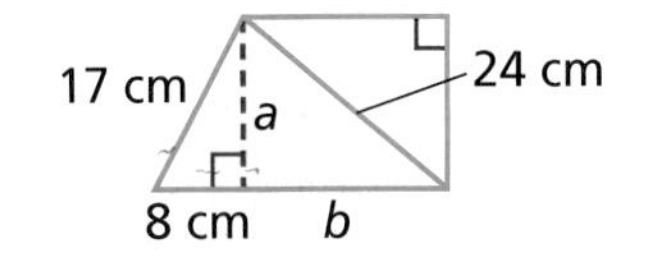

31. How far is the sailboat from the lighthouse, to the nearest kilometer?

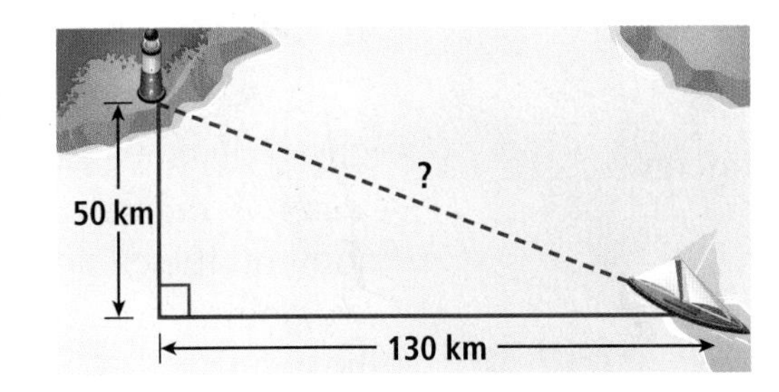

32. ***CONSTRUCTION*** A construction company is pouring a concrete foundation. The measures of two sides that meet in a corner are 33 ft and 56 ft. For the corner to be square (a right angle), what would the length of the diagonal have to be? (*Hint:* Draw a diagram.)

33. ***SOCIAL STUDIES*** The state of Colorado is shaped approximately like a rectangle. To the nearest mile, what is the distance between opposite corners of the state?

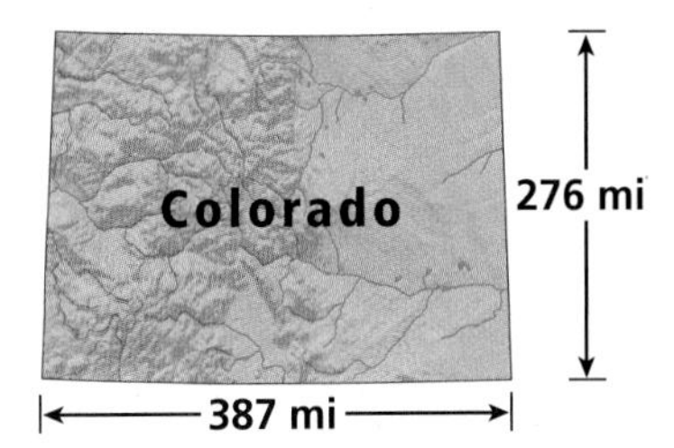

34. ***WRITE A PROBLEM*** Use a street map to write and solve a problem that requires the use of the Pythagorean Theorem.

35. ***WRITE ABOUT IT*** Explain how to use the converse of the Pythagorean Theorem to show that a triangle is a right triangle. (See Exercises 22–29.)

36. ***CHALLENGE*** A right triangle has legs of length $6x$ m and $8x$ m and hypotenuse of length 90 m. Find the lengths of the legs of the triangle.

Spiral Review

Solve. (Lesson 2-4)

37. $x + 13 = 22$ **38.** $b + 5 = -2$ **39.** $2y + 9 = 19$ **40.** $4a + 2 = -18$

41. **EOG PREP** Which real number lies between $3\frac{1}{5}$ and $3\frac{4}{7}$? (Lesson 3-10)

A 3.216 B 3.59 C 3.701 D 3.9

6-4 Circles

Learn to find the area and circumference of circles.

Vocabulary

circle

radius

diameter

circumference

A bicycle odometer uses a magnet attached to a wheel and a sensor attached to the bicycle frame. Each time the magnet passes the sensor, the odometer registers the distance traveled. This distance is the *circumference* of the wheel.

A **circle** is the set of points in a plane that are a fixed distance from a given point, called the *center*. A **radius** connects the center to any point on the circle, and a **diameter** connects two points on the circle and passes through the center.

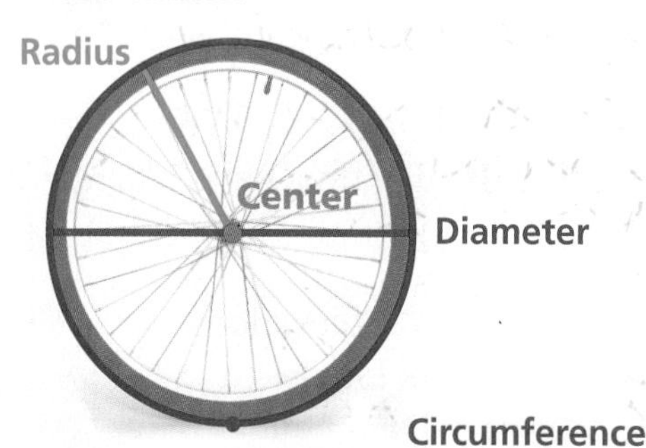

The diameter d is twice the radius r.

$$d = 2r$$

The **circumference** of a circle is the distance around the circle.

Remember!

Pi (π) is an irrational number that is often approximated by the rational numbers 3.14 and $\frac{22}{7}$.

CIRCUMFERENCE OF A CIRCLE

Words	Numbers	Formula
The circumference C of a circle is π times the diameter d, or 2π times the radius r.	3, 6; $C = \pi(6)$ $= 2\pi(3)$ ≈ 18.8 units	$C = \pi d$ or $C = 2\pi r$

EXAMPLE 1 Finding the Circumference of a Circle

Find the circumference of each circle, both in terms of π and to the nearest tenth. Use 3.14 for π.

A circle with radius 5 cm

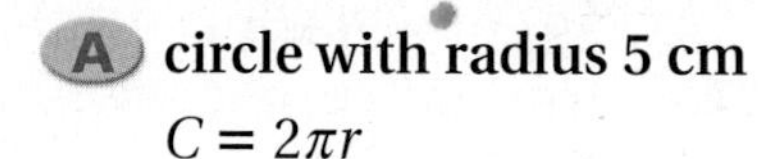

$C = 2\pi r$

$= 2\pi(5)$

$= 10\pi \text{ cm} \approx 31.4 \text{ cm}$

B circle with diameter 1.5 in.

$C = \pi d$

$= \pi(1.5)$

$= 1.5\pi \text{ in.} \approx 4.7 \text{ in.}$

AREA OF A CIRCLE

Words	Numbers	Formula
The area A of a circle is π times the square of the radius r.	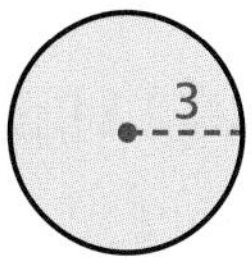 $A = \pi(3^2)$ $= 9\pi$ $\approx 28.3 \text{ units}^2$	$A = \pi r^2$

EXAMPLE 2 Finding the Area of a Circle

Find the area of each circle, both in terms of π and to the nearest tenth. Use 3.14 for π.

A circle with radius 5 cm

$A = \pi r^2 = \pi(5^2)$

$= 25\pi \text{ cm}^2 \approx 78.5 \text{ cm}^2$

B circle with diameter 1.5 in.

$A = \pi r^2 = \pi(0.75^2)$ *$\frac{d}{2} = 0.75$*

$= 0.5625\pi \text{ in}^2 \approx 1.8 \text{ in}^2$

EXAMPLE 3 Finding Area and Circumference on a Coordinate Plane

Graph the circle with center (−1, 1) that passes through (−1, 3). Find the area and circumference, both in terms of π and to the nearest tenth. Use 3.14 for π.

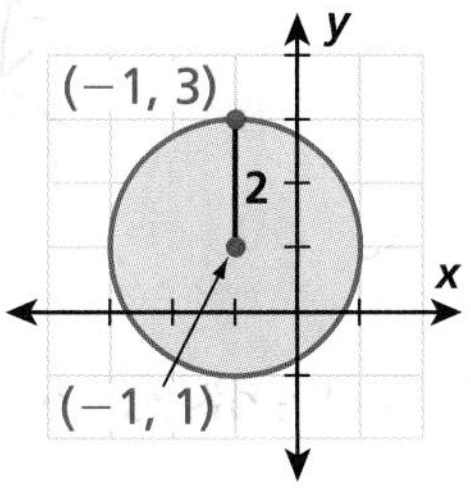

$A = \pi r^2$
$= \pi(2^2)$
$= 4\pi \text{ units}^2$
$\approx 12.6 \text{ units}^2$

$C = \pi d$
$= \pi(4)$
$= 4\pi \text{ units}$
$\approx 12.6 \text{ units}$

EXAMPLE 4 *Transportation Application*

A bicycle odometer recorded 147 revolutions of a wheel with diameter $\frac{4}{3}$ ft. How far did the bicycle travel? Use $\frac{22}{7}$ for π.

$C = \pi d = \pi\left(\frac{4}{3}\right) \approx \frac{22}{7}\left(\frac{4}{3}\right) = \frac{88}{21}$ *Find the circumference.*

The distance traveled is the circumference of the wheel times the number of revolutions, or about $\frac{88}{21} \cdot 147 = 616$ ft.

Think and Discuss

1. **Compare** the circumference of a circle with diameter x to the circumference of a circle with diameter $2x$.
2. **Give** the formula for area of a circle in terms of the diameter d.

6-4 Exercises

FOR EOG PRACTICE

see page 665

internet connect
Homework Help Online
go.hrw.com Keyword: MT4 6-4

2.01, 3.01, 3.02, 5.04

GUIDED PRACTICE

See Example 1

Find the circumference of each circle, both in terms of π and to the nearest tenth. Use 3.14 for π.

1. circle with diameter 8 cm
2. circle with radius 3.2 in.

See Example 2

Find the area of each circle, both in terms of π and to the nearest tenth. Use 3.14 for π.

3. circle with radius 1.5 ft
4. circle with diameter 15 cm

See Example 3

5. Graph a circle with center (3, −1) that passes through (0, −1). Find the area and circumference, both in terms of π and to the nearest tenth. Use 3.14 for π.

See Example 4

6. Estimate the diameter of a wheel that makes 9 revolutions and travels 50 feet. Use $\frac{22}{7}$ for π.

INDEPENDENT PRACTICE

See Example 1

Find the circumference of each circle, both in terms of π and to the nearest tenth. Use 3.14 for π.

7. circle with radius 7 in.
8. circle with diameter 11.5 m
9. circle with radius 20.2 cm
10. circle with diameter 2 ft

See Example 2

Find the area of each circle, both in terms of π and to the nearest tenth. Use 3.14 for π.

11. circle with diameter 24 cm
12. circle with radius 1.4 yd
13. circle with radius 18 in.
14. circle with diameter 17 ft

See Example 3

15. Graph a circle with center (−4, 2) that passes through (−4, −4). Find the area and circumference, both in terms of π and to the nearest tenth. Use 3.14 for π.

See Example 4

16. If the diameter of a wheel is 2 ft, about how many revolutions does the wheel make for every mile driven? Use $\frac{22}{7}$ for π. (*Hint:* 1 mi = 5280 ft.)

PRACTICE AND PROBLEM SOLVING

Find the circumference and area of each circle to the nearest tenth. Use 3.14 for π.

17.

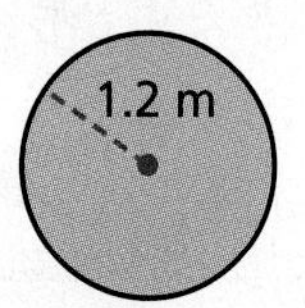

18.

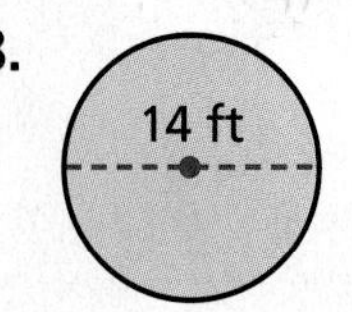

19.

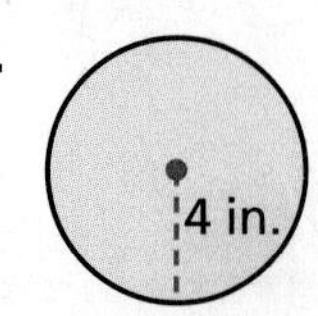

Find the radius of each circle with the given measurement.

20. $C = 18\pi$ in. **21.** $C = 12.8\pi$ cm **22.** $C = 25\pi$ ft

23. $A = 16\pi$ cm^2 **24.** $A = 169\pi$ in^2 **25.** $A = 136.89\pi$ m^2

Find the shaded area to the nearest tenth. Use 3.14 for π.

26.

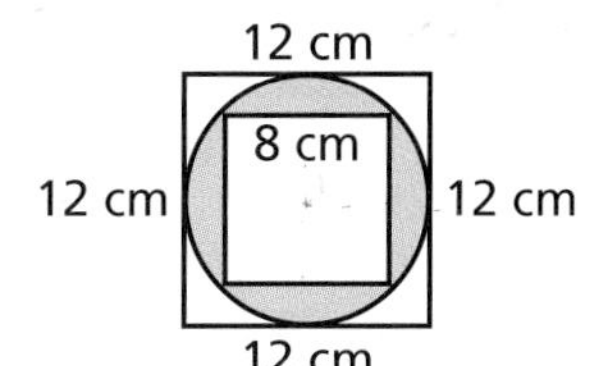

27.

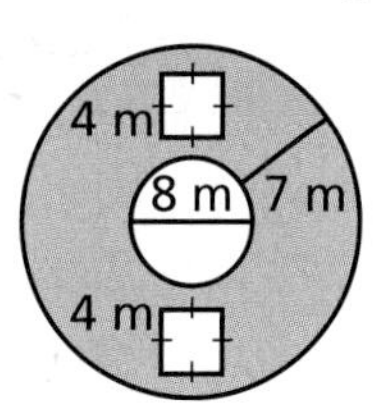

Entertainment LINK

The London Eye takes its passengers on a 30-minute flight that reaches a height of 450 feet above the River Thames.

28. ***ENTERTAINMENT*** The London Eye is an observation wheel with a diameter greater than 135 meters and less than 140 meters. Describe the range of the possible circumferences of the wheel to the nearest meter.

29. ***SPORTS*** The radius of the free-throw circle on an NBA basketball court is 6 ft. What is its circumference and area to the nearest tenth?

30. ***FOOD*** A pancake restaurant serves small silver dollar pancakes and regular-size pancakes.

a. What is the area of a silver dollar pancake to the nearest tenth?

b. What is the area of a regular pancake to the nearest tenth?

c. If 6 silver dollar pancakes are the same price as 3 regular pancakes, which is a better deal?

31. ***WHAT'S THE ERROR?*** The area of a circle is 169π in^2. A student says that this means the diameter is 13 in. What is the error?

32. ***WRITE ABOUT IT*** Explain how you would find the area of the composite figure shown. Then find the area.

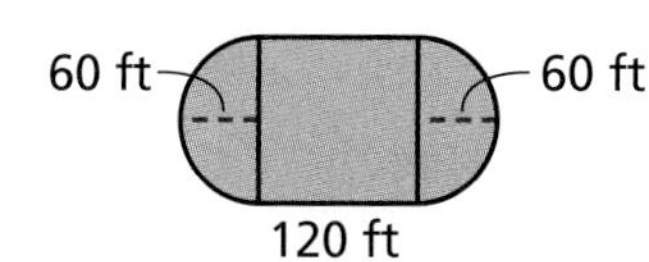

33. ***CHALLENGE*** Graph the circle with center (1, 2) that passes through the point (4, 6). Find its area and circumference, both in terms of π and to the nearest tenth.

Spiral Review

Multiply. Write each answer in simplest form. (Lesson 3-3)

34. $-8\left(3\frac{3}{4}\right)$ **35.** $\frac{6}{7}\left(\frac{7}{19}\right)$ **36.** $-\frac{5}{8}\left(-\frac{6}{15}\right)$ **37.** $-\frac{9}{10}\left(\frac{7}{12}\right)$

38. **EOG PREP** $\angle 1$ and $\angle 3$ are supplementary angles. If $m\angle 1 = 63°$, what is $m\angle 3$? (Lesson 5-1)

A 27° B 63° C 87° D 117°

Chapter 6

Mid-Chapter Quiz

LESSON 6-1 (pp. 280–284)

Find the perimeter of each figure.

1.

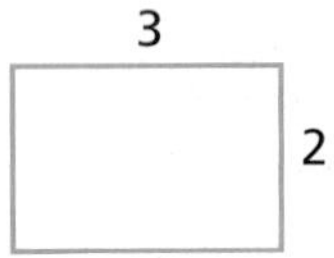

2.

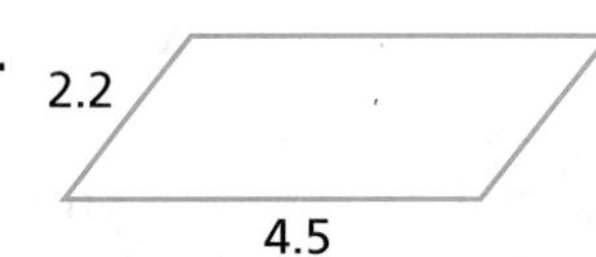

Graph and find the area of each figure with the given vertices.

3. (–3, 2), (–3, –2), (5, –2), (5, 2)

4. (–2, 4), (–2, –1), (2, –1), (2, 4)

5. (2, 4), (7, 4), (5, 0), (0, 0)

6. (7, –3), (2, –3), (–2, 3), (3, 3)

LESSON 6-2 (pp. 285–288)

Find the perimeter of each figure.

7.

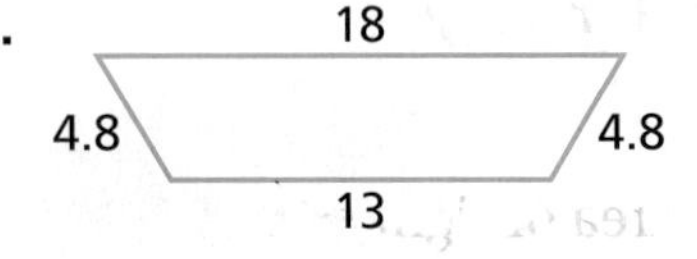

8.

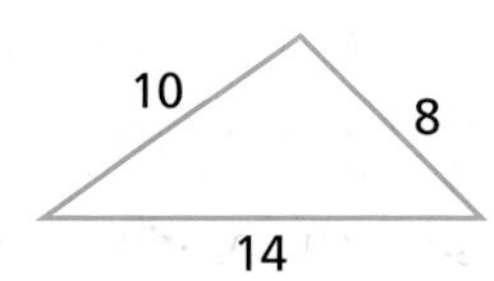

Graph and find the area of each figure with the given vertices.

9. (−6, −2), (4, −2), (−3, 3)

10. (−5, 0), (0, 0), (4, 4)

11. (2, −2), (3, 3), (−4, 3), (−3, −2)

12. (0, 4), (3, 6), (3, −3), (0, −3)

LESSON 6-3 (pp. 290–293)

Use the Pythagorean Theorem to find the height of each figure. Then find the area of each figure. If necessary, round the area to the nearest tenth of a square unit.

13.

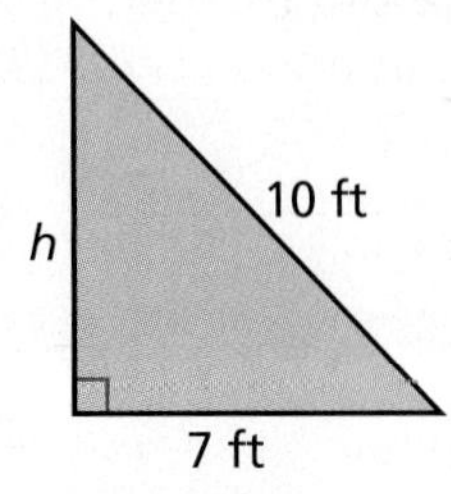

14.

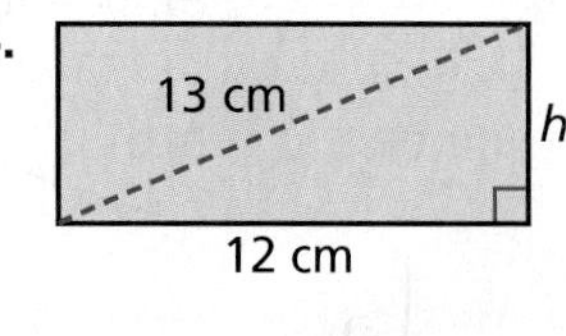

15.

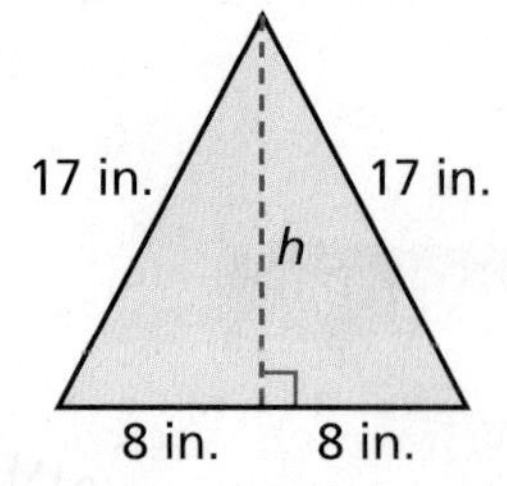

LESSON 6-4 (pp. 294–297)

Find the area and circumference of each circle, both in terms of π and to the nearest tenth. Use 3.14 for π.

16. radius = 15 cm

17. diameter = 6.5 ft

18. radius = $7\frac{1}{2}$ ft

Focus on Problem Solving

Look Back

- **Does your solution answer the question?**

When you think you have solved a problem, think again. Your answer may not really be the solution to the problem. For example, you may solve an equation to find the value of a variable, but to find the answer the problem is asking for, the value of the variable may need to be substituted into an expression.

Write and solve an equation for each problem. Check to see whether the value of the variable is the answer to the question. If not, give the answer to the question.

1. Triangle *ABC* is an isosceles triangle. Find its perimeter.

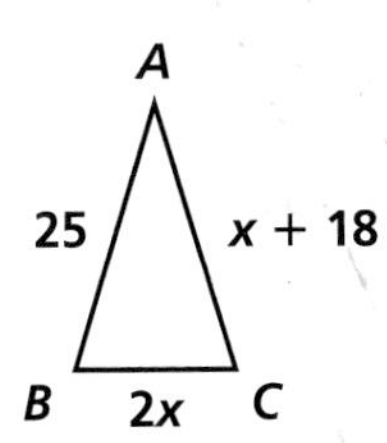

2. Find the measure of the smallest angle in triangle *DEF*.

3. Find the measure of the largest angle in triangle *DEF*.

4. Find the area of right triangle *GHI*.

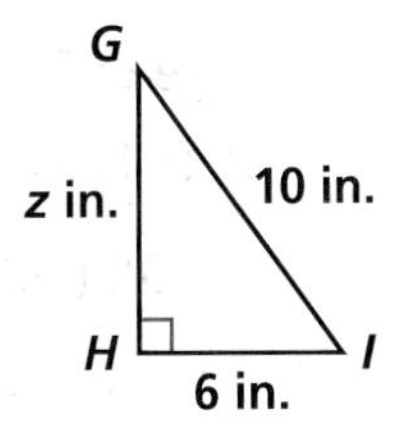

5. A *pediment* is a triangular space filled with statues on the front of a building. The approximate measurements of an isosceles triangular pediment are shown below. Find the area of the pediment.

Patterns of Solid Figures

Use with Lesson 6-5

WHAT YOU NEED	REMEMBER
• Ruler • Protractor • Tape • Paper	• A polygon is a closed plane figure formed by three or more line segments. • The faces of a regular polyhedron are congruent polygons.

A **polyhedron** is a solid figure in which every surface is a polygon. A net is a pattern of polygons used to model a regular polyhedron.

There are 5 regular polyhedra: tetrahedron, cube, octahedron, dodecahedron, and icosahedron.

Activity

1 Follow the directions to make each net. Then fold your nets into three-dimensional figures.

a. Draw a 2 in. equilateral triangle.
The measurement of each angle will be 60°.
Draw three more of them to look like Figure 1.
There will be **4** triangles.
Join the common edges and tape them together.
This is the net of a **tetrahedron.**

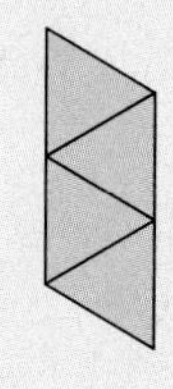

Figure 1

b. Draw a 2 in. square.
Draw 5 more of them to look like Figure 2.
There will be **6** squares.
Join the common edges and tape them together.
This is the net of a **cube.**

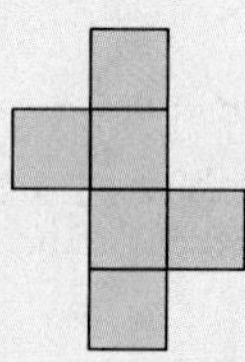

Figure 2

c. Draw a 2 in. equilateral triangle.
Draw 7 more of them to look like Figure 3.
There will be **8** triangles.
Join the common edges and tape them together.
This is the net of an **octahedron.**

Figure 3

d. Draw a regular pentagon with each side measuring 2 inches. The measurement of each angle will be 108°.
Draw 11 more of them to look like Figure 4.
There will be **12** pentagons.
Join the common edges and tape them together.
This is the net of a **dodecahedron.**

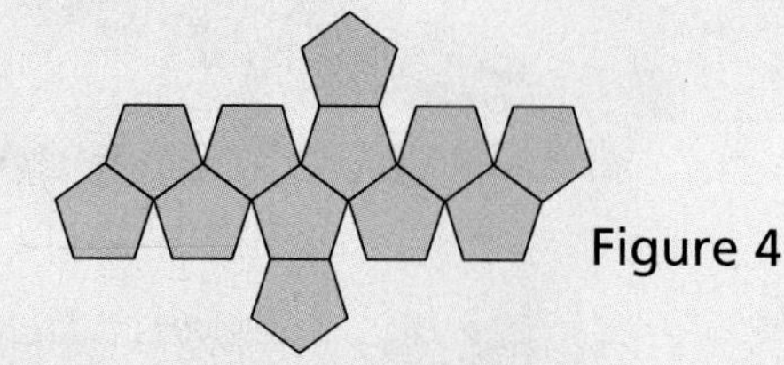

Figure 4

e. Draw a 2 in. equilateral triangle.
Draw 19 more of them to look like Figure 5.
There will be **20** triangles.
Join the common edges and tape them together.
This is the net of an **icosahedron.**

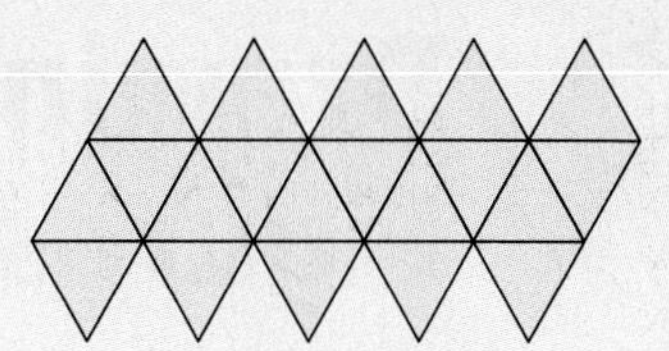
Figure 5

Copy the following table. Compare the number of vertices, faces, and edges in your polyhedra to the numbers listed in the table.

Polyhedron	Number of Vertices (V)	Number of Faces (F)	Number of Edges (E)
Tetrahedron	4	4	6
Cube	8	6	12
Octahedron	6	8	12
Dodecahedron	20	12	30
Icosahedron	12	20	30

Think and Discuss

1. Look for patterns in the table. What relationship can you find between the number of vertices, the number of faces, and the number of edges of regular polyhedra?
2. Can you make a net for an octahedron that is different from the net in Figure 3? Show the new net.

Try This

Copy and fold each net to determine whether it is the net of a polyhedron. If so, name the regular polyhedron that the net forms.

1. 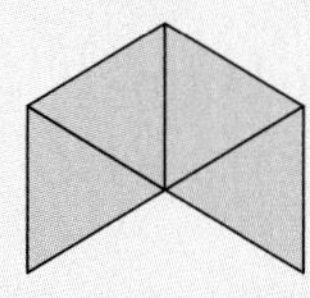

2. 

3. 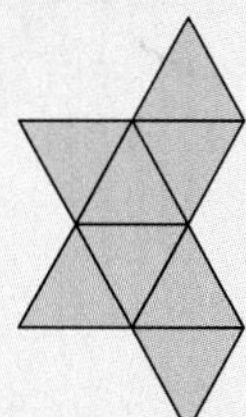

4. 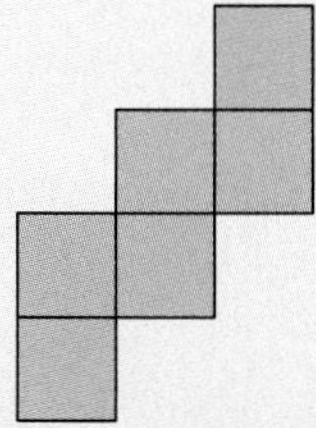

5. 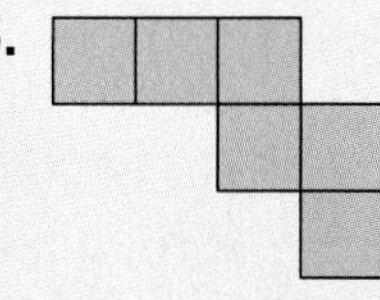

6. 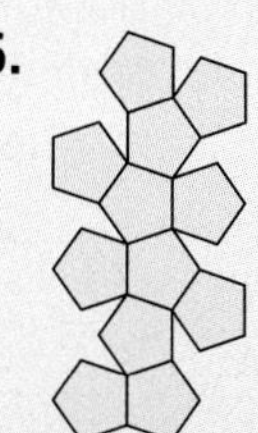

Give the missing number for each regular polygon.

7. 12 edges
   ■ vertices
   8 faces

8. ■ edges
   12 vertices
   20 faces

9. 30 edges
   20 vertices
   ■ faces

10. 6 edges
    ■ vertices
    4 faces

6-5 Drawing Three-Dimensional Figures

Learn to draw and identify parts of three-dimensional figures.

Vocabulary

face

edgc

vertex

perspective

vanishing point

horizon line

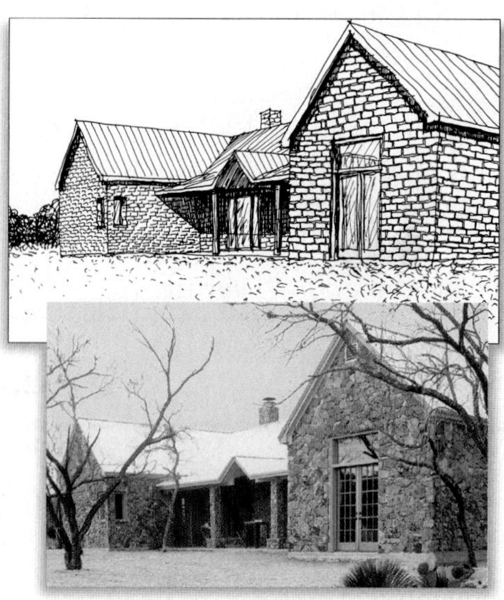

Architects use drawings to show what the exteriors of buildings will look like. Since they are drawing three-dimensional objects on two-dimensional surfaces, they must use special techniques to give the appearance of three dimensions.

Three-dimensional figures have *faces, edges,* and *vertices.* A **face** is a flat surface, an **edge** is where two faces meet, and a **vertex** is where three or more edges meet.

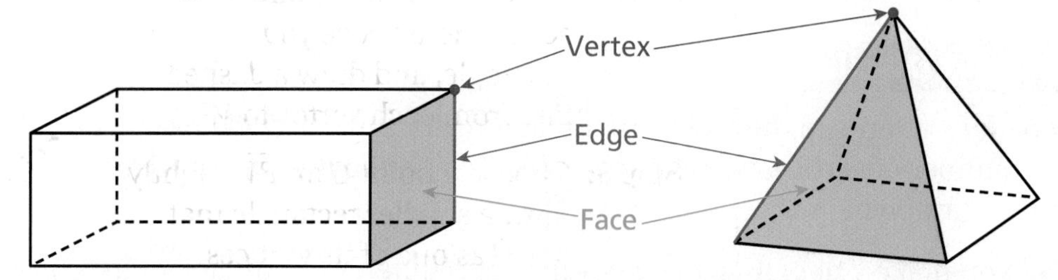

Isometric dot paper can be used to draw three-dimensional figures.

EXAMPLE 1 Drawing a Rectangular Box

Use isometric dot paper to sketch a rectangular box that is 4 units long, 2 units wide, and 3 units high.

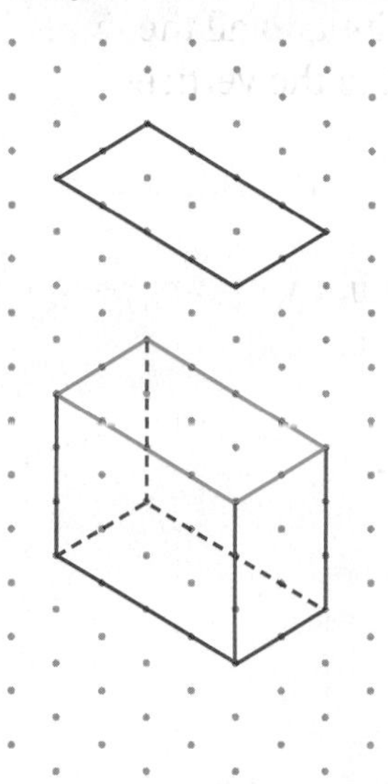

Step 1: Lightly draw the edges of the bottom face. It will look like a parallelogram.
2 units by 4 units

Step 2: Lightly draw the vertical line segments from the vertices of the base.
3 units high

Step 3: Lightly draw the top face by connecting the vertical lines to form a parallelogram.
2 units by 4 units

Step 4: Darken the lines.
Use solid lines for the edges that are visible and dashed lines for the edges that are hidden.

Perspective is a technique used to make drawings of three-dimensional objects appear to have depth and distance. In one-point perspective drawings, there is one **vanishing point**.

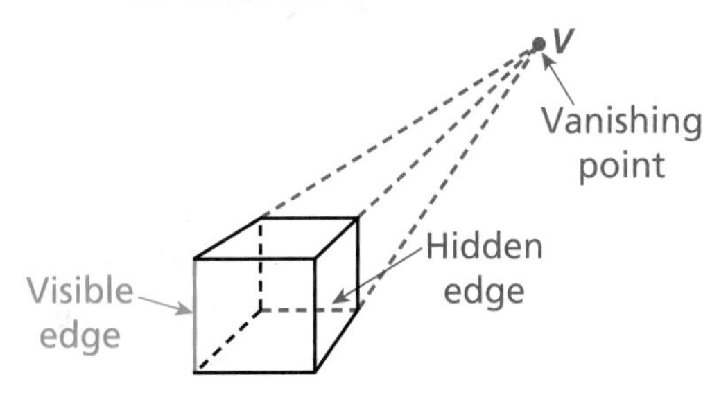

EXAMPLE 2 Sketching a One-Point Perspective Drawing

Sketch a one-point perspective drawing of a rectangular box.

Step 1: Draw a rectangle. This will be the front face.

Label the vertices A through D.

Step 2: Mark a vanishing point V somewhere above your rectangle, and draw a dashed line from each vertex to V.

Step 3: Choose a point G on $\overline{BV}$. Lightly draw a smaller rectangle that has G as one of its vertices.

Step 4: Connect the vertices of the two rectangles along the dashed lines.

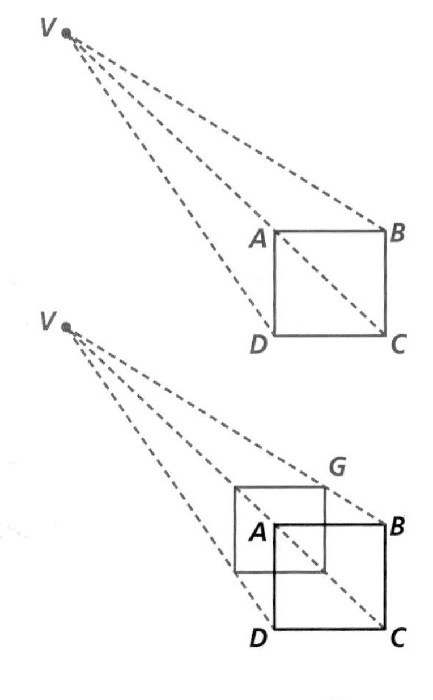

Step 5: Darken the visible edges, and draw dashed segments for the hidden edges. Erase the vanishing point and all the lines connecting it to the vertices.

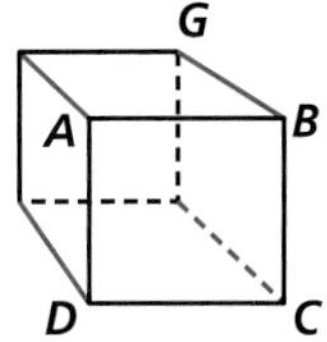

You can also draw a figure in two-point perspective by using two vanishing points and a **horizon line**.

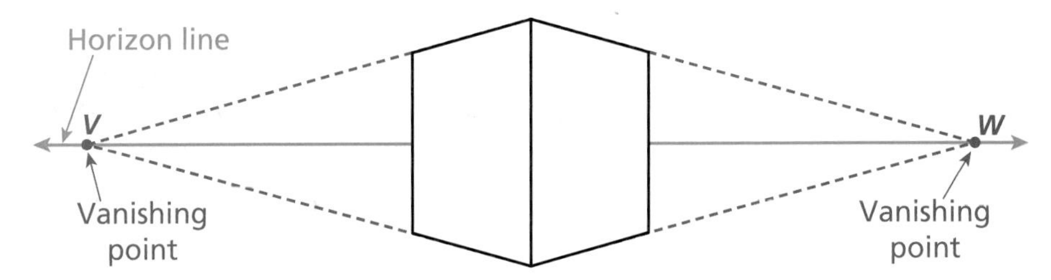

Moving the horizon line up and down gives you different views of the figure.

EXAMPLE 3 Sketching a Two-Point Perspective Drawing

Sketch a two-point perspective drawing of a rectangular box.

Step 1: Draw vertical segment $\overline{AD}$. Draw a horizontal line above $\overline{AD}$. Label vanishing points V and W on the line. Draw dashed segments $\overline{AV}$, $\overline{AW}$, $\overline{DV}$, and $\overline{DW}$.

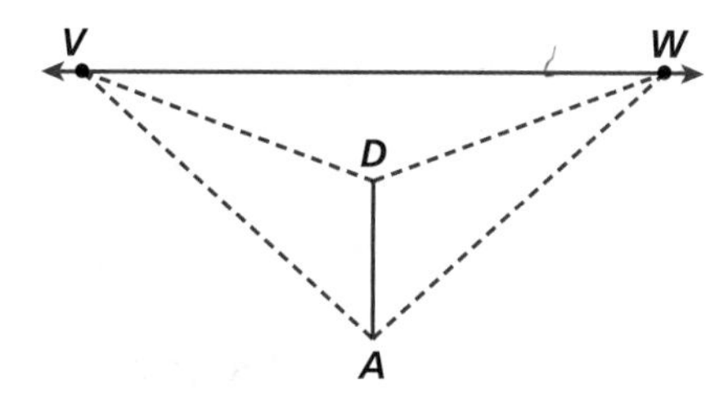

Step 2: Label points C on $\overline{DV}$ and E on $\overline{DW}$. Draw vertical segments through C and E. Draw $\overline{EV}$ and $\overline{CW}$.

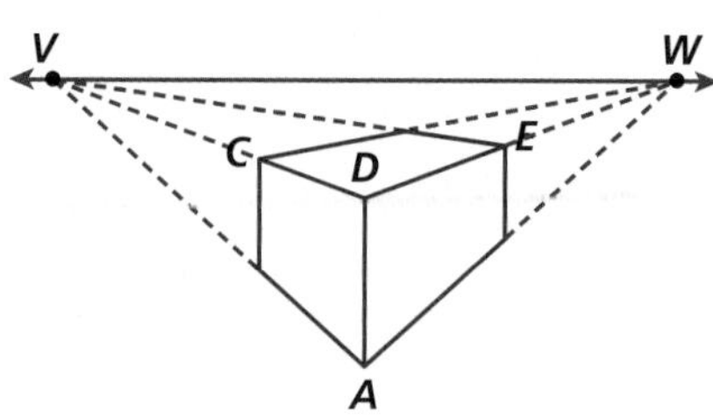

Step 3: Darken the visible edges. Erase horizon lines and dashed segments.

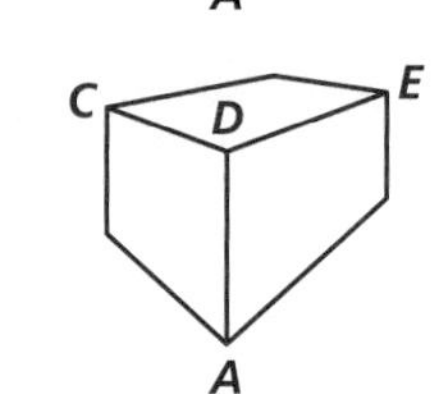

Think and Discuss

1. **Explain** whether parallel edges on a cube are always parallel on a perspective drawing of the cube.
2. **Demonstrate** your understanding of parallel and perpendicular edges, faces, and vertices of a rectangular box, using a cardboard box as a model.

6-5 Exercises

FOR EOG PRACTICE

see page 666

GUIDED PRACTICE

See Example 1

1. Use isometric dot paper to sketch a rectangular box that is 3 units long, 2 units wide, and 4 units high.

See Example 2

2. Sketch a one-point perspective drawing of a triangular box.

See Example 3

3. Sketch a two-point perspective drawing of a rectangular box.

INDEPENDENT PRACTICE

See Example 1 

4. Use isometric dot paper to sketch a rectangular box with a base 4 units long by 3 units wide and a height of 1 unit.

See Example 2

5. Sketch a one-point perspective drawing of a rectangular box.

See Example 3

6. Sketch a two-point perspective drawing of a triangular box.

PRACTICE AND PROBLEM SOLVING

Name all of the faces in each figure.

7. 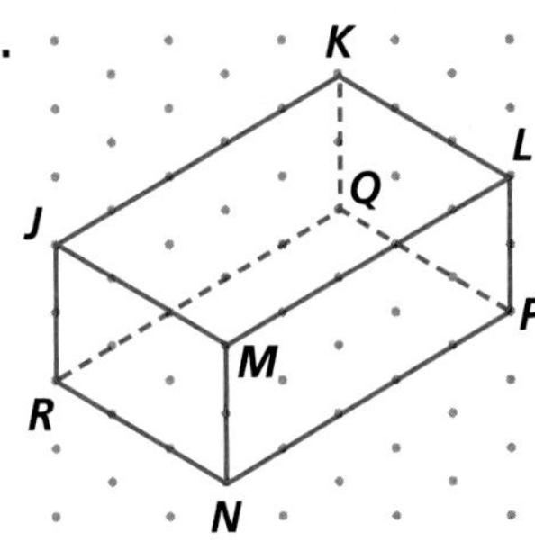

8. 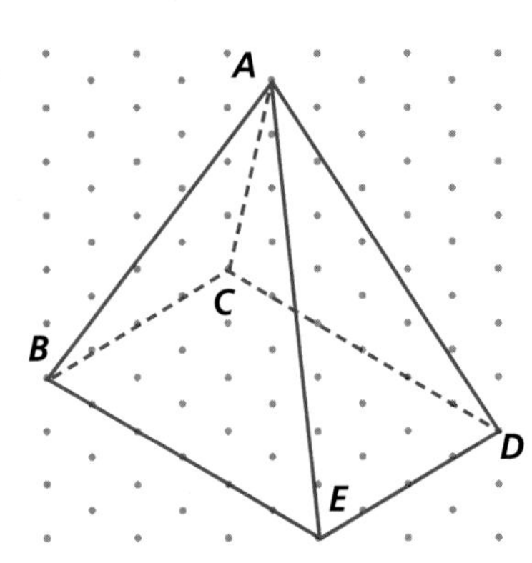

9. 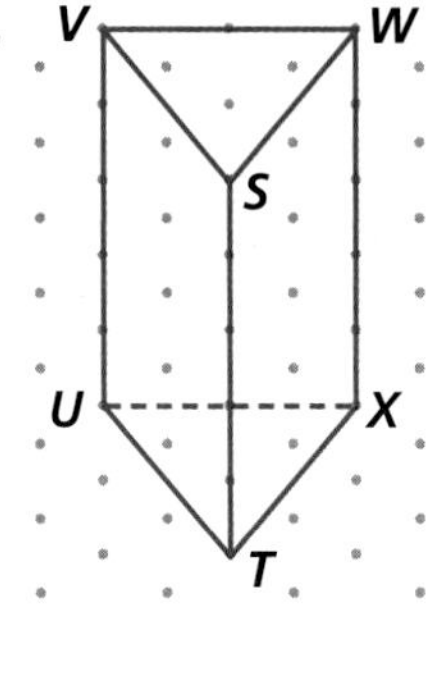

Use isometric dot paper to sketch each figure.

10. a cube 3 units on each side
11. a triangular box 5 units high
12. a rectangular box 7 units high, with base 5 units by 2 units
13. a box with parallel faces that are 3 units by 2 units and 4 units by 4 units
14. a box with parallel faces that are 2 units by 2 units and 5 units by 3 units

Use the one-point perspective drawing for Exercises 15–19.

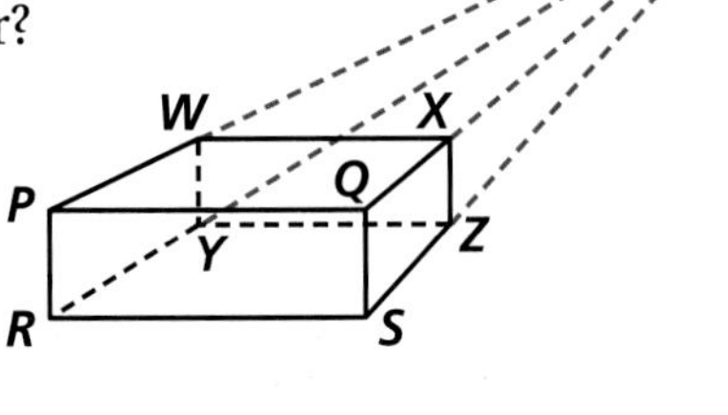

15. Name the vanishing point.
16. Which segments are parallel to each other?
17. Which face is the front face?
18. Which face is the back face?
19. Which segments are hidden edges of the figure?

Use the two-point perspective drawing for Exercises 20–24.

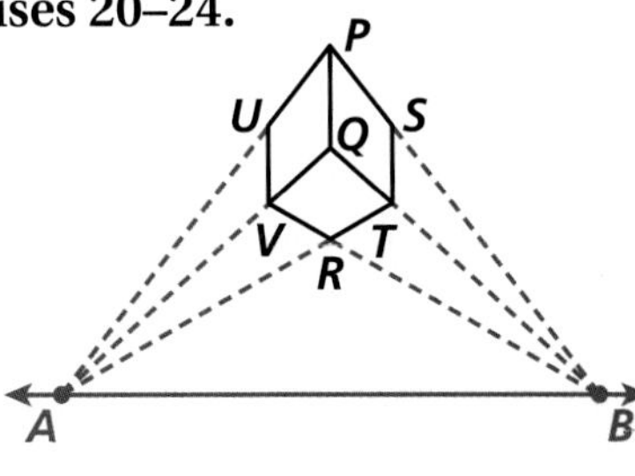

20. Name the vanishing points.
21. Which segments are parallel to each other?
22. Name the horizon line.
23. Which edge is nearest to the viewer?
24. Which segments are not edges of the figure?

The Chunnel was built from both ends at the same time. Specially designed tunneling machines completed an average of 125 m per week.

25. ***TRANSPORTATION*** Engineers long dreamed of linking England with the European mainland. In 1994, the dream became a reality with the opening of the Channel Tunnel, or Chunnel, which links Britain and France. The drawing shows the train *Eurostar* in the Chunnel. Is this an example of one-point or two-point perspective?

26. ***TECHNOLOGY*** Architects often use CADD (Computer Aided Design/Drafting) programs to create 3-D images of their ideas. Is the image an example of one-point or two-point perspective?

27. Copy the drawing below, and add another building like the one shown, with its lower front edge at $\overline{AB}$.

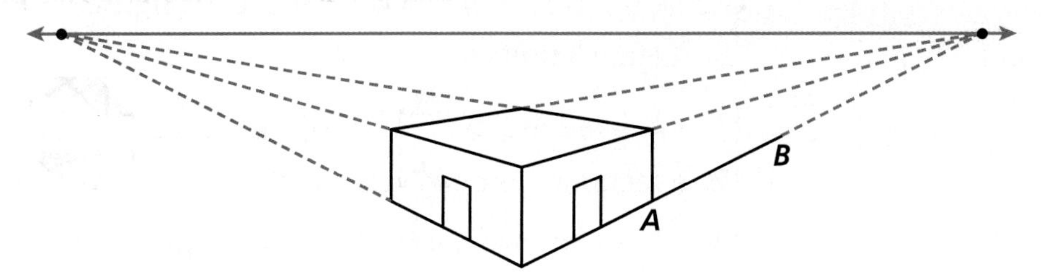

28. ***WHAT'S THE ERROR?*** A student sketched a 3-unit cube on dot paper. The student said that four faces and eight edges were visible in the sketch. What was the student's error?

29. ***WRITE ABOUT IT*** Describe the differences between a dot-paper drawing of a cube and a perspective drawing of a cube.

30. ***CHALLENGE*** Use one-point perspective to create a block-letter sign of your name.

Spiral Review

Write each number in standard notation. (Lesson 2-9)

31. 2.75×10^3 **32.** -4.2×10^2 **33.** 6.3×10^{-7} **34.** -1.9×10^{-4}

35. **EOG PREP** Which type of triangle can be constructed with a 50° angle between two 8-inch sides? (Lesson 5-3)

A Equilateral B Isosceles C Obtuse D Scalene

6-6 Volume of Prisms and Cylinders

Learn to find the volume of prisms and cylinders.

Vocabulary

prism

cylinder

Kansai International Airport, in Japan, is built on the world's largest man-made island. To find the amount of rock, gravel, and concrete needed to build the island, you need to know how to find the volume of a *rectangular prism.*

A **prism** is a three-dimensional figure named for the shape of its bases. The two bases are congruent polygons. All of the other faces are parallelograms. A **cylinder** has two circular bases.

Remember!

If all six faces of a rectangular prism are squares, it is a cube.

Triangular prism

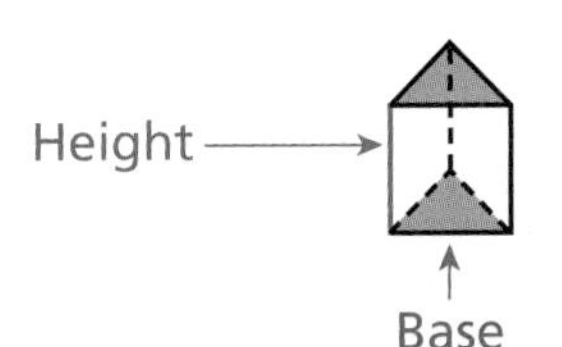

Rectangular prism

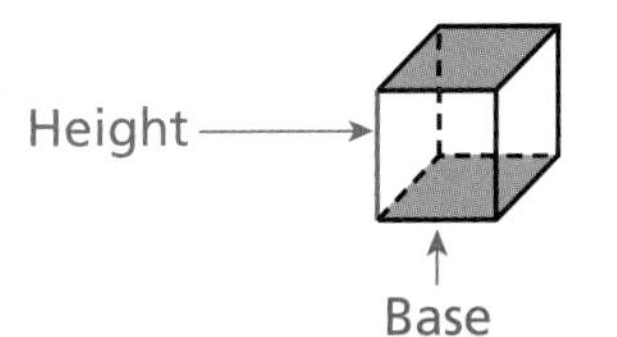

Cylinder

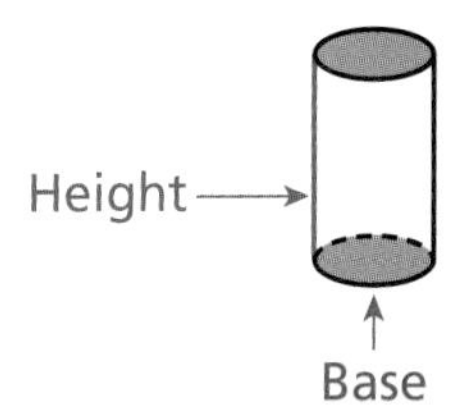

VOLUME OF PRISMS AND CYLINDERS		
Words	**Numbers**	**Formula**
Prism: The volume V of a prism is the area of the base B times the height h.	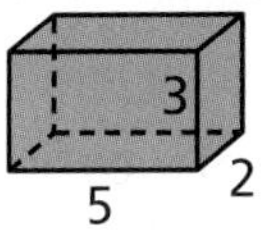 (prism 5, 2, 3) $B = 2(5) = 10 \text{ units}^2$; $V = (10)(3) = 30 \text{ units}^3$	$V = Bh$
Cylinder: The volume of a cylinder is the area of the base B times the height h.	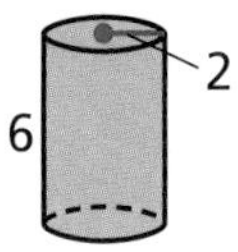 (cylinder radius 2, height 6) $B = \pi(2^2) = 4\pi \text{ units}^2$; $V = (4\pi)(6) = 24\pi \approx 75.4 \text{ units}^3$	$V = Bh = (\pi r^2)h$

EXAMPLE 1 **Finding the Volume of Prisms and Cylinders**

Helpful Hint

Area is measured in *square units.* Volume is measured in *cubic units.*

Find the volume of each figure to the nearest tenth.

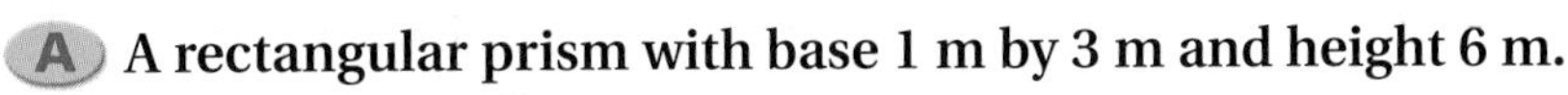

A **A rectangular prism with base 1 m by 3 m and height 6 m.**

$B = 1 \cdot 3 = 3 \text{ m}^2$ *Area of base*

$V = Bh$ *Volume of prism*

$= 3 \cdot 6 = 18 \text{ m}^3$

Find the volume of each figure to the nearest tenth.

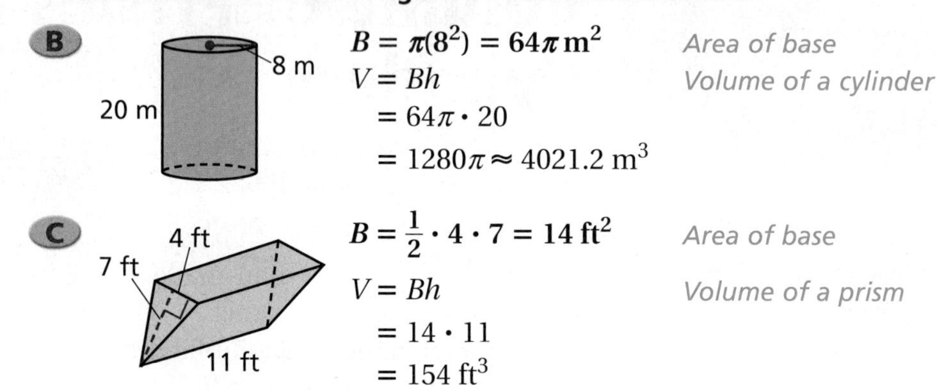

B

$B = \pi(8^2) = 64\pi \text{ m}^2$ *Area of base*

$V = Bh$ *Volume of a cylinder*

$= 64\pi \cdot 20$

$= 1280\pi \approx 4021.2 \text{ m}^3$

C

$B = \frac{1}{2} \cdot 4 \cdot 7 = 14 \text{ ft}^2$ *Area of base*

$V = Bh$ *Volume of a prism*

$= 14 \cdot 11$

$= 154 \text{ ft}^3$

The volume of a rectangular prism can be written as $V = \ell wh$, where ℓ is the length, w is the width, and h is the height.

EXAMPLE 2 Exploring the Effects of Changing Dimensions

A **A juice box measures 3 in. by 2 in. by 4 in. Explain whether doubling the length, width, or height of the box would double the amount of juice the box holds.**

Original Dimensions	Double the Length	Double the Width	Double the Height
$V = \ell wh$ $= 3 \cdot 2 \cdot 4$ $= 24 \text{ in}^3$	$V = (2\ell)wh$ $= 6 \cdot 2 \cdot 4$ $= 48 \text{ in}^3$	$V = \ell(2w)h$ $= 3 \cdot 4 \cdot 4$ $= 48 \text{ in}^3$	$V = \ell w(2h)$ $= 3 \cdot 2 \cdot 8$ $= 48 \text{ in}^3$

The original box has a volume of 24 in^3. You could double the volume to 48 in^3 by doubling any one of the dimensions. So doubling the length, width, or height would double the amount of juice the box holds.

B **A juice can has a radius of 1.5 in. and a height of 5 in. Explain whether doubling the height of the can would have the same effect on the volume as doubling the radius.**

Original Dimensions	Double the Radius	Double the Height
$V = \pi r^2 h$ $= 1.5^2\pi \cdot 5$ $= 11.25\pi \text{ in}^3$	$V = \pi(2r)^2 h$ $= 3^2\pi \cdot 5$ $= 45\pi \text{ in}^3$	$V = \pi r^2(2h)$ $= 1.5^2\pi \cdot (2 \cdot 5)$ $= 22.5\pi \text{ in}^3$

By doubling the height, you would double the volume. By doubling the radius, you would increase the volume to four times the original.

EXAMPLE 3 Construction Application

Japan
Osaka
Awaji Island
Kansai Island

Kansai International Airport is on a man-made island that is a rectangular prism measuring 60 ft deep, 4000 ft wide, and 2.5 miles long. What is the volume of rock, gravel, and concrete that was needed to build the island?

length = 2.5 mi = 2.5(5280) ft
= 13,200 ft *1 mi = 5280 ft*

width = 4000 ft

height = 60 ft

$V = 13{,}200 \cdot 4000 \cdot 60 \text{ ft}^3$ *V = lwh*

$= 3{,}168{,}000{,}000 \text{ ft}^3$

The volume of rock, gravel, and concrete needed was $3{,}168{,}000{,}000 \text{ ft}^3$, which is equivalent to nearly 24 billion gallons of water.

To find the volume of a composite three-dimensional figure, find the volume of each part and add the volumes together.

EXAMPLE 4 Finding the Volume of Composite Figures

Find the volume of the milk carton.

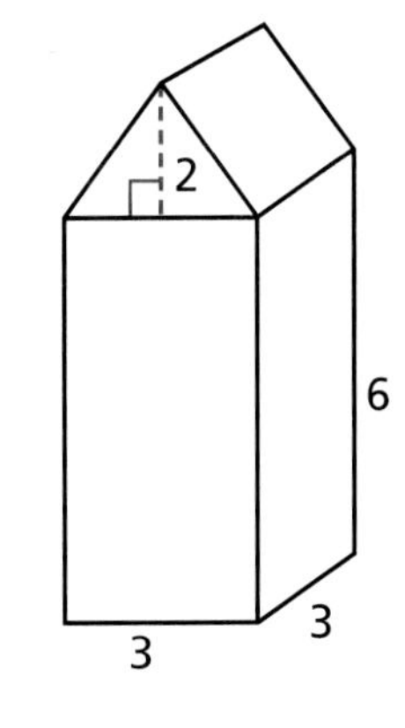

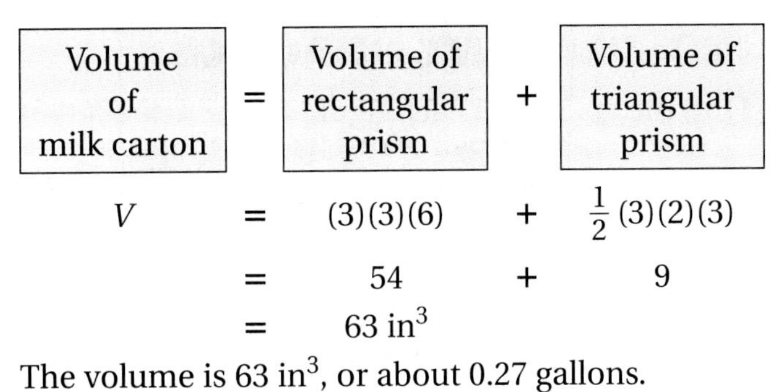

Volume of milk carton	=	Volume of rectangular prism	+	Volume of triangular prism
V	=	(3)(3)(6)	+	$\frac{1}{2}$(3)(2)(3)
	=	54	+	9
	=	63 in^3		

The volume is 63 in^3, or about 0.27 gallons.

Think and Discuss

1. **Give an example** that shows that two rectangular prisms can have different heights but the same volume.
2. **Apply** your results from Example 2 to make a conclusion about changing dimensions in a triangular prism.
3. **Describe** what happens to the volume of a cylinder when the diameter of the base is tripled.

6-6 Exercises

FOR EOG PRACTICE

see page 666

internet connect
Homework Help Online
go.hrw.com Keyword: MT4 6-6

2.01, 3.01, 3.02, 5.04

GUIDED PRACTICE

See Example 1

Find the volume of each figure to the nearest tenth. Use 3.14 for π.

1. 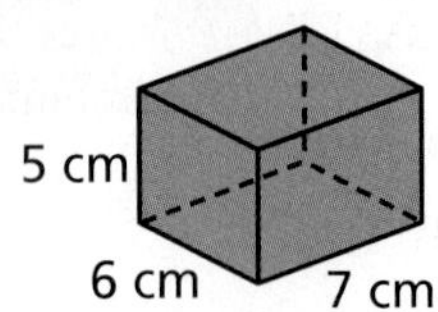

2. 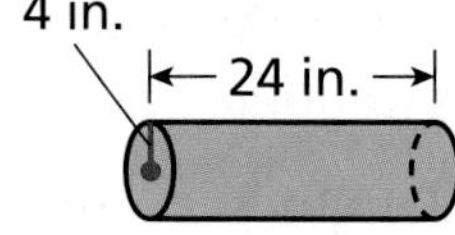

3. 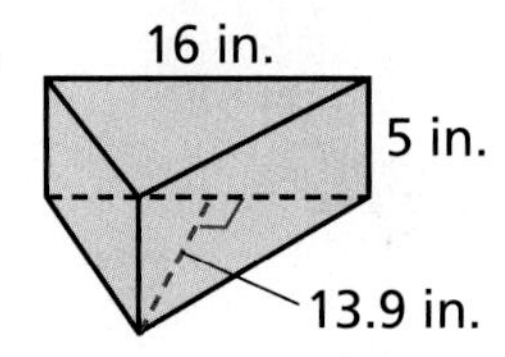

See Example 2

4. A box measures 4 in. by 3 in. by 5 in. Explain whether tripling a side from 4 in. to 12 in. would triple the volume of the box.

5. A can of vegetables has radius 2 in. and height 4 in. Explain whether tripling the radius would triple the volume of the can.

See Example 3

6. Grain is stored in cylindrical structures called *silos*. What is the volume of a silo with diameter 15 feet and height 25 feet?

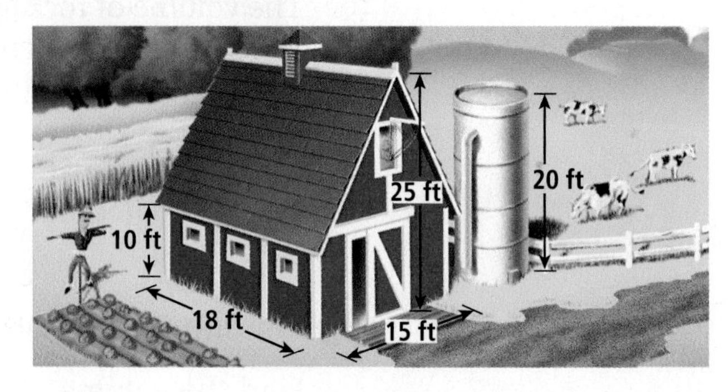

See Example 4

7. Find the volume of the barn.

INDEPENDENT PRACTICE

See Example 1

Find the volume of each figure to the nearest tenth. Use 3.14 for π.

8. 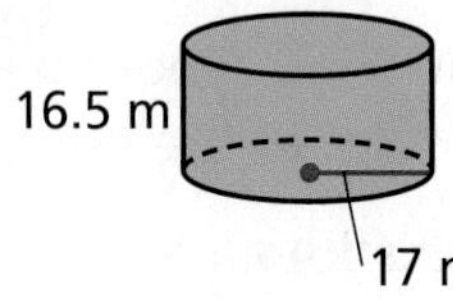

9. 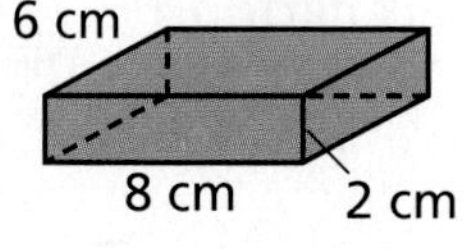

10. 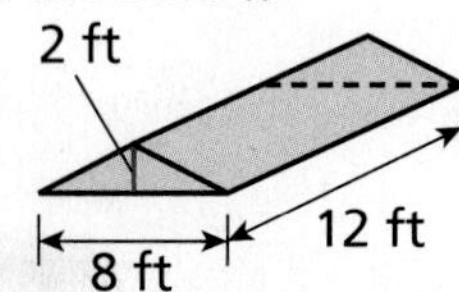

See Example 

11. A toy box measures 4 ft by 3 ft by 2 ft. Explain whether increasing the height by four times, from 2 ft to 8 ft, would increase the volume by four times.

12. A cylindrical oatmeal box has diameter 4 in. and height 7 in. Explain whether increasing the diameter by 1.5 times would increase the volume by 1.5 times.

See Example 

13. An ink cartridge for a printer is 5 cm by 3 cm by 4 cm. What is the volume of the ink cartridge?

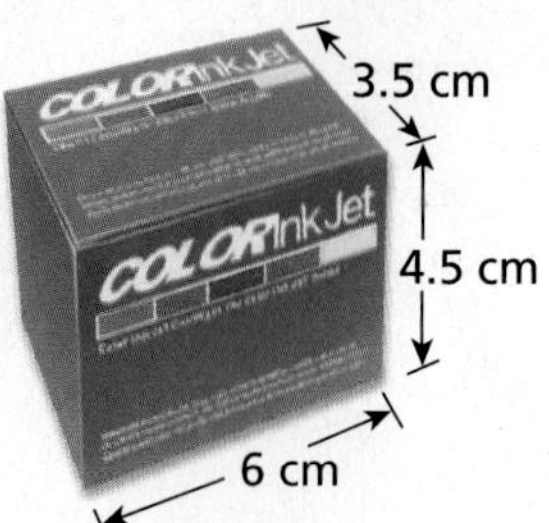

See Example 

14. Find the volume of the box containing the ink cartridge.

PRACTICE AND PROBLEM SOLVING

Through the 52 large windows of the Giant Ocean Tank, visitors can see 3000 corals and sponges as well as large sharks, sea turtles, barracudas, moray eels, and hundreds of tropical fishes.

15. ***LIFE SCIENCE*** The cylindrical Giant Ocean Tank at the New England Aquarium in Boston has a volume of 200,000 gallons.
    a. One gallon of water equals 231 cubic inches. How many cubic inches of water are in the Giant Ocean Tank?
    b. Use your answer from part **a** as the volume. The tank is 24 ft deep. Find the radius in feet of the Giant Ocean Tank.

16. ***ENTERTAINMENT*** An outdoor theater group sets up a portable stage. The stage comes in sections that are 48 in. by 96 in. by 36 in.
    a. What are the dimensions in feet of one stage section?
    b. What is the volume in cubic feet of one section?
    c. If the stage has a total volume of 864 ft^3, how many sections make up the stage?

17. ***SOCIAL STUDIES*** The tablet held by the Statue of Liberty is approximately a rectangular prism with volume 1,107,096 in^3. Estimate the thickness of the tablet.

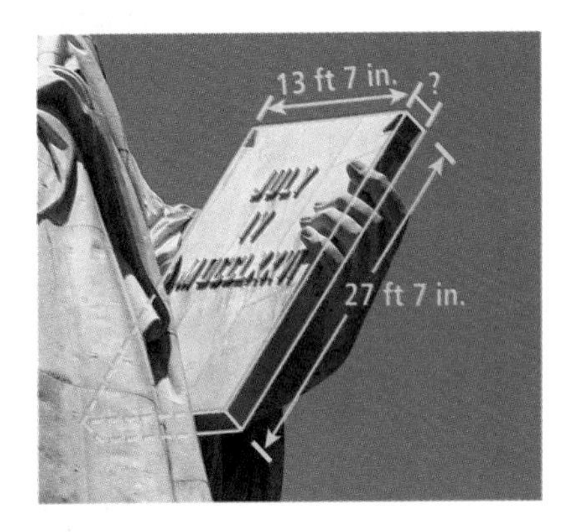

18. ***LIFE SCIENCE*** Air has about 4000 bacteria per cubic meter. There are about 120,000 bacteria in a room that is 3 m long by 4 m wide. What is the height of the room?

19. ***WHAT'S THE ERROR?*** A student read this statement in a book: "The volume of a triangular prism with height 10 cm and base area 25 cm is 250 cm^3." Correct the error in the statement.

20. ***WRITE ABOUT IT*** Explain why one cubic foot equals 1728 cubic inches.

21. ***CHALLENGE*** A 6 cm section of plastic water pipe has inner diameter 12 cm and outer diameter 15 cm. Find the volume of the plastic pipe, not the hollow interior, to the nearest tenth.

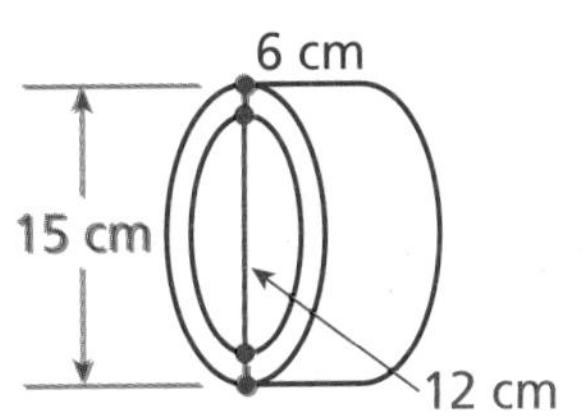

Spiral Review

Find the mean, median, and mode of each data set to the nearest tenth. (Lesson 4-3)

22. 3, 5, 5, 6, 9, 3, 5, 2, 5 **23.** 17, 15, 14, 16, 18, 13 **24.** 100, 75, 48, 75, 48, 63, 45

25. **EOG PREP** What is the sum of the angle measures of an octagon? (Lesson 5-4)

A 8° B 135° C 1080° D 1440°

6-7 Volume of Pyramids and Cones

Learn to find the volume of pyramids and cones.

Vocabulary

pyramid

cone

The Great Pyramid's height is equivalent to that of a forty-story skyscraper. The pyramid covers an area of thirteen acres.

The Great Pyramid of Giza was built using about 2.5 million blocks of stone, each weighing at least two tons. It is believed that 20,000 to 30,000 workers took about 20 years to complete the pyramid.

A **pyramid** is named for the shape of its base. The base is a polygon, and all of the other faces are triangles. A **cone** has a circular base. The height of a pyramid or cone is measured from the highest point to the base along a perpendicular line.

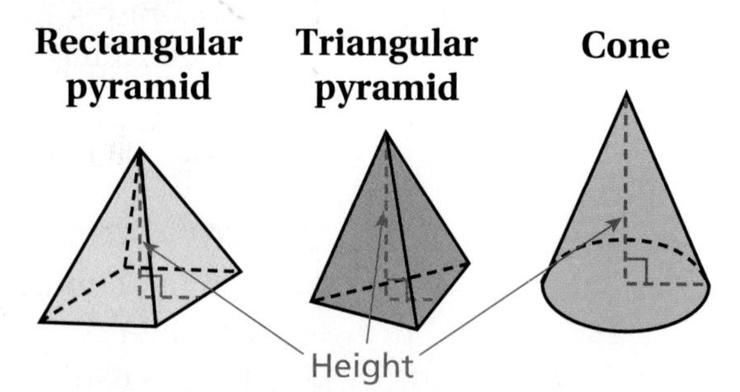

VOLUME OF PYRAMIDS AND CONES

Words	Numbers	Formula
Pyramid: The volume V of a pyramid is one-third of the area of the base B times the height h.	$B = 3(3)$ $= 9 \text{ units}^2$ $V = \frac{1}{3}(9)(4)$ $= 12 \text{ units}^3$	$V = \frac{1}{3}Bh$
Cone: The volume of a cone is one-third of the area of the circular base B times the height h.	$B = \pi(2^2)$ $= 4\pi \text{ units}^2$ $V = \frac{1}{3}(4\pi)(3)$ $= 4\pi$ $\approx 12.6 \text{ units}^3$	$V = \frac{1}{3}Bh$ or $V = \frac{1}{3}\pi r^2 h$

EXAMPLE 1 Finding the Volume of Pyramids and Cones

Find the volume of each figure.

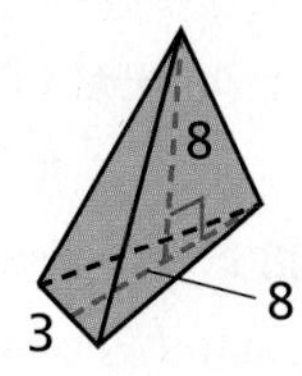

$B = \frac{1}{2}(3 \cdot 8) = 12 \text{ units}^2$

$V = \frac{1}{3} \cdot 12 \cdot 8$ $\quad V = \frac{1}{3}Bh$

$V = 32 \text{ units}^3$

Find the volume of each figure.

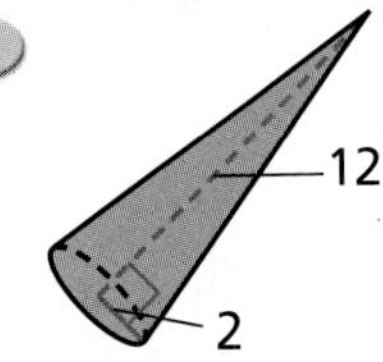

$B = \pi(2^2) = 4\pi \text{ units}^2$

$V = \frac{1}{3} \cdot 4\pi \cdot 12$ — *$V = \frac{1}{3}Bh$*

$V = 16\pi \approx 50.3 \text{ units}^3$ — *Use 3.14 for π.*

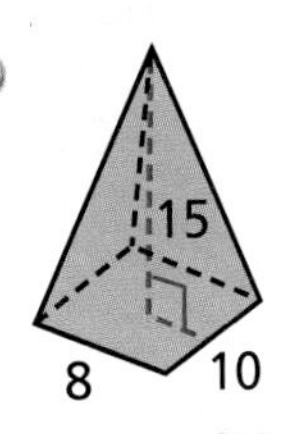

$B = 10 \cdot 8 = 80 \text{ units}^2$

$V = \frac{1}{3} \cdot 80 \cdot 15$ — *$V = \frac{1}{3}Bh$*

$V = 400 \text{ units}^3$

EXAMPLE 2 Exploring the Effects of Changing Dimensions

A cone has radius 7 ft and height 14 ft. Explain whether doubling the height would have the same effect on the volume of the cone as doubling the radius.

Original Dimensions	Double the Height	Double the Radius
$V = \frac{1}{3}\pi r^2 h$	$V = \frac{1}{3}\pi r^2(2h)$	$V = \frac{1}{3}\pi(2r)^2 h$
$= \frac{1}{3}\pi(7^2)(14)$	$= \frac{1}{3}\pi(7^2)(2 \cdot 14)$	$= \frac{1}{3}\pi(2 \cdot 7)^2(14)$
$\approx 718.01 \text{ ft}^3$	$\approx 1436.03 \text{ ft}^3$	$\approx 2872.05 \text{ ft}^3$

When the height of the cone is doubled, the volume is doubled. When the radius is doubled, the volume becomes 4 times the original volume.

EXAMPLE 3 *Social Studies Application*

The Great Pyramid of Giza is a square pyramid. Its height is 481 ft, and its base has 756 ft sides. Find the volume of the pyramid.

$B = 756^2 = 571{,}536 \text{ ft}^2$ — *$A = bh$*

$V = \frac{1}{3}(571{,}536)(481)$ — *$V = \frac{1}{3}Bh$*

$V = 91{,}636{,}272 \text{ ft}^3$

Think and Discuss

1. **Describe** two or more ways that you can change the dimensions of a rectangular pyramid to double its volume.
2. **Compare** the volume of a cube with 1 in. sides with a pyramid that is 1 in. high and has a square base with 1 in. sides.

6-7 Exercises

FOR EOG PRACTICE

see page 666

internet connect

Homework Help Online

go.hrw.com Keyword: MT4 6-7

2.01, 3.01, 3.02, 5.04

GUIDED PRACTICE

See Example 1 **Find the volume of each figure to the nearest tenth. Use 3.14 for π.**

1.
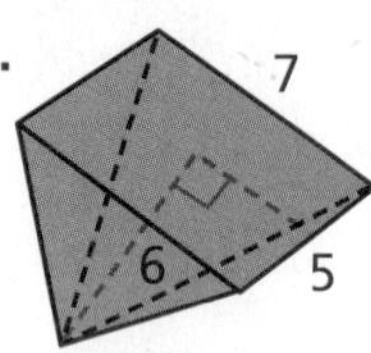

2.
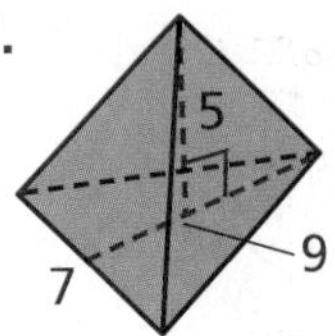

3.
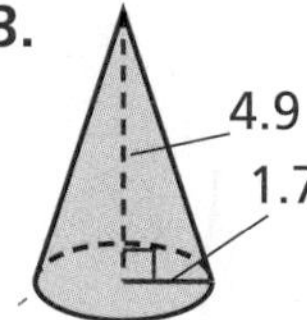

4.
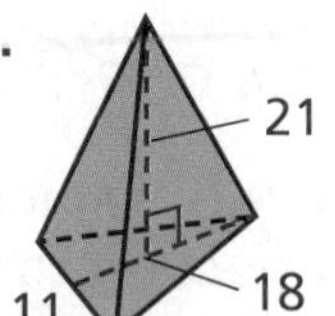

5.
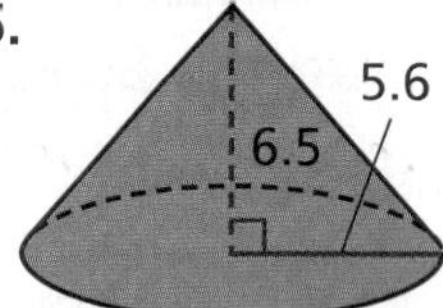

6.
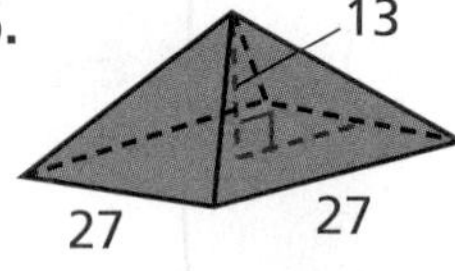

See Example 

7. A square pyramid has height 4 ft and a base that measures 3 ft on each side. Explain whether doubling the height would double the volume of the pyramid.

See Example 

8. The Transamerica Pyramid in San Francisco has a base area of 22,000 ft^2 and a height of 853 ft. What is the volume of the building?

INDEPENDENT PRACTICE

See Example  **Find the volume of each figure to the nearest tenth. Use 3.14 for π.**

9.
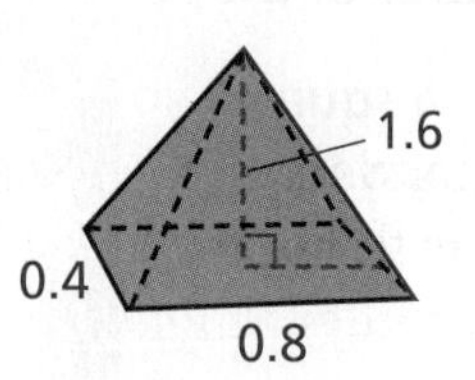

10.
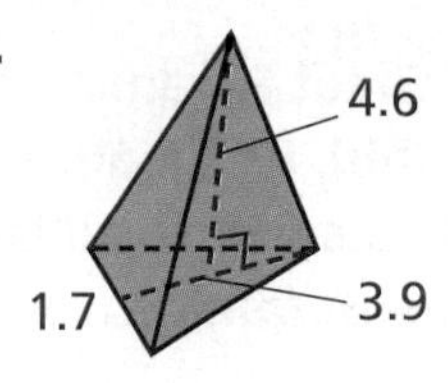

11.
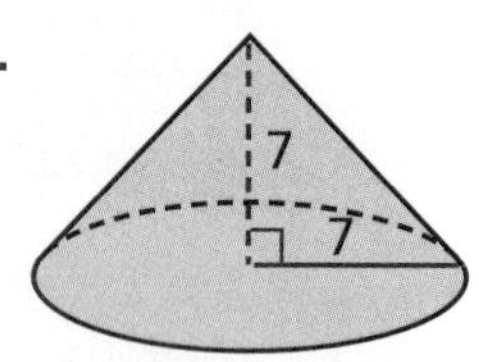

12.
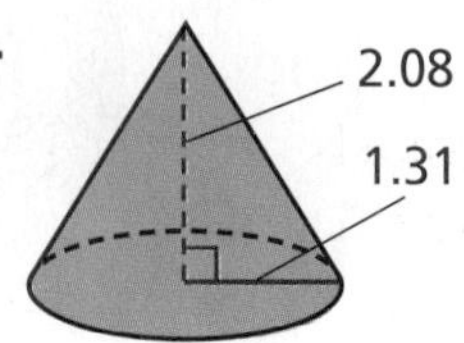

13.
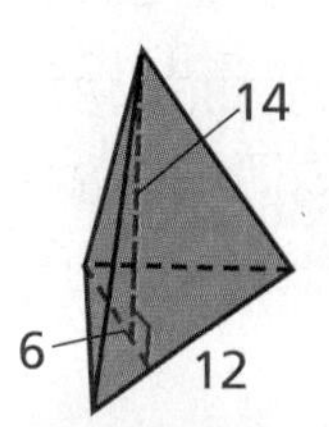

14.
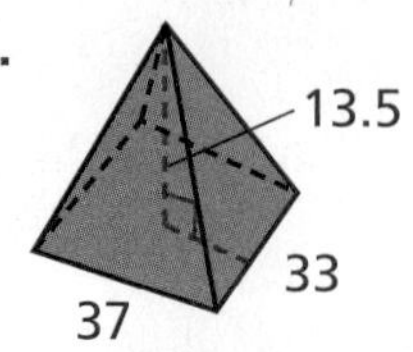

See Example 

15. A triangular pyramid has a height of 6 ft. The triangular base has a height of 6 ft and a width of 6 ft. Explain whether doubling the height of the base would double the volume of the pyramid.

See Example 

16. A cone-shaped building is commonly used to store rock salt. What would be the volume of a cone-shaped building with diameter 70 ft and height 50 ft, to the nearest hundredth?

PRACTICE AND PROBLEM SOLVING

Find the missing measure to the nearest tenth. Use 3.14 for π.

17. cone:
radius = 3 cm
height = ▒
volume = 37.7 cm^3

18. cylinder:
radius = ▒
height = 2 cm
volume = 75.36 cm^3

19. triangular pyramid:
base height = ▒
base width = 10 ft
height = 7 ft
volume = 105 ft^3

20. rectangular pyramid:
base length = 3 ft
base width = ▒
height = 7 ft
volume = 42 ft^3

Career LINK

I. M. Pei, designer of the Louvre Pyramid, has designed more than 50 buildings around the world and has won many major awards.

go.hrw.com
KEYWORD: MT4 Pei

21. ***ARCHITECTURE*** The pyramid at the entrance to the Louvre in Paris has a height of 72 feet and a square base that is 112 feet long on each side. What is the volume of this pyramid?

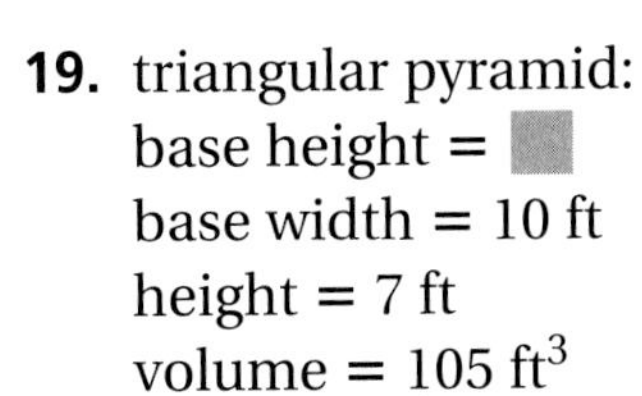

22. ***TRANSPORTATION*** Orange traffic cones, or pylons, come in a variety of sizes. What is the volume in cubic inches of a pylon with height 3 feet and diameter 9 inches?

23. ***ARCHITECTURE*** The Pyramid Arena in Memphis, Tennessee, is 321 feet tall and has a square base that is 200 yards on each side.

a. What is the volume in cubic feet of the arena?

b. How many cubic feet are in one cubic yard?

c. What is the volume in cubic yards of the arena to the nearest hundredth?

24. ***WHAT'S THE ERROR?*** A student says that the formula for the volume of a cylinder is the same as the formula for the volume of a pyramid, $\frac{1}{3}Bh$. What error did this student make?

25. ***WRITE ABOUT IT*** How would a cone's volume be affected if you doubled the height? the radius?

26. ***CHALLENGE*** The diameter of a cone is x in., the height is 12 in., and the volume is 36π in^3. What is x?

Spiral Review

Use guess and check to estimate each square root to two decimal places. (Lesson 3-9)

27. $\sqrt{35}$ **28.** $\sqrt{45}$ **29.** $\sqrt{55}$ **30.** $\sqrt{65}$

31. **EOG PREP** What is $\frac{15^3 \cdot 15^{11}}{15^{-13}}$ written as one power? (Lesson 2-7)

A 1 B 15^1 C 15^{27} D 15^{46}

Surface Area of Prisms and Cylinders

Learn to find the surface area of prisms and cylinders.

Vocabulary

surface area

lateral facc

lateral surface

An *anamorphic image* is a distorted picture that becomes recognizable when reflected onto a cylindrical mirror.

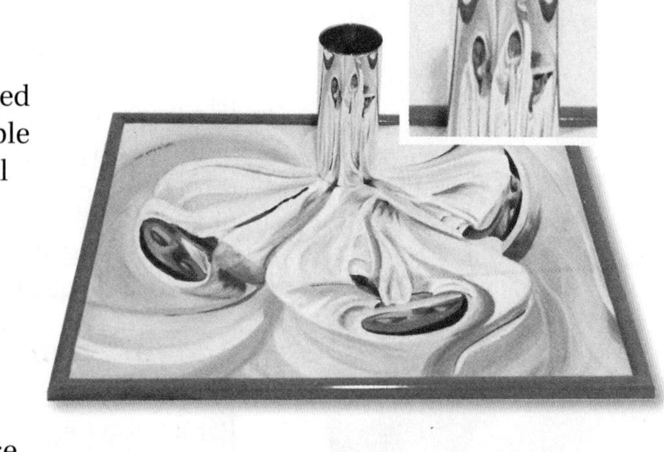

Surface area is the sum of the areas of all surfaces of a figurc. The **lateral faces** of a prism are parallelograms that connect the bases. The **lateral surface** of a cylinder is the curved surface.

SURFACE AREA OF PRISMS AND CYLINDERS

Words	Numbers	Formula
Prism: The surface area S of a prism is twice the base area B plus the lateral area F. The lateral area is the base perimeter P times the height h.	$S = 2(3 \cdot 2) + (10)(5) = 62$ units^2	$S = 2B + F$ or $S = 2B + Ph$
Cylinder: The surface area S of a cylinder is twice the base area B plus the lateral area L. The lateral area is the base circumference $2\pi r$ times the height h.	$S = 2\pi(5^2) + 2\pi(5)(6) \approx 345.4$ units^2	$S = 2B + L$ or $S = 2\pi r^2 + 2\pi rh$

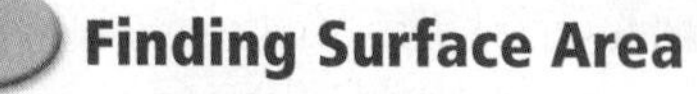

Finding Surface Area

Find the surface area of each figure.

A

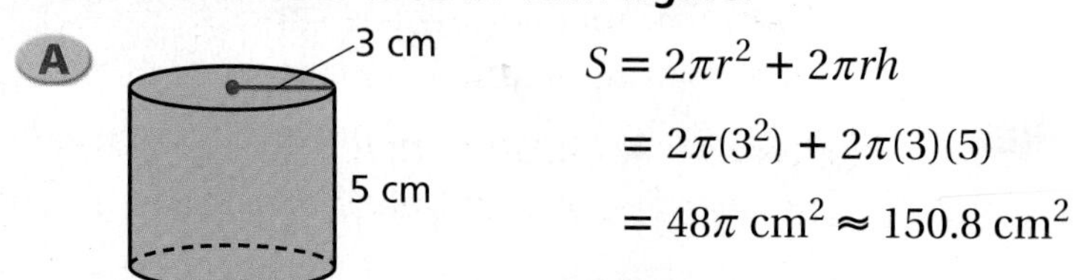

$$S = 2\pi r^2 + 2\pi rh$$
$$= 2\pi(3^2) + 2\pi(3)(5)$$
$$= 48\pi \text{ cm}^2 \approx 150.8 \text{ cm}^2$$

Find the surface area of each figure.

B

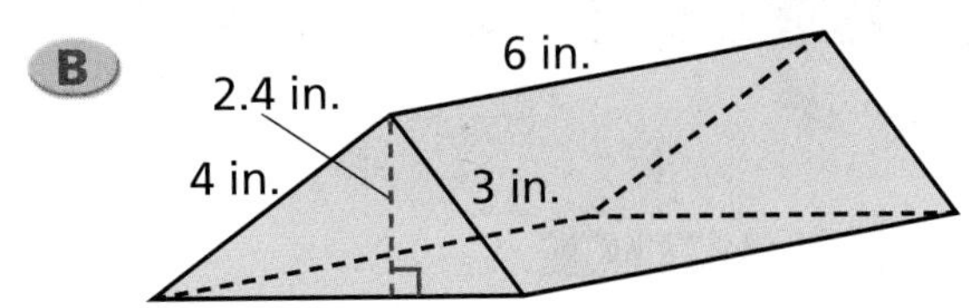

$S = 2B + Ph$

$= 2(\frac{1}{2} \cdot 5 \cdot 2.4) + (12)(6)$

$= 84 \text{ in}^2$

EXAMPLE 2 Exploring the Effects of Changing Dimensions

A cylinder has diameter 8 in. and height 3 in. Explain whether doubling the height would have the same effect on the surface area as doubling the radius.

Original Dimensions	Double the Height	Double the Radius
$S = 2\pi r^2 + 2\pi rh$ $= 2\pi(4)^2 + 2\pi(4)(3)$ $= 56\pi \text{ in}^2 \approx 175.8 \text{ in}^2$	$S = 2\pi r^2 + 2\pi rh$ $= 2\pi(4)^2 + 2\pi(4)(6)$ $= 80\pi \text{ in}^2 \approx 251.2 \text{ in}^2$	$S = 2\pi r^2 + 2\pi rh$ $= 2\pi(8)^2 + 2\pi(8)(3)$ $= 176\pi \text{ in}^2 \approx 552.6 \text{ in}^2$

They would not have the same effect. Doubling the radius would increase the surface area more than doubling the height.

EXAMPLE 3 *Art Application*

A Web site advertises that it can turn your photo into an anamorphic image. To reflect the picture, you need to cover a cylinder that is 32 mm in diameter and 100 mm tall with reflective material. How much reflective material do you need?

$L = 2\pi rh$ — *Only the lateral surface needs to be covered.*

$= 2\pi(16)(100)$ — *The diameter is 32 mm, so r = 16mm.*

$\approx 10{,}048 \text{ mm}^2$

Think and Discuss

1. **Compare** the formula for the surface area of a cylinder to the formula for the surface area of a prism.
2. **Explain** how finding the surface area of a cylindrical drinking glass would be different from finding the surface area of a cylinder.
3. **Compare** the amount of paint needed to cover a cube with 1 ft sides to the amount needed to cover a cube with 2 ft sides.

6-8 Exercises

FOR EOG PRACTICE
see page 667

internet connect
Homework Help Online
go.hrw.com Keyword: MT4 6-8

2.01, 3.01, 3.02

GUIDED PRACTICE

See Example 1

Find the surface area of each figure to the nearest tenth. Use 3.14 for π.

1. 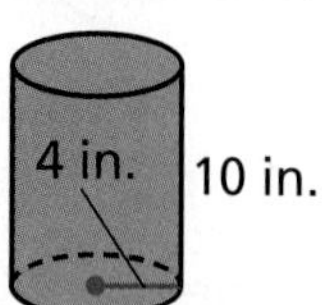

2. 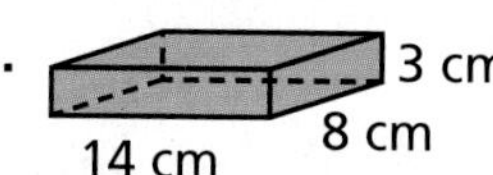

See Example 2

3. A rectangular prism is 6 cm by 8 cm by 9 cm. Explain whether doubling all of the dimensions would double the surface area.

See Example 3

4. To the nearest tenth of a square inch, how much paper is needed for a soup can label if the can is 6.4 in. tall and has a diameter of 4 in.?

INDEPENDENT PRACTICE

See Example 1

Find the surface area of each figure to the nearest tenth. Use 3.14 for π.

5. 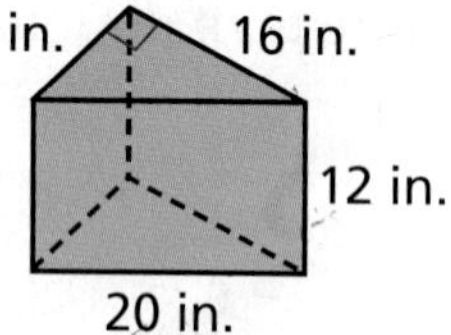

6. 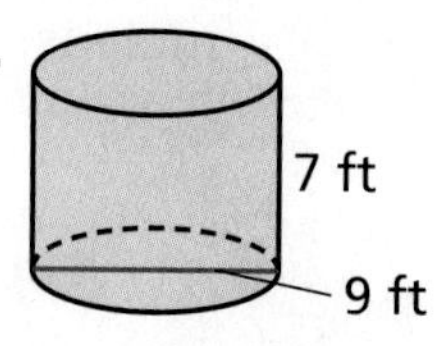

See Example 2

7. A cylinder has diameter 4 ft and height 9 ft. Explain whether halving the diameter has the same effect on the surface area as halving the height.

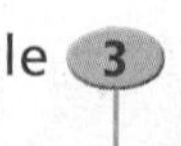
See Example 3

8. How much aluminum foil, to the nearest tenth of a square inch, would it take to cover a loaf of banana-nut bread that is a rectangular prism measuring 8.5 in. by 4 in. by 3.5 in.?

PRACTICE AND PROBLEM SOLVING

Find the surface area of each figure with the given dimensions. Use 3.14 for π.

9. rectangular prism: 9 in. by 12 in. by 15 in.
10. cylinder: $d = 20$ mm, $h = 37$ mm
11. cylinder: $r = 7.8$ cm, $h = 8.2$ cm
12. rectangular prism: $4\frac{1}{2}$ ft by 6 ft by 11 ft

Find the missing dimension in each figure with the given surface area.

13. 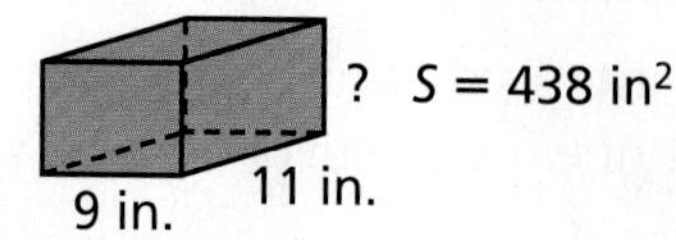

14. 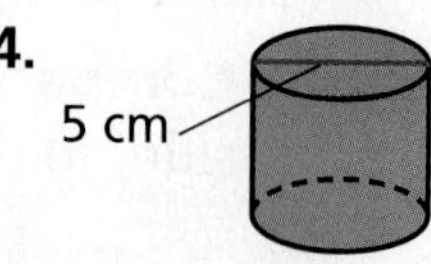

One machine used to shape the inside of a half-pipe is called the Pipe Dragon. Others are the Pipe Master, Turbo Grinder, Scorpion, and Pipe Magician.

Find the surface area of each battery to the nearest tenth of a square centimeter.

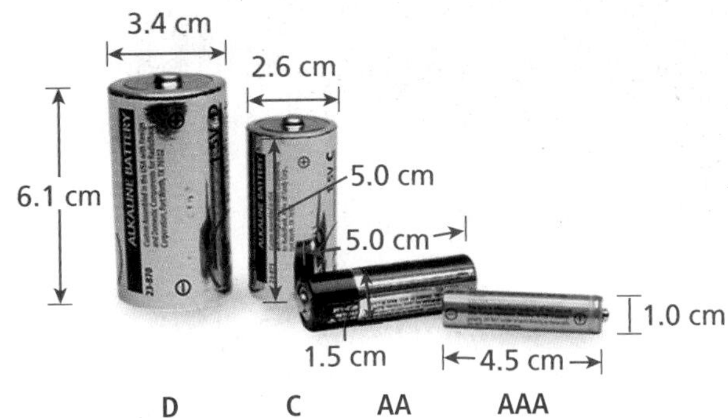

15. D

16. C

17. AA

18. AAA

19. Jesse makes rectangular tin boxes measuring 4 in. by 6 in. by 6 in. If tin costs $0.09 per in^2, how much will the tin for one box cost?

20. ***SPORTS*** In the snowboard half-pipe, competitors ride back and forth on a course shaped like a cylinder cut in half lengthwise. What is the surface area of this half-pipe course?

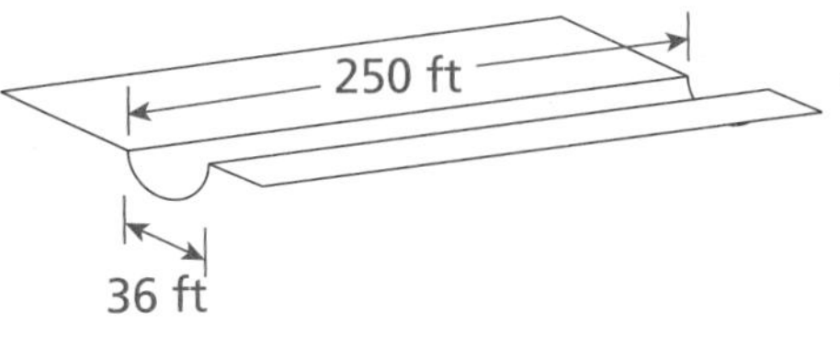

21. ***CHOOSE A STRATEGY*** Which of the following unfolded figures can be folded into the given three-dimensional figure?

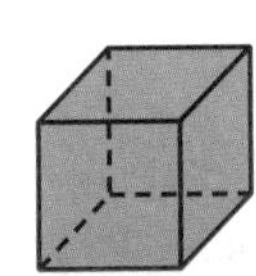

A 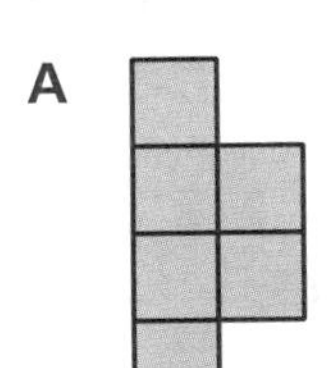

B 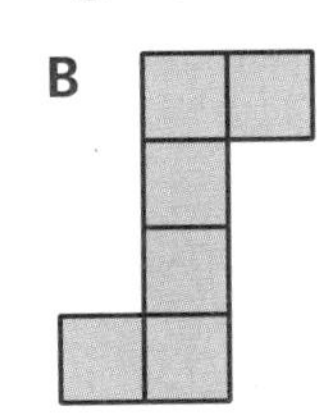

C 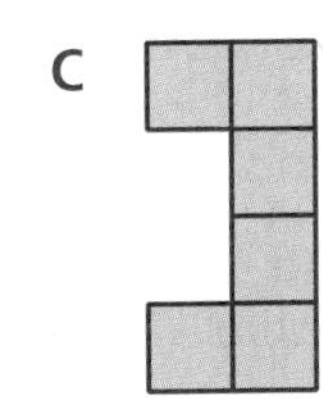

D 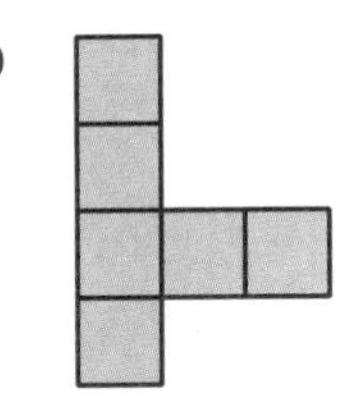

22. ***WRITE ABOUT IT*** Compare the formulas for surface area of a prism and surface area of a cylinder.

23. ***CHALLENGE*** A rectangular wood block that is 12 cm by 9 cm by 5 cm has a hole drilled through the center with diameter 4 cm. What is the total surface area of the wood block?

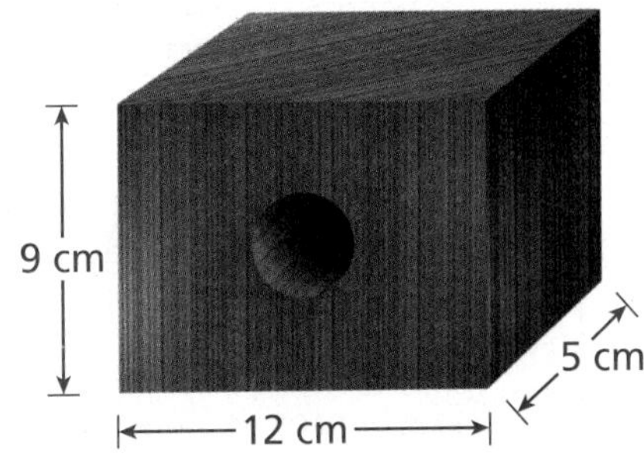

Spiral Review

Divide. Write each answer in simplest form. (Lesson 3-4)

24. $-\frac{4}{11} \div \frac{2}{7}$

25. $\frac{4}{9} \div 8$

26. $-\frac{7}{15} \div \frac{14}{45}$

27. $3\frac{1}{3} \div \frac{7}{9}$

28. In a right triangle, if $a = 9$ and $b = 12$, what is the value of the hypotenuse c? (Lesson 6-3)

29. **EOG PREP** Which ordered pair is a solution of $y = 5x - 3$? (Lesson 1-7)

A (−8, −1) B (−3, −16) C (2, 0) D (2, 7)

6-9 Surface Area of Pyramids and Cones

Learn to find the surface area of pyramids and cones.

Vocabulary

slant height

regular pyramid

right cone

The **slant height** of a pyramid or cone is measured along its lateral surface.

The base of a **regular pyramid** is a regular polygon, and the lateral faces are all congruent.

In a **right cone**, a line perpendicular to the base through the tip of the cone passes through the center of the base.

Right cone

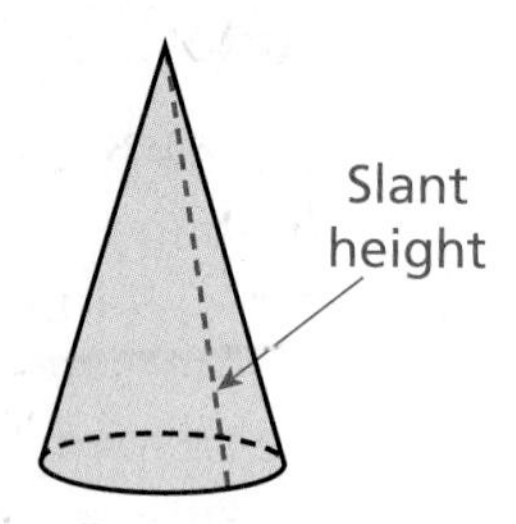

Regular pyramid

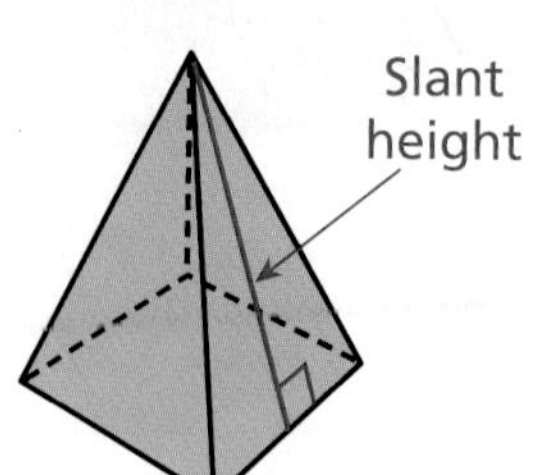

SURFACE AREA OF PYRAMIDS AND CONES

Words	Numbers	Formula
Pyramid: The surface area S of a regular pyramid is the base area B plus the lateral area F. The lateral area is one-half the base perimeter P times the slant height ℓ.	8, 12, 12; 8, 12, 12 $S = (12 \cdot 12) + \frac{1}{2}(48)(8) = 336 \text{ units}^2$	$S = B + F$ or $S = B + \frac{1}{2}P\ell$
Cone: The surface area S of a right cone is the base area B plus the lateral area L. The lateral area is one-half the base circumference $2\pi r$ times the slant height ℓ.	4, 3; 3, 4, 4 $S = \pi(3^2) + \pi(3)(4) = 21\pi \approx 65.94 \text{ units}^2$	$S = B + L$ or $S = \pi r^2 + \pi r\ell$

EXAMPLE  1 **Finding Surface Area**

Find the surface area of each figure.

A

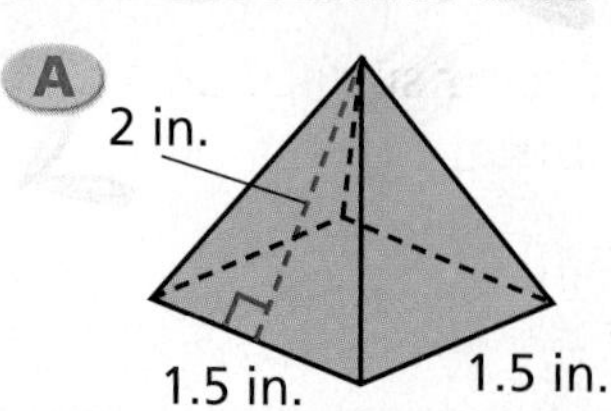

$$S = B + \frac{1}{2}P\ell$$
$$= (1.5 \cdot 1.5) + \frac{1}{2}(6)(2)$$
$$= 8.25 \text{ in}^2$$

Find the surface area of each figure.

$S = \pi r^2 + \pi r\ell$

$= \pi(2)^2 + \pi(2)(5)$

$= 14\pi \approx 44.0 \text{ m}^2$

EXAMPLE 2 Exploring the Effects of Changing Dimensions

A cone has diameter 8 in. and slant height 5 in. Explain whether doubling the slant height would have the same effect on the surface area as doubling the radius.

Original Dimensions	Double the Slant Height	Double the Radius
$S = \pi r^2 + \pi r\ell$ $= \pi(4)^2 + \pi(4)(5)$ $= 36\pi \text{ in}^2 \approx 113.1 \text{ in}^2$	$S = \pi r^2 + \pi r(2\ell)$ $= \pi(4)^2 + \pi(4)(10)$ $= 56\pi \text{ in}^2 \approx 175.9 \text{ in}^2$	$S = \pi(2r)^2 + \pi(2r)\ell$ $= \pi(8)^2 + \pi(8)(5)$ $= 104\pi \text{ in}^2 \approx 326.7 \text{ in}^2$

They would not have the same effect. Doubling the radius would increase the surface area more than doubling the slant height.

EXAMPLE 3 *Life Science Application*

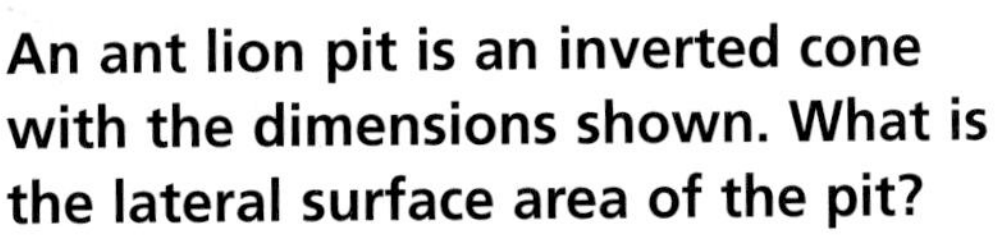

An ant lion pit is an inverted cone with the dimensions shown. What is the lateral surface area of the pit?

The slant height, radius, and depth of the pit form a right triangle.

$a^2 + b^2 = \ell^2$ *Pythagorean Theorem*

$(2.5)^2 + 2^2 = \ell^2$

$10.25 = \ell^2$

$\ell \approx 3.2$

$L = \pi r\ell$ *Lateral surface area*

$= \pi(2.5)(3.2) \approx 25.1 \text{ cm}^2$

Life Science LINK

Ant lions are the larvae of an insect similar to a dragonfly. They dig cone-shaped pits in the sand to trap ants and other crawling insects.

Think and Discuss

1. **Compare** the formula for surface area of a pyramid to the formula for surface area of a cone.
2. **Explain** how you would find the slant height of a square pyramid with base edge length 6 cm and height 4 cm.

6-9 Exercises

FOR EOG PRACTICE
see page 667

internet connect
Homework Help Online
go.hrw.com Keyword: MT4 6-9

2.01, 3.01, 3.02

GUIDED PRACTICE

See Example 1

Find the surface area of each figure to the nearest tenth. Use 3.14 for π.

1. 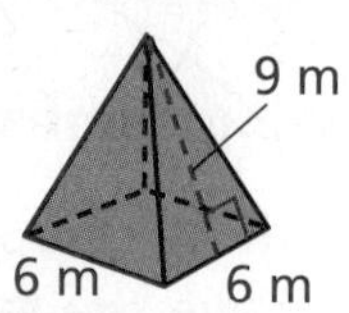

2. 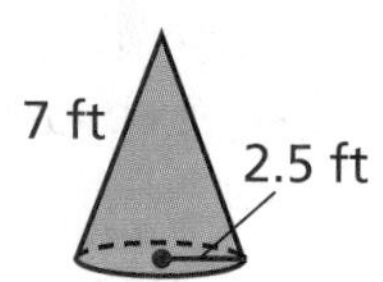

See Example 2

3. A cone has diameter 10 in. and slant height 8 in. Tell whether doubling both dimensions would double the surface area.

See Example 3

4. The cone-shaped wigwams at the Wigwam Village Motel in Cave City, Kentucky, are about 20 ft high and have a diameter of about 20 ft. Estimate the lateral surface area of a wigwam.

INDEPENDENT PRACTICE

See Example 1

Find the surface area of each figure to the nearest tenth. Use 3.14 for π.

5. 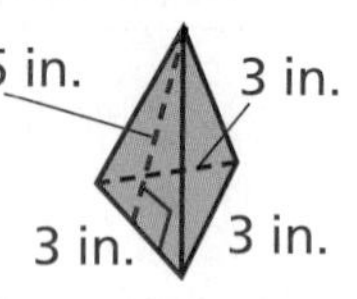

6. 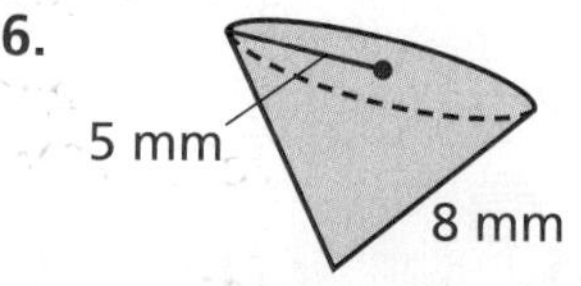

See Example 2

7. A regular square pyramid has a base with 10 yd sides and has slant height 6 yd. Tell whether doubling both dimensions would double the surface area.

See Example 3

8. In the late 1400s, Leonardo daVinci designed a parachute shaped like a pyramid. His design called for a tent-like structure made of linen, measuring 21 feet on each side and 12 feet high. How much material would be needed to make the parachute?

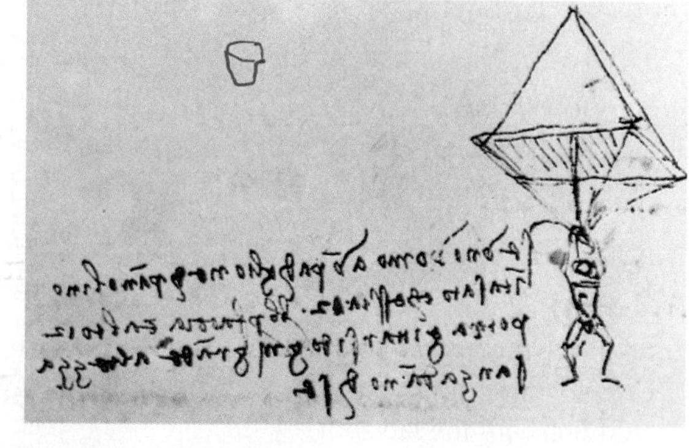

PRACTICE AND PROBLEM SOLVING

Find the surface area of each figure with the given dimensions. Use 3.14 for π.

9. regular square pyramid:
base perimeter = 60 cm
slant height = 18 cm

10. regular triangular pyramid:
base area = 0.04 km^2
base perimeter = 0.9 km
slant height = 0.2 km

11. cone: $d = 38$ ft
slant height = 53 ft

12. cone: $r = 12\frac{1}{2}$ mi
slant height = $44\frac{1}{4}$ mi

13. ***EARTH SCIENCE*** When the Moon is between the Sun and Earth, it casts a conical shadow called the *umbra.* If the shadow is 2140 mi in diameter and 260,955 mi along the edge, what is the lateral surface area of the shadow?

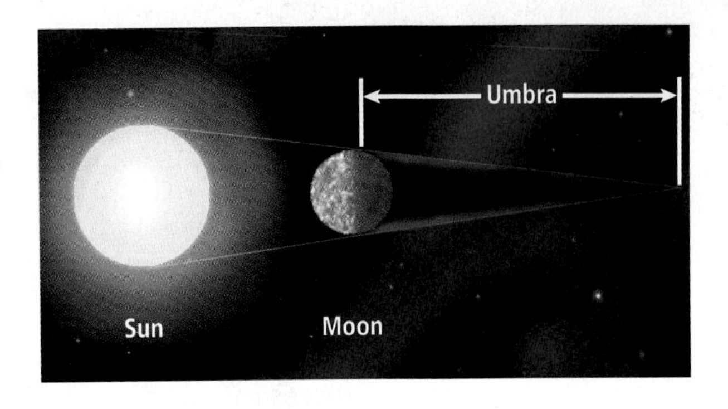

14. ***SOCIAL STUDIES*** The Pyramid of the Sun, in Teotihuacán, Mexico, is about 65 m tall and has a square base with side length 225 m. What is the lateral surface area of the pyramid?

15. The table shows the dimensions of three square pyramids.
    a. Complete the table.
    b. Which pyramid has the greatest lateral surface area? What is its lateral surface area?
    c. Which pyramid has the least volume? What is its volume?

Dimensions of Giza Pyramids (ft)

Pyramid	Height	Slant Height	Side of Base
Khufu	481	612	756
Khafre	471	■	704
Menkaure	■	277	346

16. ***WHAT'S THE ERROR?*** Correct the error in this statement: The lateral surface area of a cone is π times the radius of the base times the height of the cone.

17. ***WRITE ABOUT IT*** The dimensions of a square pyramid give its height and base dimensions. Explain how to find the slant height.

18. ***CHALLENGE*** The oldest pyramid is said to be the Step Pyramid of King Zoser, built around 2650 B.C. in Saqqara, Egypt. The base is a rectangle that measures 358 ft by 411 ft, and the height of the pyramid is 204 ft. Find the lateral surface area of the pyramid.

Spiral Review

Find each sum. (Lesson 3-2)

19. $-1.7 + 2.3$

20. $-\frac{2}{3} + \left(-\frac{1}{6}\right)$

21. $23.75 + (-25.15)$

22. $-\frac{4}{9} + \frac{2}{9}$

23. Find the length of the hypotenuse of a right triangle with legs measuring 15 m and 22 m. (Lesson 6-3)

24. **EOG PREP** Triangle $EFG \cong$ triangle JIH. What is the value of x? (Lesson 5-6)

A 5.67

B 30

C 63

D 71

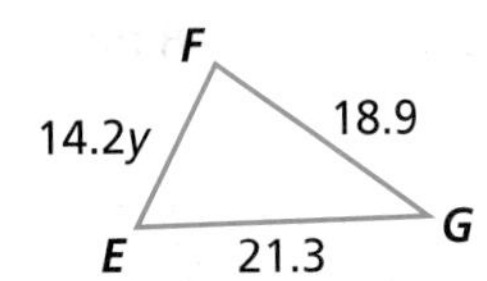

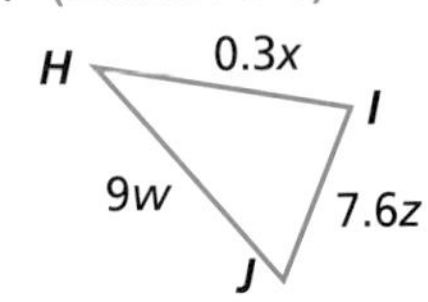

6-10 Spheres

Learn to find the volume and surface area of spheres.

Vocabulary

sphere

hemisphere

great circle

Earth is not a perfect *sphere*, but it has been molded by gravitational forces into a spherical shape. Earth has a diameter of about 7926 miles and a surface area of about 197 million square miles.

A **sphere** is the set of points in three dimensions that are a fixed distance from a given point, the center. A plane that intersects a sphere through its center divides the sphere into two halves, or **hemispheres**. The edge of a hemisphere is a **great circle**.

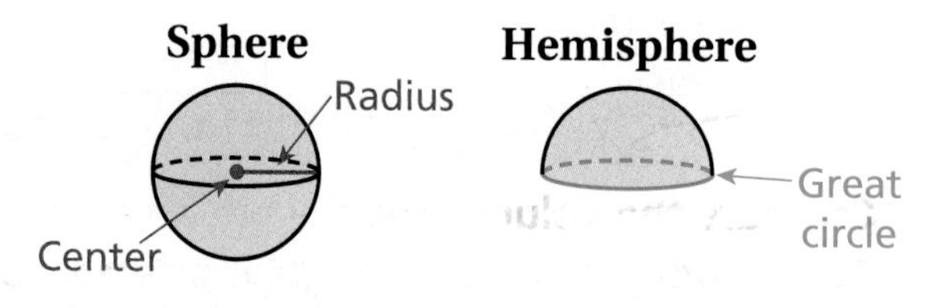

The volume of a hemisphere is exactly halfway between the volume of a cone and the volume of a cylinder with the same radius r and height equal to r.

VOLUME OF A SPHERE

Words	Numbers	Formula
The volume V of a sphere is $\frac{4}{3}\pi$ times the cube of the radius r.	3 $V = \left(\frac{4}{3}\right)\pi(3^3)$ $= \frac{108}{3}\pi$ $= 36\pi$ ≈ 113.1 units^3	$V = \left(\frac{4}{3}\right)\pi r^3$

EXAMPLE 1 Finding the Volume of a Sphere

Find the volume of a sphere with radius 6 ft, both in terms of π and to the nearest tenth.

$V = \left(\frac{4}{3}\right)\pi r^3$ *Volume of a sphere*

$= \left(\frac{4}{3}\right)\pi(6)^3$ *Substitute 6 for r.*

$= 288\pi \text{ ft}^3 \approx 904.3 \text{ ft}^3$

The surface area of a sphere is four times the area of a great circle.

SURFACE AREA OF A SPHERE		
Words	**Numbers**	**Formula**
The surface area S of a sphere is 4π times the square of the radius r.	2 $S = 4\pi(2^2)$ $= 16\pi$ $\approx 50.3 \text{ units}^2$	$S = 4\pi r^2$

EXAMPLE 2 **Finding Surface Area of a Sphere**

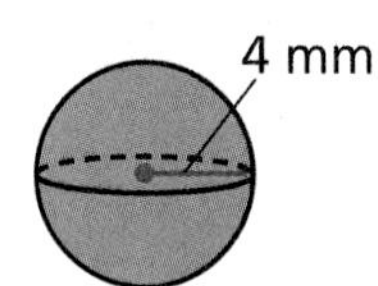

Find the surface area, both in terms of π and to the nearest tenth.

$S = 4\pi r^2$ *Surface area of a sphere*

$= 4\pi(4^2)$ *Substitute 4 for r.*

$= 64\pi \text{ mm}^2 \approx 201.1 \text{ mm}^2$

EXAMPLE 3 **Comparing Volumes and Surface Areas**

Compare the volume and surface area of a sphere with radius 21 cm with that of a rectangular prism measuring 28 × 33 × 42 cm.

Sphere:

$V = \left(\frac{4}{3}\right)\pi r^3 = \left(\frac{4}{3}\right)\pi(21^3)$

$\approx \left(\frac{4}{3}\right)\left(\frac{22}{7}\right)(9261)$

$\approx 38{,}808 \text{ cm}^3$

$S = 4\pi r^2 = 4\pi(21^2)$

$= 1764\pi$

$\approx 1764\left(\frac{22}{7}\right) \approx 5544 \text{ cm}^2$

Rectangular prism:

$V = \ell wh$

$= (28)(33)(42)$

$= 38{,}808 \text{ cm}^3$

$S = 2\ell w + 2\ell h + 2wh$

$= 2(28)(33) + 2(28)(42) + 2(33)(42)$

$= 6972 \text{ cm}^2$

The sphere and the prism have approximately the same volume, but the prism has a larger surface area.

Think and Discuss

1. **Compare** the area of a great circle with the surface area of a sphere.
2. **Explain** which would hold the most water: a bowl with radius r and height r, a cylindrical glass with radius r and height r, or a conical drinking cup with radius r and height r.

6-10 Exercises

FOR EOG PRACTICE
see page 667

2.01, 3.01, 3.02, 5.04

GUIDED PRACTICE

See Example 1

Find the volume of each sphere, both in terms of π and to the nearest tenth. Use 3.14 for π.

1. $r = 2$ cm **2.** $r = 10$ ft **3.** $d = 3.4$ m **4.** $d = 8$ mi

See Example 2

Find the surface area of each sphere, both in terms of π and to the nearest tenth. Use 3.14 for π.

5.
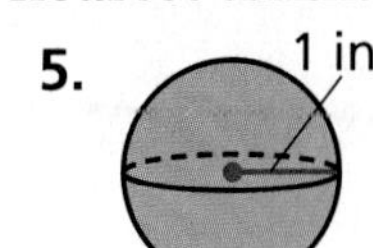

6. 6.6 mm

7.
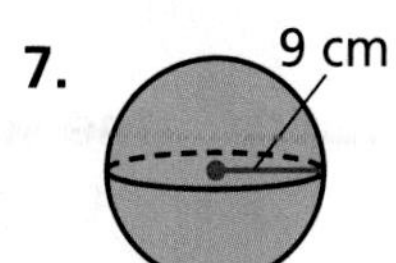

8.
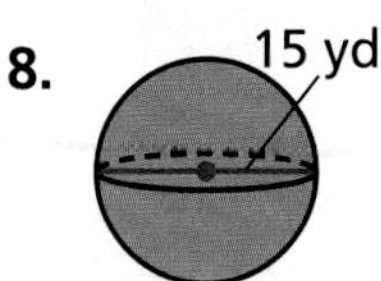

See Example 3

9. Compare the volume and surface area of a sphere with radius 4 in. with that of a cube with sides measuring 6.45 in.

INDEPENDENT PRACTICE

See Example 1

Find the volume of each sphere, both in terms of π and to the nearest tenth. Use 3.14 for π.

10. $r = 12$ ft **11.** $r = 4.8$ cm **12.** $d = 22$ mm **13.** $d = 1$ in.

See Example 2

Find the surface area of each sphere, both in terms of π and to the nearest tenth. Use 3.14 for π.

14.
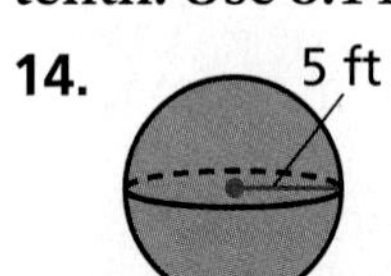

15. 7.2 m

16.

17.
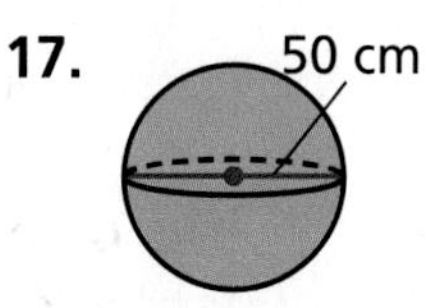

See Example 3

18. Compare the volume and surface area of a sphere with diameter 3 ft with that of a cylinder with height 1 ft and a base with radius 2 ft.

PRACTICE AND PROBLEM SOLVING

Find the missing measurements of each sphere, both in terms of π and to the nearest hundredth. Use 3.14 for π.

19. radius = 5.5 in.
volume = ■
surface area = 121π in.^{2}

20. radius = 10.8 m
volume = 1679.62π m^2
surface area = ■

21. diameter = 6.2 yd
volume = ■
surface area = ■

22. radius = ■
diameter = 18 in.
surface area = ■

23. radius = ■
volume = ■
surface area = 3600π km^2

24. radius = ■
diameter = ■
surface area = 1697.44π mi^2

Life Science LINK

Eggs come in many different shapes. The eggs of birds that live on cliffs are often extremely pointed to keep the eggs from rolling. Other birds, such as great horned owls, have eggs that are nearly spherical. Turtles and crocodiles also have nearly spherical eggs, and the eggs of many dinosaurs were spherical.

25. To lay their eggs, green turtles travel hundreds of miles to the beach where they were born. The eggs are buried on the beach in a hole about 40 cm deep. The eggs are approximately spherical, with an average diameter of 4.5 cm, and each turtle lays an average of 113 eggs at a time. Estimate the total volume of eggs laid by a green turtle at one time.

26. Fossilized embryos of dinosaurs called titanosaurid sauropods have recently been found in spherical eggs in Patagonia. The eggs were 15 cm in diameter, and the adult dinosaurs were more than 12 m in length. Find the volume of an egg.

27. The glasshouse spider mite lays spherical eggs that are translucent and about 0.1 mm in diameter. Find the surface area of an egg.

28. Hummingbirds lay eggs that are nearly spherical and about 1 cm in diameter. Find the surface area of an egg.

29. ***CHALLENGE*** An ostrich egg has about the same volume as a sphere with a diameter of 5 inches. If the shell is about $\frac{1}{12}$ inch thick, estimate the volume of just the shell, not including the interior of the egg.

Spiral Review

Multiply or divide. Write each answer in simplest form. (Lessons 3-3 and 3-4)

30. $\frac{2}{3} \cdot \frac{9}{10}$ **31.** $\frac{4}{5} \cdot \frac{3}{8}$ **32.** $\frac{1}{3} \div \frac{2}{3}$ **33.** $\frac{11}{15} \div \frac{5}{22}$

34. **EOG PREP** If two angles are complementary, what is the sum of their measures? (Lesson 5-1)

A 90° B 180° C 270° D 360°

35. **EOG PREP** If two angles are supplementary, what is the sum of their measures? (Lesson 5-1)

A 90° B 180° C 270° D 360°

EXTENSION Symmetry in Three Dimensions

Learn to identify types of symmetry in three dimensions.

Vocabulary
bilateral symmetry
cross section

Solid figures can have different kinds of symmetry.

A solid figure with *rotational symmetry* is unchanged in appearance when it is turned a specific number of degrees about a line.

A solid figure with **bilateral symmetry** has two-sided symmetry, or *reflection symmetry*, across a plane.

EXAMPLE 1 Identifying Symmetry in a Solid Figure

Identify all types of symmetry in each figure.

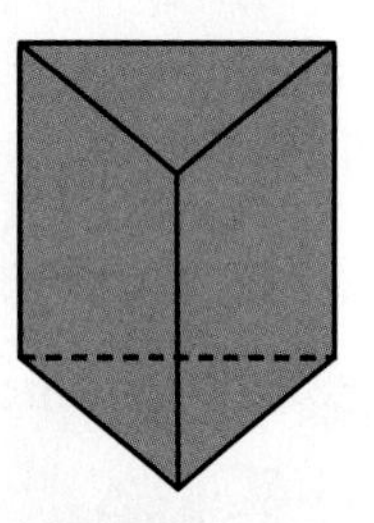

This triangular prism has both rotational symmetry and bilateral symmetry.

B

This chair has only bilateral symmetry.

When a solid and a plane intersect, the intersection is called a **cross section** .

EXAMPLE 2 Drawing a Cross Section

Draw the cross section and describe its symmetry.

The cross section is an equilateral triangle, which has three-fold rotational symmetry and line symmetry. There are three lines of symmetry, one from each vertex to the midpoint of the opposite side.

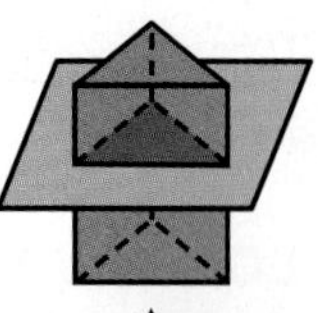

EXTENSION

Exercises

Identify all types of symmetry in each figure.

1. 

2. 

3. 

Draw the cross section and describe its symmetry.

4. 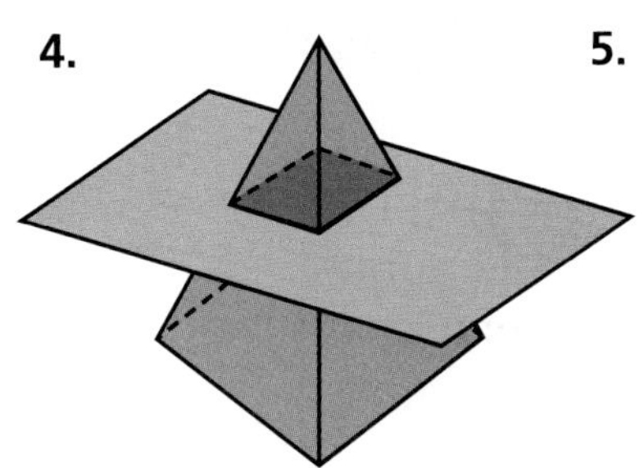

5. 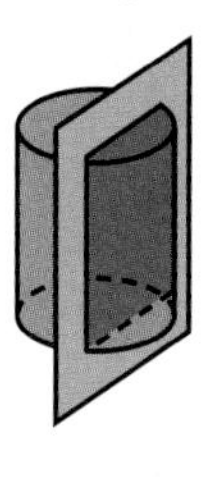

6. 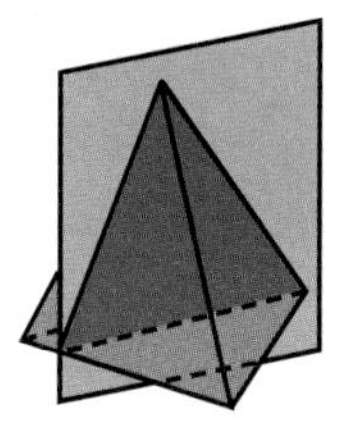

Identify all types of symmetry in each figure.

7. 

8. 

9. 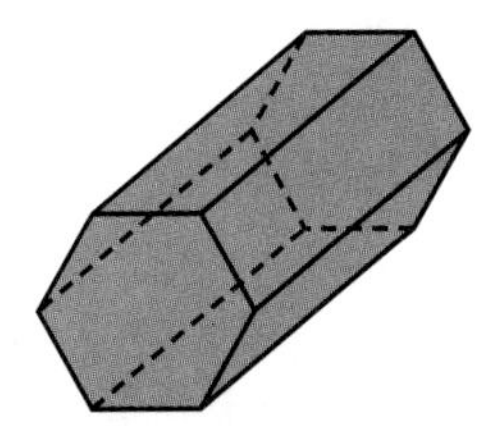

Draw the cross section and describe its symmetry.

10. 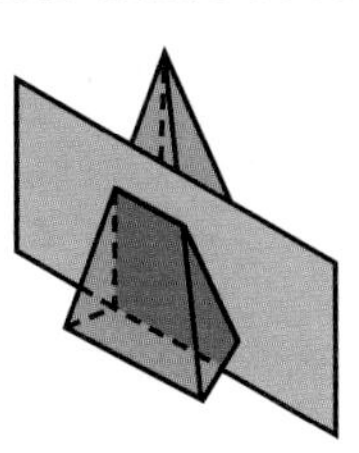

11. 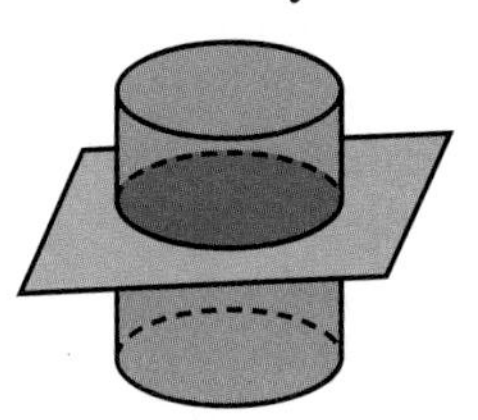

12. 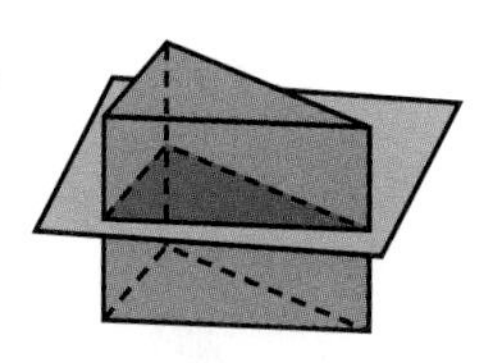

13. The Transamerica Pyramid in San Francisco is a square pyramid. Each floor is a horizontal cross section of the pyramid. What is the shape of such a cross section? How is the size of each floor related to the size of the floor below it?

14. When a plane and a cube intersect, is it possible for the cross section to be a six-sided figure? Explain.

15. Describe the possible cross sections of a sphere.

Problem Solving on Location

NORTH CAROLINA

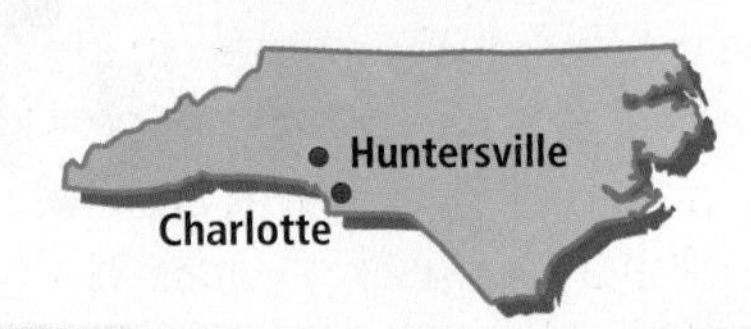

Historic Latta Plantation

Beginning around 1800, the Latta Plantation in Huntersville, North Carolina, was a thriving cotton plantation. Now it is a living-history farm. Two things to see at the farm are the chicken coop and the federal-style main house.

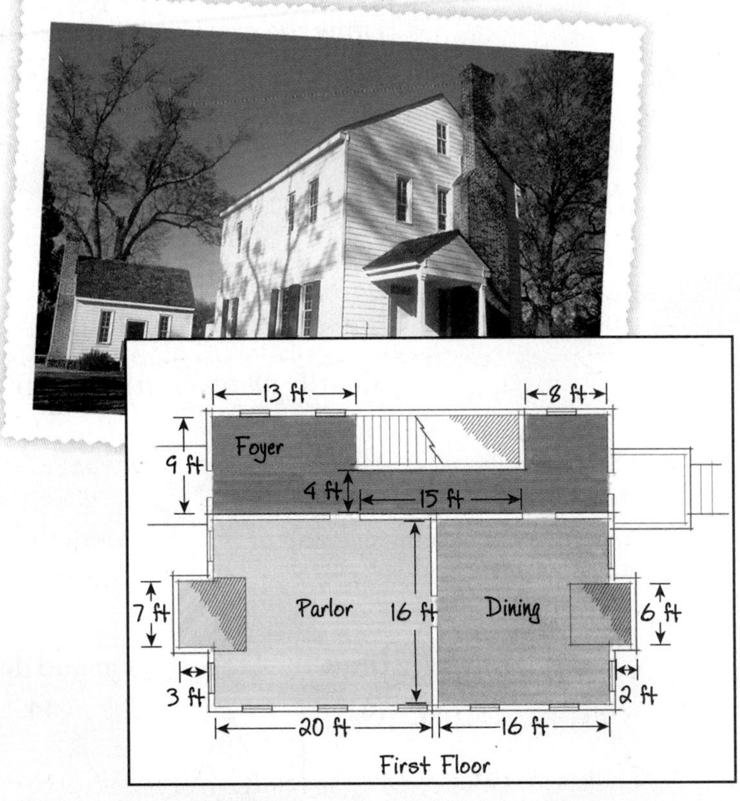

1. In the 1800s, the number 13 was considered patriotic because it was the number of original colonies. For this reason, it was often included in decorative pieces such as the secretary desk found in the parlor of the main house. The desk has two circular decorations carved into the wood above the glass doors. Each circle is divided into 13 pie-shaped wedges. Assume the diameter of each circle is 13 cm. Find the area to the nearest tenth of one pie-shaped wedge. Use 3.14 for π.

2. The chicken coop at Latta Plantation is a triangular prism. If the triangular end of the coop is an isosceles triangle with base length 10 ft and congruent side lengths 13 ft, what is the area of the triangular end?

For 3–5, use the approximate measurements shown on the floor plan.

3. Find the area of the foyer of the main house.

4. Find the perimeter of the dining room, including the fireplace.

5. The exterior walls of the main house are about 38 ft long and about 26 ft wide. If the rectangular-prism-shaped section of the house is 21 ft tall, what are the volume and lateral surface area of this section of the house?

Ericsson Stadium

Located in Charlotte, North Carolina, Ericsson Stadium is home to the Carolina Panthers football team. The stadium can seat 73,258 people and includes training facilities, practice fields, and administrative offices. The playing field is a rectangle that is 398 ft long and 280 ft wide. The *stadium footprint* is the total area covered by the stadium. Ericsson Stadium's footprint is 900 ft long and 800 ft wide.

1. Use the Pythagorean Theorem to find the length to the nearest foot of the diagonal of the playing field.

2. Ericsson Stadium contains 44,700 square yards of carpet. If this carpeting covered the floor of a single rectangular room with length 149 yards, what would the width of the room be?

3. The distance around a stadium is its *concourse length.* Ericsson Stadium has a concourse length of about 2640 ft. Estimate to the nearest whole number the radius and area of the circle whose circumference is equal to the stadium's concourse length. Use 3.14 for π.

4. Find the area of a rectangle with the same width and length as the stadium footprint.

5. The stadium footprint is actually an oval with area 653,400 ft^2. Compare the areas you found in problems **3** and **4** with the area of the stadium. Which is a better estimate of the stadium footprint?

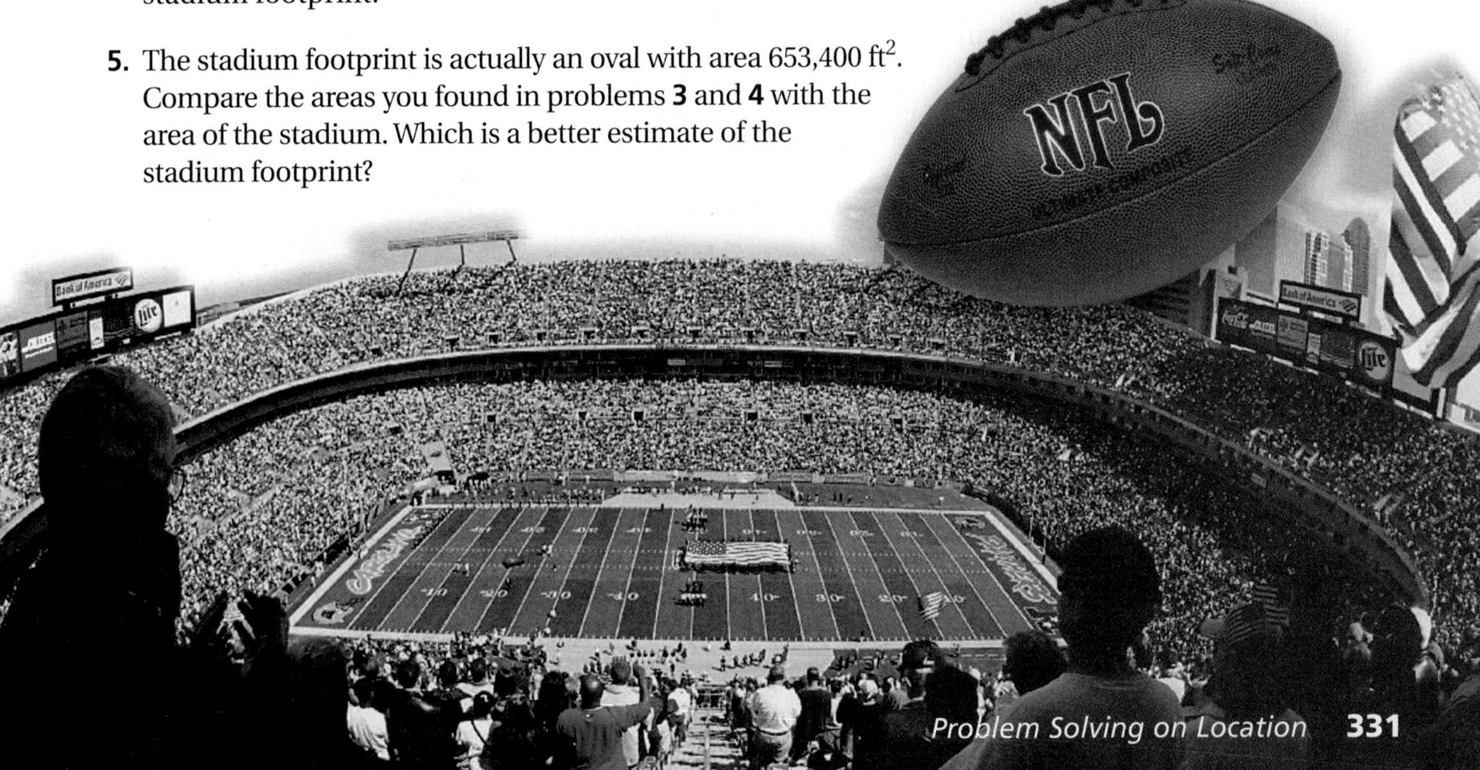

MATH-ABLES

Planes in Space

Some three-dimensional figures can be generated by plane figures.

Experiment with a circle first. Move the circle around. See if you recognize any three-dimensional shapes.

If you rotate a circle around a diameter, you get a sphere.

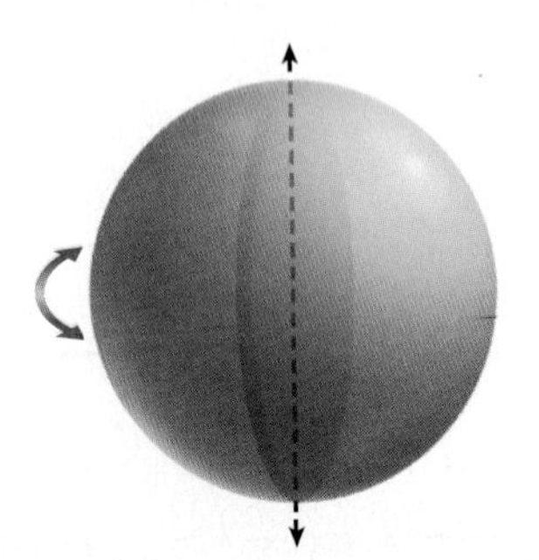

If you translate a circle up along a line perpendicular to the plane that the circle is in, you get a cylinder.

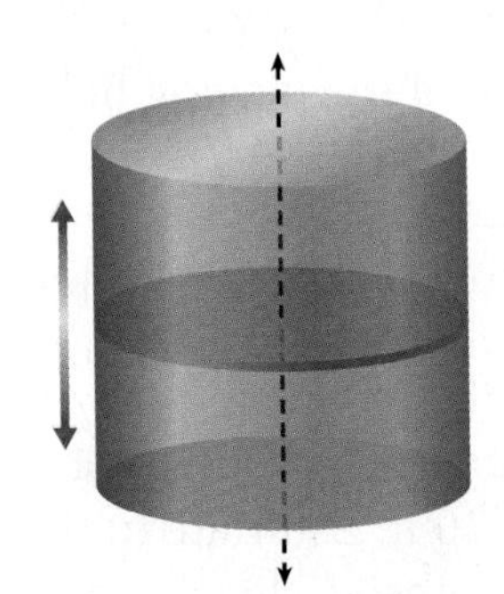

If you rotate a circle around a line outside the circle but in the same plane as the circle, you get a donut shape called a *torus*.

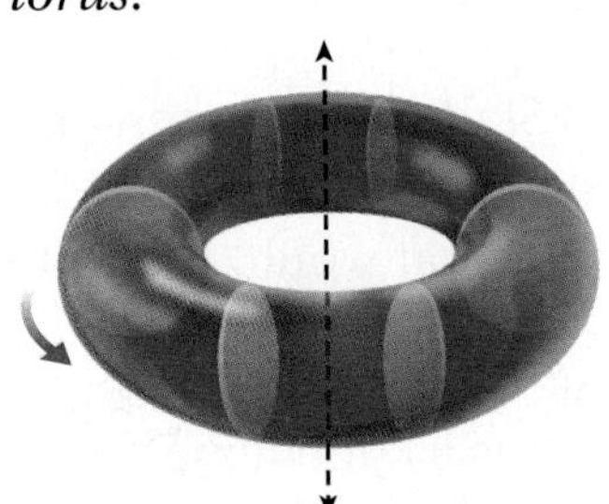

Draw or describe the three-dimensional figure generated by each plane figure.

1. a square translated along a line perpendicular to the plane it is in
2. a rectangle rotated around one of its edges
3. a right triangle rotated around one of its legs

Triple Concentration

The goal of this game is to form *Pythagorean triples,* which are sets of three whole numbers a, b, and c such that $a^2 + b^2 = c^2$. A set of cards with numbers on them are arranged face down. A turn consists of drawing 3 cards to try to form a Pythagorean triple. If the cards do not form a Pythagorean triple, they are replaced in their original positions.

internet connect

Go to ***go.hrw.com*** for a complete set of rules and cards.
KEYWORD: MT4 Game6

Pythagorean Triples

Use with Lesson 6-3

Three positive integers *a, b,* and *c* that satisfy the equation $a^2 + b^2 = c^2$ are called **Pythagorean triples.** You know that $3^2 + 4^2 = 5^2$. So 3, 4, and 5 are Pythagorean triples.

Activity

You can generate Pythagorean triples *a, b,* and *c* by starting with two different whole numbers *m* and *n*, where *m* is the larger number. The Pythagorean triple will be as follows:

$a = m^2 - n^2$
$b = 2mn$
$c = m^2 + n^2$

Example: Using $m = 2$ and $n = 1$
$a = 2^2 - 1^2 = 4 - 1 = 3$
$b = 2(2)(1) = 4$
$c = 2^2 + 1^2 = 5$

Using a spreadsheet, enter the letters *m, n, a, b,* and *c*, respectively, in cells A1 to E1.

Then enter the formula for *a* as **=A2^2−B2^2** in cell C2.

For *b*, enter **=2*A2*B2** in cell D2, and for *c*, enter **=A2^2+B2^2** in cell E2.

	A	B	C	D	E
1	m	n	a	b	c
2			=A2^2-B2^2		0

Highlight cells C2, D2, and E2, and click the **Copy** button on the toolbar. Then select cells C2, D2, and E2, and drag down to highlight 7 rows. Click the **Paste** button on the toolbar.

	A	B	C	D	E
1	m	n	a	b	c
2			0	0	0
3			0	0	0
4			0	0	0
5			0	0	0
6			0	0	0
7			0	0	0
8			0	0	0

Next, enter 5 and 4 in cells A2 and B2, 5 and 3 in cells A3 and B3, and so forth until you complete the seventh row for the patterns given.

All integers *a, b,* and *c* shown in the last three columns are Pythagorean triples.

	A	B	C	D	E
1	m	n	a	b	c
2	5	4	9	40	41
3	5	3	16	30	34
4	5	2	21	20	29
5	5	1	24	10	26
6	4	3	7	24	25
7	4	2	12	16	20
8	4	1	15	8	17

Think and Discuss

1. Is order important in a Pythagorean triple? Why?

Try This

1. Using a spreadsheet, generate 30 Pythagorean triples.

Chapter 6

Study Guide and Review

Vocabulary

Complete the sentences below with vocabulary words from the list above. Words may be used more than once.

1. In a two-dimensional figure, __?__ is the distance around the outside of the figure, while __?__ is the number of square units in the figure.
2. In a three-dimensional figure, a(n) __?__ is where two faces meet, and a(n) __?__ is where three or more edges meet.
3. A(n) __?__ divides a sphere into two halves, or __?__.

6-1 Perimeter and Area of Rectangles and Parallelograms (pp. 280–284)

EXAMPLE

■ **Find the area and perimeter of a rectangle with base 2 ft and height 5 ft.**

$A = bh$
$= 5(2)$
$= 10 \text{ ft}^2$

$P = 2l + 2w$
$= 2(5) + 2(2)$
$= 10 + 4$
$= 14 \text{ ft}$

EXERCISES

Find the area and perimeter of each figure.

4. a rectangle with base $2\frac{1}{3}$ in. and height $5\frac{2}{3}$ in.
5. a parallelogram with base 16 m, side length 24 m, and height 13 m.

6-2 Perimeter and Area of Triangles and Trapezoids (pp. 285–288)

EXAMPLE

- **Find the area of a right triangle with base 6 cm and height 3 cm.**
  $A = \frac{1}{2}bh = \frac{1}{2}(6)(3) = 9 \text{ cm}^2$

EXERCISES

Find the area and perimeter of each figure.

6. a triangle with base 8 cm, sides 4.1 cm and 8.1 cm, and height 4 cm
7. trapezoid $DEFG$ with $DE = 4.5$ in., $EF = 10.1$ in., $FG = 16.5$ in., and $DG = 2.9$ in., where $\overline{DE} \parallel \overline{FG}$ and $h = 2.0$ in.

6-3 The Pythagorean Theorem (pp. 290–293)

EXAMPLE

- **Find the length of side b in the right triangle where $a = 8$ and $c = 17$.**
  $a^2 + b^2 = c^2$
  $8^2 + b^2 = 17^2$
  $64 + b^2 = 289$
  $b^2 = 225$
  $b = \sqrt{225} = 15$

EXERCISES

Solve for the unknown side in each right triangle.

8. If $a = 6$ and $b = 8$, find c.
9. If $b = 24$ and $c = 26$, find a.

6-4 Circles (pp. 294–297)

EXAMPLE

- **Find the area and circumference of a circle with radius 3.1 cm.**
  $A = \pi r^2$ $C = 2\pi r$
  $= \pi(3.1)^2$ $= 2\pi(3.1)$
  $= 9.61\pi \approx 30.2 \text{ cm}^2$ $= 6.2\pi \approx 19.5 \text{ cm}$

EXERCISES

Find the area and circumference of each circle, both in terms of π and to the nearest tenth. Use 3.14 for π.

10. $r = 15$ in.
11. $r = 2.4$ cm
12. $d = 8$ m
13. $d = 1.2$ ft

6-5 Drawing Three-Dimensional Figures (pp. 302–306)

EXAMPLE

- **Use isometric dot paper to sketch a rectangular prism that is 3 units long, 1 unit wide, and 2 units high.**

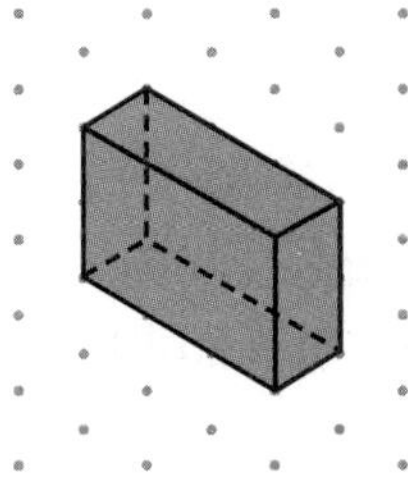

EXERCISES

Use isometric dot paper to sketch each figure.

14. a rectangular box that is 4 units long, 3 units deep, and 1 unit high
15. a cube with 3-unit sides
16. a box with a 2-unit square base and a height of 4 units

6-6 Volume of Prisms and Cylinders (pp. 307–311)

EXAMPLE

- Find the volume.

  $V = Bh = (\pi r^2)h$

  $= \pi(4^2)(6)$

  $= (16\pi)(6) = 96\pi\ \text{cm}^3$

  $\approx 301.6\ \text{cm}^3$

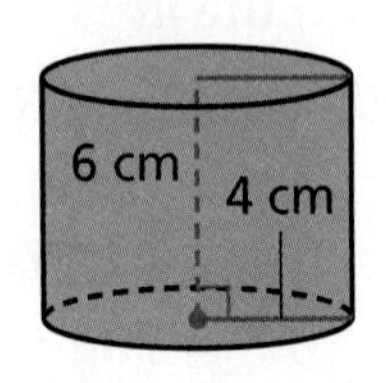

EXERCISES

Find the volume of each figure.

17.

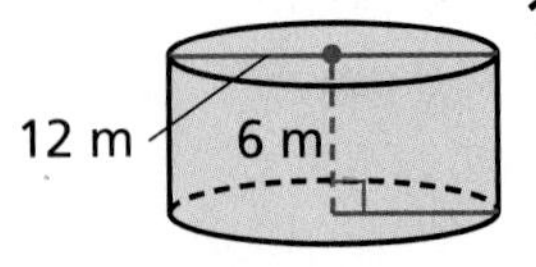

18.

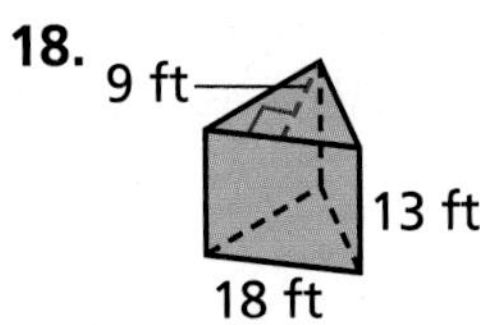

6-7 Volume of Pyramids and Cones (pp. 312–315)

EXAMPLE

- Find the volume.

  $V = \frac{1}{3}Bh = \frac{1}{3}(6)(4)(8)$

  $= \frac{1}{3}(24)(8) = 64\ \text{in}^3$

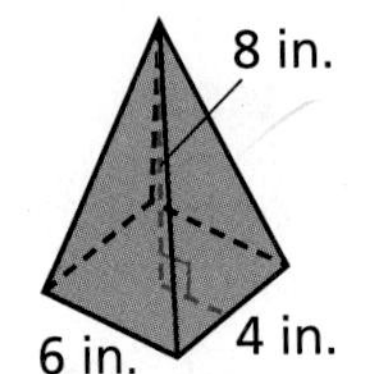

EXERCISES

Find the volume of each figure.

19.

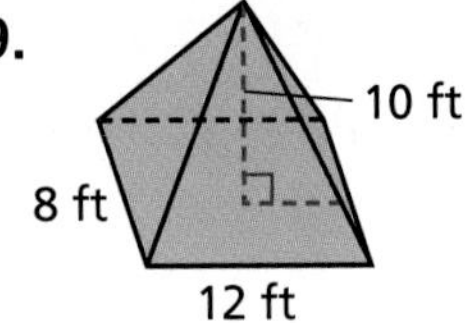

20.

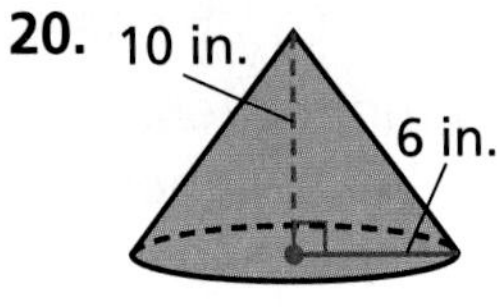

6-8 Surface Area of Prisms and Cylinders (pp. 316–319)

EXAMPLE

- Find the surface area.

  $S = 2B + Ph$

  $= 2(6) + (10)(4)$

  $= 52\ \text{in}^2$

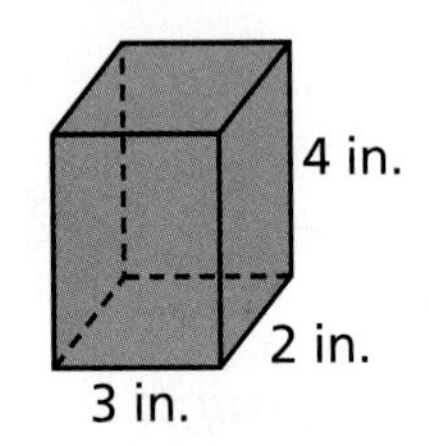

EXERCISES

Find the surface area of the figure.

21. 6.5 cm; 4 cm; 5 cm; 3 cm

6-9 Surface Area of Pyramids and Cones (pp. 320–323)

EXAMPLE

- Find the surface area.

  $S = B + \frac{1}{2}P\ell$

  $= 16 + \frac{1}{2}(16)(5)$

  $= 56\ \text{in}^2$

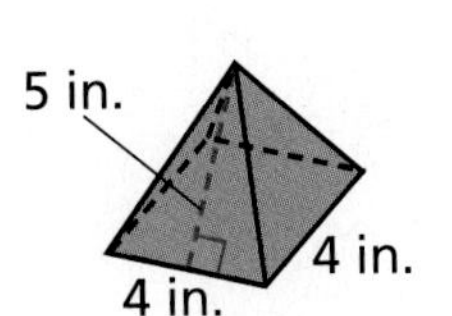

EXERCISES

Find the surface area.

22.

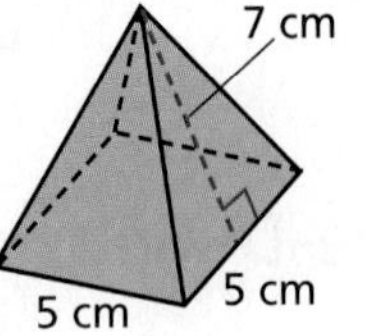

23.

6-10 Spheres (pp. 324–327)

EXAMPLE

- Find the volume of a sphere of radius 12 cm.

  $V = \frac{4}{3}\pi r^3 = \frac{4}{3}\pi(12^3)$

  $= 2304\pi\ \text{cm}^3 \approx 7234.6\ \text{cm}^3$

EXERCISES

Find the volume of each sphere, both in terms of π and to the nearest tenth. Use 3.14 for π.

24. $r = 9$ in.

25. $d = 30$ m

Chapter Test

Chapter 6

Graph and find the area of each figure with the given vertices.

1. (4, 1), (−3, 1), (−3, −4), (4, −4)

2. (0, 4), (2, 3), (2, −3), (0, −2)

3. (−3, 0), (2, 0), (4, −2)

4. (2, 3), (6, −2), (−5, −2), (−2, 3)

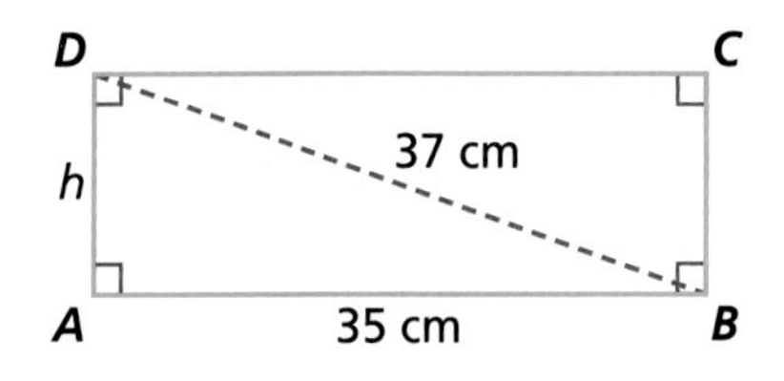

5. Use the Pythagorean Theorem to find the height of rectangle *ABCD*.

6. Find the area of rectangle *ABCD*.

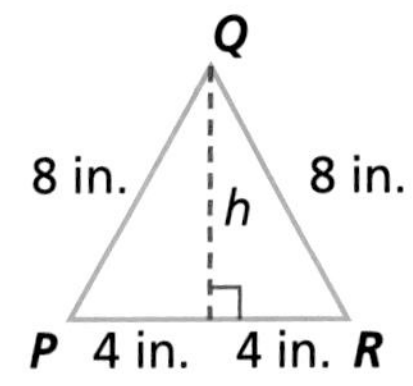

7. Use the Pythagorean Theorem to find the height of equilateral triangle *PQR* to the nearest hundredth.

8. Find the area of equilateral triangle *PQR* to the nearest tenth.

Find the area of the circle to the nearest tenth. Use 3.14 for π.

9. radius = 11 in.

10. diameter = 26 cm

Find the volume of each figure.

11. a sphere of radius 8 cm

12. a cylinder of height 10 in. and radius 6 in.

13. a pyramid with a 3 ft by 3 ft square base and height 5 ft

14. a cone of diameter 12 in. and height 18 in.

Find the surface area of each figure.

15.

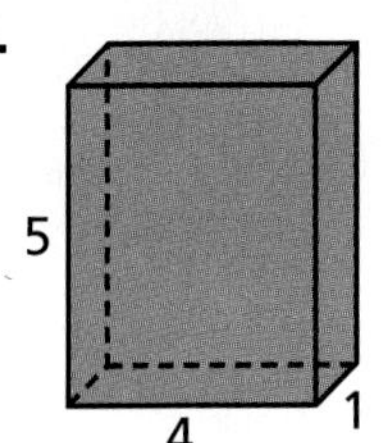

16.

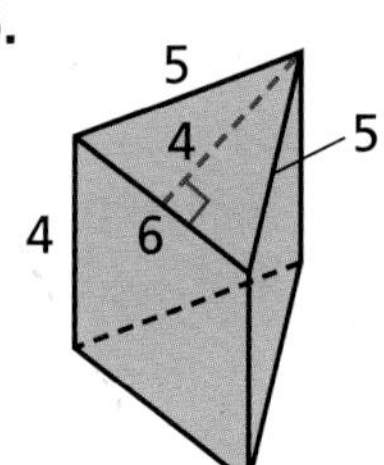

17.

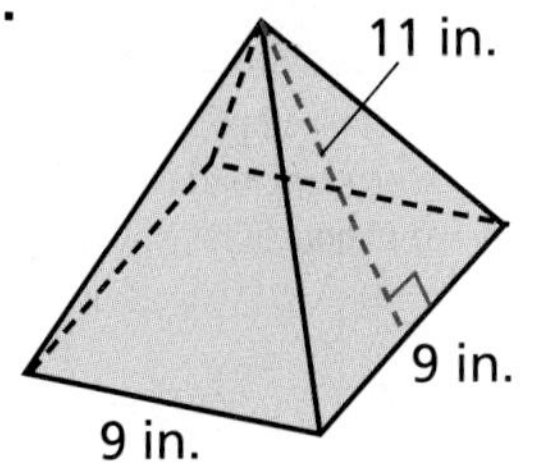

18.

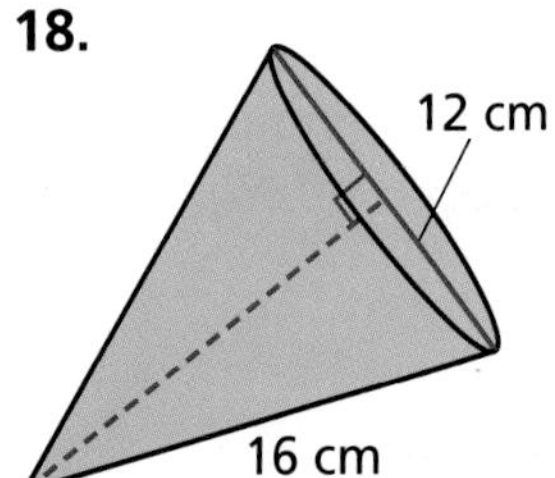

Chapter Test

Chapter 6

Performance Assessment

Show What You Know

Create a portfolio of your work from this chapter. Complete this page and include it with your four best pieces of work from Chapter 6. Choose from your homework or lab assignments, mid-chapter quiz, or any journal entries you have done. Put them together using any design you want. Make your portfolio represent what you consider your best work.

Short Response

Trace each figure, and then locate the vanishing point or horizon line.

1. Draw a rectangle with base length 7 cm and height 4 cm. Then draw a rectangle with base length 14 cm and height 1 cm. Which rectangle has the larger area? Which rectangle has the larger perimeter? Show your work or explain in words how you determined your answers.

2. A cylinder with a height of 6 in. and a diameter of 4 in. is filled with water. A cone with a height of 6 in. and a diameter of 2 in. is placed in the cylinder, point down, with its base even with the top of the cylinder. Draw a diagram to illustrate the situation described, and then determine how much water is left in the cylinder. Show your work.

Extended Problem Solving

3. A *geodesic dome* is constructed of triangles. The surface is approximately spherical.
   a. A pattern for a geodesic dome that approximates a hemisphere uses 30 triangles with base 8 ft and height 5.63 ft and 75 triangles with base 8 ft and height 7.13 ft. Find the surface area of the dome.
   b. The base of the dome is approximately a circle with diameter 41 ft. Use a hemisphere with this diameter to estimate the surface area of the dome.
   c. Compare your answer from part **a** to your estimate from part **b**. Explain the difference.

Richard Buckminster Fuller created the *geodesic dome* and designed the Dymaxion™ house, car, and map.

internet connect

EOG Test Practice Online

go.hrw.com Keyword: MT4 TestPrep

Getting Ready for EOG

Chapter 6

Cumulative Assessment, Chapters 1–6

1. The shaded figure below is a net that can be used to form a rectangular prism. What is the surface area of the prism?

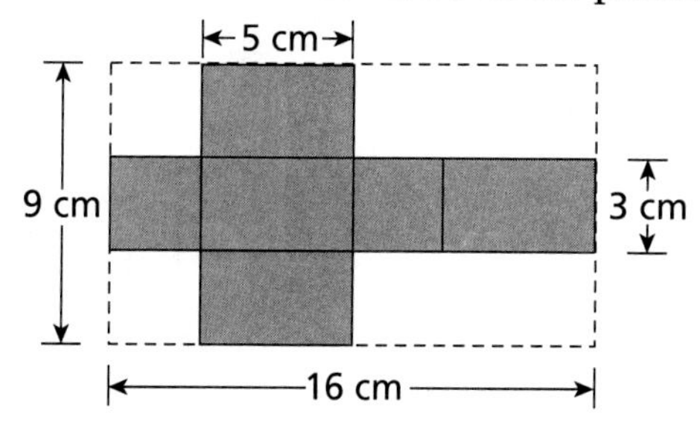

A 15 cm^2 C 144 cm^2

B 78 cm^2 D 180 cm^2

2. What is the value of x in the table below?

Number of Inches	5	10	x
Number of Centimeters	12.7	25.4	50.8

A 15 C 20

B 18 D 22

3. The quantity (3×8^{12}) is how many times the quantity (3×8^5)?

A 7 C 21

B 8 D 8^7

4. The arithmetic mean of 3 numbers is 60. If two of the numbers are 50 and 60, what is the third number?

A 55 C 65

B 60 D 70

5. What is the value of $26 - 24 \cdot 2^3$?

A 18 C -118

B 16 D -166

6. If $p = 3$, what is $4r(3 - 2p)$ in terms of r?

A $-12r$ C $-7r$

B $-8r$ D $12r - 6$

7. Point A' is formed by reflecting $A(-9, -8)$ across the y-axis. What are the coordinates of A'?

A $(9, 8)$ C $(-8, -9)$

B $(9, -8)$ D $(-9, 8)$

TEST TAKING TIP!

Look for a pattern in the data set to help you find the answer.

8. In the cylinder, point A lies on the top edge and point B on the bottom edge. If the radius of the cylinder is 2 units and the height is 5 units, what is the greatest straight-line distance between A and B?

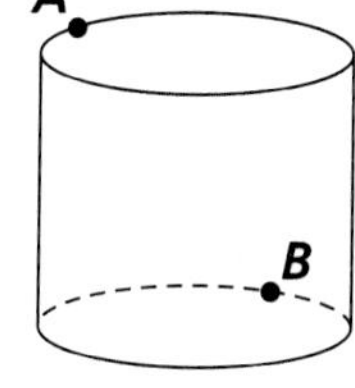

A 5 C $\sqrt{29}$

B 7 D $\sqrt{41}$

9. ***SHORT RESPONSE*** On a number line, point A has the coordinate -3 and point B has the coordinate 12. Point P is $\frac{2}{3}$ of the way from A to B. Draw and label the three points on a number line.

10. ***SHORT RESPONSE*** The tip of a blade on an electric fan is 1.5 feet from the axis of rotation. If the fan spins at a full rate of 1760 revolutions per minute, how many miles will a point at the tip of a blade travel in one hour? (1 mile = 5280 feet) Show your work.

Chapter 7

Ratios and Similarity

TABLE OF CONTENTS

Tree	Natural Height (ft)	Bonsai Height (in.)
Chinese elm	60	10
Brush cherry	50	8
Juniper	10	6
Pitch pine	200	14
Eastern hemlock	80	18

Career *Horticulturist*

Chances are that a horticulturist helped create many of the varieties of plants at your local nursery. Horticulturists work in vegetable development, fruit growing, flower growing, and landscape design. Horticulturists who are also scientists work to develop new types of plants or ways to control plant diseases.

The art of *bonsai*, or making miniature plants, began in China and became popular in Japan. Now bonsai is practiced all over the world.

internet connect

Chapter Opener Online
go.hrw.com
KEYWORD: MT4 Ch7

ARE YOU READY?

Choose the best term from the list to complete each sentence.

1. To solve an equation, you use __?__ to isolate the variable. So to solve the __?__ $3x = 18$, divide both sides by 3.
2. In the fractions $\frac{2}{3}$ and $\frac{1}{6}$, 18 is a __?__, but 6 is the __?__.
3. If two polygons are congruent, all of their __?__ sides and angles are congruent.

common denominator
corresponding
inverse operations
least common denominator
multiplication equation

Complete these exercises to review skills you will need for this chapter.

✔ Simplify Fractions

Write each fraction in simplest form.

4. $\frac{8}{24}$ **5.** $\frac{15}{50}$ **6.** $\frac{18}{72}$ **7.** $\frac{25}{125}$

✔ Use a Least Common Denominator

Find the least common denominator for each set of fractions.

8. $\frac{2}{3}$ and $\frac{1}{5}$ **9.** $\frac{3}{4}$ and $\frac{1}{8}$ **10.** $\frac{5}{7}, \frac{3}{7}$, and $\frac{1}{14}$ **11.** $\frac{1}{2}, \frac{2}{3}$, and $\frac{3}{5}$

✔ Order Decimals

Write each set of decimals in order from least to greatest.

12. 4.2, 2.24, 2.4, 0.242 **13.** 1.1, 0.1, 0.01, 1.11 **14.** 1.4, 2.53, $1.\overline{3}$, $0.\overline{9}$

✔ Solve Multiplication Equations

Solve.

15. $5x = 60$ **16.** $0.2y = 14$ **17.** $\frac{1}{2}t = 10$ **18.** $\frac{2}{3}z = 9$

✔ Identify Corresponding Parts of Congruent Figures

If $\triangle ABC \cong \triangle JRW$, complete each congruence statement.

19. $\overline{AB} \cong$ __?__ **20.** $\angle R \cong$ __?__ **21.** $\overline{AC} \cong$ __?__ **22.** $\angle C \cong$ __?__

7-1 Ratios and Proportions

Learn to find equivalent ratios to create proportions.

Vocabulary

ratio

equivalent ratio

proportion

Relative density is the ratio of the density of a substance to the density of water at 4°C. The relative density of silver is 10.5. This means that silver is 10.5 times as heavy as an equal volume of water.

The comparisons of water to silver in the table are *ratios* that are all equivalent.

Comparisons of Mass of Equal Volumes of Water and Silver				
Water	1 g	2 g	3 g	4 g
Silver	10.5 g	21 g	31.5 g	42 g

Mexico and Peru are the world's largest silver producers.

Reading Math

Ratios can be written in several ways. A colon is often used. 90:3 and $\frac{90}{3}$ name the same ratio.

A **ratio** is a comparison of two quantities by division. In one rectangle, the ratio of shaded squares to unshaded squares is 7:5. In the other rectangle, the ratio is 28:20. Both rectangles have equivalent shaded areas. Ratios that make the same comparison are **equivalent ratios**.

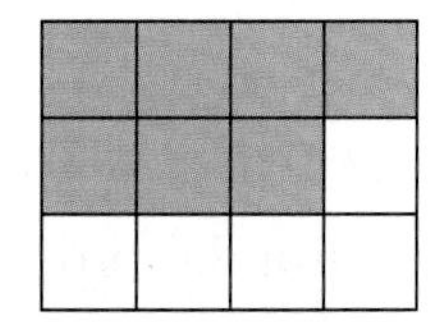

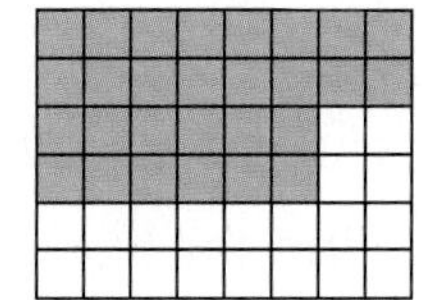

EXAMPLE 1 Finding Equivalent Ratios

Find two ratios that are equivalent to each given ratio.

A $\frac{6}{8}$

$\frac{6}{8} = \frac{6 \cdot 2}{8 \cdot 2} = \frac{12}{16}$

$\frac{6}{8} = \frac{6 \div 2}{8 \div 2} = \frac{3}{4}$

Multiply or divide the numerator and denominator by the same nonzero number.

Two ratios equivalent to $\frac{6}{8}$ are $\frac{12}{16}$ and $\frac{3}{4}$.

B $\frac{48}{27}$

$\frac{48}{27} = \frac{48 \cdot 2}{27 \cdot 2} = \frac{96}{54}$

$\frac{48}{27} = \frac{48 \div 3}{27 \div 3} = \frac{16}{9}$

Two ratios equivalent to $\frac{48}{27}$ are $\frac{96}{54}$ and $\frac{16}{9}$.

Ratios that are equivalent are said to be *proportional*, or in **proportion**. Equivalent ratios are identical when they are written in simplest form.

EXAMPLE 2 Determining Whether Two Ratios are in Proportion

Simplify to tell whether the ratios form a proportion.

A $\frac{7}{21}$ **and** $\frac{2}{6}$

$\frac{7}{21} = \frac{7 \div 7}{21 \div 7} = \frac{1}{3}$

$\frac{2}{6} = \frac{2 \div 2}{6 \div 2} = \frac{1}{3}$

Since $\frac{1}{3} = \frac{1}{3}$, the ratios are in proportion.

B $\frac{9}{12}$ **and** $\frac{16}{24}$

$\frac{9}{12} = \frac{9 \div 3}{12 \div 3} = \frac{3}{4}$

$\frac{16}{24} = \frac{16 \div 8}{24 \div 8} = \frac{2}{3}$

Since $\frac{3}{4} \neq \frac{2}{3}$, the ratios are *not* in proportion.

EXAMPLE 3 *Earth Science Application*

At 4°C, two cubic feet of silver has the same mass as 21 cubic feet of water. At 4°C, would 126 cubic feet of water have the same mass as 6 cubic feet of silver?

$\frac{2}{21} \stackrel{?}{=} \frac{6}{126}$

$\frac{2}{21} \stackrel{?}{=} \frac{6 \div 6}{126 \div 6}$ *Simplify.*

$\frac{2}{21} \neq \frac{1}{21}$

Since $\frac{2}{21}$ is not equal to $\frac{1}{21}$, 126 cubic feet of water would not have the same mass at 4°C as 6 cubic feet of silver.

Earth Science LINK

Silver is a rare mineral usually mined along with lead, copper, and zinc.

Think and Discuss

1. **Describe** how two ratios can form a proportion.
2. **Give** three ratios equivalent to 12:24.
3. **Explain** why the ratios 2:4 and 6:10 do not form a proportion.
4. **Give an example** of two ratios that are proportional and have numerators with different signs.

7-1 Exercises

FOR EOG PRACTICE

see page 668

internet connect

Homework Help Online

go.hrw.com Keyword: MT4 7-1

GUIDED PRACTICE

See Example 1 **Find two ratios that are equivalent to each given ratio.**

1. $\frac{4}{10}$ **2.** $\frac{3}{9}$ **3.** $\frac{21}{7}$ **4.** $\frac{40}{32}$

See Example 2 **Simplify to tell whether the ratios form a proportion.**

5. $\frac{6}{30}$ and $\frac{3}{15}$ **6.** $\frac{6}{9}$ and $\frac{10}{18}$ **7.** $\frac{35}{21}$ and $\frac{20}{12}$

See Example 3 **8.** A recipe calls for 1.5 cups of mix to make 8 pancakes. Mike wants to make 12 pancakes and uses 2 cups of mix. Does Mike have the correct ratio for the recipe? Explain.

INDEPENDENT PRACTICE

See Example 1 **Find two ratios that are equivalent to each given ratio.**

9. $\frac{1}{7}$ **10.** $\frac{5}{11}$ **11.** $\frac{16}{14}$ **12.** $\frac{65}{15}$

See Example 2 **Simplify to tell whether the ratios form a proportion.**

13. $\frac{7}{14}$ and $\frac{13}{28}$ **14.** $\frac{80}{100}$ and $\frac{4}{5}$ **15.** $\frac{1}{3}$ and $\frac{15}{45}$

See Example 3 **16.** A molecule of carbonic acid contains 3 atoms of oxygen for every 2 atoms of hydrogen. Could a compound containing 81 hydrogen atoms and 54 oxygen atoms be carbonic acid? Explain.

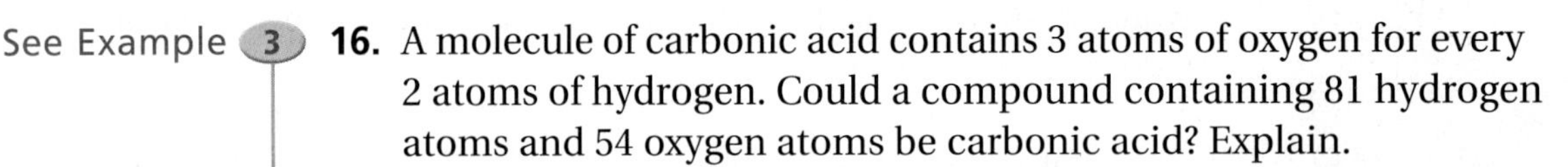

PRACTICE AND PROBLEM SOLVING

Tell whether the ratios form a proportion. If not, find a ratio that would form a proportion with the first ratio.

17. $\frac{8}{14}$ and $\frac{6}{21}$ **18.** $\frac{7}{9}$ and $\frac{140}{180}$ **19.** $\frac{4}{7}$ and $\frac{12}{49}$

20. $\frac{30}{36}$ and $\frac{15}{16}$ **21.** $\frac{13}{12}$ and $\frac{39}{36}$ **22.** $\frac{11}{20}$ and $\frac{22}{40}$

23. $\frac{16}{84}$ and $\frac{6}{62}$ **24.** $\frac{24}{10}$ and $\frac{44}{18}$ **25.** $\frac{11}{121}$ and $\frac{33}{363}$

26. ***BUSINESS*** Cal pays his employees weekly. He would like to start paying them four times the weekly amount on a monthly basis. Is a month equivalent to four weeks? Explain.

27. ***TRANSPORTATION*** Aaron's truck has a 12-gallon gas tank. He just put 3 gallons of gas into the tank. Is this equivalent to a third of a tank? If not, what amount of gas is equivalent to a third of a tank?

28. ***ENTERTAINMENT*** The table lists prices for movie tickets.

a. Are the ticket prices proportional?

b. How much do 6 movie tickets cost?

c. If Suzie paid $57.75 for movie tickets, how many did she buy?

Movie Ticket Prices

Number of Tickets	1	2	3
Price	$8.25	$16.50	$24.75

29. ***HOBBIES*** A bicycle chain moves between two sprockets when you shift gears. The number of teeth on the front sprocket and the number of teeth on the rear sprocket form a ratio. Equivalent ratios provide equal pedaling power. Find a ratio equivalent to the ratio shown, $\frac{52}{24}$.

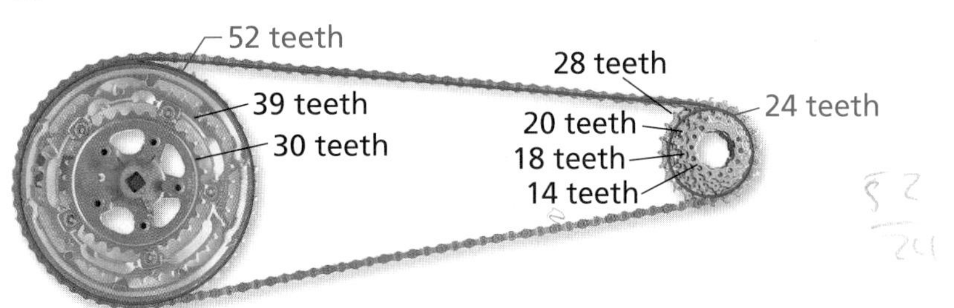

30. ***COMPUTERS*** While a file downloads, a computer displays the total number of kilobytes downloaded and the number of seconds that have passed. If the display shows 42 kilobytes after 7 seconds, is the file downloading at about 6 kilobytes per second? Explain.

31. ***WRITE A PROBLEM*** The ratio of the number of bones in the human skull to the number of bones in the ears is 11:3. There are 22 bones in the skull and 6 in the ears. Use this information to write a problem using equivalent ratios. Explain your solution.

32. ***WRITE ABOUT IT*** Describe at least two ways, given a ratio, to create an equivalent ratio.

33. ***CHALLENGE*** Write all possible proportions using each of the numbers 2, 4, 8, and 16 once.

Spiral Review

Add or subtract. (Lesson 3-5)

34. $\frac{5}{7} + \frac{2}{3}$ **35.** $\frac{4}{9} + \left(-1\frac{3}{4}\right)$ **36.** $\frac{3}{5} - \frac{7}{10}$ **37.** $2\frac{7}{9} - 1\frac{8}{11}$

Find the two square roots of each number. (Lesson 3-8)

38. 49 **39.** 9 **40.** 81 **41.** 169

42. **EOG PREP** Which two integers does $-\sqrt{74}$ lie between? (Lesson 3-9)

A -7 and -6 B -8 and -7 C -9 and -8 D -11 and -10

7-2 Ratios, Rates, and Unit Rates

Learn to work with rates and ratios.

Vocabulary

rate

unit rate

unit price

Movie and television screens range in shape from almost perfect squares to wide rectangles. An *aspect ratio* describes a screen by comparing its width to its height. Common aspect ratios are 4:3, 37:20, 16:9, and 47:20.

Most high-definition TV screens have an aspect ratio of 16:9.

EXAMPLE 1 ***Entertainment Application***

By design, movies can be viewed on screens with varying aspect ratios. The most common ones are 4:3, 37:20, 16:9, and 47:20.

A **Order the width-to-height ratios from least (standard TV) to greatest (wide-screen).**

$4{:}3 = \frac{4}{3} = 1.\overline{3}$ *Divide.* $\frac{4}{3} = \frac{1.\overline{3}}{1}$

$37{:}20 = \frac{37}{20} = 1.85$

$16{:}9 = \frac{16}{9} = 1.\overline{7}$

$47{:}20 = \frac{47}{20} = 2.35$

The decimals in order are $1.\overline{3}$, $1.\overline{7}$, 1.85, and 2.35.
The width-to-height ratios in order from least to greatest are 4:3, 16:9, 37:20, and 47:20.

B **A wide-screen television has screen width 32 in. and height 18 in. What is the aspect ratio of this screen?**

The ratio of the width to the height is 32:18.

The ratio $\frac{32}{18}$ can be simplified: $\frac{32}{18} = \frac{2(16)}{2(9)} = \frac{16}{9}$.

The screen has the aspect ratio 16:9.

A ratio is a comparison of two quantities. A **rate** is a comparison of two quantities that have different units.

ratio: $\frac{90}{3}$ rate: $\frac{90 \text{ miles}}{3 \text{ hours}}$ ← *Read as "90 miles per 3 hours."*

Unit rates are rates in which the second quantity is 1. The ratio $\frac{90}{3}$ can be simplified by dividing: $\frac{90}{3} = \frac{30}{1}$.

unit rate: $\frac{30 \text{ miles}}{1 \text{ hour}}$, or 30 mi/h

EXAMPLE 2 Using a Bar Graph to Determine Rates

The number of acres destroyed by wildfires in 2000 is shown for the states with the highest totals. Use the bar graph to find the number of acres, to the nearest acre, destroyed in each state per day.

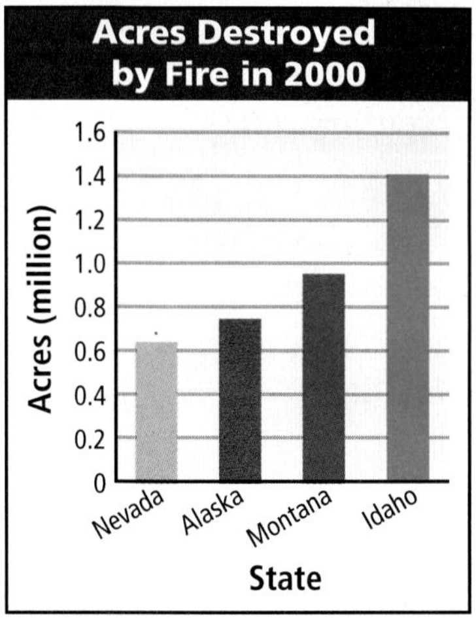

Acres Destroyed by Fire in 2000

Source: National Interagency Fire Center

$$\text{Nevada} = \frac{640{,}000 \text{ acres}}{366 \text{ days}} \approx \frac{1749 \text{ acres}}{1 \text{ day}}$$

$$\text{Alaska} = \frac{750{,}000 \text{ acres}}{366 \text{ days}} \approx \frac{2049 \text{ acres}}{1 \text{ day}}$$

$$\text{Montana} = \frac{950{,}000 \text{ acres}}{366 \text{ days}} \approx \frac{2596 \text{ acres}}{1 \text{ day}}$$

$$\text{Idaho} = \frac{1{,}400{,}000 \text{ acres}}{366 \text{ days}} \approx \frac{3825 \text{ acres}}{1 \text{ day}}$$

Nevada: 1749 acres/day; Alaska: 2049 acres/day; Montana: 2596 acres/day; Idaho: 3825 acres/day

Unit price is a unit rate used to compare costs per item.

EXAMPLE 3 Finding Unit Prices to Compare Costs

A **Blank videotapes can be purchased in packages of 3 for \$4.99, or 10 for \$15.49. Which is the better buy?**

$$\frac{\text{price for package}}{\text{number of videotapes}} = \frac{\$4.99}{3} \approx \$1.66$$

$$\frac{\text{price for package}}{\text{number of videotapes}} = \frac{\$15.49}{10} \approx \$1.55$$

Divide the price by the number of tapes.

The better buy is the package of 10 for \$15.49.

B **Leron can buy a 64 oz carton of orange juice for \$2.49 or a 96 oz carton for \$3.99. Which is the better buy?**

$$\frac{\text{price for carton}}{\text{number of ounces}} = \frac{\$2.49}{64} \approx \$0.0389$$

$$\frac{\text{price for carton}}{\text{number of ounces}} = \frac{\$3.99}{96} \approx \$0.0416$$

Divide the price by the number of ounces.

The better buy is the 64 oz carton for \$2.49.

Think and Discuss

1. Choose the quantity that has a lower unit price: 6 oz for \$1.29 or 15 oz for \$3.00. Explain your answer.

2. Explain why an aspect ratio is not considered a rate.

3. Determine two different units of measurement for speed.

7-2 Exercises

FOR EOG PRACTICE

see page 668

internet connect
Homework Help Online
go.hrw.com Keyword: MT4 7-2

GUIDED PRACTICE

See Example 1

1. The height of a bridge is 68 ft, and its length is 340 ft. Find the ratio of its height to its length in simplest form.

See Example 2

For Exercises 2 and 3, use the bar graph to find each unit rate.

2. Ellen's words per minute
3. Yoshiko's words per minute

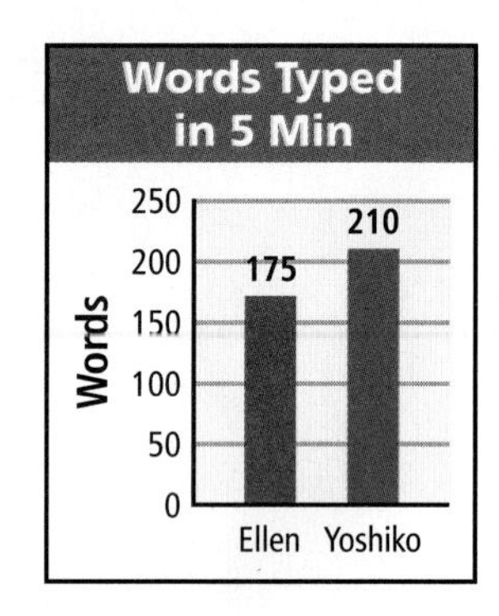

See Example 3

Determine the better buy.

4. a 15 oz can of corn for $1.39 or a 22 oz can for $1.85
5. a dozen golf balls for $22.99 or 20 golf balls for $39.50

INDEPENDENT PRACTICE

See Example 1

6. A child's basketball hoop is 6 ft tall. Find the ratio of its height to the height of a regulation basketball hoop, which is 10 ft tall. Express the ratio in simplest form.

See Example 2

For Exercises 7 and 8, use the bar graph to find each unit rate.

7. gallons per hour for machine A
8. gallons per hour for machine B

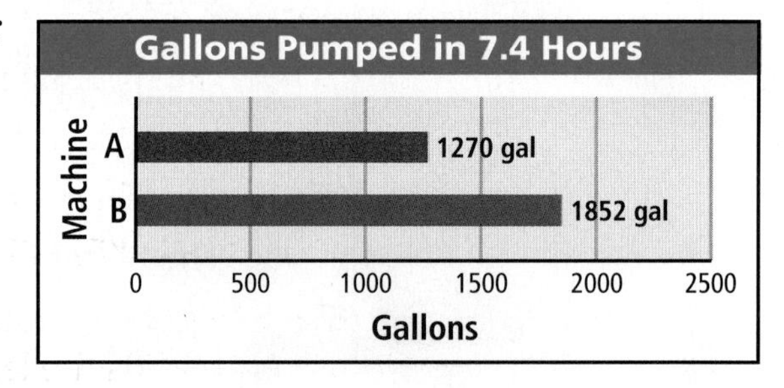

See Example 3

Determine the better buy.

9. 4 boxes of cereal for $9.56; 2 boxes of cereal for $4.98
10. 8 oz jar of soup for $2.39; 10 oz jar of soup for $2.69

PRACTICE AND PROBLEM SOLVING

Find each unit rate.

11. $525 for 20 hours of work
12. 96 chairs in 8 rows
13. 12 slices of pizza for $9.25
14. 64 beats in 4 measures of music

Find each unit price and tell which is the better buy.

15. $7.47 for 3 yards of fabric; $11.29 for 5 yards of fabric
16. A $\frac{1}{2}$-pound hamburger for $3.50; a $\frac{1}{3}$-pound hamburger for $3.25
17. 10 gallons of gasoline for $13.70; 12.5 gallons of gasoline for $17.75
18. $1.65 for 5 pounds of bananas; $3.15 for 10 pounds of bananas

19. ***COMMUNICATIONS*** Super-Cell offers a wireless phone plan that includes 250 base minutes for $24.99 a month. Easy-Phone has a plan that includes 325 base minutes for $34.99.

   a. Find the unit rate for the base minutes for each plan.

   b. Which company offers a lower rate for base minutes?

20. ***BUSINESS*** A cereal company pays $59,969 to have its new cereal placed in a grocery store display for one week. Find the daily rate for this display.

21. ***ENTERTAINMENT*** Tom, Cherise, and Tina work as film animators. The circle graph shows the number of frames they each rendered in an 8-hour day.

   a. Find the hourly unit rendering rate for each employee.

   b. Who was the most efficient employee?

   c. How many more frames per hour did Cherise render than Tom?

   d. How many more frames per hour did Tom and Cherise together render than Tina?

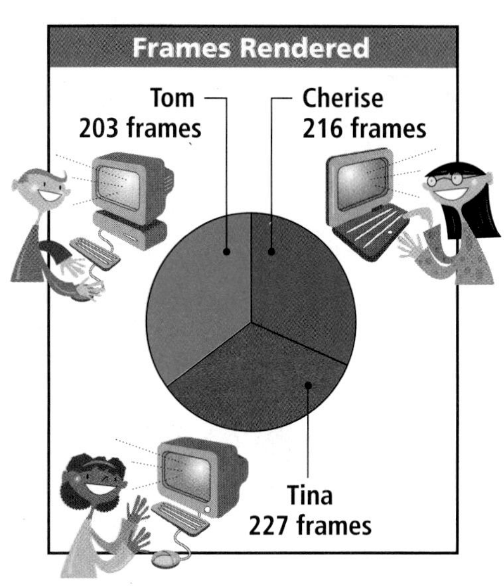

22. ***WHAT'S THE ERROR?*** A clothing store charges $30 for 12 pairs of socks. A student says that the unit price is $0.40 per pair. What is the error? What is the correct unit price?

23. ***WRITE ABOUT IT*** Explain how to find unit rates. Give an example and explain how consumers can use unit rates to save money.

24. ***CHALLENGE*** The size of a television (13 in., 25 in., 32 in., and so on) represents the length of the diagonal of the television screen. A 25 in. television has an aspect ratio of 4:3. What is the width and height of the screen?

Spiral Review

Evaluate each expression for the given value of the variable. (Lesson 2-1)

25. $c + 4$ for $c = -8$ **26.** $m - 2$ for $m = 13$ **27.** $5 + d$ for $d = -10$

Evaluate each expression for the given value of the variable. (Lesson 3-2)

28. $45.6 + x$ for $x = -11.1$ **29.** $17.9 - b$ for $b = 22.3$ **30.** $r + (-4.9)$ for $r = 31.8$

31. **EOG PREP** How much fencing, to the nearest foot, is needed to enclose a square lot with an area of 350 ft^2? (Lesson 3-9)

A 65 ft B 68 ft C 74 ft D 75 ft

7-3 Analyze Units

 Problem Solving Skill

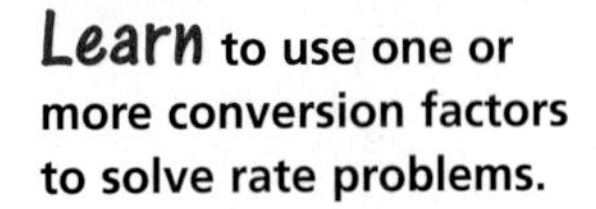

Learn to use one or more conversion factors to solve rate problems.

Vocabulary
conversion factor

You can measure the speed of an object using a strobe lamp and a camera in a dark room. Each time the lamp flashes, the camera records the object's position.

Problems often require *dimensional analysis,* also called *unit analysis,* to convert from one unit to another unit.

To convert units, multiply by one or more ratios of equal quantities called **conversion factors**.

For example, to convert inches to feet you would use the ratio at right as a conversion factor. $\frac{1 \text{ ft}}{12 \text{ in.}}$

Multiplying by a conversion factor is like multiplying by a fraction that reduces to 1, such as $\frac{5}{5}$. $\frac{1 \text{ ft}}{12 \text{ in.}} = \frac{12 \text{ in.}}{12 \text{ in.}}$, or $\frac{1 \text{ ft}}{1 \text{ ft}}$, $= 1$

EXAMPLE 1 Finding Conversion Factors

Helpful Hint

The conversion factor
- must introduce the unit desired in the answer and
- must cancel the original unit so that the unit desired is all that remains.

Find the appropriate factor for each conversion.

A quarts to gallons

There are 4 quarts in 1 gallon. To convert quarts to gallons, multiply the number of **quarts** by $\frac{1 \text{ gal}}{4 \text{ qt}}$.

B meters to centimeters

There are 100 centimeters in 1 meter. To convert meters to centimeters, multiply the number of **meters** by $\frac{100 \text{ cm}}{1 \text{ m}}$.

EXAMPLE 2 Using Conversion Factors to Solve Problems

The average American eats 23 pounds of pizza per year. Find the number of ounces of pizza the average American eats per year.

The problem gives the ratio 23 *pounds* to 1 year and asks for an answer in *ounces* per year.

$\frac{23 \text{ lb}}{1 \text{ yr}} \cdot \frac{16 \text{ oz}}{1 \text{ lb}}$ *Multiply the ratio by the conversion factor.*

$= \frac{23 \cdot 16 \text{ oz}}{1 \text{ yr}}$ *Cancel lb units.* $\frac{\cancel{lb}}{yr} \cdot \frac{oz}{\cancel{lb}} = \frac{oz}{yr}$

$= 368$ oz per year *Multiply 23 by 16 oz.*

The average American eats 368 ounces of pizza per year.

EXAMPLE 3 PROBLEM SOLVING APPLICATION

A car traveled 990 feet down a road in 15 seconds. How many miles per hour was the car traveling?

1 Understand the Problem

The problem is stated in units of feet and seconds. The question asks for the **answer** in units of miles and hours. You will need to use several conversion factors.

List the important information:

- Feet to miles ⟶ $\frac{1 \text{ mi}}{5280 \text{ ft}}$
- Seconds to minutes ⟶ $\frac{60 \text{ s}}{1 \text{ min}}$; minutes to hours ⟶ $\frac{60 \text{ min}}{1 \text{ h}}$

2 Make a Plan

Multiply by each conversion factor separately, or **simplify the problem** and multiply by several conversion factors at once.

3 Solve

First, convert 990 feet in 15 seconds into a unit rate.

$$\frac{990 \text{ ft}}{15 \text{ s}} = \frac{(990 \div 15) \text{ ft}}{(15 \div 15) \text{ s}} = \frac{66 \text{ ft}}{1 \text{ s}}$$

Create a single conversion factor to convert seconds directly to hours:

seconds to minutes ⟶ $\frac{60 \text{ s}}{1 \text{ min}}$; minutes to hours ⟶ $\frac{60 \text{ min}}{1 \text{ h}}$

seconds to hours $= \frac{60 \text{ s}}{1 \cancel{\text{min}}} \cdot \frac{60 \cancel{\text{min}}}{1 \text{ h}} = \frac{3600 \text{ s}}{1 \text{ h}}$

$\frac{66 \text{ ft}}{1 \text{ s}} \cdot \frac{1 \text{ mi}}{5280 \text{ ft}} \cdot \frac{3600 \text{ s}}{1 \text{ h}}$ *Set up the conversion factors.*

Do not include the numbers yet. Notice what happens to the units.

$\frac{\cancel{\text{ft}}}{\cancel{\text{s}}} \cdot \frac{\text{mi}}{\cancel{\text{ft}}} \cdot \frac{\cancel{\text{s}}}{\text{h}}$ *Simplify. Only $\frac{mi}{h}$ remain.*

$\frac{66 \text{ ft}}{1 \text{ s}} \cdot \frac{1 \text{ mi}}{5280 \text{ ft}} \cdot \frac{3600 \text{ s}}{1 \text{ h}}$ *Multiply.*

$$\frac{66 \cdot 1 \text{ mi} \cdot 3600}{1 \cdot 5280 \cdot 1 \text{ h}} = \frac{237{,}600 \text{ mi}}{5280 \text{ h}} = \frac{45 \text{ mi}}{1 \text{ h}}$$

The car was traveling 45 miles per hour.

4 Look Back

A rate of 45 mi/h is less than 1 mi/min. 15 seconds is $\frac{1}{4}$ min. A car traveling 45 mi/h would go less than $\frac{1}{4}$ of 5280 ft in 15 seconds. It goes 990 ft, so 45 mi/h is a reasonable speed.

EXAMPLE 4 *Physical Science Application*

Physical Science LINK

A strobe light flashing on dripping liquid can make droplets appear to stand still or even move upward.

A strobe lamp can be used to measure the speed of an object. The lamp flashes every $\frac{1}{1000}$ s. A camera records the object moving 7.5 cm between flashes. How fast is the object moving in m/s?

$$\frac{7.5 \text{ cm}}{\frac{1}{1000}\text{ s}}$$ *Use rate = $\frac{\text{distance}}{\text{time}}$.*

It may help to eliminate the fraction $\frac{1}{1000}$ first.

$$\frac{7.5 \text{ cm}}{\frac{1}{1000}\text{ s}} = \frac{1000 \cdot 7.5 \text{ cm}}{1000 \cdot \frac{1}{1000}\text{ s}}$$ *Multiply top and bottom by 1000.*

$$= \frac{7500 \text{ cm}}{1 \text{ s}}$$

Now convert centimeters to meters.

$$\frac{7500 \text{ cm}}{1 \text{ s}}$$

$$= \frac{7500 \text{ cm}}{1 \text{ s}} \cdot \frac{1 \text{ m}}{100 \text{ cm}}$$ *Multiply by the conversion factor.*

$$= \frac{7500 \text{ m}}{100 \text{ s}} = \frac{75 \text{ m}}{1 \text{ s}}$$

The object is traveling 75 m/s.

EXAMPLE 5 *Transportation Application*

The rate of one knot equals one nautical mile per hour. One nautical mile is 1852 meters. What is the speed in meters per second of a ship traveling at 20 knots?

20 knots = 20 nautical mi/h

Set up the units to obtain m/s in your answer.

$$\frac{\cancel{\text{nautical mi}}}{\cancel{\text{h}}} \cdot \frac{\text{m}}{\cancel{\text{nautical mi}}} \cdot \frac{\cancel{\text{h}}}{\text{s}}$$ *Examine the units.*

$$\frac{20 \text{ nautical mi}}{\text{h}} \cdot \frac{1852 \text{ m}}{\text{nautical mi}} \cdot \frac{1 \text{ h}}{3600 \text{ s}}$$

$$\frac{20 \cdot 1852}{3600} \approx 10.3$$

The ship is traveling about 10.3 m/s.

Think and Discuss

1. **Give** the conversion factor for converting $\frac{\text{lb}}{\text{yr}}$ to $\frac{\text{lb}}{\text{mo}}$.
2. **Explain** how to find whether 10 miles per hour is faster than 15 feet per second.
3. **Give an example** of a conversion between units that includes ounces as a unit in the conversion.

7-3 Exercises

GUIDED PRACTICE

See Example 1

Find the appropriate factor for each conversion.

1. feet to inches
2. gallons to pints
3. centimeters to meters

See Example 2

4. Aihua drinks 4 cups of water a day. Find the total number of gallons of water she drinks in a year.

See Example 3

5. A model airplane flies 22 feet in 2 seconds. What is the airplane's speed in miles per hour?

See Example 4

6. If a fish swims 0.09 centimeter every hundredth of a second, how fast in meters per second is it swimming?

See Example 5

7. There are about 400 cocoa beans in a pound. There are 2.2 pounds in a kilogram. About how many grams does a cocoa bean weigh?

INDEPENDENT PRACTICE

See Example 1

Find the appropriate factor for each conversion.

8. kilometers to meters
9. inches to yards
10. days to weeks

See Example 2

11. A theme park sells 71,175 yards of licorice each year. How many feet per day does the park sell?

See Example 3

12. A yellow jacket can fly 4.5 meters in 9 seconds. How fast in kilometers per hour can a yellow jacket fly?

See Example 4

13. Brilco Manufacturing produces 0.2 of a brick every tenth of a second. How many bricks can be produced in an 8-hour day?

See Example 5

14. Assume that one dollar is equal to 1.14 euros. If 500 g of an item is selling for 25 euros, what is its price in dollars per kg?

PRACTICE AND PROBLEM SOLVING

Use conversion factors to find each specified amount.

15. radios produced in 5 hours at a rate of 3 radios per minute
16. distance traveled (in feet) after 12 seconds at 87 miles per hour
17. hot dogs eaten in a month at a rate of 48 hot dogs eaten each year
18. umbrellas sold in a year at a rate of 5 umbrellas sold per day
19. miles jogged in 1 hour at an average rate of 7.3 feet per second
20. states visited in a two-week political campaign at a rate of 2 states per day

21. ***SPORTS*** Use the graph to find each world-record speed in miles per hour. (*Hint:* 1 mile ≈ 1609 m.)

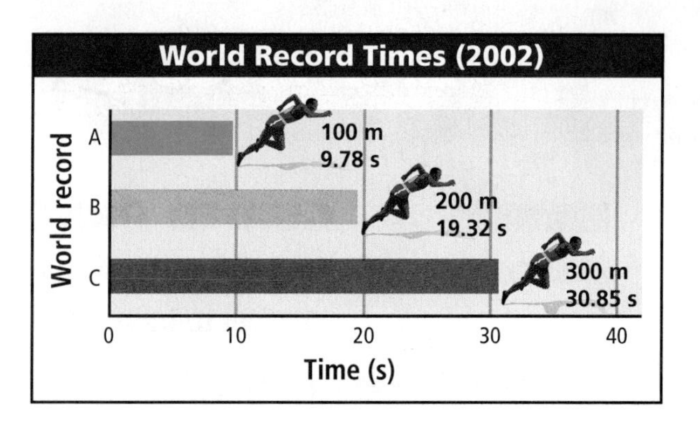

22. ***LIFE SCIENCE*** The Kelp Forest exhibit at the Monterey Bay Aquarium holds 335,000 gallons. How many days would it take to fill it at a rate of 1 gallon per second?

23. ***TRANSPORTATION*** An automobile engine is turning at 3000 revolutions per minute. During each revolution, each of the four spark plugs fires. How many times do the spark plugs fire in one second?

24. ***CHOOSE A STRATEGY*** The label on John's bottle of cough syrup says a person should take 3 teaspoons. Which spoon could John use to take the cough medicine? (*Hint:* 1 teaspoon = $\frac{1}{6}$ oz.)

A A 1.5 oz spoon

B A 0.5 oz spoon

C A 1 oz spoon

D None of these

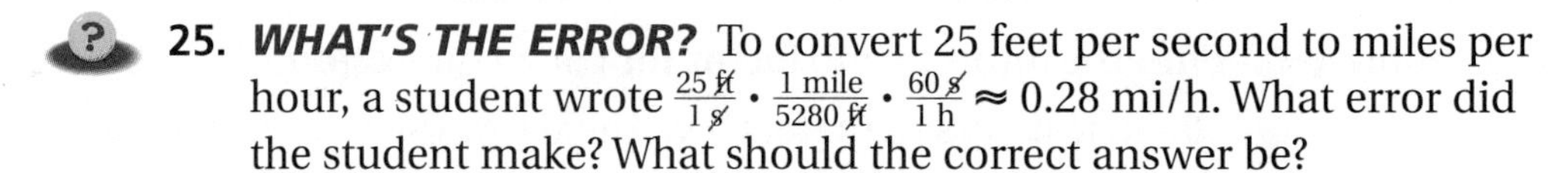

25. ***WHAT'S THE ERROR?*** To convert 25 feet per second to miles per hour, a student wrote $\frac{25\text{ ft}}{1\text{ s}} \cdot \frac{1\text{ mile}}{5280\text{ ft}} \cdot \frac{60\text{ s}}{1\text{ h}} \approx 0.28$ mi/h. What error did the student make? What should the correct answer be?

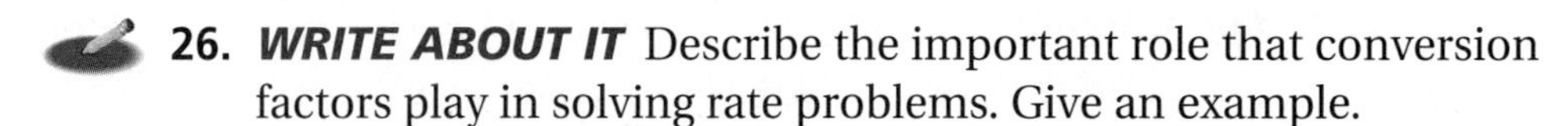

26. ***WRITE ABOUT IT*** Describe the important role that conversion factors play in solving rate problems. Give an example.

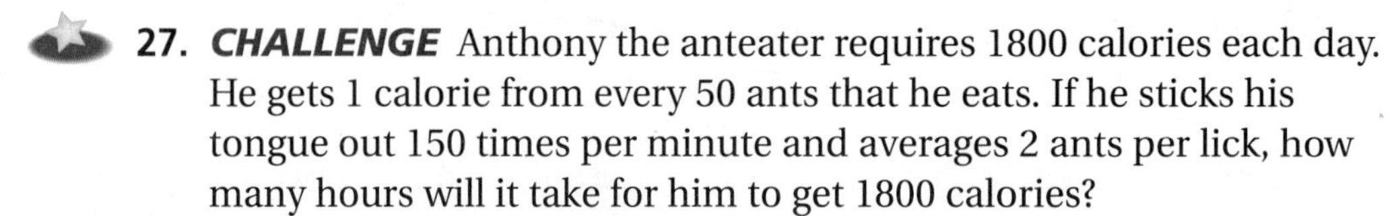

27. ***CHALLENGE*** Anthony the anteater requires 1800 calories each day. He gets 1 calorie from every 50 ants that he eats. If he sticks his tongue out 150 times per minute and averages 2 ants per lick, how many hours will it take for him to get 1800 calories?

Spiral Review

Find the area of the quadrilateral with the given vertices. (Lesson 6-1)

28. (0, 0), (0, 9), (5, 9), (5, 0)

29. (–3, 1), (4, 1), (6, 3), (–1, 3)

Find the area of each circle to the nearest tenth. Use 3.14 for π. (Lesson 6-4)

30. circle with radius 7 ft

31. circle with diameter 17 in.

32. circle with radius 3.5 cm

33. circle with diameter 2.2 mi

34. **EOG PREP** A cylinder has radius 6 cm and height 14 cm. If the radius were cut in half, what would the volume of the cylinder be? Use 3.14 for π and round to the nearest tenth. (Lesson 6-6)

A 791.3 cm^3 B 422.3 cm^3 C 395.6 cm^3 D 393.5 cm^3

Model Proportions

Use with Lesson 7-4

WHAT YOU NEED:
- Ruler
- Pattern blocks

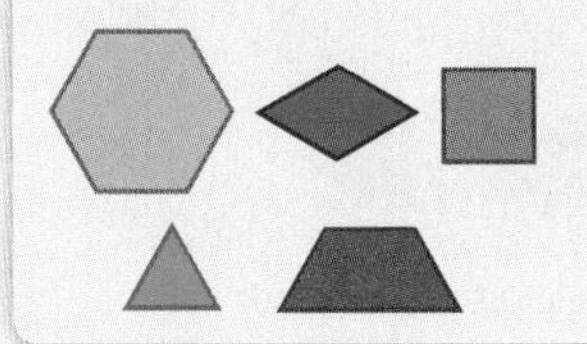

REMEMBER
- Use the area formulas to find the area of each pattern block except the hexagon.

To find the area of the hexagon, think of the pieces that can fit together to make the hexagon.

Activity

1 Measure each type of pattern block to the nearest eighth of an inch to determine its area. Use pattern blocks to find several area relationships that represent fractions equivalent to one-half. For example,

$$\frac{\triangle}{\diamond} = \frac{1}{2}.$$

2 Above, you related area of pattern blocks to a ratio. Now make a proportion based upon area that uses only pattern blocks on both sides of the equal sign. Then write these proportions using numbers based on your measurements for area. Use cross products to check your work.

Think and Discuss

1. Which pattern-block area relationships equal $\frac{5}{6}$?

2. What area relationships can you make with a triangle and a trapezoid?

3. What area relationships can you make with only a triangle?

Try This

Use pattern blocks to complete each proportion based on area. Then write these proportions using numbers based on your measurements for area.

1.

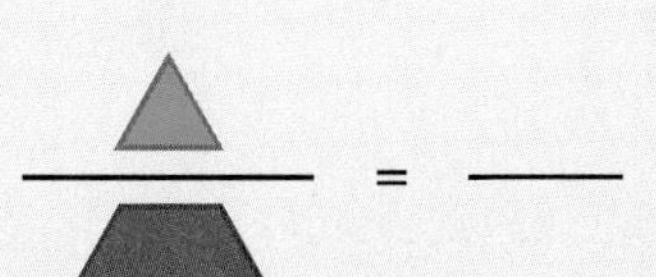

2.

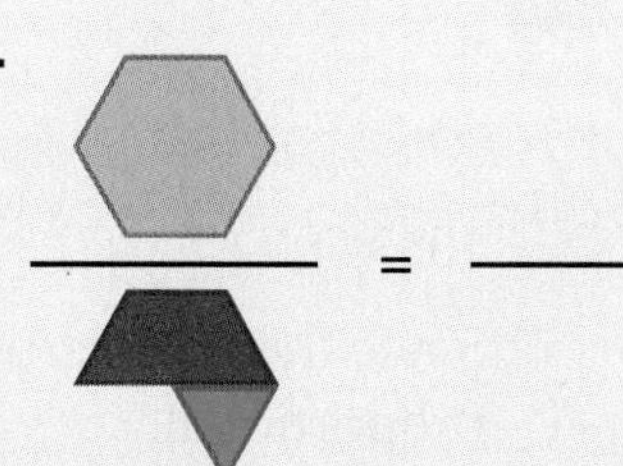

7-4 Solving Proportions

Learn to solve proportions.

Vocabulary

cross product

Unequal masses will not balance on a *fulcrum* if they are an equal distance from it; one side will go up and the other side will go down.

Unequal masses will balance when the following proportion is true:

$$\frac{\text{mass 1}}{\text{length 2}} = \frac{\text{mass 2}}{\text{length 1}}$$

Alexander Calder's sculpture *Totem* stands in Paris. Calder is known as the father of the mobile.

One way to find whether ratios, such as those above, are equal is to find a common denominator. The ratios are equal if their numerators are equal after the fractions have been rewritten with a common denominator.

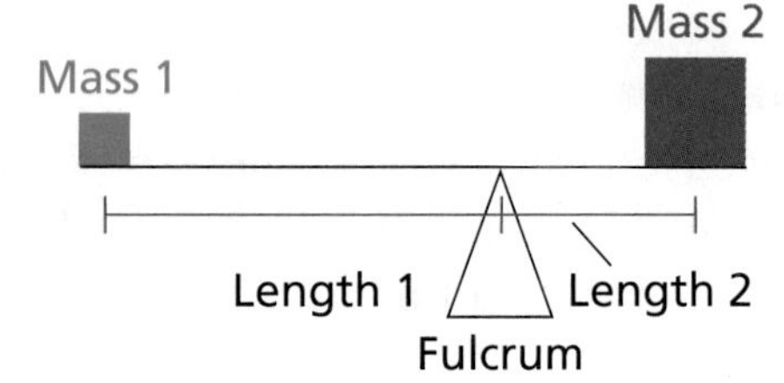

$$\frac{6}{8} = \frac{72}{96} \quad \frac{9}{12} = \frac{72}{96} \qquad \frac{6}{8} = \frac{9}{12}$$

CROSS PRODUCTS

Cross products in proportions are equal. If the ratios are *not* in proportion, the cross products are not equal.

Proportions		*Not* Proportions	
$\frac{6}{8} = \frac{9}{12}$	$\frac{5}{2} = \frac{15}{6}$	$\frac{1}{6} \neq \frac{2}{7}$	$\frac{5}{12} \neq \frac{2}{5}$
$6 \cdot 12 = 8 \cdot 9$	$5 \cdot 6 = 2 \cdot 15$	$1 \cdot 7 \neq 6 \cdot 2$	$5 \cdot 5 \neq 12 \cdot 2$
$72 = 72$	$30 = 30$	$7 \neq 12$	$25 \neq 24$

Helpful Hint

The cross product represents the numerator of the fraction when a common denominator is found by multiplying the denominators.

EXAMPLE  Using Cross Products to Identify Proportions

Tell whether the ratios are proportional.

A $\frac{5}{6} \stackrel{?}{=} \frac{15}{21}$

$\frac{5}{6} \times \frac{15}{21}$ → 90, 105   *Find cross products.*

$105 \neq 90$

Since the cross products are not equal, the ratios are not proportional.

B **A shade of paint is made by mixing 5 parts red paint with 7 parts blue paint. If you mix 12 quarts of blue paint with 8 quarts of red paint, will you get the correct shade?**

$\frac{5 \text{ parts red}}{7 \text{ parts blue}} \stackrel{?}{=} \frac{8 \text{ quarts red}}{12 \text{ quarts blue}}$ *Set up ratios.*

$5 \cdot 12 = 60 \quad 7 \cdot 8 = 56$ *Find the cross products.*

$60 \neq 56$

The ratios are not equal. You will not get the correct shade of paint.

When you do not know one of the four numbers in a proportion, set the cross products equal to each other and solve.

EXAMPLE 2 Solving Proportions

Solve the proportion.

$\frac{12}{d} = \frac{4}{14}$

$\frac{12}{d} = \frac{4}{14}$

$12 \cdot 14 = 4d$ *Find the cross products.*

$168 = 4d$ *Solve.*

$42 = d$ $\frac{12}{42} = \frac{4}{14}$ ✔; *the proportion checks.*

EXAMPLE 3 *Physical Science Application*

Two masses can be balanced on a fulcrum when $\frac{\text{mass 1}}{\text{length 2}} = \frac{\text{mass 2}}{\text{length 1}}$. The green box and the blue box are balanced. What is the mass of the blue box?

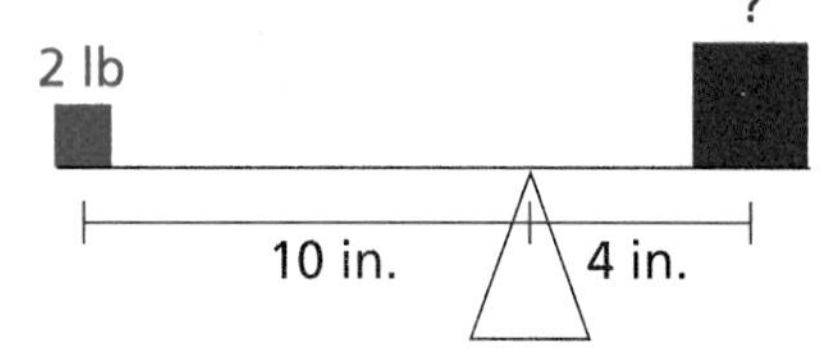

$\frac{2}{4} = \frac{m}{10}$ *Set up the proportion.*

$2 \cdot 10 = 4m$ *Find the cross products.*

$\frac{20}{4} = \frac{4m}{4}$ *Solve for m.*

$5 = m$

The mass of the blue box is 5 lb.

Think and Discuss

1. **Explain** what the cross products of two ratios represent.
2. **Tell** what it means if the cross products are not equal.
3. **Describe** how to solve a proportion when one of the four numbers is a variable.

7-4 Exercises

FOR EOG PRACTICE

see page 669

internet connect

Homework Help Online

go.hrw.com Keyword: MT4 7-4

2.02

GUIDED PRACTICE

See Example 1

Tell whether the ratios in each pair are proportional.

1. $\frac{7}{14} \stackrel{?}{=} \frac{14}{28}$
2. $\frac{2}{9} \stackrel{?}{=} \frac{6}{27}$
3. $\frac{3}{7} \stackrel{?}{=} \frac{6}{15}$
4. $\frac{15}{25} \stackrel{?}{=} \frac{9}{15}$
5. A bubble solution can be made with a ratio of one part detergent to eight parts water. Would a mixture of 56 oz water and 8 oz detergent be proportional to this ratio? Explain.

See Example 2

Solve each proportion.

6. $\frac{x}{5} = \frac{2}{10}$
7. $\frac{4}{9} = \frac{n}{18}$
8. $\frac{11}{d} = \frac{66}{12}$
9. $\frac{21}{7} = \frac{h}{2}$
10. $\frac{12}{f} = \frac{16}{13}$
11. $\frac{t}{7} = \frac{8}{28}$
12. $\frac{1}{2} = \frac{s}{18}$
13. $\frac{28}{7} = \frac{50}{q}$

See Example 3

14. A 10 kg weight is positioned 5 cm from a fulcrum. At what distance from the fulcrum must a 15 kg weight be positioned to keep the scale balanced?

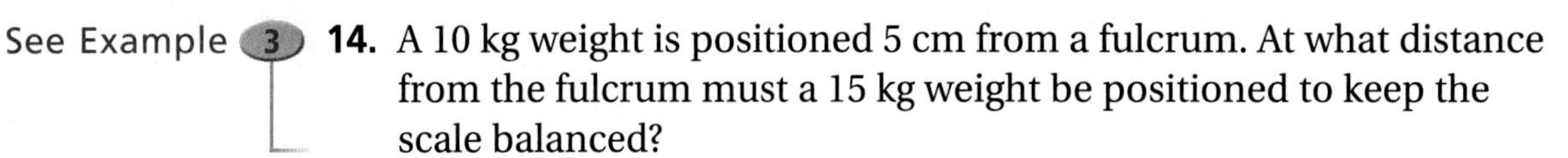

INDEPENDENT PRACTICE

See Example 1

Tell whether the ratios in each pair are proportional.

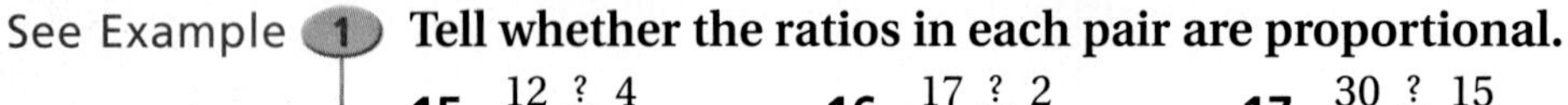

15. $\frac{12}{49} \stackrel{?}{=} \frac{4}{7}$
16. $\frac{17}{51} \stackrel{?}{=} \frac{2}{6}$
17. $\frac{30}{36} \stackrel{?}{=} \frac{15}{16}$
18. $\frac{7}{8} \stackrel{?}{=} \frac{35}{40}$
19. A class had 18 girls and 12 boys. Then 2 boys and 3 girls transferred out of the class. Did the ratio of girls to boys stay the same? Explain.

See Example 2

Solve each proportion.

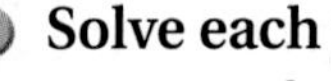

20. $\frac{3}{9} = \frac{b}{21}$
21. $\frac{27}{90} = \frac{b}{10}$
22. $\frac{4}{1} = \frac{0.56}{m}$
23. $\frac{y}{5} = \frac{42}{35}$
24. $\frac{r}{7} = \frac{3}{2}$
25. $\frac{48}{16} = \frac{12}{n}$
26. $\frac{p}{9} = \frac{2}{12}$
27. $\frac{2}{d} = \frac{6}{1.5}$

See Example 3

28. Jo weighs 65 lb and Tim weighs 78 lb. If Tim is seated 6 ft from the center of a balanced seesaw, how far is Jo seated from the center?

PRACTICE AND PROBLEM SOLVING

For each set of ratios, find the two that are proportional.

29. $\frac{6}{3}, \frac{18}{9}, \frac{51}{25}$
30. $\frac{1}{4}, \frac{11}{44}, \frac{111}{440}$
31. $\frac{30}{14}, \frac{66}{21}, \frac{22}{7}$
32. $\frac{54}{168}, \frac{9}{28}, \frac{52}{142}$
33. $\frac{0.25}{4}, \frac{0.125}{6}, \frac{1}{16}$
34. $\frac{a}{c}, \frac{a}{b}, \frac{4a}{4b}$
35. ***PHYSICAL SCIENCE*** Each molecule of sulfuric acid reacts with 2 molecules of ammonia. How many molecules of sulfuric acid react with 24 molecules of ammonia?

Health LINK

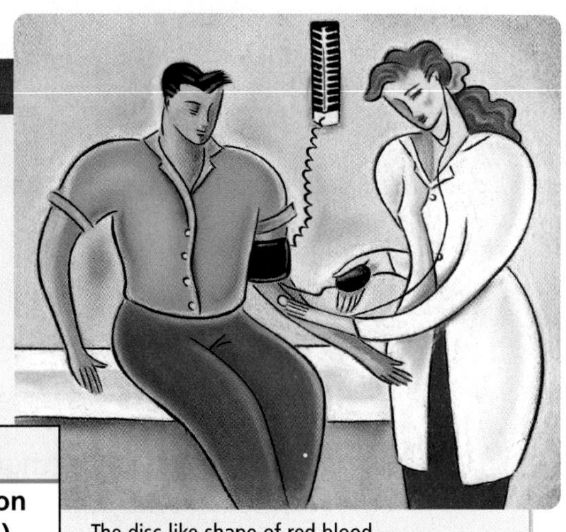

A doctor reports blood pressure in millimeters of mercury (mm Hg) as a ratio of *systolic* blood pressure to *diastolic* blood pressure (such as 140 over 80). Systolic pressure is measured when the heart beats, and diastolic pressure is measured when it rests. Refer to the table of blood pressure ranges for adults for Exercises 36–39.

Blood Pressure Ranges

	Optimal	Normal–High	Hypertension (very high)
Systolic	under 120 mm Hg	120–140 mm Hg	over 140 mm Hg
Diastolic	under 80 mm Hg	80–90 mm Hg	over 90 mm Hg

The disc-like shape of red blood cells allows them to pass through tiny capillaries.

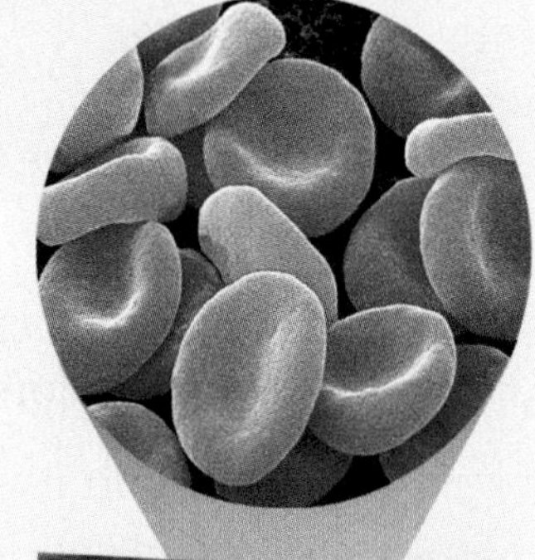

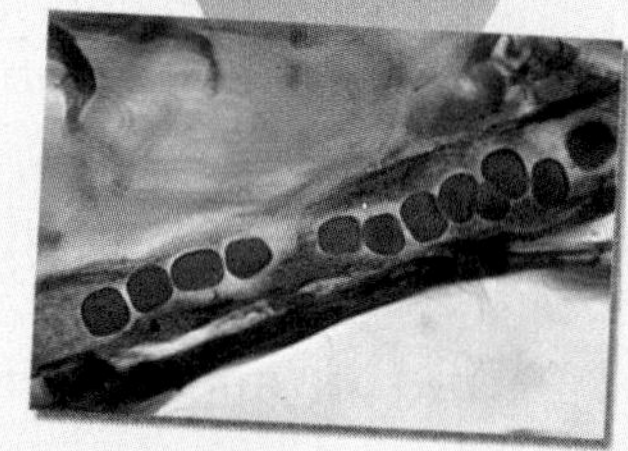

About $\frac{9}{20}$ of your blood is made up of cells; the rest is plasma.

36. Eduardo is a healthy 37-year-old man whose blood pressure is in the optimal category.

a. Calculate an approximate ratio of systolic to diastolic blood pressure in the optimal range.

b. If Eduardo's systolic blood pressure is 102 mm Hg, use the ratio from part **a** to predict his diastolic blood pressure.

37. The midpoint of a range of values can be found by adding the highest and lowest numbers together and dividing by 2.

a. Calculate an approximate ratio of systolic to diastolic blood pressure for the normal–high category.

b. Tyra's diastolic blood pressure is 88 mm Hg. Use the ratio from part **a** to predict her systolic blood pressure.

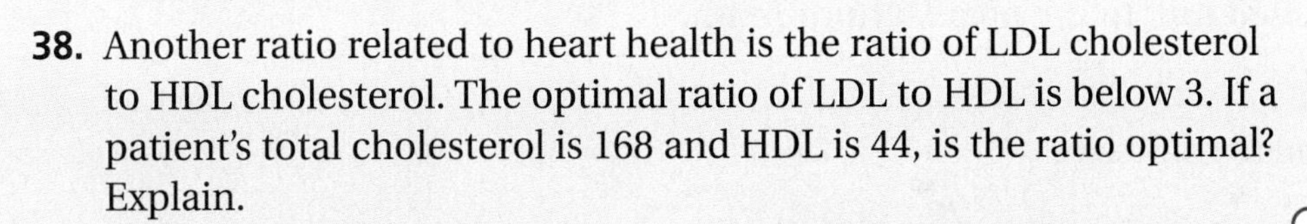

38. Another ratio related to heart health is the ratio of LDL cholesterol to HDL cholesterol. The optimal ratio of LDL to HDL is below 3. If a patient's total cholesterol is 168 and HDL is 44, is the ratio optimal? Explain.

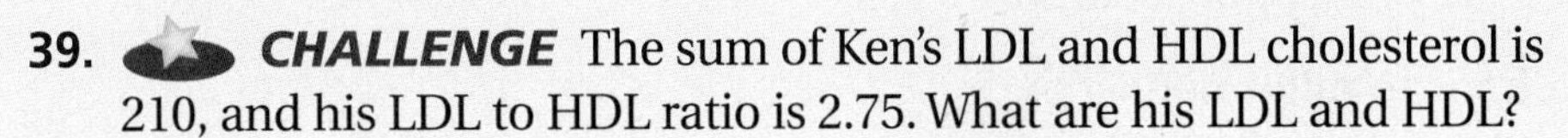

39. ***CHALLENGE*** The sum of Ken's LDL and HDL cholesterol is 210, and his LDL to HDL ratio is 2.75. What are his LDL and HDL?

go.hrw.com
KEYWORD: MT4 Health
CNN Student News

Spiral Review

Write each decimal as a fraction in simplest form. (Lesson 3-1)

40. 0.65 **41.** −1.25 **42.** 0.723 **43.** 11.17

44. **EOG PREP** A $4\frac{5}{8}$ ft section of wood is cut from a $7\frac{1}{2}$ ft board. How much of the original board remains? (Section 3-5)

A $3\frac{9}{16}$ ft B $3\frac{5}{8}$ ft C $2\frac{7}{8}$ ft D $2\frac{3}{8}$ ft

Chapter 7

Mid-Chapter Quiz

LESSON 7-1 (pp. 342–345)

Simplify to tell whether the ratios form a proportion.

1. $\frac{4}{5}$ and $\frac{16}{20}$
2. $\frac{33}{60}$ and $\frac{11}{21}$
3. $\frac{12}{42}$ and $\frac{6}{21}$
4. $\frac{8}{20}$ and $\frac{4}{25}$
5. Josh is following a recipe that calls for 2.5 cups of sugar to make 2 dozen cookies. He uses 3.5 cups of sugar to make 3 dozen cookies. Has he followed the recipe? Explain.

LESSON  (pp. 346–349)

Find the unit price for each offer and tell which is the better buy.

6. a long distance phone charge of $1.40 for 10 min or $4.50 for 45 min
7. Buy one 10 pack of AAA batteries for $5.49 and get one free, or buy two 4 packs for $2.98.
8. A 64 oz bottle of juice costs $2.39, and a 20 oz bottle costs $0.79. You can use a 20-cents-off coupon if you buy four 20 oz bottles or a 15-cents-off coupon if you buy a 64 oz bottle. Which is the better buy?

LESSON  (pp. 350–354)

Find the appropriate factor for each conversion.

9. gallons to quarts
10. millimeters to centimeters
11. minutes to days

Convert to the indicated unit to the nearest hundredth.

12. Change 60 ounces to pounds.
13. Change 25 pounds to ounces.
14. Change 5 feet per minute to feet per second.
15. Change 40 miles per hour to miles per second.
16. Driving at a constant rate, Noah covered 140 miles in 3.5 hours. Express his driving rate in feet per minute.

LESSON  (pp. 356–359)

Solve.

17. $\frac{6}{9} = \frac{n}{72}$
18. $\frac{18}{12} = \frac{3}{x}$
19. $\frac{0.7}{1.4} = \frac{z}{28}$
20. $\frac{12}{y} = \frac{32}{16}$
21. $\frac{c}{5} = \frac{9}{24}$
22. $\frac{5}{3} = \frac{g}{27}$
23. $\frac{0.5}{h} = \frac{2}{3}$
24. $\frac{9}{0.9} = \frac{72}{b}$
25. Tim can input 110 data items in 2.5 minutes. Typing at the same rate, how many data items can he input in 7 minutes?

Mid-Chapter Quiz

Focus on Problem Solving

Solve

- **Choose an operation: Multiplication or division**

When you are converting units, think about whether the number in the answer will be greater or less than the number given in the question. This will help you to decide whether to multiply or divide to convert the units.

For example, if you are converting feet to inches, you know that the number of inches will be greater than the number of feet because each foot is 12 inches. So you know that you should multiply by 12 to get a greater number.

In general, if you are converting to smaller units, the number of units will have to be greater to represent the same quantity.

For each problem, determine whether the number in the answer will be greater or less than the number given in the question. Use your answer to decide whether to multiply or divide by the conversion factor. Then solve the problem.

1. The speed a boat travels is usually measured in nautical miles, or knots. The Golden Gate–Sausalito ferry in California, which provides service between Sausalito and San Francisco, can travel at 20.5 knots. Find the speed in miles per hour. (*Hint:* 1 knot = 1.15 miles per hour)

2. When it is finished, the Crazy Horse Memorial in the Black Hills of South Dakota will be the world's largest sculpture. The sculpture's height will be 563 feet. Find the height in meters. (*Hint:* 1 meter = 3.28 feet)

3. The amounts of water typically used for common household tasks are given in the table below. Find the number of liters needed for each task. (*Hint:* 1 gallon = 3.79 liters)

Task	Water Used (gal)
Laundry (1 load)	40
5-minute shower	12.5
Washing hands	0.5
Flushing toilet	3.5

4. Lake Baikal, in Siberia, is so large that it would take all of the rivers on Earth combined an entire year to fill it. At 1.62 kilometers deep, it is the deepest lake in the world. Find the depth of Lake Baikal in miles. (1 mile = 1.61 kilometers)

7-5 Exercises

FOR EOG PRACTICE
see page 670

3.03

GUIDED PRACTICE

See Example 1 Tell whether each transformation is a dilation.

1. 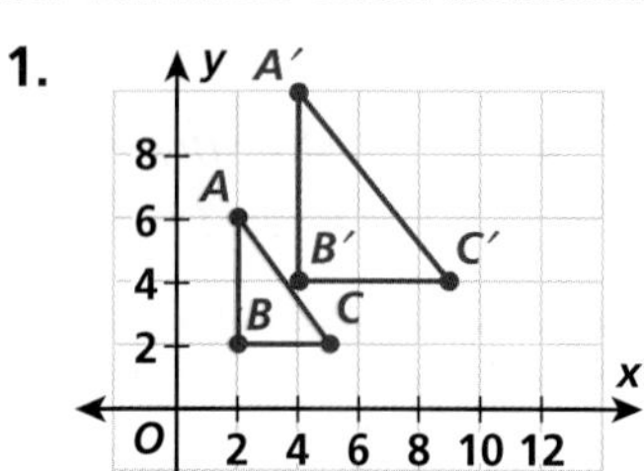

2. 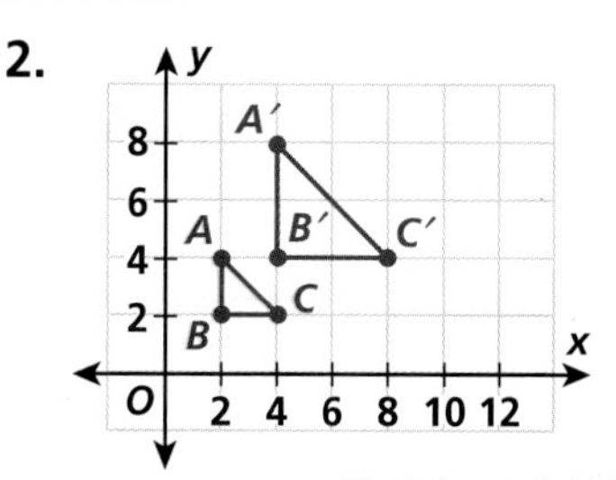

See Example 2 Dilate each figure by the given scale factor with P as the center of dilation.

3. 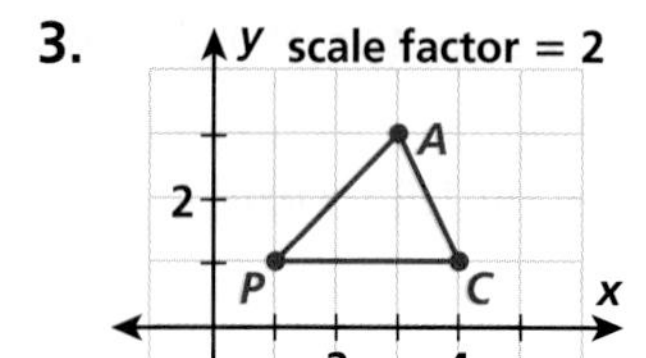

4. 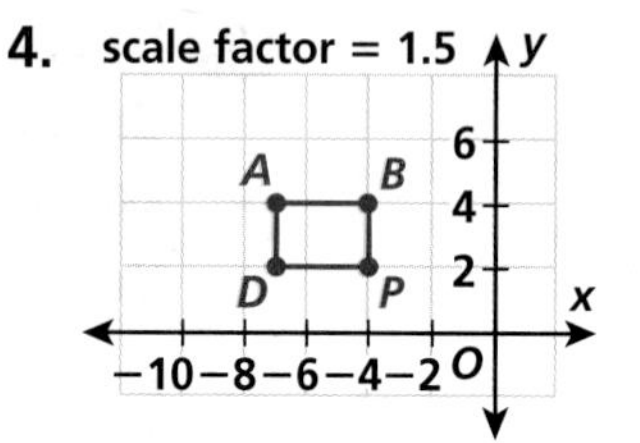

See Example 3 Dilate each figure by the given scale factor with the origin as the center of dilation. What are the vertices of the image?

5. 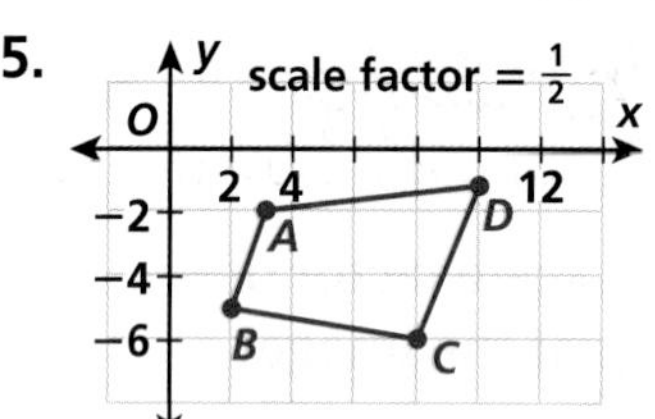

6. 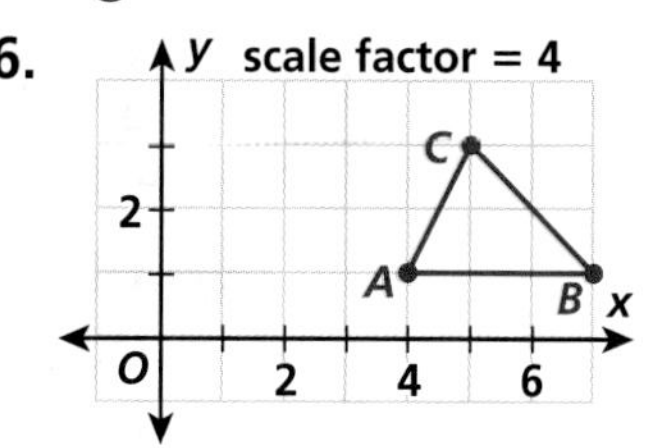

INDEPENDENT PRACTICE

See Example 1 Tell whether each transformation is a dilation.

7. 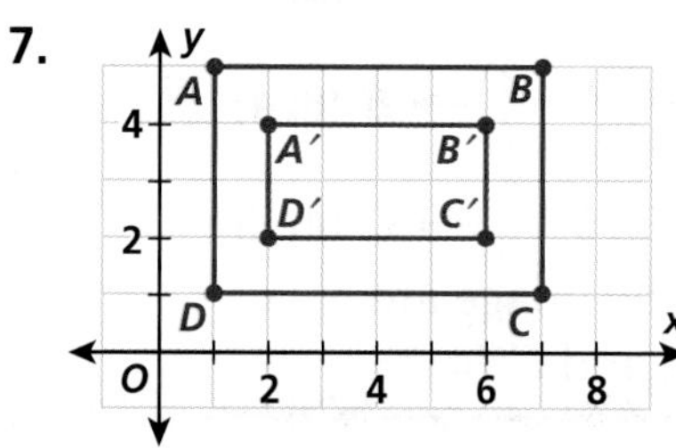

8. 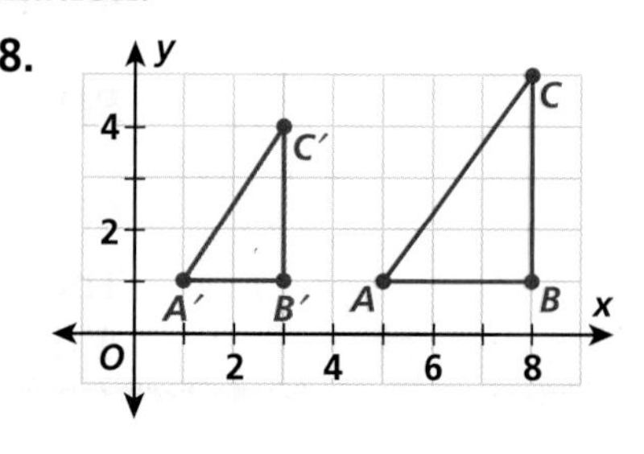

See Example 2 Dilate each figure by the given scale factor with P as the center of dilation.

9. 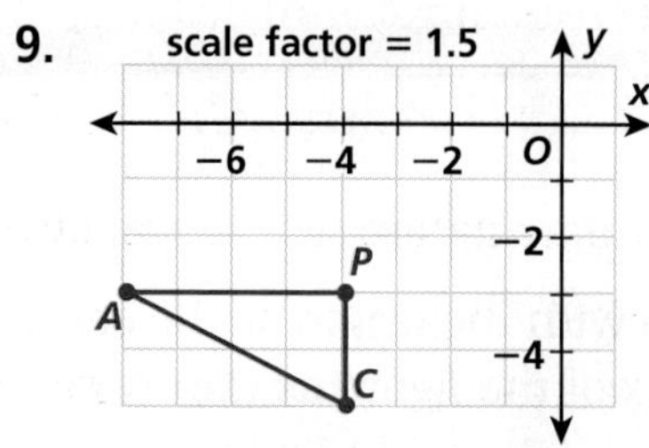

10. 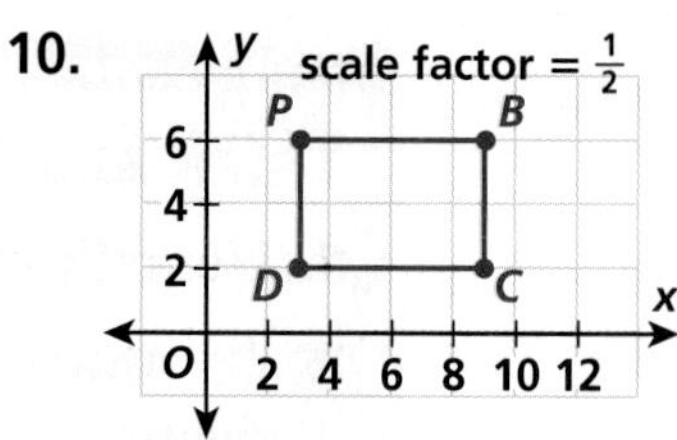

See Example 3 Dilate each figure by the given scale factor with the origin as the center of dilation. What are the vertices of the image?

11.

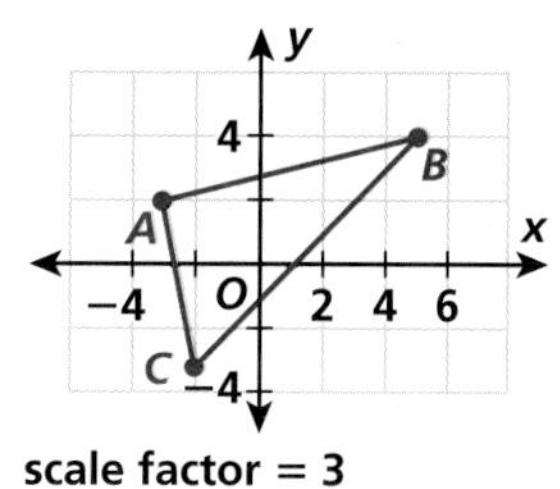

scale factor = 3

12.

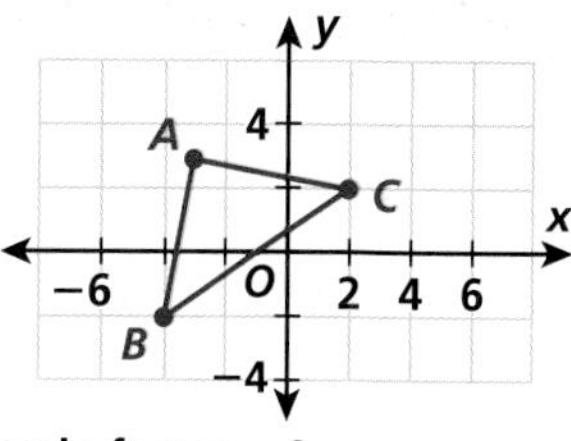

scale factor = 2

PRACTICE AND PROBLEM SOLVING

Identify the scale factor used in each dilation.

13.

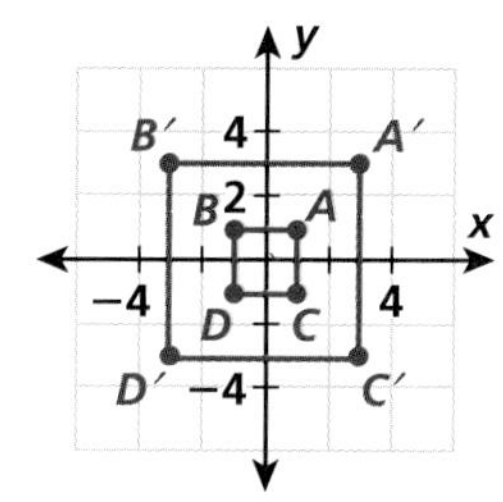

14.

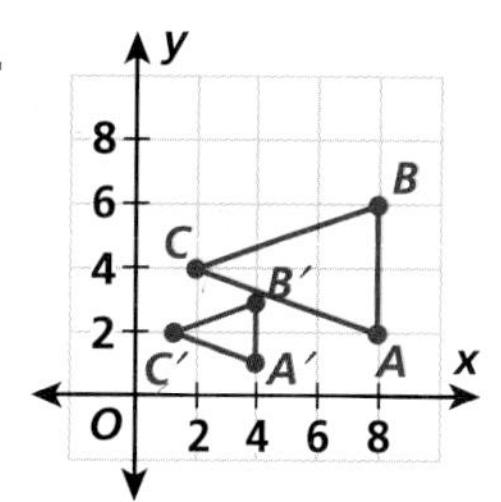

Photography LINK

In a camera lens, a larger aperture lets in more light than a smaller one.

15. ***PHOTOGRAPHY*** The *aperture* is the polygonal opening in a camera lens when a picture is taken. The aperture can be small or large. Is an aperture a dilation? Why or why not?

16. A rectangle has vertices $A(4, 4)$, $B(9, 4)$, $C(9, 0)$, and $D(4, 0)$. Give the coordinates after dilating from the origin by a scale factor of 2.5.

17. ***CHOOSE A STRATEGY*** The perimeter of an equilateral triangle is 48 cm. If the triangle is dilated by a scale factor of 0.25, what is the length of each side of the new triangle?

A 3 cm **B** 4 cm **C** 16 cm **D** 8 cm

18. ***WRITE ABOUT IT*** Explain how you can check the drawing of a dilation for accuracy.

19. ***CHALLENGE*** What scale factor was used in the dilation of a triangle with vertices $A(6, -2)$, $B(8, 3)$, and $C(-12, 10)$ to the triangle with vertices $A'\left(-2, \frac{2}{3}\right)$, $B'\left(-2\frac{2}{3}, -1\right)$, and $C'\left(4, -3\frac{1}{3}\right)$?

Spiral Review

Find the area of each figure with the given vertices. (Lesson 6-2)

20. (1, 0), (10, 0), (1, −6)

21. (5, 5), (2, 1), (11, 1), (8, 5)

22. (−8, −8), (8, −8), (4, 4), (−4, 4)

23. (−12, 4), (−6, 4), (−7, 11)

24. **EOG PREP** A pyramid has a rectangular base measuring 12 cm by 9 cm and height 15 cm. What is the volume of the pyramid? (Lesson 6-7)

A 540 cm^3 **B** 405 cm^3 **C** 315 cm^3 **D** 270 cm^3

Explore Similarity

Use with Lesson 7-6

WHAT YOU NEED:

- Two pieces of graph paper with different-sized boxes, such as 1 cm graph paper and $\frac{1}{4}$ in. graph paper
- Number cube
- Metric ruler
- Protractor

internet connect
Lab Resources Online
go.hrw.com
KEYWORD: MT4 Lab7B

Triangles that have the same shape have some interesting relationships.

Activity

1 Follow the steps below to draw two triangles.

a. On a sheet of graph paper, plot a point below and to the left of the center of the paper. Label the point *A*. On the other sheet of paper, plot a point below and to the left of the center and label this point *D*.

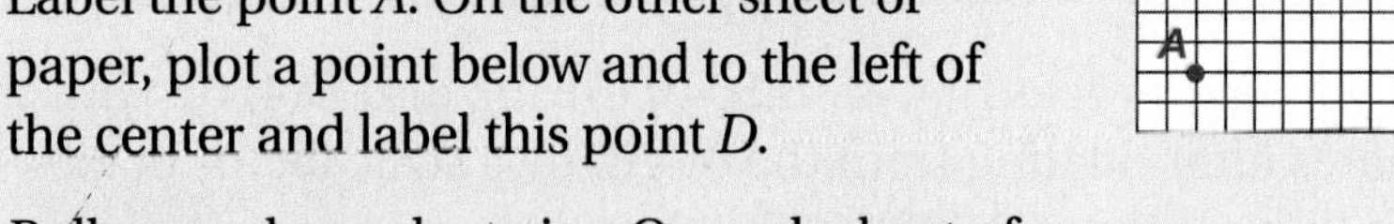

b. Roll a number cube twice. On each sheet of graph paper, move up the number on the first roll, move right the number on the second roll, and plot this location as point *B* on the first sheet and point *E* on the second sheet.

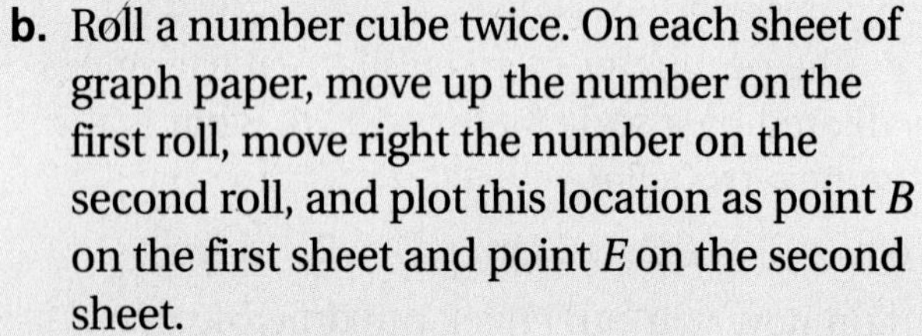

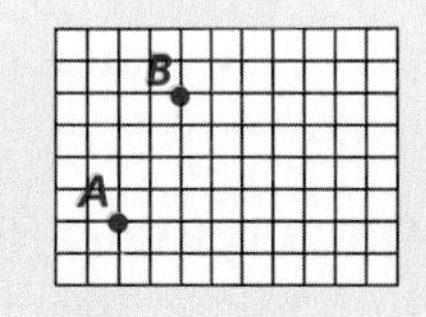

c. Roll the number cube twice again. On each sheet of graph paper, move down the number on the first roll, move right the number on the second roll, and plot point *C* on the first sheet and point *F* on the second sheet.

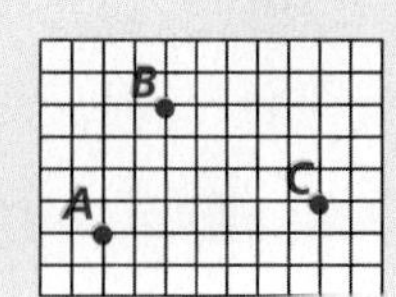

d. Connect the three points on each sheet of graph paper to form triangles *ABC* and *DEF*.

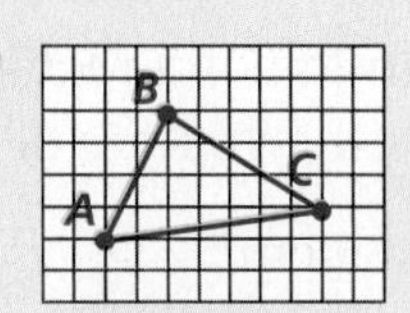

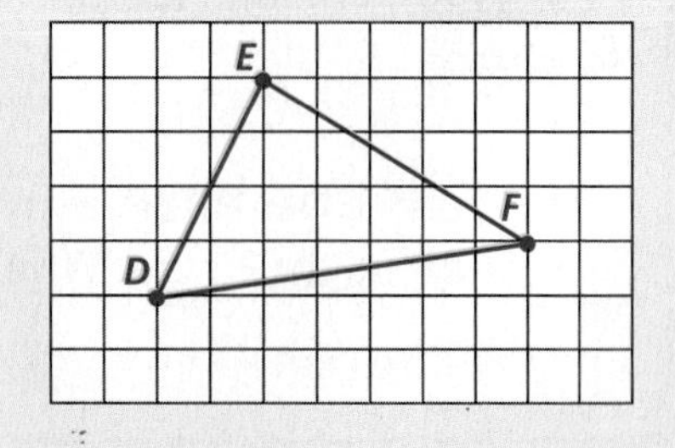

e. Measure the angles of each triangle. Measure the side lengths of each triangle to the nearest millimeter. Find the following:

$m\angle A$	$m\angle D$	$m\angle B$	$m\angle E$	$m\angle C$	$m\angle F$
AB	DE	$\frac{AB}{DE}$	BC	EF	$\frac{BC}{EF}$
AC	DF	$\frac{AC}{DF}$			

2 Follow the steps below to draw two triangles.

a. On one sheet of graph paper, plot a point below and to the left of the center of the paper. Label the point *A*.

b. Roll a number cube twice. Move up the number on the first roll, move right the number on the second roll, and plot this location as point *B*. From *B*, move up the number on the first roll, move right the number on the second roll, and label this point *D*.

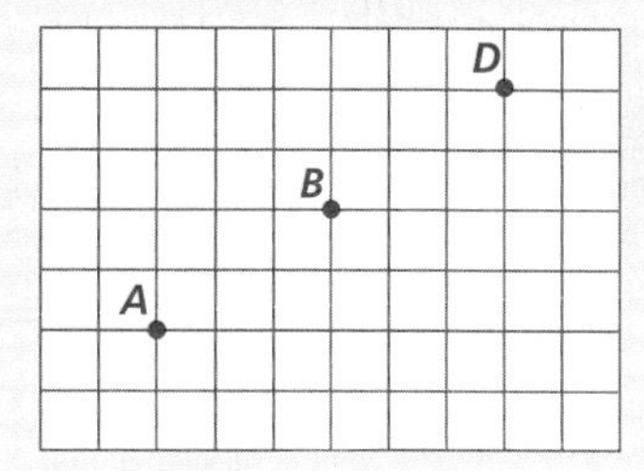

c. Roll a number cube twice. From *B*, move down the number on the first roll, move right the number on the second roll, and plot this location as point *C*.

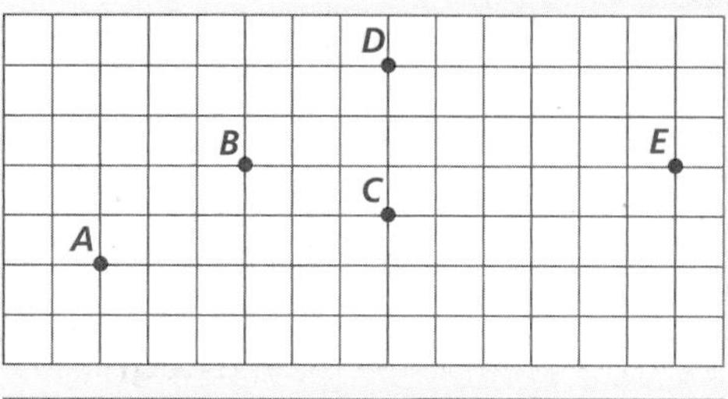

d. From *D*, move down twice the number on the first roll, move right twice the number on the second roll, and label this point *E*.

e. Connect points to form triangles *ABC* and *ADE*.

f. Measure the angles of each triangle. Measure the side lengths of each triangle to the nearest millimeter.

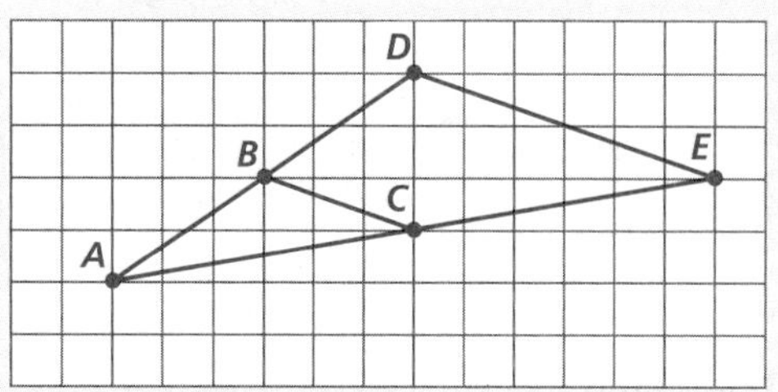

Think and Discuss

1. How do corresponding angles of triangles with the same shape compare?
2. How do corresponding side lengths of triangles with the same shape compare?
3. Suppose you enlarge a triangle on a copier machine. What measurements or values would be the same on the enlargement?

Try This

1. Make a small trapezoid on graph paper and triple the length of each side. Compare the angle measures and side lengths of the trapezoids.
2. Make a large polygon on graph paper. Use a copier to reduce the size of the polygon. Compare the angle measures and side lengths of the polygons.

7-6 Similar Figures

Learn to determine whether figures are similar, to use scale factors, and to find missing dimensions in similar figures.

Vocabulary

similar

The heights of letters in newspapers and on billboards are measured using *points* and *picas.* There are 12 points in 1 pica and 6 picas in one inch.

A letter 36 inches tall on a billboard would be 216 picas, or 2592 points. The first letter in this paragraph is 12 points.

12 points	24 points	48 points	72 points
1 pica	2 picas	4 picas	6 picas
A	A	A	A

Congruent figures have the same size and shape. **Similar** figures have the same shape, but not necessarily the same size. The *A*'s in the table are similar. They have the same shape, but they are not the same size.

For polygons to be similar,

- corresponding angles must be congruent, and
- corresponding sides must have lengths that form equivalent ratios.

The ratio formed by the corresponding sides is the scale factor.

EXAMPLE 1 Using Scale Factors to Find Missing Dimensions

A picture 4 in. tall and 9 in. wide is to be scaled to 2.5 in. tall to be displayed on a Web page. How wide should the picture be on the Web page for the two pictures to be similar?

To find the scale factor, divide the known height of the scaled picture by the corresponding height of the original picture.

0.625 $\quad \frac{2.5}{4} = 0.625$

Then multiply the width of the original picture by the scale factor.

5.625 $\quad 9 \cdot 0.625$

The picture should be 5.625 in. wide.

EXAMPLE 2 Using Equivalent Ratios to Find Missing Dimensions

A company's logo is in the shape of an isosceles triangle with two sides that are each 2.4 in. long and one side that is 1.8 in. long. On a billboard, the triangle in the logo has two sides that are each 8 ft long. What is the length of the third side of the triangle on the billboard?

Set up a proportion.

$$\frac{2.4 \text{ in.}}{8 \text{ ft}} = \frac{1.8 \text{ in.}}{x \text{ ft}}$$

$2.4 \text{ in.} \cdot x \text{ ft} = 8 \text{ ft} \cdot 1.8 \text{ in.}$ *Find the cross products.*

$2.4 \cancel{\text{in.}} \cdot x \cancel{\text{ft}} = 8 \cancel{\text{ft}} \cdot 1.8 \cancel{\text{in.}}$ *in.· ft is on both sides*

$2.4x = 8 \cdot 1.8$ *Cancel the units.*

$2.4x = 14.4$ *Multiply.*

$x = \frac{14.4}{2.4} = 6$ *Solve for x.*

The third side of the triangle is 6 ft long.

EXAMPLE 3 Identifying Similar Figures

Which rectangles are similar?

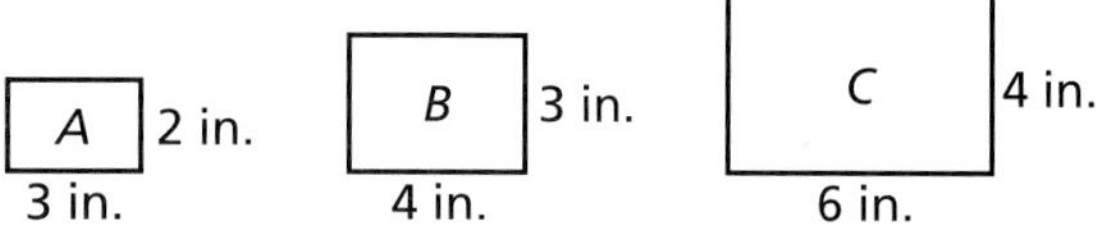

Since the three figures are all rectangles, all the angles are right angles. So the corresponding angles are congruent.

Compare the ratios of corresponding sides to see if they are equal.

$$\frac{\text{length of rectangle } A}{\text{length of rectangle } B} \rightarrow \frac{3}{4} \stackrel{?}{=} \frac{2}{3} \leftarrow \frac{\text{width of rectangle } A}{\text{width of rectangle } B}$$

$$9 \neq 8$$

The ratios are not equal. Rectangle A is not similar to rectangle B.

$$\frac{\text{length of rectangle } A}{\text{length of rectangle } C} \rightarrow \frac{3}{6} = \frac{2}{4} \leftarrow \frac{\text{width of rectangle } A}{\text{width of rectangle } C}$$

$$12 = 12$$

The ratios are equal. Rectangle A is similar to rectangle C. The notation $A \sim C$ shows similarity.

Remember!

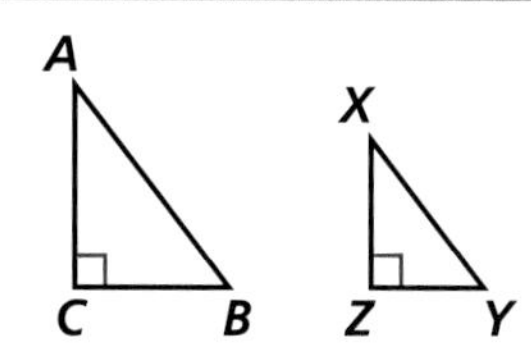

The following are matching, or corresponding:
$\angle A$ and $\angle X$
$\angle B$ and $\angle Y$
$\angle C$ and $\angle Z$
$\overline{AB}$ and $\overline{XY}$.
$\overline{BC}$ and $\overline{YZ}$.
$\overline{AC}$ and $\overline{XZ}$.

Think and Discuss

1. **Compare** an image formed by a scale factor greater than 1 to an image formed by a scale factor less than 1.
2. **Describe** one way for two figures not to be similar.
3. **Explain** whether two congruent figures are similar.

7-6 Exercises

FOR EOG PRACTICE

see page 670

internet connect

Homework Help Online
go.hrw.com Keyword: MT4 7-6

3.02

GUIDED PRACTICE

See Example 1

1. Fran scans a document that is 8.5 in. wide by 11 in. long into her computer. If she scales the length down to 7 in., how wide should the similar document be?

See Example 2

2. An isosceles triangle has a base of 12 cm and legs measuring 18 cm. How wide is the base of a similar triangle with legs measuring 22 cm?

See Example 3

3. Which rectangles are similar?

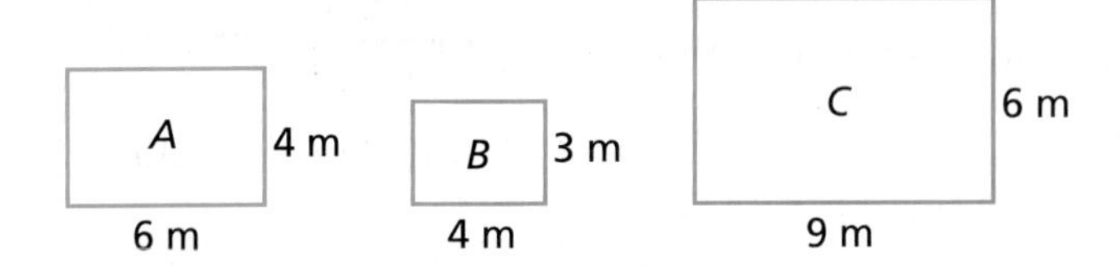

INDEPENDENT PRACTICE

See Example 1

4. A rectangular airfield measures 4.3 mi wide and 7.5 mi long. On a map, the width of the airfield is 3.75 in. How long is the airport on the map?

See Example 2

5. Rich drew a 7 in. wide by 4 in. tall picture that will be turned into a 40 ft wide billboard. How tall will the billboard be?

See Example 3

6. Which rectangles are similar?

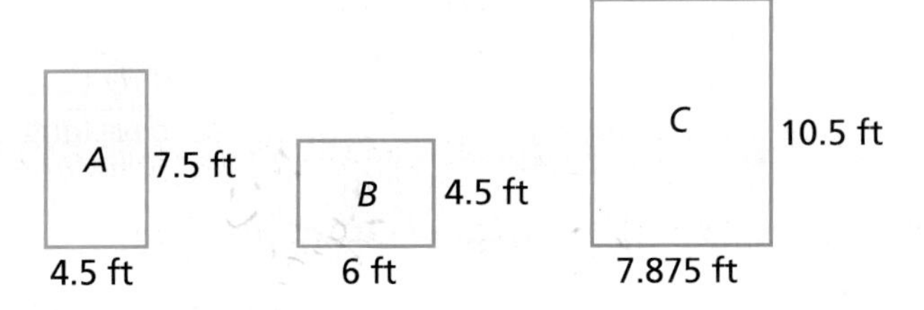

PRACTICE AND PROBLEM SOLVING

Tell whether the figures are similar. If they are not similar, explain.

7.

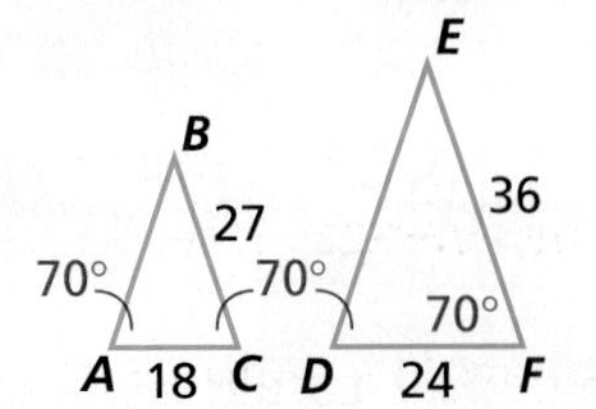

8.

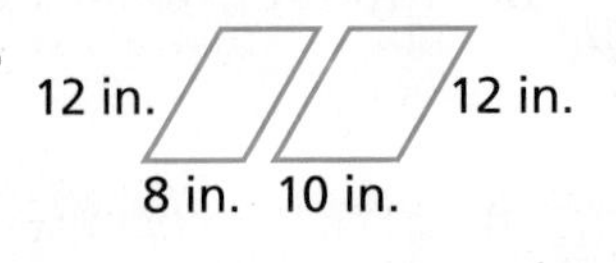

9.

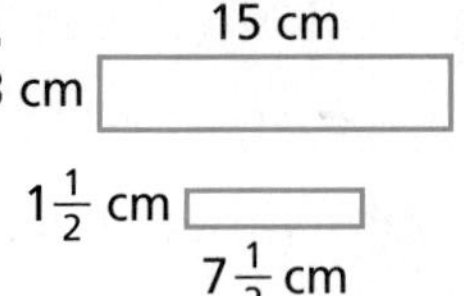

10. Draw a right triangle with vertices (0,0), (4,0), and (4,6) on a coordinate plane. Extend the hypotenuse to (6, 9), and form a new triangle with vertices (0, 0) and (6, 0). Are the triangles similar? Explain.

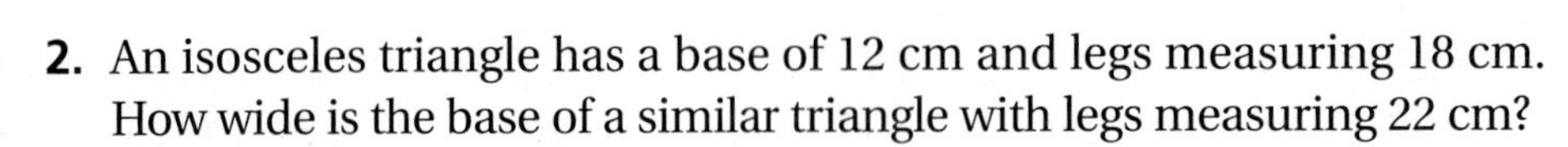

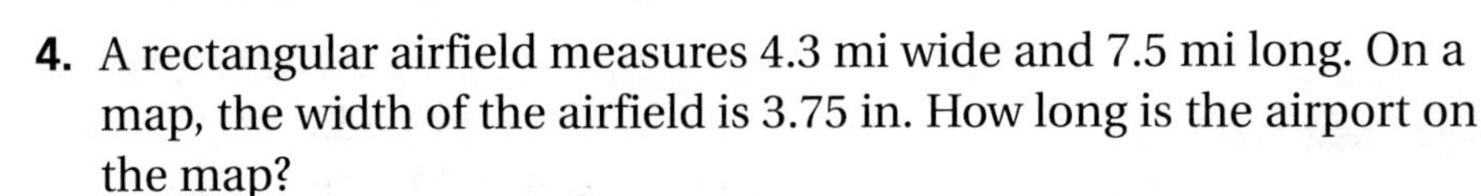

Art LINK

Many reproductions of artwork have been enlarged to fit unusual surfaces.

The figures in each pair are similar. Find the scale factor to solve for *x*.

11.

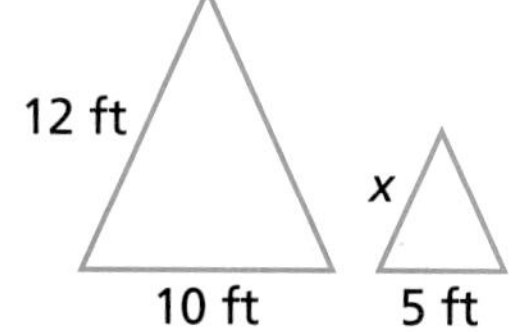

12.

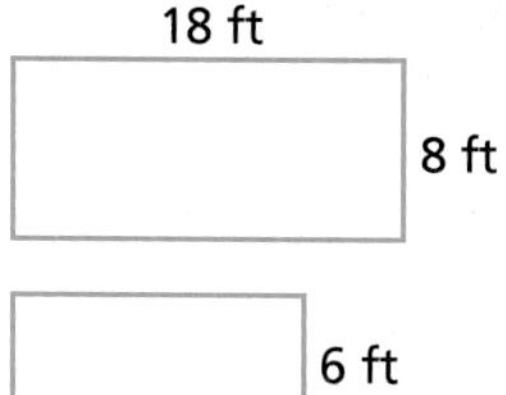

13.

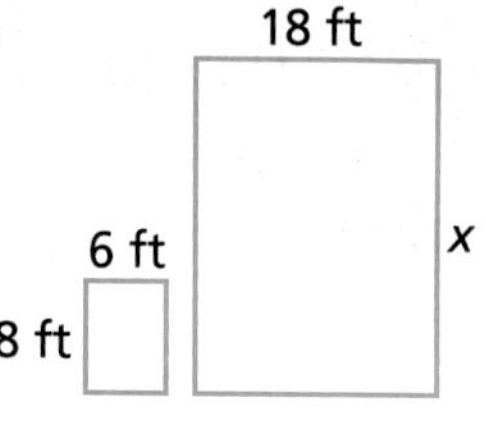

14. ***ART*** Helen is copying a printed reproduction of the *Mona Lisa.* The print is 24 in. wide and 36 in. tall. If Helen's canvas is 12 in. wide, how tall should her canvas be?

15. Ann's room is 10 ft by 12 ft 6 in. Her sketch of the room is 8 in. by 10 in. Is Ann's sketch a scale drawing? If so, what scale factor did she use?

16. A rectangle is 14 cm long and 9 cm wide. A similar rectangle is 4.5 cm wide and x cm long. Find x.

17. ***PHYSICAL SCIENCE*** Bill is 6 ft tall. He casts a 4 ft shadow at the same time that a tree casts a 16 ft shadow. Use similar triangles to find the height of the tree.

18. ***WRITE A PROBLEM*** A drawing on a sheet of graph paper shows a kite 8 cm wide and 10 cm long. The width of the kite is labeled 2 ft. Write and solve a problem about the kite.

19. ***WRITE ABOUT IT*** Consider the statement "All similar figures are congruent." Is this statement true or false? Explain.

20. ***CHALLENGE*** In right triangle ABC, $\angle B$ is the right angle, $AB =$ 21 cm, and $BC =$ 15 cm. Right triangle ABC is similar to triangle DEF, which has length $DE =$ 7 cm. Find the area of triangle DEF.

Spiral Review

Find the volume of each cone to the nearest tenth cubic unit. Use 3.14 for π. (Lesson 6-7)

21. radius 10 mm; height 12 mm

22. diameter 4 ft; height 5.7 ft

23. radius and height 12.5 cm

24. diameter 15 in.; height 35 in.

25. **EOG PREP** A data set contains 10 numbers in order. What is the median? (Lesson 4-3)

A the fifth number

B the number occurring most often

C the average of the numbers

D the average of the fifth and sixth numbers

26. **EOG PREP** Which of the following describes how the volume of a sphere changes when the radius is doubled? (Lesson 6-10)

A The volume is tripled.

B The volume is 9 times greater.

C The volume is $\frac{1}{9}$ the original volume.

D The volume is 8 times greater.

7-7 Scale Drawings

Learn to make comparisons between and find dimensions of scale drawings and actual objects.

Vocabulary

scale drawing

scale

reduction

enlargement

Stan Herd is a crop artist and farmer who has created works of art that are as large as 160 square acres. Herd first makes a *scale drawing* of each piece, and then he determines the actual lengths of the parts that make up the art piece.

To get an idea of scale, notice the red tractor at the lower right.

A **scale drawing** is a two-dimensional drawing that accurately represents an object. The scale drawing is mathematically similar to the object.

A **scale** gives the ratio of the dimensions in the drawing to the dimensions of the object. All dimensions are reduced or enlarged using the same scale. Scales can use the same units or different units.

Reading Math

The scale *a*:*b* is read "*a* to *b*." For example, the scale 1 cm:3 ft is read "one centimeter to three feet."

Scale	Interpretation
1:20	1 unit on the drawing is 20 units.
1 cm:1 m	1 cm on the drawing is 1 m.
$\frac{1}{4}$ in. = 1 ft	$\frac{1}{4}$ in. on the drawing is 1 ft.

EXAMPLE 1 Using Proportions to Find Unknown Scales or Lengths

A The length of an object on a scale drawing is 5 cm, and its actual length is 15 m. The scale is 1 cm:▢ m. What is the scale?

$\frac{1 \text{ cm}}{x \text{ m}} = \frac{5 \text{ cm}}{15 \text{ m}}$ *Set up proportion using* $\frac{\text{scale length}}{\text{actual length}}$.

$1 \cdot 15 = x \cdot 5$ *Find the cross products.*

$x = 3$ *Solve the proportion.*

The scale is 1 cm:3 m.

B The length of an object on a scale drawing is 3.5 in. The scale is 1 in:12 ft. What is the actual length of the object?

$\frac{1 \text{ in.}}{12 \text{ ft}} = \frac{3.5 \text{ in.}}{x \text{ ft}}$ *Set up proportion using* $\frac{\text{scale length}}{\text{actual length}}$.

$1 \cdot x = 3.5 \cdot 12$ *Find the cross products.*

$x = 42$ *Solve the proportion.*

The actual length is 42 ft.

A scale drawing that is smaller than the actual object is called a **reduction**. A scale drawing can also be larger than the object. In this case, the drawing is referred to as an **enlargement**.

EXAMPLE 2 *Life Science Application*

Under a 1000:1 microscope view, a paramecium appears to have length 39 mm. What is its actual length?

$$\frac{1000}{1} = \frac{39 \text{ mm}}{x \text{ mm}} \begin{array}{l} \leftarrow \text{scale length} \\ \leftarrow \text{actual length} \end{array}$$

$1000 \cdot x = 1 \cdot 39$ — *Find the cross products.*

$x = 0.039$ — *Solve the proportion.*

The actual length of the paramecium is 0.039 mm.

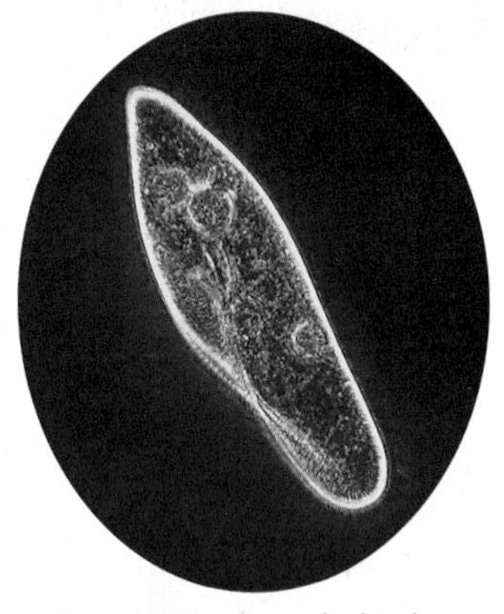

A paramecium is a cylindrical or foot-shaped microorganism.

A drawing that uses the scale $\frac{1}{4}$ in. = 1 ft is said to be in $\frac{1}{4}$ in. scale. Similarly, a drawing that uses the scale $\frac{1}{2}$ in. = 1 ft is in $\frac{1}{2}$ in. scale.

EXAMPLE 3 Using Scales and Scale Drawings to Find Heights

A **If a wall in a $\frac{1}{4}$ in. scale drawing is 3 in. tall, how tall is the actual wall?**

$$\frac{0.25 \text{ in.}}{1 \text{ ft}} = \frac{3 \text{ in.}}{x \text{ ft}} \begin{array}{l} \leftarrow \text{scale length} \\ \leftarrow \text{actual length} \end{array}$$ *Length ratios are equal.*

$0.25 \cdot x = 1 \cdot 3$ — *Find the cross products.*

$x = 12$ — *Solve the proportion.*

The wall is 12 ft tall.

B **How tall is the wall if a $\frac{1}{2}$ in. scale is used?**

$$\frac{0.5 \text{ in.}}{1 \text{ ft}} = \frac{3 \text{ in.}}{x \text{ ft}} \begin{array}{l} \leftarrow \text{scale length} \\ \leftarrow \text{actual length} \end{array}$$ *Length ratios are equal.*

$0.5 \cdot x = 1 \cdot 3$ — *Cross multiply.*

$x = 6$ — *Solve the proportion.*

The wall is 6 ft tall.

Think and Discuss

1. **Describe** which scale would produce the largest drawing of an object: 1:20, 1 in. = 1 ft, or $\frac{1}{4}$ in. = 1 ft.
2. **Describe** which scale would produce the smallest drawing of an object: 1:10, 1 cm = 10 cm, or 1 mm:1 m.

7-7 Exercises

FOR EOG PRACTICE

see page 671

internet connect
Homework Help Online
go.hrw.com Keyword: MT4 7-7

2.02, 3.02

GUIDED PRACTICE

See Example 1

1. A 10 ft fence is 8 in. long on a scale drawing. What is the scale?
2. Using a scale of 2 cm:9 m, how long is an object that is 4.5 cm long in a drawing?

See Example 2

3. Under a 100:1 microscope view, a microorganism appears to have a length of 0.85 in. How long is the microorganism?
4. Using the microscope from Exercise 3, how long would a 0.075 mm microorganism appear to be under the microscope?

See Example 3

5. On a $\frac{1}{4}$ in. scale, a tree is 13 in. tall. How tall is the actual tree?
6. How high is a 54 ft bridge on a $\frac{1}{2}$ in. scale drawing?

INDEPENDENT PRACTICE

See Example 1

7. What is the scale of a drawing where a 6 m wall is 4 cm long?
8. If a scale of 2 in:10 ft is used, how long is an object that is 14 in. long in a drawing?

See Example 2

9. Using a 1000:1 magnification microscope, a paramecium has length 23 mm. What is the actual length of the paramecium?
10. If a 0.27 cm long crystal appears to be 13.5 cm long under a microscope, what is the power of the microscope?

See Example 3

11. Using a $\frac{1}{2}$ in. scale, how tall would a 40 ft statue be in a drawing?
12. How wide is a 3 ft doorway in a $\frac{1}{4}$ in. scale drawing?

PRACTICE AND PROBLEM SOLVING

The scale of a map is 1 in. = 15 mi. Find each length on the map.

13. 30 mi
14. 45 mi
15. 7.5 mi
16. 153.75 mi

The scale of a drawing is 3 in. = 27 ft. Find each actual measurement.

17. 2 in.
18. 5 in.
19. 6.5 in.
20. 11.25 in.

21. Use the scale of the map and a ruler to find the distance in miles between Two Egg, Florida, and Gnaw Bone, Indiana.

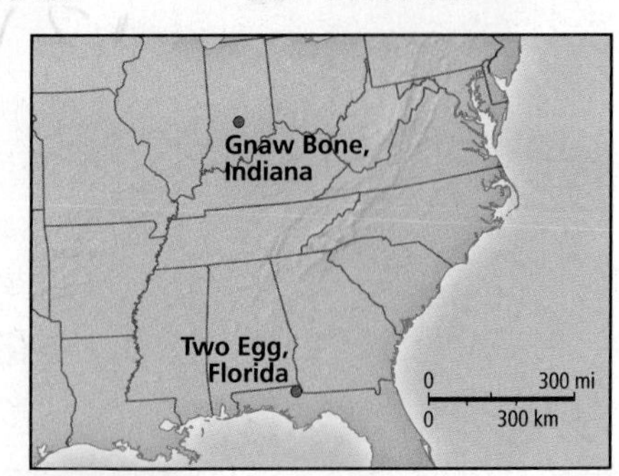

Architecture LINK

Use a metric ruler to measure the width of the 36-inch-wide door on the blueprint of the family room below.

For Exercises 22–28, indicate the scale that you used.

22. How wide are the pocket doors (shown by the red line)?

23. What is the distance s between two interior studs?

24. How long is the oak mantle? (The right side ends just above the *B* in the word *BRICK.*)

25. Could a 4 ft wide bookcase fit along the right-hand wall without blocking the pocket doors? Explain.

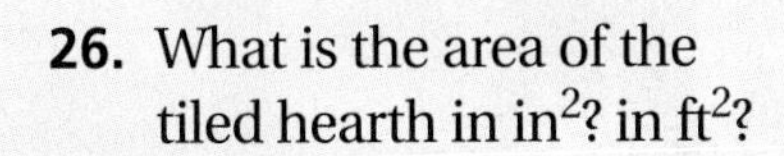

26. What is the area of the tiled hearth in in^2? in ft^2?

27. What is the area of the entire family room in ft^2?

28. Blueprint paper has a maximum width of 36 in., or about 91.4 cm. What does this width represent in the real world corresponding to the scale that you used?

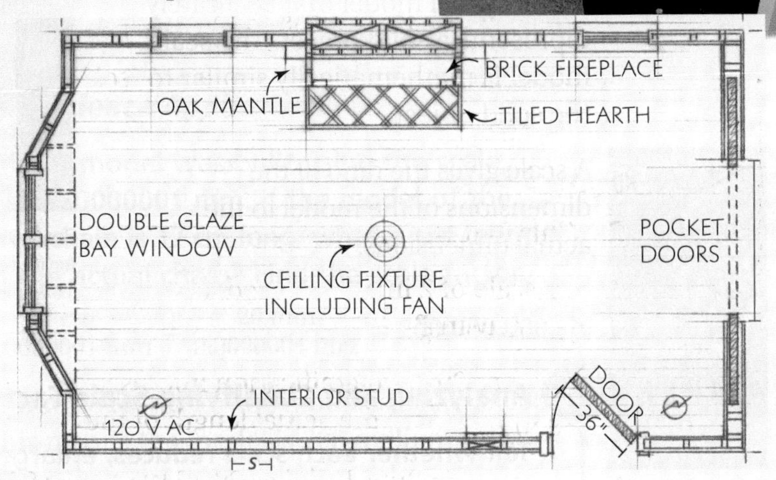

29. **CHALLENGE** Suppose the architect used a $\frac{1}{8}$ in. = 1 ft scale.

a. What would the dimensions of the family room be?

b. Use the result from part **a** to find the area of the family room.

c. If the carpet the Andersons want costs \$4.99 per square foot, how much would it cost to carpet the family room?

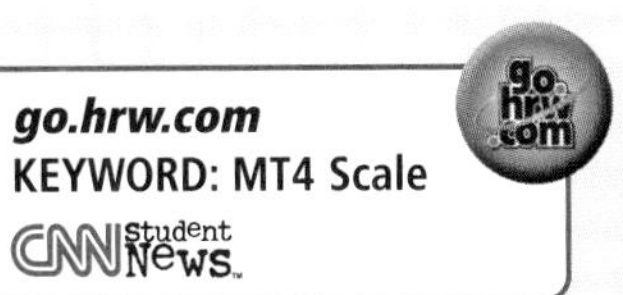

Spiral Review

State whether the ratios in each pair are in proportion. (Lesson 7-1)

30. $\frac{3}{7}$ and $\frac{6}{14}$ **31.** $\frac{5}{8}$ and $\frac{10}{4}$ **32.** $\frac{13}{4}$ and $\frac{52}{16}$ **33.** $\frac{22}{7}$ and $\frac{11}{3}$

34. **EOG PREP** A tree was 3.5 ft tall after 2 years and 8.75 ft tall after 5 years. If the tree grew at a constant rate, how tall was it after 3 years? (Lesson 7-4)

A 5 ft B 5.25 ft C 5.75 ft D 6.5 ft

7-8 Exercises

FOR EOG PRACTICE
see page 671

internet connect
Homework Help Online
go.hrw.com Keyword: MT4 7-8

2.02, 3.02

GUIDED PRACTICE

See Example 1

Tell whether each scale reduces, enlarges, or preserves the size of the actual object.

1. 1 in:18 in.
2. 4 ft:15 in.
3. 1 m:1000 mm
4. 1 cm:10 mm
5. 6 in:100 ft
6. 80 ft:20 in.

See Example 2

7. What scale factor relates a 15 in. tall model boat to a 30 ft tall yacht?

See Example 3

8. A model of a 42 ft tall shopping mall was built using the scale 1 in:3 ft. What is the height of the model?

See Example 4

9. A molecular model uses the scale 2.5 cm:0.00001 mm. If the model is 7 cm long, how long is the molecule?

INDEPENDENT PRACTICE

See Example 1

Tell whether each scale reduces, enlarges, or preserves the size of the actual object.

10. 10 ft:24 in.
11. 1 mi:5280 ft
12. 6 in:100 ft
13. 0.25 in:1 ft
14. 50 ft:1 in.
15. 250 cm:1 km

See Example 2

16. What scale factor was used to build a 55 ft wide billboard from a 25 in. wide model?

See Example 3

17. A model of a house was built using the scale 5 in:25 ft. If a window in the model is 1.5 in. wide, how wide is the actual window?

See Example 4

18. To create a model of an artery, a health teacher uses the scale 2.5 cm:0.75 mm. If the diameter of the artery is 2.7 mm, what is the diameter on the model?

PRACTICE AND PROBLEM SOLVING

Change both measurements to the same unit of measure, and find the scale factor.

19. 1 ft model of a 1 in. fossil
20. 8 cm model of a 24 m rocket
21. 2 ft model of a 30 yd sports field
22. 4 ft model of a 6 yd whale
23. 40 cm model of a 5 m tree
24. 6 in. model of a 6 ft sofa

25. ***LIFE SCIENCE*** Wally has an 18 in. model of a 42 ft dinosaur, the *Tyrannosaurus rex.* What scale factor does this represent?

The Gateway Arch in St. Louis, Missouri, consists of 143 triangular sections, each about 12 feet tall. The sections decrease in cross section from 54 feet at the base to 17 feet at the top.

go.hrw.com
KEYWORD: MT4 Arch

CNN Student News

26. ***BUSINESS*** Engineers designed a theme park by creating a model using the scale 0.5 in:32 ft.
    a. If the dimensions of the model are 41.25 in. by 82.5 in., what are the dimensions of the park?
    b. What is the area of the park in square feet?
    c. If the builders estimate that it will cost $250 million to build the park, how much will it cost per square foot?

27. ***ARCHITECTURE*** Maurice is building a 2 ft high model of the Gateway Arch in St. Louis, Missouri. If he is using a 3 in:78.75 ft scale, how high is the actual arch?

28. ***ENTERTAINMENT*** At Tobu World Square, a theme park in Japan, there are more than 100 scale models of world-famous landmarks, $\frac{1}{25}$ the size of the originals. Using this scale factor,
    a. how tall in inches would a scale model of Big Ben's 320 ft clock tower be?
    b. how tall would a 5 ft tall person be in the model?

The models in Tobu World Square are often seen in movies and television.

29. ***WHAT'S THE ERROR?*** A student is asked to find the scale factor that relates a 10 in. scale model to a 45 ft building. She solves the problem by writing $\frac{10 \text{ in.}}{45 \text{ ft}} = \frac{2}{9} = \frac{1}{4.5}$. What error did the student make? What is the correct scale factor?

30. ***WRITE ABOUT IT*** Explain how you can tell whether a scale factor will make an enlarged scale model or a reduced scale model.

31. ***CHALLENGE*** A scientist wants to build a model, reduced 11,000,000 times, of the Moon revolving around Earth. Will the scale 48 ft:100,000 mi give the desired reduction?

Spiral Review

Find the surface area of each sphere. Use 3.14 for π. (Lesson 6-10)

32. radius 5 mm **33.** radius 12.2 ft **34.** diameter 4 in. **35.** diameter 20 cm

Find each unit rate. (Lesson 7-2)

36. $90 for 8 hours of work **37.** 5 apples for $0.85 **38.** 24 players on 2 teams

39. **EOG PREP** How long would it take to drain a 750-gallon hot tub at a rate of 12.5 gallons per minute? (Lesson 7-3)

A 1 hour B 45 minutes C 55 minutes D 80 minutes

Make a Scale Model

Use with Lesson 7-9

WHAT YOU NEED	REMEMBER
• Card stock • Scissors • Ruler • Tape	A scale such as 1 in. = 200 ft results in a smaller-scale model than a scale of 1 in. = 20 feet.

You can make a scale model of a solid object, such as a rectangular prism, in many ways; you can make a net and fold it, or you can cut card stock and tape the pieces together. The most important thing is to find a good scale.

Activity 1

The Trump Tower in New York City is a rectangular prism with these approximate dimensions: height, 880 feet; base length, 160 feet; base width, 80 feet.

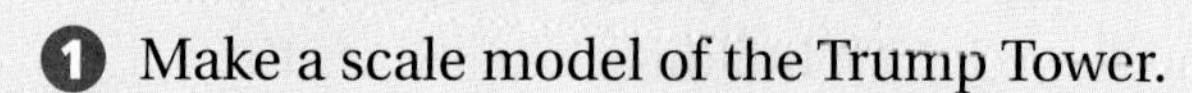

1 Make a scale model of the Trump Tower.

First determine the appropriate height for your model and find a good scale.

To use $8\frac{1}{2}$ in. by 11 in. card stock, divide the longest dimension by 11 to find a scale.

$$\frac{880 \text{ ft}}{11 \text{ in.}} = \frac{80 \text{ ft}}{1 \text{ in.}}$$

Let 1 in. = 80 ft.

The dimensions of the model using this scale are $\frac{880}{80} = 11$ in., $\frac{160}{80} = 2$ in., and $\frac{80}{80} = 1$ in.

So you will need to cut the following:

Two 11 in. × 2 in. rectangles

Two 11 in. × 1 in. rectangles

Two 2 in. × 1 in. rectangles

Tape the pieces together to form the model.

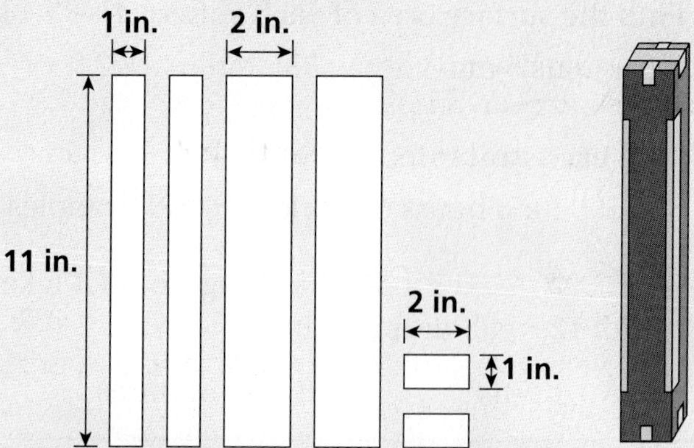

Think and Discuss

1. How tall would a model of a 500 ft tall building be if the same scale were used?
2. Why would a building stand more solidly than your model?
3. What could be another scale of the model if the numbers were without units?

Try This

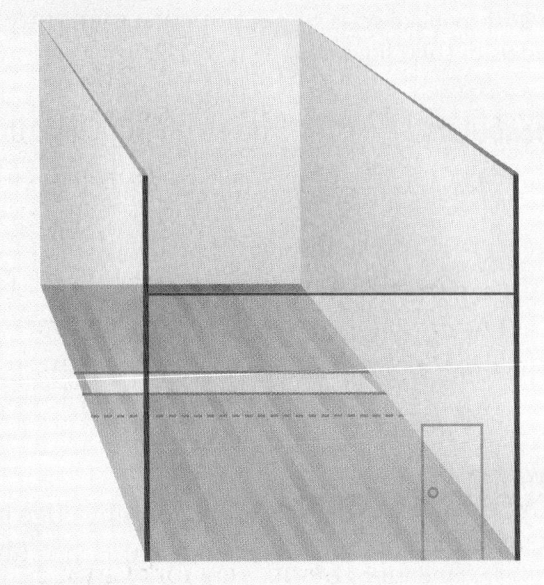

1. Build a scale model of a four-wall handball court. The court is an open-topped rectangular prism 20 feet wide and 40 feet long. Three of the walls are 20 feet tall, and the back wall is 14 feet tall.

A scale model can also be used to make a model that is larger than the original object.

Activity 2

1 A size-AA battery has a diameter of about 0.57 inches and a height of about 2 inches. Make a scale model of a AA battery.

You can roll up paper or card stock to create a cylinder. Find the circumference of the battery: $0.57\pi \approx 1.8$ in.

Note that the height is greater than the circumference, so use the height to find a scale.

$$\frac{11 \text{ in.}}{2 \text{ in.}} = 5.5$$

To use $8\frac{1}{2}$ in. by 11 in. paper or card stock, try multiplying the dimensions of the battery by 5.5.

$2(5.5) = 11$ in. $\qquad$ $1.8(5.5) = 9.9$ in.

Note that 9.9 in. by 11 in. is larger than an 8.5 in. by 11 in. piece of paper. Divide the width of the paper by the height of the battery to find a smaller scale. $8.5 \div 2 = 4.25$. Use the scale to find the new dimensions: diameter $\approx$ 2.4 in., circumference $\approx$ 7.7 in., and height = 8.5 in. The pieces for the scale model are shown.

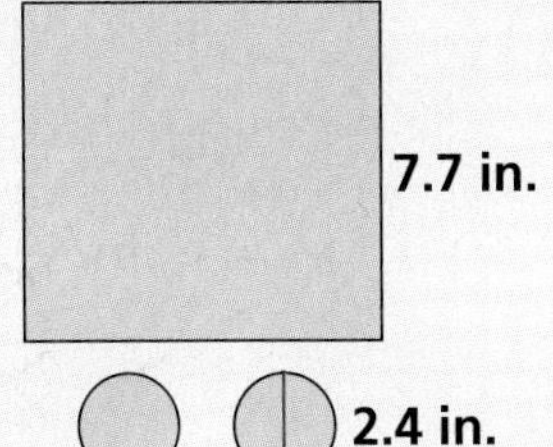

Think and Discuss

1. A salt crystal is one-sixteenth inch long on each side. What would a good scale be for a model of the crystal?

Try This

1. Measure the diameter of the terminal at the top of the battery. Make a scale model of the terminal using the same scale used to make a model of the battery.

7-9 Scaling Three-Dimensional Figures

Learn to make scale models of solid figures.

Vocabulary
capacity

A popcorn company sells a small box of popcorn that measures 1 ft × 1 ft × 1 ft. They also sell a large box that measures 3 ft × 3 ft × 3 ft. It takes 5 seconds for a machine to fill the smaller box with popcorn. It takes quite a bit longer to fill the larger box.

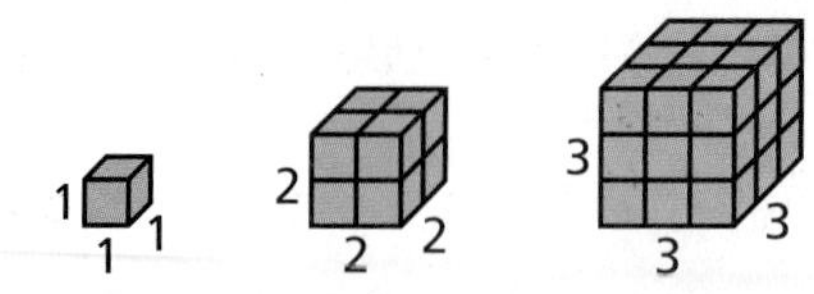

Edge Length	1 ft	2 ft	3 ft
Volume	$1 \times 1 \times 1 = 1\text{ ft}^3$	$2 \times 2 \times 2 = 8\text{ ft}^3$	$3 \times 3 \times 3 = 27\text{ ft}^3$
Surface Area	$6 \cdot 1 \times 1 = 6\text{ ft}^2$	$6 \cdot 2 \times 2 = 24\text{ ft}^2$	$6 \cdot 3 \times 3 = 54\text{ ft}^2$

Helpful Hint

Multiplying the linear dimensions of a solid by n creates n^2 as much surface area and n^3 as much volume.

Corresponding edge lengths of any two cubes are in proportion to each other because the cubes are similar. However, volumes and surface areas do not have the same scale factor as edge lengths.

Each edge of the 2 ft cube is 2 times as long as each edge of the 1 ft cube. However, the cube's volume, or **capacity**, is 8 times as large, and its surface area is 4 times as large as the 1 ft cube's.

EXAMPLE 1 Scaling Models That Are Cubes

A 5 cm cube is built from small cubes, each 1 cm on an edge. Compare the following values.

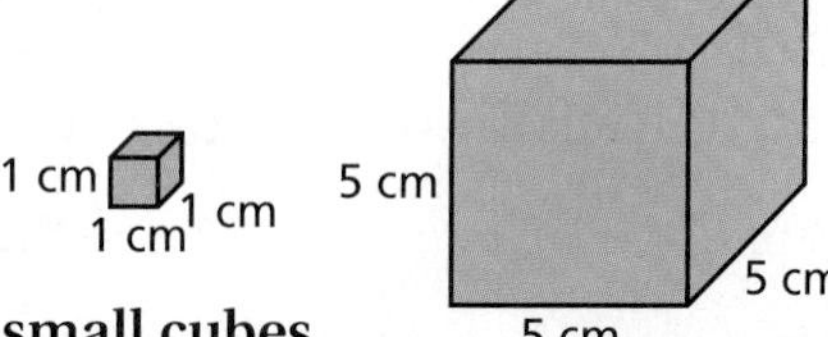

A **the edge lengths of the large and small cubes**

$$\frac{5\text{ cm cube}}{1\text{ cm cube}} \longrightarrow \frac{5\text{ cm}}{1\text{ cm}} = 5$$ *Ratio of corresponding edges*

The edges of the large cube are 5 times as long as those of the small cube.

B **the surface areas of the two cubes**

$$\frac{5\text{ cm cube}}{1\text{ cm cube}} \longrightarrow \frac{150\text{ cm}^2}{6\text{ cm}^2} = 25$$ *Ratio of corresponding areas*

The surface area of the large cube is 25 times that of the small cube.

C **the volumes of the two cubes**

$$\frac{5\text{ cm cube}}{1\text{ cm cube}} \longrightarrow \frac{125\text{ cm}^3}{1\text{ cm}^3} = 125$$ *Ratio of corresponding volumes*

The volume of the large cube is 125 times that of the small cube.

EXAMPLE 2 Scaling Models That Are Other Solid Figures

The Fuller Building in New York, also known as the Flatiron Building, can be modeled as a trapezoidal prism with the approximate dimensions shown. For a 10 cm tall model of the Fuller Building, find the following.

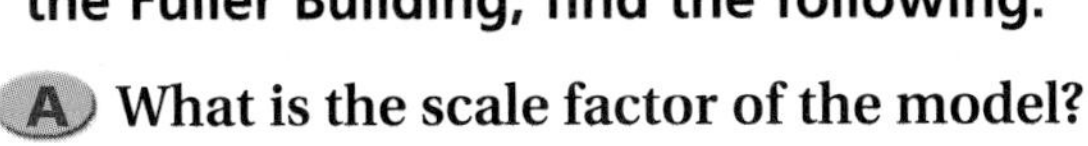

A **What is the scale factor of the model?**

$\frac{10 \text{ cm}}{93 \text{ m}} = \frac{10 \text{ cm}}{9300 \text{ cm}} = \frac{1}{930}$ *Convert and simplify.*

The scale factor of the model is 1:930.

B **What are the other dimensions of the model?**

left side: $\frac{1}{930} \cdot 65 \text{ m} = \frac{6500}{930} \text{ cm} \approx 6.99 \text{ cm}$

back: $\frac{1}{930} \cdot 30 \text{ m} = \frac{3000}{930} \text{ cm} \approx 3.23 \text{ cm}$

right side: $\frac{1}{930} \cdot 60 \text{ m} = \frac{6000}{930} \text{ cm} \approx 6.45 \text{ cm}$

front: $\frac{1}{930} \cdot 2 \text{ m} = \frac{200}{930} \text{ cm} \approx 0.22 \text{ cm}$

The trapezoidal base has side lengths 6.99 cm, 3.23 cm, 6.45 cm, and 0.22 cm.

EXAMPLE 3 *Business Application*

A machine fills a cubic box that has edge lengths of 1 ft with popcorn in 5 seconds. How long does it take the machine to fill a cubic box that has edge lengths of 3 ft?

$V = 3 \text{ ft} \cdot 3 \text{ ft} \cdot 3 \text{ ft} = 27 \text{ ft}^3$ *Find the volume of the larger box.*

Set up a proportion and solve.

$\frac{5}{1 \text{ ft}^3} = \frac{x}{27 \text{ ft}^3}$ *Cancel units.*

$5 \cdot 27 = x$ *Multiply.*

$135 = x$ *Calculate the fill time.*

It takes 135 seconds to fill the larger box.

Think and Discuss

1. **Describe** how the volume of a model compares to the original object if the linear scale factor of the model is 1:2.
2. **Explain** one possible way to double the surface area of a rectangular prism.

7-9 Exercises

FOR EOG PRACTICE
see page 671

internet connect
Homework Help Online
go.hrw.com Keyword: MT4 7-9

2.02, 3.02

GUIDED PRACTICE

See Example 1

A 4 in. cube is built from small cubes, each 1 in. on a side. Compare the following values.

1. the side lengths of the large and small cubes
2. the surface areas of the two cubes
3. the volumes of the two cubes

See Example 2

4. The dimensions of a basketball arena are 500 ft long, 375 ft wide, and 125 ft high. The scale model used to build the arena is 40 in. long. Find the width and height of the model.

See Example 3

5. A 2 ft by 1 ft by 1 ft fish tank in the shape of a rectangular prism drains in 2 min. How long would it take an 8 ft by 3 ft by 3 ft fish tank to drain at the same rate?

INDEPENDENT PRACTICE

See Example 1

A 7 m cube is built from small cubes, each 1 m on a side. Compare the following values.

6. the side lengths of the large and small cubes
7. the surface areas of the two cubes
8. the volumes of the two cubes

See Example 2

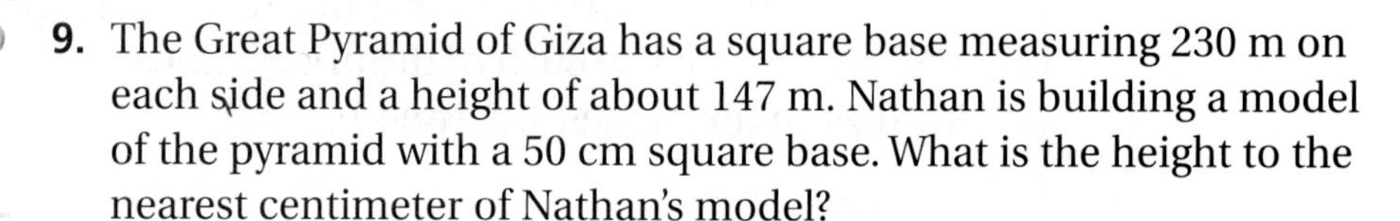

9. The Great Pyramid of Giza has a square base measuring 230 m on each side and a height of about 147 m. Nathan is building a model of the pyramid with a 50 cm square base. What is the height to the nearest centimeter of Nathan's model?

See Example 3

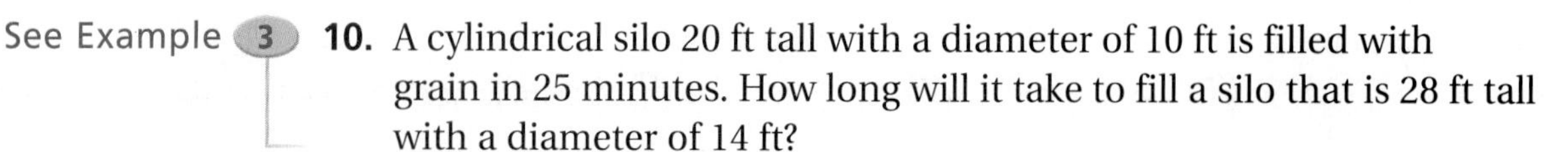

10. A cylindrical silo 20 ft tall with a diameter of 10 ft is filled with grain in 25 minutes. How long will it take to fill a silo that is 28 ft tall with a diameter of 14 ft?

PRACTICE AND PROBLEM SOLVING

For each cube, a reduced scale model is built using a scale factor of $\frac{1}{2}$. Find the length of the model and the number of 1 cm cubes used to build it.

11. a 4 cm cube
12. a 6 cm cube
13. an 8 cm cube
14. a 2 cm cube
15. a 10 cm cube
16. a 12 cm cube
17. What is the volume in cm^3 of a 1 m cube?

18. ***ART*** A piece of pottery requires 2 pounds of modeling clay. How much clay would be required to double all the dimensions of the piece?

19. If it took 100,000 Lego® blocks to build a cylindrical monument with a 5 m diameter, about how many Legos would be needed to build a monument with an 8 m diameter and the same height?

20. ***PHYSICAL SCIENCE*** For a model of the solar system to be accurate, the Sun's diameter would need to be about 612 times the diameter of Pluto. How would their volumes be related?

21. A cereal box that holds 20 oz of cereal is reduced using a linear factor of 0.9. About how many ounces does the new box hold?

Legoland, in Billund, Denmark, contains Lego models of the Taj Mahal, Mount Rushmore, other monuments, and visitors, too.

22. ***CHOOSE A STRATEGY*** Five 1 cm cubes are used to build a solid. How many cubes are used to build a scale model of the solid with a linear scale factor of 2 to 1?

A. 10 cubes
B. 20 cubes
C. 40 cubes
D. 100 cubes

23. ***WRITE ABOUT IT*** If the linear scale factor of a model is $\frac{1}{4}$, what is the relationship between the volume of the original object and the volume of the model?

24. ***CHALLENGE*** To double the volume of a rectangular prism, what number is multiplied by each of the prism's linear dimensions? Give your answer to the nearest hundredth.

Spiral Review

Find two ratios that are equivalent to each given ratio. (Lesson 7-1)

25. $\frac{3}{5}$ **26.** $\frac{13}{26}$ **27.** $\frac{4}{11}$ **28.** $\frac{10}{9}$

The scale of a drawing is 2 in. = 3 ft. Find the actual measurement for each length in the drawing. (Lesson 7-7)

29. 1 in. **30.** 5 in. **31.** 12 in. **32.** 8.5 in.

33. **EOG PREP** What scale factor was used to create a 10 in. tall model from a 15 ft tall statue? (Lesson 7-8)

A $\frac{1}{1.5}$ B $\frac{1}{3}$ C $\frac{1}{15}$ D $\frac{1}{18}$

EXTENSION Trigonometric Ratios

Learn to find the three basic trigonometric ratios for a right triangle and to use them to find missing lengths.

Vocabulary

trigonometric ratios

sine

cosine

tangent

Look at the ratios of the side lengths in the two similar right triangles, ABC and DEF.

The ratios of corresponding sides are equal.

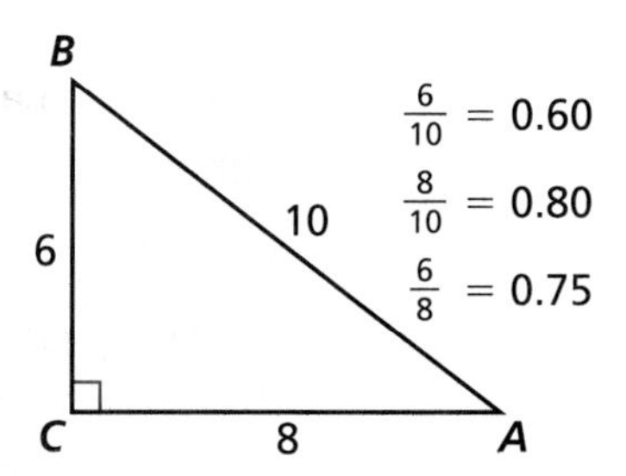

Special ratios called **trigonometric ratios** compare the lengths of the side *opposite* an acute angle in a right triangle, the side *adjacent* (next to) the acute angle, and the length of the hypotenuse. The hypotenuse is never the adjacent side.

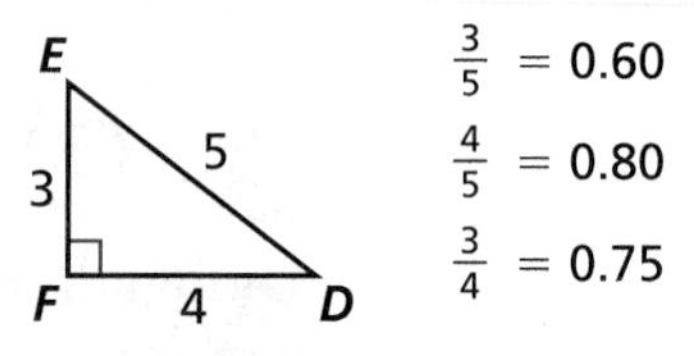

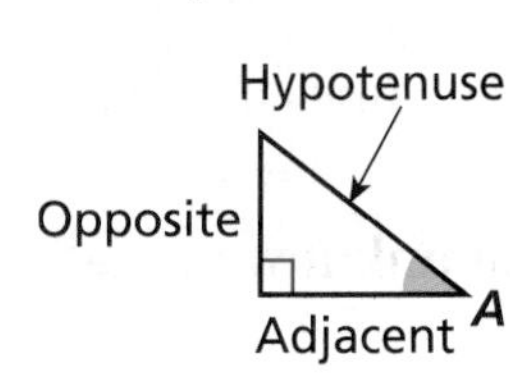

sine of $\angle A = \sin A = \dfrac{\text{length of side opposite } \angle A}{\text{hypotenuse}}$

cosine of $\angle A = \cos A = \dfrac{\text{length of side adjacent to } \angle A}{\text{hypotenuse}}$

tangent of $\angle A = \tan A = \dfrac{\text{length of side opposite } \angle A}{\text{length of side adjacent to } \angle A}$

Trigonometric ratios are constant for a given angle measure.

EXAMPLE 1 Finding the Value of a Trigonometric Ratio

Find the cosine of 50°.

In triangle ABC: $\cos A = \frac{AC}{AB} = \frac{54}{84} \approx 0.64$

On a calculator: COS 50 = 0.64278761

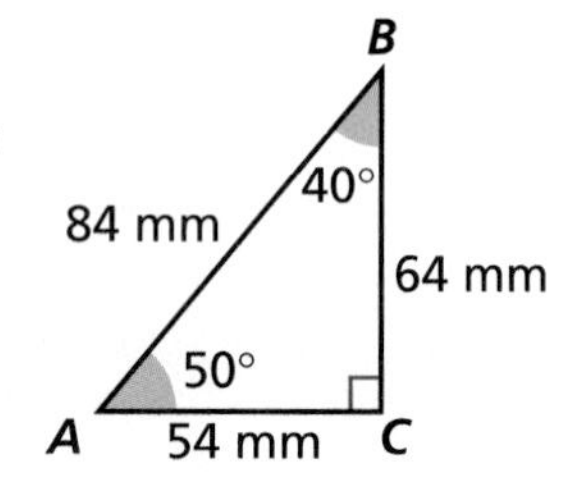

EXAMPLE 2 Using Trigonometric Ratios to Find Missing Lengths

Find the height of the Washington Monument to the nearest foot.

$\tan 70° = \frac{x}{202}$ *Write the tangent ratio for a 70° angle.*

$2.75 \approx \frac{x}{202}$ *Use a calculator to find the value of tan 70°.*

$x \approx 2.75(202) \approx 555.5$ *Solve the equation.*

The height of the Washington Monument is about 556 ft.

EXTENSION Exercises

2.02, 3.02

Find the value of each trigonometric ratio to the nearest thousandth.

1. sin 51° **2.** tan 72° **3.** cos 89°

Find each indicated height to the nearest foot.

4.

5.

Use trigonometric ratios to find each unknown length *x* to the nearest tenth.

6.

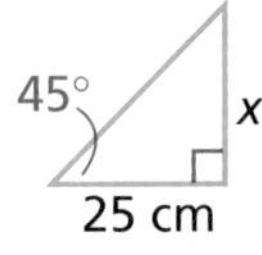

7.

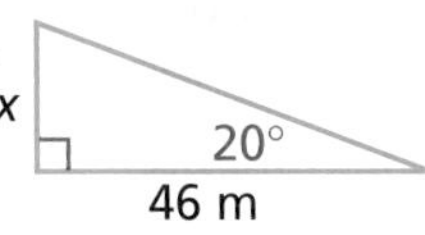

8.

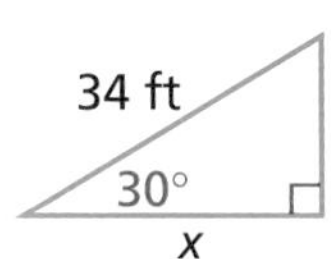

9.

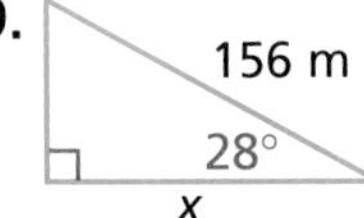

10.

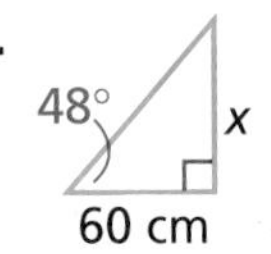

11.

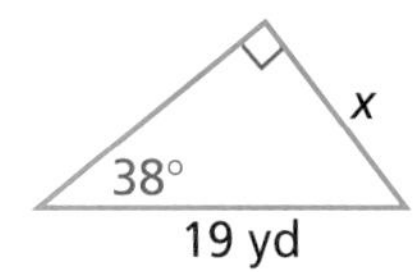

12. Joaquim puts a flagpole on his front porch. He attaches a support wire from the house to hold the flagpole in place. The wire attaches to the house at a right angle. Find, to the nearest tenth of a foot, the length of the support wire.

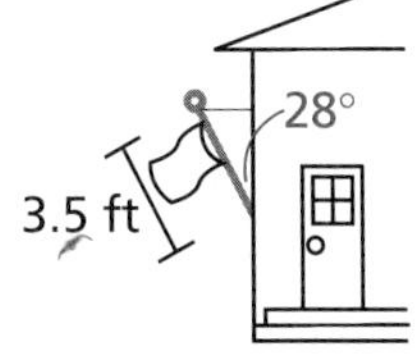

13. Samantha is building a shed. She wants the pitch of the roof to be 36°. Find, to the nearest foot, how high from the ground the peak of the roof will be.

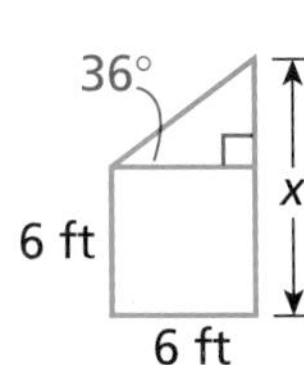

14. Since the hypotenuse is always the longest side of a right triangle, which trigonometric ratio(s) cannot be greater than 1?

15. What angle has a tangent of 1? Explain why this is true.

16. A right triangle has acute angles A and B, where $m\angle A = 36°$ and $m\angle B = 54°$. Compare sin A to cos B. Compare cos A to sin B. Explain your findings.

Problem Solving on Location

NORTH CAROLINA

Topographic Maps

Detailed maps illustrating the landforms of a geographical area are called *topographic maps.* Topographic maps of North Carolina are drawn to a large variety of scales. These scales range from 1:2000 to 1:1,000,000. Although each map displays only a section of the state, some scales create maps that, when put together to display the whole state, would measure more than 200 ft across.

1. Two of the scales used to draw topographic maps of North Carolina are 1:24,000 and 1:100,000. Which of these scales will produce a greater map size? Explain.

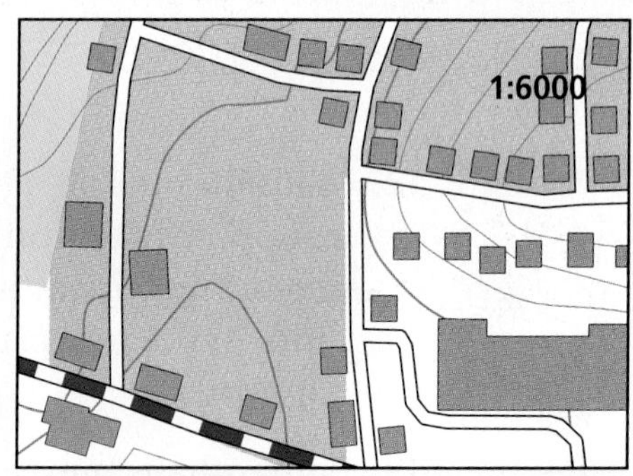

2. If the distance between Royal Pines and Sawmill is about 78 inches on a topographic map that was drawn to a 1:62,500 scale, what would the distance between the two cities be on a map that was drawn to a scale of 1:24,000? What is the actual distance in miles between the two cities?

3. On a topographic map that was drawn to a 1:50,000 scale, the distance between Salisbury and Lexington is about 58 centimeters. What would the distance between the two cities be on a map that was drawn to a scale of 1:125,000? What is the actual distance in miles between the two cities?

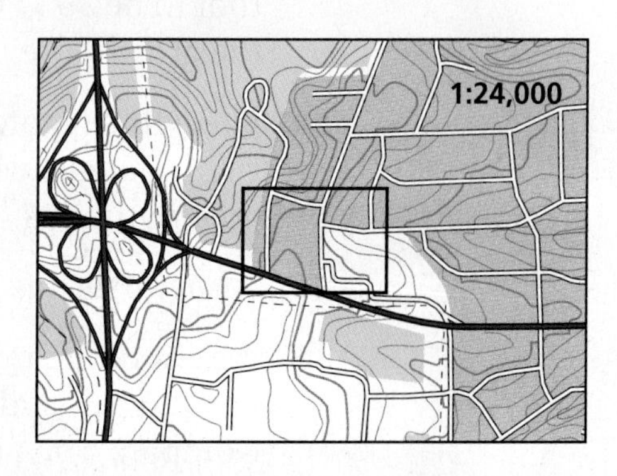

4. The actual distance between Asheville and Greensboro is about 173 miles. Give the distance in feet between the two cities on a topographic map that is drawn to a 1:500,000 scale.

Grapes

Between the mountains and the coast of North Carolina, you can find more than 220 vineyards. These vineyards produce more than 500,000 gallons of grape products annually.

1. The following proportion is true for the 16 American hybrid table grape varieties that are grown in North Carolina.

$$\frac{\text{yellow seedless varieties}}{\text{American hybrid varieties}} = \frac{\text{yellow seeded varieties}}{\text{blue seeded varieties}}$$

If there are 5 seeded varieties, use the table at right to determine how many yellow seeded and blue seeded varieties there are.

The slope of a vineyard site is the ratio of the vineyard's vertical height to its horizontal length. The greater the slope is, the less the vines are affected by cold weather. When choosing land for a vineyard, however, a grower should choose land with a slope less than or equivalent to $\frac{3}{20}$ in order to avoid soil erosion and conditions that make operating machinery dangerous.

Number of Varieties of American Hybrid Table Grapes			
	Yellow	Blue	Red
Seeded	■	■	0
Seedless	4	3	4

2. Tell which of the following vineyard slopes are proportional: $\frac{12}{106}$, $\frac{16}{108}$, $\frac{6}{54}$, $\frac{15}{135}$, $\frac{28}{189}$.

3. If you were choosing a vineyard site from the three described below, which would you choose, and why?
   - a site with a vertical height of 200 ft and a horizontal length of 30 ft
   - a site with a vertical height of 54 ft and a horizontal length of 360 ft
   - a site with a vertical height of 16 ft and a horizontal length of 102 ft

MATH-ABLES

Copy-Cat

You can use this method to copy a well-known work of art or any drawing. First, draw a grid over the work you want to copy, or draw a grid on tracing paper and tape it over the picture.

Next, on a separate sheet of paper draw a blank grid with the same number of squares. The squares do not have to be the same size. Copy each square from the original exactly onto the blank grid. Do not look at the overall picture as you copy. When you have copied all of the squares, the drawing on your finished grid should look just like the original work.

Suppose you are copying an image from a 12 in. by 18 in. print, and that you use 1-inch squares on the first grid.

1. If you use 3-inch squares on the blank grid, what size will your finished copy be?
2. If you want to make a copy that is 10 inches tall, what size should you make the squares on your blank grid? How wide will the copy be?
3. Choose a painting, drawing, or cartoon, and copy it using the method above.

Tic-Frac-Toe

Draw a large tic-tac-toe board. In each square, draw a blank proportion, $\frac{\square}{\square} = \frac{\square}{\square}$. Players take turns using a spinner with 12 sections or a 12-sided die. A player's turn consists of placing a number anywhere in one of the proportions. The player who correctly completes the proportion can claim that square. A square may also be blocked by filling in three parts of a proportion that cannot be completed with a number from 1 to 12. The first player to claim three squares in a row wins.

internet connect go.hrw.com

Go to ***go.hrw.com*** for a copy of the game board.
KEYWORD: MT4 Game7

Dilations of Geometric Figures

Use with Lesson 7-5

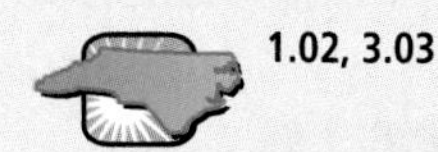

A **dilation** is a geometric transformation that changes the size but not the shape of a figure.

Activity

1. Construct a triangle similar to the one shown below. Label the vertices *A, B,* and *C.*

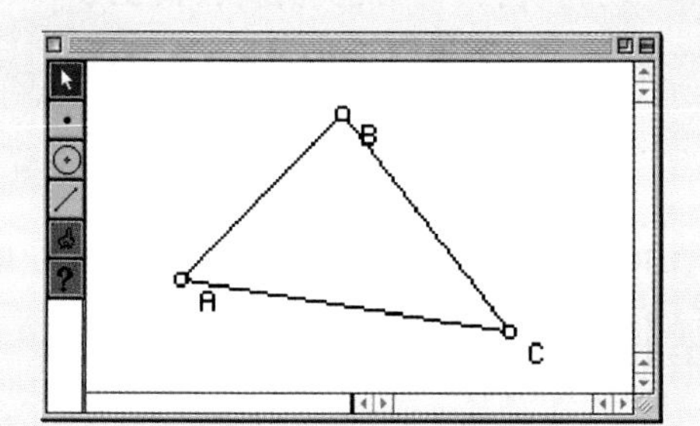

2. Next pick a center of dilation inside triangle *ABC* and label it point *D.*

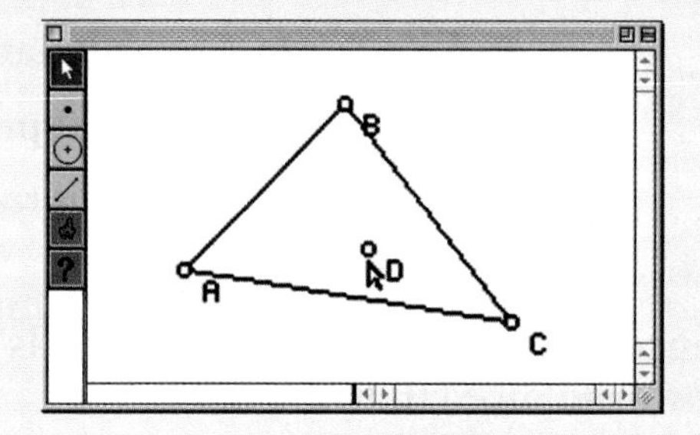

3. Use the dilation tool on your software to shrink the triangle by a ratio of 1 to 2.

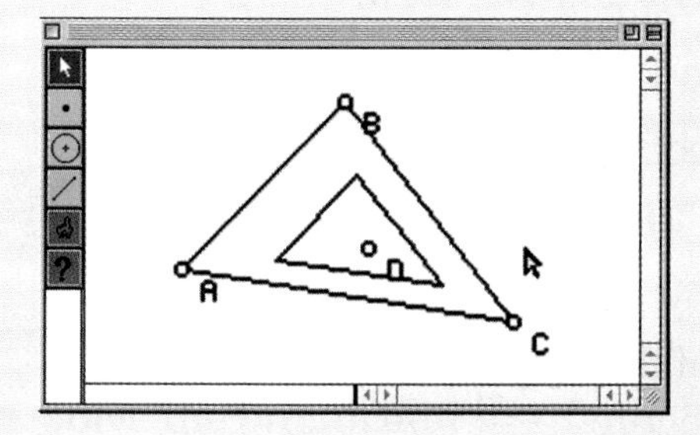

4. Use the dilation tool again to stretch the original triangle by a ratio of 4 to 3.

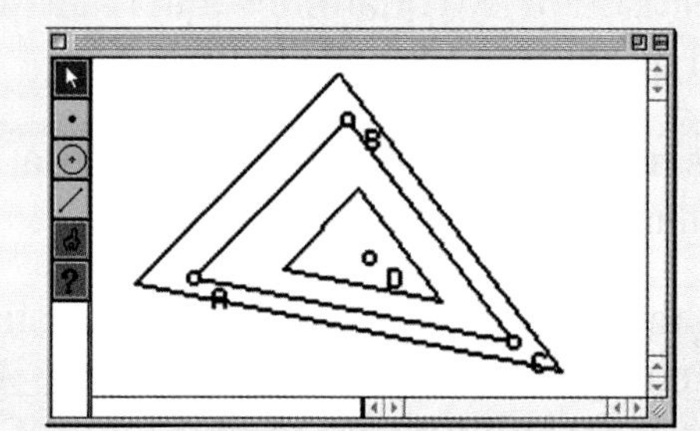

Notice that the dilations of triangle *ABC* are exactly the same *shape* as the original triangle, but they are different *sizes.*

Think and Discuss

1. Are all of the triangles shown in the last figure similar?
2. If the center of dilation is inside the triangle, and the dilated triangle is shrunk, is the smaller triangle always completely inside the original triangle?

Try This

1. Use geometry software to construct a quadrilateral *ABCD.*
   a. Choose a center of dilation inside *ABCD.* Shrink *ABCD* by a factor of 1 to 3.
   b. Choose a center of dilation outside *ABCD.* Stretch *ABCD* by a factor of 3 to 2.

7-6 Similar Figures (pp. 368–371)

EXAMPLE

■ **A stamp 1.2 in. tall and 1.75 in. wide is to be scaled to 4.2 in. tall. How wide should the new stamp be for the two stamps to be similar?**

$\frac{\text{scaled height}}{\text{original height}} = \frac{4.2}{1.2} = 3.5 = \text{scale factor}$

scaled width = original width · scale factor
= 1.75(3.5) = 6.125

The larger stamp should be 6.125 in. wide.

EXERCISES

27. A picture 3 in. wide by 5 in. tall is to be scaled to 7.5 in. wide to be put on a flyer. How tall should the flyer picture be?

28. A picture 8 in. wide by 10 in. tall is to be scaled to 2.5 in. wide to be put on an invitation. How tall should the invitation picture be?

7-7 Scale Drawings (pp. 372–375)

EXAMPLE

■ **A length on a map is 4.2 in. The scale is 1 in:100 mi. Find the actual distance.**

$\frac{1 \text{ in.}}{100 \text{ mi}} = \frac{4.2 \text{ in.}}{x \text{ mi}}$ *Proportion using* $\frac{\text{scale length}}{\text{actual length}}$

$1 \cdot x = 100 \cdot 4.2 = 420$ mi

The actual distance is 420 mi.

EXERCISES

29. A length on a scale drawing is 5.4 cm. The scale is 1 cm:12 m. Find the actual length.

30. A 79.2 ft length is to be scaled on a drawing with the scale 1 in:12 ft. Find the scaled length.

7-8 Scale Models (pp. 376–379)

EXAMPLE

■ **Tell whether the scale 1000 m:1 km reduces, enlarges, or preserves the size of the actual object.**

$\frac{1000 \text{ m}}{1 \text{ km}} = \frac{1000 \text{ m}}{1000 \text{ m}} = 1$ *Convert 1 km = 1000 m and simplify.*

The scale preserves the size since the scale factor is 1.

EXERCISES

Find each scale factor, and tell whether the scale reduces, enlarges, or preserves the size of the actual object.

31. 100 in:1 yd

32. 5 in:2 in.

33. 10 m:1 km

34. 1 km:100,000 cm

7-9 Scaling Three-Dimensional Figures (pp. 382–385)

EXAMPLE

■ **A 4 in. cube is built from small cubes, each 2 in. on a side. Compare the volumes of the large cube and the small cube.**

$\frac{\text{vol. of large cube}}{\text{vol. of small cube}} = \frac{4^3 \text{ in}^3}{2^3 \text{ in}^3} = \frac{64 \text{ in}^3}{8 \text{ in}^3} = 8$

The volume of the large cube is 8 times that of the small cube.

EXERCISES

A 3 ft cube is built from small cubes, each 1 ft on a side. Compare the indicated measures of the large cube and the small cube.

35. side lengths

36. surface areas

37. volumes

Chapter Test

Simplify to tell whether the ratios form a proportion.

1. $\frac{24}{72}$ and $\frac{36}{108}$
2. $\frac{15}{20}$ and $\frac{9}{16}$

Use conversion factors.

3. Change 15 quarts to gallons.
4. Change 40 kilometers per hour to meters per hour.
5. Change 45 miles per hour to feet per second.

Solve each proportion.

6. $\frac{3}{5} = \frac{18}{n}$
7. $\frac{x}{15} = \frac{7}{35}$
8. $\frac{10}{y} = \frac{35}{63}$
9. Use the scale 10 in:50 ft. Find the scale factor. Tell whether the scale factor reduces, enlarges, or preserves the size of an object.
10. Use the scale 1000 mm:100 cm. Find the scale factor. Tell whether the scale factor reduces, enlarges, or preserves the size of an object.

A 9 cm cube is built from small cubes, each 1 cm on a side. Compare the indicated measures of the large cube and the small cube.

11. the side lengths
12. the surface areas
13. the volumes

Dilate each triangle *ABC* by the given scale factor with the origin as the center of dilation.

14. $A(1, 1)$, $B(3, 1)$, $C(1, 3)$; scale factor = 3

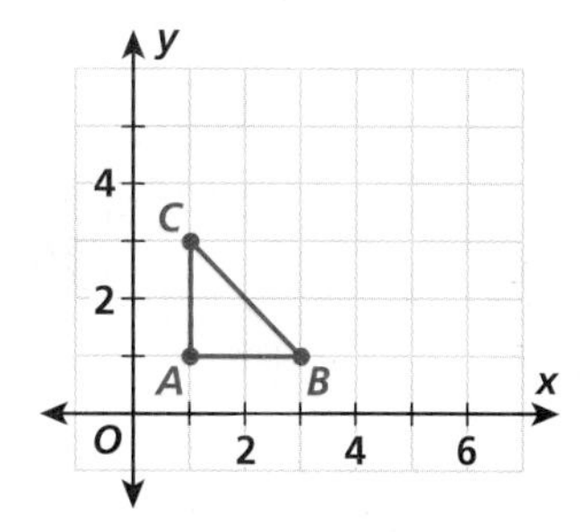

15. $A(2, 2)$, $B(4, 6)$, $C(8, 4)$; scale factor = 0.5

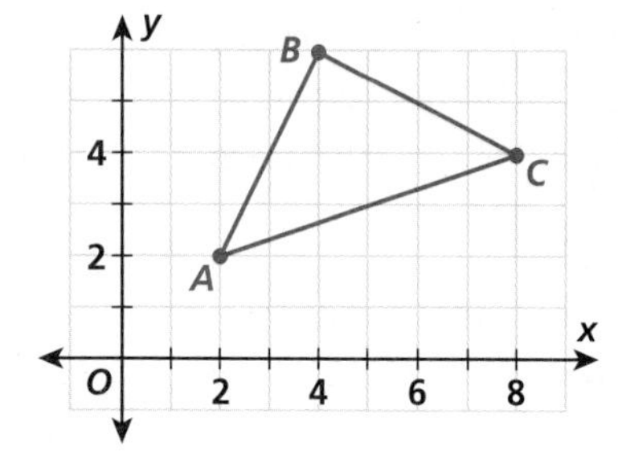

16. Dina earned $28 for working 2 hours. At the same rate of pay, how many hours must she work to earn $49?
17. The ratio of the length of a rectangular field to its width is 10:7. If the width of the field is 70 meters, find the perimeter of the field.
18. A company sells snack-size boxes of raisins with dimensions 5 in. by 2.5 in. by 1.5 in., each weighing 2 oz. They want to make a family-size box that would be 7.5 in. by 3.75 in. by 2.25 in. What would be the weight of the family-size box of raisins?

Chapter 7

Performance Assessment

Show What You Know

Create a portfolio of your work from this chapter. Complete this page and include it with your four best pieces of work from Chapter 7. Choose from your homework assignments, mid-chapter quiz, or any journal entries you have done. Put them together using any design you want. Make your portfolio represent what you consider your best work.

Short Response

1. At the school cafeteria the ratio of pints of chocolate milk sold to pints of plain milk sold is 4 to 7. At this rate, how many pints of chocolate milk will be sold if 168 pints of plain milk are sold? Show your work.

2. While shopping for school supplies Sara finds boxes of pencils in two sizes. One box has 8 pencils for \$0.89, and the other box has 12 pencils for \$1.25.

   a. Which box is the better bargain? Why? Round your answer to the nearest cent.

   b. How much would it save to buy 48 pencils at the better rate? Show your work.

Extended Problem Solving

3. To build an accurate model of the solar system, choose a diameter for the model of the Sun. Then all distances and sizes of the planets can be calculated proportionally using the table below.

   Suppose the Sun in the model has a 1 in. diameter.

   a. What is the diameter of Pluto in the model?

   b. What is Pluto's distance from the Sun in the model?

   c. What would Pluto's distance from the Sun be in the model if the Sun's diameter were changed to 2 ft?

There were only six known planets when this mechanical model was created in the early 1700's.

	Sun	Mars	Jupiter	Pluto
Diameter (mi)	864,000	4200	88,640	1410
Distance from Sun (million mi)	n/a	141	483	3670

internet connect
EOG Test Practice Online
go.hrw.com Keyword: MT4 TestPrep

Getting Ready for EOG

Chapter 7

Cumulative Assessment, Chapters 1–7

1. Joan paid \$6.40 for 80 copies of a flyer. What is the unit price?

   A 8 copies per dollar

   B 16 copies per dollar

   C \$0.80 per copy

   D \$0.08 per copy

2. A 9-inch model is made of a 15-foot boat. What is the scale factor?

   A 1:20 C 5:3

   B 3:5 D 20:1

3. If $x = yz$, which of the following must be equal to xy?

   A yz C y^2z

   B yz^2 D $\frac{z^2}{y}$

4. Each of these fractions is in its simplest form: $\frac{4}{n}, \frac{5}{n}, \frac{7}{n}$. Which of the following could be the value of n?

   A 28 C 26

   B 27 D 25

5. In the equation $A = \pi r^2$, if r is doubled, by what number is A multiplied?

   A 2 C 4

   B $\frac{1}{2}$ D $\frac{1}{4}$

6. Which of the following is true for the data set 20, 30, 50, 70, 80, 80, 90?

   I. The mean is greater than 70.
   II. The median is greater than 70.
   III. The mode is greater than 70.

   A I and II only C III only

   B II and III only D I, II, and III

7. What is the next number in the sequence $-27, 9, -3, 1, \ldots$?

   A -3 C $-\frac{1}{3}$

   B -1 D 0

TEST TAKING TIP!

Redraw a figure: Answers to geometry problems may become apparent as you redraw the figure.

8. How many edges are in the prism below?

   A 5 C 10

   B 7 D 15

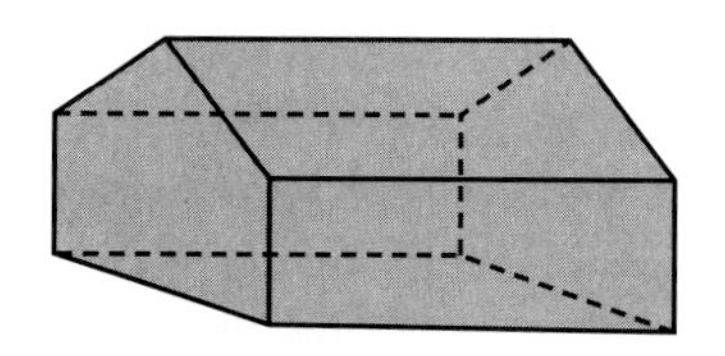

9. ***SHORT RESPONSE*** What is the value of $(-1 - 2)^3 + 2.5^1$? Use the order of operations, and show each step.

10. ***SHORT RESPONSE*** Use the map to estimate to the nearest 10 km the distance that the Steward family will travel as they sail from St. Petersburg to Pensacola, Florida. Explain in words how you determined your answer.

Chapter

8

Percents

TABLE OF CONTENTS

Player	Age	Home Runs	At Bats/Home Run
Barry Bonds	37	576	14.0
Sammy Sosa	33	450	14.4
Ken Griffey Jr.	32	460	14.6
Alex Rodriguez	26	241	15.6

Career *Sports Statistician*

Statisticians are mathematicians who work with data, creating statistics, graphs, and tables that describe and explain the real world. Sports statisticians combine their love of sports with their ability to use mathematics.

Statistics not only explain what has happened, but can help you predict what may happen in the future. The table describes the home run hitting of some active Major League baseball players.

Chapter Opener Online
go.hrw.com
KEYWORD: MT4 Ch8

ARE YOU READY?

Choose the best term from the list to complete each sentence.

cross multiply
equivalent ratios
proportion
ratio

1. A __?__ is a comparison of two quantities by division.
2. Ratios that make the same comparison are __?__.
3. Two ratios that are equivalent are in __?__.
4. To solve a proportion, you __?__.

Complete these exercises to review skills you will need for this chapter.

✓ Write Fractions as Decimals

Write each fraction as a decimal.

5. $\frac{3}{4}$ **6.** $\frac{5}{8}$ **7.** $\frac{2}{5}$ **8.** $\frac{2}{3}$

✓ Write Decimals as Fractions

Write each decimal as a fraction in simplest form.

9. 0.7 **10.** 0.6 **11.** 0.25 **12.** 0.375

13. 0.2 **14.** 0.9 **15.** 0.86 **16.** 0.99

✓ Solve Proportions

Solve each proportion.

17. $\frac{x}{3} = \frac{9}{27}$ **18.** $\frac{7}{8} = \frac{h}{4}$ **19.** $\frac{9}{n} = \frac{2}{3}$

20. $\frac{3}{8} = \frac{12}{t}$ **21.** $\frac{4}{5} = \frac{28}{z}$ **22.** $\frac{100}{p} = \frac{90}{45}$

✓ Read Circle Graphs

Refer to the graph to answer each question.

23. Which item accounts for nearly half the budget?

24. What dollar amount is spent on computer equipment?

25. What dollar amount is spent on new books and programs?

26. What dollar amount is spent on other expenses?

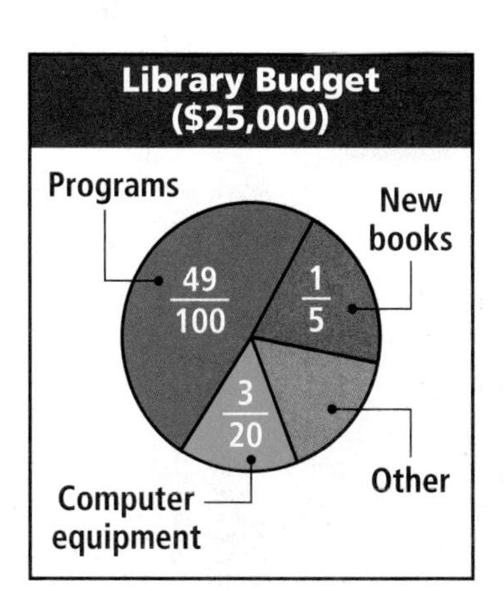

8-1 Relating Decimals, Fractions, and Percents

Learn to relate decimals, fractions, and percents.

Vocabulary

percent

In an average day, a typical koala sleeps 20 out of 24 hours. The part of a day the koala sleeps can be shown in several ways:

$$\frac{20}{24} = 0.83\overline{3} = 83.\overline{3}\%$$

So koalas sleep over 80% of the time.

Koalas usually sleep in the fork of a tree. They are most active after sunset.

Percents are ratios that compare a number to 100.

Reading Math

Think of the % symbol as meaning /100.
0.75 = 75% = 75/100

Ratio	Equivalent Ratio with Denominator of 100	Percent
$\frac{3}{10}$	$\frac{30}{100}$	30%
$\frac{1}{2}$	$\frac{50}{100}$	50%
$\frac{3}{4}$	$\frac{75}{100}$	75%

To convert a fraction to a decimal, divide the numerator by the denominator.

$$\frac{1}{8} = 1 \div 8 = 0.125$$

$$\begin{array}{r} 0.125 \\ 8\overline{)1.000} \\ \underline{8} \\ 20 \\ \underline{16} \\ 40 \\ \underline{40} \\ 0 \end{array}$$

To convert a decimal to a percent, multiply by 100 and insert the percent symbol.

$$0.125 \cdot 100 \rightarrow 12.5\%$$

EXAMPLE 1 Finding Equivalent Ratios and Percents

Find the missing ratio or percent equivalent for each letter *a–g* on the number line.

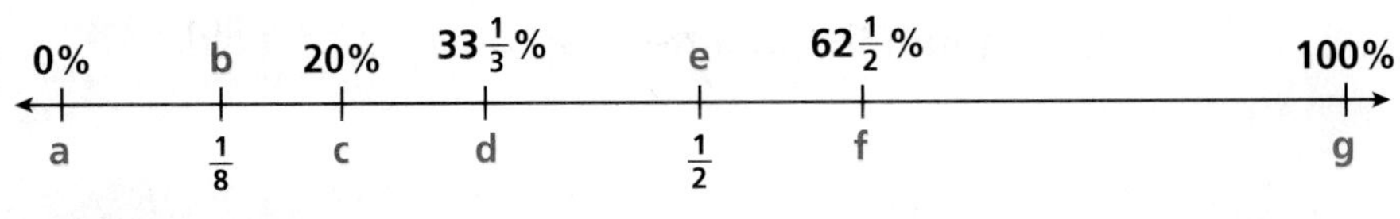

a: $0\% = \frac{0}{100} = 0$

b: $\frac{1}{8} = 0.125 = 12.5\% = 12\frac{1}{2}\%$

c: $20\% = \frac{20}{100} = \frac{2}{10} = \frac{1}{5}$

d: $33\frac{1}{3}\% = 0.33\overline{3} = \frac{1}{3}$

e: $\frac{1}{2} = 0.5 = 50\%$

f: $62\frac{1}{2}\% = 0.625 = \frac{625}{1000} = \frac{5}{8}$

g: $100\% = \frac{100}{100} = 1$

Remember!

Here are some percents and their equivalent ratios:

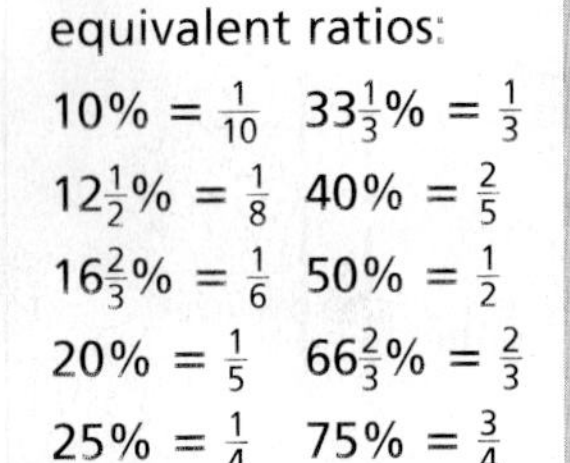

$10\% = \frac{1}{10}$ $33\frac{1}{3}\% = \frac{1}{3}$
$12\frac{1}{2}\% = \frac{1}{8}$ $40\% = \frac{2}{5}$
$16\frac{2}{3}\% = \frac{1}{6}$ $50\% = \frac{1}{2}$
$20\% = \frac{1}{5}$ $66\frac{2}{3}\% = \frac{2}{3}$
$25\% = \frac{1}{4}$ $75\% = \frac{3}{4}$

EXAMPLE 2 Finding Equivalent Fractions, Decimals, and Percents

Find the equivalent value missing from the table for each value given on the circle graph.

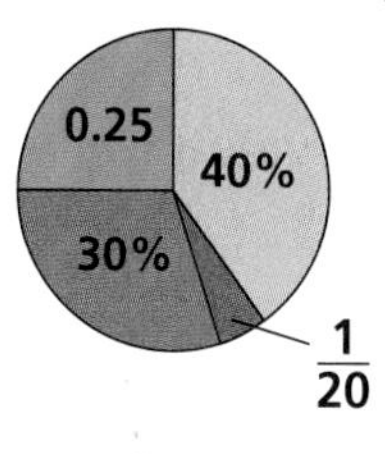

Fraction	Decimal	Percent
$\frac{25}{100} = \frac{1}{4}$	**0.25**	$0.25(100) = 25\%$
$\frac{40}{100} = \frac{2}{5}$	$\frac{2}{5} = 0.4$	**40%**
$\mathbf{\frac{1}{20}}$	$\frac{1}{20} = 0.05$	$0.05(100) = 5\%$
$\frac{30}{100} = \frac{3}{10}$	$\frac{3}{10} = 0.3$	**30%**

You can use the information in each column of Example 2 to make three equivalent circle graphs. One shows the breakdown by fractions, one shows the breakdown by decimals, and one shows the breakdown by percents.

Fraction

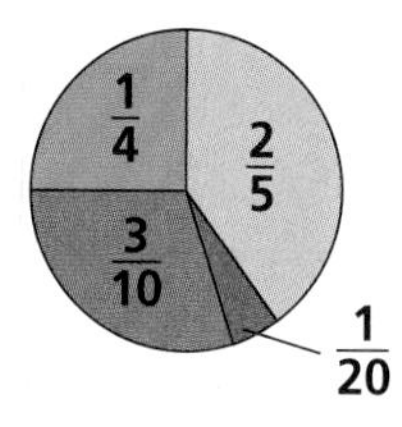

The sum of the fractions should be 1.

Decimal

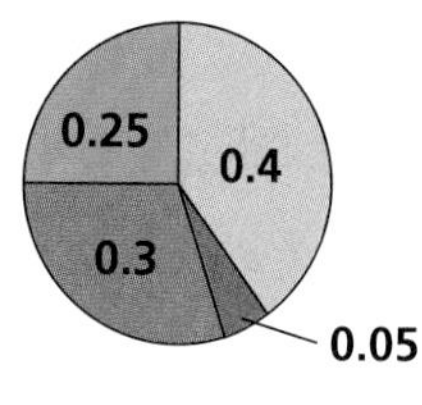

The sum of the decimals should be 1.

Percent

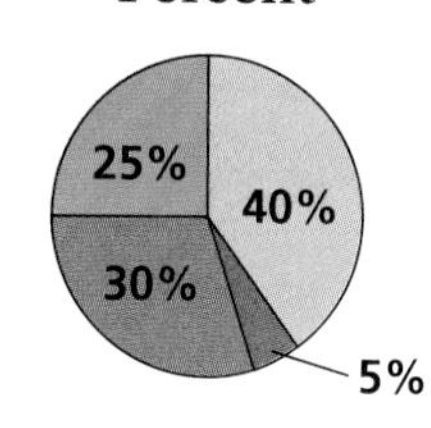

The sum of the percents should be 100%.

EXAMPLE 3 *Physical Science Application*

Gold that is 24 karat is 100% pure gold. Gold that is 18 karat is 18 parts pure gold and 6 parts another metal, such as copper, zinc, silver, or nickel.

What percent of 18-karat gold is pure gold?

$\frac{\text{parts pure gold}}{\text{total parts}} \rightarrow \frac{18}{24} = \frac{3}{4}$ *Set up a ratio and reduce.*

$\frac{3}{4} = 3 \div 4 = 0.75 = 75\%$ *Find the percent.*

So 18-karat gold is 75% pure gold.

Think and Discuss

1. **Give an example** of a real-world situation in which you would use (1) decimals (2) fractions, and (3) percents.
2. **Show** 25 cents as a part of a dollar in terms of (1) a reduced fraction (2) a percent, and (3) a decimal. Which is most common?

8-1 Exercises

FOR EOG PRACTICE
see page 672

internet connect
Homework Help Online
go.hrw.com Keyword: MT4 8-1

GUIDED PRACTICE

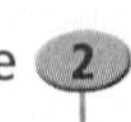

See Example 1 Find the missing ratio or percent equivalent for each letter on the number line.

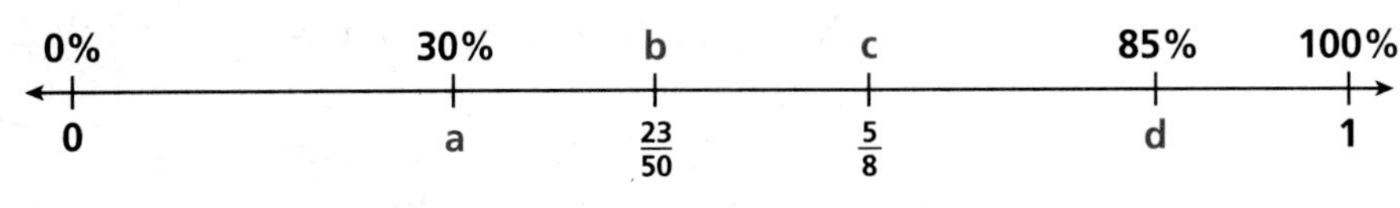

1. a 2. b 3. c 4. d

See Example 2 Find each equivalent value.

5. $\frac{2}{5}$ as a percent 6. 32% as a fraction 7. $\frac{7}{8}$ as a decimal

See Example 3

8. A molecule of water is made up of 2 atoms of hydrogen and 1 atom of oxygen. What percent of the atoms of a water molecule is oxygen?

INDEPENDENT PRACTICE

See Example 1 Find the missing ratio or percent equivalent for each letter on the number line.

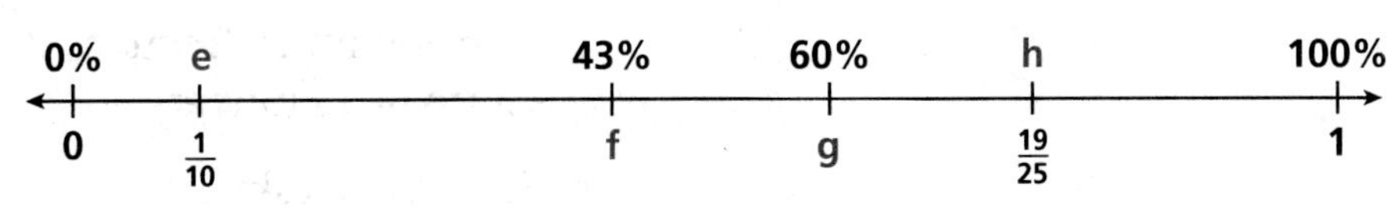

9. e 10. f 11. g 12. h

See Example 2 Find each equivalent value as indicated.

13. 32% as a decimal 14. $\frac{23}{25}$ as a percent 15. 0.545 as a fraction

See Example 3

16. Sterling silver is an alloy combining 925 parts pure silver and 75 parts of another metal, such as copper. What percent of sterling silver is not pure silver?

PRACTICE AND PROBLEM SOLVING

Write the labels from each circle graph as percents.

17. 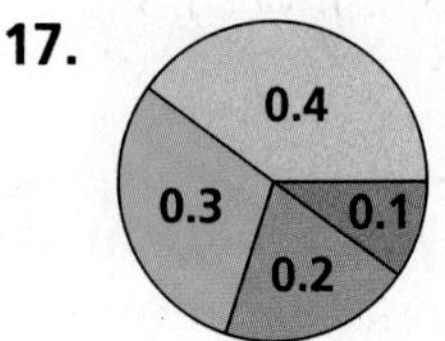

18. 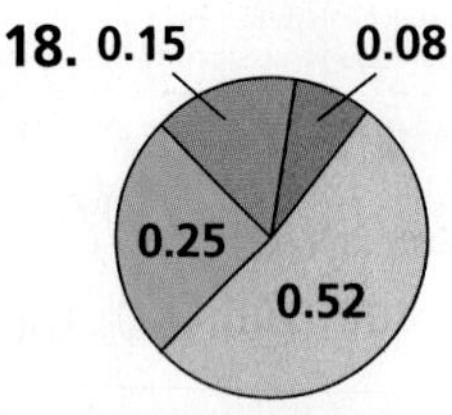

19. 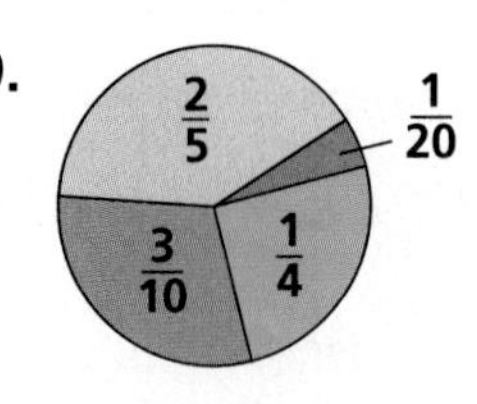

20. A nickel is 5% of a dollar. Write the value of a nickel as a decimal and as a fraction.

21. ***PHYSICAL SCIENCE*** Of the 20 highest mountains in the United States, 17 are located in Alaska. What percent of the highest mountains in the United States are in Alaska?

22. ***LIFE SCIENCE*** When collecting plant specimens, it is a good idea to remove no more than 5% of a population of plants. A botanist wants to collect plants from an area with 60 plants. What is the greatest number of plants she should remove?

23. The graph shows the percents of the total U.S. land area taken up by the five largest states. The sixth section of the graph represents the area of the remaining 45 states.

a. Alaska is the largest state in total land area. Write Alaska's portion of the total U.S. land area as a fraction and as a decimal.

b. What percent of the total U.S. land area is Alaska and Texas combined? How might you describe this percent?

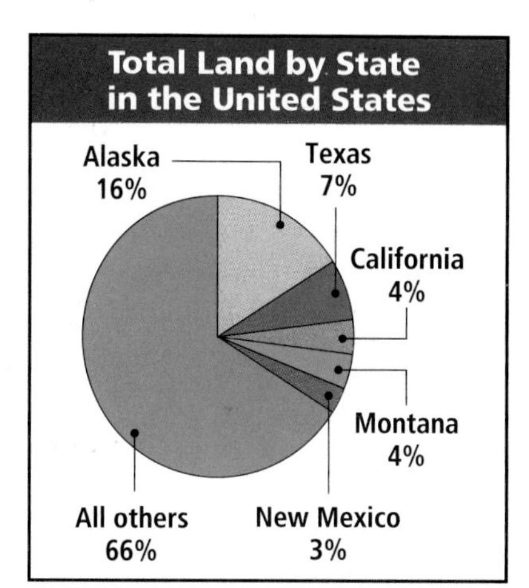

24. ***WHAT'S THE ERROR?*** An analysis showed that 0.03% of the video games produced by one company were defective. Wynn says this is 3 out of every 100. What is Wynn's error?

25. ***WRITE ABOUT IT*** How can you find a fraction, decimal, or percent when you have only one form of a number?

26. ***CHALLENGE*** Luke and Lissa were asked to solve a percent problem using the numbers 17 and 45. Luke found 17% of 45, but Lissa found 45% of 17. Explain why they both got the same answer. Would this work for other numbers as well? Why or why not?

Spiral Review

Tell whether the two lines described in each exercise are parallel, perpendicular, or neither. (Lesson 5-5)

27. $\overleftrightarrow{PQ}$ has slope $\frac{3}{2}$. $\overleftrightarrow{EF}$ has slope $-\frac{2}{3}$.

28. $\overleftrightarrow{AB}$ has slope $\frac{9}{11}$. $\overleftrightarrow{CD}$ has slope $-\frac{3}{4}$.

29. $\overleftrightarrow{XY}$ has slope $\frac{13}{25}$. $\overleftrightarrow{QR}$ has slope $\frac{13}{25}$.

30. $\overleftrightarrow{MN}$ has slope $-\frac{1}{8}$. $\overleftrightarrow{OP}$ has slope 8.

31. **EOG PREP** A cone has diameter 12 cm and height 9 cm. What is the volume of the cone to the nearest tenth? Use 3.14 for π. (Lesson 6-7)

A 56.5 cm^3 B 118.3 cm^3 C 1356.5 cm^3 D 339.1 cm^3

32. **EOG PREP** What is the value of $Q - 1\frac{2}{3}$ for $Q = 4\frac{3}{4}$? (Lesson 3-5)

A $3\frac{1}{12}$ B $5\frac{1}{12}$ C $1\frac{1}{6}$ D $3\frac{5}{12}$

Make a Circle Graph

Use with Lesson 8-1

WHAT YOU NEED:
- Compass
- Protractor
- Ruler
- Paper

REMEMBER
- A circle measures 360°.
- Percent compares a number to 100.

Lab Resources Online
go.hrw.com
KEYWORD: MT4 Lab8A

Activity

1 Skunks are legal pets in some states but not in most. Use the information from the table to make a circle graph showing the percents for each category.

Skunks as Pets by State	
Legality	**Number of States**
Legal (no restrictions)	6
Legal with permit	12
Legal in some areas	2
Illegal	27
Other conditions	3

a. Use a compass to draw a large circle. Use a ruler to draw a vertical radius.

b. Extend the table to show the percent of states with each category of legality.

c. Use the percents to determine the angle measure of each sector of the graph.

d. Use a protractor to draw each angle clockwise from the radius.

e. Label the graph and each sector. Color the sectors.

Legality	Number of States	Percent of States	Angle of Section
Legal (no restrictions)	6	$\frac{6}{50} = 12\%$	$\frac{12}{100} \cdot 360 = 43.2°$
Legal with permit	12	$\frac{12}{50} = 24\%$	$\frac{24}{100} \cdot 360 = 86.4°$
Legal in some areas	2	$\frac{2}{50} = 4\%$	$\frac{4}{100} \cdot 360 = 14.4°$
Illegal	27	$\frac{27}{50} = 54\%$	$\frac{54}{100} \cdot 360 = 194.4°$
Other conditions	3	$\frac{3}{50} = 6\%$	$\frac{6}{100} \cdot 360 = 21.6°$

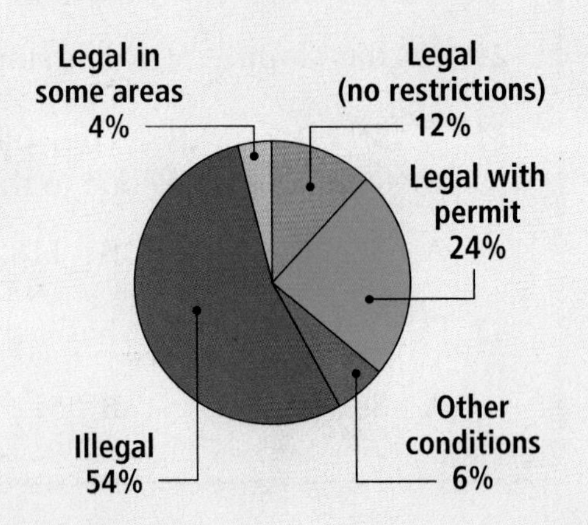

Think and Discuss

1. How many states would need to legalize skunks for the largest sector to be 180°?

Try This

1. Make a circle graph to show only the states where skunks are not illegal.

8-2 Finding Percents

Learn to find percents.

Relative humidity is a measure of the amount of water vapor in the air. When the relative humidity is 100%, the air has the maximum amount of water vapor. At this point, any additional water vapor would cause precipitation. To find the relative humidity on a given day, you would need to find a percent.

The rainy season in some parts of Indochina extends from March to November, and the average humidity is close to 90%.

EXAMPLE 1 Finding the Percent One Number Is of Another

A What percent of 162 is 90?

Method 1: Set up an equation to find the percent.

$p \cdot 162 = 90$ *Set up an equation.*

$p = \frac{90}{162}$ *Solve for p.*

$p = 0.\overline{5}$, or approximately 0.56. *0.56 is 56%*

So 90 is approximately 56% of 162.

B Earth has a surface area of approximately 197 million square miles. About 58 million square miles of that surface area is land. Find the percent of Earth's surface area that is land.

Method 2: Set up a proportion to find the percent.

Think: **What number** is to 100 as 58 is to 197?

$\frac{\text{number}}{100} = \frac{\text{part}}{\text{whole}}$ *Set up a proportion.*

$\frac{n}{100} = \frac{58}{197}$ *Substitute.*

$n \cdot 197 = 100 \cdot 58$ *Find the cross products.*

$197n = 5800$

$n = \frac{5800}{197}$ *Solve for n.*

$n \approx 29.44$, or approximately 29.

$\frac{29}{100} \approx \frac{58}{197}$ *The proportion is reasonable.*

So approximately 29% of Earth's surface area is land.

20. ***LANGUAGE ARTS*** The Hawaiian words shown contain all of the letters of the Hawaiian alphabet. The ` is actually a consonant!

   a. What percent of the Hawaiian alphabet are vowels?

   b. To the nearest tenth, what percent of the letters in the English alphabet are also in the Hawaiian alphabet?

21. ***EARTH SCIENCE*** If there are 3.87 cm^3 of oxygen in an 18 cm^3 sample of air, what percent of the sample is oxygen?

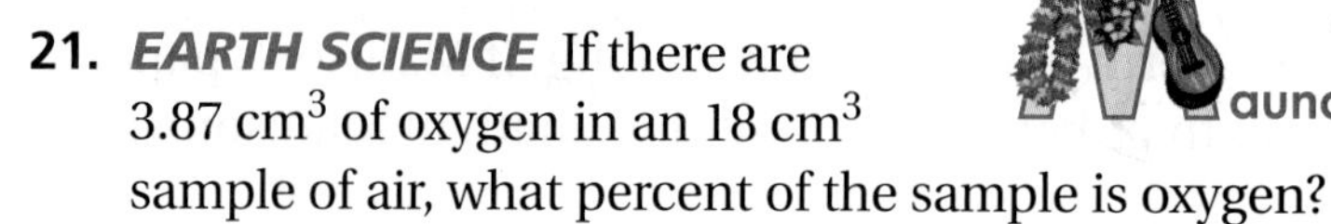

22. ***SOCIAL STUDIES*** According to the 2000 U.S. Census, approximately 2.5 million Americans spend $12\frac{1}{2}$% of the 24-hour day commuting. How many hours a day does a person in this group spend commuting?

23. ***SOCIAL STUDIES*** Of the 50 states in the Union, 32% have names that begin with either *M* or *N*. How many states have names beginning with either *M* or *N*?

24. ***LIFE SCIENCE*** The General Sherman sequoia tree, in California, is thought to be the largest living thing on Earth by volume. It has a height of 275 ft. Its lowest large branch is at a height of 130 ft. What percent of the height of the tree would you need to climb to reach that branch?

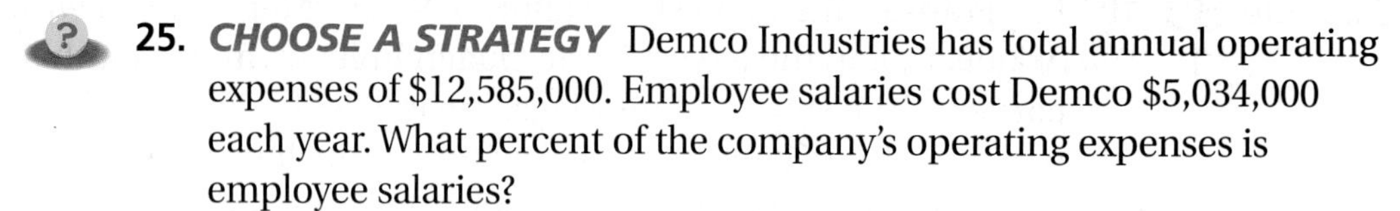

25. ***CHOOSE A STRATEGY*** Demco Industries has total annual operating expenses of $12,585,000. Employee salaries cost Demco $5,034,000 each year. What percent of the company's operating expenses is employee salaries?

   A 4% **B** 40% **C** 25% **D** 250%

26. ***WRITE ABOUT IT*** A question on a math quiz asks, "What is 150% of 88?" Mark calculates 13.2 as the answer. Is this a reasonable answer? Explain why or why not.

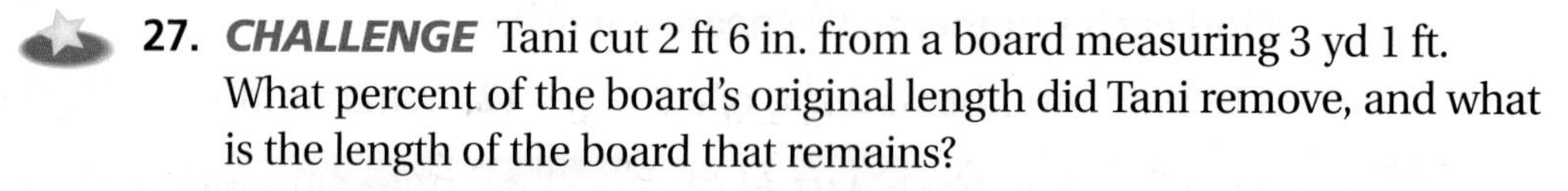

27. ***CHALLENGE*** Tani cut 2 ft 6 in. from a board measuring 3 yd 1 ft. What percent of the board's original length did Tani remove, and what is the length of the board that remains?

Spiral Review

State if each number is rational, irrational, or not a real number. (Lesson 3-10)

28. -14 **29.** $\sqrt{13}$ **30.** $\frac{127}{46{,}191}$ **31.** $\sqrt{-\frac{5}{6}}$

32. **EOG PREP** Each edge of a gift box is 4 in. long. How much wrapping paper would it take to cover the surface of the gift box? (Lesson 6-8)

A 96 in^2 B 64 in^2 C 32 in^2 D 128 in^2

Find Percent Error

Use with Lesson 8-2

2.01

A measurement is only as precise as the device that is used to measure. There is often a difference between a measured value and an accepted or actual value. When the difference is given as a percent of the accepted value, this is called the *percent error.*

Percent error is always nonnegative, so use absolute value.

$$\text{percent error} = \frac{|\text{measured value} - \text{accepted value}|}{\text{accepted value}} \cdot 100$$

Activity

1. A student uses an 8 oz cup and finds the volume of a container to the nearest 8 oz as 64 oz. The actual volume of the container is 67.6 oz. Find the percent error of the measurement to the nearest tenth of a percent.

   a. Store the measured volume on your calculator as M and the actual volume as A. Type 64 STO▶ ALPHA M ENTER and 67.6 STO▶ ALPHA A ENTER .

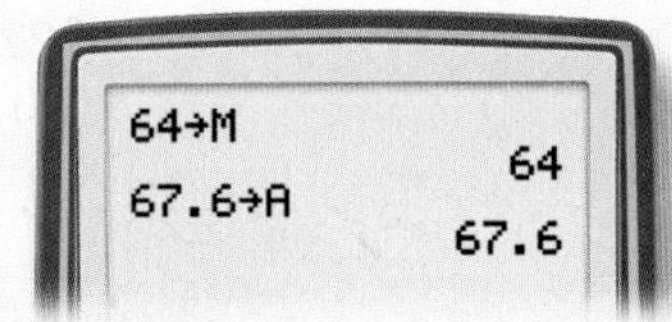

   b. Find the percent error by using the following keystrokes:

(MATH NUM 1: ABS (ALPHA M − ALPHA A) ÷ ALPHA A) × 100

To the nearest tenth of a percent, the percent error is 5.3%.

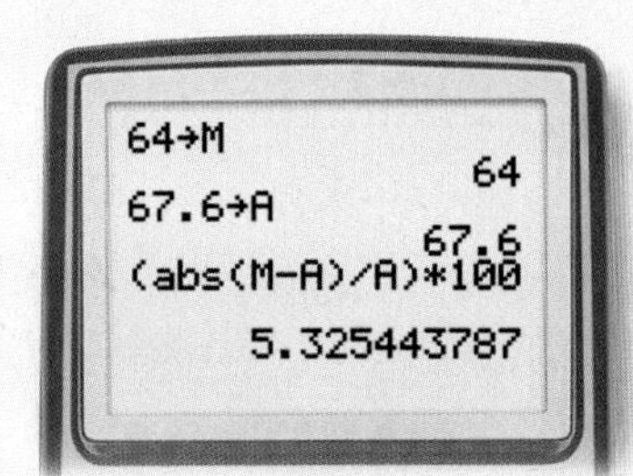

Think and Discuss

1. Can percent error exceed 100%? Explain.
2. Tell why one measurement that is 0.1 cm from an actual length may have a larger percent error than another measurement that is 25 cm from a different actual length.
3. Describe why a ruler with centimeter markings can only measure accurately to within $\frac{1}{2}$ cm of an actual length.

Try This

Find the percent error to the nearest tenth of a percent.

1. measured length 3 cm; actual length 3.4 cm
2. measured length 250 ft; actual length 246.9 ft

8-3 Finding a Number When the Percent Is Known

Learn to find a number when the percent is known.

The Pacific giant squid can grow to a weight of 2000 pounds. This is 1250% of the maximum weight of the Pacific giant octopus. When one number is known, and its relationship to another number is given by a percent, the other number can be found.

In studies, the Pacific giant octopus has been able to travel through mazes and unscrew jar lids for food.

EXAMPLE 1 Finding a Number When the Percent Is Known

36 is 4% of what number?

Set up an equation to find the number.

$36 = 4\% \cdot n$ *Set up an equation.*

$36 = 0.04n$ *$4\% = \frac{4}{100}$*

$\frac{36}{0.04} = \frac{0.04}{0.04}n$ *Divide both sides by 0.04.*

$900 = n$

36 is 4% of 900.

EXAMPLE 2 *Physical Science Application*

In a science lab, a sample of a compound contains 16.5 grams of sodium. If 82.5% of the sample is sodium, find the number of grams the entire sample weighs.

Choose a method: Set up a proportion to find the number.

Think: 82.5 is to 100 as 16.5 is to **what number?**

$\frac{82.5}{100} = \frac{16.5}{n}$ *Set up a proportion.*

$82.5 \cdot n = 100 \cdot 16.5$ *Find the cross products.*

$82.5n = 1650$ *Solve for n.*

$n = \frac{1650}{82.5}$

$n = 20$

The entire sample weighs 20 grams.

EXAMPLE 3 Life Science Application

A **The Pacific giant squid can grow to a weight of 2000 pounds. This is 1250% of the maximum weight of the Pacific giant octopus. To the nearest pound, find the maximum weight of the octopus.**

Choose a method: Set up an equation.

Think: 2000 is 1250% of what number?

$2000 = 1250\% \cdot n$ *Set up an equation.*

$2000 = 12.50 \cdot n$ *1250% = 12.50*

$\frac{2000}{12.50} = n$ *Solve for n.*

$160 = n$

The maximum weight of the Pacific giant octopus is about 160 lb.

B **The king cobra, the world's largest venomous snake, can reach a length of 18 feet. This is only about 60% of the length of the largest reticulated python. Find the length of the largest reticulated python.**

Choose a method: Set up a proportion.

Think: 60 is to 100 as 18 is to what number?

$\frac{60}{100} = \frac{18}{n}$ *Set up a proportion.*

$60 \cdot n = 100 \cdot 18$ *Find the cross products.*

$60n = 1800$

$n = \frac{1800}{60}$ *Solve for n.*

$n = 30$

The largest reticulated python is 30 feet long.

Life Science LINK

Reticulated means "net-like" or "forming a network." The reticulated python is named for the pattern on its skin.

You have now seen all three types of percent problems.

Three Types of Percent Problems	
1. Finding the percent of a number	15% of 120 = n
2. Finding the percent one number is of another	p% of 120 = 18
3. Finding a number when the percent is known	15% of n = 18

Think and Discuss

1. Compare finding a number when a percent is known to finding the percent one number is of another number.

2. Explain whether a number is greater than or less than 36 if 22% of the number is 36.

8-3 Exercises

FOR EOG PRACTICE

see page 673

internet connect
Homework Help Online
go.hrw.com Keyword: MT4 8-3

GUIDED PRACTICE

See Example 1

Find each number to the nearest tenth.

1. 4.3 is $12\frac{1}{2}$% of what number?

2. 56 is $33\frac{1}{3}$% of what number?

3. 18% of what number is 30?

4. 30% of what number is 96?

See Example 2

5. The only kind of rock that floats in water is pumice. Chalk, although denser, absorbs more water than pumice does. How much water can a 5.2 oz piece of chalk absorb if it can absorb 32% of its weight?

See Example 3

6. At 3 P.M., a chimney casts a shadow that is 135% of its actual height. If the shadow is 37.8 ft, what is the actual height of the chimney?

INDEPENDENT PRACTICE

See Example 1

Find each number to the nearest tenth.

7. 105 is $33\frac{1}{3}$% of what number?

8. 77 is 25% of what number?

9. 51 is 6% of what number?

10. 24 is 15% of what number?

11. 84% of what number is 14?

12. 56% of what number is 39.2?

13. 10% of what number is 57?

14. 180% of what number is 6?

See Example 2

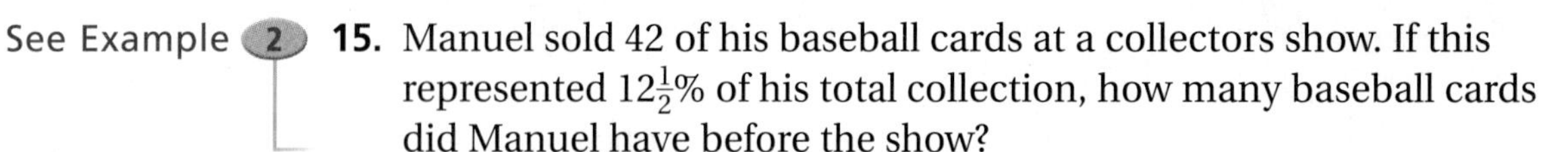

15. Manuel sold 42 of his baseball cards at a collectors show. If this represented $12\frac{1}{2}$% of his total collection, how many baseball cards did Manuel have before the show?

See Example 3

16. When a tire is labeled "185/70/14," that means it is 185 mm wide, the sidewall height (from the rim to the road) is 70% of its width, and the wheel has a diameter of 14 in. What is the tire's sidewall height?

PRACTICE AND PROBLEM SOLVING

Complete each statement.

17. Since 1% of 600 is 6,

a. 2% of ____ is 6.

b. 4% of ____ is 6.

c. 8% of ____ is 6.

18. Since 100% of 8 is 8,

a. 50% of ____ is 8.

b. 25% of ____ is 8.

c. 10% of ____ is 8.

19. Since 5% of 80 is 4,

a. 10% of ____ is 4.

b. 20% of ____ is 4.

c. 40% of ____ is 4.

20. In a poll of 225 students, 36 said that their favorite Thanksgiving food was turkey, and 56 said that their favorite was stuffing. Give the percent of students who said that each food was their favorite.

Social Studies 

The U.S. census collects information about state populations, economics, income and poverty levels, births and deaths, and so on. This information can be used to study trends and patterns. For Exercises 21–23, round answers to the nearest tenth.

2000 U.S. Census Data

	Population	Male	Female
Alaska	626,932	324,112	302,820
New York	18,976,457	9,146,748	9,829,709
Age 34 and Under	139,328,990	71,053,554	68,275,436
Age 35 and Over	142,092,916	67,000,009	75,092,907
Total U.S.	281,421,906	138,053,563	143,368,343

21. What percent of New York's population is male?

22. What percent of the entire country's population, to the nearest tenth of a percent, is made up of people in New York?

23. Tell what percent of the U.S. population each represents.

a. people 34 and under **b.** people 35 and over

c. male **d.** female

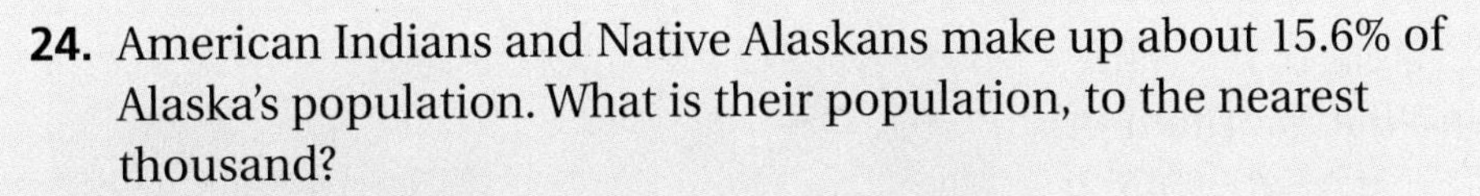

24. American Indians and Native Alaskans make up about 15.6% of Alaska's population. What is their population, to the nearest thousand?

25. ***CHALLENGE*** About 71% of the U.S. population age 85 and over is female. Of the fractions that round to 71% when rounded to the nearest percent, which has the least denominator?

The New York counties with the greatest populations are Kings (Brooklyn) and Queens.

go.hrw.com
KEYWORD: MT4 Census
CNN Student News

Spiral Review

Find the range of each set of data. (Lesson 4-4)

26. 16, 32, 1, 54, 30, 28

27. 105, 969, 350, 87, 410

28. 0.2, 0.8, 0.65, 0.7, 1.6, 1.1

Find the first and third quartiles of the data set. (Lesson 4-4)

29. 55, 60, 40, 45, 70, 65, 35, 40, 75, 50, 60, 80, 45, 55

30. **EOG PREP** A triangle has vertices $A(4, 4)$, $B(6, -2)$, and $C(-4, -12)$. What are the vertices after dilating the triangle by a scale factor of 2 with the origin as the center of dilation? (Lesson 7-5)

A $A'(2, 2)$, $B'(3, -1)$, $C'(-2, -6)$

B $A'(-8, -8)$, $B'(-12, 4)$, $C'(8, 24)$

C $A'(8, 8)$, $B'(12, -4)$, $C'(-8, -24)$

D $A'(16, 16)$, $B'(36, 4)$, $C'(16, 144)$

Chapter 8

Mid-Chapter Quiz

LESSON 8-1 (pp. 400–403)

Find the equivalent value missing from the table for each value given on the circle graph.

Fraction	Decimal	Percent
$\frac{1}{8}$	1. ■	2. ■
3. ■	0.25	4. ■
5. ■	6. ■	$37\frac{1}{2}\%$
$\frac{1}{4}$	7. ■	8. ■

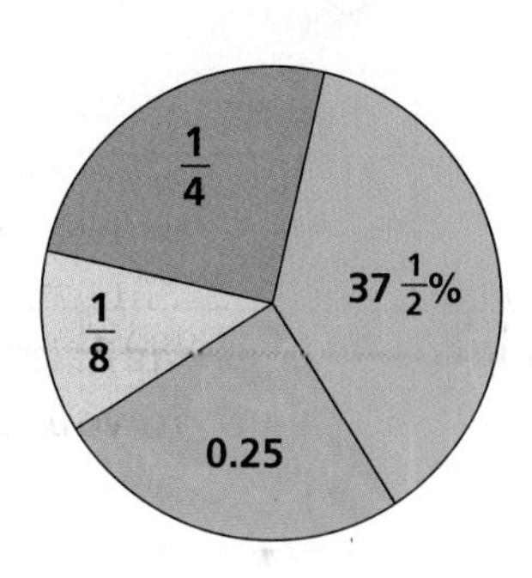

LESSON 8-2 (pp. 405–408)

9. What is 27% of 16?

10. 48 is what percent of 384?

11. In the November 2001 election, only 191,411 of the 509,719 voters registered in Westchester County, New York, cast a ballot. This was the lowest turnout in at least a century. To the nearest tenth, what percent of registered voters actually voted in the election?

12. Use the height of the 88-story Jin Mao Tower in Shanghai and the information shown at right to find the heights of the Eiffel Tower and Russia's Motherland Statue.

13. Of Canada's total area of 9,976,140 km^2, 755,170 km^2 is water. To the nearest tenth of a percent, what part of Canada is water?

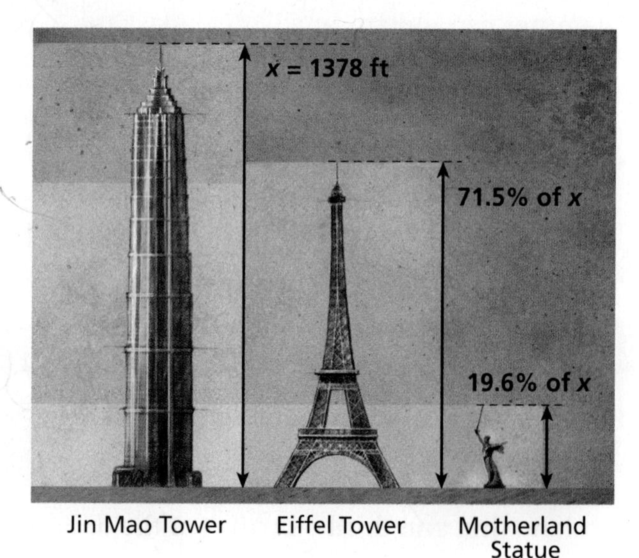

LESSON 8-3 (pp. 410–413)

14. 30 is 12.5% of what number?

15. 244 is 250% of what number?

16. The speed of sound in air at sea level at 32°F is 1088 ft/s. If that represents only 22.04% of the speed of sound in ice-cold water, what is the speed of sound in ice-cold water, to the nearest whole number?

17. In 2000, U.S. imports from Canada totaled $230,838.3 million. This was about 129% of the total dollar value of the U.S. exports to Canada. To the nearest ten million dollars, what was the value of U.S. exports to Canada?

1.02

Focus on Problem Solving

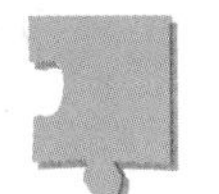

Make a Plan

- **Do you need an estimate or an exact answer?**

When you are solving a word problem, ask yourself whether you need an exact answer or whether an estimate is sufficient. For example, if the amounts given in the problem are approximate, only an approximate answer can be given. If an estimate is sufficient, you may wish to use estimation techniques to save time in your calculations.

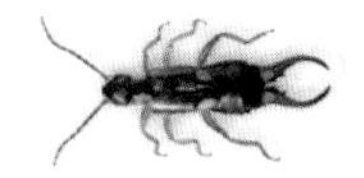

For each problem below, explain whether an exact answer is needed or whether an estimate is sufficient. Then find the answer.

1. In a poll of 3000 registered voters in a certain district, 1800 favored a proposed school bond package. What percent favored the bond package?

2. George needs to score 76% on his final exam to get a B in his math class. If the final is worth 200 points, how many points does he need?

3. Karou is trying to save about $3500 for a trip to Japan. If she has $1000 in an account that earns 8% interest and puts $100 per month in the account, will she have enough in 2 years?

4. Erik makes $7.60 per hour at his job. If he receives a 5% raise, how much will he be making per hour?

5. Jamie is planning to tile her kitchen floor. The room is 330 square feet. It is recommended that she buy enough tile for an area 15% greater than the actual kitchen floor in case of breakage. How many square feet of tile should she buy?

6. There are about 1,032,000 known species of animals on Earth. Of these, about 751,000 are insects. What percent of known species are insects?

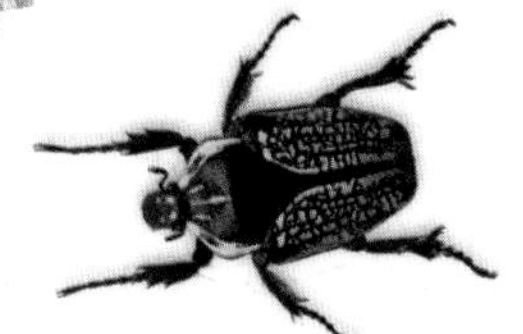

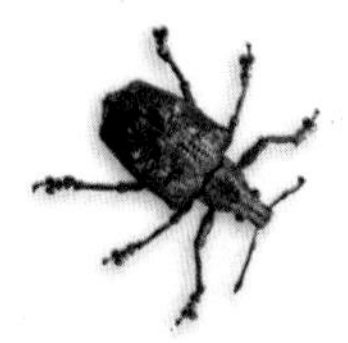

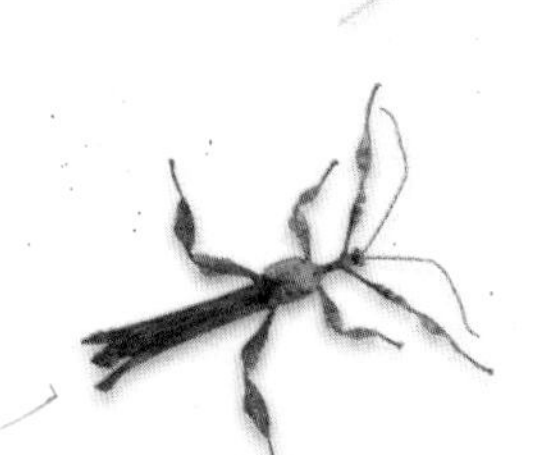

8-4 Percent Increase and Decrease

Learn to find percent increase and decrease.

Vocabulary

percent change

percent increase

percent decrease

Many animals hibernate during the winter to survive harsh conditions and food shortages. While they sleep, their body temperatures drop, their breathing rates decrease, and their heart rates slow. They may even appear to be dead.

"He hums in his sleep."

Percents can be used to describe a change. **Percent change** is the ratio of the *amount of change* to the *original amount.*

$$\text{percent change} = \frac{\text{amount of change}}{\text{original amount}}$$

Percent increase describes how much the original amount increases. **Percent decrease** describes how much the original amount decreases.

EXAMPLE 1 Finding Percent Increase or Decrease

Find the percent increase or decrease from 20 to 24.

This is percent increase.

$24 - 20 = 4$ — *First find the amount of change.*

Think: What percent is 4 of 20?

$\frac{\text{amount of increase}}{\text{original amount}} \rightarrow \frac{4}{20}$ — *Set up the ratio.*

$\frac{4}{20} = 0.2$ — *Find the decimal form.*

$= 20\%$ — *Write as a percent.*

From 20 to 24 is a 20% increase.

EXAMPLE 2 Life Science Application

A The heart rate of a hibernating woodchuck slows from 80 to 4 beats per minute. What is the percent decrease?

$80 - 4 = 76$ — *First find the amount of change.*

Think: What percent is 76 of 80?

$\frac{\text{amount of decrease}}{\text{original amount}} \rightarrow \frac{76}{80}$ — *Set up the ratio.*

$\frac{76}{80} = 0.95$ — *Find the decimal form.*

$= 95\%$ — *76 is 95% of 80.*

The woodchuck's heart rate decreases by 95% during hibernation.

B **According to the U.S. Census Bureau, 69.9 million children lived in the United States in 1998. It is estimated that there will be 77.6 million children in 2020. What is the percent increase, to the nearest percent?**

$77.6 - 69.9 = 7.7$ — *First find the amount of change.*

Think: What percent is 7.7 of 69.9?

$\frac{\text{amount of increase}}{\text{original amount}} = \frac{7.7}{69.9}$ — *Set up the ratio.*

$\frac{7.7}{69.9} \approx 0.1102$ — *Find the decimal form.*

$\approx 11.02\%$ — *Write as a percent.*

The number of children in the United States is estimated to increase 11%.

EXAMPLE 3 Using Percent Increase or Decrease to Find Prices

A **Anthony bought an LCD monitor originally priced at $750 that was reduced in price by 35%. What was the reduced price?**

$750 • 35% — *First find 35% of $750.*

$750 • 0.35 = $262.50 — *35% = 0.35*

The amount of decrease is $262.50.

Think: The reduced price is $262.50 *less than* $750.

$750 − $262.50 — *Subtract the amount of decrease.*

= $487.50

The reduced price of the monitor was $487.50.

B **Mr. Salazar received a shipment of sofas that cost him $366 each. He marks the price of each sofa up $33\frac{1}{3}\%$ to find the *retail price*. What is the retail price of each sofa?**

$\$366 \cdot 33\frac{1}{3}\%$ — *First find $33\frac{1}{3}\%$ of $366.*

$\$366 \cdot \frac{1}{3} = \$122$ — $33\frac{1}{3}\% = \frac{1}{3}$

The amount of increase is $122.

Think: The retail price is $122 *more than* $366.

$366 + $122 = $488 — *Add the amount of increase.*

The retail price of each sofa is $488.

Think and Discuss

1. **Explain** whether a 150% increase or a 150% decrease is possible.
2. **Compare** finding a 20% increase to finding 120% of a number.
3. **Explain** how you could find the percent of change if you knew the U.S. populations in 1990 and 2000.

8-4 Exercises

FOR EOG PRACTICE

see page 674

GUIDED PRACTICE

See Example 1 **Find each percent increase or decrease to the nearest percent.**

1. from 40 to 55
2. from 85 to 30
3. from 75 to 150
4. from 55 to 90
5. from 110 to 82
6. from 82 to 110

See Example 2

7. A population of geese rose from 234 to 460 over a period of two years. What is the percent increase, to the nearest tenth of a percent?

See Example 3

8. An automobile dealer agrees to cut 5% off the $10,288 sticker price of a new car for a customer. What is the price of the car for the customer?

INDEPENDENT PRACTICE

See Example 1 **Find each percent increase or decrease to the nearest percent.**

9. from 55 to 60
10. from 111 to 200
11. from 9 to 5
12. from 800 to 1500
13. from 0.84 to 0.67
14. from 45 to 20

See Example 2

15. The boiling point of water is lower at higher altitudes. Water boils at 212°F at sea level and 193.7°F at 10,000 ft. What is the percent decrease in the temperatures, to the nearest tenth of a percent?

See Example 3

16. Mr. Simmons owns a hardware store and typically marks up merchandise by 28% over warehouse cost. How much would he charge for a hammer that costs him $13.50?

PRACTICE AND PROBLEM SOLVING

Find each percent increase or decrease to the nearest percent.

17. from $49.60 to $38.10
18. from $67 to $104
19. from $575 to $405
20. from $822 to $766
21. from $0.23 to $0.19
22. from $12.50 to $14.75

Find each missing number.

23. originally: $500
new price: ■
20% increase
24. originally: 140
new amount: ■
50% increase
25. originally: ■
new amount: 230
15% increase
26. originally: ■
new price: $4.20
5% decrease
27. originally: 32
new amount: 48
■% increase
28. originally: $65
new price: $52
■% decrease

29. Maria purchased a CD burner for $199. Six months later, the same burner was selling for $119. By what percent had the price decreased, to the nearest percent?

30. ***LIFE SCIENCE*** The *Carcharodon megaladon* shark of the Miocene era is believed to have been about 12 m long. The modern great white shark is about 6 m long. Write the change in length of these longest sharks over time as a percent increase or decrease.

31. A sale ad shows a \$240 winter coat discounted 35%.
    a. How much is the price decrease?
    b. What is the sale price of the coat?
    c. If the coat is reduced in price by an additional $33\frac{1}{3}$%, what will be the new sale price?
    d. What percent decrease does this final sale price represent?

32. Is the percent change the same when a blouse is marked up from \$15 to \$20 as when it is marked down from \$20 to \$15? Explain.

33. ***EARTH SCIENCE*** After the Mount St. Helens volcano erupted in 1980, the elevation of the mountain decreased by about 13.6%. Its elevation had been 9677 ft. What was its elevation after the eruption?

34. ***CHOOSE A STRATEGY*** A printer originally sold for \$199. Six months later, the price was reduced 45%. During a sale, the printer was discounted an additional 20% off the reduced price. What was the final price of the printer?

    A \$17.91 **B** \$87.56 **C** \$101.89 **D** \$98.97

35. ***WRITE ABOUT IT*** Describe how you can use mental math to find the percent increase from 80 to 100 and the percent decrease from 100 to 80.

36. ***CHALLENGE*** During a sale, the price of a computer game was decreased by 40%. By what percent must the sale price be increased to restore the original price?

Spiral Review

Find the surface area of each figure to the nearest tenth. Use 3.14 for π. (Lesson 6-9)

37. a square pyramid with base 13 m by 13 m and slant height 7.5 m
38. a cone with a diameter 90 cm and slant height 125 cm
39. a square pyramid with base length 6 yd and slant height 4 yd
40. **EOG PREP** A 1 lb 8 oz package of corn sells for \$5.76. What is the unit price? (Lesson 7-2)

    A \$0.34 per oz B \$0.32 per oz C \$0.24 per oz D \$0.64 per oz

8-5 Estimating with Percents

Learn to estimate with percents.

Vocabulary

estimate

compatible numbers

Waiters, waitresses, and other restaurant employees depend upon tips for much of their income. Typically, a tip is 15% to 20% of the bill. Tips do not have to be calculated exactly, so estimation is often used. When the sales tax is about 8%, doubling the tax gives a good estimate for a tip.

Some problems require only an **estimate**. Estimates involving percents and fractions can be found by using **compatible numbers**, numbers that go well together because they have common factors.

$\frac{13}{24}$ The numbers 13 and 24 are not compatible numbers.

Change 13 to 12. $\frac{13}{24}$ is nearly equivalent to $\frac{12}{24}$.

$\approx \frac{12}{24}$ 12 and 24 are compatible numbers. 12 is a common factor.

The fraction $\frac{12}{24}$ simplifies to $\frac{1}{2}$. *$\frac{13}{24} \approx \frac{1}{2}$*

EXAMPLE 1 Estimating with Percents

Helpful Hint

Methods of estimating:
1. Use compatible numbers.
2. Round to common percents. (10%, 25%, $33\frac{1}{3}$%)
3. Break percents into smaller parts. (1%, 5%, 10%)

Estimate.

A 26% of 48

Instead of computing the exact answer of 26% · 48, estimate.

$26\% = \frac{26}{100} \approx \frac{25}{100}$ *Use compatible numbers, 25 and 100.*

$\approx \frac{1}{4}$ *Simplify.*

$\frac{1}{4} \cdot 48 = 12$ *Use mental math: 48 ÷ 4.*

So 26% of 48 is about 12.

B 14% of 20

Instead of computing the exact answer of 14% · 20, estimate.

$14\% \approx 15\%$ *Round.*

$\approx 10\% + 5\%$ *Break down the percent into smaller parts.*

$15\% \cdot 20 = (10\% + 5\%) \cdot 20$ *Set up an equation.*

$= 10\% \cdot 20 + 5\% \cdot 20$ *Use Distributive Property.*

$= 2 + 1$ *10% of 20 is 2, so 5% of 20 is 1.*

So 14% of 20 is about 3.

You can calculate compound interest using a formula.

$A = P\left(1 + \frac{r}{k}\right)^{n \cdot k}$, where A = amount (new balance),

P = principal (original amount of account),

r = rate of annual interest,

n = number of years, and

k = number of compounding periods per year.

EXAMPLE 2 Calculating Compound Interest Using a Formula

Use the formula to find the amount after 3 years if $5000 is invested at 3% annual interest that is compounded semiannually.

$A = 5000\left(1 + \frac{0.03}{2}\right)^{3 \cdot 2}$ *Substitute P = 5000, r = 0.03, k = 2, n = 3.*

$A = 5000(1.015)^6$ *Evaluate in the parentheses and the exponent.*

$A = 5000(1.093443264)$ *Evaluate the power. Use a calculator.*

$A = \$5467.22$ *Evaluate the product, and round.*

There would be a total of $5467.22 at the end of 3 years.

EXTENSION Exercises

Use a spreadsheet or calculator to find the value of each investment after 3 years, compounded annually.

1. $10,000 at 8% annual interest
2. $1000 at 6% annual interest

Use the compound interest formula to find the value of each investment after 5 years, compounded semiannually.

3. $10,000 at 8% annual interest
4. $1000 at 6% annual interest

Use the compound interest formula to find the value of the investment.

5. $12,500 at 4% annual interest, compounded annually, for 5 years
6. $800 at $5\frac{1}{2}$% annual interest, compounded semiannually, for 7 years
7. $2000 at 7% annual interest, compounded quarterly, for 3 years
8. Determine the value of a $20,000 inheritance after 20 years if it is invested at a 4% annual rate of interest that is compounded annually, semiannually, and quarterly.
9. Determine the value of a $5000 savings account paying 6% interest, compounded monthly, over a 5-year period, assuming that no additional deposits or withdrawals are made during that time.
10. Explain whether money earns more compounded annually or quarterly.

Problem Solving on Location

NORTH CAROLINA

University of North Carolina

Students planning to attend the University of North Carolina can choose from six different campuses. The main campus is located in Chapel Hill, but the total enrollment for all six campuses is about 74,246.

For 1–2, use the table.

1. What percent of the total enrollment does the enrollment at each of the six campuses represent?

2. Approximately 40% of the students at the Asheville campus are involved in *intramural* (on-campus) sports. The percent of students involved in intramural sports at the other campuses are as follows: Chapel Hill, 50%; Charlotte, 40%; Greensboro, 9%; Pembroke, 10%; and Wilmington, 60%.
   - **a.** Find the number of students involved in intramural sports at each campus.
   - **b.** What percent of the total University of North Carolina enrollment is made up of students involved in intramural sports?

3. If the new freshman class at Chapel Hill has about 3650 students and the campus admits about 37% of the students who apply, about how many freshman applied?

4. About 24,231 University of North Carolina students live on campus. What percent of students live off campus?

University of North Carolina Enrollment

Campus	Enrollment
Asheville	3247
Chapel Hill	24,892
Charlotte	18,916
Greensboro	13,861
Pembroke	3445
Wilmington	9885

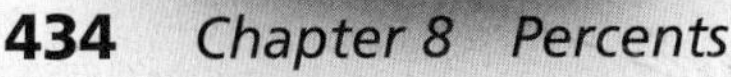

Highlands Cove

Mountain golf courses provide a challenging and engaging alternative to traditional courses. The 18-hole Highlands Cove mountain golf course is located 6 miles east of Highlands, North Carolina, and offers panoramic views of the Blue Ridge Mountains. The course contains several holes with large vertical drops, including one hole that drops 300 feet from the tee to the fairway.

For 1–5, use the circle graph.

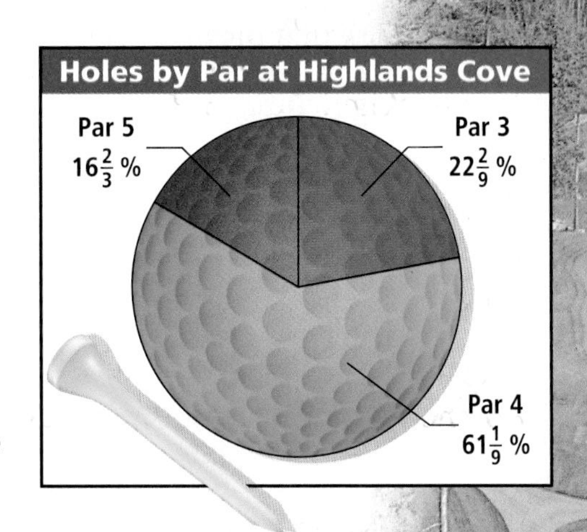

1. *Par* is the estimated number of strokes needed to complete a hole or a course. How many par-3 holes are on the Highlands Cove golf course? par-4 holes? par-5 holes?

2. Find the total par for the 18 holes at the Highlands Cove golf course.

3. If the number of par-3 holes were reduced by 25%, how many par-3 holes would remain on the course?

4. If the number of par-4 holes on the course were changed to 5 holes, what would the approximate percent decrease in the number of par-4 holes be?

5. If the number of par-5 holes on the course were changed to 5 holes, what would the percent increase in the number of par-5 holes be?

MATH-ABLES

Percent Puzzlers

Prove your precision with these perplexing percent puzzlers!

1. A farmer is dividing his sheep among four pens. He puts 20% of the sheep in the first pen, 30% in the second pen, 37.5% in the third pen, and the rest in the fourth pen. What is the smallest number of sheep he could have?

2. Karen and Tina are on the same baseball team. Karen has hit in 35% of her 200 times at bat. Tina has hit in 30% of her 20 times at bat. If Karen hits in 100% of her next five times at bat and Tina hits in 80% of her next five times at bat, who will have the higher percentage of hits?

3. Joe was doing such a great job at work that his boss gave him a 10% raise! Then he made such a huge mistake that his boss gave him a 10% pay cut. What percent of his original salary does Joe make now?

4. Suppose you have 100 pounds of saltwater that is 99% water (by weight) and 1% salt. Some of the water evaporates so that the remaining liquid is 98% water and 2% salt. How much does the remaining liquid weigh?

Percent Tiles

Use cardboard or heavy paper to make 100 tiles with a digit from 0 through 9 (10 of each) on each tile, and print out a set of cards. Each player draws seven tiles. Lay four cards out on the table as shown. The object of the game is to collect as many cards as possible. To collect a card, use numbered tiles to correctly complete the statement on the card.

internet connect go.hrw.com

Go to ***go.hrw.com*** for cards and a complete set of rules.
KEYWORD: MT4 Game8

Compute Compound Interest

Use with Chapter 8 Extension

1.02
internet connect
Lab Resources Online
go.hrw.com
KEYWORD: MT4 TechLab8

The formula for compound interest is $A = P\left(1 + \frac{r}{k}\right)^{nk}$, where A is the final dollar value, P is the initial dollar investment, r is the rate for each interest period, n is the number of interest periods, and k is the number of compounding periods per year.

Activity

1 Use a calculator to find the value after 9 years of $1500 invested in a savings bank that pays 3% interest compounded annually.

The initial investment P is $1500. The rate r is 3% = 0.03. The interest period is one year. The number of interest periods n is 9, and $k = 1$.

$$A = 1500\left(1 + \frac{0.03}{1}\right)^{9 \cdot 1} = 1500(1.03)^9$$

On your graphing calculator, press

1500 [×] [(] 1.03 [)] [^] 9 [ENTER].

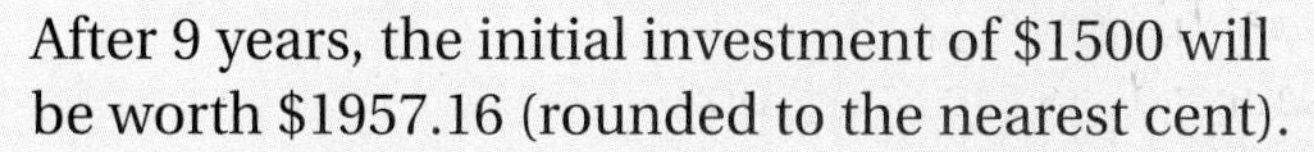

After 9 years, the initial investment of $1500 will be worth $1957.16 (rounded to the nearest cent).

2 Use a calculator to find the value after 9 years of $1500 invested in a savings bank that pays 6% interest compounded semi-annually (twice a year).

The initial investment P is $1500. Since interest is compounded twice a year, there are **18** interest periods in 9 years, and $n = 9$. The interest rate for each period r is 6% divided by 2, or 3% = **0.03**.

$$A = 1500 \times \left(1 + \frac{0.06}{2}\right)^{9 \cdot 2} = 1500 \times (\mathbf{1.03})^{\mathbf{18}}$$

On your calculator, press 1500 [×] [(] 1.03 [)] [^] 18 [ENTER]. You should find that A = $2553.65.

Think and Discuss

1. Compare the value of an initial deposit of $1000 at 6% simple interest for 10 years with the same initial deposit at 6% annual compound interest for 10 years. Which is greater? Why?

Try This

1. Find the value of an initial investment of $2500 for the specified term and interest rate.

a. 8 years, 5% compounded annually

b. 20 years, 5% compounded monthly

Chapter 8

Study Guide and Review

Vocabulary

Complete the sentences below with vocabulary words from the list above. Words may be used more than once.

1. A ratio that compares a number to 100 is called a(n) ___?___.
2. The ratio $\frac{\text{amount of change}}{\text{original amount}}$ is called the ___?___.
3. Percent is used to calculate ___?___, a fee paid to a person who makes a sale.
4. The formula $I = Prt$ is used to calculate ___?___. In the formula, P represents the amount borrowed or invested, which is called the ___?___, r is the ___?___, and t is the period of time that the money is borrowed or invested.

8-1 Relating Decimals, Fractions, and Percents (pp. 400–403)

EXAMPLE

■ Complete the table.

Fraction	Decimal	Percent
$\frac{3}{4}$	0.75	0.75(100) = 75%
$\frac{625}{1000} = \frac{5}{8}$	0.625	0.625(100) = 62.5%
$\frac{80}{100} = \frac{4}{5}$	$\frac{80}{100} = 0.80$	80%

EXERCISES

Complete the table.

Fraction	Decimal	Percent
$\frac{7}{16}$	5. ■	6. ■
7. ■	1.125	8. ■
9. ■	10. ■	70%
11. ■	0.004	12. ■

8-2 Finding Percents (pp. 405–408)

EXAMPLE

- **A raw apple weighing 5.3 oz contains about 4.45 oz of water. What percent of an apple is water?**

$\frac{\text{number}}{100} = \frac{\text{part}}{\text{whole}}$ *Set up a proportion.*

$\frac{n}{100} = \frac{4.45}{5.3}$ *Substitute.*

$5.3n = 445$ *Cross multiply.*

$n = \frac{445}{5.3} \approx 83.96 \approx 84\%$

An apple is about 84% water.

EXERCISES

13. The length of a year on Mercury is about 88 Earth days. The length of a year on Venus is about 225 Earth days. About what percent of the length of Venus's year is Mercury's year?
14. The main span of the Brooklyn Bridge is 1595 feet long. The Golden Gate Bridge is about 263% the length of the Brooklyn Bridge. To the nearest hundred feet, how long is the Golden Gate Bridge?

8-3 Finding a Number When the Percent Is Known (pp. 410–413)

EXAMPLE

- **The population of Fairbanks, Alaska, is 30,224. This is about 477% of the population of Kodiak, Alaska. To the nearest ten people, find the population of Kodiak.**

$\frac{477}{100} = \frac{30{,}224}{n}$ *Set up a proportion.*

$477n = 3{,}022{,}400$ *Cross multiply.*

$n = \frac{3{,}022{,}400}{477} \approx 6336.2683 \approx 6340$

The population of Kodiak is about 6340.

EXERCISES

15. The diameter at the equator of the planet Jupiter is 88,846 miles. This is about 2930% of the diameter of Mercury at its equator. To the nearest ten miles, find the diameter of Mercury at its equator.
16. At the age of 12 weeks, Rachel weighed 8 lb 2 oz. Her birth weight was about $66\frac{2}{3}\%$ of her 12-week weight. To the nearest ounce, what was her birth weight?

8-4 Percent Increase and Decrease (pp. 416–419)

EXAMPLE

- **In 1990, there were 639,270 robberies reported in the United States. This number decreased to 409,670 in 1999. What was the percent decrease?**

$639{,}270 - 409{,}670 = 229{,}600$ *Amount of decrease*

$\frac{\text{amount of decrease}}{\text{original amount}} = \frac{229{,}600}{639{,}270}$

$\approx 0.3592 \approx 35.92\%$

From 1990 to 1999, the number of reported robberies in the United States decreased by 35.92%.

EXERCISES

17. On sale, a shirt was reduced from $20 to $16. Find the percent decrease.
18. In 1900, the U.S. public debt was $1.2 billion dollars. This number increased to $5674.2 billion dollars in the year 2000. Find the percent increase.
19. At the beginning of a 10-week medically supervised diet, Ken weighed 202 lb. After the diet, Ken weighed 177 lb. Find the percent decrease.

8-5 Estimating with Percents (pp. 420–423)

EXAMPLE

- **Estimate the percent that 5 is of 17.**

$\frac{5}{17} \approx \frac{5}{15}$ *Use compatible numbers.*

$\frac{1}{3} = 33\frac{1}{3}\%$ *Simplify; change to %.*

So 5 is about $33\frac{1}{3}\%$ of 17.

EXERCISES

Use compatible numbers to estimate.

20. the percent that 6 is of 25

21. the percent that 7 is of 33

22. 23% of 64

23. 78% of 19

24. 14% of 40

25. 16% of 30

8-6 Applications of Percents (pp. 424–427)

EXAMPLE

- **As an appliance salesman, Jim earns a base pay of $450 per week plus an 8% commission on his weekly sales. Last week, his sales totaled $2750. How much did he earn for the week?**

Find the amount of commission.

$8\% \cdot \$2750 = 0.08 \cdot \$2750 = \$220$

Add the commission amount to his base pay.

$\$220 + \$450 = \$670$

Last week Jim earned $670.

EXERCISES

26. As a real-estate agent, Hal earns $3\frac{1}{2}\%$ commission on the houses he sells. In the first quarter of this year, he sold two houses, one for $125,000 and the other for $189,000. How much was Hal's commission for this quarter?

27. If the sales tax is $6\frac{3}{4}\%$, how much tax would Raymond pay if he bought a radio for $19.99 and a camera for $24.99?

8-7 More Applications of Percents (pp. 428–431)

EXAMPLE

- **For home improvements, the Walters borrowed $10,000 for 3 years at simple interest. They repaid a total of $11,050. What was the interest rate of the loan?**

Find the amount of interest.

$P + I = A$

$10{,}000 + I = 11{,}050$

$I = 11{,}050 - 10{,}000 = 1050$

Substitute into the simple interest formula.

$I = P \cdot r \cdot t$

$1050 = 10{,}000 \cdot r \cdot 3$

$1050 = 30{,}000r$

$\frac{1050}{30{,}000} = r$

$0.035 = r$

The interest rate of the loan was 3.5%.

EXERCISES

Using the simple interest formula, find the missing number.

28. interest = ▒; principal = $12,500; rate = $5\frac{3}{4}\%$ per year; time = $2\frac{1}{2}$ years

29. interest = $90; principal = ▒; rate = 3% per year; time = 6 years

30. interest = $367.50; principal = $1500; rate per year = ▒; time = $3\frac{1}{2}$ years

31. interest = $1237.50; principal = $45,000; rate = $5\frac{1}{2}\%$ per year; time = ▒

Which simple-interest loan would cost the borrower less? How much less?

32. $2000 at 4% for 3 years or $2000 at 4.75% for 2 years

Chapter Test

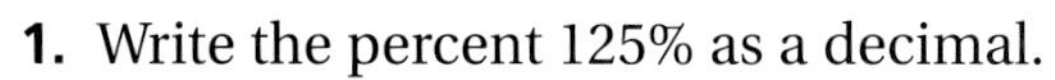

1. Write the percent 125% as a decimal.
2. Write the fraction $\frac{7}{20}$ as a percent.
3. Write the decimal 0.0375 as a percent.
4. Write the percent $87\frac{1}{2}\%$ as a fraction in simplest form.

Calculate.

5. What percent of 72 is 9?
6. What is 25% of 48?
7. 15.9 is $33\frac{1}{3}\%$ of what number?
8. What percent of 19 is 61.75?

Use compatible numbers to estimate.

9. the percent that 7 is of 23
10. the percent that 110 is of 48
11. 83% of 197

Using the simple interest formula, find the missing number.

12. interest = ■; principal = $15,500; rate = $4\frac{1}{2}\%$ per year; time = 3 years
13. interest = $87.50; principal = ■; rate = $3\frac{1}{2}\%$ per year; time = 6 months
14. interest = $401.63; principal = $2550; rate per year = ■; time = $3\frac{1}{2}$ years
15. interest = $562.50; principal = $20,000; rate = $3\frac{3}{4}\%$ per year, time = ■

Solve each problem. Give percents to the nearest hundredth.

16. The mean distance of Earth from the Sun is 92,960,000 miles. This is about 258% of the mean distance of Mercury from the Sun. To the nearest ten million miles, what is Mercury's mean distance from the Sun?
17. In 2000, U.S. trade with Saudi Arabia totaled $20.6 billion. Of this total, $6.2 billion were U.S. exports to Saudi Arabia. What percent of its total trade with Saudi Arabia were U.S. exports?
18. In the third quarter of 2001, the median sale price for a single-family home in Putnam County, New York, was $259,970. A year earlier, the price was $242,555. What was the percent increase?
19. As a real-estate agent, Walter Jordan earns $3\frac{3}{4}\%$ commission on the houses he sells. In the last quarter of this year, he sold two houses, one for $225,000 and the other for $199,000. How much was Walter Jordan's commission for this quarter?
20. If the sales tax rate is $7\frac{1}{4}\%$ and Jessica paid $1.45 in sales tax for a sweater, what was the price of the sweater?
21. Determine the amount of simple interest on a $1250 loan at $6\frac{1}{2}\%$ simple interest for 3 years.

Chapter 8

Performance Assessment

Show What You Know

Create a portfolio of your work from this chapter. Complete this page and include it with your four best pieces of work from Chapter 8. Choose from your homework or lab assignments, mid-chapter quiz, or any journal entries you have done. Put them together using any design you want. Make your portfolio represent what you consider your best work.

Short Response

1. If 10 kg of pure acid are added to 15 kg of pure water, what percent of the resulting solution is acid? Show your work.
2. In the chemistry laboratory, Jim is working with six large jars of capacities 5 L, 4 L, 3 L, 2 L, 1 L, and 10 L. The 5 L jar is filled with an acid mix, and the rest of the jars are empty. Jim uses the 5 L jar to fill the 4 L jar and pours the excess into the 10 L jar. Then he uses the 4 L jar to fill the 3 L jar and pours the excess into the 10 L jar. He repeats the process until all but the 1 L and 10 L jars are empty. What percent of the 10 L jar is filled? Show your work.

Extended Problem Solving

3. The 60 students in a physical education class were asked to choose an elective. The results of the selection are shown in the table. Make a circle graph to display the results as percents. When making your graph, use a protractor to draw the angles at the center of the circle for each sector of the circle.
   a. Which two groups make up 50% of the graph?
   b. Could you make all the activities have an equal percent of participation? Explain.

Activity	Number of Students
Badminton	15
Basketball	12
Gymnastics	6
Volleyball	15
Wrestling	12

internet connect
EOG Test Practice Online
go.hrw.com Keyword: MT4 TestPrep

Getting Ready for EOG

Cumulative Assessment, Chapters 1–8

1. A club with 30 girls and 40 boys sponsored a boat ride. If 60% of the girls and 25% of the boys went on the boat ride, what percent of the club went on the ride?

   A 30% C 40%
   B 35% D 60%

2. For every 1000 m^3 of air that goes through the filtering system in Ken's bedroom, 0.05 g of dust is removed. How many grams of dust are removed when 10^7 m^3 of air are filtered?

   A 50,000 g C 500 g
   B 5,000 g D 50 g

3. Let operation ◆ be defined as $x \blacklozenge y = x^y$. What is the value of $3 \blacklozenge (-1)$?

   A -3 C -1
   B $-\frac{1}{3}$ D $\frac{1}{3}$

4. If $\frac{1}{2}$ of a number is 2 more than $\frac{1}{3}$ of the number, what is the number?

   A 6 C 20
   B 12 D 24

5. When a certain rectangle is divided in half, two squares are formed, each with perimeter 48 inches. What is the perimeter of the original rectangle?

   A 24 inches C 48 inches
   B 36 inches D 72 inches

6. Consider this equation: $\frac{20}{x} = \frac{4}{x-5}$. Which of the following is equivalent to the given equation?

   A $x(x - 5) = 80$
   B $20x = 4(x - 5)$
   C $20(x - 5) = 4x$
   D $24 = x + (x - 5)$

7. A 1000-ton load is increased by 1%. What is the weight of the adjusted load?

   A 1001 tons C 1100 tons
   B 1010 tons D 1110 tons

TEST TAKING TIP!

Using diagrams: Remember that no inferences are to be taken from a diagram that are not stated as given.

8. In the figure, what is y in terms of x?

   A $90 + x$
   B $90 + 2x$
   C $180 - x$
   D $180 - 2x$

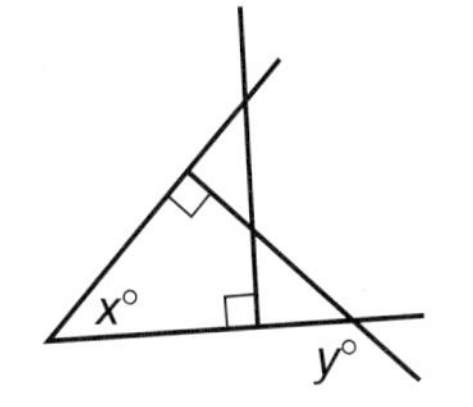

9. ***SHORT RESPONSE*** If $(x + 3)(9 - 5) = 16$, what is the value of x? Show your work or explain in words how you determined your answer.

10. ***SHORT RESPONSE*** The table shows the test scores of four students.

	Test 1	Test 2	Test 3	Test 4
Ann	80	100	100	90
Dan	60	90	90	100
Juan	100	80	100	60
Leon	100	100	100	65

Find the students' mean scores. What is the mode of the mean scores?

Chapter 9

Probability

TABLE OF CONTENTS

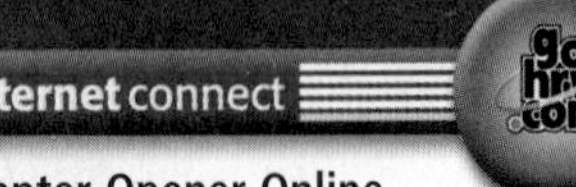

internet connect

Chapter Opener Online
go.hrw.com
KEYWORD: MT4 Ch9

Letter	Code
A	1000001
E	1000101
H	1001000
I	1001001
L	1001100
M	1001101
O	1001111
T	1010100
V	1010110

Career *Cryptographer*

1001001100110010011111010110
10001011001101100000110101001001000

Is this pattern of zeros and ones some kind of message or secret code? A cryptographer could find out. Cryptographers create and break codes by assigning number values to letters of the alphabet.

Almost all text sent over the Internet is encrypted to ensure security for the sender. Codes made up of zeros and ones, or *binary codes*, are frequently used in computer applications.

Use the table to break the code above.

ARE YOU READY?

Choose the best term from the list to complete each sentence.

fraction
percent
range
ratio

1. The term __?__ means "per hundred."
2. A __?__ is a comparison of two numbers.
3. In a set of data, the __?__ is the largest number minus the smallest number.
4. A __?__ is in simplest form when its numerator and denominator have no common factors other than 1.

Complete these exercises to review skills you will need for this chapter.

✔ Simplify Ratios

Write each ratio in simplest form.

5. 5:50 **6.** 95 to 19 **7.** $\frac{20}{100}$ **8.** $\frac{192}{80}$

✔ Write Fractions as Decimals

Express each fraction as a decimal.

9. $\frac{52}{100}$ **10.** $\frac{7}{1000}$ **11.** $\frac{3}{5}$ **12.** $\frac{2}{9}$

✔ Write Fractions as Percents

Express each fraction as a percent.

13. $\frac{19}{100}$ **14.** $\frac{1}{8}$ **15.** $\frac{5}{2}$ **16.** $\frac{2}{3}$

17. $\frac{3}{4}$ **18.** $\frac{9}{20}$ **19.** $\frac{7}{10}$ **20.** $\frac{2}{5}$

✔ Operations with Fractions

Add. Write each answer in simplest form.

21. $\frac{3}{8} + \frac{1}{4} + \frac{1}{6}$ **22.** $\frac{1}{6} + \frac{2}{3} + \frac{1}{9}$ **23.** $\frac{1}{8} + \frac{1}{4} + \frac{1}{8} + \frac{1}{2}$ **24.** $\frac{1}{3} + \frac{1}{4} + \frac{2}{5}$

Multiply. Write each answer in simplest form.

25. $\frac{3}{8} \cdot \frac{1}{5}$ **26.** $\frac{2}{3} \cdot \frac{6}{7}$ **27.** $\frac{3}{7} \cdot \frac{14}{27}$ **28.** $\frac{13}{52} \cdot \frac{3}{51}$

29. $\frac{4}{5} \cdot \frac{11}{4}$ **30.** $\frac{5}{2} \cdot \frac{3}{4}$ **31.** $\frac{27}{8} \cdot \frac{4}{9}$ **32.** $\frac{1}{15} \cdot \frac{30}{9}$

9-1 Probability

Learn to find the probability of an event by using the definition of probability.

Vocabulary

experiment
trial
outcome
sample space
event
probability
impossible
certain

An **experiment** is an activity in which results are observed. Each observation is called a **trial**, and each result is called an **outcome**. The **sample space** is the set of all possible outcomes of an experiment.

Experiment	Sample space
• flipping a coin	• heads, tails
• rolling a number cube	• 1, 2, 3, 4, 5, 6
• guessing the number of jelly beans in a jar	• whole numbers

An **event** is any set of one or more outcomes. The **probability** of an event, written P(event), is a number from 0 (or 0%) to 1 (or 100%) that tells you how likely the event is to happen.

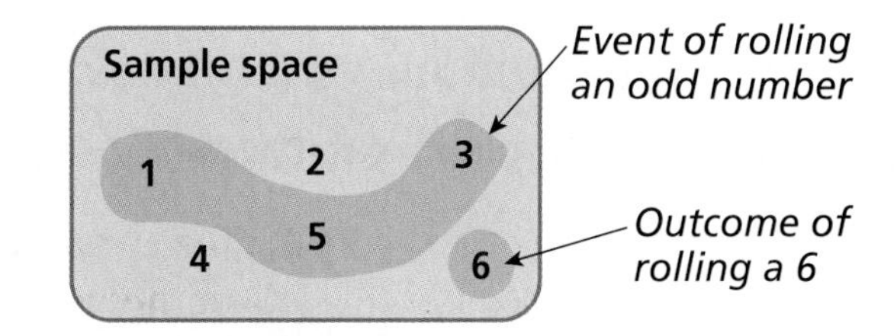

- A probability of 0 means the event is **impossible**, or can never happen.
- A probability of 1 means the event is **certain**, or has to happen.
- The probabilities of all the outcomes in the sample space add up to 1.

Never happens		Happens about half the time		Always happens
0	$\frac{1}{4}$	$\frac{1}{2}$	$\frac{3}{4}$	1
0	0.25	0.5	0.75	1
0%	25%	50%	75%	100%

EXAMPLE 1 Finding Probabilities of Outcomes in a Sample Space

Give the probability for each outcome.

A **The weather forecast shows a 40% chance of rain.**

Outcome	Rain	No rain
Probability		

The probability of rain is $P(\text{rain}) = 40\% = 0.4$. The probabilities must add to 1, so the probability of no rain is $P(\text{no rain}) = 1 - 0.4 = 0.6$, or 60%.

Give the probability for each outcome.

B

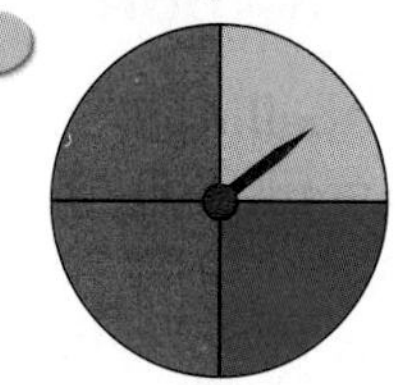

Outcome	Red	Yellow	Blue
Probability	■	■	■

Half of the spinner is red, so a reasonable estimate of the probability that the spinner lands on red is $P(\text{red}) = \frac{1}{2}$.
One-fourth of the spinner is yellow, so a reasonable estimate of the probability that the spinner lands on yellow is $P(\text{yellow}) = \frac{1}{4}$.
One-fourth of the spinner is blue, so a reasonable estimate of the probability that the spinner lands on blue is $P(\text{blue}) = \frac{1}{4}$.

Check The probabilities of all the outcomes must add to 1.

$\frac{1}{2} + \frac{1}{4} + \frac{1}{4} = 1$ ✓

To find the probability of an event, add the probabilities of all the outcomes included in the event.

EXAMPLE 2 Finding Probabilities of Events

A quiz contains 5 multiple-choice questions. Suppose you guess randomly on every question. The table below gives the probability of each score.

Score	0	1	2	3	4	5
Probability	0.237	0.396	0.264	0.088	0.014	0.001

A What is the probability of guessing one or more correct?

The event "one or more correct" consists of the outcomes 1, 2, 3, 4, 5.

$P(\text{one or more correct}) = 0.396 + 0.264 + 0.088 + 0.014 + 0.001$
$= 0.763$, or 76.3%

B What is the probability of guessing fewer than 2 correct?

The event "fewer than 2 correct" consists of the outcomes 0 and 1.

$P(\text{fewer than 2 correct}) = 0.237 + 0.396$
$= 0.633$, or 63.3%

C What is the probability of passing the quiz (getting 4 or 5 correct) by guessing?

The event "passing the quiz" consists of the outcomes 4 and 5.

$P(\text{passing the quiz}) = 0.014 + 0.001$
$= 0.015$, or 1.5%

EXAMPLE 3 PROBLEM SOLVING APPLICATION

Six students remain in a spelling bee. Amy's probability of winning is $\frac{1}{3}$. Amy is twice as likely to win as Kim. Bob has the same chance as Kim. Pat, Ani, and Jo all have the same chance of winning. Create a table of probabilities for the sample space.

1 Understand the Problem

The **answer** will be a table of probabilities. Each probability will be a number from 0 to 1. The probabilities of all outcomes add to 1.

List the **important information:**

- $P(\text{Amy}) = \frac{1}{3}$
- $P(\text{Kim}) = \frac{1}{2} \cdot P(\text{Amy}) = \frac{1}{2} \cdot \frac{1}{3} = \frac{1}{6}$
- $P(\text{Kim}) = P(\text{Bob}) = \frac{1}{6}$
- $P(\text{Pat}) = P(\text{Ani}) = P(\text{Jo})$

2 Make a Plan

You know the probabilities add to 1, so use the strategy **write an equation**. Let p represent the probability for Pat, Ani, and Jo.

$$P(\text{Amy}) + P(\text{Kim}) + P(\text{Bob}) + P(\text{Pat}) + P(\text{Ani}) + P(\text{Jo}) = 1$$

$$\frac{1}{3} + \frac{1}{6} + \frac{1}{6} + p + p + p = 1$$

$$\frac{2}{3} + 3p = 1$$

3 Solve

$$\frac{2}{3} + 3p = 1$$

$$\underline{-\frac{2}{3}} \qquad \underline{-\frac{2}{3}}$$ *Subtract $\frac{2}{3}$ from both sides.*

$$3p = \frac{1}{3}$$

$$\frac{1}{3} \cdot 3p = \frac{1}{3} \cdot \frac{1}{3}$$ *Multiply both sides by $\frac{1}{3}$.*

$$p = \frac{1}{9}$$

Outcome	Amy	Kim	Bob	Pat	Ani	Jo
Probability	$\frac{1}{3}$	$\frac{1}{6}$	$\frac{1}{6}$	$\frac{1}{9}$	$\frac{1}{9}$	$\frac{1}{9}$

4 Look Back

Check that the probabilities add to 1.

$$\frac{1}{3} + \frac{1}{6} + \frac{1}{6} + \frac{1}{9} + \frac{1}{9} + \frac{1}{9} = 1 \checkmark$$

Think and Discuss

1. **Give** a probability for each of the following: usually, sometimes, always, never. Compare your values with the rest of your class.
2. **Explain** the difference between an outcome and an event.

9-1 Exercises

FOR EOG PRACTICE

see page 676

internet connect

Homework Help Online
go.hrw.com Keyword: MT4 9-1

GUIDED PRACTICE

See Example 

1. The weather forecast calls for a 55% chance of snow. Give the probability for each outcome.

Outcome	Snow	No snow
Probability		

See Example 2

An experiment consists of drawing 4 marbles from a bag and counting the number of blue marbles. The table gives the probability of each outcome.

Number of Blue Marbles	0	1	2	3	4
Probability	0.024	0.238	0.476	0.238	0.024

2. What is the probability of drawing at least 3 blue marbles?
3. What is the probability of drawing fewer than 3 blue marbles?

See Example 

4. There are 4 teams in a school tournament. Team A has a 25% chance of winning. Team B has the same chance as Team D. Team C has half the chance of winning as Team B. Create a table of probabilities for the sample space.

INDEPENDENT PRACTICE

See Example 

5. Give the probability for each outcome.

Outcome	Red	Blue	Yellow	Green
Probability				

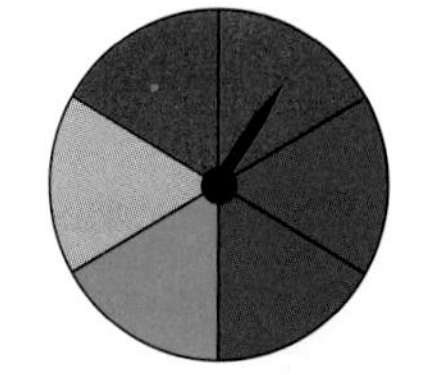

See Example 

Raul needs 3 more classes to graduate from college. He registers late, so he may not get all the classes he needs. The table gives the probabilities for the number of courses he will be able to register for.

Number of Classes Available	0	1	2	3
Probability	0.015	0.140	0.505	0.340

6. What is the probability that at least 1 of the classes will be available?
7. What is the probability that fewer than 2 of the classes will be available?

See Example 

8. There are 5 candidates for class president. Makyla and Jacob have the same chance of winning. Daniel has a 20% chance of winning, and Samantha and Maria are both half as likely to win as Daniel. Create a table of probabilities for the sample space.

PRACTICE AND PROBLEM SOLVING

Use the table to find the probability of each event.

Outcome	A	B	C	D	E
Probability	0.204	0.115	0	0.535	0.146

9. A, B, or C occurring

10. A or E occurring

11. A, B, D, or E occurring

12. C not occurring

13. D not occurring

14. C or D occurring

15. Jamal has a 10% chance of winning a contest, Elroy has the same chance as Tina and Mel, and Gina is three times as likely as Jamal to win. Create a table of probabilities for the sample space.

16. ***BUSINESS*** Community planners have decided that a new strip mall has a 32% chance of being built in Zone A, 20% in Zone B, and 48% in Zone C. What is the probability that it will not be built in Zone C?

17. ***ENTERTAINMENT*** Contestants in a festival game have a 2% chance of winning \$5, a 7% chance of winning \$1, a 15% chance of winning \$0.50, and a 20% chance of winning \$0.25. What is the probability of not winning anything?

18. ***WHAT'S THE ERROR?*** Two people are playing a game. One of them says, "Either I will win or you will. The sample space contains two outcomes, so we each have a probability of one-half." What is the error?

19. ***WRITE ABOUT IT*** Suppose an event has a probability of p. What can you say about the value of p? What is the probability that the event will not occur? Explain.

20. ***CHALLENGE*** List all possible events in the sample space with outcomes A, B, and C.

Spiral Review

Find the surface area of each figure. Use 3.14 for π. (Lesson 6-8)

21. a rectangular prism with base 4 in. by 3 in. and height 2.5 in.

22. a cylinder with radius 10 cm and height 7 cm

23. a cylinder with diameter 7.5 yd and height 11.3 yd

24. a cube with side length 3.2 ft

25. **EOG PREP** The surface area of a sphere is 50.24 cm^2. What is its diameter? Use 3.14 for π. (Lesson 6-10)

A 2 cm   B 4 cm   C 1 cm   D 2.5 cm

9-2 Experimental Probability

Learn to estimate probability using experimental methods.

Vocabulary

experimental probability

Car insurance rates are typically lower for teenage girls than for teenage boys. Rates for adults over age 25 are lower than for adults under age 25. Rates may also be lower if you are married, a nonsmoker, or a student with good grades. Insurance companies estimate the probability that you will have an accident by studying accident rates for different groups of people.

In **experimental probability**, the likelihood of an event is estimated by repeating an experiment many times and observing the number of times the event happens. That number is divided by the total number of trials. The more the experiment is repeated, the more accurate the estimate is likely to be.

$$\text{probability} \approx \frac{\text{number of times the event occurs}}{\text{total number of trials}}$$

EXAMPLE 1 Estimating the Probability of an Event

A **After 1000 spins of the spinner, the following information was recorded. Estimate the probability of the spinner landing on red.**

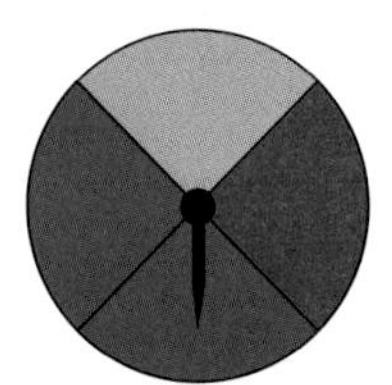

Outcome	Blue	Red	Yellow
Spins	448	267	285

$$\text{probability} \approx \frac{\text{number of spins that landed on red}}{\text{total number of spins}} = \frac{267}{1000} = 0.267$$

The probability of landing on red is about 0.267, or 26.7%.

B **A marble is randomly drawn out of a bag and then replaced. The table shows the results after 100 draws. Estimate the probability of drawing a yellow marble.**

Outcome	Green	Red	Yellow	Blue	White
Draws	30	18	18	21	13

$$\text{probability} \approx \frac{\text{number of yellow marbles drawn}}{\text{total number of draws}} = \frac{18}{100} = 0.18$$

The probability of drawing a yellow marble is about 0.18 or 18%.

C **A researcher has been observing cars passing through an intersection where there is heavy traffic. Of the last 50 cars, 21 turned left, 15 turned right, and 14 went straight. Estimate the probability that a car will turn right.**

Outcome	Left turn	Right turn	Straight
Observations	21	15	14

$$\text{probability} \approx \frac{\text{number of right turns}}{\text{total number of cars}} = \frac{15}{50} = 0.30 = 30\%$$

The probability that a car will turn right is about 0.30 or 30%.

EXAMPLE 2 *Safety Application*

Use the table to compare the probabilities of being involved in an accident for a driver between ages 16 and 25 and a driver between ages 26 and 35.

Auto Accidents in Ohio, 1999		
Age	Number of Licensed Drivers Involved in Accidents	Total Number of Licensed Drivers
16–25	186,026	1,354,729
26–35	133,451	1,584,345
36–45	124,347	1,779,620
46–55	84,715	1,480,101
56–65	45,525	731,118
66–75	29,527	724,530
76 and over	17,820	499,167

Source: Ohio Department of Public Safety

$$\text{probability} \approx \frac{\text{number of licensed drivers in accidents}}{\text{total number of licensed drivers}}$$

probability for a driver between 16 and 25 $\approx \frac{186{,}026}{1{,}354{,}729} \approx 0.137$

probability for a driver between 26 and 35 $\approx \frac{133{,}451}{1{,}584{,}345} \approx 0.084$

A driver between ages 16 and 25 is more likely to be in an accident than a driver between ages 26 and 35.

Think and Discuss

1. **Compare** the probability in Example 1A of the spinner landing on red to what you think the probability should be.
2. **Give** a possible number of marbles of each color in the bag in Example 1B.

9-2 Exercises

FOR EOG PRACTICE

see page 677

internet connect
Homework Help Online
go.hrw.com Keyword: MT4 9-2

GUIDED PRACTICE

See Example 

1. A game spinner was spun 500 times. It was found that A was spun 170 times, B was spun 244 times, and C was spun 86 times. Estimate the probability of the spinner landing on A.

2. A coin was randomly drawn from a bag and then replaced. After 200 draws, it was found that 22 pennies, 53 nickels, 87 dimes, and 38 quarters had been drawn. Estimate the probability of drawing a penny.

See Example 

3. Use the table to compare the probability that a person is listening to talk radio to the probability that the person is listening to a rock station.

Favorite Station	Number of Listeners
Talk/News	12,115
Rock	18,230
Country	11,455
Other	23,160

INDEPENDENT PRACTICE

See Example 

4. A researcher polled 230 freshmen at a university and found that 110 of them were enrolled in a history class. Estimate the probability that a randomly selected freshman is enrolled in a history class.

5. Tyler has made 65 out of his last 150 free throw attempts. Estimate the probability that he will make his next free throw.

See Example 

6. Ed polled 128 students about their favorite hobbies. Use the table to compare the probability that a student's favorite hobby is sports to the probability that it is reading.

Favorite Hobby	Number of Students
Movies	36
Sports	32
Reading	32
Video games	28

PRACTICE AND PROBLEM SOLVING

Use the table for Exercises 7–11. Estimate the probability of each event.

7. A batter hits a single.

8. A batter hits a double.

9. A batter hits a triple.

10. A batter hits a home run.

11. A batter makes an out.

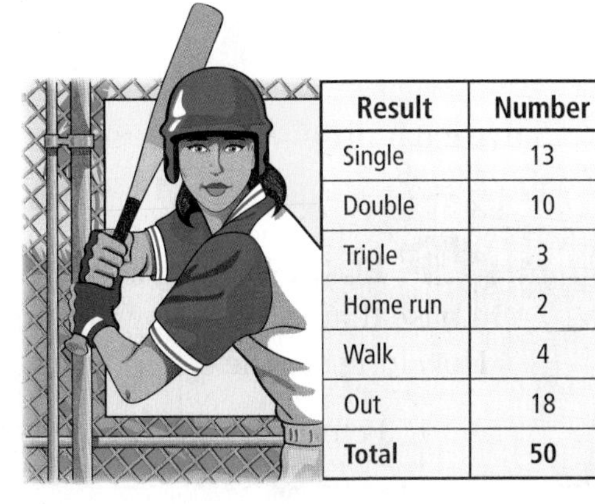

Result	Number
Single	13
Double	10
Triple	3
Home run	2
Walk	4
Out	18
Total	**50**

Earth Science LINK

The strength of an earthquake is measured on the Richter scale. A *major* earthquake measures between 7 and 7.9 on the Richter scale, and a *great* earthquake measures 8 or higher. The table shows the number of major and great earthquakes per year worldwide from 1970 to 1995.

12. Estimate the probability that there will be more than 15 major earthquakes next year.

13. Estimate the probability that there will be fewer than 12 major earthquakes next year.

14. Estimate the probability that there will be no great earthquakes next year.

15. ***WRITE ABOUT IT*** Suppose you want to know the probability that there will be more than five earthquakes next year in a certain country. What would you need to know, and how would you estimate the probability?

16. ***CHALLENGE*** Estimate the probability that there will be more than one major earthquake in the next month.

Number of Earthquakes Worldwide

Year	Major	Great	Year	Major	Great
1970	20	0	1983	14	0
1971	19	1	1984	8	0
1972	15	0	1985	13	1
1973	13	0	1986	5	1
1974	14	0	1987	11	0
1975	14	1	1988	8	0
1976	15	2	1989	6	1
1977	11	2	1990	12	0
1978	16	1	1991	11	0
1979	13	0	1992	23	0
1980	13	1	1993	15	1
1981	13	0	1994	13	2
1982	10	1	1995	22	3

go.hrw.com
KEYWORD: MT4 Quake
CNN Student News

Spiral Review

Solve each proportion. (Lesson 7-4)

17. $\frac{x}{3} = \frac{8}{12}$ **18.** $\frac{7}{y} = \frac{49}{98}$ **19.** $\frac{10}{12} = \frac{b}{6}$ **20.** $\frac{12}{36} = \frac{4}{c}$

21. **EOG PREP** An isosceles triangle has two sides that are 4.5 cm long and a base that is 3 cm long. A similar triangle has a base that is 1.5 cm long. How long are the other two legs of the similar triangle? (Lesson 7-6)

A 150 cm B 3.75 cm C 4.5 cm D 2.25 cm

Technology LAB 9A

Use with Lesson 9-3

Generate Random Numbers

A spreadsheet can be used to generate random decimal numbers that are greater than or equal to 0 but less than 1. By using formulas, you can shift these numbers into a useful range.

Activity

1 Use a spreadsheet to generate five random decimal numbers that are between 0 and 1. Then convert these numbers to integers from 1 to 10.

	A
1	0.063515
2	

a. Type =**RAND()** into cell A1 and press **ENTER**. A random decimal number appears.

	A
1	0.20589
2	0.837083
3	0.445334
4	0.939134
5	0.993354
6	

b. Click to highlight cell A1. Go to the **Edit** menu and **Copy** the contents of A1. Then click and drag to highlight cells A2 through A5. Go to the **Edit** menu and use **Paste** to fill cells A2 through A5.

Notice that the random number in cell A1 changed when you filled the other cells.

RAND() gives a decimal number greater than or equal to 0, but less than 1. To generate random integers from 1 to 10, you need to do the following:

- Multiply **RAND()** by 10 (to give a number greater than or equal to 0 but less than 10).
- Use the **INT** function to drop the decimal part of the result (to give an integer from 0 to 9).
- Add 1 (to give an integer from 1 to 10).

c. Change the formula in A1 to =**INT(10*RAND()) + 1** and press **ENTER.** Repeat the process in part **b** to fill cells A2 through A5.

A2 = =INT(10*RAND()) + 1

	A	B	C	D
1	9			
2	1			
3	7			
4	7			
5	6			
6				

The formula =**INT(10*RAND()) + 1** generates random integers from 1 to 10.

Think and Discuss

1. Explain how **INT(10*RAND()) + 1** generates random integers from 1 to 10.

Try This

1. Use a spreadsheet to simulate three rolls of a number cube.

Chapter 9

Mid-Chapter Quiz

LESSON 9-1 (pp. 446–450)

Use the table of probabilities for the sample space to find the probability of each event.

Outcome	A	B	C	D
Probability	0.4	0.3	0.2	0.1

1. *P*(D)
2. *P*(not C)
3. *P*(A or B)
4. There are 4 students in a race. Jennifer has a 30% chance of winning. Anjelica has the same chance as Jennifer. Debra and Yolanda have equal chances. Create a table of probabilities for the sample space.

LESSON 9-2 (pp. 451–454)

An experiment consists of drawing a marble from a bag and putting it back. The experiment is repeated 100 times, with the following results.

Outcome	Red	Green	Blue	Yellow
Draws	23	18	47	12

5. Estimate the probability of each outcome. Create a table of probabilities for the sample space.
6. Estimate *P*(red or blue).
7. Estimate *P*(not green).

LESSON 9-3 (pp. 456–459)

Use the table of random numbers to simulate each situation. Use at least 10 trials for each simulation.

93840	03363	31168	57602	19464	52245	98744	61040
68395	76832	56386	45060	57512	38816	51623	23252
16805	92120	74443	49176	49898	62042	65847	15380
85178	78842	16598	28335	84837	76406	53436	45043

8. At a local school, 68% of the tenth grade students are studying geometry. Estimate the probability that at least 6 out of 8 randomly selected tenth grade students are studying geometry.
9. Kayla has a package of 100 multicolored beads that contains 15 purple beads. If she randomly selects 8 beads to make a friendship bracelet, estimate the probability that she will get more than 1 purple bead.

Focus on Problem Solving

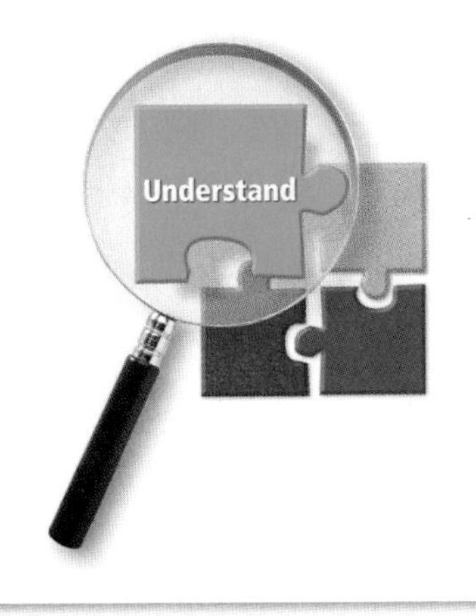

Understand the Problem

- **Understand the words in the problem**

Words that you don't understand can make a simple problem seem difficult. Before you try to solve a problem, you will need to know the meaning of the words in it.

If a problem gives a name of a person, place, or thing that is difficult to understand, such as *Eulalia*, you can use another name or a pronoun in its place. You could replace *Eulalia* with *she*.

Read the problems so that you can hear yourself saying the words.

Copy each problem, and circle any words that you do not understand. Look up each word and write its definition, or use context clues to replace the word with a similar word that is easier to understand.

1. A point in the circle is chosen randomly. What is the probability that the point is in the inscribed triangle?

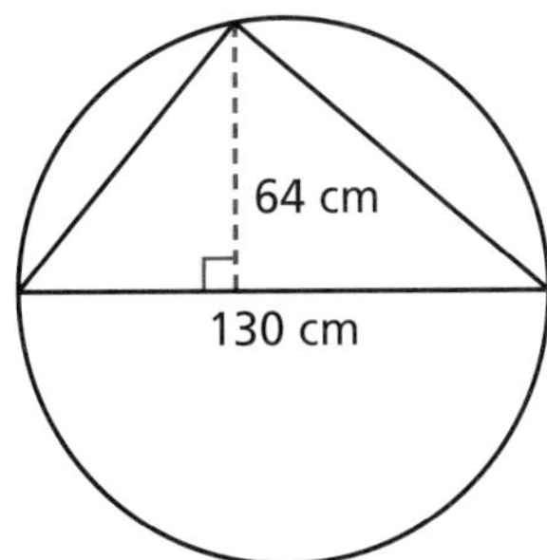

2. A chef has observed the number of people ordering each entrée from the evening's specials. Estimate the probability that the next customer will order Boeuf Bourguignon.

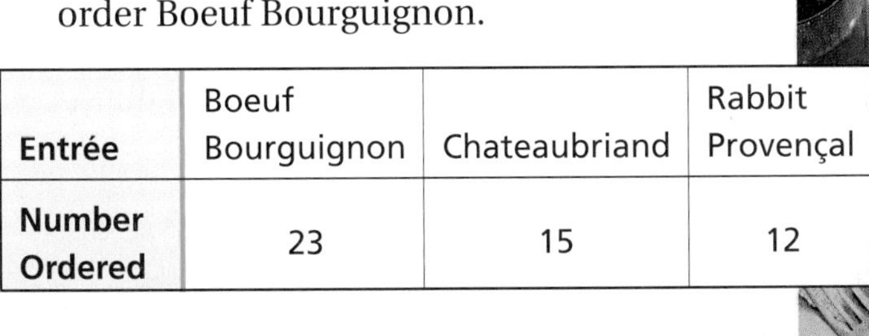

Entrée	Boeuf Bourguignon	Chateaubriand	Rabbit Provençal
Number Ordered	23	15	12

3. Eulalia and Nunzio play cribbage 5 times a week. Eulalia skunked Nunzio 3 times in the last 12 weeks. Estimate the probability that Eulalia will skunk Nunzio the next time they play cribbage.

4. A pula has a coat of arms on the obverse and a running zebra on the reverse. If a pula is tossed 150 times and lands with the coat of arms facing up 70 times, estimate the probability of its landing with the zebra facing up.

9-4 Theoretical Probability

Learn to estimate probability using theoretical methods.

Vocabulary

theoretical probability

equally likely

fair

mutually exclusive

In the game of Monopoly®, you can get out of jail if you roll doubles, but if you roll doubles three times in a row, you have to go to jail. Your turn is decided by the probability that both dice will be the same number.

Theoretical probability is used to estimate probabilities by making certain assumptions about an experiment. Suppose a sample space has 5 outcomes that are **equally likely**, that is, they all have the same probability, x. The probabilities must add to 1.

$$x + x + x + x + x = 1$$
$$5x = 1$$
$$x = \frac{1}{5}$$

THEORETICAL PROBABILITY FOR EQUALLY LIKELY OUTCOMES

Suppose there are n equally likely outcomes in the sample space of an experiment.

- The probability of each outcome is $\frac{1}{n}$.
- The probability of an event is $\frac{\text{number of outcomes in the event}}{n}$.

A coin, die, or other object is called **fair** if all outcomes are equally likely.

EXAMPLE 1 Calculating Theoretical Probability

An experiment consists of rolling a fair die. There are 6 possible outcomes: 1, 2, 3, 4, 5, and 6.

A What is the probability of rolling a 3?

The die is fair, so all 6 outcomes are equally likely. The probability of the outcome of rolling a 3 is $P(3) = \frac{1}{6}$.

B What is the probability of rolling an odd number?

There are 3 outcomes in the event of rolling an odd number: 1, 3, and 5.

$$P(\text{rolling an odd number}) = \frac{\text{number of possible odd numbers}}{6} = \frac{3}{6} = \frac{1}{2}$$

An experiment consists of rolling a fair die. There are 6 possible outcomes: 1, 2, 3, 4, 5, and 6.

C What is the probability of rolling a number less than 5?

There are 4 outcomes in the event of rolling a number less than 5: 1, 2, 3, and 4.

$P(\text{rolling a number less than } 5) = \frac{4}{6} = \frac{2}{3}$

Suppose you roll two fair dice. Are all outcomes equally likely? It depends on how you consider the outcomes. You could look at the number on each die or at the total shown on the dice.

If you look at the total, all outcomes are not equally likely. For example, there is only one way to get a total of 2, 1 + 1, but a total of 5 can be 1 + 4, 2 + 3, 3 + 2, or 4 + 1.

EXAMPLE 2 Calculating Theoretical Probability for Two Fair Dice

An experiment consists of rolling two fair dice.

A Show a sample space that has all outcomes equally likely.

Suppose the dice are two different colors, red and blue.

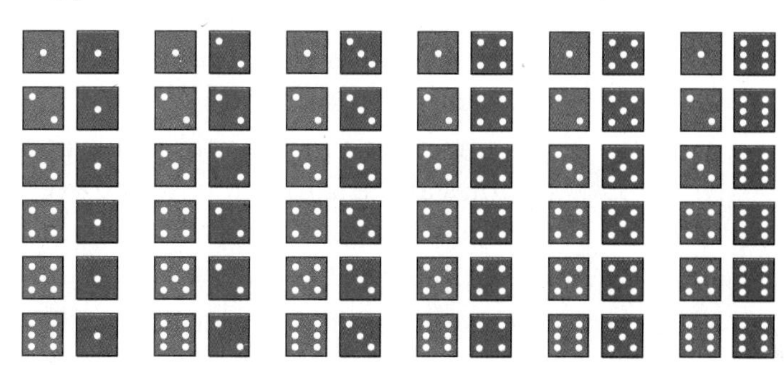

The outcome of a red 3 and a blue 6 can be written as the ordered pair (3, 6). There are 36 possible outcomes in the sample space.

B What is the probability of rolling doubles?

There are 6 outcomes in the event "rolling doubles": (1, 1), (2, 2), (3, 3), (4, 4), (5, 5) and (6, 6).

$P(\text{doubles}) = \frac{6}{36} = \frac{1}{6}$

C What is the probability that the total shown on both dice is 10?

There are 3 outcomes in the event "a total of 10": (4, 6), (5, 5), and (6, 4).

$P(\text{total} = 10) = \frac{3}{36} = \frac{1}{12}$

D What is the probability that the total shown is less than 5?

There are 6 outcomes in the event "a total less than 5": (1, 1), (1, 2), (1, 3), (2, 1), (2, 2), and (3, 1).

$P(\text{total} < 5) = \frac{6}{36} = \frac{1}{6}$

Two events are **mutually exclusive** if they cannot both occur in the same trial of an experiment. Suppose A and B are two mutually exclusive events.

- $P(\text{both } A \text{ and } B \text{ will occur}) = 0$
- $P(\text{either } A \text{ or } B \text{ will occur}) = P(A) + P(B)$

Examples 2C and 2D are mutually exclusive, because the total cannot be less than 5 and equal to 10 at the same time. Examples 2B and 2C are *not* mutually exclusive, because the outcome (5, 5) is a double *and* has a sum of 10.

EXAMPLE 3 Finding the Probability of Mutually Exclusive Events

Suppose you are playing a game of Monopoly and have just rolled doubles two times in a row. If you roll doubles again, you will go to jail. You will also go to jail if you roll a total of 3, because you are 3 spaces away from the "Go to Jail" square. What is the probability that you will go to jail?

It is impossible to roll a total of 3 and doubles at the same time, so the events are mutually exclusive. Add the probabilities to find the probability of going to jail on the next roll.

The event "total = 3" consists of two outcomes, (1, 2) and (2, 1), so $P(\text{total of } 3) = \frac{2}{36}$. From Example 2B, $P(\text{doubles}) = \frac{6}{36}$.

$$P(\text{going to jail}) = P(\text{doubles}) + P(\text{total} = 3)$$
$$= \frac{6}{36} + \frac{2}{36}$$
$$= \frac{8}{36}$$

The probability of going to jail is $\frac{8}{36} = \frac{2}{9}$, or about 22.2%.

Think and Discuss

1. **Describe** a sample space for tossing two coins that has all outcomes equally likely.
2. **Give an example** of an experiment in which it would not be reasonable to assume that all outcomes are equally likely.

9-4 Exercises

FOR EOG PRACTICE
see page 678

internet connect
Homework Help Online
go.hrw.com Keyword: MT4 9-4

GUIDED PRACTICE

See Example 1

An experiment consists of rolling a fair die.

1. What is the probability of rolling an even number?
2. What is the probability of rolling a 3 or a 5?

See Example 2

An experiment consists of rolling two fair dice. Find each probability.

3. *P*(total shown = 7)
4. *P*(rolling two 5's)
5. *P*(rolling two even numbers)
6. *P*(total shown > 8)

See Example 3

7. Suppose you are playing a game in which two fair dice are rolled. To make the first move, you need to roll doubles or a sum of 3 or 11. What is the probability that you will be able to make the first move?

INDEPENDENT PRACTICE

See Example 1

An experiment consists of rolling a fair die.

8. What is the probability of rolling a 7?
9. What is the probability of not rolling a 6?
10. What is the probability of rolling a number greater than 2?

See Example 2

An experiment consists of rolling two fair dice. Find each probability.

11. *P*(total shown = 12)
12. *P*(not rolling doubles)
13. *P*(total shown > 0)
14. *P*(total shown < 4)

See Example 3

15. Suppose you are playing a game in which two fair dice are rolled. You need 7 to land on the finish by an exact count, or 4 to land on a "roll again" space. What is the probability of landing on the finish or rolling again?

PRACTICE AND PROBLEM SOLVING

Three fair coins are tossed: a penny, a dime, and a quarter. The table shows a sample space with all outcomes equally likely. Find each probability.

16. *P*(HTT)
17. *P*(THT)
18. *P*(TTT)
19. *P*(2 heads)
20. *P*(0 tails)
21. *P*(at least 1 head)
22. *P*(1 tail)
23. *P*(all the same)

Penny	Dime	Quarter	Outcome
H	H	H	HHH
H	H	T	HHT
H	T	H	HTH
H	T	T	HTT
T	H	H	THH
T	H	T	THT
T	T	H	TTH
T	T	T	TTT

Life Science 

What color are your eyes? Can you roll your tongue? These traits are determined by the genes you inherited from your parents before you were born. A *Punnett square* shows all possible gene combinations for two parents whose genes are known.

To make a Punnett square, draw a two-by-two grid. Write the genes of one parent above the top row, and the other parent along the side. Then fill in the grid as shown.

	B	*b*
b	*Bb*	*bb*
b	*Bb*	*bb*

24. In the Punnett square above, one parent has the gene combination *Bb*, which represents one gene for brown eyes and one gene for blue eyes. The other parent has the gene combination *bb*, which represents two genes for blue eyes. If all outcomes in the Punnett square are equally likely, what is the probability of a child with the gene combination *bb*?

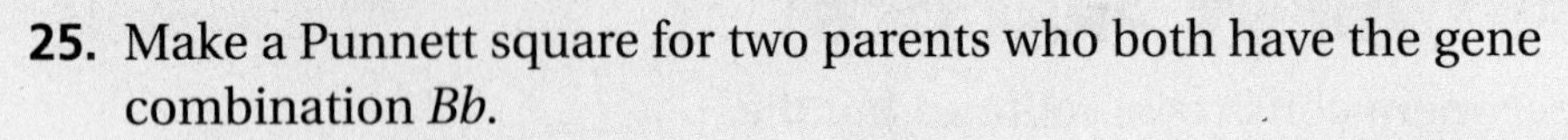

25. Make a Punnett square for two parents who both have the gene combination *Bb*.

a. If all outcomes in the Punnett square are equally likely, what is the probability of a child with the gene combination *BB*?

b. The gene combinations *BB* and *Bb* will result in brown eyes, and the gene combination *bb* will result in blue eyes. What is the probability that the couple will have a child with brown eyes?

26. ***CHALLENGE*** The combinations *Tt* and *TT* represent the ability to roll your tongue, and the combination *tt* means you cannot roll your tongue. Draw a Punnett square that results in a probability of $\frac{1}{2}$ that the child can roll his or her tongue. What can you say about whether the parents can roll their tongues?

Spiral Review

Write each value as indicated. (Lesson 8-1)

27. $\frac{9}{10}$ as a percent **28.** 46% as a fraction **29.** $\frac{3}{8}$ as a decimal

30. $\frac{7}{14}$ as a decimal **31.** 0.78 as a fraction **32.** 52.5% as a decimal

33. **EOG PREP** Last year, a factory produced 1,235,600 parts. If the company expects a 12% increase in production this year, how many parts will the factory produce? (Lesson 8-4)

A 1,383,872 B 14,827,200 C 12,625,400 D 1,482,720

34. **EOG PREP** Angles 1 and 2 are supplementary, and m∠1 = 50°. What is m∠2? (Lesson 5-1)

A 40° B 50° C 130° D 140°

9-5 The Fundamental Counting Principle

Learn to find the number of possible outcomes in an experiment.

Vocabulary

Fundamental Counting Principle

tree diagram

Computers can generate random passwords that are hard to guess because of the many possible arrangements of letters, numbers, and symbols.

If you tried to guess another person's password, you might have to try over one billion different codes!

"Your logon password is XB#2D940. Write it down and don't lose it again."

THE FUNDAMENTAL COUNTING PRINCIPLE

If there are m ways to choose a first item and n ways to choose a second item after the first item has been chosen, then there are $m \cdot n$ ways to choose all the items.

EXAMPLE 1 Using the Fundamental Counting Principle

A computer randomly generates a 5-character password of 2 letters followed by 3 digits. All passwords are equally likely.

A **Find the number of possible passwords.**

Use the Fundamental Counting Principle.

first letter	second letter	first digit	second digit	third digit
?	?	?	?	?
26 choices	26 choices	10 choices	10 choices	10 choices

$26 \cdot 26 \cdot 10 \cdot 10 \cdot 10 = 676{,}000$

The number of possible 2-letter, 3-digit passwords is 676,000.

B **Find the probability of being assigned the password MQ836.**

$$P(\text{MQ836}) = \frac{1}{\text{number of possible passwords}} = \frac{1}{676{,}000} \approx 0.0000015$$

C **Find the probability of a password that does not contain an *A*.**

First use the Fundamental Counting Principle to find the number of passwords that do not contain an *A*.

$25 \cdot 25 \cdot 10 \cdot 10 \cdot 10 = 625{,}000$ possible passwords without an *A*

There are 25 choices for any letter except A.

$$P(\text{no A}) = \frac{625{,}000}{676{,}000} = \frac{625}{676} \approx 0.925$$

A computer randomly generates a 5-character password of 2 letters followed by 3 digits. All passwords are equally likely.

D **Find the probability that a password contains exactly one 4.**

Only one of the digits can be a 4. The other two can be any of the 9 other digits. The 4 could be in one of three positions.

One digit must be a 4. *Other digits can be any digit but 4.*

$26 \cdot 26 \cdot 1 \cdot 9 \cdot 9 =$ 54,756 possible passwords with 4 as 1st digit
$26 \cdot 26 \cdot 9 \cdot 1 \cdot 9 =$ 54,756 possible passwords with 4 as 2nd digit
$26 \cdot 26 \cdot 9 \cdot 9 \cdot 1 =$ 54,756 possible passwords with 4 as 3rd digit
164,268 containing exactly one 4

$$P(\text{exactly one } 4) = \frac{164{,}268}{676{,}000} = \frac{243}{1000} = 0.243$$

The Fundamental Counting Principle tells you only the *number* of outcomes in some experiments, not what the outcomes are. A **tree diagram** is a way to show all of the possible outcomes.

EXAMPLE 2 Using a Tree Diagram

You pack 2 pairs of pants, 3 shirts, and 2 sweaters for your vacation. Describe all of the outfits you can make if each outfit consists of a pair of pants, a shirt, and a sweater.

You can find all of the possible outcomes by making a tree diagram. There should be $2 \cdot 3 \cdot 2 = 12$ different outfits.

Each "branch" of the tree diagram represents a different outfit. The outfit shown in the circled branch could be written as (black, red, gray). The other outfits are as follows:
(black, red, tan), (black, green, gray), (black, green, tan),
(black, yellow, gray), (black, yellow, tan),
(blue, red, gray), (blue, red, tan), (blue, green, gray),
(blue, green, tan), (blue, yellow, gray), (blue, yellow, tan)

Think and Discuss

1. **Suppose** in Example 2 you could pack one more item. Which would you bring, another shirt or another pair of pants? Explain.

9-5 Exercises

FOR EOG PRACTICE

see page 678

internet connect
Homework Help Online
go.hrw.com Keyword: MT4 9-5

GUIDED PRACTICE

See Example 1

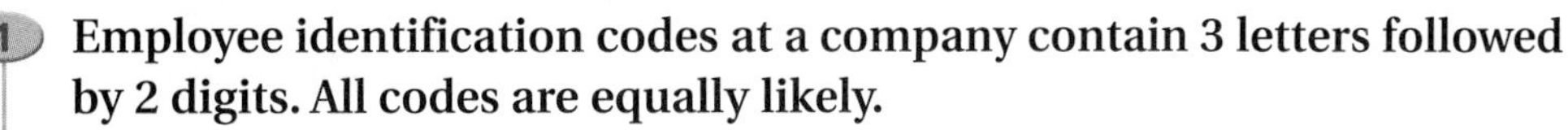

Employee identification codes at a company contain 3 letters followed by 2 digits. All codes are equally likely.

1. Find the number of possible identification codes.
2. Find the probability of being assigned the ID ABC35.
3. Find the probability that an ID code does not contain the number 7.
4. Find the probability that an ID code contains exactly one *F*.

See Example 2

5. There are 3 ways to travel from Los Angeles to San Francisco (car, train, or plane) and 2 ways to travel from San Francisco to Honolulu (plane or boat). Describe all the ways a person can travel from Los Angeles to Honolulu with a stopover in San Francisco.
6. The soup choices at a restaurant are chicken, bean, and vegetable. The sandwich choices are cheese, ham, and turkey. Describe all of the different soup and sandwich options available.

INDEPENDENT PRACTICE

See Example 

License plates in a certain state contain 3 letters followed by 3 digits. All license plates are equally likely.

7. Find the number of possible license plates.
8. Find the probability of not being assigned a plate containing *A* or *B*.
9. Find the probability of receiving a plate containing no vowels (*A, E, I, O, U*).
10. Find the probability of receiving a plate with all odd numbers.

See Example 

11. An interior-decorating catalog offers a chair in a dark, light, or oak finish, with a choice of a tan, black, or cream seat cover, and in regular or tall height. Describe all of the different chairs that are available.
12. A washing machine has regular, delicate, and permanent press cycles with hot, warm, or cold wash. Describe all of the washing options available.

PRACTICE AND PROBLEM SOLVING

Find the number of possible outcomes.

13. birds: parrot, parakeet, cockatiel
    cages: round, square
14. bagels: sesame, sourdough, plain
    spreads: plain, chive, veggie
15. colors: purple, red, blue, orange
    sizes: small, medium, large
16. destinations: Paris, London, Rome
    months: May, June, July, August

Technology LINK

In the process of spin-coating a CD-ROM, the disc is rotated at high speeds. This process is used to apply layers as thin as $\frac{1}{8}$ of a micron, which is 640 times thinner than a human hair.

17. Mario needs to register for one course in each of six subject areas. The school offers 2 math courses, 3 foreign languages courses, 4 science courses, 4 English courses, 4 social studies courses, and 5 elective courses. In how many ways can he register?

18. ***TECHNOLOGY*** Tim is buying a new computer from an online store. His options are shown at right. He can choose a color, one software package, and one hardware option.

    a. How many computer choices are available?

    b. Tim decides he wants a red computer. Describe all of the choices available to him.

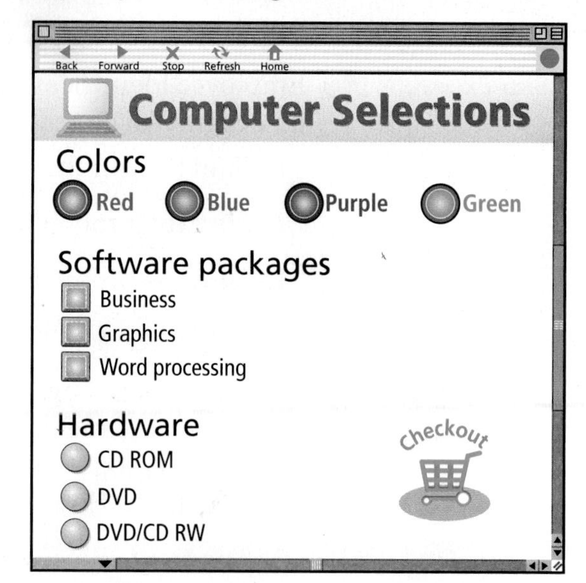

19. ***WHAT'S THE ERROR?*** To find the total number of outfits that can be made with 5 tops, 3 pants, and 2 jackets, a student answers, "5 + 3 + 2 = 10 different outfits." What error has the student made, and what is the correct answer?

20. ***WRITE ABOUT IT*** Describe when you would want to use the Fundamental Counting Principle instead of a tree diagram. Describe when a tree diagram would be more useful than the Fundamental Counting Principle.

21. ***CHALLENGE*** A password can have letters, numerals, or 32 other keyboard symbols in each of its 5 character spaces. There are two restrictions. The password cannot start with an *A* or a 1 and it may not end with a 0. Find the total number of possible passwords.

Spiral Review

Find each number. (Lesson 8-3)

22. 60% of what number is 12?
23. 112 is 80% of what number?
24. 30 is 2% of what number?
25. 90% of what number is 18?
26. 75% of what number is 200?
27. 18 is 45% of what number?

28. **EOG PREP** Last year, Tyrone earned $45,672. Of this amount, $6,622.44 was withheld for income taxes. What percent of Tyrone's income was withheld? (Lesson 8-6)

A 12% B 17.8% C 14.5% D 13%

9-6 Permutations and Combinations

Learn to find permutations and combinations.

Vocabulary
factorial
permutation
combination

Some pizza restaurants have many choices of toppings. You can use *factorials* to find out how many different pizzas you could order.

The **factorial** of a number is the product of all the whole numbers from the number down to 1. The factorial of 0 is defined to be 1.

$$5! = 5 \cdot 4 \cdot 3 \cdot 2 \cdot 1 = 120$$

EXAMPLE 1 Evaluating Expressions Containing Factorials

Reading Math
Read 5! as "five factorial."

Evaluate each expression.

A $9!$

$9 \cdot 8 \cdot 7 \cdot 6 \cdot 5 \cdot 4 \cdot 3 \cdot 2 \cdot 1 = 362{,}880$

B $\frac{6!}{3!}$

$\frac{6 \cdot 5 \cdot 4 \cdot \cancel{3} \cdot \cancel{2} \cdot \cancel{1}}{\cancel{3} \cdot \cancel{2} \cdot \cancel{1}}$ *Write out each factorial and simplify.*

$6 \cdot 5 \cdot 4 = 120$ *Multiply remaining factors.*

C $\frac{14!}{(11 - 4)!}$ *Subtract within parentheses.*

$\frac{14!}{7!}$

$\frac{14 \cdot 13 \cdot 12 \cdot 11 \cdot 10 \cdot 9 \cdot 8 \cdot \cancel{7} \cdot \cancel{6} \cdot \cancel{5} \cdot \cancel{4} \cdot \cancel{3} \cdot \cancel{2} \cdot \cancel{1}}{\cancel{7} \cdot \cancel{6} \cdot \cancel{5} \cdot \cancel{4} \cdot \cancel{3} \cdot \cancel{2} \cdot \cancel{1}}$

$14 \cdot 13 \cdot 12 \cdot 11 \cdot 10 \cdot 9 \cdot 8 = 17{,}297{,}280$

A **permutation** is an arrangement of things in a certain order.

If no letter can be used more than once, there are 6 permutations of the first 3 letters of the alphabet: ABC, ACB, BAC, BCA, CAB, and CBA.

first letter		second letter		third letter
?		?		?
3 choices	·	2 choices	·	1 choice

The product can be written as a factorial.

$3 \cdot 2 \cdot 1 = 3! = 6$

If no letter can be used more than once, there are 60 permutations of the first 5 letters of the alphabet, when taken 3 at a time: ABC, ABD, ABE, ACD, ACE, ADB, ADC, ADE, and so on.

first letter		second letter		third letter
?		?		?
5 choices	·	4 choices	·	3 choices = 60 permutations

Notice that the product can be written as a quotient of factorials.

$$60 = 5 \cdot 4 \cdot 3 = \frac{5 \cdot 4 \cdot 3 \cdot 2 \cdot 1}{2 \cdot 1} = \frac{5!}{2!}$$

PERMUTATIONS

The number of permutations of n things taken r at a time is

$$_nP_r = \frac{n!}{(n-r)!}.$$

EXAMPLE 2 Finding Permutations

Helpful Hint

By definition, $0! = 1$.

There are 8 runners in a race.

A Find the number of orders in which all 8 runners can finish.

The number of runners is 8.

$$_8P_8 = \frac{8!}{(8-8)!} = \frac{8!}{0!} = \frac{8 \cdot 7 \cdot 6 \cdot 5 \cdot 4 \cdot 3 \cdot 2 \cdot 1}{1} = 40{,}320$$

All 8 runners are taken at a time.

There are 40,320 permutations. This means there are 40,320 orders in which the 8 runners can finish.

B Find the number of ways the 8 runners can finish first, second, and third.

The number of runners is 8.

$$_8P_3 = \frac{8!}{(8-3)!} = \frac{8!}{5!} = \frac{8 \cdot 7 \cdot 6 \cdot \cancel{5} \cdot \cancel{4} \cdot \cancel{3} \cdot \cancel{2} \cdot \cancel{1}}{\cancel{5} \cdot \cancel{4} \cdot \cancel{3} \cdot \cancel{2} \cdot \cancel{1}} = 8 \cdot 7 \cdot 6 = 336$$

The top 3 places are taken at a time.

There are 336 permutations. This means that the 8 runners can finish in first, second, and third in 336 ways.

A **combination** is a selection of things in any order.

If no letter can be used more than once, there is only 1 combination of the first 3 letters of the alphabet. ABC, ACB, BAC, BCA, CAB, and CBA are considered to be the same combination of A, B, and C because the order does not matter.

If no letter is used more than once, there are 10 combinations of the first 5 letters of the alphabet, when taken 3 at a time. To see this, look at the list of permutations below.

ABC	ABD	ABE	ACD	ACE	ADE	BCD	BCE	BDE	CDE
ACB	ADB	AEB	ADC	AEC	AED	BDC	BEC	BED	CED
BAC	BAD	BAE	CAD	CAE	DAE	CBD	CBE	DBE	DCE
BCA	BDA	BEA	CDA	CEA	DEA	CDB	CEB	DEB	DEC
CAB	DAB	EAB	DAC	EAC	EAD	DCB	EBC	EBD	ECD
CBA	DBA	EBA	DCA	ECA	EDA	DBC	ECB	EDB	EDC

These 6 permutations are all the same combination.

In the list of 60 permutations, each combination is repeated 6 times. The number of combinations is $\frac{60}{6} = 10$.

COMBINATIONS

The number of combinations of n things taken r at a time is

$$_nC_r = \frac{_nP_r}{r!} = \frac{n!}{r!(n-r)!}.$$

EXAMPLE 3 Finding Combinations

A gourmet pizza restaurant offers 9 topping choices.

A Find the number of 2-topping pizzas that can be ordered.

9 possible toppings

$$_9C_2 = \frac{9!}{2!(9-2)!} = \frac{9!}{2!7!} = \frac{9 \cdot 8 \cdot \cancel{7} \cdot \cancel{6} \cdot \cancel{5} \cdot \cancel{4} \cdot \cancel{3} \cdot \cancel{2} \cdot \cancel{1}}{(2 \cdot 1)(\cancel{7} \cdot \cancel{6} \cdot \cancel{5} \cdot \cancel{4} \cdot \cancel{3} \cdot \cancel{2} \cdot \cancel{1})} = 36$$

2 toppings chosen at a time

There are 36 combinations. This means that there are 36 different 2-topping pizzas that can be ordered.

B Find the number of 5-topping pizzas that can be ordered.

9 possible toppings

$$_9C_5 = \frac{9!}{5!(9-5)!} = \frac{9!}{5!4!} = \frac{9 \cdot 8 \cdot 7 \cdot 6 \cdot \cancel{5} \cdot \cancel{4} \cdot \cancel{3} \cdot \cancel{2} \cdot \cancel{1}}{(\cancel{5} \cdot \cancel{4} \cdot \cancel{3} \cdot \cancel{2} \cdot \cancel{1})(4 \cdot 3 \cdot 2 \cdot 1)} = 126$$

5 toppings chosen at a time

There are 126 combinations. This means that there are 126 different 5-topping pizzas.

Think and Discuss

1. **Explain** the difference between a combination and a permutation.
2. **Give an example** of an experiment where order is important and one where order is not important.

9-6 Exercises

FOR EOG PRACTICE

see page 679

internet connect

Homework Help Online

go.hrw.com Keyword: MT4 9-6

GUIDED PRACTICE

See Example 1

Evaluate each expression.

1. $7!$
2. $\frac{6!}{2!}$
3. $\frac{8!}{(6-4)!}$
4. $\frac{5!}{(4-1)!}$

See Example 2

There are 10 cyclists in a race.

5. In how many possible orders can all 10 cyclists finish the race?
6. How many ways can the 10 cyclists finish first, second, and third?

See Example 3

A group of 8 people is forming several committees.

7. Find the number of different 3-person committees that can be formed.
8. Find the number of different 5-person committees that can be formed.

INDEPENDENT PRACTICE

See Example 1

Evaluate each expression.

9. $3!$
10. $\frac{7!}{3!}$
11. $\frac{4!}{(3-2)!}$
12. $\frac{10!}{(6-3)!}$

See Example 2

Ann has 7 books she wants to put on her bookshelf.

13. How many possible arrangments of books are there?
14. Suppose Ann has room on the shelf for only 3 of the 7 books. In how many ways can she arrange the books now?

See Example 3

If Diane joins a CD club, she gets 6 free CDs.

15. If Diane can select from a list of 40 CDs, how many groups of 6 different CDs are possible?
16. If Diane can select from a list of 55 CDs, how many groups of 6 different CDs are possible?

PRACTICE AND PROBLEM SOLVING

Evaluate each expression.

17. $\frac{9!}{(9-2)!}$
18. $\frac{12!}{5!(12-5)!}$
19. ${}_{11}P_{11}$
20. ${}_{7}C_{2}$
21. ${}_{15}C_{15}$
22. ${}_{9}C_{6}$
23. $\frac{10!}{9!}$
24. ${}_{8}P_{4}$

Simplify each expression.

25. $\frac{n!}{(n-1)!}$
26. ${}_{n}C_{n}$
27. ${}_{n}P_{n}$
28. ${}_{n}C_{1}$
29. ${}_{n}P_{1}$
30. ${}_{n}C_{0}$
31. ${}_{n}P_{0}$
32. ${}_{n}C_{n-1}$

Art LINK

Josef Albers used the simple design of nested squares to investigate color relationships. He did not mix colors, but instead created hundreds of variations using paint straight from the tube.

33. How many ways can a softball coach choose the first, second, and third batters in the lineup for a team of 9 players?

34. How many teams of 3 people can be made from 6 employees?

35. In how many ways can 6 people line up to get on a bus?

36. If 10 students go hiking in pairs, how many different pairs of students are possible?

37. Levi is making a salad. He can choose from the following toppings: carrots, cheese, radishes, cauliflower, broccoli, mushrooms, and hard-boiled eggs. If he wants to have 5 different toppings, how many possible salads can he make?

38. ***ART*** An artist is making a painting of three nested squares. He has 12 different colors to choose from. How many different paintings could he make if the squares are all different colors?

39. ***SPORTS*** At a track meet, there are 7 runners competing in the 100 m dash.

a. Find the number of orders in which all 7 runners can finish.

b. Find the number of orders in which the 7 runners can finish in first, second, and third places.

40. ***LIFE SCIENCE*** There are 12 different species of fish in a lake. In how many ways can researchers capture, tag, and release fish of 5 different species?

41. ***WHAT'S THE QUESTION?*** There are 11 items available at a buffet. Customers can choose up to 5 of these items. If the answer is 462, what is the question?

42. ***WRITE ABOUT IT*** Explain how you could use combinations and permutations to find the probability of an event.

43. ***CHALLENGE*** How many ways can a local chapter of the American Mathematical Society schedule 3 speakers for 3 different meetings in one day if all of the speakers are available on any of 5 dates?

Spiral Review

Find the interest and the total amount to the nearest cent. (Lesson 8-7)

44. \$300 at 5% per year for 2 years

45. \$750 at 4.5% per year for 4 years

46. \$1250 at 7% per year for 10 years

47. \$410 at 2.6% per year for 1.5 years

48. \$1000 at 6% per year for 5 years

49. \$90 at 8% per year for 3 years

50. **EOG PREP** A spinner was spun 220 times. It stopped on red 58 times. Which is the *best* estimate of the probability of the spinner stopping on red? (Lesson 9-2)

A 0.225

B 0.264

C 0.126

D 0.32

Pascal's Triangle

Use with Lesson 9-6

Pascal's Triangle is a triangular array of counting numbers. The first row of the triangle is a 1, and every other number in the triangle is the sum of the two numbers diagonal to it in the row above it. On the outer edge of the triangle, each number has only one number diagonal to it, so each row of the triangle begins and ends with the number 1.

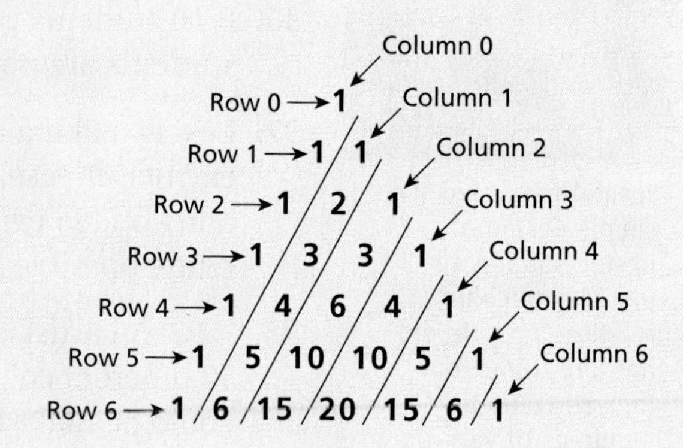

Activity

1. Copy Pascal's Triangle onto your paper.
2. Add two more rows to Pascal's Triangle.
3. Look at row 5 of Pascal's Triangle.

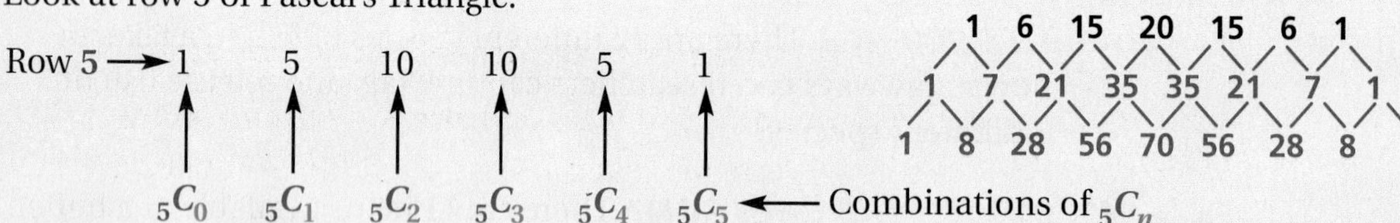

This row shows all possible combinations of 5 items taken 1, 2, 3, 4, or 5 at a time ($_5C_n$).

If there are 5 people in a club, how many different combinations of 2 club members can bring refreshments to a meeting?

$$_5C_2 = 10$$

Think and Discuss

1. Can you use Pascal's Triangle to find $_{31}C_8$? Would this be easier than using the combination formula? Explain your answer.

Try This

1. Find a pattern for the numbers in column 1 of Pascal's Triangle. What will the number in row 7 column 1 be?
2. Find the sum for each row of numbers. Write each sum as a power of 2.
3. Misha has 7 pens that are each a different color. He brings 2 to school each day. How many days can he have a different combination of pen colors before he must start repeating combinations? Explain how Pascal's triangle can help you answer this question.

9-7 Independent and Dependent Events

Learn to find the probabilities of independent and dependent events.

Vocabulary

independent events

dependent events

It is critical that the engine of a single-engine airplane not fail during flight. These planes often have two *independent* electrical systems. In the event that one electrical system fails, for example, due to a faulty spark plug, the second system will still be able to keep the plane in flight.

Events are **independent events** if the occurrence of one event does not affect the probability of the other. Events are **dependent events** if the occurrence of one does affect the probability of the other.

EXAMPLE 1 Classifying Events as Independent or Dependent

Determine if the events are dependent or independent.

A **a coin landing heads on one toss and tails on another toss**

The result of one toss does not affect the result of the other, so the events are independent.

B **drawing a heart and a spade from a deck at the same time**

The cards drawn cannot be the same card, so the events are dependent.

FINDING THE PROBABILITY OF INDEPENDENT EVENTS

If A and B are independent events, then $P(A \text{ and } B) = P(A) \cdot P(B)$.

EXAMPLE 2 Finding the Probability of Independent Events

An experiment consists of spinning the spinner 3 times. For each spin, all outcomes are equally likely.

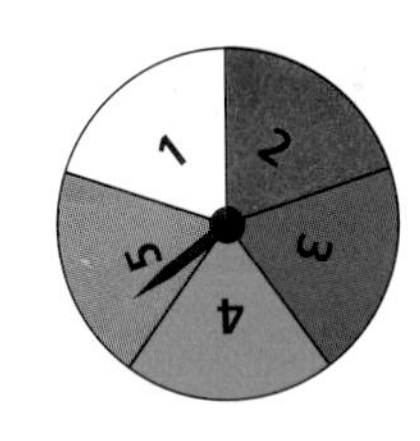

A **What is the probability of spinning a 5 all 3 times?**

The result of each spin does not affect the results of the other spins, so the spin results are independent.

For each spin, $P(5) = \frac{1}{5}$.

$P(5, 5, 5) = \frac{1}{5} \cdot \frac{1}{5} \cdot \frac{1}{5} = \frac{1}{125} = 0.008$ *Multiply.*

9-8 Exercises

FOR EOG PRACTICE

see page 679

internet connect

Homework Help Online

go.hrw.com Keyword: MT4 9-8

GUIDED PRACTICE

See Example 1

A family collects 63 game pieces for a contest. Three pieces win a prize.

1. Estimate the odds in favor of winning a prize in the contest.
2. Estimate the odds against winning a prize in the contest.

See Example 2

3. If the odds in favor of winning a new set of golf clubs are 1:999, what is the probability of winning the golf clubs?
4. If the odds against winning a game system are 2249:1, what is the probability of winning the game system?

See Example 3

5. The probability of winning a DVD is $\frac{1}{75}$. What are the odds in favor of winning the DVD?
6. The probability of winning a vacation is $\frac{1}{22{,}750}$. What are the odds against winning the vacation?

INDEPENDENT PRACTICE

See Example 1

Of the 1260 visitors to a convention, 70 win door prizes.

7. Estimate the odds in favor of winning a door prize at the convention.
8. Estimate the odds against winning a door prize at the convention.

See Example 2

9. If the odds in favor of winning a new computer are 1:9999, what is the probability of winning a new computer?
10. If the odds against being randomly selected for a committee are 19:1, what is the probability of being selected?

See Example 3

11. The probability of winning a shopping spree is $\frac{1}{845}$. What are the odds in favor of winning?
12. The probability of winning a new car is $\frac{1}{500{,}000}$. What are the odds against winning the car?

PRACTICE AND PROBLEM SOLVING

You roll two fair number cubes. Find the odds in favor of and against each event.

13. rolling two 6's
14. rolling a total of 7
15. rolling a total of 11
16. rolling doubles
17. rolling two odd numbers
18. rolling a 1 and a 4

Passenger Train Routes

The 6 passenger routes of North Carolina serve a total of 16 cities. Each route has stops in at least two North Carolina cities, and all but one have stops in at least one other state. There are two trains per route, one northbound and one southbound, for a total of 12 trains.

1. In how many different orders could a person ride all 12 trains?

2. No routes have stops in both Fayetteville and Charlotte. If 3 routes have stops in Charlotte and 2 routes have stops in Fayetteville, what are the odds of a passenger's choosing a route at random that stops in neither Fayetteville nor Charlotte?

3. Only 1 route has stops in both Raleigh and Selma. Is it impossible for more than 1 route to have stops in Selma? Explain.

4. The probability of choosing a route at random that has a stop in Raleigh is $\frac{1}{2}$, and the probability of choosing a route at random that has a stop in Rocky Mount is $\frac{2}{3}$. Two routes have stops in both Raleigh and Rocky Mount. How many routes have a stop in Raleigh but not Rocky Mount? in Rocky Mount but not Raleigh?

5. The *Carolinian* makes 12 stops, and it stops in every city where the *Piedmont* stops. The *Piedmont* stops in only $\frac{3}{4}$ of the cities where the *Carolinian* stops. What is the probability of randomly choosing a city where the *Carolinian* stops but the *Piedmont* does not from the 16 cities serviced by passenger trains?

MATH-ABLES

The Paper Chase

Stephen's desk has 8 drawers. When he receives a paper, he usually chooses a drawer at random to put it in. However, 2 out of 10 times he forgets to put the paper away, and it gets lost.

The probability that a paper will get lost is $\frac{2}{10}$, or $\frac{1}{5}$.

- What is the probability that a paper will get put into a drawer?
- If all drawers are equally likely to be chosen, what is the probability that a paper will get put in drawer 3?

When Stephen needs a document, he looks first in drawer 1 and then checks each drawer in order until the paper is found or until he has looked in all the drawers.

1. If Stephen checked drawer 1 and didn't find the paper he was looking for, what is the probability that the paper will be found in one of the remaining 7 drawers?
2. If Stephen checked drawers 1, 2, and 3, and didn't find the paper he was looking for, what is the probability that the paper will be found in one of the remaining 5 drawers?
3. If Stephen checked drawers 1–7 and didn't find the paper he was looking for, what is the probability that the paper will be found in the last drawer?

Try to write a formula for the probability of finding a paper.

Permutations

Use a set of Scrabble™ tiles, or make a similar set of lettered cards. Draw 2 vowels and 3 consonants, and place them face up in the center of the table. Each player tries to write as many permutations as possible in 60 seconds. Score 1 point per permutation, with a bonus point for each permutation that forms an English word.

internet connect go.hrw.com

Go to ***go.hrw.com*** for a complete set of rules and game pieces.
KEYWORD: MT4 Game9

Permutations and Combinations

Use with Lesson 9-6

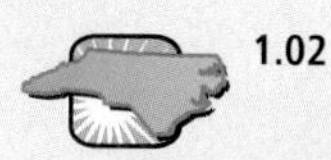

1.02

internet connect
Lab Resources Online
go.hrw.com
KEYWORD: MT4 TechLab9

Graphing calculators have features to help with computing factorials, permutations, and combinations.

Activity

1 In a stock-car race, 11 cars finish the race. The number of different orders in which they can finish is 11! A calculator can help you do the computation. Both ways are shown—the direct way, using the definition of *factorial,* and the calculator factorial command.

To compute 11! on a graphing calculator, enter 11 MATH, press ▶ to go to the **PRB** menu, and select **4:!** ENTER.

The number of ways the 11 cars can finish first, second, third, and fourth is given by 11 · 10 · 9 · 8, or in *permutation* notation, ${}_{11}P_4$, 11 things taken 4 at a time. Both the direct and calculator *nPr* command methods are shown. The *nPr* command is also found in the **PRB** menu.

To compute ${}_{11}P_4$, enter 11 MATH, press ▶ to go to the **PRB** menu, select **2:nPr**, type 4, and press ENTER.

2 Twenty girls try out for 5 open places on a hockey team. Since order is not considered, the number of different *combinations* of these girls that can be chosen is given by ${}_{20}C_5$, the number of combinations of 20 things taken 5 at a time. Both the direct and calculator *nCr* command computations are shown.

To compute ${}_{20}C_5$, press 20 MATH, press ▶ to go to the **PRB** menu, select **3:nCr**, and press 5 ENTER.

Think and Discuss

1. Explain why *nPr* is usually greater than *nCr* for the same values of *n* and *r*.

2. Can *nPr* ever equal *nCr*?

Try This

Compute each value by direct calculator multiplication and division and by using the calculator permutation and combination commands.

1. ${}_{14}P_6$ **2.** ${}_{25}P_{17}$ **3.** ${}_{8}P_3$ **4.** ${}_{8}C_3$ **5.** ${}_{16}C_4$ **6.** ${}_{40}C_6$

Chapter 9

Study Guide and Review

Vocabulary

Complete the sentences below with vocabulary words from the list above. Words may be used more than once.

1. The __?__ of an event tells you how likely the event is to happen.
   - A probability of 0 means it is __?__ for the event to occur.
   - A probability of 1 means it is __?__ that the event will occur.

2. The set of all possible outcomes of an experiment is called the __?__.

3. A(n) __?__ is an arrangement where order is important.
   A(n) __?__ is an arrangement where order is not important.

9-1 Probability (pp. 446–450)

EXAMPLE

■ **Of the raw diamonds received by a diamond cutter, it is expected that about $\frac{1}{8}$ of them will be acceptable.**

Outcome	Acceptable	Unacceptable
Probability	▇	▇

$P(\text{acceptable}) = \frac{1}{8} = 0.125 = 12.5\%$

$P(\text{unacceptable}) = 1 - \frac{1}{8} = \frac{7}{8} = 0.875 = 87.5\%$

EXERCISES

Give the probability for each outcome.

4. About 75% of the people attending a book signing have already read the book.

Outcome	Read	Not read
Probability	▇	▇

9-2 Experimental Probability (pp. 451–454)

EXAMPLE

■ **The table shows the results of spinning a spinner 80 times. Estimate the probability of the spinner landing on blue.**

Outcome	White	Red	Blue	Black
Spins	32	17	24	7

$$\text{probability} \approx \frac{\text{spins that landed on blue}}{\text{total number of spins}}$$
$$= \frac{24}{80} = \frac{3}{10} = 0.3$$

The probability of the spinner landing on blue is about 0.3, or 30%.

EXERCISES

5. The table shows the results of spinning a spinner 100 times. Estimate the probability of the spinner landing on 5.

Outcome	1	2	3	4	5	6
Spins	17	22	11	18	17	15

6. The table shows the results of a survey of 500 students. Estimate the probability that a randomly selected student's favorite subject is math.

Favorite Subject	Math	Science	Art	Other
Number of Students	140	105	75	180

9-3 Use a Simulation (pp. 456–459)

EXAMPLE

■ **At a local school, 75% of the students study a foreign language. If 5 students are chosen randomly, estimate the probability that at least 4 study a foreign language. Use the random number table to make a simulation with at least 10 trials.**

08	57	09	92	75	27	37	87	52	36
16	73	29	39	73	78	65	88	02	42
53	19	18	65	79	64	46	47	60	51
73	16	79	89	12	63	84	60	59	57
13	89	68	35	51	22	56	51	23	81

The probability is about $\frac{8}{10}$, or 80%.

EXERCISES

08570	99275	27378	75236	16732
93973	78658	80242	53191	86579
64464	76051	73167	98912	63846
05957	13896	83551	22565	12381
93861	72073	87891	19845	71302

7. On an assembly line, 25% of the items are rejected. Estimate the probability that at least 2 of the next 6 items will be rejected. Use the random number table to make a simulation with at least 10 trials.

9-4 Theoretical Probability (pp. 462–466)

EXAMPLE

■ **A fair die is rolled once. Find the probability of getting an odd number or a 4.**

$$P(\text{odd or } 4) = P(\text{odd}) + P(4)$$
$$= \frac{3}{6} + \frac{1}{6} = \frac{4}{6} = \frac{2}{3}$$

EXERCISES

8. A marble is drawn at random from a box that contains 7 red, 12 blue, and 5 white marbles. What is the probability of getting a red or a white marble?

9-5 The Fundamental Counting Principle (pp. 467–470)

EXAMPLE

- **A code contains 4 letters. How many possible codes are there?**

  $26 \cdot 26 \cdot 26 \cdot 26 = 456{,}976$ codes

EXERCISES

A building has 6 doors to the outside.

9. How many ways can you enter and leave the building?

10. How many ways can you enter by one door and leave by a different door?

9-6 Permutations and Combinations (pp. 471–475)

EXAMPLE

- **Blaire has 5 plants to arrange on a shelf that will hold 3 plants. How many ways are there to arrange the plants, if the order is important? if the order is not important?**

  order important: ${}_5P_3 = \frac{5!}{(5-3)!} = \frac{5!}{2!} = 60$ ways

  order not important: ${}_5C_3 = \frac{5!}{3!\,(5-3)!} = 10$ ways

EXERCISES

11. Seven people are arranged in a row of 3 seats. How many different arrangements are possible?

12. A school's debate club has 9 members. A team of 4 students will be chosen to represent the school at a competition. How many different teams are possible?

9-7 Independent and Dependent Events (pp. 477–481)

EXAMPLE

- **Two marbles are drawn from a jar containing 3 red marbles and 4 black marbles. What is *P*(red, black) if the first marble is replaced? if the first marble is not replaced?**

	P(red)	*P*(black)	*P*(red, black)
Replaced	$\frac{3}{7}$	$\frac{4}{7}$	$\frac{12}{49} \approx 0.24$
Not replaced	$\frac{3}{7}$	$\frac{4}{6}$	$\frac{12}{42} \approx 0.29$

EXERCISES

13. A number cube is rolled three times. What is the probability of getting a 4 all three times?

14. Two cards are drawn at random from a deck that has 26 red and 26 black cards. What is the probability that the first card is red and the second is black?

9-8 Odds (pp. 482–485)

EXAMPLE

- **A digit from 1 to 9 is selected at random. What are the odds in favor of selecting an even number?**

  favorable ⟶ 4:5 ⟵ *unfavorable*

EXERCISES

15. A letter is selected at random from the alphabet. What are the odds in favor of getting a vowel (*A*, *E*, *I*, *O*, *U*)?

Chapter Test

1. Outcomes A, C, D, and F have the same probability. Complete the probability table.

Outcome	A	B	C	D	E	F
Probability		$\frac{1}{6}$			$\frac{1}{3}$	

2. Madeline is choosing 3 of her 10 best flower displays to be entered in a competition. How many different selections are possible?

3. Jim wants to hang 4 pictures in a row on his wall. If he has 6 pictures to choose from, how many different arrangements are possible?

4. A fair coin is tossed three times. What is the probability of getting heads all three times?

5. In the Westcreek neighborhood, 37% of the families have a dog. Each block has 16 families, 8 on each side. Estimate the probability that 3 or more families on one side of a given block have a dog. Use the random number table to make a simulation with at least 10 trials.

97120 08320 17871 21826 74838 37240 36810 20423

12562 45677 88983 94930 31599 76585 61429 05379

34628 46304 66531 96270 21309 31567 30762 47240

30883 71946 25948 97988 26267 21350 59356 43952

6. Jill has 6 cans of food without labels. She knows there are 2 cans of fruit, 3 of corn, and 1 of beans. If she chooses a can at random, what is the probability that it will not be fruit?

7. Julio's parents write down 10 different chores on slips of paper and put them in a box. If Julio has to draw 2 different chores from the box, what is the probability that he will draw his 2 least favorite chores, vacuuming and pulling weeds?

8. The table shows the results of an interview in which 1000 college students were asked whether they went home during spring and winter breaks. Estimate the probability a student will go home during winter break.

	Spring (yes)	Spring (no)
Winter (yes)	170	520
Winter (no)	233	77

9. A frame shop has a special offer. Pictures can be framed in gold, silver, or brass, the mat can be any one of 16 colors, and the glass can be regular or nonglare. How many ways can a picture be framed with this offer?

10. At a bazaar, the odds in favor of winning a door prize is 1:15. What is the probability of winning a door prize?

Chapter 9

Performance Assessment

Show What You Know

Create a portfolio of your work from this chapter. Complete this page and include it with your four best pieces of work from Chapter 9. Choose from your homework or lab assignments, mid-chapter quiz, or any journal entries you have done. Put them together using any design you want. Make your portfolio represent what you consider your best work.

Short Response

1. A dart thrown at the square board shown lands in a random spot on the board. What is the probability it lands in the blue square? Show your work.

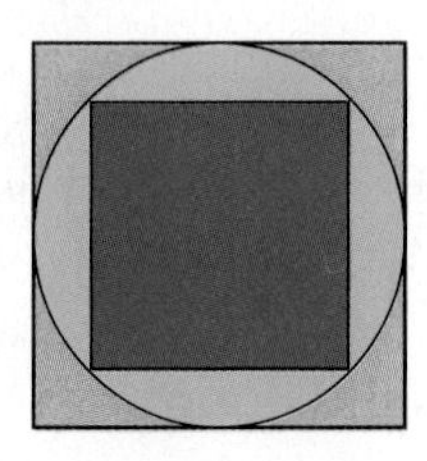

2. The pilot of a hot air balloon is trying to land in a 2 km square field. The field has a large tree in each corner. The balloon's ropes will tangle in a tree if it lands within $\frac{1}{7}$ km of its trunk. What is the probability the balloon will land in the field without getting caught in a tree? Express your answer to the nearest tenth of a percent. Show your work.

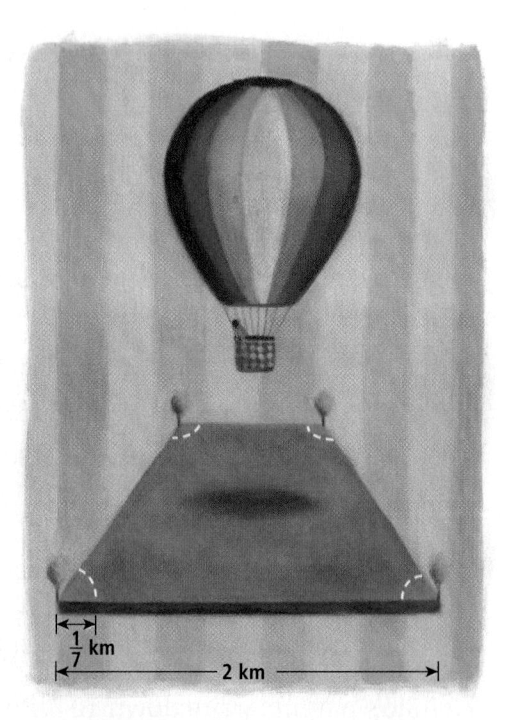

Extended Problem Solving

3. The students at a new high school are choosing a school mascot and school color. The mascot choices are a bear, a lion, a jaguar, or a tiger. The color choices are red, orange, or blue.
   a. How many different combinations do the students have to choose from? Show your work.
   b. If a second school color is added, either gold or silver, how many different combinations do the students have to choose from? Show your work.
   c. How would adding a choice from among n names change the number of combinations to choose from?

Performance Assessment

internet connect
EOG Test Practice Online
go.hrw.com Keyword: MT4 TestPrep

Getting Ready for EOG

Chapter 9

Cumulative Assessment, Chapters 1–9

1. In a box containing gumdrops, 78 are red, 24 are green, and the rest are yellow. If the probability of selecting a yellow gumdrop is $\frac{1}{3}$, how many yellow gumdrops are in the box?

   A 34 C 54
   B 51 D 102

2. $P(-2, -3)$ is reflected across the y-axis to P'. What are the coordinates of P'?

   A $(-3, -2)$ C $(-2, 3)$
   B $(-3, 2)$ D $(2, -3)$

3. If 125% of x is equal to 80% of y, and $y \neq 0$, what is the value of $\frac{x}{y}$?

   A $\frac{16}{25}$ C $\frac{25}{16}$
   B $\frac{4}{5}$ D $\frac{5}{4}$

4. Mia made 5 payments on a loan, with each payment being twice the amount of the one before it. If the total of all 5 payments was $465, how much was the first payment?

   A $5 C $31
   B $15 D $93

5. An electric pump can fill a 45-gallon tub in half an hour. At this rate, how long would it take to fill a 60-gallon tub?

   A 35.0 minutes C 40.0 minutes
   B 37.5 minutes D 42.5 minutes

6. Which of the following ratios is equivalent to the ratio 1.2:1?

   A 1:2 C 5:6
   B 12:1 D 6:5

7. If $20 \cdot 3000 = 6 \cdot 100^x$, what is the value of x?

   A 2 C 4
   B 3 D 5

8. In the chart below, the amount represented by each shaded square is twice that represented by each unshaded square.

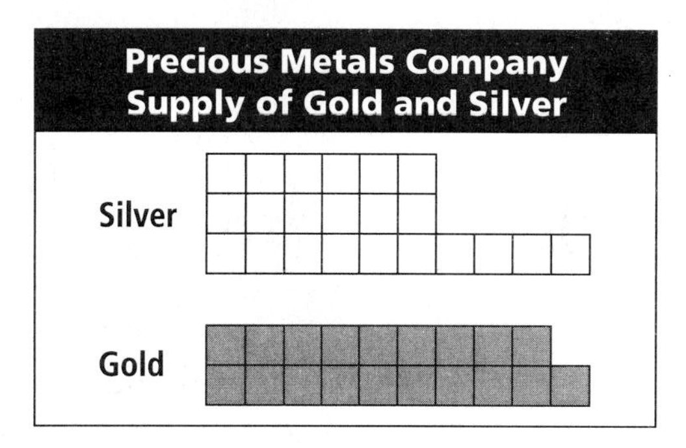

   What is the ratio of the amount of gold to the amount of silver?

   A $\frac{19}{22}$ C $\frac{22}{19}$
   B $\frac{13}{19}$ D $\frac{19}{11}$

TEST TAKING TIP!

To check that answers are reasonable, you can sometimes draw a graph or a diagram on graph paper.

9. ***SHORT RESPONSE*** Which type of quadrilateral is a figure with vertices $(-3, 4)$, $(3, 4)$, $(3, -2)$, and $(-3, -2)$? What is its area? Show your work.

10. ***SHORT RESPONSE*** The heights of two similar triangles are 4 in. and 5 in. What percent of the height of the smaller triangle is the height of the larger triangle? Show your work.

Chapter

10

More Equations and Inequalities

TABLE OF CONTENTS

River	Location	Discharge (m^3/s)
Colorado	Glen Canyon Dam, CO	314.6
Snake	Hells Canyon Dam, ID	726.04
Missouri	St. Joseph, MO	1751.4
Columbia	The Dalles, OR	6331.65

internet connect

Chapter Opener Online
go.hrw.com
KEYWORD: MT4 Ch10

Career *Hydrologist*

Hydrologists measure water flow between rivers, streams, lakes, and oceans. They map their results to record locations and movement of water above and below the earth's surface.

Hydrologists are involved in projects such as water-resource studies, field irrigation, flood management, soil-erosion prevention, and the study of water discharge from creeks, streams, and rivers. The table shows the rate of water discharge for four U.S. rivers.

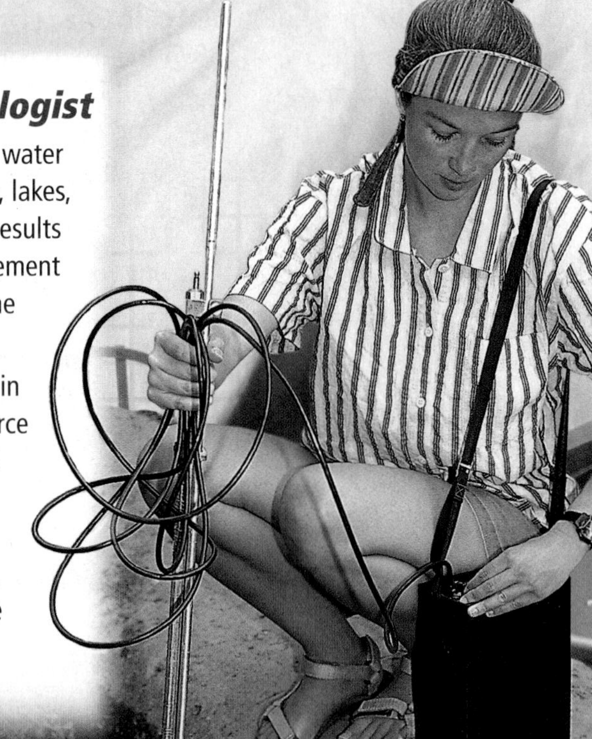

ARE YOU READY?

Choose the best term from the list to complete each sentence.

1. A letter that represents a value that can change is called a(n) __?__.
2. A(n) __?__ has one or more variables.
3. The algebraic expression $5x^2 - 3y + 4x^2 + 7$ has four __?__. Since they have the same variable raised to the same power, $5x^2$ and $4x^2$ are __?__.
4. When you individually multiply the numbers inside parentheses by the factor outside the parentheses, you are applying the __?__.

algebraic expression
Distributive Property
like terms
terms
variable

Complete these exercises to review skills you will need for this chapter.

✔ Distribute Multiplication

Replace each ■ with a number so that each equation illustrates the Distributive Property.

5. $6 \cdot (11 + 8) = 6 \cdot 11 + 6 \cdot$ ■
6. $7 \cdot (14 + 12) =$ ■ $\cdot\ 14 +$ ■ $\cdot 12$
7. $9 \cdot (6 -$ ■$) = 9 \cdot 6 - 9 \cdot 2$
8. $14 \cdot ($■ $- 7) = 14 \cdot 20 - 14 \cdot 7$

✔ Simplify Algebraic Expressions

Simplify each expression by applying the Distributive Property and combining like terms.

9. $3(x + 2) + 7x$
10. $4(y - 3) + 8y$
11. $2(z - 1) - 3z$
12. $-4(t - 6) - t$
13. $-(r - 3) - 8r$
14. $-5(4 - 2m) + 7$

✔ Connect Words and Equations

Write an equation to represent each situation.

15. The perimeter P of a rectangle is the sum of twice the length ℓ and twice the width w.
16. The volume V of a rectangular prism is the product of its three dimensions: length ℓ, width w, and height h.
17. The surface area S of a sphere is the product of 4π and the square of the radius r.
18. The cost c of a telegram of 18 words is the cost f of the first 10 words plus the cost a of each additional word.

10-1 Solving Two-Step Equations

Learn to solve two-step equations.

Sometimes more than one inverse operation is needed to solve an equation. Before solving, ask yourself, "What is being done to the variable, and in what order?" Then work backward to undo the operations.

Landscapers charge an hourly rate for labor, plus the cost of the plants. The number of hours a landscaper worked can be found by solving a two-step equation.

EXAMPLE 1 PROBLEM SOLVING APPLICATION

Chris's landscaping bill is \$380. The plants cost \$212, and the labor cost \$48 per hour. How many hours did the landscaper work?

1 Understand the Problem

The **answer** is the number of hours the landscaper worked on the yard. List the **important information:** The plants cost \$212, the labor cost \$48 per hour, and the total bill is \$380.

Let h represent the hours the landscaper worked.

Total bill	=	Plants	+	Labor
380	=	212	+	$48h$

2 Make a Plan

Think: First the variable is multiplied by 48, and then 212 is added to the result. Work backward to solve the equation. Undo the operations in reverse order: First subtract 212 from both sides of the equation, and then divide both sides of the new equation by 48.

3 Solve

$$380 = 212 + 48h$$
$$\underline{-212} \quad \underline{-212} \qquad \textit{Subtract to undo addition.}$$
$$168 = 48h$$
$$\frac{168}{48} = \frac{48h}{48} \qquad \textit{Divide to undo multiplication.}$$
$$3.5 = h$$

The landscaper worked 3.5 hours.

4 Look Back

If the landscaper worked 3.5 hours, the labor would be \$48(3.5) = \$168. The sum of the plants and the labor would be \$212 + \$168 = \$380.

EXAMPLE 2 Solving Two-Step Equations

Solve.

A $\frac{p}{4} + 5 = 13$

Think: First the variable is **divided by 4,** and then **5 is added.** To isolate the variable, **subtract 5,** and then **multiply by 4.**

$$\begin{aligned} \frac{p}{4} + 5 &= 13 \\ -5 &\quad -5 \\ \frac{p}{4} &= 8 \end{aligned}$$ *Subtract to undo addition.*

$$4 \cdot \frac{p}{4} = 4 \cdot 8$$ *Multiply to undo division.*

$$p = 32$$

Check $\frac{p}{4} + 5 \stackrel{?}{=} 13$

$\frac{32}{4} + 5 \stackrel{?}{=} 13$ *Substitute 32 into the original equation.*

$8 + 5 \stackrel{?}{=} 13$ ✔

B $1.8 = -2.5m - 1.7$

Think: First the variable is **multiplied by −2.5,** and then **1.7 is subtracted.** To isolate the variable, **add 1.7,** and then **divide by −2.5.**

$$\begin{aligned} 1.8 &= -2.5m - 1.7 \\ +1.7 &\qquad\quad +1.7 \\ 3.5 &= -2.5m \end{aligned}$$ *Add to undo subtraction.*

$$\frac{3.5}{-2.5} = \frac{-2.5m}{-2.5}$$ *Divide to undo multiplication.*

$$-1.4 = m$$

C $\frac{k+4}{9} = 6$

Think: First **4 is added** to the variable, and then the result is **divided by 9.** To isolate the variable, **multiply by 9,** and then **subtract 4.**

$$\frac{k+4}{9} = 6$$

$$9 \cdot \frac{k+4}{9} = 9 \cdot 6$$ *Multiply to undo division.*

$$\begin{aligned} k + 4 &= 54 \\ -4 &\quad -4 \\ k &= 50 \end{aligned}$$ *Subtract to undo addition.*

Think and Discuss

1. Describe how you would solve $4(x - 2) = 16$.

10-1 Exercises

FOR EOG PRACTICE
see page 680

internet connect
Homework Help Online
go.hrw.com Keyword: MT4 10-1

5.03, 5.04

GUIDED PRACTICE

See Example 1

1. Joe is paid a weekly salary of $520. He is paid an additional $21 for every hour of overtime he works. This week his total pay, including regular salary and overtime pay, was $604. How many hours of overtime did Joe work this week?

See Example 2

Solve.

2. $9t + 12 = 75$
3. $-2.4 = -1.2x + 1.8$
4. $\frac{r}{7} + 11 = 25$
5. $\frac{b + 24}{2} = 13$
6. $14q - 17 = 39$
7. $\frac{a - 3}{28} = 3$

INDEPENDENT PRACTICE

See Example 1

8. The cost of a family membership at a health club is $58 per month plus a one-time $129 start-up fee. If a family spent $651, how many months is their membership?

See Example 2

Solve.

9. $\frac{m}{-3} - 2 = 8$
10. $\frac{c - 1}{2} = 12$
11. $15g - 4 = 46$
12. $\frac{h + 19}{19} = 2$
13. $6y + 3 = -27$
14. $9.2 = 4.4z - 4$

PRACTICE AND PROBLEM SOLVING

Solve.

15. $5w + 3.8 = 16.3$
16. $15 - 3x = -6$
17. $\frac{m}{5} + 6 = 9$
18. $2.3a + 8.6 = -5.2$
19. $\frac{q + 4}{7} = 1$
20. $9 = -5g - 23$
21. $6z - 2 = 0$
22. $\frac{5}{2}d - \frac{3}{2} = -\frac{1}{2}$
23. $47k + 83 = 318$
24. $8 = 6 + \frac{p}{4}$
25. $46 - 3n = -23$
26. $\frac{7 + s}{5} = -4$
27. $9y - 7.2 = 4.5$
28. $\frac{2}{3} - 6h = -\frac{11}{6}$
29. $-1 = \frac{3}{5}b + \frac{1}{5}$

Write an equation for each sentence, and then solve it.

30. The quotient of a number and 2, minus 9, is 14.
31. A number increased by 5 and then divided by 7 is 12.
32. The sum of 10 and 5 times a number is 25.

Life Science LINK

About 20% of the more than 2500 species of snakes are venomous. The United States has 20 domestic venomous snake species, including coral snakes, rattlesnakes, copperheads, and cottonmouths.

Venom is collected from snakes and injected into horses, which develop antibodies. The horses' blood is sterilized to make antivenom.

33. The inland taipan of central Australia is the world's most toxic venomous snake. Just 1 mg of its venom is enough to kill 1000 mice. One bite contains up to 110 mg of venom. About how many mice could be killed with the venom contained in just one inland taipan bite?

34. A rattlesnake grows a new rattle segment each time it sheds its skin. Rattlesnakes shed their skin an average of three times per year. However, segments often break off. If a rattlesnake had 44 rattle segments break off in its lifetime and it had 10 rattles when it died, approximately how many years did the rattlesnake live?

35. All snakes shed their skin as they grow. The shed skin of a snake is an average of 10% longer than the actual snake. If the shed skin of a coral snake is 27.5 inches long, estimate the length of the coral snake.

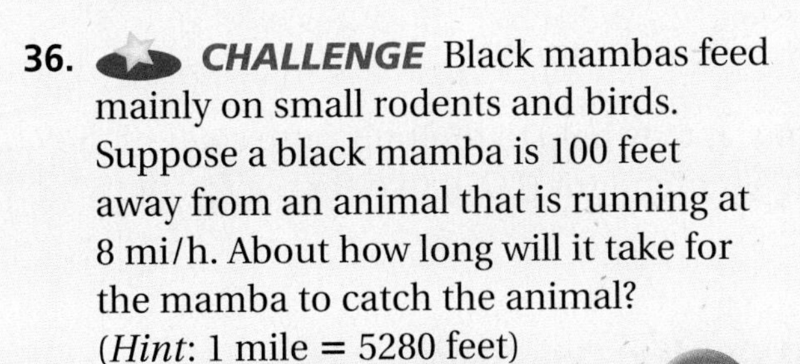

36. ***CHALLENGE*** Black mambas feed mainly on small rodents and birds. Suppose a black mamba is 100 feet away from an animal that is running at 8 mi/h. About how long will it take for the mamba to catch the animal? (*Hint*: 1 mile = 5280 feet)

Records of World's Most Venomous Snakes

Category	Record	Type of Snake
Fastest	12 mi/h	Black mamba
Longest	18 ft 9 in.	King cobra
Heaviest	34 lb	Eastern diamondback rattlesnake
Longest fangs	2 in.	Gaboon viper

Spiral Review

Simplify. (Lesson 1-6)

37. $x + 4x + 3 + 7x$ **38.** $-2m + 4 + 2m$ **39.** $w - 17 + 2$ **40.** $5s + 3r + s - 5r$

41. **EOG PREP** What is the area of the parallelogram? (Lesson 6-1)

24 cm
14 cm
12 cm

A $38\ cm^2$ B $76\ cm^2$ C $288\ cm^2$ D $336\ cm^2$

10-2 Solving Multistep Equations

Learn to solve multistep equations.

To solve a complicated equation, you may have to simplify the equation first by combining like terms.

EXAMPLE 1 Solving Equations That Contain Like Terms

Solve.

$2x + 4 + 5x - 8 = 24$

$$\begin{aligned} 2x + 4 + 5x - 8 &= 24 \\ 7x - 4 &= 24 && \textit{Combine like terms.} \\ +4 \quad &\;\;\, +4 && \textit{Add to undo subtraction.} \\ 7x \quad &= 28 \\ \frac{7x}{7} &= \frac{28}{7} && \textit{Divide to undo multiplication.} \\ x &= 4 \end{aligned}$$

Check

$$\begin{aligned} 2x + 4 + 5x - 8 &= 24 \\ 2(4) + 4 + 5(4) - 8 &\stackrel{?}{=} 24 && \textit{Substitute 4 for x.} \\ 8 + 4 + 20 - 8 &\stackrel{?}{=} 24 \\ 24 &\stackrel{?}{=} 24 \checkmark \end{aligned}$$

If an equation contains fractions, it may help to multiply both sides of the equation by the least common denominator (LCD) to clear the fractions before you isolate the variable.

EXAMPLE 2 Solving Equations That Contain Fractions

Solve.

A $\frac{3y}{7} + \frac{5}{7} = -\frac{1}{7}$

Multiply both sides by 7 to clear fractions, and then solve.

$$\begin{aligned} 7\left(\frac{3y}{7} + \frac{5}{7}\right) &= 7\left(-\frac{1}{7}\right) \\ 7\left(\frac{3y}{7}\right) + 7\left(\frac{5}{7}\right) &= 7\left(-\frac{1}{7}\right) && \textit{Distributive Property} \\ 3y + 5 &= -1 \\ -5 \quad &\;\;\, -5 && \textit{Subtract to undo addition.} \\ 3y \quad &= -6 \\ \frac{3y}{3} &= \frac{-6}{3} && \textit{Divide to undo multiplication.} \\ y &= -2 \end{aligned}$$

Solve.

B $\frac{2p}{3} + \frac{p}{4} - \frac{1}{6} = \frac{7}{2}$

The LCD is 12.

$$\begin{aligned}
12\left(\frac{2p}{3} + \frac{p}{4} - \frac{1}{6}\right) &= 12\left(\frac{7}{2}\right) && \textit{Multiply both sides by the LCD.}\\
12\left(\frac{2p}{3}\right) + 12\left(\frac{p}{4}\right) - 12\left(\frac{1}{6}\right) &= 12\left(\frac{7}{2}\right) && \textit{Distributive Property}\\
8p + 3p - 2 &= 42\\
11p - 2 &= 42 && \textit{Combine like terms.}\\
+2 \quad &\quad +2 && \textit{Add to undo subtraction.}\\
11p &= 44\\
\frac{11p}{11} &= \frac{44}{11} && \textit{Divide to undo multiplication.}\\
p &= 4
\end{aligned}$$

Check

$$\begin{aligned}
\frac{2p}{3} + \frac{p}{4} - \frac{1}{6} &= \frac{7}{2}\\
\frac{2(4)}{3} + \frac{4}{4} - \frac{1}{6} &\stackrel{?}{=} \frac{7}{2} && \textit{Substitute 4 for p.}\\
\frac{8}{3} + 1 - \frac{1}{6} &\stackrel{?}{=} \frac{7}{2}\\
\frac{16}{6} + \frac{6}{6} - \frac{1}{6} &\stackrel{?}{=} \frac{21}{6} && \textit{The LCD is 6.}\\
\frac{21}{6} &\stackrel{?}{=} \frac{21}{6} \checkmark
\end{aligned}$$

Remember!

The least common denominator (LCD) is the smallest number that each of the denominators will divide into.

EXAMPLE 3 *Money Application*

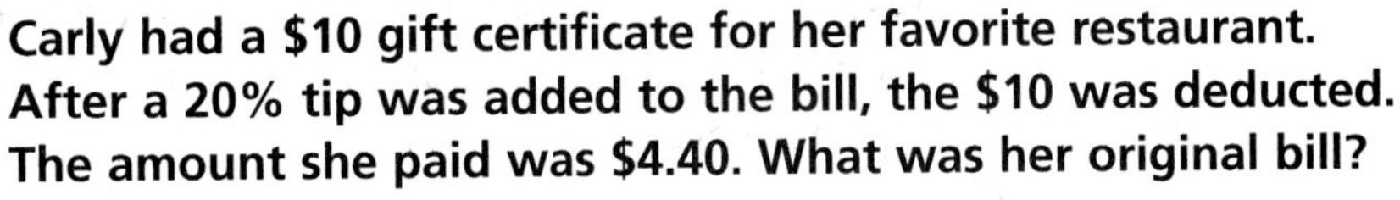

Carly had a \$10 gift certificate for her favorite restaurant. After a 20% tip was added to the bill, the \$10 was deducted. The amount she paid was \$4.40. What was her original bill?

Let b represent the amount of the original bill.

$$\begin{aligned}
b + 0.20b - 10 &= 4.40 && \textit{bill + tip − gift certificate = amount paid}\\
1.20b - 10 &= 4.40 && \textit{Combine like terms.}\\
+10 \quad &\quad +10 && \textit{Add 10 to both sides.}\\
1.20b &= 14.40\\
\frac{1.20b}{1.20} &= \frac{14.40}{1.20} && \textit{Divide both sides by 1.20.}\\
b &= 12 && \text{Her original bill was \$12.}
\end{aligned}$$

Think and Discuss

1. List the steps required to solve $3x - 4 + 2x = 7$.

2. Tell how you would clear the fractions in the equation $\frac{3x}{4} - \frac{2x}{3} + \frac{5}{8} = 1$.

10-2 Exercises

FOR EOG PRACTICE

see page 681

internet connect

Homework Help Online
go.hrw.com Keyword: MT4 10-2

5.03, 5.04

GUIDED PRACTICE

See Example 1

Solve.

1. $8d - 11 + 3d + 2 = 13$
2. $2y + 5y + 4 = 25$
3. $10e - 2e - 9 = 39$
4. $3c - 7 + 12c = 53$
5. $4h + 8 + 7h - 2h = 89$
6. $8x - 3x + 2 = -33$

See Example 2

7. $\frac{5x}{11} + \frac{4}{11} = -\frac{1}{11}$
8. $\frac{y}{2} - \frac{3y}{8} + \frac{1}{4} = \frac{1}{2}$
9. $\frac{4}{5} - \frac{2p}{5} = \frac{6}{5}$
10. $\frac{9}{4}z + \frac{1}{2} = 2$

See Example 3

11. Joley used a \$20 gift certificate to help pay for dinner for herself and a friend. After an 18% tip was added to the bill, the \$20 was deducted. The amount she paid was \$8.90. What was the original bill?

INDEPENDENT PRACTICE

See Example 1

Solve.

12. $6n + 4n - n + 5 = 23$
13. $-83 = 6k + 17 + 4k$
14. $36 - 4c - 3c = 22$
15. $10 + 4w - 3w = 13$
16. $28 = 10a - 5a - 2$
17. $30 = 7y - 35 + 6y$

See Example 2

18. $\frac{3}{8} + \frac{p}{8} = 3\frac{1}{8}$
19. $\frac{9h}{10} - \frac{3h}{10} = \frac{18}{10}$
20. $\frac{4g}{14} - \frac{3}{7} - \frac{g}{14} = \frac{3}{14}$
21. $\frac{5}{18} = \frac{4m}{9} - \frac{m}{3} + \frac{1}{2}$
22. $\frac{5}{11} = -\frac{3b}{11} + \frac{8b}{22}$
23. $\frac{3x}{4} - \frac{11x}{24} = -1\frac{1}{6}$

See Example 3

24. Pat bought 6 shirts that were all the same price. He used a traveler's check for \$25, and then paid the difference of \$86. What was the price of each shirt?

PRACTICE AND PROBLEM SOLVING

Solve and check.

25. $\frac{5n}{6} - \frac{1}{4} = \frac{3}{8}$
26. $5n + 12 - 9n = -16$
27. $6b - 1 - 10b = 51$
28. $\frac{x}{2} + \frac{2}{3} = \frac{5}{6}$
29. $-2x - 7 + 3x = 10$
30. $\frac{3r}{4} - \frac{2}{3} = \frac{5}{6}$
31. $5y - 2 - 8y = 31$
32. $7n - 10 - 9n = -13$
33. $\frac{h}{6} + \frac{h}{8} = 1\frac{1}{6}$
34. $2a + 7 + 3a = 32$
35. $\frac{b}{6} + \frac{3b}{8} = \frac{5}{12}$
36. $-10 = 9m - 13 - 7m$

You can estimate the weight in pounds of a fish that is L inches long and G inches around at the thickest part by using the formula $W \approx \frac{LG^2}{800}$.

37. Gina is paid 1.5 times her normal hourly rate for each hour she works over 40 hours in a week. Last week she worked 48 hours and earned \$634.40. What is her normal hourly rate?

38. ***SPORTS*** The average weight of the top 5 fish at a fishing tournament was 12.3 pounds. The weights of the second-, third-, fourth-, and fifth-place fish are shown in the table. What was the weight of the heaviest fish?

Winning Entries

Caught By	Weight (lb)
Wayne S.	■
Carla P.	12.8
Deb N.	12.6
Virgil W.	11.8
Brian B.	9.7

39. ***PHYSICAL SCIENCE*** The formula $C = \frac{5}{9}(F - 32)$ is used to convert a temperature from degrees Fahrenheit to degrees Celsius. Water boils at 100°C. Use the formula to find the boiling point of water in degrees Fahrenheit.

40. At a bulk food store, Kerry bought $\frac{2}{3}$ lb of coffee that cost \$4.50/lb, $\frac{3}{4}$ lb of coffee that cost \$5.20/lb, and $\frac{1}{5}$ lb of coffee that did not have a price marked. If her total cost was \$8.18, what was the price per pound of the third type of coffee?

41. ***WHAT'S THE ERROR?*** A student's work in solving an equation is shown. What error has the student made, and what is the correct answer?

$$\frac{1}{3}x + 3x = 7$$
$$x + 3x = 21$$
$$4x = 21$$
$$x = \frac{21}{4}$$

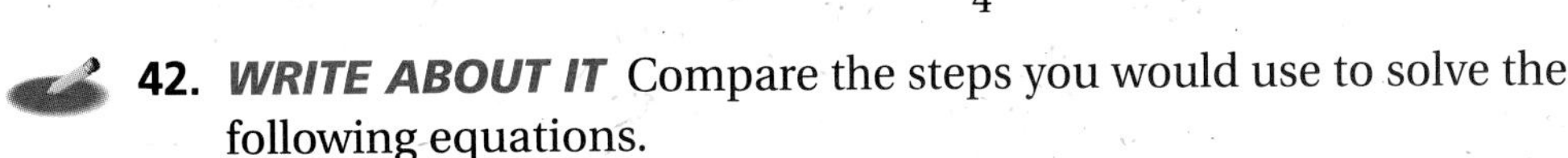

42. ***WRITE ABOUT IT*** Compare the steps you would use to solve the following equations.

$4x - 8 = 16$ $\qquad$ $4(x - 2) = 16$

43. ***CHALLENGE*** List the steps you would use to solve the following equation.

$$\frac{5\left(\frac{1}{2}x - \frac{1}{3}\right) + \frac{7}{6}x}{2} + 2 = 3$$

Spiral Review

Evaluate each expression for the given value of the variable. (Lesson 3-2)

44. $19.4 - x$ for $x = -5.6$ **45.** $11 - r$ for $r = 13.5$ **46.** $p + 65.1$ for $p = -42.3$

47. $-\frac{3}{7} - t$ for $t = 1\frac{5}{7}$ **48.** $3\frac{5}{11} + y$ for $y = -2\frac{4}{11}$ **49.** $-\frac{1}{19} + g$ for $g = \frac{18}{19}$

50. **EOG PREP** $\overleftrightarrow{AB}$ has a slope of $\frac{2}{5}$. What is the slope of a line perpendicular to $\overleftrightarrow{AB}$? (Lesson 5-5)

A $-\frac{2}{5}$ B $\frac{5}{2}$ C $-\frac{5}{2}$ D $\frac{7}{5}$

Model Equations with Variables on Both Sides

Use with Lesson 10-3

1.02, 5.04

KEY

Algebra tiles

[+] $= x$ [−] $= -x$

[−] $= 1$ [+] $= -1$

REMEMBER

It will not change the value of an expression if you add or remove zero.

[+] + [−] $= 0$ [+] + [−] $= 0$

internet connect
Lab Resources Online
go.hrw.com
KEYWORD: MT4 Lab10A

To solve an equation with the same variable on both sides of the equal sign, you must first add or subtract to eliminate the variable term from one side of the equation.

Activity

1 Model and solve the equation $-x + 2 = 2x - 4$.

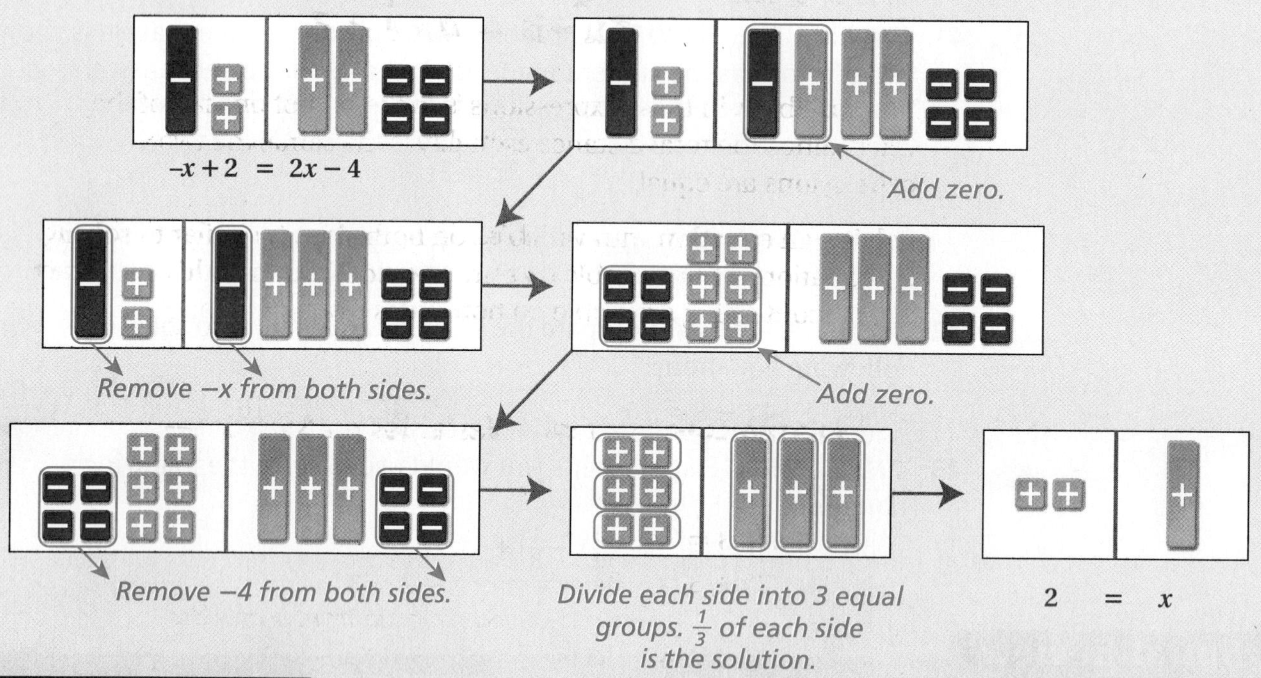

Think and Discuss

1. How would you check the solution to $-x + 2 = 2x - 4$ using algebra tiles?
2. Why must you isolate the variable terms by having them on only one side of the equation?

Try This

Model and solve each equation.

1. $x + 1 = -x - 1$ **2.** $3x = -3x + 18$ **3.** $4 - 2x = -5x + 7$ **4.** $2x + 2x + 1 = x + 10$

10-3 Solving Equations with Variables on Both Sides

Learn to solve equations with variables on both sides of the equal sign.

Some problems produce equations that have variables on both sides of the equal sign. For instance, Elaine runs the same distance each day. On Mondays, Fridays, and Saturdays, she runs 3 laps on the track and an additional 5 miles off the track. On Tuesdays and Thursdays, she runs 4 laps on the track and 2.5 miles off the track.

Expression for Mondays, Fridays, and Saturdays $3x+5$ $4x+2.5$ *Expression for Tuesdays and Thursdays*

$$3x+5 = 4x+2.5$$

The variable x in these expressions is the length of one lap of the track. Since the total distance each day is the same, the two expressions are equal.

Solving an equation with variables on both sides is similar to solving an equation with a variable on only one side. You can add or subtract a term containing a variable on both sides of an equation.

EXAMPLE 1 Solving Equations with Variables on Both Sides

Solve.

A $2a + 3 = 3a$

$$2a + 3 = 3a$$
$$-2a \qquad -2a \qquad \text{Subtract 2a from both sides.}$$
$$3 = a$$

B $4v - 7 = 5 + 7v$

$$4v - 7 = 5 + 7v$$
$$-4v \qquad -4v \qquad \text{Subtract 4v from both sides.}$$
$$-7 = 5 + 3v$$
$$-5 \quad -5 \qquad \text{Subtract 5 from both sides.}$$
$$-12 = 3v$$
$$\frac{-12}{3} = \frac{3v}{3} \qquad \text{Divide both sides by 3.}$$
$$-4 = v$$

Helpful Hint

If the variables in an equation are eliminated and the resulting statement is false, the equation has no solution.

Solve.

C $g + 5 = g - 2$

$$\begin{aligned} g + 5 &= g - 2 \\ -g \quad &\quad -g \quad && \text{Subtract } g \text{ from both sides.} \\ 5 &\neq -2 \end{aligned}$$

No solution. There is no number that can be substituted for the variable g to make the equation true.

To solve multistep equations with variables on both sides, first combine like terms and clear fractions. Then add or subtract variable terms to both sides so that the variable occurs on only one side of the equation. Then use properties of equality to isolate the variable.

EXAMPLE 2 Solving Multistep Equations with Variables on Both Sides

Solve.

A $2c + 4 - 3c = -9 + c + 5$

$$\begin{aligned} 2c + 4 - 3c &= -9 + c + 5 \\ -c + 4 &= -4 + c && \textit{Combine like terms.} \\ +c \quad &\quad +c && \textit{Add c to both sides.} \\ 4 &= -4 + 2c \\ +4 &\quad +4 && \textit{Add to undo subtraction.} \\ 8 &= 2c \\ \frac{8}{2} &= \frac{2c}{2} && \textit{Divide to undo multiplication.} \\ 4 &= c \end{aligned}$$

B $\frac{w}{2} - \frac{3w}{4} + \frac{1}{3} = w + \frac{7}{6}$

$$\begin{aligned} \frac{w}{2} - \frac{3w}{4} + \frac{1}{3} &= w + \frac{7}{6} \\ 12\left(\frac{w}{2} - \frac{3w}{4} + \frac{1}{3}\right) &= 12\left(w + \frac{7}{6}\right) && \textit{Multiply by LCD, 12.} \\ 12\left(\frac{w}{2}\right) - 12\left(\frac{3w}{4}\right) + 12\left(\frac{1}{3}\right) &= 12(w) + 12\left(\frac{7}{6}\right) \\ 6w - 9w + 4 &= 12w + 14 \\ -3w + 4 &= 12w + 14 && \textit{Combine like terms.} \\ +3w \quad &\quad +3w && \textit{Add 3w to both sides.} \\ 4 &= 15w + 14 \\ -14 &\quad -14 && \textit{Subtract 14 from both sides.} \\ -10 &= 15w \\ \frac{-10}{15} &= \frac{15w}{15} && \textit{Divide both sides by 15.} \\ -\frac{2}{3} &= w \end{aligned}$$

EXAMPLE 3 Sports Application

Elaine runs the same distance every day. On Mondays, Fridays, and Saturdays, she runs 3 laps on the track, and then runs 5 more miles. On Tuesdays and Thursdays, she runs 4 laps on the track, and then runs 2.5 more miles. On Wednesdays, she just runs laps. How many laps does she run on Wednesdays?

First solve for the distance around the track.

$$3x + 5 = 4x + 2.5$$ *Let x represent the distance around the track.*

$$\underline{-3x} \quad = \underline{-3x}$$ *Subtract 3x from both sides.*

$$5 = x + 2.5$$

$$\underline{-2.5} \quad \underline{-2.5}$$ *Subtract 2.5 from both sides.*

$$2.5 = x$$ *The track is 2.5 miles around.*

Now find the total distance Elaine runs each day.

$$3x + 5$$ *Choose one of the original expressions.*

$$3(2.5) + 5 = 12.5$$ *Elaine runs 12.5 miles each day.*

Find the number of laps Elaine runs on Wednesdays.

$$2.5n = 12.5$$ *Let n represent the number of 2.5-mile laps.*

$$\frac{2.5n}{2.5} = \frac{12.5}{2.5}$$ *Divide both sides by 2.5.*

$$n = 5$$

Elaine runs 5 laps on Wednesdays.

Helpful Hint

The value of the variable is not necessarily the answer to the question.

Think and Discuss

1. Give an example of an equation that has no solution.

10-3 Exercises

5.03, 5.04

FOR EOG PRACTICE
see page 681

internet connect
Homework Help Online
go.hrw.com Keyword: MT4 10-3

GUIDED PRACTICE

See Example 1

Solve.

1. $5x + 2 = x + 6$
2. $6a - 6 = 8 + 4a$
3. $3x + 9 = 10x - 5$
4. $4y - 2 = 6y + 6$

See Example 2

5. $4x - 5 + 2x = 13 + 9x - 21$
6. $\frac{2n}{5} + \frac{n}{10} - 4 = 6 + 3n - 15$
7. $\frac{3}{10} + \frac{9d}{10} - 2 = 2d + 4 - 3d$
8. $4(x - 5) + 2 = x + 3$

See Example 3

9. June has a set of folding chairs for her flute students. If she arranges them in 5 rows for a recital, she has 2 chairs left over. If she arranges them in 3 rows of the same length, she has 14 left over. How many chairs does she have?

INDEPENDENT PRACTICE

See Example 1

Solve.

10. $2n + 12 = 5n$
11. $9x - 2 = 10 - 3x$
12. $5n + 3 = 14 - 6n$
13. $9y - 6 = 7y + 8$
14. $5x + 2 = x + 6$
15. $2(4x + 15) = 8x + 3$

See Example 2

16. $\frac{2p}{9} + \frac{5p}{18} - \frac{5}{6} = \frac{2}{3} + \frac{p}{12} + \frac{1}{6}$
17. $3(x - 4) - 4 = 5x + 6.9 - 3x$
18. $\frac{1}{2}(2n + 6) = 5n - 12 - n$
19. $\frac{a}{22} - 4.5 + 2a = \frac{7}{11} + \frac{17a}{11} + \frac{4}{11}$

See Example 3

20. Sean and Laura have the same number of action figures in their collections. Sean has 6 complete sets plus 2 individual figures, and Laura has 3 complete sets plus 20 individual figures. How many figures are in a complete set?

PRACTICE AND PROBLEM SOLVING

Solve and check.

21. $8y - 3 = 17 - 2y$
22. $2n + 6 = 7n - 9$
23. $2n + 12n = 2(n + 12)$
24. $3(4x - 2) = 12x$
25. $100(x - 3) = 450 - 50x$
26. $5p - 15 = 15 - 5p$
27. $\frac{1}{2} - \frac{3m}{4} + 7 = 4m - 9 - \frac{m}{28}$
28. $7(x - 1) = 3\left(x + \frac{1}{3}\right)$
29. $4(x - 5) + 2 = \frac{1}{3}(x + 9) + \frac{2x}{3}$
30. $12\left(4r - \frac{5r}{6}\right) + 20 = 19r - 15 + \frac{45r}{2}$

Both figures have the same perimeter. Find each perimeter.

31.

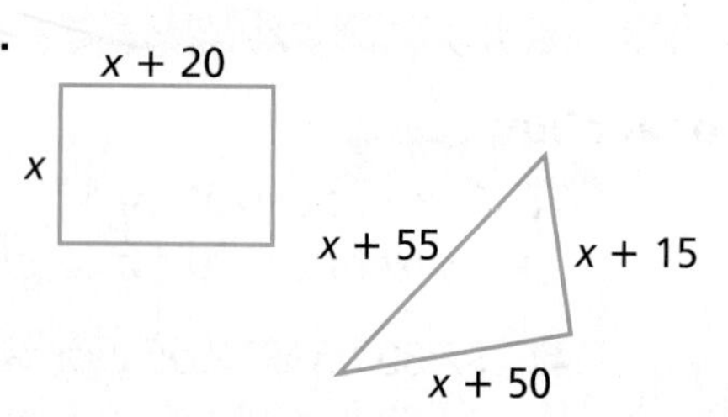

32.

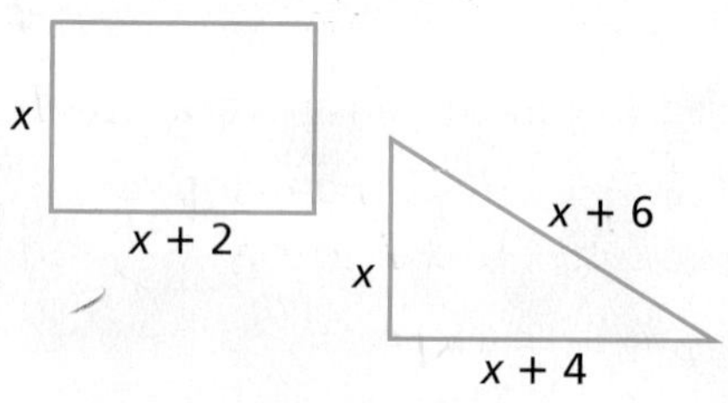

33. Find two consecutive whole numbers such that one-fourth of the first number is one more than one-fifth of the second number. (*Hint:* Let n represent the first number. Then $n + 1$ represents the next consecutive whole number.)

34. Find three consecutive whole numbers such that the sum of the first two numbers equals the third number. (*Hint:* Let n represent the first number. Then $n + 1$ and $n + 2$ represent the next two whole numbers.)

Physical Science LINK

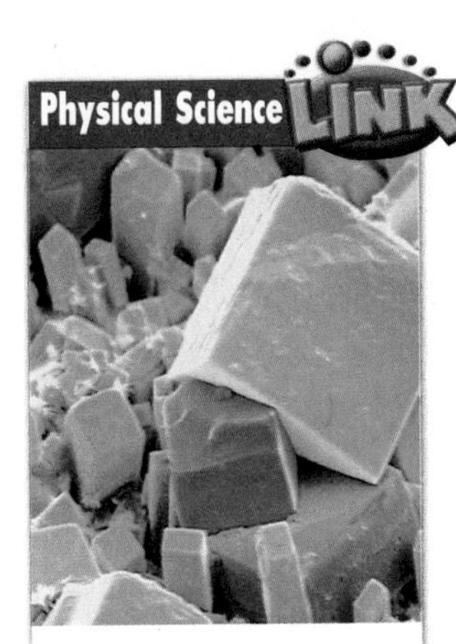

Sodium and chlorine bond together to form sodium chloride, or salt. The atomic structure of sodium chloride causes it to form cubes.

35. ***PHYSICAL SCIENCE*** An atom of chlorine (Cl) has 6 more protons than an atom of sodium (Na). The atomic number of chlorine is 5 less than twice the atomic number of sodium. The atomic number of an element is equal to the number of protons per atom.

a. How many protons are in an atom of chlorine?

b. What is the atomic number of sodium?

36. ***EARTH SCIENCE*** *Specific gravity* compares the density of a mineral with the density of water. The following equation relates a mineral's specific gravity s, its weight in air a, and its weight in water w.

Mineral	Specific Gravity	Weight in Air	Weight in Water
Granite	■	152.3 g	97.2 g
Gold	19.3	10 g	■
Quartz	2.65	■	6.5 g

$s(a - w) = a$

a. Find the specific gravity of a piece of granite.

b. Find the weight in water of a piece of gold that weighs 10 g in air.

c. Find the weight in air of a piece of quartz that weighs 6.5 g in water.

37. ***CHOOSE A STRATEGY*** Solve the following equation for t. How can you determine the solution once you have combined like terms?

$2(t - 24) = 5t - 3(t + 16)$

38. ***WRITE ABOUT IT*** Two cars are traveling in the same direction. The first car is going 45 mi/h, and the second car is going 60 mi/h. The first car left 2 hours before the second car. Explain how you could solve an equation with variables on both sides to find how long it will take the second car to catch up to the first car.

39. ***CHALLENGE*** Solve the equation $\frac{x+1}{7} = \frac{3}{4} + \frac{x-3}{5}$.

Spiral Review

Find both unit prices and tell which is the better buy. (Lesson 7-2)

40. $11.99 for 2 yd of fencing
$25 for 10 ft of fencing

41. 20 oz of cereal for $3.49
16 oz of cereal for $2.99

42. 4 tickets for $110
6 tickets for $180

43. $2.39 for a 12 oz can of carrots
$3.68 for a 20 oz can of carrots

44. $5.47 for a box of 100 nails
$13.12 for a box of 250 nails

45. $747 for 3 computer monitors
$550 for 2 computer monitors

46. **EOG PREP** A square has a perimeter of 56 cm. If the square is dilated by a scale factor of 0.2, what is the length of each side of the new square? (Lesson 7-5)

A 11.2 cm B 2.8 cm C 5.6 cm D 14 cm

Chapter 10 Mid-Chapter Quiz

LESSON 10-1 (pp. 498–501)

Solve.

1. $5x + 17 = 47$
2. $4y + 1 = -15$
3. $16 - z = 12$
4. $\frac{1}{2}t + 9 = 25$
5. $-32 = \frac{7}{3}w - 11$
6. $\frac{2}{3}q - 9 = -1$
7. $\frac{x + 8}{4} = -10$
8. $5 = \frac{21 - z}{3}$
9. $\frac{a - 4}{3} = 5$
10. A car rental company charges \$39.99 per day plus \$0.20 per mile. Jill rented a car for one day and the charges were \$47.39, before tax. How many miles did Jill drive?

LESSON 10-2 (pp. 502–505)

Solve.

11. $4c + 2c + 6 = 24$
12. $\frac{2x}{5} - \frac{3}{5} = \frac{11}{5}$
13. $\frac{t}{5} + \frac{t}{3} = \frac{8}{15}$
14. $\frac{4m}{3} - \frac{m}{6} = \frac{7}{2}$
15. $8 - 6g + 15 = 19$
16. $\frac{2}{5}b - \frac{1}{4}b = 3$
17. $\frac{r}{3} + 7 - \frac{r}{5} = -3$
18. $5k + 9.3 = 21.8$
19. $\frac{x}{4} - \frac{x}{5} - \frac{1}{3} = \frac{16}{15}$
20. On his last three math tests, Mark scored 85, 95, and 80. What grade must he get on his next test to have an average of 90 for all four tests?

LESSON 10-3 (pp. 507–511)

Solve.

21. $3x + 13 = x + 1$
22. $q + 7 = 2q + 5$
23. $8n + 24 = 3n + 59$
24. $m + 5 = m + 3$
25. $9w - 2w + 8 = 4w + 38$
26. $-2a + a + 9 = 3a - 9$
27. $\frac{5c}{4} = \frac{2c}{3} + 7$
28. $\frac{3z}{2} - \frac{17}{3} = \frac{2z}{3} - \frac{3}{2}$
29. $\frac{7}{12}y - \frac{1}{4} = 2y - \frac{5}{3}$
30. The rectangle and the triangle have the same perimeter. Find the perimeter of each figure.

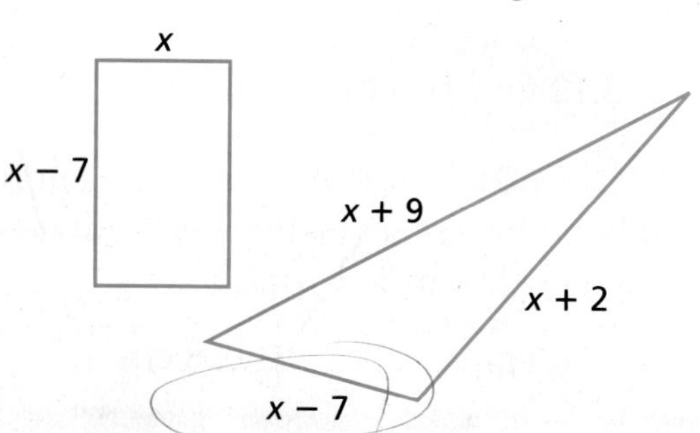

Focus on Problem Solving

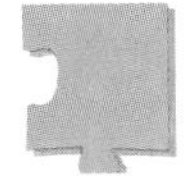

Make a Plan

- **Write an equation**

Several steps may be needed to solve a problem. It often helps to write an equation that represents the steps.

Example:

Juan's first 3 exam scores are 85, 93, and 87. What does he need to score on his next exam to average 90 for the 4 exams?

Let x be the score on his next exam. The average of the exam scores is the sum of the 4 scores, divided by 4. This amount must equal 90.

$$\begin{aligned} \text{Average of exam scores} &= 90 \\ \frac{85 + 93 + 87 + x}{4} &= 90 \\ \frac{265 + x}{4} &= 90 \\ 4\left(\frac{265 + x}{4}\right) &= 4(90) \\ 265 + x &= 360 \\ -265 \quad & \quad -265 \\ x &= 95 \end{aligned}$$

Juan needs a 95 on his next exam.

Read each problem and write an equation that could be used to solve it.

1. The average of two numbers is 27. The first number is twice the second number. What are the two numbers?

2. Nancy spends $\frac{1}{3}$ of her monthly salary on rent, $\frac{1}{10}$ on her car payment, $\frac{1}{12}$ on food, $\frac{1}{5}$ on other bills, and has $680 left for other expenses. What is Nancy's monthly salary?

3. A vendor at a concert sells caps and T-shirts. The T-shirts cost 1.5 times as much as the caps. If 5 caps and 7 T-shirts cost $248, what is the price of each item?

4. Amanda and Rick have the same amount to spend on school supplies. Amanda buys 4 notebooks and has $8.60 left. Rick buys 7 notebooks and has $7.55 left. How much does each notebook cost?

Solve and graph.

C $2x + 3 > 5x - 6$

$$\begin{aligned} 2x + 3 &> 5x - 6 \\ \underline{-2x} \quad &\quad \underline{-2x} \\ 3 &> 3x - 6 \\ \underline{+6} \quad &\quad \underline{+6} \\ 9 &> 3x \\ \frac{9}{3} &> \frac{3x}{3} \\ 3 &> x \end{aligned}$$

Subtract 2x from both sides.

Add 6 to both sides.

Divide both sides by 3.

−2 0 2 4 6

EXAMPLE 3 *Business Application*

The student council sells T-shirts with the school logo on them. The unit cost is $10.50 for the shirt and the ink. They have a fixed cost of $60 for silk screen equipment. If they sell the shirts for $12 each, how many must they sell to make a profit?

Let R represent the revenue and C represent the cost. In order for the student council to make a profit, the revenue must be greater than the cost.

$$R > C$$

The revenue from selling x shirts at $12 each is $12x$. The cost of producing x shirts is the fixed cost plus the unit cost times the number of shirts produced, or $60 + 10.50x$. Substitute the expressions for R and C.

$$12x > 60 + 10.50x$$ *Let x represent the number of shirts sold.*

$$\underline{-10.50x} \qquad \underline{-10.50x}$$ *Subtract 10.50x from both sides.*

$$1.5x > 60$$

$$\frac{1.5x}{1.5} > \frac{60}{1.5}$$ *Divide both sides by 1.5.*

$$x > 40$$

The student council must sell more than 40 shirts to make a profit.

Think and Discuss

1. **Compare** solving a multistep equation with solving a multistep inequality.
2. **Describe** two situations in which you would have to reverse the inequality symbol when solving a multistep inequality.

10-4 Exercises

FOR EOG PRACTICE see page 682

internet connect
Homework Help Online
go.hrw.com Keyword: MT4 10-4

5.03, 5.04

GUIDED PRACTICE

See Example 1 — **Solve and graph.**

1. $2k + 4 > 10$
2. $\frac{1}{2}z - 5.5 \leq 4.5$
3. $5y + 10 < -25$
4. $-4x + 6 \geq 14$
5. $4y + 1.5 \geq 13.5$
6. $3k - 2 > 13$

See Example 2

7. $4x - 3 + x < 12$
8. $\frac{4b}{5} + \frac{7}{10} \geq \frac{1}{2}$
9. $4 + 9h - 7 \leq 3h + 3$
10. $14c + 2 - 3c > 8 + 8c$
11. $\frac{1}{9} + \frac{d}{3} < \frac{1}{2} - \frac{2d}{3}$
12. $\frac{5}{6} \geq \frac{4m}{9} - \frac{1}{3} + \frac{2m}{9}$

See Example 3

13. A school's Spanish club is selling printed caps to raise money for a trip. The printer charges $150 in advance plus $3 for every cap ordered. If the club sells caps for $12.50 each, at least how many caps do they need to sell to make a profit?

INDEPENDENT PRACTICE

See Example 1 — **Solve and graph.**

14. $6k - 8 > 22$
15. $10x + 2 > 42$
16. $5p - 5 \leq 45$
17. $14 \geq 13q - 12$
18. $3.6 + 7.2n < 25.2$
19. $-8x - 12 \geq 52$

See Example 2

20. $7p + 5 < 6p - 12$
21. $11 + 17a \geq 13a - 1$
22. $\frac{11}{13} + \frac{n}{2} > \frac{25}{26}$
23. $\frac{2}{3} \leq \frac{1}{2}k - \frac{5}{6}$
24. $\frac{n}{7} + \frac{11}{14} \leq -\frac{17}{14}$
25. $3r - 16 + 7r < 14$

See Example 3

26. Josef is on the planning committee for the eighth-grade holiday party. The food, decoration, and entertainment costs total $350. The committee has $75 in the treasury. If the committee expects to sell the tickets for $5 each, at least how many tickets must be sold to cover the remaining cost of the party?

PRACTICE AND PROBLEM SOLVING

Solve and graph.

27. $3p - 3 \leq 19$
28. $12n + 26 > -10$
29. $4 - 9w < 13$
30. $-8x - 18 \geq 14$
31. $16a + 3 > 11$
32. $-2y + 1 \geq 8$
33. $3q - 5q > -12$
34. $\frac{3m}{4} + \frac{2}{3} > \frac{m}{2} + \frac{7}{8}$
35. $7b - 4.6 < 3b + 6.2$
36. $6k + 4 - 3k \geq 2$
37. $26 - \frac{33}{4} \leq -\frac{2}{3}f - \frac{1}{4}$
38. $\frac{7}{9}v + \frac{5}{12} - \frac{3}{18}v \geq \frac{3}{4}v + \frac{1}{3}$
39. ***ENTERTAINMENT*** A concert is being held in a gymnasium that can hold no more than 550 people. A permanent bleacher will seat 30 people. The event organizers are setting up 20 rows of chairs. At most, how many chairs can be in each row?

40. Katie and April are making a string of pi beads for pi day (March 14). They use 10 colors of beads that represent the digits 0–9, and the beads are strung in the order of the digits of π. The string already has 70 beads. If they have 30 days to string the beads, and they want to string 1000 beads by π day, at least how many beads do they have to string each day?

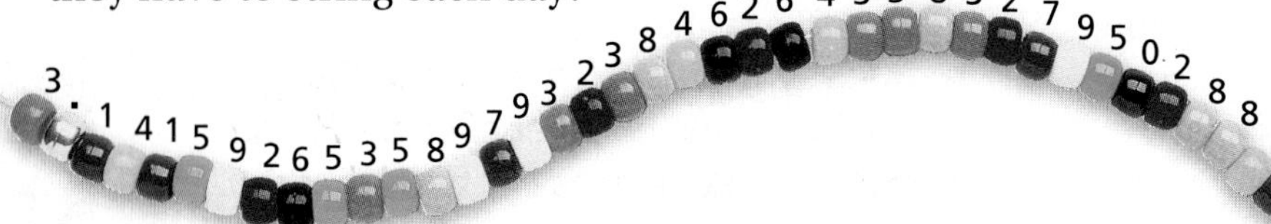

41. *SPORTS* The Cubs have won 44 baseball games and have lost 65 games. They have 53 games remaining. At least how many of the remaining 53 games must the Cubs win to have a winning season? (A winning season means they win more than 50% of their games.)

42. *ECONOMICS* Satellite TV customers can either purchase a dish and receiver for \$249 or pay a \$50 fee and rent the equipment for \$12 a month.

a. How much would it cost to rent the equipment for 9 months?

b. How many months would it take for the rental charges to exceed the purchase price?

43. *WRITE A PROBLEM* Write and solve an inequality using the following shipping rates for orders from a mail-order catalog.

Mail-Order Shipping Rates				
Merchandise Amount	\$0.01–\$20	\$20.01–\$30	\$30.01–\$45	\$45.01–\$60
Shipping Cost	\$4.95	\$5.95	\$7.95	\$8.95

44. *WRITE ABOUT IT* Describe two ways to solve the inequality below. In one way, you must reverse the inequality symbol, but in the other way, you do not need to reverse the symbol.

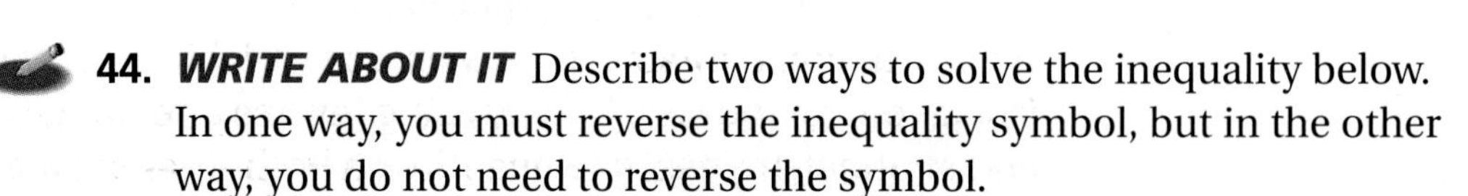

$$-2x - 3 < x + 4$$

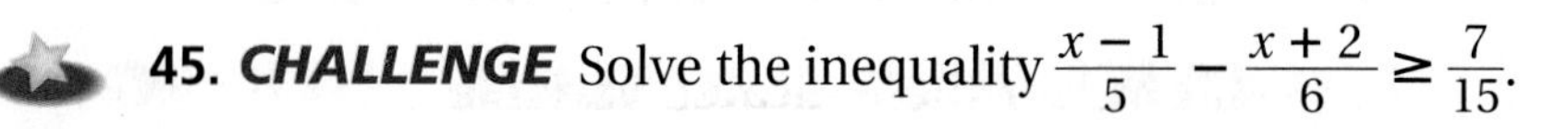

45. *CHALLENGE* Solve the inequality $\frac{x-1}{5} - \frac{x+2}{6} \geq \frac{7}{15}$.

Spiral Review

Find each number. (Lesson 8-3)

46. 19 is 20% of what number?

47. 74% of what number is 481?

48. 32% of what number is 58.88?

49. 0.7488 is 52% of what number?

50. **EOG PREP** What is the probability of rolling an odd number on a fair number cube? (Lesson 9-4)

A $\frac{1}{2}$ B $\frac{2}{3}$ C $\frac{1}{6}$ D $\frac{1}{3}$

10-5 Solving for a Variable

Learn to solve an equation for a variable.

Euler's formula relates the number of vertices V, the number of edges E, and the number of faces F of a polyhedron.

$$V - E + F = 2$$

Tetrahedron:
4 faces **6 edges** **4 vertices**
$4 - 6 + 4 = 2$

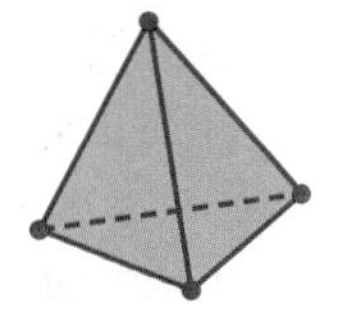

Leonard Euler (1707–1783) made major contributions to nearly every area of mathematics, including algebra, geometry, and calculus.

Suppose a polyhedron has 8 vertices and 12 edges. How many faces does it have? One way to find the answer is to substitute values into the formula and solve. Another way to find the answer is to solve for the variable first and then substitute the values.

Substitute, then solve:

$$\begin{aligned} V - E + F &= 2 \\ 8 - 12 + F &= 2 \\ -4 + F &= 2 \\ \underline{+4\quad} & \quad \underline{+4} \\ F &= 6 \end{aligned}$$

Solve, then substitute:

$$\begin{aligned} V - E + F &= 2 \\ \underline{-V + E\quad} & \quad \underline{-V + E} \\ F &= 2 - V + E \\ F &= 2 - 8 + 12 \\ F &= 6 \end{aligned}$$

If an equation contains more than one variable, you can sometimes isolate one of the variables by using inverse operations. You can add and subtract any variable quantity on both sides of an equation.

EXAMPLE 1 Solving for a Variable by Addition or Subtraction

Solve for the indicated variable.

A Solve $V - E + F = 2$ for V.

$$\begin{aligned} V - E + F &= 2 \\ \underline{+E - F\quad} & \quad \underline{+E - F} \\ V \quad &= 2 + E - F \end{aligned}$$

Add E and subtract F from both sides.

Isolate V.

B Solve $V - E + F = 2$ for E.

$$\begin{aligned} V - E + F &= 2 \\ \underline{-V \quad -F} & \quad \underline{-V - F} \\ -E &= 2 - V - F \\ -1 \cdot (-E) &= -1 \cdot (2 - V - F) \\ E &= -2 + V + F \end{aligned}$$

Subtract V and F from both sides.

Multiply both sides by −1.

Isolate E.

To isolate a variable, you can multiply or divide both sides of an equation by a variable if it can never be equal to 0. You can also take the square root of both sides of an equation that cannot have negative values.

EXAMPLE 2 Solving for a Variable by Division or Square Roots

Helpful Hint

In the geometry formulas $A = \frac{1}{2}bh$ and $a^2 + b^2 = c^2$, the variables represent distances or areas, so they cannot be 0 or negative.

Solve for the indicated variable. Assume all values are positive.

A Solve $A = \frac{1}{2}bh$ for h.

$$A = \frac{1}{2}bh$$
$$2 \cdot A = 2 \cdot \frac{1}{2}bh$$
$$\frac{2A}{b} = \frac{bh}{b}$$
$$\frac{2A}{b} = h$$

B Solve $a^2 + b^2 = c^2$ for a.

$$a^2 + b^2 = c^2$$
$$\underline{-b^2 \qquad -b^2}$$
$$a^2 = c^2 - b^2$$
$$\sqrt{a^2} = \sqrt{c^2 - b^2}$$
$$a = \sqrt{c^2 - b^2}$$

C Solve the formula for the surface area of a cylinder for h.

$$S = 2\pi r^2 + 2\pi rh \qquad \textit{Write the formula.}$$
$$\underline{-2\pi r^2 \qquad -2\pi r^2} \qquad \textit{Subtract } 2\pi r^2 \textit{ from each side.}$$
$$S - 2\pi r^2 = 2\pi rh$$
$$\frac{S}{2\pi r} - \frac{2\pi r^2}{2\pi r} = \frac{2\pi rh}{2\pi r} \qquad \textit{The radius r cannot be 0.}$$
$$\frac{S}{2\pi r} - r = h \qquad \textit{Isolate h.}$$

When graphing on a coordinate plane, it is helpful to solve for y. Most graphing calculators will graph only equations that are solved for y.

EXAMPLE 3 Solving for y and Graphing

Remember!

To find solutions (x, y), choose values for x and substitute to find y.

Solve for y and graph $2x + 3y = 6$.

$$2x + 3y = 6$$
$$\underline{-2x \qquad -2x}$$
$$3y = -2x + 6$$
$$\frac{3y}{3} = \frac{-2x + 6}{3}$$
$$y = \frac{-2x}{3} + 2$$

x	y
−3	4
0	2
3	0
6	−2

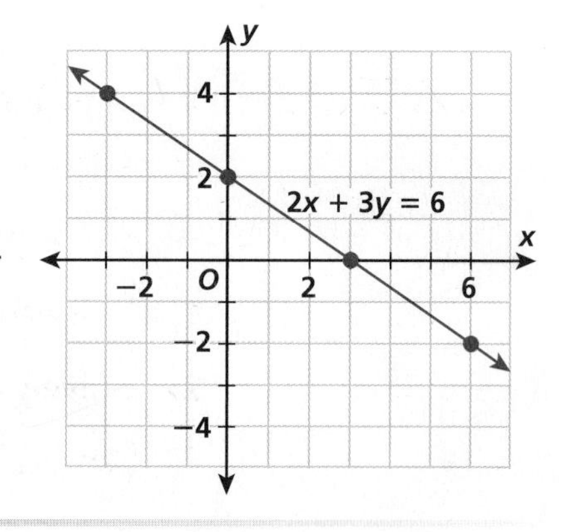

Think and Discuss

1. **List** the steps you would use to solve $P = 2b + 2h$ for h.
2. **Describe** how to graph the equation $\frac{1}{2}x + y = 4$.

10-5 Exercises

FOR EOG PRACTICE

see page 682

internet connect
Homework Help Online
go.hrw.com Keyword: MT4 10-5

5.04

GUIDED PRACTICE

See Example 1

Solve for the indicated variable. Assume all values are positive.

1. Solve $\ell_1 + \ell_2 + \ell_3 = P$ for ℓ_2.
2. Solve $\ell_1 + \ell_2 + \ell_3 = P$ for ℓ_1.
3. Solve $A - B + 2 = C$ for A.
4. Solve $A - B + 2 = C$ for B.

See Example 2

5. Solve $A = \frac{1}{2}d_1d_2$ for d_1.
6. Solve $a^2 + b^2 = c^2$ for b.
7. Solve $S = (n - 2)180$ for n.
8. Solve $F = \frac{9}{5}C + 32$ for C.

See Example 3

Solve each equation for y and graph the equation.

9. $y + 3x = 15$
10. $2y - 9x = 14$
11. $6x - 3y - 3 = 0$

INDEPENDENT PRACTICE

See Example 1

Solve for the indicated variable. Assume all values are positive.

12. Solve $A_1 + A_2 + A_3 = 180$ for A_1.
13. Solve $A_1 + A_2 + A_3 = 180$ for A_3.
14. Solve $p - c = 100 + a$ for a.
15. Solve $p - c = 100 + a$ for c.

See Example 2

16. Solve $E = mc^2$ for m.
17. Solve $E = mc^2$ for c.
18. Solve $p = \frac{w}{t}$ for time t.
19. Solve $A = \frac{1}{2}(b_1 + b_2)h$ for b_1.

See Example 3

Solve each equation for y and graph the equation.

20. $3y + 6x = 24$
21. $-2y - 9x = 10$
22. $5 = 4y - 3x$

PRACTICE AND PROBLEM SOLVING

Solve for the indicated variable. Assume all values are positive.

23. Solve $\frac{1}{2}x - 2 = y$ for x.
24. Solve $7y + 7x = 21x + 35$ for y.
25. Solve $\frac{3}{2}k + \ell^2 = 6\ell^2$ for k.
26. Solve $4m - 2n^2 = 2m - 72$ for m.
27. Solve $y = x - 21y$ for y.
28. Solve $9g + 7h = 9g - 7h + g$ for g.
29. Solve $z^2 + y = 5y$ for z.
30. Solve $\frac{3}{4}r - \frac{1}{2} = 1\frac{1}{4}r + 8s$ for r.

In Lesson 11-3, you will learn an important equation in algebra, $y = mx + b$. Solve $y = mx + b$ for the given variable.

31. Solve for m.
32. Solve for x.
33. Solve for b.

Solve each equation for y. Then substitute the given values, and graph the equation.

34. $ax + 5y = c$; $a = -10$ and $c = 15$
35. $ax + 2y = c$; $a = 12$ and $c = 16$
36. $-6x + by = c$; $b = 1$ and $c = 16$

Physical Science LINK

Electric power companies bill customers based on the amount of electrical energy (in kilowatt-hours) they use. The amount of electrical energy E used to run a household appliance is the power P in kilowatts multiplied by the time T in hours that the appliance is used, or $E = P \cdot T$. Use the table for Exercises 37–39.

Power Ratings of Household Appliances	
Appliance	**Power (kilowatts)**
Clothes dryer	4
Hair dryer	1
Color television	0.2
Radio	0.1
Alarm clock	

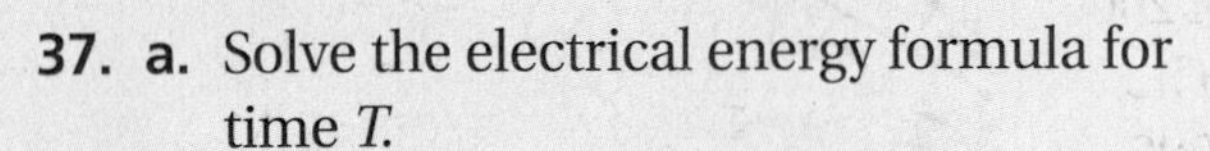

37. **a.** Solve the electrical energy formula for time T.

b. How long could you run your clothes dryer and use only 6 kilowatt-hours of energy?

38. **a.** Solve the electrical energy formula for power P.

b. Suppose your alarm clock used 2.16 kilowatt-hours in 30 days. If your clock is plugged in 24 hours a day, determine the power used in kilowatts each hour.

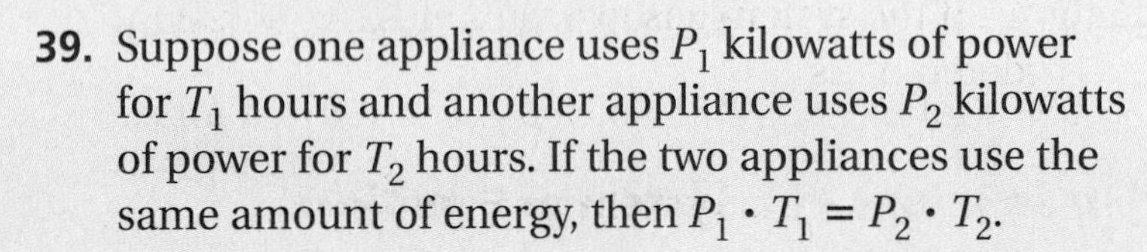

39. Suppose one appliance uses P_1 kilowatts of power for T_1 hours and another appliance uses P_2 kilowatts of power for T_2 hours. If the two appliances use the same amount of energy, then $P_1 \cdot T_1 = P_2 \cdot T_2$.

a. Solve the equation $P_1 \cdot T_1 = P_2 \cdot T_2$ for T_1.

b. Use your result from part **a** to determine how many hours of listening to the radio is equivalent to 15 minutes, or $\frac{1}{4}$ hour, of using a hair dryer.

40. *Ohm's Law,* $V = I \cdot R$, relates current I, voltage V, and resistance R. Solve Ohm's Law for I, the current of a circuit.

41. ***CHALLENGE*** The current in a series circuit with two different resistances R_1 and R_2 is given by the formula $I = \frac{V}{R_1 + R_2}$. Solve this formula for R_1.

Spiral Review

Solve and check. (Lesson 10-2)

42. $6x - 3 + x = 4$ **43.** $32 = 13 - 4x + 21$ **44.** $5x + 14 - 2x = 23$

45. **EOG PREP** Which of the following are three consecutive integers such that the sum of the first two integers is 10 more than the third integer? (Lesson 10-3)

A 35, 36, 37 B 11, 12, 13 C 4, 5, 6 D −7, −6, −5

10-6 Systems of Equations

Learn to solve systems of equations.

Vocabulary

system of equations

solution of a system of equations

Tickets for a concert are \$40 for main-floor seats and \$25 for upper-level seats. A total of 2000 concert tickets were sold. The total ticket sales were \$62,000. How many main-floor tickets were sold and how many upper-level tickets were sold? You can solve this problem using two equations.

A **system of equations** is a set of two or more equations that contain two or more variables. A **solution of a system of equations** is a set of values that are solutions of all of the equations. If the system has two variables, the solutions can be written as ordered pairs.

EXAMPLE 1 Identifying Solutions of a System of Equations

Determine if each ordered pair is a solution of the system of equations below.

$$2x + 3y = 8$$
$$x - 4y = 15$$

A (−2, 4)

$2x + 3y = 8$	$x - 4y = 15$	
$2(-2) + 3(4) \stackrel{?}{=} 8$	$-2 - 4(4) \stackrel{?}{=} 15$	*Substitute for x and y.*
$8 = 8$ ✔	$-18 \neq 15$ ✘	

The ordered pair (−2, 4) is not a solution of the system of equations.

B (7, −2)

$2x + 3y = 8$	$x - 4y = 15$	
$2(7) + 3(-2) \stackrel{?}{=} 8$	$7 - 4(-2) \stackrel{?}{=} 15$	*Substitute for x and y.*
$8 = 8$ ✔	$15 = 15$ ✔	

The ordered pair (7, −2) is a solution of the system of equations.

C (11, −1)

$2x + 3y = 8$	$x - 4y = 15$	
$2(11) + 3(-1) \stackrel{?}{=} 8$	$11 - 4(-1) \stackrel{?}{=} 15$	*Substitute for x and y.*
$19 \neq 8$ ✘	$15 = 15$ ✔	

The ordered pair (11, −1) is not a solution of the system of equations.

EXAMPLE 2 Solving Systems of Equations

Solve the system of equations.

$$y = x + 3$$
$$y = 2x + 5$$

The expressions $x + 3$ and $2x + 5$ both equal y, so they equal each other.

$$y = y$$
$$y = x + 3 \qquad y = 2x + 5$$
$$x + 3 = 2x + 5$$

Solve the equation to find x.

$$
\begin{aligned}
x + 3 &= 2x + 5 \\
-x \quad &\quad -x && \textit{Subtract x from both sides.} \\
3 &= x + 5 \\
-5 \quad &\quad -5 && \textit{Subtract 5 from both sides.} \\
-2 &= x
\end{aligned}
$$

Helpful Hint

When solving systems of equations, remember to find values for all of the variables.

To find y, substitute -2 for x in one of the original equations.

$y = x + 3 = -2 + 3 = 1$

The solution is $(-2, 1)$.

Check: Substitute -2 for x and 1 for y in each equation.

$y = x + 3$	$y = 2x + 5$
$1 \stackrel{?}{=} -2 + 3$	$1 \stackrel{?}{=} 2(-2) + 5$
$1 = 1$ ✔	$1 = 1$ ✔

To solve a general system of two equations with two variables, you can solve both equations for x or both for y.

EXAMPLE 3 Solving Systems of Equations

Solve the system of equations.

A

$$x + y = 5$$
$$x - 2y = -4$$

Solve both equations for x.

$$
\begin{aligned}
x + y &= 5 \\
-y \quad &\quad -y \\
x &= 5 - y
\end{aligned}
\qquad
\begin{aligned}
x - 2y &= -4 \\
+2y \quad &\quad +2y \\
x &= -4 + 2y
\end{aligned}
$$

$$
\begin{aligned}
5 - y &= -4 + 2y \\
+y \quad &\quad +y && \textit{Add y to both sides.} \\
5 &= -4 + 3y \\
+4 \quad &\quad +4 && \textit{Add 4 to both sides.} \\
9 &= 3y \\
3 &= y && \textit{Divide both sides by 3.}
\end{aligned}
$$

$$
\begin{aligned}
x &= 5 - y \\
&= 5 - 3 = 2 && \textit{Substitute 3 for y.}
\end{aligned}
$$

The solution is $(2, 3)$.

Solve the system of equations.

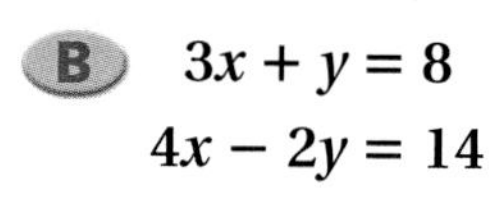

B $3x + y = 8$
$4x - 2y = 14$

Solve both equations for y.

$$\begin{aligned} 3x + y &= 8 \\ -3x \quad & \quad -3x \\ y &= 8 - 3x \end{aligned}$$

$$\begin{aligned} 4x - 2y &= 14 \\ -4x \quad & \quad -4x \\ -2y &= 14 - 4x \\ \frac{-2y}{-2} &= \frac{14}{-2} - \frac{4x}{-2} \\ y &= -7 + 2x \end{aligned}$$

$$\begin{aligned} 8 - 3x &= -7 + 2x \\ +3x \quad & \quad +3x && \text{Add } 3x \text{ to both sides.} \\ 8 &= -7 + 5x \\ +7 \quad & \quad +7 && \text{Add 7 to both sides.} \\ 15 &= 5x \\ 3 &= x && \text{Divide both sides by 5.} \end{aligned}$$

$$\begin{aligned} y &= 8 - 3x \\ &= 8 - 3(3) = -1 && \text{Substitute 3 for } x. \end{aligned}$$

The solution is $(3, -1)$.

Helpful Hint

You can choose either variable to solve for. It is usually easiest to solve for a variable that has a coefficient of 1.

Think and Discuss

1. **Compare** an equation to a system of equations.
2. **Describe** how you would know whether $(-1, 0)$ is a solution of the system of equations below.

$$x + 2y = -1$$
$$-3x + 4y = 3$$

10-6 Exercises

5.03, 5.04

FOR EOG PRACTICE
see page 683

internet connect
Homework Help Online
go.hrw.com Keyword: MT4 10-6

GUIDED PRACTICE

See Example 1

Determine if the ordered pair is a solution of each system of equations.

1. $(2, 3)$ $y = 2x - 1$; $y = x + 1$

2. $(2, 7)$ $y = 5x - 3$; $y = 3x + 1$

3. $(2, 4)$ $y = 4x - 4$; $y = 2x$

4. $(2, 2)$ $y = 2x + 1$; $y = 3x - 2$

See Example 2 Solve each system of equations.

5. $y = x + 1$
$y = 2x - 1$

6. $y = -3x + 2$
$y = 4x - 5$

7. $y = 5x - 3$
$y = 2x + 6$

8. $y = 4x - 3$
$y = 2x + 5$

9. $y = -2x + 6$
$y = 3x - 9$

10. $y = 5x + 7$
$y = -3x + 7$

See Example 3

11. $x + y = 8$
$x + 3y = 14$

12. $x + y = 20$
$x = y - 4$

13. $2x + y = 12$
$3x - y = 13$

14. $4x - 3y = 33$
$x = -4y - 25$

15. $5x - 2y = 4$
$11x + 4y = -8$

16. $x = -3y$
$7x - 2y = -69$

INDEPENDENT PRACTICE

See Example 1 Determine if the ordered pair is a solution of the system of equations.

17. (0, 1) $y = -2x - 1$
$y = 2x + 1$

18. (5, 11) $y = 3x - 4$
$y = 2x + 1$

19. (−1, 5) $y = 4x + 1$
$y = 3x$

20. (−6, −9) $y = x - 3$
$y = 2x + 3$

See Example 2 Solve each system of equations.

21. $y = -x - 2$
$y = 3x + 2$

22. $y = 3x - 6$
$y = x + 2$

23. $y = -3x + 5$
$y = x - 3$

24. $y = 2x - 3$
$y = 4x - 3$

25. $y = x + 6$
$y = -2x - 12$

26. $y = 3x - 1$
$y = -2x + 9$

See Example 3

27. $x + y = 5$
$x - 2y = -4$

28. $x + 2y = 4$
$2x - y = 3$

29. $y = 5x - 2$
$4x + 3y = 13$

30. $2x + 3y = 1$
$4x - 3y = -7$

31. $5x - 9y = 11$
$3x + 7y = 19$

32. $12x + 18y = 30$
$4x - 13y = 67$

PRACTICE AND PROBLEM SOLVING

Solve each system of equations.

33. $y = 3x - 2$
$y = x + 2$

34. $y = 5x - 11$
$y = -2x + 10$

35. $x + y = -1$
$x - y = 5$

36. $y = 2x + 7$
$x + y = 4$

37. $4x - 3y = 0$
$-7x + 9y = 0$

38. $10x + 15y = 74$
$30x - 5y = -68$

39. $3x - y = 5$
$x - 4y = -2$

40. $x = 9y - 100$
$x = -5y + 54$

41. $2x + 6y = 1$
$4x - 3y = 0$

42. $3x - 4y = -5$
$x + 6y = 35$

43. $\frac{1}{3}x + \frac{1}{4}y = 6$
$-\frac{1}{2}x + y = 2$

44. $y = 2x - 2$
$y = -2$

45. $9.7x - 1.5y = 62.7$
$-2.3x - 7.4y = 8.4$

46. $-1.2x + 2.7y = 9.9$
$4.2x + 6.8y = 40.1$

47. $\frac{5}{6}x - 4y = -\frac{5}{2}$
$\frac{10}{3}x + \frac{1}{4}y = \frac{5}{6}$

Entertainment LINK

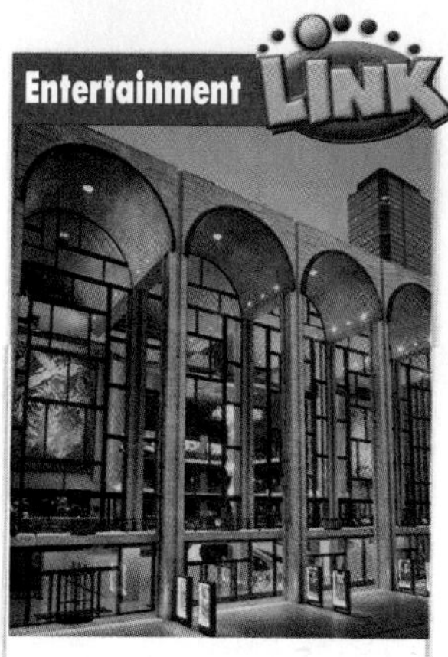

The Metropolitan Opera House in New York has 6 levels and 3500 seats.

go.hrw.com
KEYWORD: MT4 Music

CNN Student News

Write and solve a system of equations for Exercises 48–50.

48. Two numbers have a sum of 23 and a difference of 9. Find the two numbers.

49. Two numbers have a sum of 18. The first number is 2 more than 3 times the second number. Find the two numbers.

50. Two numbers have a difference of 6. The first number is 9 more than 2 times the second number. Find the two numbers.

51. ***ENTERTAINMENT*** Tickets for a concert are $40 for main-floor seats and $25 for upper-level seats. A total of 2000 concert tickets were sold. The ticket sales were $62,000. Let m represent the number of main-floor tickets and u the number of upper-level tickets.

a. Write an equation about the total number of tickets sold.

b. Write an equation about the total ticket sales.

c. Solve the system of equations to find how many main-floor tickets were sold and how many upper-level tickets were sold.

52. ***CHOOSE A STRATEGY*** Jan invested some money at 7% interest and $500 more than that at 9% interest. The total interest earned in 1 year was $141. How much did she invest at each rate?

A $350 at 7%, $850 at 9%
B $800 at 7%, $1300 at 9%
C $575 at 7%, $1075 at 9%
D $600 at 7%, $1100 at 9%

53. ***WRITE ABOUT IT*** List the steps you would use to solve the system of equations below. Explain which variable you would solve for and why.

$$x + 2y = 7$$
$$2x + y = 8$$

54. ***CHALLENGE*** Solve the following system of equations.

$$\frac{x-2}{4} + \frac{y+3}{8} = 1$$
$$\frac{2x-1}{12} + \frac{y+3}{6} = \frac{5}{4}$$

Spiral Review

Use the Fundamental Counting Principle to find the number of possible outcomes. (Lesson 9-5)

55. toppings: mayo, onion, lettuce, tomato
sandwich: burger, fish, chicken

56. stain: oak, redwood, pine, amber, rosewood
finish: glossy, matte, clear

57. distances: 50 m, 100 m, 400 m
races: freestyle, backstroke, butterfly

58. snacks: nachos, candy, hot dog, pizza
drinks: water, soda

59. **EOG PREP** If A and B are independent events such that $P(A) = 0.14$ and $P(B) = 0.28$, what is the probability that both A and B will occur? (Lesson 9-7)

A 0.42 B 0.24 C 0.0784 D 0.0392

NORTH CAROLINA

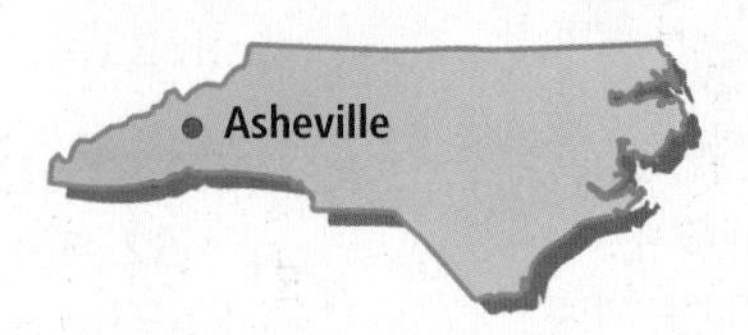

The Blue Ridge Parkway

The 469-mile Blue Ridge Parkway is a recreational highway connecting Great Smoky Mountains National Park and Shenandoah National Park. The parkway's construction began in September 1935 and was finished in September 1987 with the completion of the 7.5-mile missing link. The completion of the missing link included the difficult construction of the Linn Cove Viaduct.

Mile markers can be found along the parkway, starting with mile 0 just south of Shenandoah National Park.

For 1–3, use the formula $d = rt$ (distance = rate × time) and the table.

1. A tourist drove from Cumberland Knob to the Linn Cove Viaduct in 2.5 hours. Solve the distance formula for *r* and then find the average rate (speed) to the nearest mile per hour that the tourist traveled.

2. The Perez family traveled from the Virginia–North Carolina state line to Brinegar Cabin at an average speed of 27 mi/h. In the same amount of time, the Lewis family traveled from Hare Mill Pond to Alligator Back parking overlook. At what speed in miles per hour did the Lewis family travel?

3. Solve the distance formula $d = rt$ for *t*, and then find the time it takes to travel at an average speed of 35 mi/h from Brinegar Cabin to the Daniel Boone Wilderness Trail.

Points of Interest Along the Blue Ridge Parkway

Mile Marker	Point of Interest
216.9	Virginia–North Carolina state line
217.5	Cumberland Knob
225.2	Hare Mill Pond
238.5	Brinegar Cabin
242.4	Alligator Back parking overlook
261.2	Horse Gap
285.1	Daniel Boone Wilderness Trail
304.4	Linn Cove Viaduct

Biltmore Estate

The Biltmore Estate in Asheville, North Carolina, is the site of the largest home in the United States. The house was commissioned by George Vanderbilt and designed by Richard Hunt. The grounds of the estate were designed by Frederick Olmstead, the designer of New York City's Central Park. Many movies have been filmed at the estate, including *Patch Adams* and *Richie Rich*. The estate has over 800,000 visitors yearly.

1. The White House in Washington, D.C., has 132 rooms. This is 18 less than $\frac{3}{5}$ the number of rooms in Biltmore House. How many rooms are there in Biltmore House?

2. The number of bathrooms plus the number of bedrooms in Biltmore House is 77. There are 9 more bathrooms than bedrooms. How many bedrooms are there? How many bathrooms are there?

3. The rectangular banquet-hall floor has a width that is $\frac{7}{12}$ its length and a perimeter of 228 feet. What are the dimensions of the banquet-hall floor?

4. The length of the rectangular-prism-shaped pool in Biltmore House is 53 ft, and the width is 27 ft. If the volume in cubic feet is equal to 595 less than 1501 times the height in feet of the pool, what is the volume of water the pool can hold?

MATH-ABLES

Trans-Plants

Solve each equation below. Then use the values of the variables to decode the answer to the question.

$3a + 17 = -25$

$2b - 25 + 5b = 7 - 32$

$2.7c - 4.5 = 3.6c - 9$

$\frac{5}{12}d + \frac{1}{6}d + \frac{1}{3}d + \frac{1}{12}d = 6$

$4e - 6e - 5 = 15$

$420 = 29f - 73$

$2(g + 6) = -20$

$2h + 7 = -3h + 52$

$96i + 245 = 53$

$3j + 7 = 46$

$\frac{1}{2}k = \frac{3}{4}k - \frac{1}{2}$

$30l + 240 = 50l - 160$

$4m + \frac{3}{8} = \frac{67}{8}$

$24 - 6n = 54$

$8.4o - 6.8 = 14.2 + 6.3o$

$4p - p + 8 = 2p + 5$

$16 - 3q = 3q + 40$

$4 + \frac{1}{3}r = r - 8$

$\frac{2}{3}s - \frac{5}{6}s + \frac{1}{2} = -\frac{3}{2}$

$4 - 15 = 4t + 17$

$45 + 36u = 66 + 23u + 31$

$6v + 8 = -4 - 6v$

$4w + 3w - 6w = w + 15 + 2w - 3w$

$x + 2x + 3x + 4x + 5 = 75$

$\frac{4 - y}{5} = \frac{2 - 2y}{8}$

$-11 = 25 - 4.5z$

What happens to plants that live in a math classroom?

−7, 9, −10, −11   −16, 18, 10, 15   12, −4, 4, −14, 18, −10   18, 10, 10, −7, 12

24 Points

This traditional Chinese game is played using a deck of 52 cards numbered 1–13, with four of each number. The cards are shuffled, and four cards are placed face-up in the center. The winner is the first player who comes up with an expression that equals 24, using each of the numbers on the four cards once. Players can use grouping symbols and the operations addition, subtraction, multiplication, and division.

internet connect

Go to ***go.hrw.com*** for a complete set of rules and game cards.
KEYWORD: MT4 Game10

Solve Two-Step Equations by Graphing

Use with Lesson 10-3

1.02, 5.03

A graphing calculator is another tool used to solve equations.

internet connect
Lab Resources Online
go.hrw.com
KEYWORD: MT4 TechLab10

Activity

To solve the equation $2x - 3 = 4x + 1$, use the Y= menu.

Enter the left side of the equation in **Y1** and the right side in **Y2.**

1. Press Y= 2 X,T,θ,n − 3 ENTER 4 X,T,θ,n + 1.

2. To select the standard viewing window and graph the two equations, press ZOOM **6:ZStandard.**

On the figure, the two graphs *appear* to intersect at the point $(-2, -7)$. The coordinates -2 and -7 are *approximations.*

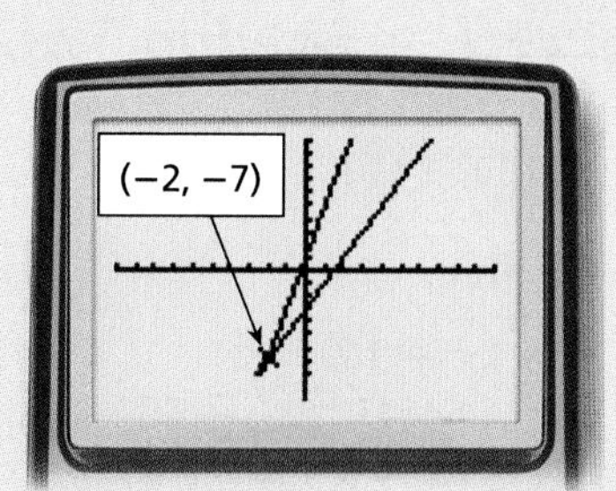

The solution of the original equation is the x-coordinate of the point of intersection of the two lines. The solution is approximately $x = -2$.

If the equation is solved algebraically by adding $-4x$ and 3 to both sides, the result is $2x - 3 + (-4x) + 3 = 4x + 1 + (-4x) + 3$. This becomes $-2x = 4$, or $x = -2$, thus confirming the estimated graphical solution.

Another way to estimate the solution is to use the TRACE key.

3. Press TRACE and the left arrow key 9 times to get the screen shown. As you press the arrow keys, observe how the coordinates change. This is as close to $x = -2$ as you can get using this window. The x-value –1.914… is only an estimate of the exact solution, -2.

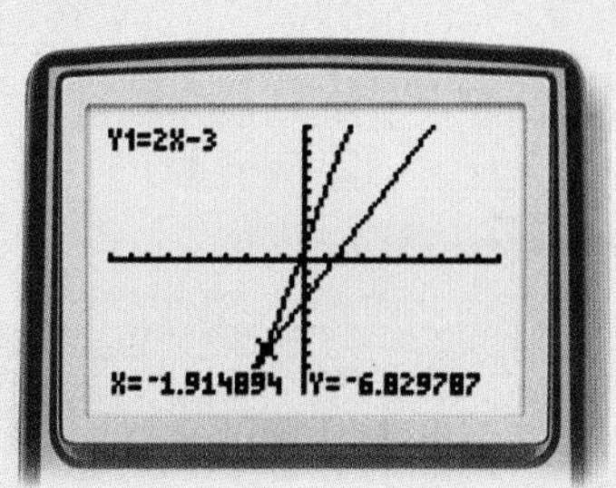

Think and Discuss

1. Explain why using the TRACE key may show only estimates.
2. Graph both sides of the equation $x + 4 = -2x + 7$ in the standard viewing window. What must be done to solve the equation graphically?

Try This

Use a graphing calculator to find an approximate solution to each equation. Specify the window used. Confirm your estimate by solving algebraically.

1. $2x + 1 = x - 4$
2. $\frac{1}{2}x - 3 = 2x + 4$
3. $3x - 5 = 2x + 6$
4. $3x + 5 = 4 - 2x$

Chapter 10

Study Guide and Review

Vocabulary

Complete the sentences below with vocabulary words from the list above.

1. Two or more equations that contain two or more variables is called a(n) ___?___.

2. A set of values that are solutions of all of the equations of a system is the ___?___.

10-1 Solving Two-Step Equations (pp. 498–501)

EXAMPLE

Solve.

■ $7x + 12 = 33$

Think: First the variable is **multiplied by 7**, and then **12 is added.** To isolate the variable, **subtract 12**, and then **divide by 7.**

$$\begin{aligned} 7x + 12 &= 33 \\ -12 \quad &\;\; -12 \\ 7x &= 21 \end{aligned}$$ *Subtract to undo addition.*

$$\frac{7x}{7} = \frac{21}{7}$$ *Divide to undo multiplication.*

$$x = 3$$

■ $\frac{z}{3} - 8 = 5$

Think: First the variable is **divided by 3**, and then **8 is subtracted.** To isolate the variable, **add 8**, and then **multiply by 3.**

$$\begin{aligned} \frac{z}{3} - 8 &= 5 \\ +8 \quad &\;\; +8 \\ \frac{z}{3} &= 13 \end{aligned}$$ *Add to undo subtraction.*

$$3 \cdot \frac{z}{3} = 3 \cdot 13$$ *Multiply to undo division.*

$$z = 39$$

EXERCISES

Solve.

3. $3m + 5 = 35$
4. $55 = 7 - 6y$
5. $2c + 1 = -31$
6. $5r + 15 = 0$
7. $\frac{t}{2} + 7 = 15$
8. $\frac{w}{4} - 5 = 11$
9. $-25 = \frac{7r}{3} - 11$
10. $\frac{2h}{5} - 9 = -19$
11. $\frac{x+2}{3} = 18$
12. $\frac{d-3}{4} = -9$
13. $21 = \frac{a-4}{3}$
14. $14 = \frac{c+8}{7}$

10-2 Solving Multistep Equations (pp. 502–505)

EXAMPLE

■ **Solve.**

$\frac{5x}{9} - \frac{x}{6} + \frac{1}{3} = \frac{3}{2}$

$18\left(\frac{5x}{9} - \frac{x}{6} + \frac{1}{3}\right) = 18\left(\frac{3}{2}\right)$

$18\left(\frac{5x}{9}\right) - 18\left(\frac{x}{6}\right) + 18\left(\frac{1}{3}\right) = 18\left(\frac{3}{2}\right)$

$10x - 3x + 6 = 27$

$7x + 6 = 27$ *Combine like terms.*

$\underline{-6 \quad -6}$ *Subtract to undo addition.*

$7x = 21$

$\frac{7x}{7} = \frac{21}{7}$ *Divide to undo multiplication.*

$x = 3$

EXERCISES

Solve.

15. $5y + 3 + 2y - 9 = 8$

16. $3h - 4 - h + 8 = 10$

17. $\frac{4t}{5} + \frac{3}{5} = -\frac{1}{5}$

18. $\frac{2r}{9} - \frac{4}{9} = \frac{2}{9}$

19. $\frac{3z}{4} - \frac{2z}{3} + \frac{1}{2} = \frac{5}{6}$

20. $\frac{3a}{8} - \frac{a}{12} + \frac{7}{2} = 7$

10-3 Solving Equations with Variables on Both Sides (pp. 507–511)

EXAMPLE

■ **Solve.**

$3x + 5 - 5x = -12 + x + 2$

$-2x + 5 = -10 + x$ *Combine like terms.*

$\underline{+2x \qquad +2x}$

$5 = -10 + 3x$

$\underline{+10 \quad +10}$

$15 = 3x$

$\frac{15}{3} = \frac{3x}{3}$

$5 = x$

EXERCISES

Solve.

21. $22s = 16 + 3(5s + 4)$

22. $\frac{5c}{8} - \frac{c}{3} = \frac{5c}{6} - 13$

23. $2 + x = 5 - 3x$

24. $6 - 4y = 6y$

10-4 Solving Multistep Inequalities (pp. 514–518)

EXAMPLE

■ **Solve and graph.**

$5x - 3 - 8x < 9$

$-3x - 3 < 9$ *Combine like terms.*

$\underline{+3 \quad +3}$

$-3x < 12$

$\frac{-3x}{-3} > \frac{12}{-3}$ *Change $<$ to $>$.*

$x > -4$

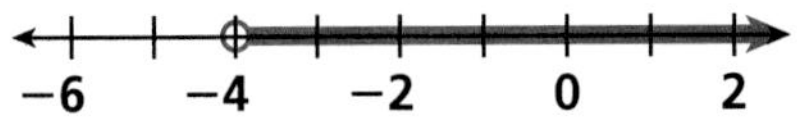

EXERCISES

Solve and graph.

25. $5z + 3z - 4 > 4$

26. $2h + 7 \le 3h + 1$

27. $\frac{a}{2} + \frac{a}{3} + \frac{a}{4} < 26$

28. $1 + \frac{2x}{3} \ge \frac{x}{2}$

10-5 Solving for a Variable (pp. 519–522)

EXAMPLE

Solve for the indicated variable.

■ **Solve $A = 3b - 4c$ for b.**

$$A = 3b - 4c$$
$$\underline{+4c} \quad \underline{+4c} \qquad \textit{Add 4c to both sides.}$$
$$A + 4c = 3b$$
$$\frac{A+4c}{3} = \frac{3b}{3} \qquad \textit{Divide by 3.}$$
$$\frac{A}{3} + \frac{4c}{3} = b$$

■ **Solve $m = \frac{100y}{x}$ for y.**

$$m = \frac{100y}{x}$$
$$x \cdot m = x \cdot \left(\frac{100y}{x}\right) \qquad \textit{Multiply by x.}$$
$$xm = 100y$$
$$\frac{xm}{100} = \frac{100y}{100} \qquad \textit{Divide by 100.}$$
$$\frac{xm}{100} = y$$

EXERCISES

Solve for the indicated variable.

29. Solve $P = 2w + 2\ell$ for ℓ.
30. Solve $A = P + Prt$ for r.
31. Solve $F = \frac{9}{5}C + 32$ for C.
32. Solve $2x + 3y = 9$ for y.
33. Solve $x + 3y = 7$ for y.
34. Solve $4x - 12y = 8$ for x.

10-6 Systems of Equations (pp. 523–527)

EXAMPLE

■ **Solve the system of equations.**

$$4x + y = 3$$
$$x + y = 12$$

Solve both equations for y.

$$4x + y = 3 \qquad x + y = 12$$
$$\underline{-4x \quad -4x} \qquad \underline{-x \quad -x}$$
$$y = -4x + 3 \qquad y = -x + 12$$

$$-4x + 3 = -x + 12$$
$$\underline{+4x} \quad \underline{+4x} \qquad \textit{Add 4x.}$$
$$3 = 3x + 12$$
$$\underline{-12} \quad \underline{-12} \qquad \textit{Subtract 12.}$$
$$-9 = 3x$$
$$\frac{-9}{3} = \frac{3x}{3} \qquad \textit{Divide by 3.}$$
$$-3 = x$$

$$y = -4x + 3$$
$$= -4(-3) + 3 \qquad \textit{Substitute −3 for x.}$$
$$= 12 + 3$$
$$= 15$$

The solution is $(-3, 15)$.

EXERCISES

Solve each system of equations.

35. $y = x + 7$
    $y = 2x + 5$
36. $x - y = -2$
    $x + y = 18$
37. $4x + 3y = 27$
    $2x - y = 1$
38. $4x + y = 10$
    $x - 2y = 7$
39. $3x - 4y = 26$
    $x + 2y = 2$
40. $4x - 3y = 4$
    $2x - y = 1$

Study Guide and Review

Chapter Test

Solve each equation.

1. $3t - 1 = 92$
2. $\frac{2}{5}y - 9 = 1$
3. $\frac{z-3}{5} = -4$
4. $\frac{7x}{9} - \frac{2}{9} = \frac{19}{9}$
5. $\frac{2v}{5} + \frac{v}{4} = \frac{9}{5} + \frac{3}{20}$
6. $\frac{r}{3} - \frac{2r}{5} - \frac{1}{4} = \frac{5}{12}$
7. $16z - 3z + 9 = 2z + 86$
8. $\frac{1}{4}w = 2w + \frac{35}{2}$
9. $15n = 29 + 2(3n - 1)$
10. $3(s + 1) - (s - 2) = 22s$
11. $\frac{3}{5}(15k + 10) = 12k - 9$
12. $\frac{3m}{4} - \frac{1}{9} = \frac{5m}{12} + \frac{14}{9}$
13. A delivery service charges $2.50 for the first pound and $0.75 for each additional pound. If Carl paid $7.75 for his package, how many pounds did the package weigh?
14. A wireless phone company offers two plans. In one plan, there is a monthly fee of $40 and a charge of $0.25 per minute. In the other plan, there is a monthly fee of $25 and a charge of $0.40 per minute. For what number of minutes are the costs equal?
15. The rectangle and the triangle shown have the same perimeter. Find the perimeter of each figure.

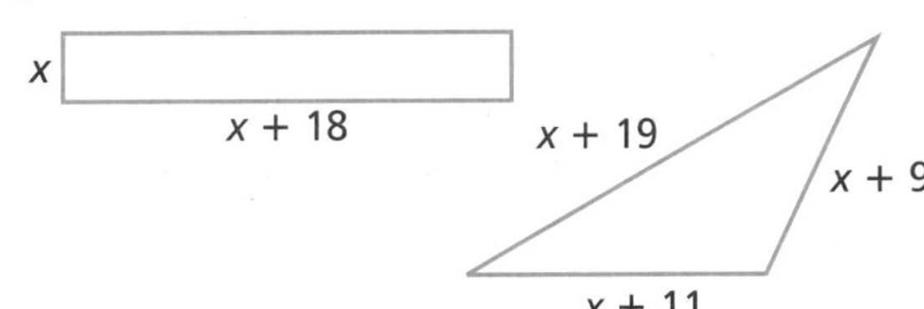

Solve and graph each inequality.

16. $6m + 4 > 2$
17. $z + 3z + 4 \geq -8$
18. $3x - 5x - 4 > 2$
19. $8 - 3p > 14$

Solve for the indicated variable.

20. Solve for w. $P = 2(\ell + w)$
21. Solve for r. $s = c + rc$
22. Solve for b. $A = \pi ab$
23. Solve for d. $x^2 + d^2 = 1$

Solve each equation for y.

24. $x + 2y = 12$
25. $10 - x + y = 0$
26. $3y - x = -6$
27. $4x + 3y = 6$

Solve each system of equations.

28. $x - 2y = 16$
$4x + y = 1$
29. $x + 2y = 6$
$4x + 3y = 4$
30. $3x - 2y = -3$
$3x + y = 3$
31. $x + 5y = 11$
$4x - y = 2$

Chapter 10

Performance Assessment

Show What You Know

Create a portfolio of your work from this chapter. Complete this page and include it with your four best pieces of work from Chapter 10. Choose from your homework or lab assignments, mid-chapter quiz, or any journal entries you have done. Put them together using any design you want. Make your portfolio represent what you consider your best work.

Short Response

1. Solve the inequality: $7x - 4 < 9x + 14$. Show your work or explain in words how you determined your answer.

2. Solve the system of equations. Show your work.

$$x - y = -3$$
$$2x - 4y = 22$$

3. Alfred and Eugene each spent \$62 on campsite and gasoline expenses during their camping trip. Each campsite they stayed at had the same per-night charge. Alfred paid for 4 nights of campsites and \$30 for gasoline. Eugene paid for 2 nights of campsites and \$46 for gasoline. Write an equation that could be used to determine the cost of one night's stay at a campsite. What was the cost of one night's stay at a campsite?

Extended Problem Solving

4. You are designing a house to fit on a rectangular lot that has 90 feet of lake frontage and is 162 feet deep. The building codes require that the house not be built closer than 10 feet to the lot boundary lines.

   a. Write an inequality and solve it to find how long the front of the house facing the lake may be.

   b. If you want the house to cover no more than 20% of the lot, what would be the maximum square footage of the house?

   c. If you want to spend a maximum of \$100,000 building the house, to the nearest whole dollar, what would be the maximum you could spend per square foot for a 1988-square-foot house?

Performance Assessment

internet connect

EOG Test Practice Online
go.hrw.com Keyword: MT4 TestPrep

Getting Ready for EOG

Chapter 10

Cumulative Assessment, Chapters 1–10

1. Which of the following is the solution of the equation $5x = 4(x + 2)$?

   A $x = 2$
   B $x = -2$
   C $x = 8$
   D $x = -8$

2. If $x + y = 3 + k$ and $2x + 2y = 10$, what is the value of k?

   A 7
   B 6
   C 3
   D 2

3. If $3 = b^x$, what is the value of $3b$?

   A b^{x+1}
   B b^{x+2}
   C b^{2x}
   D b^{3x}

4. Tim is rolling a number cube with the numbers 1 through 6 on it. He rolls the cube twice. What is the probability that the two rolls will have a sum of 10?

   A $\frac{1}{36}$
   B $\frac{1}{12}$
   C $\frac{1}{10}$
   D $\frac{1}{9}$

5. The formula $M = \frac{P(rt + 1)}{12t}$ gives the monthly payment M on a loan with principal P, annual interest rate r, and length t years. What is the monthly payment on a 2-year loan for \$3000 at an annual rate of 8%?

   A \$605
   B \$480
   C \$145
   D \$125

6. Mia earns a monthly base salary of \$640 plus a 12% commission on her total monthly sales. At the end of last month, Mia earned \$2380. What were her total monthly sales?

   A \$145
   B \$285.60
   C \$1740
   D \$14,500

7. Ten years from now, Cal will be x years old. How old was he 5 years ago?

   A $x - 5$
   B $x - 10$
   C $x - 15$
   D $x + 10$

TEST TAKING TIP!

One way to find an answer is to substitute the choices into the problem. Be sure to examine all the choices before deciding.

8. If n is the least positive integer for which $3n$ is both an even integer and the square of an integer, what is the value of n?

   A 3
   B 4
   C 6
   D 12

9. ***SHORT RESPONSE*** Between which two consecutive positive integers does $\sqrt{213}$ lie? Explain in words how you determined your answer.

10. ***SHORT RESPONSE*** The graph shows Richie's weekly budget.

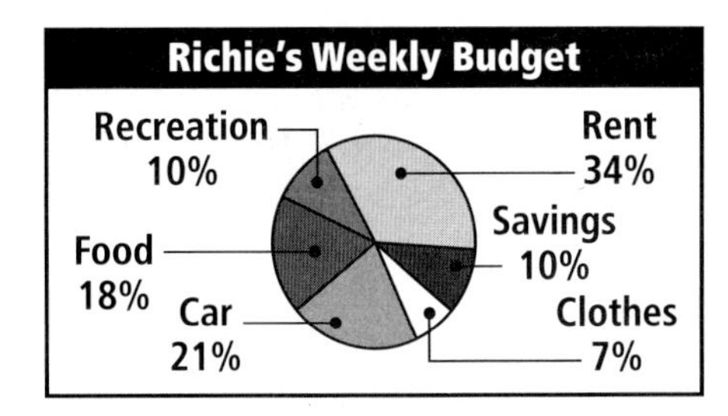

Write an equation that could be used to find how much more money m Richie allots for food than for recreation with a budget of d dollars. If Richie allots \$6.40 more for food than for recreation, what is his total monthly budget?

Getting Ready for EOG

Chapter 11

Graphing Lines

TABLE OF CONTENTS

Whooping Crane Population				
Year	1940	1960	1980	2000
Cranes	15	36	79	202

Career Wildlife Ecologist

Whatever happened to the Carolina parakeet and the passenger pigeon, two species of birds that once inhabited the United States? They are now as extinct as *Tyrannosaurus rex.* The primary focus of wildlife ecologists is to keep other animals from becoming extinct.

They have been successful with the whooping crane, the largest wild bird in North America. The table shows how the whooping crane has come back from the brink of extinction.

Chapter Opener Online
go.hrw.com
KEYWORD: MT4 Ch11

ARE YOU READY?

Choose the best term from the list to complete each sentence.

1. The expression $4 - 3$ is an example of a(n) __?__ expression.
2. When you divide both sides of the equation $2x = 20$ by 2, you are __?__.
3. An example of a(n) __?__ is $3x > 12$.
4. The expression $7 - 6$ can be rewritten as the __?__ expression $7 + (-6)$.

addition
inequality in one variable
solving for the variable
subtraction

Complete these exercises to review skills you will need for this chapter.

✔ Operations with Integers

Simplify.

5. $\frac{7 - 5}{-2}$ **6.** $\frac{-3 - 5}{-2 - 3}$ **7.** $\frac{-8 + 2}{-2 + 8}$

8. $\frac{-16}{-2}$ **9.** $\frac{-22}{2}$ **10.** $-12 + 9$

✔ Equations

Solve.

11. $3p - 4 = 8$ **12.** $2(a + 3) = 4$ **13.** $9 = -2k + 27$

14. $3s - 4 = 1 - 3s$ **15.** $7x + 1 = x$ **16.** $4m - 5(m + 2) = 1$

Determine whether each ordered pair is a solution to $-\frac{1}{2}x + 3 = y$.

17. $(4, 1)$ **18.** $\left(-\frac{8}{2}, 2\right)$ **19.** $(0, 5)$ **20.** $(-4, 5)$

21. $(8, 1)$ **22.** $(2, 2)$ **23.** $(-2, 4)$ **24.** $(0, 1)$

✔ Solve for One Variable

Solve each equation for the indicated variable.

25. Solve for x: $5y - x = 4$. **26.** Solve for y: $3y + 9 = 2x$.

27. Solve for y: $2y + 3x = 6$. **28.** Solve for x: $ax + by = c$.

✔ Solve Inequalities in One Variable

Solve and graph each inequality.

29. $x + 4 > 2$ **30.** $-3x < 9$ **31.** $x - 1 \leq -5$

11-1 Graphing Linear Equations

Learn to identify and graph linear equations.

Vocabulary

linear equation

In most bowling leagues, bowlers have a handicap added to their scores to make the game more competitive. For some leagues, the *linear equation* $h = 160 - 0.8s$ expresses the handicap h of a bowler who has an average score of s.

A **linear equation** is an equation whose solutions fall on a line on the coordinate plane. All solutions of a particular linear equation fall on the line, and all the points on the line are solutions of the equation. To find a solution that lies between two points (x_1, y_1) and (x_2, y_2), choose an x-value between x_1 and x_2 and find the corresponding y-value.

Reading Math

Read x_1 as "x sub one" or "x one."

If an equation is linear, a constant change in the x-value corresponds to a constant change in the y-value. The graph shows an example where each time the x-value increases by 3, the y-value increases by 2.

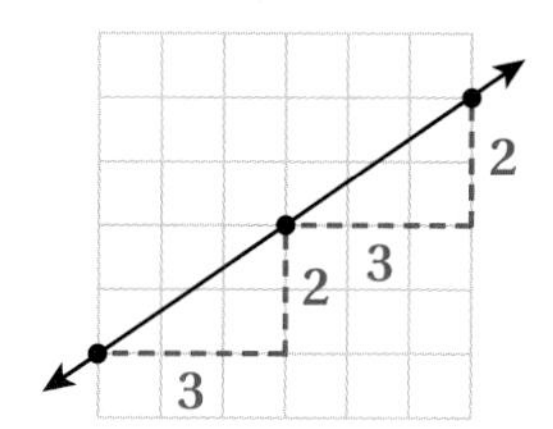

EXAMPLE 1 Graphing Equations

Graph each equation and tell whether it is linear.

A $y = 2x - 3$

x	$2x - 3$	y	(x, y)
−2	2(−2) − 3	−7	(−2, −7)
−1	2(−1) − 3	−5	(−1, −5)
0	2(0) − 3	−3	(0, −3)
1	2(1) − 3	−1	(1, −1)
2	2(2) − 3	1	(2, 1)
3	2(3) − 3	3	(3, 3)

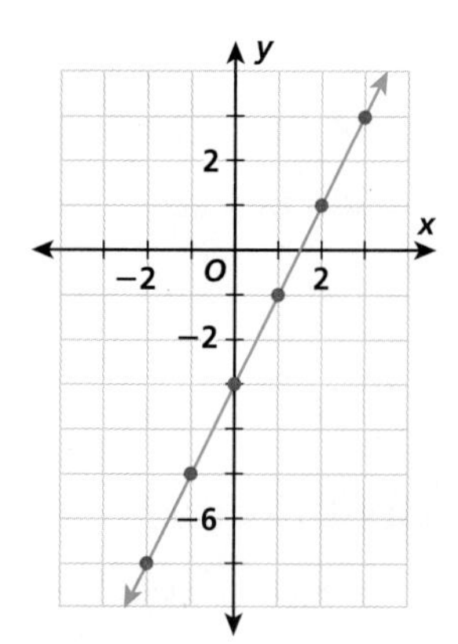

The equation $y = 2x - 3$ is a linear equation because it is the graph of a straight line and each time x increases by 1 unit, y increases by 2 units.

Graph each equation and tell whether it is linear.

B $y = x^2$

x	x^2	y	(x, y)
−2	$(-2)^2$	4	(−2, 4)
−1	$(-1)^2$	1	(−1, 1)
0	$(0)^2$	0	(0, 0)
1	$(1)^2$	1	(1, 1)
2	$(2)^2$	4	(2, 4)

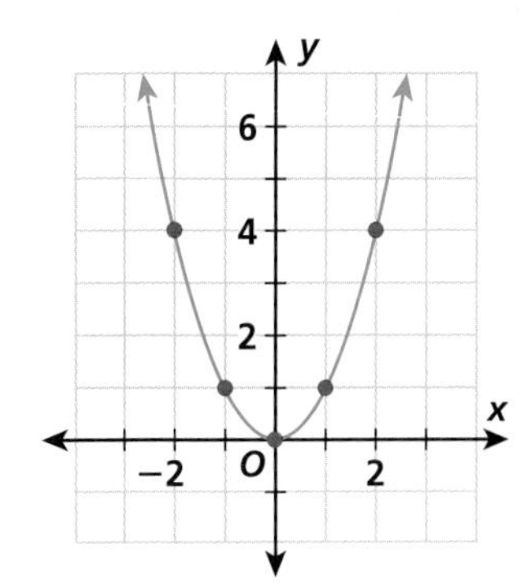

The equation $y = x^2$ is not a linear equation because its graph is not a straight line. Also notice that as x increases by a constant of 1, the change in y is not constant.

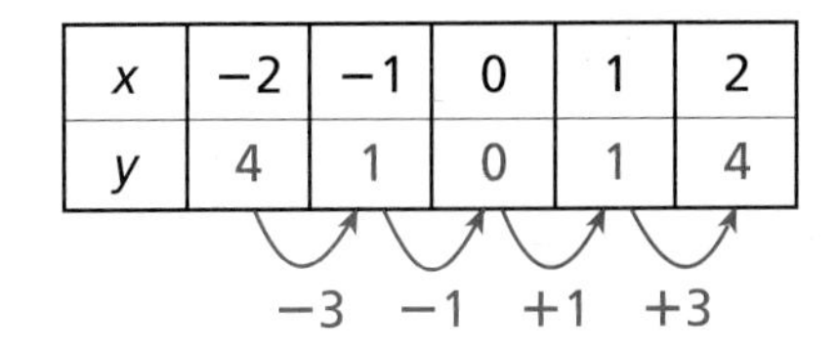

x	−2	−1	0	1	2
y	4	1	0	1	4

C $y = \frac{2x}{3}$

x	$\frac{2x}{3}$	y	(x, y)
−2	$\frac{2(-2)}{3}$	$-\frac{4}{3}$	$(-2, -\frac{4}{3})$
−1	$\frac{2(-1)}{3}$	$-\frac{2}{3}$	$(-1, -\frac{2}{3})$
0	$\frac{2(0)}{3}$	0	(0, 0)
1	$\frac{2(1)}{3}$	$\frac{2}{3}$	$(1, \frac{2}{3})$
2	$\frac{2(2)}{3}$	$\frac{4}{3}$	$(2, \frac{4}{3})$

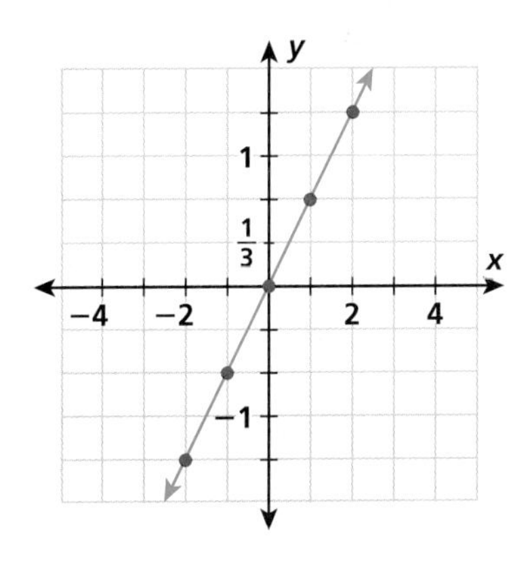

The equation $y = \frac{2x}{3}$ is a linear equation because the points form a straight line. Each time the value of x increases by 1, the value of y increases by $\frac{2}{3}$, or y increases by 2 each time x increases by 3.

D $y = -3$

x	−3	y	(x, y)
−2	−3	−3	(−2, −3)
−1	−3	−3	(−1, −3)
0	−3	−3	(0, −3)
1	−3	−3	(1, −3)
2	−3	−3	(2, −3)

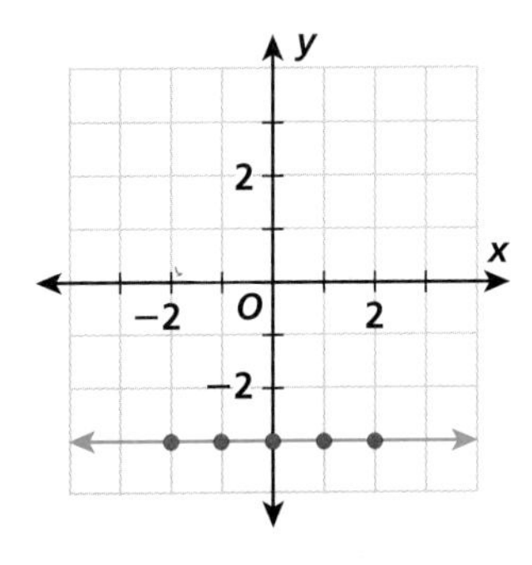

For any value of x, y = −3.

The equation $y = -3$ is a linear equation because the points form a straight line. As the value of x increases, the value of y has a constant change of 0.

EXAMPLE 2 Sports Application

In bowling, the equation $h = 160 - 0.8s$ represents the handicap h calculated for a bowler with average score s. How much will the handicap be for each bowler listed in the table? Draw a graph that represents the relationship between the average score and the handicap.

Bowler	Average Score
Sandi	145
Dominic	125
Leo	160
Sheila	140
Tawana	175

s	$h = 160 - 0.8s$	h	(s, h)
145	$h = 160 - 0.8(145)$	44	(145, 44)
125	$h = 160 - 0.8(125)$	60	(125, 60)
160	$h = 160 - 0.8(160)$	32	(160, 32)
140	$h = 160 - 0.8(140)$	48	(140, 48)
175	$h = 160 - 0.8(175)$	20	(175, 20)

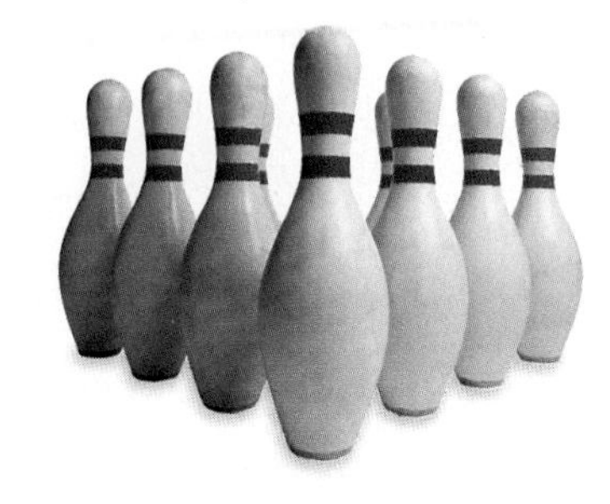

The handicaps are: Sandi, 44 pins; Dominic, 60 pins; Leo, 32 pins; Sheila, 48 pins; and Tawana, 20 pins. This is a linear equation because when s increases by 10 units, h decreases by 8 units. Note that a bowler with an average score of over 200 is given a handicap of 0.

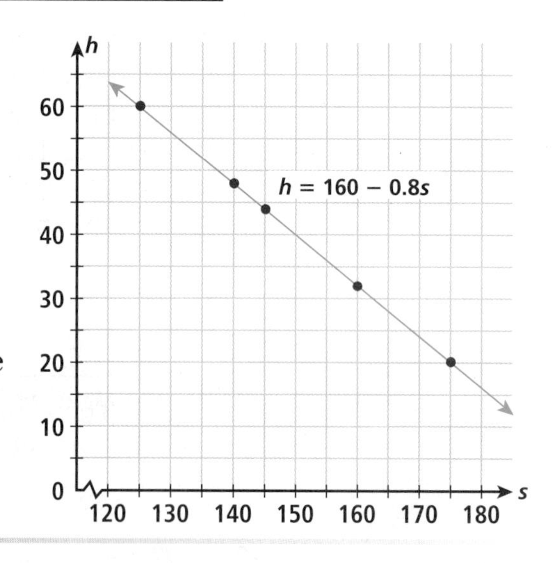

Think and Discuss

1. **Explain** whether an equation is linear if three ordered-pair solutions lie on a straight line but a fourth does not.
2. **Compare** the equations $y = 3x + 2$ and $y = 3x^2$. Without graphing, explain why one of the equations is not linear.
3. **Describe** why the ordered pair for a bowler with an average score of 210 would not fall on the line in Example 2.

11-1 Exercises

FOR EOG PRACTICE
see page 684

internet connect
Homework Help Online
go.hrw.com Keyword: MT4 11-1

5.01a, 5.01b, 5.03

GUIDED PRACTICE

See Example 1

Graph each equation and tell whether it is linear.

1. $y = x + 2$
2. $y = -2x$
3. $y = x^3$

See Example 2 

4. Kelp is one of the fastest-growing plants in the world. It grows about 2 ft every day. If you found a kelp plant that was 124 ft long, the equation $\ell = 2d + 124$ would represent the length ℓ of the plant d days later. How long would the plant be after 3 days? after 4.5 days? after 6 days? Graph the equation. Is this a linear equation?

INDEPENDENT PRACTICE

See Example 1

Graph each equation and tell whether it is linear.

5. $y = \frac{1}{3}x - 2$
6. $y = -6$
7. $y = \frac{1}{2}x^2$
8. $x = 3$
9. $y = x^2 - 12$
10. $y = 2x + 1$

See Example 2 

11. A catering service charges a $150 setup fee plus $7.50 for each guest at a reception. This is represented by the equation $C = 7.5g + 150$, where C is the total cost based on g guests. Find the total cost of catering for the following numbers of guests: 100, 150, 200, 250, 300. Is this a linear equation? Draw a graph that represents the relationship between the total cost and the number of guests.

PRACTICE AND PROBLEM SOLVING

Evaluate each equation for $x = -1, 0,$ and 1. Then graph the equation.

12. $y = 4x$
13. $y = 2x + 5$
14. $y = 6x - 3$
15. $y = x - 10$
16. $y = 4x - 2$
17. $y = 4x + 3$
18. $y = 2x - 4$
19. $y = x + 7$
20. $y = 3x + 2.5$

21. ***PHYSICAL SCIENCE*** The force exerted on an object by Earth's gravity is given by the formula $F = 9.8m$, where F is the force in newtons and m is the mass of the object in kilograms. How many newtons of gravitational force are exerted on a student with mass 52 kg?

22. At a rate of $0.08 per kilowatt-hour, the equation $C = 0.08t$ gives the cost of a customer's electric bill for using t kilowatt-hours of energy. Complete the table of values and graph the energy cost equation for t ranging from 0 to 1000.

Kilowatt-hours (*t*)	540	580	620	660	700	740
Cost in Dollars (*C*)	■	■	■	■	■	■

23. The minute hand of a clock moves $\frac{1}{10}$ degree every second. If you look at the clock when the minute hand is 10 degrees past the 12, you can use the equation $y = \frac{1}{10}x + 10$ to find how many degrees past the 12 the minute hand is after x seconds. Graph the equation and tell whether it is linear.

24. ***ENTERTAINMENT*** A bowling alley charges \$4 for shoe rental plus \$1.75 per game bowled. Write an equation that shows the total cost of bowling g games. Graph the equation. Is it linear?

25. ***BUSINESS*** A car wash pays d dollars an hour. The table shows how much employees make based on the number of hours they work.

Car Wash Wages				
Hours Worked (h)	20	25	30	40
Earnings (E)	\$150.00	\$187.50	\$225.00	\$300.00

a. Write and solve an equation to find the hourly wage.

b. Write an equation that gives an employee's earnings E for h hours of work.

c. Graph the equation for h between 0 and 50 hours.

d. Is the equation linear?

26. ***WHAT'S THE QUESTION?*** The equation $C = 9.5n + 1350$ gives the total cost of producing n trailer hitches. If the answer is \$10,850, what is the question?

27. ***WRITE ABOUT IT*** Explain how you could show that $y = 5x + 1$ is a linear equation.

28. ***CHALLENGE*** Three solutions of an equation are (1, 1), (3, 3), and (5, 5). Draw one possible graph that would show that the equation is not a linear equation.

Spiral Review

Two fair dice are rolled. Find each probability. (Lessons 9-4 and 9-7)

29. rolling two odd numbers

30. rolling a two and a prime number

31. rolling a pair of ones

32. rolling a six and a seven

33. **EOG PREP** The probability of winning a raffle is $\frac{1}{1200}$. What are the odds in favor of winning the raffle? (Lesson 9-8)

A 1:1200

B 1:1199

C 1199:1

D 1200:1

34. **EOG PREP** A bag of 9 marbles has 3 red marbles and 6 blue marbles in it. What is the probability of drawing a red marble? (Lesson 9-4)

A 1

B $\frac{2}{3}$

C $\frac{1}{3}$

D $\frac{1}{2}$

11-2 Slope of a Line

Learn to find the slope of a line and use slope to understand and draw graphs.

Remember!

You looked at slope on the coordinate plane in Lesson 5-5 (p. 244).

In skiing, the term *slope* refers to a slanted mountainside. The steeper a slope is, the higher its difficulty rating will be. In math, slope defines the "slant" of a line. The larger the absolute value of the slope of a line is, the "steeper," or more vertical, the line will be.

Linear equations have constant slope.
For a line on the coordinate plane, slope is the following ratio:

$$\frac{\text{vertical change}}{\text{horizontal change}} = \frac{\text{change in } y}{\text{change in } x}$$

This ratio is often referred to as $\frac{\text{rise}}{\text{run}}$, or "rise over run,"

where *rise* indicates the number of units moved up or down and *run* indicates the number of units moved to the left or right. Slope can be positive, negative, zero, or undefined. A line with positive slope goes up from left to right. A line with negative slope goes down from left to right.

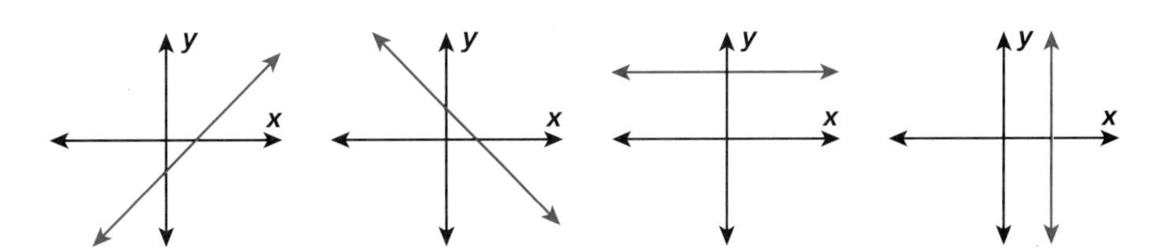

Positive slope **Negative slope** **Zero slope** **Undefined slope**

If you know any two points on a line, or two solutions of a linear equation, you can find the slope of the line without graphing.
The slope of a line through the points (x_1, y_1) and (x_2, y_2) is as follows:

$$\frac{y_2 - y_1}{x_2 - x_1}$$

EXAMPLE 1 Finding Slope, Given Two Points

Find the slope of the line that passes through (2, 5) and (8, 1).

Let (x_1, y_1) be (2, 5) and (x_2, y_2) be (8, 1).

$$\frac{y_2 - y_1}{x_2 - x_1} = \frac{1 - 5}{8 - 2}$$ *Substitute 1 for y_2, 5 for y_1, 8 for x_2, and 2 for x_1.*

$$= \frac{-4}{6} = -\frac{2}{3}$$

The slope of the line that passes through (2, 5) and (8, 1) is $-\frac{2}{3}$.

When choosing two points to evaluate the slope of a line, you can choose any two points on the line because slope is constant.

Below are two graphs of the same line.

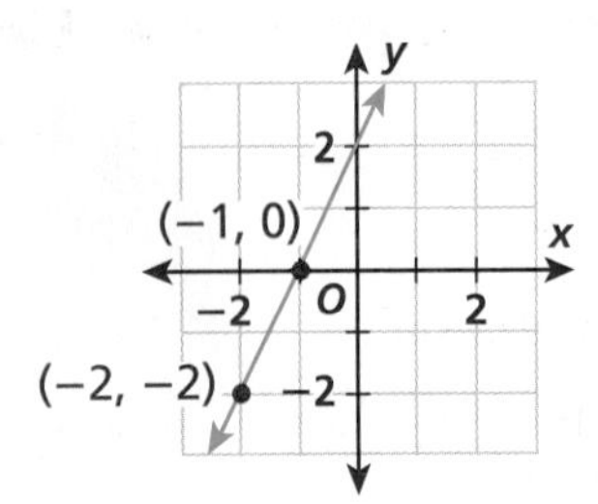

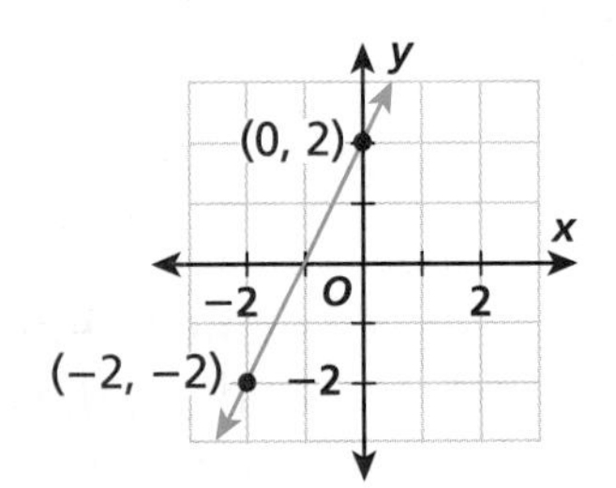

$$\frac{y_2 - y_1}{x_2 - x_1} = \frac{0 - (-2)}{-1 - (-2)} = \frac{2}{1} = 2$$

$$\frac{y_2 - y_1}{x_2 - x_1} = \frac{2 - (-2)}{0 - (-2)} = \frac{4}{2} = \frac{2}{1} = 2$$

The slope of the line is 2. Notice that although different points were chosen in each case, the slope formula still results in the same slope for the line.

EXAMPLE 2 Finding Slope from a Graph

Use the graph of the line to determine its slope.

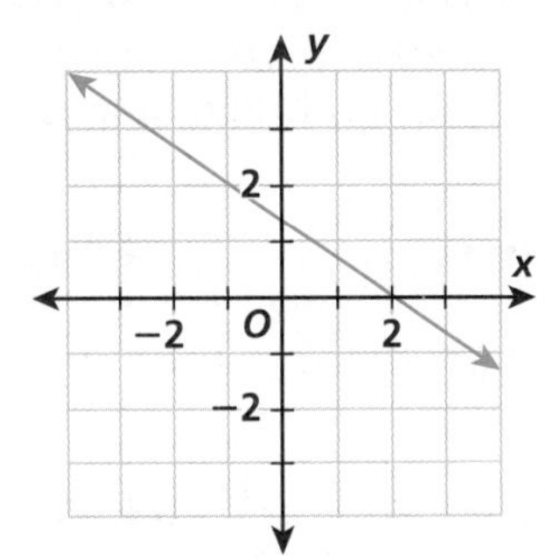

Choose two points on the line: **(−1, 2)** and **(2, 0)**.

Guess by looking at the graph:

$$\frac{\text{rise}}{\text{run}} = \frac{-2}{3} = -\frac{2}{3}$$

Use the slope formula.

Let (2, 0) be (x_1, y_1) and (−1, 2) be (x_2, y_2).

$$\frac{y_2 - y_1}{x_2 - x_1} = \frac{2 - 0}{-1 - 2} = \frac{2}{-3} = -\frac{2}{3}$$

Notice that if you switch (x_1, y_1) and (x_2, y_2), you get the same slope:

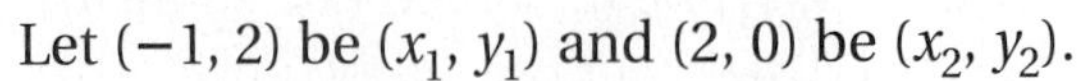

Let (−1, 2) be (x_1, y_1) and (2, 0) be (x_2, y_2).

$$\frac{y_2 - y_1}{x_2 - x_1} = \frac{0 - 2}{2 - (-1)} = \frac{-2}{3} = -\frac{2}{3}$$

The slope of the given line is $-\frac{2}{3}$.

> **Helpful Hint**
>
> It does not matter which point is chosen as (x_1, y_1) and which point is chosen as (x_2, y_2).

Recall that two parallel lines have the same slope. The slopes of two perpendicular lines are negative reciprocals of each other.

EXAMPLE 3 Identifying Parallel and Perpendicular Lines by Slope

Tell whether the lines passing through the given points are parallel or perpendicular.

A **line 1: (1, 9) and (−1, 5); line 2: (−3, −5) and (4, 9)**

$$\text{slope of line 1: } \frac{y_2 - y_1}{x_2 - x_1} = \frac{5 - 9}{-1 - 1} = \frac{-4}{-2} = 2$$

$$\text{slope of line 2: } \frac{y_2 - y_1}{x_2 - x_1} = \frac{9 - (-5)}{4 - (-3)} = \frac{14}{7} = 2$$

Both lines have a slope equal to 2, so the lines are parallel.

Remember!

The product of the slopes of perpendicular lines is −1.

B **line 1: (−10, 0) and (20, 6); line 2: (−1, 4) and (2, −11)**

$$\text{slope of line 1: } \frac{y_2 - y_1}{x_2 - x_1} = \frac{6 - 0}{20 - (-10)} = \frac{6}{30} = \frac{1}{5}$$

$$\text{slope of line 2: } \frac{y_2 - y_1}{x_2 - x_1} = \frac{-11 - 4}{2 - (-1)} = \frac{-15}{3} = -5$$

Line 1 has a slope equal to $\frac{1}{5}$ and line 2 has a slope equal to −5. $\frac{1}{5}$ and −5 are negative reciprocals of each other, so the lines are perpendicular.

You can graph a line if you know one point on the line and the slope.

EXAMPLE 4 Graphing a Line Using a Point and the Slope

Graph the line passing through (1, 1) with slope $-\frac{1}{3}$.

The slope is $-\frac{1}{3}$. So for every 1 unit down, you will move 3 units to the right, and for every 1 unit up, you will move 3 units to the left.

Plot the point (1, 1). Then move 1 unit down, and right 3 units and plot the point (4, 0). Use a straightedge to connect the two points.

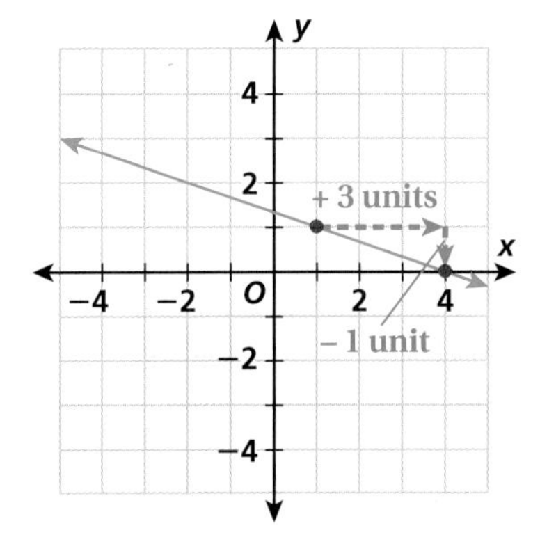

Think and Discuss

1. **Explain** why it does not matter which point you choose as (x_1, y_1) and which point you choose as (x_2, y_2) when finding slope.
2. **Give an example** of two pairs of points from each of two parallel lines.

11-2 Exercises

FOR EOG PRACTICE
see page 684

internet connect
Homework Help Online
go.hrw.com Keyword: MT4 11-2

5.01c, 5.02, 5.03

GUIDED PRACTICE

See Example 1 Find the slope of the line that passes through each pair of points.

1. (1, 3) and (2, 4) **2.** (2, 6) and (0, 2) **3.** (−1, 2) and (5, 5)

See Example 2 Use the graph of each line to determine its slope.

4.

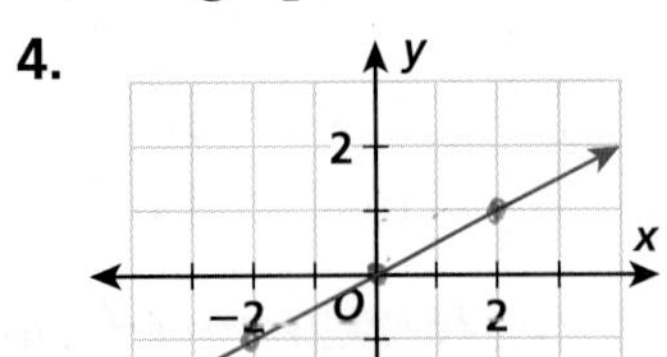

5.

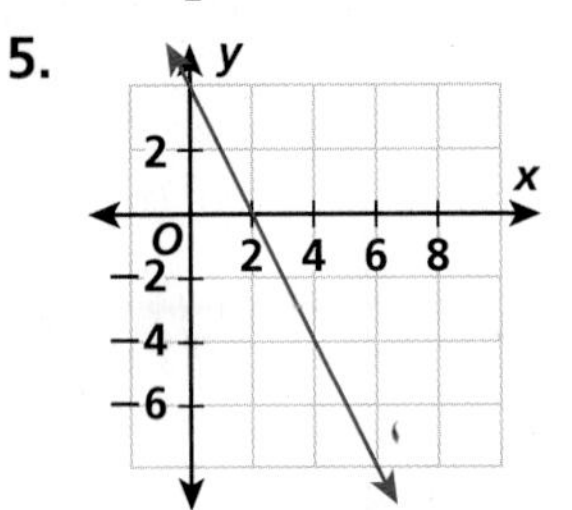

See Example 3 Tell whether the lines passing through the given points are parallel or perpendicular.

6. line 1: (2, 3) and (4, 7)
line 2: (5, 2) and (9, 0)

7. line 1: (−4, 1) and (0, 29)
line 2: (3, 3) and (5, 17)

See Example 4

8. Graph the line passing through (0, 2) with slope $-\frac{1}{2}$.

9. Graph the line passing through (−2, 0) with slope $\frac{2}{3}$.

INDEPENDENT PRACTICE

See Example 1 Find the slope of the line that passes through each pair of points.

10. (−1, −1) and (−3, 2) **11.** (0, 0) and (6, −3) **12.** (2, −5) and (1, −2)

13. (3, 1) and (0, 3) **14.** (−2, −3) and (2, 4) **15.** (0, −2) and (−6, 3)

See Example 2 Use the graph of each line to determine its slope.

16.

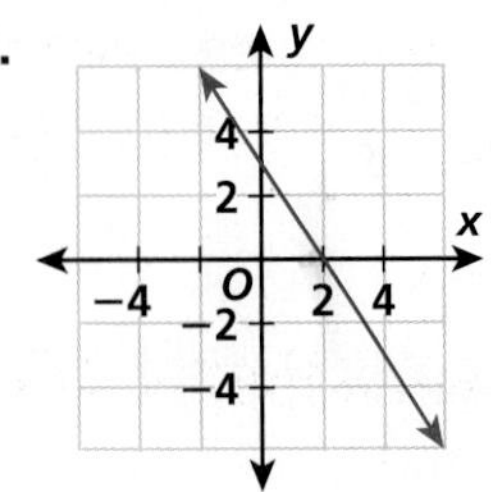

17.

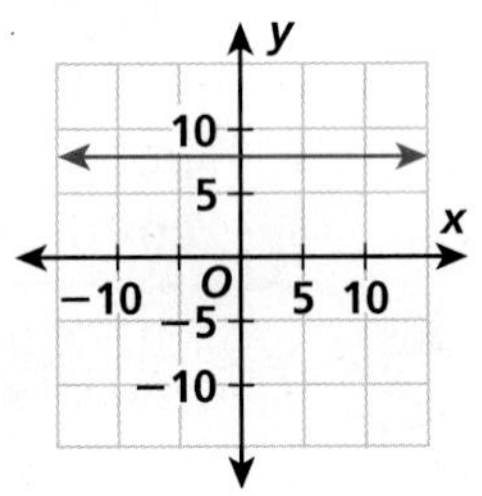

See Example 3 Tell whether the lines passing through the given points are parallel or perpendicular.

18. line 1: (1, 4) and (6, 6)
line 2: (−1, −6) and (4, −4)

19. line 1: (−1, −1) and (−3, 2)
line 2: (7, −3) and (13, 1)

See Example 4

20. Graph the line passing through (−1, 3) with slope $\frac{1}{4}$.

21. Graph the line passing through (4, 2) with slope $-\frac{4}{5}$.

PRACTICE AND PROBLEM SOLVING

22. ***SAFETY*** To accommodate a 2.5 foot vertical rise, a wheelchair ramp extends horizontally for 30 feet. Find the slope of the ramp.

For Exercises 23–26, find the slopes of each pair of lines. Use the slopes to determine whether the lines are perpendicular, parallel, or neither.

23.

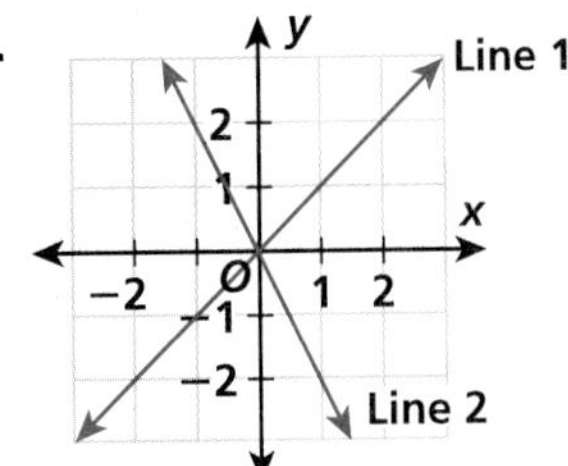

24.

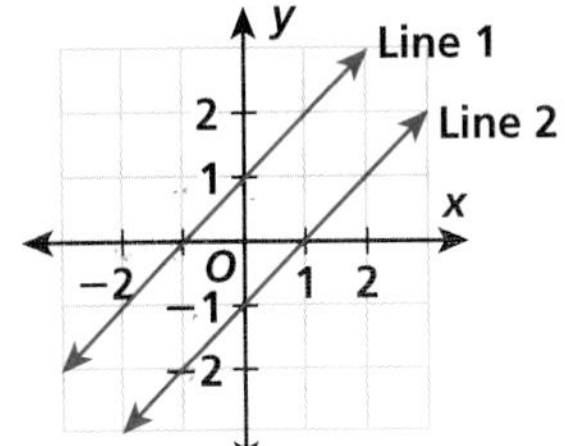

25.

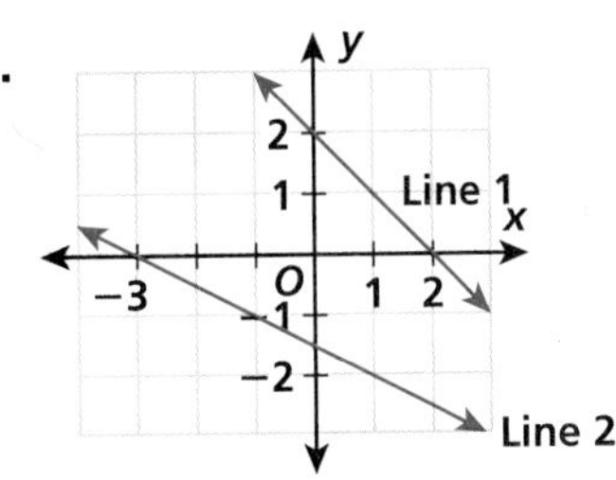

26.

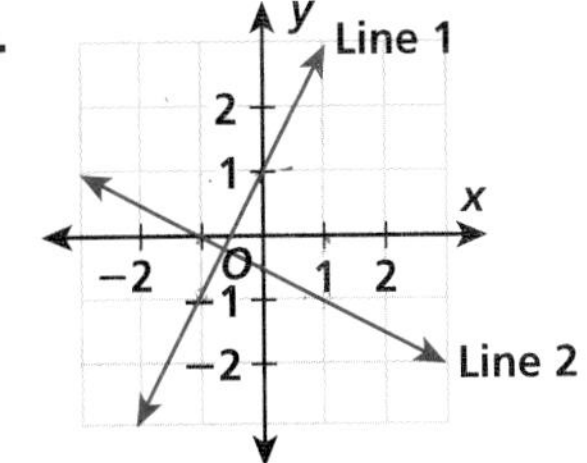

27. The Luxor Hotel in Las Vegas, Nevada, has a 350 ft tall glass pyramid. The elevator of the pyramid moves at an incline, which has a slope of $-\frac{4}{5}$. Graph the line that describes the path it travels along. (*Hint:* The point (0, 350) is the top of the pyramid.)

28. ***WHAT'S THE ERROR?*** The slope of the line through the points (1, 4) and (−1, −4) is $\frac{1-(-1)}{4-(-4)} = \frac{1}{4}$. What is the error in this statement?

29. ***WRITE ABOUT IT*** The equation of a vertical line is $x = a$ where a is any number. Explain why the slope of a vertical line is undefined, using a specific vertical line.

30. ***CHALLENGE*** Graph the equations $y = 2x - 3$, $y = -\frac{1}{2}x$ and $y = 2x + 4$ on one coordinate plane. Find the slope of each line and determine whether each combination of two lines is parallel, perpendicular, or neither. Explain how to tell whether two lines are parallel, perpendicular, or neither by their equations.

Spiral Review

Find the area of each figure with the given dimensions. (Lesson 6-2)

31. triangle: $b = 4$, $h = 6$

32. triangle: $b = 3$, $h = 14$

33. trapezoid: $b_1 = 9$, $b_2 = 11$, $h = 12$

34. trapezoid: $b_1 = 3.4$, $b_2 = 6.6$, $h = 1.8$

35. **EOG PREP** A circular flower bed has radius 22 in. What is the circumference of the bed to the nearest tenth of an inch? Use 3.14 for π. (Lesson 6-4)

A 1519.8 in. B 69.1 in. C 103.7 in. D 138.2 in.

EXAMPLE 3 Entertainment Application

Helpful Hint

The y-intercept represents the initial number of points (50). The slope represents the rate of change (−3.5 points per game).

An arcade deducts 3.5 points from your 50-point game card for each Skittle-ball game you play. The linear equation $y = -3.5x + 50$ represents the number of points y on your card after x games. Graph the equation using the slope and y-intercept.

$y = -3.5x + 50$ *The equation is in slope-intercept form.*

$m = -3.5$ $b = 50$

The slope of the line is −3.5, and the y-intercept is 50. The line crosses the y-axis at the point (0, 50) and moves down 3.5 units for every 1 unit it moves to the right.

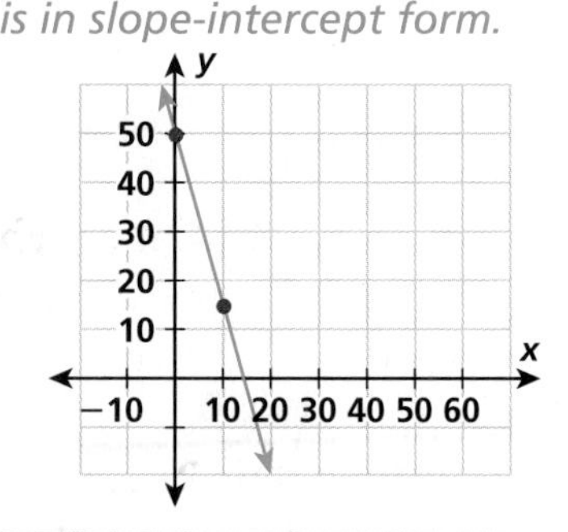

EXAMPLE 4 Writing Slope-Intercept Form

Write the equation of the line that passes through (−3, 1) and (2, −1) in slope-intercept form.

Find the slope.

$$\frac{y_2 - y_1}{x_2 - x_1} = \frac{-1 - 1}{2-(-3)} = \frac{-2}{5} = -\frac{2}{5}$$ *The slope is $-\frac{2}{5}$.*

Choose either point and substitute it along with the slope into the slope-intercept form.

$$y = mx + b$$

$$-1 = -\frac{2}{5}(2) + b$$ *Substitute 2 for x, −1 for y, and $-\frac{2}{5}$ for m.*

$$-1 = -\frac{4}{5} + b$$ *Simplify.*

Solve for b.

$$-1 = -\frac{4}{5} + b$$

$$\underline{+\frac{4}{5} \quad +\frac{4}{5}}$$ *Add $\frac{4}{5}$ to both sides.*

$$-\frac{1}{5} = b$$

Write the equation of the line, using $-\frac{2}{5}$ for m and $-\frac{1}{5}$ for b.

$$y = -\frac{2}{5}x + \left(-\frac{1}{5}\right), \text{ or } y = -\frac{2}{5}x - \frac{1}{5}$$

Think and Discuss

1. **Describe** the line represented by the equation $y = -5x + 3$.
2. **Give** a real-life example with a graph that has a slope of 5 and a y-intercept of 30.

11-3 Exercises

FOR EOG PRACTICE
see page 685

internet connect
Homework Help Online
go.hrw.com Keyword: MT4 11-3

5.01a, 5.01c, 5.01d, 5.02, 5.03

GUIDED PRACTICE

See Example 1 **Find the x-intercept and y-intercept of each line. Use the intercepts to graph the equation.**

1. $x - y = 5$ **2.** $2x + 3y = 12$ **3.** $3x + 5y = -15$ **4.** $-5x + 2y = -10$

See Example 2 **Write each equation in slope-intercept form, and then find the slope and y-intercept.**

5. $2x = 4y$ **6.** $3x - y = 14$ **7.** $3x - 9y = 27$ **8.** $x + 2y = 8$

See Example 3 **9.** A freight company charges $22 plus $3.50 per pound to ship an item that weighs n pounds. The total shipping charges are given by the equation $C = 3.5n + 22$. Identify the slope and y-intercept, and use them to graph the equation for n between 0 and 100 pounds.

See Example 4 **Write the equation of the line that passes through each pair of points in slope-intercept form.**

10. $(-1, -6)$ and $(2, 6)$ **11.** $(0, 5)$ and $(3, -1)$ **12.** $(3, 5)$ and $(6, 6)$

INDEPENDENT PRACTICE

See Example 1 **Find the x-intercept and y-intercept of each line. Use the intercepts to graph the equation.**

13. $2y = 20 - 4x$ **14.** $4x = 12 + 3y$ **15.** $-y = 18 - 6x$ **16.** $2x + y = 7$

See Example 2 **Write each equation in slope-intercept form, and then find the slope and y-intercept.**

17. $-y = 2x$ **18.** $5y + 2x = 15$ **19.** $-4y - 8x = 8$ **20.** $2y + 6x = -14$

See Example 3 **21.** A salesperson receives a weekly salary of $300 plus a commission of $15 for each TV sold. Total weekly pay is given by the equation $P = 15n + 300$. Identify the slope and y-intercept, and use them to graph the equation for n between 0 and 40 TVs.

See Example 4 **Write the equation of the line that passes through each pair of points in slope-intercept form.**

22. $(0, -7)$ and $(4, 25)$ **23.** $(-1, 1)$ and $(3, -3)$ **24.** $(-6, -3)$ and $(12, 0)$

PRACTICE AND PROBLEM SOLVING

Use the x-intercept and y-intercept of each line to graph the equation.

25. $y = 2x - 10$ **26.** $y = \frac{1}{3}x + 2$ **27.** $y = 4x - 2.5$ **28.** $y = -\frac{4}{5}x + 15$

Life Science LINK

Acute Mountain Sickness (AMS) occurs if you ascend in altitude too quickly without giving your body time to adjust. It usually occurs at altitudes over 10,000 feet above sea level. To prevent AMS you should not ascend more than 1000 feet per day. And every time you climb a total of 3000 feet, your body needs two nights to adjust.

Often people will get sick at high altitudes because there is less oxygen and lower atmospheric pressure.

29. The map shows a team's plan for climbing Long's Peak in Rocky Mountain National Park.

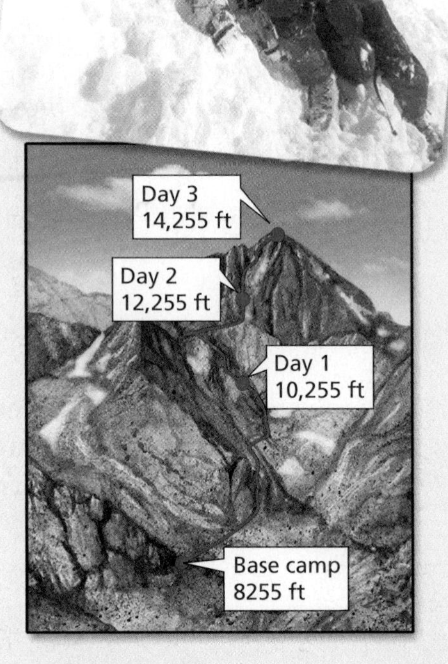

a. Make a graph of the team's plan of ascent and find the slope of the line. (Day number should be your x-value, and altitude should be your y-value.)

b. Find the y-intercept and explain what it means.

c. Write the equation of the line in slope-intercept form.

d. Does the team run a high risk of getting AMS?

30. An expedition starts at an altitude of 9056 ft and climbs at an average rate of 544 ft of elevation a day. Write an equation in slope-intercept form that describes the expedition's climb. Are the climbers likely to suffer from AMS at their present climbing rate? On what day of their climb will they be at risk?

31. The equation that describes a mountain climber's ascent up Mount McKinley in Alaska is $y = 955x + 16{,}500$, where x is the day number and y is the altitude at the end of the day. What are the slope and y-intercept? What do they mean in terms of the climb?

32. ***CHALLENGE*** Make a graph of the ascent of a team that follows the rules to avoid AMS exactly and spends the minimum number of days climbing from base camp (17,600 ft) to the summit of Mount Everest (29,035 ft). Can you write a linear equation describing this trip? Explain your answer.

Spiral Review

Estimate the number or percent. (Lesson 8-5)

33. 25% of 398 is about what number?

34. 202 is about 50% of what number?

35. About what percent of 99 is 39?

36. About what percent of 989 is 746?

37. **EOG PREP** Carlos has \$3.35 in dimes and quarters. If he has a total of 23 coins, how many dimes does he have? (Lesson 10-6)

A 16 B 11 C 18 D 9

Graph Equations in Slope-Intercept Form

Use with Lesson 11-4

1.02, 5.01c, 5.01d

internet connect
Lab Resources Online
go.hrw.com
KEYWORD: MT4 Lab11A

To graph $y = x + 1$, a linear equation in slope-intercept form, in the standard graphing calculator window, press Y= ; enter the right side of the equation, X,T,θ,n + 1; and press ZOOM **6:ZStandard.**

From the slope-intercept equation, you know that the slope of the line is 1. Notice that the standard window distorts the screen, and the line does not appear to have a great enough slope.

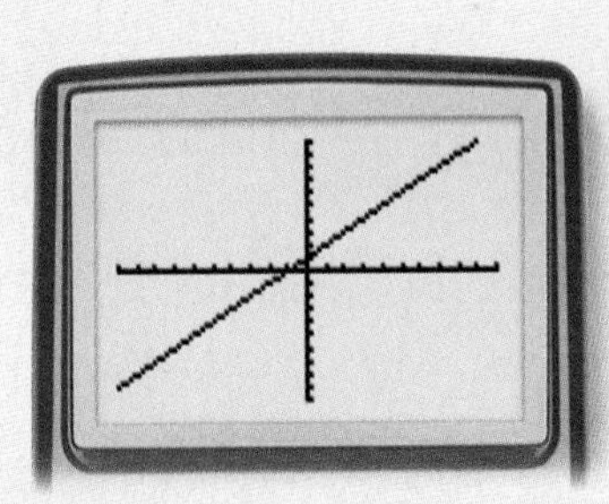

Press ZOOM **5:ZSquare**. This changes the scale for x from -10 to 10 to -15.16 to 15.16. The graph is shown at right. Or press ZOOM **8:ZInteger** ENTER . This changes the scale for x to -47 to 47 and the scale for y to -31 to 31.

Activity

1 Graph $2x + 3y = 36$ in the integer window. Find the x- and y-intercepts of the graph.

First solve $3y = -2x + 36$ for y.

$y = \frac{-2x + 36}{3}$, so $y = \frac{-2}{3}x + 12$.

Press Y= ; enter the right side of the equation,

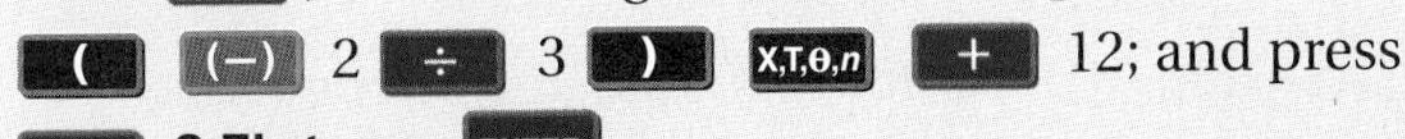

((−) 2 ÷ 3) X,T,θ,n + 12; and press ZOOM **8:ZInteger** ENTER .

Press TRACE to see the equation of the line and the y-intercept. The graph in the **ZInteger** window is shown.

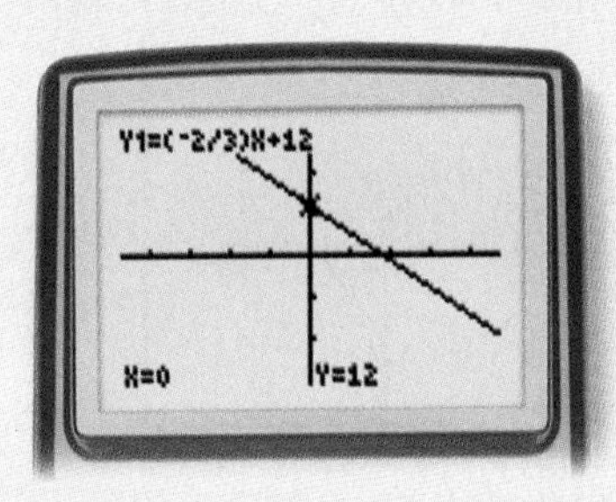

Think and Discuss

1. How do the ratios of the range of y to the range of x in the **ZSquare** and **ZInteger** windows compare?

Try This

Graph each equation in a square window.

1. $y = 2x$ **2.** $2y = x$ **3.** $2y - 4x = 12$ **4.** $2x + 5y = 40$

11-4 Point-Slope Form

Learn to find the equation of a line given one point and the slope.

Vocabulary

point-slope form

Lasers aim light along a straight path. If you know the destination of the light beam (a point on the line) and the slant of the beam (the slope), you can write an equation in *point-slope form* to calculate the height at which the laser is positioned.

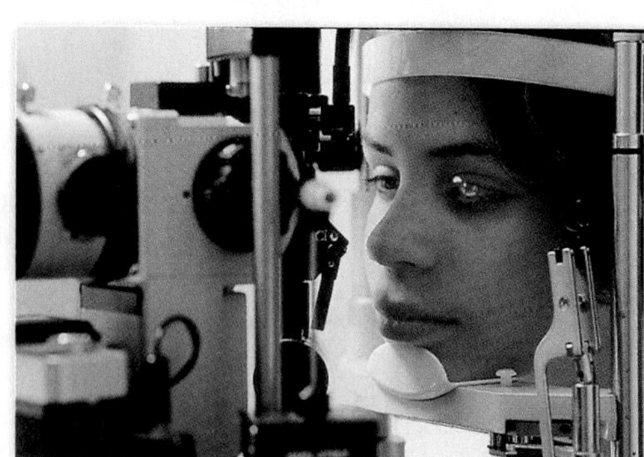

The **point-slope form** of an equation of a line with slope m passing through (x_1, y_1) is $y - y_1 = m(x - x_1)$.

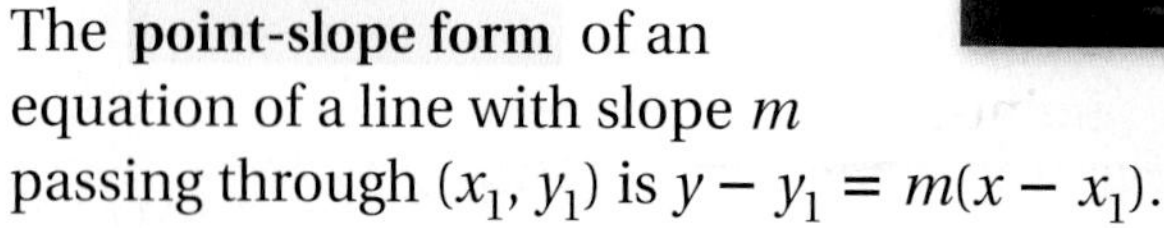

EXAMPLE 1 Using Point-Slope Form to Identify Information About a Line

Use the point-slope form of each equation to identify a point the line passes through and the slope of the line.

A $y - 9 = -\frac{2}{3}(x - 21)$

$y - y_1 = m(x - x_1)$

$y - 9 = -\frac{2}{3}(x - 21)$ — *The equation is in point-slope form.*

$m = -\frac{2}{3}$ — *Read the value of m from the equation.*

$(x_1, y_1) = (21, 9)$ — *Read the point from the equation.*

The line defined by $y - 9 = -\frac{2}{3}(x - 21)$ has slope $-\frac{2}{3}$, and passes through the point (21, 9).

B $y - 3 = 4(x + 7)$

$y - y_1 = m(x - x_1)$

$y - 3 = 4(x + 7)$

$y - 3 = 4[x - (-7)]$ — *Rewrite using subtraction instead of addition.*

$m = 4$

$(x_1, y_1) = (-7, 3)$

The line defined by $y - 3 = 4(x + 7)$ has slope 4, and passes through the point (−7, 3).

EXAMPLE 2 Writing the Point-Slope Form of an Equation

Write the point-slope form of the equation with the given slope that passes through the indicated point.

A **the line with slope –2 passing through (4, 1)**

$y - y_1 = m(x - x_1)$

$y - 1 = -2(x - 4)$ *Substitute 4 for x_1, 1 for y_1 and −2 for m.*

The equation of the line with slope −2 that passes through (4, 1) in point-slope form is $y - 1 = -2(x - 4)$.

B **the line with slope 7 passing through (−1, 3)**

$y - y_1 = m(x - x_1)$

$y - 3 = 7[x - (-1)]$ *Substitute −1 for x_1, 3 for y_1, and 7 for m.*

$y - 3 = 7(x + 1)$

The equation of the line with slope 7 that passes through (–1, 3) in point-slope form is $y - 3 = 7(x + 1)$.

EXAMPLE 3 *Medical Application*

Suppose that laser eye surgery is modeled on a coordinate grid. The laser is positioned at the *y*-intercept so that the light shifts down 1 mm for each 40 mm it shifts to the right. The light reaches the center of the cornea of the eye at (125, 0). Write the equation of the light beam in point-slope form, and find the height of the laser.

As x increases by 40, y decreases by 1, so the slope of the line is $-\frac{1}{40}$. The line must pass through the point (125, 0).

$y - y_1 = m(x - x_1)$

$y - 0 = -\frac{1}{40}(x - 125)$ *Substitute 125 for x_1, 0 for y_1, and $-\frac{1}{40}$ for m.*

The equation of the line the laser beam travels along, in point-slope form, is $y = -\frac{1}{40}(x - 125)$. Substitute 0 for x to find the y-intercept.

$y = -\frac{1}{40}(0 - 125)$

$y = -\frac{1}{40}(-125)$

$y = 3.125$

The y-intercept is 3.125, so the laser is at a height of 3.125 mm.

Think and Discuss

1. **Describe** the line, using the point-slope equation, that has a slope of 2 and passes through (−3, 4).
2. **Tell** how you find the point-slope form of the line when you know the coordinates of two points.

11-4 Exercises

FOR EOG PRACTICE
see page 685

internet connect
Homework Help Online
go.hrw.com Keyword: MT4 11-4

5.01c, 5.01d, 5.02, 5.03

GUIDED PRACTICE

See Example 1 **Use the point-slope form of each equation to identify a point the line passes through and the slope of the line.**

1. $y - 4 = -2(x + 7)$
2. $y - 9 = 5(x - 12)$
3. $y + 2.4 = 2.1(x - 1.8)$
4. $y + 1 = 11(x - 1)$
5. $y + 8 = -6(x - 9)$
6. $y - 7 = 4(x + 3)$

See Example 2 **Write the point-slope form of the equation with the given slope that passes through the indicated point.**

7. the line with slope 3 passing through (0, 4)
8. the line with slope −10 passing through (−13, 8)

See Example 3

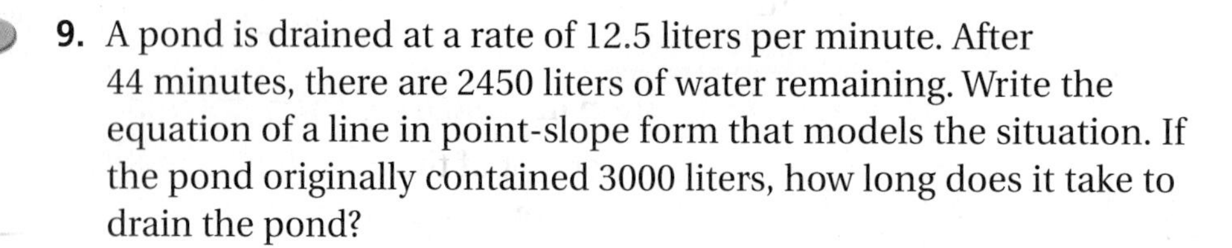

9. A pond is drained at a rate of 12.5 liters per minute. After 44 minutes, there are 2450 liters of water remaining. Write the equation of a line in point-slope form that models the situation. If the pond originally contained 3000 liters, how long does it take to drain the pond?

INDEPENDENT PRACTICE

See Example 1

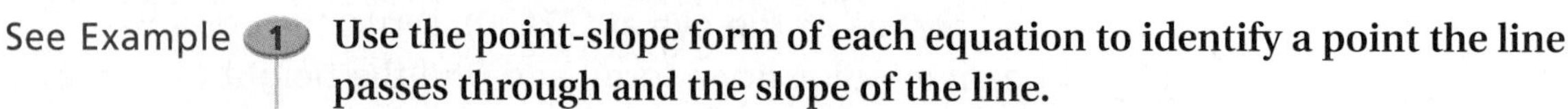

Use the point-slope form of each equation to identify a point the line passes through and the slope of the line.

10. $y - 1 = \frac{2}{3}(x + 7)$
11. $y + 7 = 3(x + 4)$
12. $y - 2 = -\frac{1}{6}(x - 11)$
13. $y - 11 = 14(x - 8)$
14. $y - 3 = -1.8(x - 5.6)$
15. $y + 7 = 1(x - 5)$

See Example 2 **Write the point-slope form of the equation with the given slope that passes through the indicated point.**

16. the line with slope −5 passing through (−3, −5)
17. the line with slope 4 passing through (−1, 0)

See Example 3

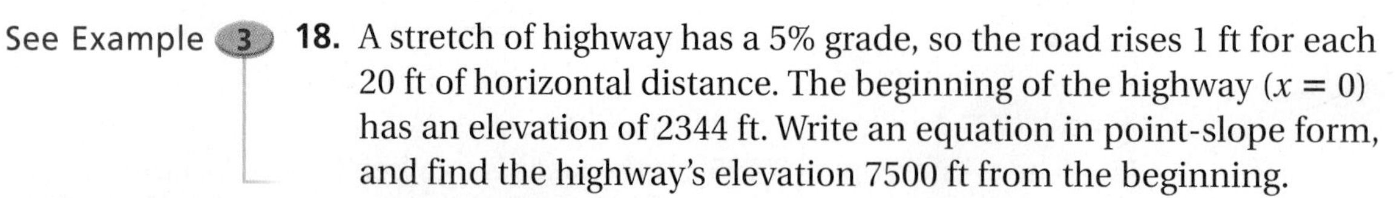

18. A stretch of highway has a 5% grade, so the road rises 1 ft for each 20 ft of horizontal distance. The beginning of the highway ($x = 0$) has an elevation of 2344 ft. Write an equation in point-slope form, and find the highway's elevation 7500 ft from the beginning.

PRACTICE AND PROBLEM SOLVING

Write the point-slope form of each line described below.

19. the line parallel to $y = 3x - 4$ that passes through (−1, 4)
20. the line perpendicular to $y = -2x$ that passes through (7, −3)
21. the line perpendicular to $y = x + 1$ that passes through (−6, −8)
22. the line parallel to $y = -10x - 5$ that passes through (−3, 0)

Earth Science LINK

Mount Etna, a volcano in Sicily, Italy, has been erupting for over half a million years. It is one of the world's most active volcanoes. When it erupted in 1669 it almost completely destroyed the city of Catania.

go.hrw.com
KEYWORD: MT4 Etna

CNN Student News

23. ***EARTH SCIENCE*** Jorullo is a cinder cone volcano in Mexico. Suppose Jorullo is 315 m tall, 50 m from the center of its base. Use the slope of a cinder cone to write a possible equation in point-slope form that approximately models the height of the volcano, x meters from the center of its base.

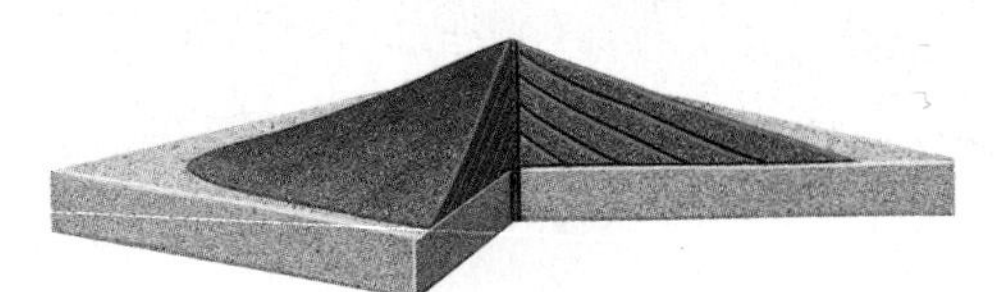

Shield volcano typical slope: 0.03–0.17

Composite volcano typical slope: 0.17–0.5

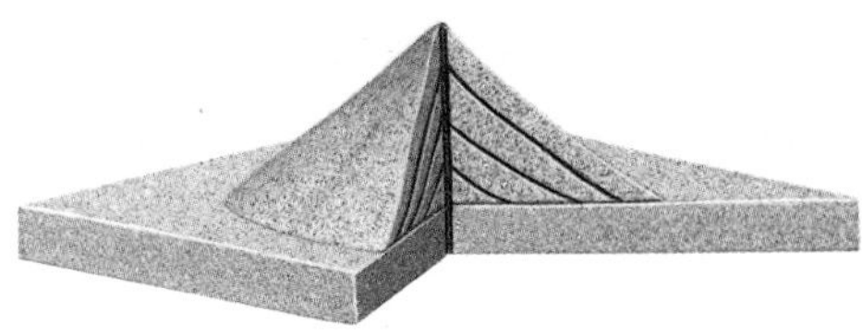

Cinder cone volcano typical slope: 0.5–0.65

24. ***LIFE SCIENCE*** Since a breed of finch was introduced to the United States, the population of the breed has increased by about 600 birds per year. After 4 years, there are roughly 2730 finches.
    a. Write an equation in point-slope form to model the finch population.
    b. What is the y-intercept of the equation in part **a**, and what does the y-intercept tell you about the finch population?

25. ***LIFE SCIENCE*** Moose antlers grow at the fastest rate of any animal bone. Each day, a moose antler grows about 1 in. Suppose you started observing a moose when its antlers were 15 in long. Write an equation in point-slope form that describes the length of the moose's antlers after d days of observation.

26. ***WRITE A PROBLEM*** Write a problem about the point-slope form of an equation using the data on a car's fuel economy.

Fuel Economy		
Gas Tank Capacity	City Efficiency	Highway Efficiency
16 gal	28 mi/gal	36 mi/gal

27. ***WRITE ABOUT IT*** Explain how you could convert an equation in point-slope form to slope-intercept form.

28. ***CHALLENGE*** The value of one line's x-intercept is the opposite of the value of its y-intercept. The line contains the point $(10, -5)$. Find the point-slope form of the equation.

Spiral Review

Solve each inequality. (Lesson 10-4)

29. $4x + 3 - x > 15$
30. $3 - 7x \leq 24$
31. $3x + 9 < 2x - 4$
32. $1 - x \geq 11 + x$

33. **EOG PREP** A landscaping company charges a $35 consultation fee, plus $50 per hour. How much would it cost to hire the company for 3 hours? (Lesson 11-1)

A $225   B $150   C $185   D $135

Chapter 11

Mid-Chapter Quiz

LESSON 11-1 (pp. 540–544)

Graph each equation and tell whether it is linear.

1. $y = 1 - 3x$ **2.** $x = 2$ **3.** $y = 2x^2$

Draw a graph that represents the relationship.

4. At Bob's Books, the equation $u = \frac{2}{3}n + 3$ represents the price for a used book u with a selling price n when the book was new. How much will a used copy cost for each of the listed new prices?

New Price	Used Price
$12	■
$15	■
$24	■
$36	■

LESSON  11-2 (pp. 545–549)

Find the slope of the line that passes through each pair of points.

5. (5, 2) and (1, 3) **6.** (1, 4) and (−1, −3) **7.** (0, −2) and (−5, 0)

Tell whether the lines passing through the given points are parallel or perpendicular.

8. line 1: (−1, −3) and (3, −11)
line 2: (−8, −3) and (6, 4)

9. line 1: (0, −1) and (−2, −9)
line 2: (2, 15) and (−1, 3)

LESSON 11-3 (pp. 550–554)

Given two points through which a line passes, write the equation of each line in slope-intercept form.

10. (−4, 3) and (−2, 1) **11.** (2, 7) and (5, 3) **12.** (4, 0) and (2, −5)

Identify the slope and y-intercept, and use them to graph the equation.

13. An airline frequent-flyer plan offers a bonus of 5000 mi to new members plus 1.5 mi for every dollar charged on a credit card endorsed by the airline. The linear equation $y = 1.5x + 5000$ represents the number of miles earned after charging x dollars on the credit card.

LESSON  11-4 (pp. 556–559)

Use the point-slope form of each equation to identify a point the line passes through and the slope of the line.

14. $y + 4 = -2(x - 1)$ **15.** $y = -(x + 4)$ **16.** $y - 7 = -3x$

Write the point-slope form of each line with the given conditions.

17. slope −3, passing through (7, 2) **18.** slope 4, passing through (−4, 1)

Focus on Problem Solving

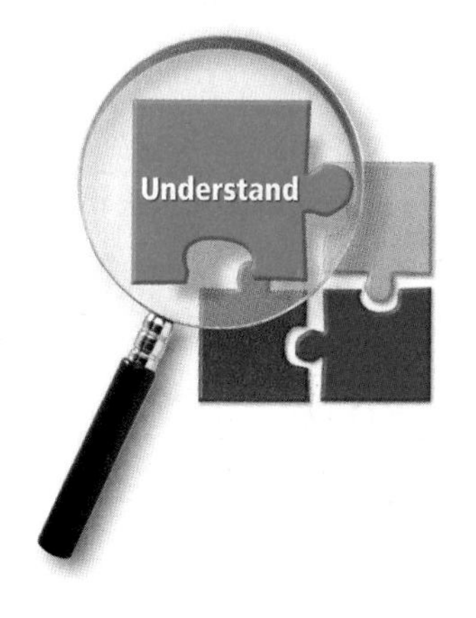

Understand the Problem

- **Identify important details in the problem**

When you are solving word problems, you need to find the information that is important to the problem.

You can write the equation of a line if you know the slope and one point on the line or if you know two points on the line.

Example:

A school bus carrying 40 students is traveling toward the school at **30 mi/hr**. After **15 minutes**, it has **20 miles to go**. How far away from the school was the bus when it started?

You can write the equation of the line in point-slope form.

$$\begin{aligned} y - y_1 &= m(x - x_1) \\ y - (-20) &= 30(x - 0.25) && \textit{The slope is the rate of change, or 30.} \\ y + 20 &= 30x - 7.5 && \textit{15 minutes = 0.25 hours} \\ -20 & \quad\;\; -20 && \textit{(0.25, −20) is a point on the line.} \\ y &= 30x - 27.5 \end{aligned}$$

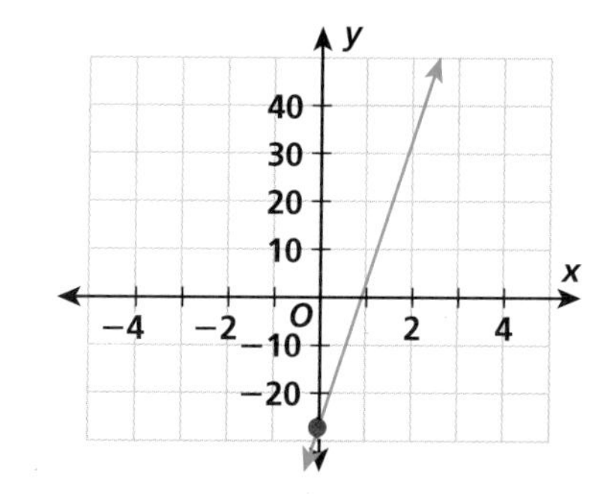

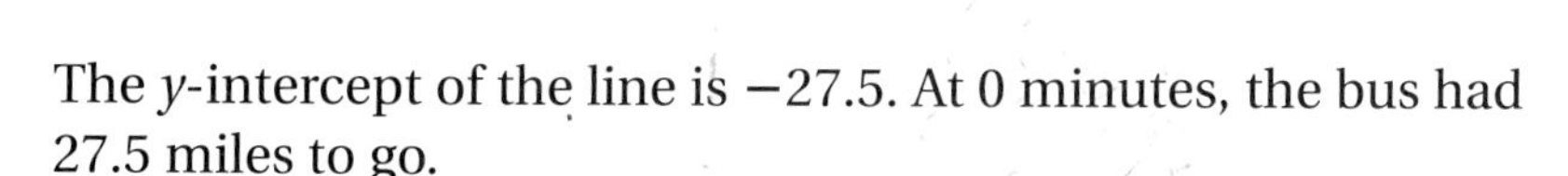

The y-intercept of the line is −27.5. At 0 minutes, the bus had 27.5 miles to go.

Read each problem, and identify the information needed to write the equation of a line. Give the slope and one point on the line, or give two points on the line.

1. At sea level, water boils at 212°F. At an altitude of 2000 ft, water boils at 208°F. If the relationship is linear, estimate the temperature that water would boil at an altitude of 5000 ft.

2. Don earns a weekly salary of $480, plus a commission of 5% of his total sales. How many dollars in merchandise does he have to sell to make $500 in one week?

3. An environmental group has a goal of planting 10,000 trees. On Arbor Day, volunteers planted 4500 trees. If the group can plant 500 trees per week, how long will it take them to plant the remaining trees to reach their goal?

4. Kayla rents a booth at a craft fair. If she sells 50 bracelets, her profit is $25. If she sells 80 bracelets, her profit is $85. What would her profit be if she sold 100 bracelets?

11-5 Direct Variation

Learn to recognize direct variation by graphing tables of data and checking for constant ratios.

Vocabulary

direct variation

constant of proportionality

A satellite in orbit travels 8 miles in 1 second, 16 miles in 2 seconds, 24 miles in 3 seconds, and so on.

The ratio of distance to time is constant. The satellite travels 8 miles every 1 second.

$$\frac{\text{distance}}{\text{time}} = \frac{8 \text{ mi}}{1 \text{ s}} = \frac{16 \text{ mi}}{2 \text{ s}} = \frac{24 \text{ mi}}{3 \text{ s}}$$

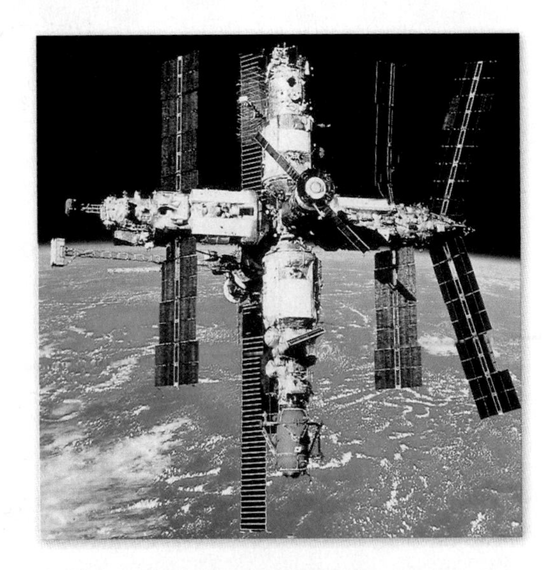

DIRECT VARIATION

Words	Numbers	Algebra
For **direct variation**, two variable quantities are related proportionally by a constant positive ratio. The ratio is called the **constant of proportionality**.	$8 = k$ $16 = 2k$ $24 = 3k$	$y = kx$ $k = \frac{y}{x}$

The distance the satellite travels *varies directly* with time and is represented by the equation $y = kx$. The constant ratio k is 8.

EXAMPLE 1 Determining Whether a Data Set Varies Directly

Helpful Hint

The graph of a direct-variation equation is always linear *and* always contains the point (0, 0). The variables x and y either increase together or decrease together.

Determine whether the data set shows direct variation.

A

Shoe Sizes					
U.S. Size	7	8	9	10	11
European Size	39	41	43	44	45

Make a graph that shows the relationship between the U.S. sizes and the European sizes. The graph is not linear.

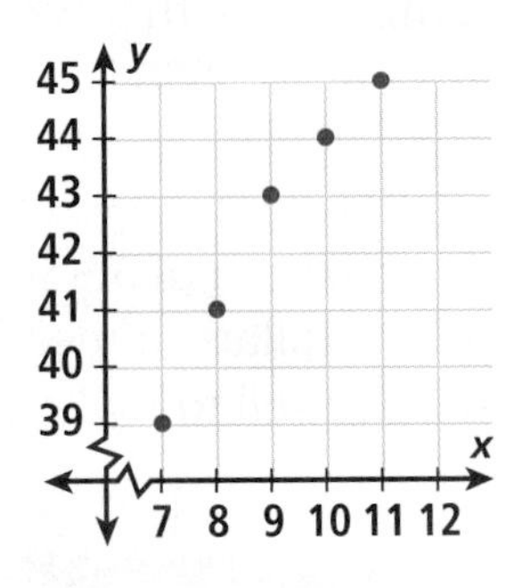

You can also compare ratios to see if a direct variation occurs.

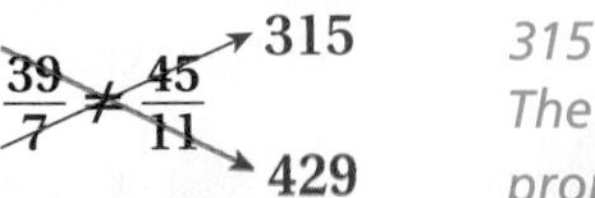

$\frac{39}{7} \neq \frac{45}{11}$ → 315, 429

$315 \neq 429$ The ratios are not proportional.

The relationship of the data is not a direct variation.

Determine whether the data set shows direct variation.

B

Distance Sound Travels at 20°C (m)					
Time (s)	0	1	2	3	4
Distance (m)	0	350	700	1050	1400

Make a graph that shows the relationship between the number of seconds and the distance sound travels.

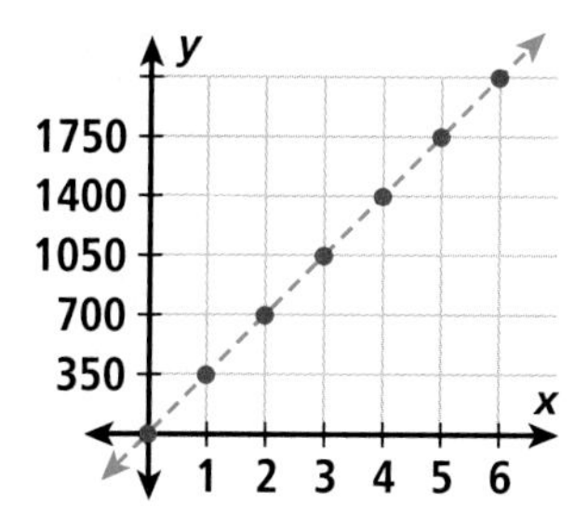

Plot the points.

The points lie in a straight line.

(0, 0) is included.

You can also compare ratios to see if a direct variation occurs.

$\frac{350}{1} = \frac{700}{2} = \frac{1050}{3} = \frac{1400}{4}$ *Compare ratios. The ratio is constant.*

The ratios are proportional. The relationship is a direct variation.

EXAMPLE 2 Finding Equations of Direct Variation

Find each equation of direct variation, given that y varies directly with x.

A y **is 52 when** x **is 4**

$y = kx$ *y varies directly with x.*

$52 = k \cdot 4$ *Substitute for x and y.*

$13 = k$ *Solve for k.*

$y = 13x$ *Substitute 13 for k in the original equation.*

B x **is 10 when** y **is 15**

$y = kx$ *y varies directly with x.*

$15 = k \cdot 10$ *Substitute for x and y.*

$\frac{3}{2} = k$ *Solve for k.*

$y = \frac{3}{2}x$ *Substitute $\frac{3}{2}$ for k in the original equation.*

C y **is 5 when** x **is 2**

$y = kx$ *y varies directly with x.*

$5 = k \cdot 2$ *Substitute for x and y.*

$\frac{5}{2} = k$ *Solve for k.*

$y = \frac{5}{2}x$ *Substitute $\frac{5}{2}$ for k in the original equation.*

EXAMPLE 3 *Physical Science Application*

When a driver applies the brakes, a car's total stopping distance is the sum of the reaction distance and the braking distance. The reaction distance is the distance the car travels before the driver presses the brake pedal. The braking distance is the distance the car travels after the brakes have been applied.

Determine whether there is a direct variation between either data set and speed. If so, find the equation of direct variation.

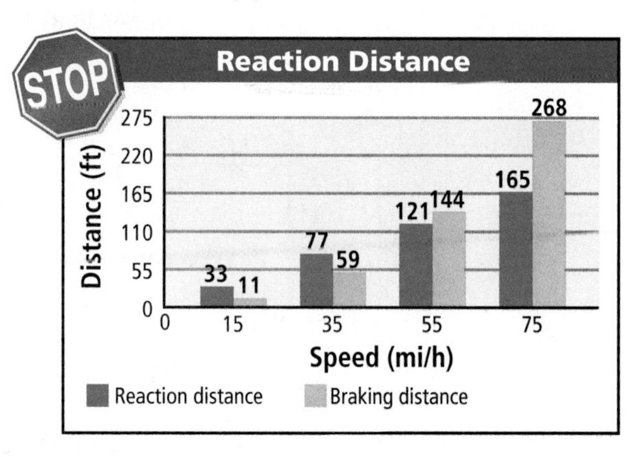

A **reaction distance and speed**

$\frac{\text{reaction distance}}{\text{speed}} = \frac{33}{15} = 2.2$ $\qquad$ $\frac{\text{reaction distance}}{\text{speed}} = \frac{77}{35} = 2.2$

The first two pairs of data result in a common ratio. In fact, all of the reaction distance to speed ratios are equivalent to 2.2.

$$\frac{\text{reaction distance}}{\text{speed}} = \frac{33}{15} = \frac{77}{35} = \frac{121}{55} = \frac{165}{75} = 2.2$$

The variables are related by a constant ratio of 2.2 to 1, and (0, 0) is included. The equation of direct variation is $y = 2.2x$, where x is the speed, y is the reaction distance, and 2.2 is the constant of proportionality.

B **braking distance and speed**

$\frac{\text{breaking distance}}{\text{speed}} = \frac{11}{15} = 0.7\overline{3}$ $\qquad$ $\frac{\text{breaking distance}}{\text{speed}} = \frac{59}{35} = 1.69$

$0.7\overline{3} \neq 1.69$

If any of the ratios are not equal, then there is no direct variation. It is not necessary to compute additional ratios or to determine whether (0, 0) is included.

Think and Discuss

1. **Describe** the slope and the y-intercept of a direct variation equation.
2. **Tell** whether two variables that do not vary directly can result in a linear graph.

11-5 Exercises

FOR EOG PRACTICE

see page 686

internet connect

Homework Help Online
go.hrw.com Keyword: MT4 11-5

5.01a, 5.01d

GUIDED PRACTICE

See Example 1

Make a graph to determine whether the data sets show direct variation.

1. The table shows an employee's pay per number of hours worked.

Hours Worked	0	1	2	3	4	5	6
Pay ($)	0	8.50	17.00	25.50	34.00	42.50	51.00

See Example 2

Find each equation of direct variation, given that y varies directly with x.

2. y is 10 when x is 2
3. y is 16 when x is 4
4. y is 12 when x is 15
5. y is 3 when x is 6
6. y is 220 when x is 2
7. y is 5 when x is 40

See Example 3

8. The following table shows how many hours it takes to travel 300 miles, depending on your speed in miles per hour. Determine whether there is direct variation between the two data sets. If so, find the equation of direct variation.

Speed (mi/h)	5	6	7.5	10	15	30	60
Time (hr)	60	50	40	30	20	10	5

INDEPENDENT PRACTICE

See Example 1

Make a graph to determine whether the data sets show direct variation.

9. The table shows the amount of current flowing through a 12-volt circuit with various resistances.

Resistance (ohms)	48	24	12	6	4	3	2
Current (amps)	0.25	0.5	1	2	3	4	6

See Example 2

Find each equation of direct variation, given that y varies directly with x.

10. y is 2.5 when x is 2.5
11. y is 2 when x is 8
12. y is 93 when x is 3
13. y is 8 when x is 22
14. y is 52 when x is 4
15. y is 10 when x is 100

See Example 3

16. The following table shows how many hours it takes to drive certain distances at a speed of 60 miles per hour. Determine whether there is direct variation between the two data sets. If so, find the equation of direct variation.

Distance (mi)	15	30	60	90	120	150	180
Time (hr)	0.25	0.5	1	1.5	2	2.5	3

PRACTICE AND PROBLEM SOLVING

Tell whether each equation represents direct variation between x and y.

17. $y = 133x$ **18.** $y = -4x^2$ **19.** $y = \frac{k}{x}$ **20.** $y = 2\pi x$

Life Science LINK

Most reptiles have a thick, scaly skin, which prevents them from drying out. As they grow, the outermost layer of this skin is shed. Although snakes shed their skins all in one piece, most reptiles shed their skins in much smaller pieces.

21. ***LIFE SCIENCE*** The weight of a person's skin is related to body weight by the equation $s = \frac{1}{16}w$, where s is skin weight and w is body weight.

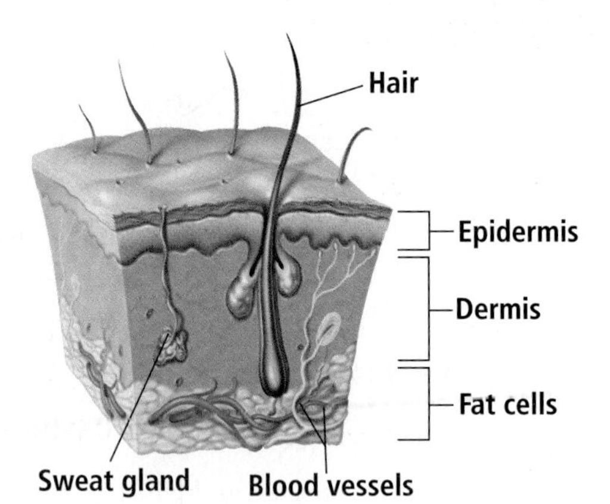

a. Does this equation show direct variation between body weight and skin weight?

b. If a person calculates skin weight as $9\frac{3}{4}$ lb, what is the person's body weight?

22. ***PHYSICAL SCIENCE*** Boyle's law states that for a fixed amount at a constant temperature, the volume of a gas increases as its pressure decreases. Explain whether the relationship between volume and pressure described by Boyle's law is a direct variation.

23. ***COOKING*** A waffle recipe calls for different amounts of mix, depending on the number of servings. Graph the data set and determine whether it shows direct variation.

Number of Servings	2	4	6	8	10	12	14
Waffle Mix (c)	1.5	3	4.5	6	7.5	9	10.5

24. ***WRITE A PROBLEM*** In physical science, Charles's law states that for a fixed amount at a constant pressure, the volume of a gas increases as the temperature increases. Write a direct variation problem about Charles's law.

25. ***WRITE ABOUT IT*** Describe how the constant of proportionality k affects the appearance of the graph of a direct variation equation.

26. ***CHALLENGE*** Bananas are sold at 39¢ a pound. Determine what condition would need to be satisfied if the price paid and the number of bananas purchased represented a direct variation.

Spiral Review

Solve. (Lesson 10-1)

27. $5x + 2 = -18$ **28.** $\frac{b}{-6} + 12 = 5$ **29.** $\frac{a+4}{11} = -3$ **30.** $\frac{1}{3}x - \frac{1}{4} = \frac{5}{12}$

31. **EOG PREP** The area of a trapezoid is given by the formula $A = \frac{1}{2}(b_1 + b_2)h$. What is the measure of b_1 if $A = 60\ \text{cm}^2$, $b_2 = 5$ m, and $h = 6$ m? (Lesson 10-5)

A 7 m B 15 m C 14.5 m D 12 m

11-6 Graphing Inequalities in Two Variables

Learn to graph inequalities on the coordinate plane.

Vocabulary

boundary line

linear inequality

Graphing can help you visualize the relationship between the maximum distance a Mars rover can travel and the number of Martian days.

Solar-powered rovers landing on Mars in 2004 will have a range of up to 330 feet per Martian day.

A graph of a linear equation separates the coordinate plane into three parts: the points on one side of the line, the points on the **boundary line**, and the points on the other side of the line.

Each point in the coordinate plane makes one of these three statements true:

Equality → $y = x + 2$

Inequality → $y > x + 2$, $y < x + 2$

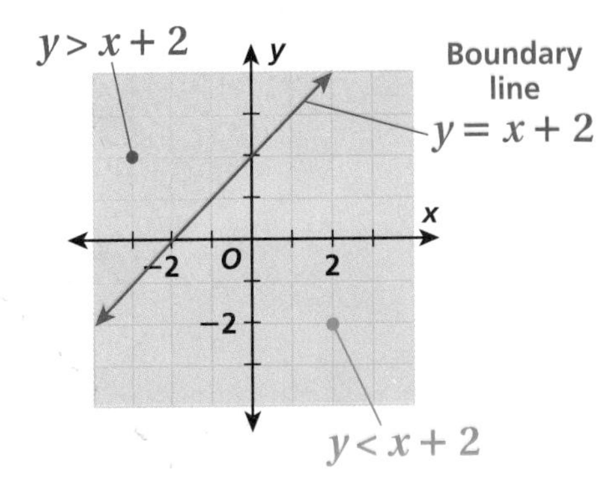

When the equality symbol is replaced in a linear equation by an inequality symbol, the statement is a **linear inequality**. Any ordered pair that makes the linear inequality true is a solution.

EXAMPLE 1 Graphing Inequalities

Helpful Hint

Any point on the line $y = x + 1$ is not a solution of $y > x + 1$ because the inequality symbol $>$ means only "greater than" and does not include "equal to."

Graph each inequality.

A $y > x + 1$

First graph the boundary line $y = x + 1$. Since no points that are on the line are solutions of $y > x + 1$, make the line *dashed*. Then determine on which side of the line the solutions lie.

$(0, 0)$ *Test a point not on the line.*

$y > x + 1$

$0 \overset{?}{>} 0 + 1$ *Substitute 0 for x and 0 for y.*

$0 \overset{?}{>} 1$

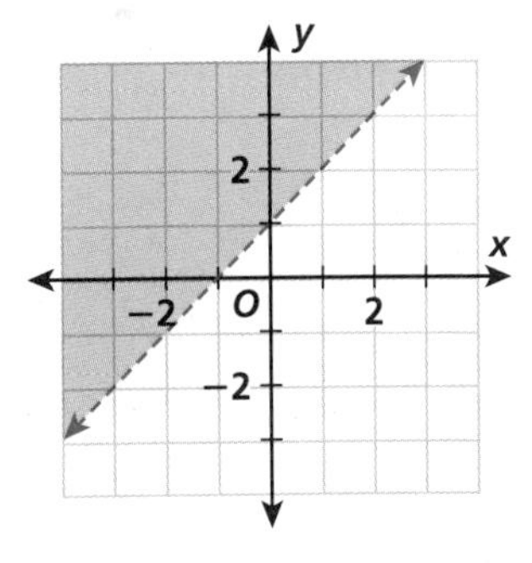

Since $0 > 1$ is not true, $(0, 0)$ is not a solution of $y > x + 1$. Shade the side of the line that does not include $(0, 0)$.

> **Helpful Hint**
>
> Any point on the line $y = x + 1$ is a solution of $y \leq x + 1$. This is because the inequality symbol $\leq$ means "less than or equal to."

Graph each inequality.

B $y \leq x + 1$

First graph the boundary line $y = x + 1$. Since points that are on the line are solutions of $y \leq x + 1$, make the line ***solid.***

Then shade the part of the coordinate plane in which the rest of the solutions of $y \leq x + 1$ lie.

$(2, 1)$ *Choose any point not on the line.*

$y \leq x + 1$

$1 \overset{?}{\leq} 2 + 1$ *Substitute 2 for x and 1 for y.*

$1 \overset{?}{\leq} 3$

Since $1 \leq 3$ is true, $(2, 1)$ is a solution of $y \leq x + 1$. Shade the side of the line that includes the point $(2, 1)$.

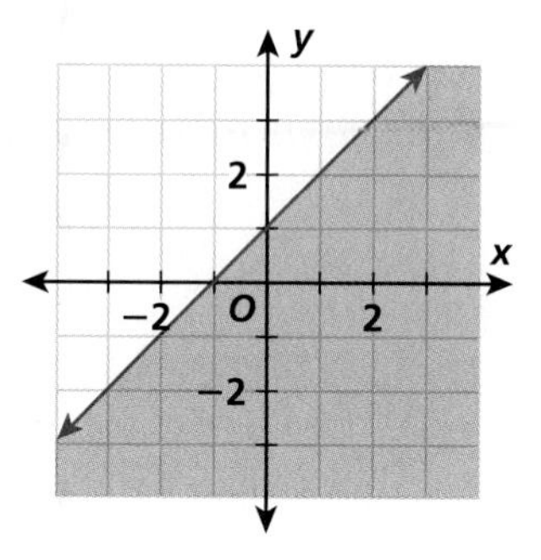

C $3y + 4x \leq 12$

First write the equation in slope-intercept form.

$3y + 4x \leq 12$

$3y \leq -4x + 12$ *Subtract 4x from both sides.*

$y \leq -\frac{4}{3}x + 4$ *Divide both sides by 3.*

Then graph the line $y = -\frac{4}{3}x + 4$. Since points that are on the line are solutions of $y \leq -\frac{4}{3}x + 4$, make the line solid. Then shade the part of the coordinate plane in which the rest of the solutions of $y \leq -\frac{4}{3}x + 4$ lie.

$(0, 0)$ *Choose any point not on the line.*

$y \leq -\frac{4}{3}x + 4$

$0 \overset{?}{\leq} 0 + 4$ *Substitute 0 for x and 0 for y.*

$0 \overset{?}{\leq} 4$

Since $0 \leq 4$ is true, $(0, 0)$ is a solution of $y \leq -\frac{4}{3}x + 1$. Shade the side of the line that includes the point $(0, 0)$.

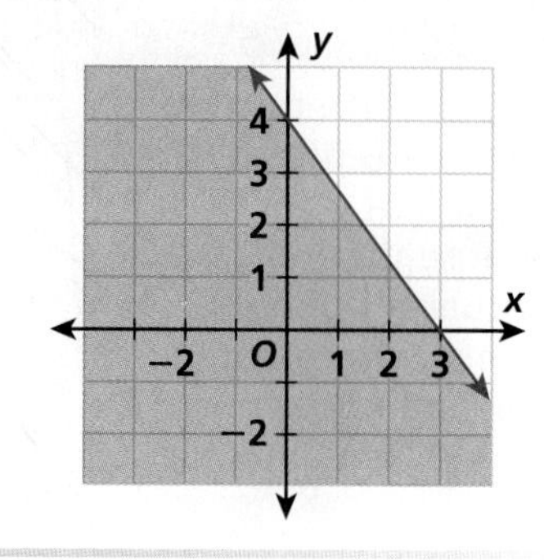

EXAMPLE 2 *Science Application*

Helpful Hint

The phrase "up to 330 ft" can be translated as "less than or equal to 330 ft."

Solar-powered rovers landing on Mars in 2004 will have a range of up to 330 feet per Martian day. Graph the relationship between the distance a rover can travel and the number of Martian days. Can a rover travel 3000 feet in 8 days?

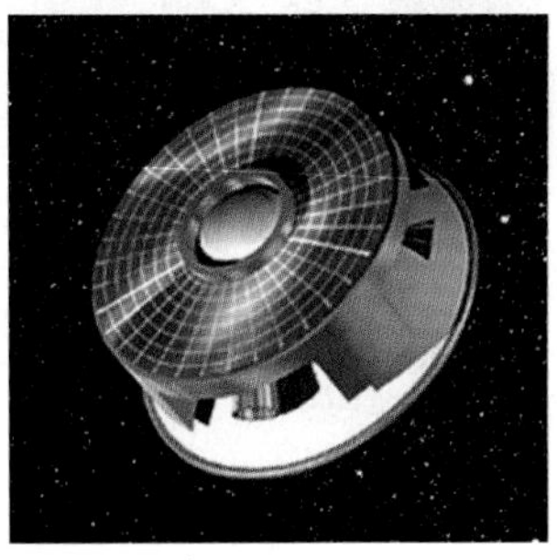

Mars rover in space.

First find the equation of the line that corresponds to the inequality.

In 0 days the rover travels 0 feet. ⟶ point (0, 0)

In 1 day the rover can travel up to 330 feet. ⟶ point (1, 330)

$m = \frac{330 - 0}{1 - 0} = \frac{330}{1} = 330$ *With two known points, find the slope.*

$y = 330x + 0$ *The y-intercept is 0.*

Graph the boundary line $y = 330x$. Since points on the line are solutions of $y \leq 330x$, make the line solid.

Shade the part of the coordinate plane in which the rest of the solutions of $y \leq 330x$ lie.

(5, 0) *Choose any point not on the line.*

$y \leq 330x$

$0 \leq 330 \cdot 5$ *Substitute 5 for x and 0 for y.*

$0 \leq 1650$

Since $0 \leq 1650$ is true, (5, 0) is a solution of $y \leq 330x$. Shade the part on the side of the line that includes point (5, 0).

The point (8, 3000) is not included in the shaded area, so the rover cannot travel 3000 feet in 8 days.

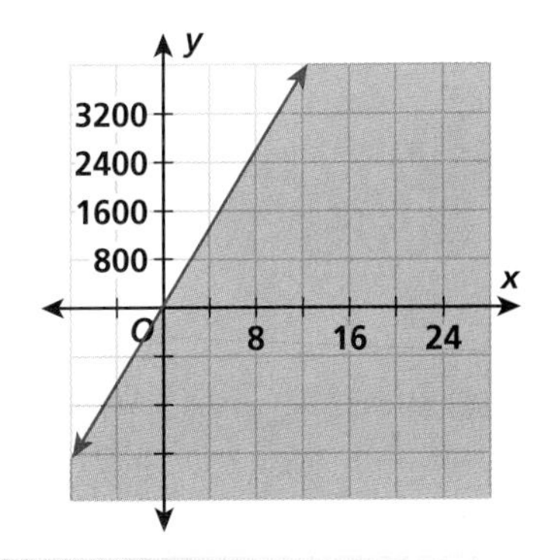

Think and Discuss

1. **Describe** the graph of $5x + y < 15$. Tell how it would change if $<$ were changed to $\geq$.
2. **Compare and contrast** the use of an open circle, a closed circle, a dashed line, and a solid line when graphing inequalities.
3. **Explain** how you can tell if a point on the line is a solution of the inequality.
4. **Name** a linear inequality for which the graph is a horizontal dashed line and all points below it.

Exercises

FOR EOG PRACTICE

see page 687

internet connect
Homework Help Online
go.hrw.com Keyword: MT4 11-6

5.03

GUIDED PRACTICE

See Example 1 **Graph each inequality.**

1. $y < x + 3$
2. $y \geq 2x - 1$
3. $y > -3x + 2$
4. $4x + y \leq 1$
5. $y \leq \frac{2}{3}x + 3$
6. $\frac{1}{2}x - \frac{1}{4}y < -1$

See Example 2

7. **a.** The organizers of a golf outing have a prize budget of \$150 to buy golf gloves and hats for the players. They can buy golf gloves for \$10 each and hats for \$12 each. Write and graph an inequality showing the different ways the organizers can spend their prize budget.
   b. Can the organizers of the golf outing purchase 7 hats and 6 golf gloves and still be within their prize budget?

INDEPENDENT PRACTICE

See Example 1 **Graph each inequality.**

8. $y \leq -\frac{1}{2}x - 4$
9. $y < -1.5x + 2.5$
10. $-4(2x + y) \geq -8$
11. $3x - \frac{3}{4}y > -2$
12. $6x - 9y > 15$
13. $3\left(\frac{2}{3}x + \frac{1}{3}y\right) \leq -3$

See Example 2

14. **a.** To avoid suffering from the bends, a diver should ascend no faster than 30 feet per minute. Write and graph an inequality showing the relationship between the depth of a diver and the time required to ascend to the surface.
    b. If a diver initially at a depth of 77 ft ascends to the surface in 2.6 minutes, is the diver in danger of developing the bends?

PRACTICE AND PROBLEM SOLVING

Tell whether the given ordered pair is a solution of each inequality shown.

15. $y \leq 2x + 4$, (2, 1)
16. $y > -6x + 1$, (−3, 19)
17. $y \geq 3x - 3$, (5, 14)
18. $y > -x + 12$, (0, 14)
19. $y \geq 3.4x + 1.9$, (4, 22)
20. $y \leq 7(x - 3)$, (3, 3)
21. **a.** Graph the inequality $y \geq x + 5$.
    b. Name an ordered pair that is a solution of the inequality.
    c. Is (3, 5) a solution of $y \geq x + 5$? Explain how to check your answer.
    d. Which side of the line $y = x + 5$ is shaded?
    e. Name an ordered pair that is a solution of $y < x + 5$.
22. ***FOOD*** The school cafeteria needs to buy no more than 30 pounds of potatoes. A supermarket sells 3-pound and 5-pound bags of potatoes. Write and graph an inequality showing the number of 3-pound and 5-pound bags of potatoes the cafeteria can buy.

23. ***SPORTS*** A basketball player scored 18 points in a game. Some of her points may have been from free throws, so her points from 2-point and 3-point field goals could be at most 18. Write and graph an inequality showing the possible numbers of 2-point and 3-point field goals she scored.

24. ***BUSINESS*** It costs a manufacturing company \$35 an hour to operate machine A and \$25 an hour to operate machine B. The total cost of operating both machines can be no more than \$250 each day.
    a. Write and graph an inequality showing the number of hours each machine can be used each day.
    b. If machine A is used for 4 hours, for how many hours can machine B be used without going over \$250?

25. ***EARTH SCIENCE*** A weather balloon can ascend at a rate of up to 800 feet per minute.
    a. Write an inequality showing the relationship between the distance the balloon can ascend and the number of minutes.
    b. Graph the inequality for time between 0 and 30 minutes.
    c. Can the balloon ascend to a height of 2 miles within 15 minutes? (One mile is equal to 5280 feet.)

26. ***CHOOSE A STRATEGY*** Which of the following ordered pairs is NOT a solution of the inequality $4x + 9y \leq 108$?
    A (0, 0) **B** (−6, 15) **C** (−4, −12) **D** (7, 8)

27. ***WRITE ABOUT IT*** When you graph a linear inequality that is solved for y, when do you shade above the boundary line and when do you shade below it? When do you use a dashed line?

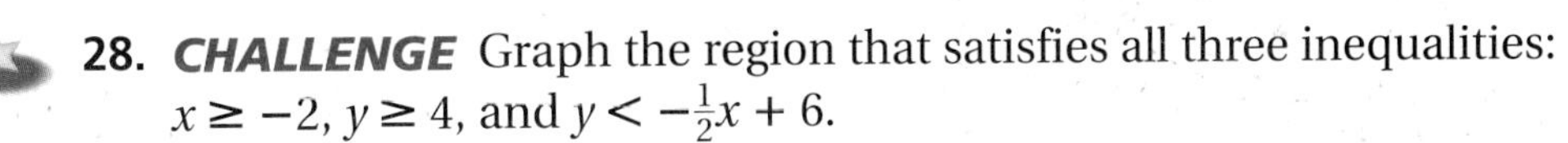

28. ***CHALLENGE*** Graph the region that satisfies all three inequalities: $x \geq -2$, $y \geq 4$, and $y < -\frac{1}{2}x + 6$.

Spiral Review

Solve for the indicated variable. (Lesson 10-5)

29. Solve $A = \frac{1}{2}bh$ for h.

30. Solve $2a + 2b + 2c = 2d$ for b.

31. Solve $A = \frac{1}{2}(b_1 + b_2)h$ for b_2.

32. Solve $W = X - 2Y + 4Z$ for Y.

33. **EOG PREP** What is the equation in slope-intercept form of the line that passes through points (1, 6) and (−1, −2)? (Lesson 11-3)
    A $y = 4x + 2$ B $y = -3x + 6$ C $y = 4x - 2$ D $y = 2x + 4$

11-7 Lines of Best Fit

Learn to recognize relationships in data and find the equation of a line of best fit.

The graph shows the winning times for the women's 3000 meter Olympic speed skating event. As is the case with many Olympic sports, the athletes keep improving and setting new records, so there is a correlation between the year and the winning time.

Winning Times for Women's 3000-Meter Olympic Speed Skating

Time (min): 0, 3, 3.5, 4, 4.5, 5, 5.5

Years since 1960: 0, 10, 20, 30, 40

When data show a correlation, you can estimate and draw a *line of best fit* that approximates a trend for a set of data and use it to make predictions.

To estimate the equation of a line of best fit:

- calculate the means of the x-coordinates and y-coordinates: (x_m, y_m).
- draw the line through (x_m, y_m) that appears to best fit the data.
- estimate the coordinates of another point on the line.
- find the equation of the line.

EXAMPLE 1 Finding a Line of Best Fit

Plot the data and find a line of best fit.

x	2	4	5	1	3	8	6	7
y	4	8	7	3	4	8	5	9

Remember!

The line of best fit is the line that comes closest to all the points on a scatter plot. Try to draw the line so that about the same number of points are above the line as below the line.

Plot the data points and find the mean of the x- and y-coordinates.

$$x_m = \frac{2+4+5+1+3+8+6+7}{8} = 4.5 \qquad y_m = \frac{4+8+7+3+4+8+5+9}{8} = 6$$

$$(x_m, y_m) = (4.5, 6)$$

Draw a line through (4.5, 6) that best represents the data. Estimate and plot the coordinates of another point on that line, such as (7, 8). Find the equation of the line.

$m = \frac{8-6}{7-4.5} = \frac{2}{2.5} = 0.8$ *Find the slope.*

$y - y_1 = m(x - x_1)$ *Use point-slope form.*

$y - 6 = 0.8(x - 4.5)$ *Substitute.*

$y - 6 = 0.8x - 3.6$

$y = 0.8x + 2.4$

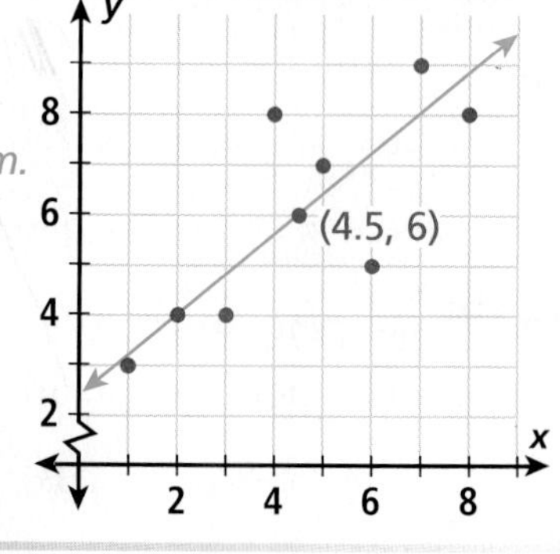

The equation of a line of best fit is $y = 0.8x + 2.4$.

EXAMPLE 2 *Sports Application*

Find a line of best fit for the women's 3000-meter speed skating. Use the equation of the line to predict the winning time in 2006.

Year	1964	1968	1972	1976	1980	1984	1988	1992	1994	1998	2002
Winning Time (min)	5.25	4.94	4.87	4.75	4.54	4.41	4.20	4.33	4.29	4.12	3.96

Let 1960 represent year 0. The first point is then (4, 5.25), and the last point is (42, 3.96).

Plot the data points and find the mean of the x- and y-coordinates.

$$x_m = \frac{4 + 8 + 12 + 16 + 20 + 24 + 28 + 32 + 34 + 38 + 42}{11} \approx 23.5$$

$$y_m = \frac{5.25 + 4.94 + 4.87 + 4.75 + 4.54 + 4.41 + 4.20 + 4.33 + 4.29 + 4.12 + 3.96}{11} \approx 4.5$$

$$(x_m, y_m) = (23.5, 4.5)$$

Draw a line through (23.5, 4.5) that best represents the data.

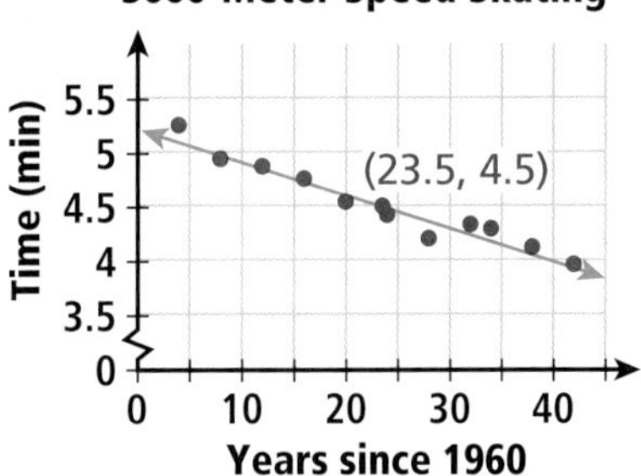

Estimate and plot the coordinates of another point on that line, (8, 5).

Find the equation of that line.

$$m = \frac{5 - 4.5}{8 - 23.5} = \frac{0.5}{-15.5} \approx -0.03$$

$y - y_1 = m(x - x_1)$

$y - 4.5 = -0.03(x - 23.5)$

$y - 4.5 = -0.03x + 0.7$ *Round 0.705 to 0.7.*

$y = -0.03x + 5.2$

The equation of a line of best fit is $y = -0.03x + 5.2$.

Helpful Hint

If you substitute 2006 instead of 46 for the year, you get a negative value for y. The answer would not be reasonable.

Since 1960 represents year 0, 2006 represents year 46.

$y = -0.03(46) + 5.2$ *Substitute.*

$y = -1.38 + 5.2$

$y = 3.82$

The equation predicts a winning time of 3.82 minutes for the year 2006.

Think and Discuss

1. **Explain** why selecting a different second point may result in a different equation.
2. **Describe** what a line of best fit can tell you.
3. **Tell** whether a line of best fit must include one or more points in the data.

11-7 Exercises

FOR EOG PRACTICE
see page 687

internet connect
Homework Help Online
go.hrw.com Keyword: MT4 11-7

4.01, 4.02, 5.01, 5.02, 5.03

GUIDED PRACTICE

See Example 1

Plot the data and find a line of best fit.

1.

x	2	7	3	4	6	1	9	5
y	4	13	7	8	11	2	17	10

2.

x	22	32	28	20	26	30	24	34
y	11	7	9	12	10	8	10	6

See Example 2

3. Ten students each did a different number of jumping jacks and then recorded their heart rates. Find and graph a line of best fit for the data. How is heart rate related to exercise?

Jumping Jacks	0	5	10	15	20	25	30	35	40	45
Heart Rate (beats/min)	78	76	84	86	93	90	96	92	100	107

INDEPENDENT PRACTICE

See Example 1

Plot the data and find a line of best fit.

4.

x	10	25	5	40	30	20	15	35
y	25	62	13	100	75	48	39	88

5.

x	0.4	0.5	0.3	0.7	0.2	0.8	0.1	0.6
y	5	5	6	2	8	1	8	3

See Example 2

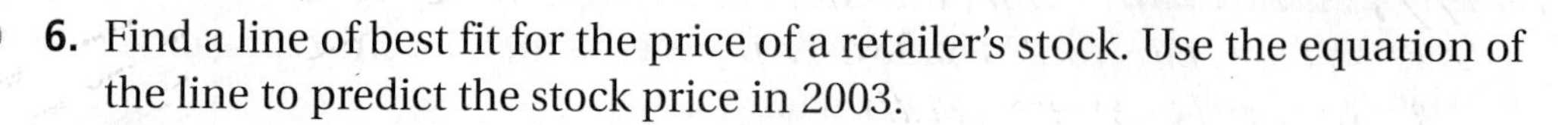

6. Find a line of best fit for the price of a retailer's stock. Use the equation of the line to predict the stock price in 2003.

Year	1994	1995	1996	1997	1998	1999	2000
Stock Price	11.70	11.95	12.28	12.54	12.77	13.00	13.26

PRACTICE AND PROBLEM SOLVING

Tell whether a line of best fit for each scatter plot would have a positive or negative slope. If a line of best fit would not be appropriate for the data, write *neither*.

7.

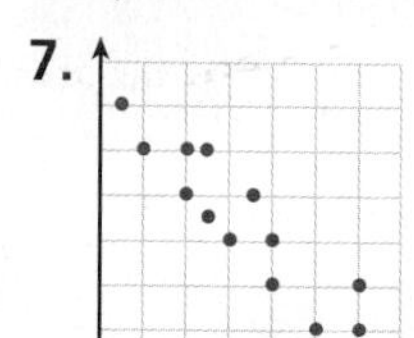

8.

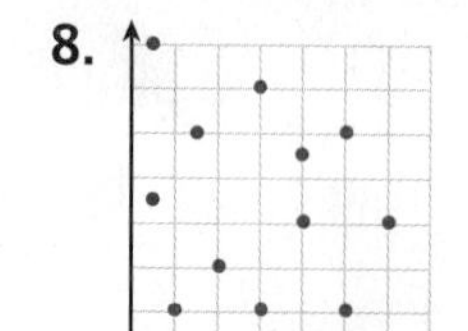

9.

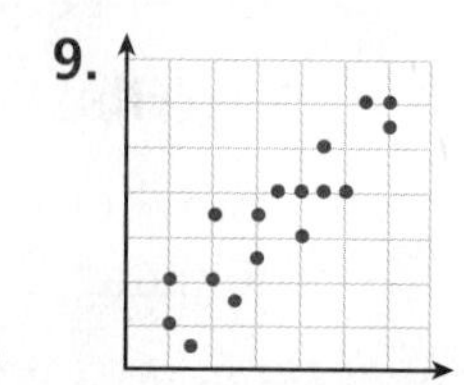

10.

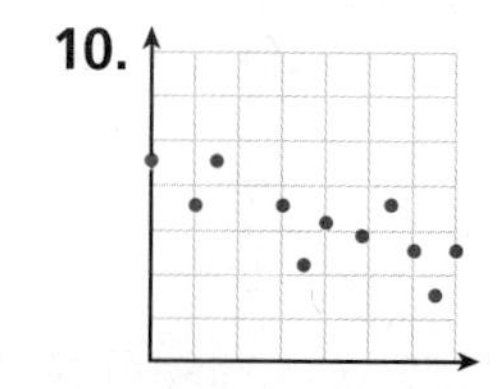

Economics LINK

Economic analysts study trends in data dealing with consumer purchases and ownership. Analysts often make predictions about future markets based on these economic trends. The table shows data on how many American households owned a computer during the past several years.

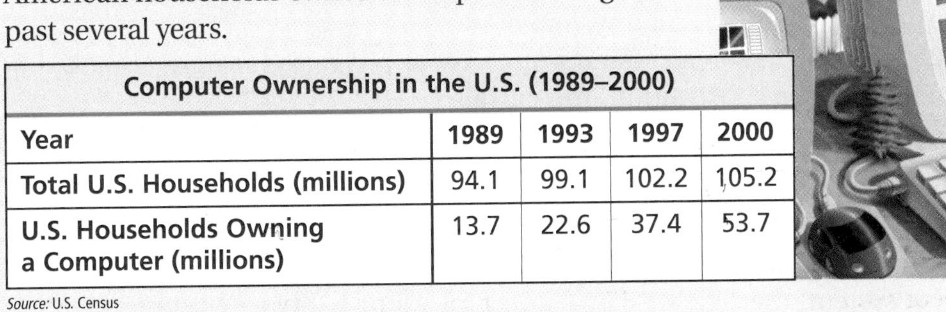

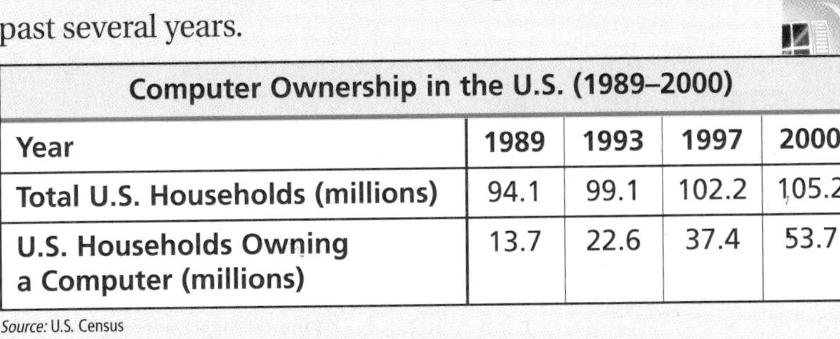

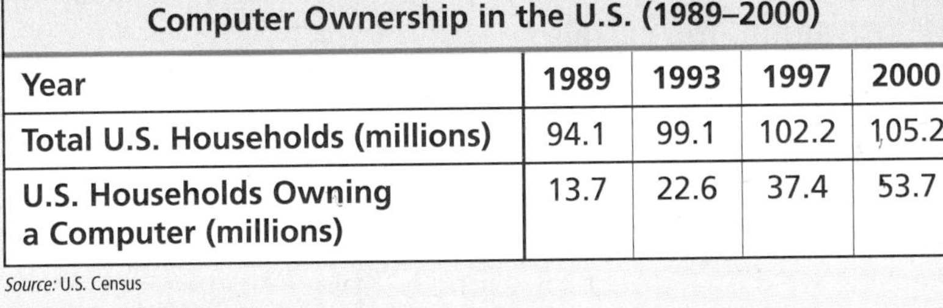

Computer Ownership in the U.S. (1989–2000)				
Year	1989	1993	1997	2000
Total U.S. Households (millions)	94.1	99.1	102.2	105.2
U.S. Households Owning a Computer (millions)	13.7	22.6	37.4	53.7

Source: U.S. Census

11. Let 1989 represent year 0 along the x-axis.

a. What is the mean number of years for the data shown?

b. Find the *percent* of U.S. households owning a computer for each year shown in the table, to the nearest tenth. Then find the mean.

12. Let y represent the percent of U.S. households that owned a computer between 1989 and 2000. Find a line of best fit, and plot it on the same graph as the data points. Use the point (6, 32) to write the equation of the line of best fit.

13. In 1998 about 42.1% of U.S. households owned a computer. What percent of households owned a computer in 1998 according to the line of best fit?

14. Use the equation of the line of best fit to predict the percent of U.S. households owning a computer in the year 2005. Do you think the actual value will be higher or lower than the predicted value?

15. ***CHALLENGE*** What information does the slope of the line of best fit give you? What would it mean to an economic analyst if the slope were negative?

go.hrw.com
KEYWORD: MT4 Economy
CNN Student News

Spiral Review

Tell whether the lines passing through the given points are parallel, perpendicular, or neither. (Lesson 11-2)

16. l: (2, 3), (4, 8)
m: (2, 3), (7, 1)

17. l: (3, −1), (7, 4)
m: (5, 5), (0, 9)

18. l: (−6, 1), (−7, 7)
m: (−3,−3), (−4, 3)

19. l: (5, 4), (−11, 0)
m: (1, −2), (0, 6)

20. **EOG PREP** Given that y varies directly with x, what is the equation of direct variation if y is 16 when x is 20? (Lesson 11-5)

A $y = 1\frac{1}{5}x$ B $y = \frac{5}{4}x$ C $y = \frac{4}{5}x$ D $y = 0.6x$

EXTENSION Systems of Equations

Learn to solve a system of equations by graphing.

Recall that two or more equations considered together form a system of equations. You've solved systems of equations using substitution. You can also use graphing to help you solve a system.

When you graph a system of linear equations in the same coordinate plane, their point of intersection is the solution of the system.

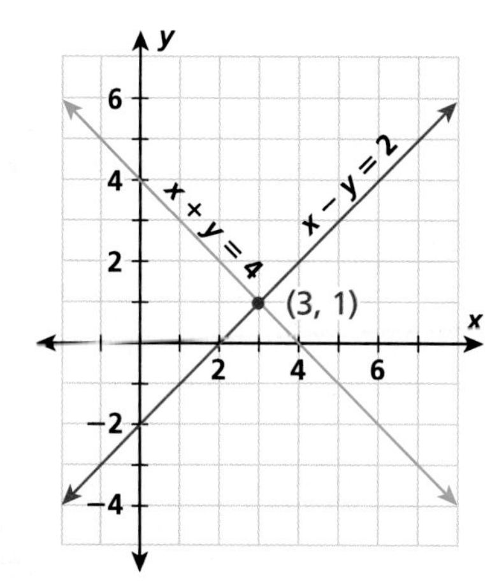

EXAMPLE 1 Using a Graph to Solve a System of Linear Equations

Solve the system graphically, and check your answer algebraically.

$$2x + y = 8$$
$$y - x = 2$$

Write each equation in slope-intercept form.

$2x + y = 8$

$y = -2x + 8$

slope = -2, y-intercept = 8

$y - x = 2$

$y = x + 2$

slope = 1, y-intercept = 2

Use each slope and y-intercept to graph. The point of intersection of the graphs, (2, 4), appears to be the solution of the system.

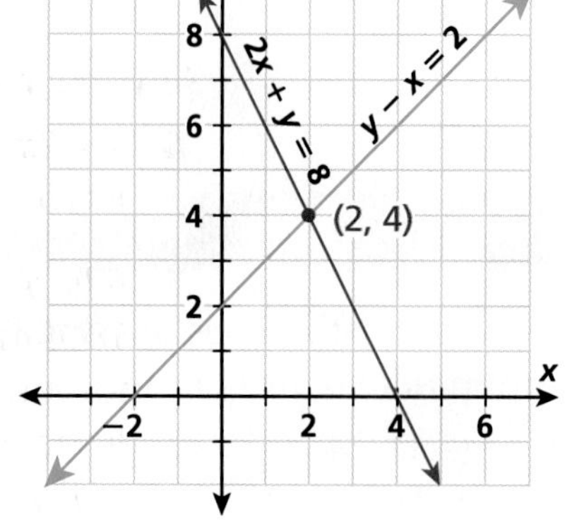

Check by substituting $x = 2$ and $y = 4$ into each of the *original* equations in the system.

Check

$2x + y = 8$

$2(2) + 4 \stackrel{?}{=} 8$

$4 + 4 \stackrel{?}{=} 8$

$8 \stackrel{?}{=} 8$ ✔

$y - x = 2$

$4 - 2 \stackrel{?}{=} 2$

$2 \stackrel{?}{=} 2$ ✔

The ordered pair (2, 4) checks in the original system of equations, so (2, 4) is the solution.

EXAMPLE 2 Graphing a System of Linear Equations to Solve a Problem

A plane left Los Angeles at 525 mi/h on a trans-Pacific flight. After the plane had traveled 1500 miles, a second plane started along the same route, flying at 600 mi/h. How many hours after the second plane leaves Los Angeles will it catch up with the first plane?

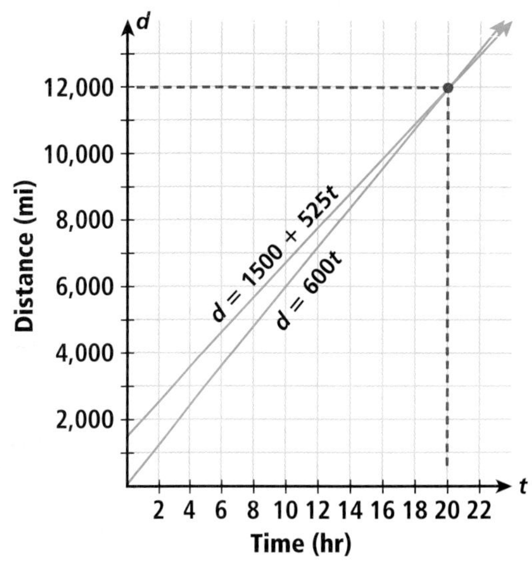

Let t = the number of hours and d = the distance in miles.

For plane 1, $d = 1500 + 525t$.

For plane 2, $d = 600t$.

Graph each equation. The point of intersection is (20, 12,000).

Check

$12{,}000 = 1500 + 525(20)$ $\quad$ $12{,}000 = 600(20)$

$12{,}000 = 12{,}000$ ✔ $\quad$ $12{,}000 = 12{,}000$ ✔

Plane 2 will catch up with plane 1 after 20 hours in flight, 12,000 miles from Los Angeles.

EXTENSION Exercises

1.02, 5.03

Tell whether the ordered pair is the solution of each given system.

1. (5, 11) $\quad y = 3x - 4$, $\; y = 2x + 1$

2. (0, 1) $\quad y = 4x + 1$, $\; y = 3x$

3. (2, −5) $\quad 3x + y = 1$, $\; -5x + y = -7$

Solve each system graphically, and check your answer algebraically.

4. $y = 2x$, $\; y = 3x - 3$

5. $y = -2x + 3$, $\; y = \frac{1}{2}x + 3$

6. $y - x = -2$, $\; x - 2y = 4$

7. A lion cub is running toward the rim of a deep gorge. The gorge is 1800 meters from his mother. The cub is running at 480 meters per minute, and the lioness races after him at 660 meters per minute. If the cub had a 450-meter head start, will his mother catch him in time?

8. Lillian has a choice of two long-distance telephone plans. The first plan has a monthly fee of $3.95 plus 5 cents per minute. The second plan has no monthly fee, but charges 7 cents per minute. If Lillian averages about 300 minutes of long-distance calls per month, which plan is better for her?

Problem Solving on Location

NORTH CAROLINA

Whitewater Falls

Although the entrance to Whitewater Falls is in North Carolina, the falls actually occupy parts of both North and South Carolina. Upper Whitewater Falls, in North Carolina, drops a total of 411 ft and is the highest waterfall east of the Rocky Mountains. Lower Whitewater Falls, in South Carolina, drops an additional 400 ft.

1. The top of Whitewater Falls is about 2700 ft above sea level. Using the information you have about the heights of both sections of Whitewater Falls, estimate the elevation of the Whitewater River at the bottom of the falls. Is it likely that the waterfalls have a constant slope? Explain.

2. The trail that leads from the lower overlook to the Whitewater River and Foothills Trail is $\frac{1}{2}$ mile long and drops 600 ft in elevation. Find the slope of the trail, assuming that the trail is a straight path with constant slope.

3. Suppose a hiker began at an elevation of about 2500 ft and walked x feet along the trail from the lower overlook to the Whitewater River. Use your answer from problem **2** to write an equation to find the new elevation y of the hiker.

4. To reach the lower overlook of the Whitewater Falls, visitors must descend 154 steps. Suppose that the stairs descend along a linear path. Would the slope of the entire set of stairs be the same as the slope of each individual stair? Explain.

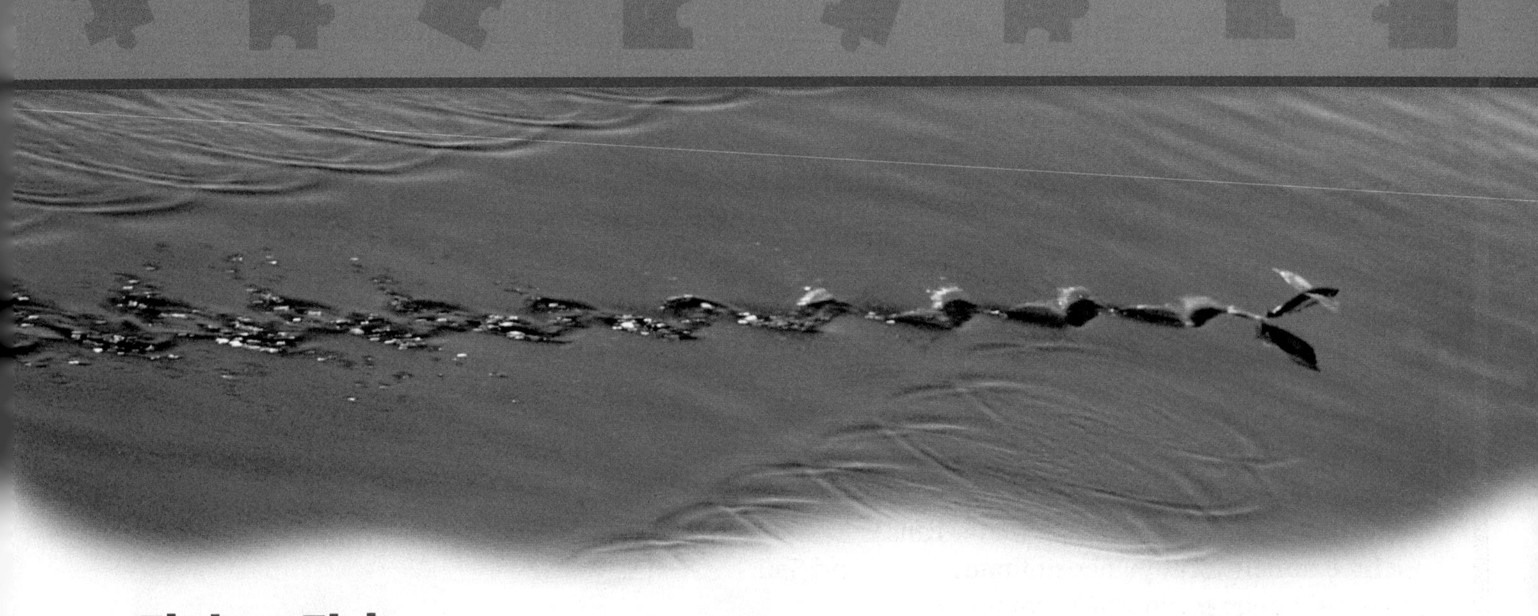

Flying Fish

Flying fish can be found in the ocean along the eastern coast of North Carolina and, in spite of their name, cannot actually fly. Instead, when they are chased by a predator, they race toward the surface of the ocean, flap their lower tail lobe rapidly to build speed, and catch an air current. They then spread their winglike fins and glide through the air to safety. If necessary, a flying fish can continue to glide horizontally for up to a quarter of a mile by repeating the rapid beating of its tail each time it returns to the water's surface.

1. A flying fish can beat its lower tail lobe up to 70 times per second. Write and graph the linear inequality that expresses the number of beats b of the tail lobe that occur in t seconds.

2. You observe a flying fish as it glides 400 ft horizontally through the air in about 10 seconds.

a. Write and graph the equation that expresses the horizontal distance d traveled by the fish after t seconds.

b. What does the slope of your line represent?

c. Use your graph from part **a** to estimate the amount of time it will take for the fish to reach its horizontal limit of $\frac{1}{4}$ mile in the air.

3. Write a linear inequality that represents a flying fish soaring through the air at speeds of up to 35 miles per hour. Graph this inequality on the same coordinate plane used for problem **2.** Be sure to change the slope of your line to the same units as those used in the slope in problem **2.** Does the equation from problem **2** seem reasonable? Explain.

4. A flying fish will usually glide at a vertical height of at least 1 ft above the water's surface, and it can reach heights of up to 35 ft above the water. Write and graph two linear inequalities on the same coordinate plane to show the possible range of heights above the water at which a flying fish might be seen.

MATH-ABLES

Graphing in Space

You can graph a point in two dimensions using a coordinate plane with an x- and a y-axis. Each point is located using an ordered pair (x, y). In three dimensions, you need three coordinate axes, and each point is located using an ordered triple (x, y, z).

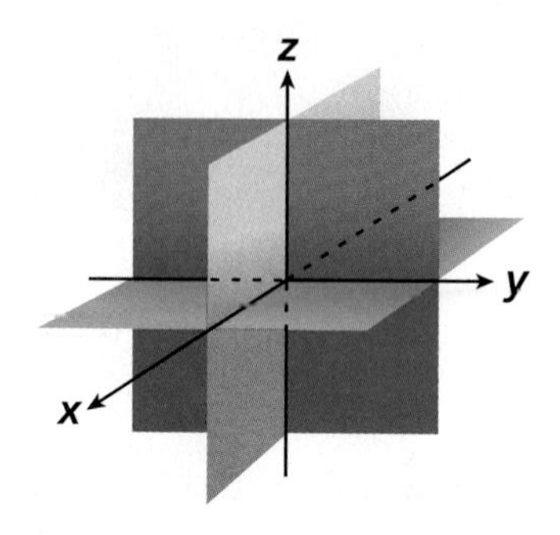

To graph a point, move along the x-axis the number of units of the x-coordinate. Then move left or right the number of units of the y-coordinate. Then move up or down the number of units of the z-coordinate.

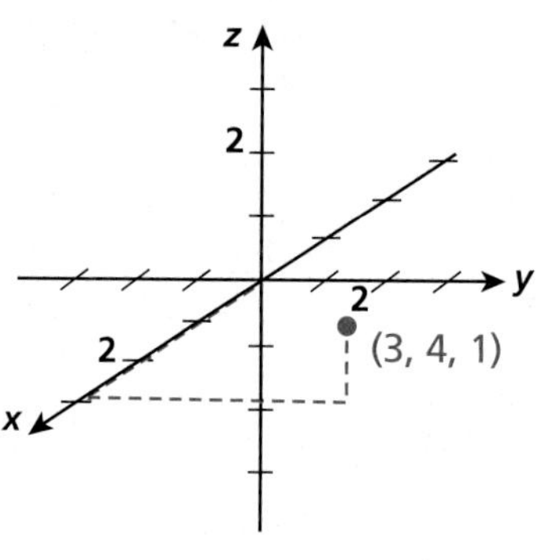

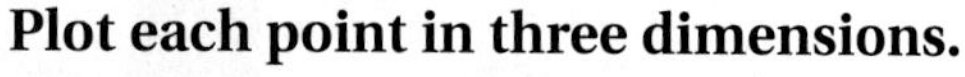

Plot each point in three dimensions.

1. $(1, 2, 5)$ **2.** $(-2, 3, -2)$

3. $(4, 0, 2)$

The graph of the equation $y = 2$ in three dimensions is a plane that is perpendicular to the y-axis and is two units to the right of the origin.

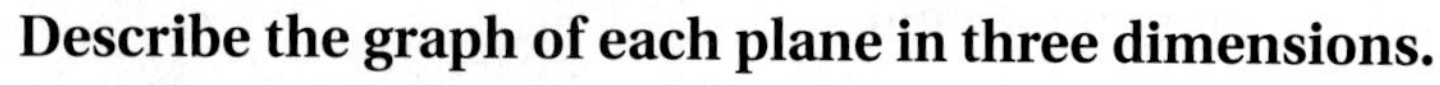

Describe the graph of each plane in three dimensions.

4. $x = 3$ **5.** $z = 1$ **6.** $y = -1$

Line Solitaire

Use a red and a blue number cube and a coordinate plane. Roll the number cubes to generate the coordinates of points on the coordinate plane. The x-coordinate of each point is the number on the red cube, and the y-coordinate is the number on the blue cube. Generate seven ordered pairs and plot the points on the coordinate plane. Then try to write the equations of three lines that divide the plane into seven regions so that each point is in a different region.

Graph Inequalities in Two Variables

1.02, 5.03

Use with Lesson 11-6

internet connect
Lab Resources Online
go.hrw.com
KEYWORD: MT4 TechLab11

A graphing calculator can be used to graph the solution of an inequality in two variables.

Activity

1 To graph the inequality $y > 2x - 4$ using a graphing calculator, use the **Y=** menu, and enter the equation $y = 2x - 4$.

Press Y= 2 X,T,θ,n − 4 GRAPH .

The line representing the graph of the equation represents the *boundary* of the solution region of the inequality. The graph of the inequality is either the region above the line or the region below the line. Use a test point to decide which region represents the graph of the inequality.

The point (0, 0) is a good test point if it is not on the line.

Substituting 0 for both x and y, $0 > 2 \cdot 0 - 4$, or $0 > -4$, which is *true*. The solution graph is the region above the line.

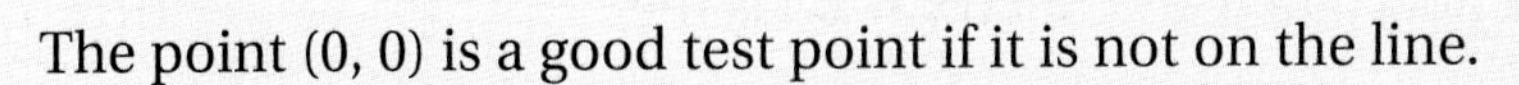

To graph this region, press Y= ◀ ◀ and notice that the edit cursor moves to the left of **Y1** onto an icon that looks like a small line segment, ⟍.

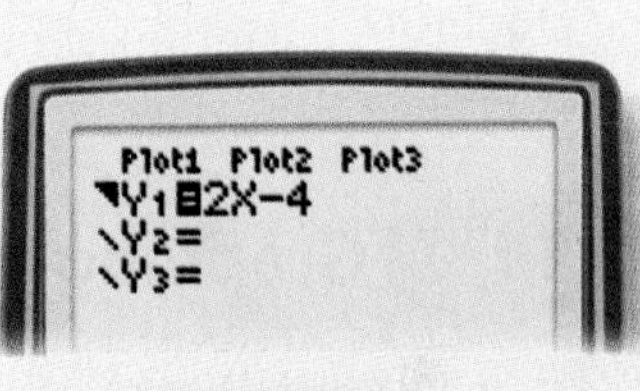

Now press the ENTER key several times and notice the different icons that are displayed. Choose the icon that looks like a shaded region above a line. Press GRAPH to display the shaded region. Any point (x, y) not on the line that is in the shaded region is a solution of $y > 2x - 4$.

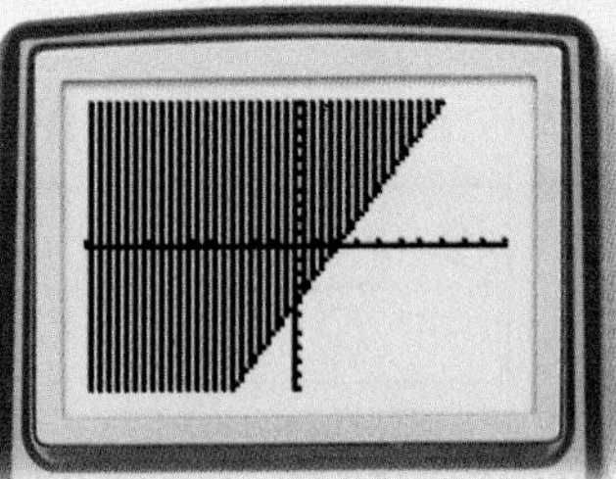

Think and Discuss

1. What inequality would the graph with all points below the x-axis shaded represent?

2. How would you use your calculator to display a graph of the region that is the intersection of the solution graphs of **both** $y > x - 2$ and $y < x + 3$?

Try This

Use a graphing calculator to graph each inequality.

1. $y < x - 4$ **2.** $y > 4 - x$ **3.** $y < 2x - 5$

4. $2x - 5y < 10$ **5.** $x + y < 4$ **6.** $3x + y > 6$

Chapter 11

Study Guide and Review

Vocabulary

boundary line	567	point-slope form	556
constant of proportionality	562	slope-intercept form	551
direct variation	562	*x*-intercept	550
linear equation	540	*y*-intercept	550
linear inequality	567		

Complete the sentences below with vocabulary words from the list above. Words may be used more than once.

1. The *x*-coordinate of the point where a line crosses the *x*-axis is its ___?___, and the *y*-coordinate of the point where the line crosses the *y*-axis is its ___?___.

2. $y = mx + b$ is the ___?___ of a line, and $y - y_1 = m(x - x_1)$ is the ___?___ .

3. Two variables related by a constant ratio are in ___?___.

11-1 Graphing Linear Equations (pp. 540–544)

EXAMPLE

■ Graph $y = x - 2$. Tell whether it is linear.

x	$x - 2$	y	(x, y)
−1	−1 − 2	−3	(−1, −3)
0	0 − 2	−2	(0, −2)
1	1 − 2	−1	(1, −1)
2	2 − 2	0	(2, 0)

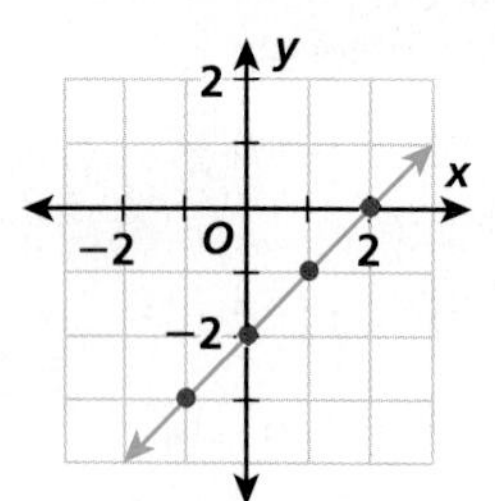

$y = x - 2$ is linear; its graph is a straight line.

EXERCISES

Graph each equation and tell whether it is linear.

4. $y = 3x - 1$
5. $y = 3 - 2x$
6. $y = -x^2$
7. $y = x^3$
8. $y = -x^3$
9. $y = 3x$
10. $y = \frac{12}{x}$ for $x \neq 0$
11. $y = -\frac{12}{x}$ for $x \neq 0$

11-2 Slope of a Line (pp. 545–549)

EXAMPLE

■ **Find the slope of the line that passes through (−1, 2) and (1, 3).**

Let (x_1, y_1) be (−1, 2) and (x_2, y_2) be (1, 3).

$$\frac{y_2 - y_1}{x_2 - x_1} = \frac{3 - 2}{1 - (-1)}$$
$$= \frac{1}{2}$$

The slope of the line that passes through (−1, 2) and (1, 3) is $\frac{1}{2}$.

EXERCISES

Find the slope of the line that passes through each pair of points.

12. (3, 1) and (6, 3)

13. (3, 2) and (4, −2)

14. (4, 4) and (−1, −2)

15. (−1, 5) and (6, −2)

16. (−3, −3) and (−4, −2)

17. (0, 0) and (−5, −7)

18. (−5, 7) and (−1, −2)

11-3 Using Slopes and Intercepts (pp. 550–554)

EXAMPLE

■ **Write $2x + 3y = 6$ in slope-intercept form. Identify the slope and y-intercept.**

$2x + 3y = 6$

$3y = -2x + 6$ *Subtract 2x from both sides.*

$\frac{3y}{3} = \frac{-2x}{3} + \frac{6}{3}$ *Divide both sides by 3.*

$y = -\frac{2}{3}x + 2$ *slope-intercept form*

$m = -\frac{2}{3}$ and $b = 2$

EXERCISES

Write each equation in slope-intercept form. Identify the slope and y-intercept.

19. $2y = 3x + 8$

20. $3y = 5x - 9$

21. $4x + 5y = 10$

22. $4y - 7x = 12$

Given two points that a line passes through, write the equation of the line in slope-intercept form.

23. (0, 4) and (−1, 1)

24. (−2, 5) and (3, −5)

25. (4, 3) and (−2, 6)

26. (3, −1) and (−1, −3)

11-4 Point-Slope Form (pp. 556–559)

EXAMPLE

■ **Write the point-slope form of the line with slope −3 that passes through (2, −1).**

$y - y_1 = m(x - x_1)$

$y - (-1) = -3(x - 2)$ *Substitute 2 for x_1,*

$y + 1 = -3(x - 2)$ *−1 for y_1, −3 for m.*

In point-slope form, the equation of the line with slope −3 that passes through (2, −1) is $y + 1 = -3(x - 2)$.

EXERCISES

Write the point-slope form of each line with the given conditions.

27. slope 4, passes through (1, 3)

28. slope −2, passes through (−3, 4)

29. slope $-\frac{3}{5}$, passes through (0, −2)

30. slope $\frac{2}{7}$, passes through (0, 0)

11-5 Direct Variation (pp. 562–566)

EXAMPLE

■ **y varies directly with x, and y is 27 when x is 3. Write the equation of direct variation.**

$y = kx$	*y varies directly with x.*
$27 = k \cdot 3$	*Substitute 3 for x and 27 for y.*
$9 = k$	*Solve for k.*
$y = 9x$	*Substitute 9 for k in the original equation.*

EXERCISES

y varies directly with x. Write the equation of direct variation for each set of conditions.

31. y is 54 when x is 9

32. x is 8 when y is 96

33. y is 9 when x is 63

11-6 Graphing Inequalities in Two Variables (pp. 567–571)

EXAMPLE

■ **Graph the inequality $y > x - 2$.**

Graph $y = x - 2$ as a dashed line. Test (0, 0) in the inequality; $0 > -2$ is true, so shade the side of the line that contains (0, 0).

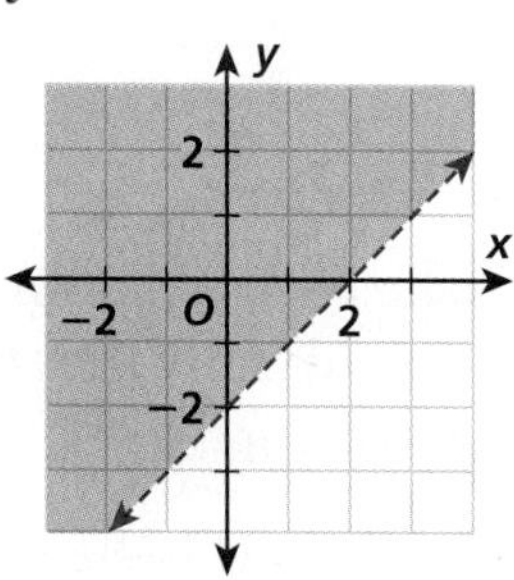

EXERCISES

Graph each inequality.

34. $y \le x + 4$

35. $2y \ge 3x + 6$

36. $2x + 5y > 10$

37. $4y - 3x < 12$

38. Jon can input up to 55 data items per minute. Graph the relationship between the number of minutes and the number of data items he inputs.

11-7 Lines of Best Fit (pp. 572–575)

EXAMPLE

■ **Plot the data and find a line of best fit.**

x	3	4	5	5	6	7
y	4	2	4	5	7	5

Calculate the means of x and y.

$x_m = \frac{30}{6} = 5$ $\quad y_m = \frac{27}{6} = 4.5$

Draw a line through (5, 4.5) to fit the data. Estimate another point on the line, (3, 3). Find the slope, 0.75, and use point-slope form to write an equation of the line.

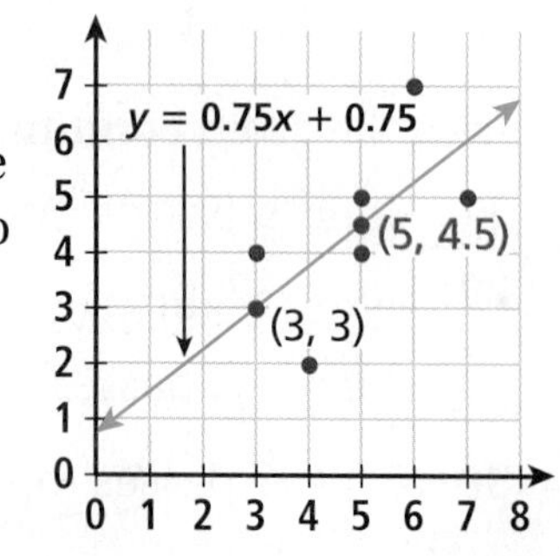

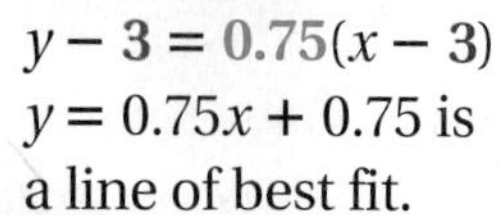

$y - 3 = 0.75(x - 3)$

$y = 0.75x + 0.75$ is a line of best fit.

EXERCISES

Plot each data set. Find a line of best fit.

39.

x	1	2	2	4	4	5
y	1	4	6	4	7	5

40.

x	1	3	4	4	6	7
y	2	1	4	7	6	7

41.

x	10	20	30	40	50	60
y	6	17	33	39	55	62

42.

x	10	25	40	55	70	85
y	67	58	41	29	28	20

Chapter Test

Chapter 11

Find the slope of the line that passes through each pair of points.

1. (2, 5) and (4, 9)

2. (7, 9) and (1, 12)

3. (0, −8) and (−1, −10)

Tell whether the lines passing through the given points are parallel or perpendicular.

4. line 1: (0, 8) and (2, 2)
line 2: (−2, 4) and (4, −14)

5. line 1: (0, −1) and (−2, −9)
line 2: (2, 15) and (−1, 3)

Given two points through which a line passes, write the equation of each line in slope-intercept form.

6. (1, 4) and (0, −3)

7. (−3, 0) and (2, −4)

8. (−1, 5) and (2, 0)

Use the point-slope form of each equation to identify a point the line passes through and the slope of the line.

9. $y - 6 = 3(x - 5)$

10. $y + 2 = -5(x - 9)$

11. $y - 1 = 7x$

Write the point-slope form of each line with the given conditions.

12. slope −2, passing through (−4, 1)

13. slope 3, passing through (2, 0)

Given that y varies directly with x, write the equation of direct variation for each set of conditions.

14. y is 225 when x is 25

15. y is 0.1875 when x is 0.25

16. x is 13 when y is 91

Graph each inequality.

17. $y > x + 3$

18. $3y \leq x - 6$

19. $2y + 3x \geq 12$

Find a line of best fit for each data set.

20.

x	0	1	1	3	5	5	6
y	1	1	2	2	3	2	3

21.

x	0	2	2	3	4	7
y	6	6	5	2	1	1

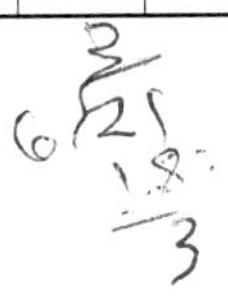

Marge made a down payment of \$200 for a computer and is making weekly payments of \$25. The equation $y = 25x + 200$ represents the amount paid after x weeks.

22. Use the slope and y-intercept to graph the equation.

23. She completed the payments in 8 weeks. How much did she pay?

24. A dragonfly beats its wings up to 30 times per second. Graph the relationship between flying time and the number of times the dragonfly beats its wings. Is it possible for a dragonfly to beat its wings 1000 times in half a minute?

Chapter Test

Chapter 11

Performance Assessment

Show What You Know

Create a portfolio of your work from this chapter. Complete this page and include it with your four best pieces of work from Chapter 11. Choose from your homework or lab assignments, mid-chapter quiz, or any journal entries you have done. Put them together using any design you want. Make your portfolio represent what you consider your best work.

Short Response

1. Graph the equation $y = |x|$, and tell whether it is linear. For x-values, use the integers from -5 through 5.

2. Scientists have found that a linear equation can be used to model the relation between the outdoor temperature and the number of chirps per minute crickets make. If a snowy tree cricket makes 100 chirps/min at 63°F and 178 chirps/min at 77°F, what is the approximate temperature when the cricket makes 126 chirps/min? Show your work.

3. Plot the points $A(-5, -4)$, $B(1, -2)$, $C(2, 3)$, and $D(-4, 1)$. Use straight segments to connect the four points in order. Then find the slope of each line segment. What special kind of quadrilateral is $ABCD$? Explain.

Extended Problem Solving

4. Tara's house is on a line between José's house and a tree that was hit by a lightning bolt. José heard the thunder six seconds after the lightning struck. Tara heard it 1.5 seconds before José did.

   a. What is the rate to the nearest foot per second at which the thunder was traveling?

   b. To the nearest 10 feet, what is the distance between José's house and Tara's house?

   c. Write a linear equation that could be used to find the distance y along a straight path the thunder traveled in x seconds.

   d. Graph your equation from part **c** on a coordinate plane.

Performance Assessment

internet connect
EOG Test Practice Online
go.hrw.com Keyword: MT4 TestPrep

Getting Ready for EOG

Chapter 11

Cumulative Assessment, Chapters 1–11

1. A savings plan requires \$1000 to start plus a monthly deposit, as shown on the graph.

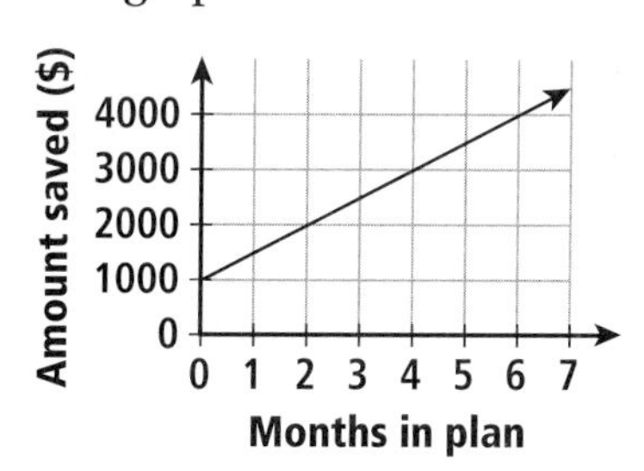

What does the slope of the line joining these points represent?

A The plan is for 500 weeks.

B Members will make 500 deposits.

C Each successive deposit is \$500 more.

D The monthly deposit is \$500.

2. What is the value of k such that the slope of the line joining the points $(k, -3)$ and $(4, 2)$ is $\frac{1}{2}$?

A 6 C 14

B -6 D -14

3. If 75% of a group of 96 graduates are older than 25 and, of those over 25, $\frac{1}{3}$ are business majors, how many are business majors?

A 72 C 48

B 64 D 24

4. What is the volume of the cube whose surface area is $150e^2$?

A $25e^3$ C $125e^3$

B $50e^3$ D $625e^3$

5. Which of the following is *not* a real number?

A $-\sqrt{5}$ C $\sqrt{-5}$

B $\sqrt[3]{-8}$ D -8

6. Playing with blocks, a child named Luke places the letters K, U, E, and L together at random. What is the probability that they spell his name?

A $\frac{1}{4}$ C $\frac{1}{12}$

B $\frac{1}{8}$ D $\frac{1}{24}$

7. One can of paint covers an area of 10 ft by 50 ft. Which is an expression for the number of cans of paint needed to paint an area l ft by w ft?

A $\frac{500}{lw}$ C $\frac{lw}{500}$

B $\frac{l+w}{500}$ D $500lw$

8. If $x \bigstar y$ means $x^2 < y^2$, then which of the following statements is true?

A $\frac{1}{4} \bigstar \frac{1}{3}$ C $-2 \bigstar \frac{1}{2}$

B $-3 \bigstar 2$ D $-4 \bigstar -2$

TEST TAKING TIP!

Read the requirement for the problem: Be sure you know what you have to find.

9. ***SHORT RESPONSE*** If $7 + x + y = 50$ and $x + y = c$, what is the value of $50 - c$? Show your work.

10. ***SHORT RESPONSE*** A factory recycled 5 of every 25 machine parts earmarked for scrap. What was the ratio of nonrecycled parts to recycled parts? Explain in words how you determined your answer.

Getting Ready for EOG

Chapter 12

Sequences and Functions

TABLE OF CONTENTS

internet connect

Chapter Opener Online
go.hrw.com
KEYWORD: MT4 Ch12

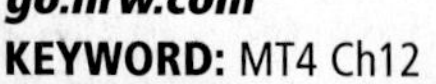

Growth Rates of *E. coli* Bacteria

Conditions	Doubling Time (min)
Optimum temperature (30°C) and growth medium	20
Low temperature (below 30°C)	40
Low nutrient growth medium	60
Low temperature and low nutrient growth medium	120

Career *Bacteriologist*

Bacteriologists study the growth and characteristics of microorganisms. They generally work in the fields of medicine and public health.

Bacteria colonies grow very quickly. The rate at which bacteria multiply depends upon temperature, nutrient supply, and other factors. The table shows growth rates of an *E. coli* bacteria colony under different conditions.

ARE YOU READY?

Choose the best term from the list to complete each sentence.

1. An equation whose solutions fall on a line on a coordinate plane is called a(n) __?__.
2. When the equation of a line is written in the form $y = mx + b$, m represents the __?__ and b represents the __?__.
3. To write an equation of the line that passes through (1, 3) and has slope 2, you might use the __?__ of the equation of a line.

linear equation
point-slope form
slope
***x*-intercept**
***y*-intercept**

Complete these exercises to review skills you will need for this chapter.

✔ Number Patterns

Find the next three numbers in the pattern.

4. $\frac{1}{-3}, \frac{3}{-4}, \frac{5}{-5}, \ldots$
5. 2, 3, 6, 11, 18, . . .
6. −11, −8, −5, . . .
7. $4, 2\frac{1}{2}, 1, \ldots$

✔ Evaluate Expressions

Evaluate each expression for the given values of the variables.

8. $a + (b - 1)c$ for $a = 6, b = 3, c = -4$
9. $a \cdot b^c$ for $a = -2, b = 4, c = 2$
10. $(ab)^c$ for $a = 3, b = -2, c = 2$
11. $-(a + b) + c$ for $a = -1, b = -4, c = -10$

✔ Graph Linear Equations

Use the slope and the *y*-intercept to graph each line.

12. $y = \frac{2}{3}x + 4$
13. $y = -\frac{1}{2}x - 2$
14. $y = 3x + 1$
15. $2y = 3x - 8$
16. $3y + 2x = 6$
17. $x - 5y = 5$

✔ Simplify Ratios

Write each ratio in simplest form.

18. $\frac{3}{9}$
19. $\frac{21}{5}$
20. $\frac{-12}{4}$
21. $\frac{27}{45}$
22. $\frac{3}{-45}$
23. $\frac{20}{-8}$

12-1 Arithmetic Sequences

Learn to find terms in an arithmetic sequence.

Vocabulary

sequence

term

arithmetic sequence

common difference

Joaquín received 5000 bonus miles for joining a frequent-flier program. Each time he flies to visit his grandparents, he earns 1250 miles.

The number of miles Joaquín has in his account is 6250 after 1 trip, 7500 after 2 trips, 8750 after 3 trips, and so on.

After 1 trip	After 2 trips	After 3 trips	After 4 trips
6250	7500	8750	10,000

Difference $7500 - 6250 = 1250$

Difference $8750 - 7500 = 1250$

Difference $10{,}000 - 8750 = 1250$

A **sequence** is a list of numbers or objects, called **terms**, in a certain order. In an **arithmetic sequence**, the difference between one term and the next is always the same. This difference is called the **common difference**. The common difference is added to each term to get the next term.

EXAMPLE 1 Identifying Arithmetic Sequences

Determine if each sequence could be arithmetic. If so, give the common difference.

Helpful Hint

You cannot tell if a sequence is arithmetic by looking at a finite number of terms, because the next term might not fit the pattern.

A **8, 13, 18, 23, 28, . . .**

8 13 18 23 28, . . .

5 5 5 5

Find the difference of each term and the term before it.

The sequence could be arithmetic with a common difference of 5.

B **1, 2, 4, 8, 16, . . .**

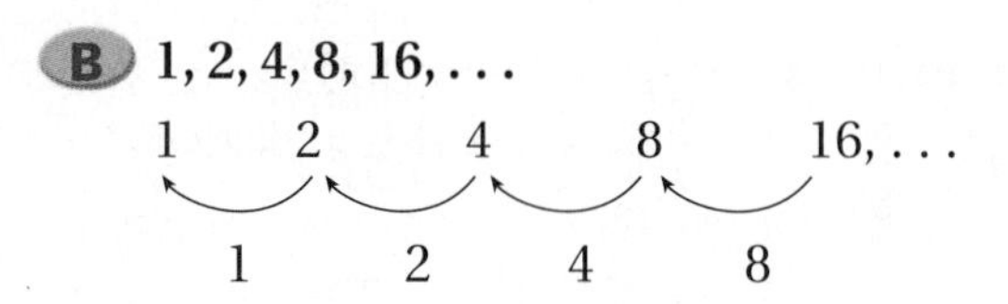

1 2 4 8 16, . . .

1 2 4 8

Find the difference of each term and the term before it.

The sequence is not arithmetic.

Determine if each sequence could be arithmetic. If so, give the common difference.

C **100, 93, 86, 79, 72, . . .**

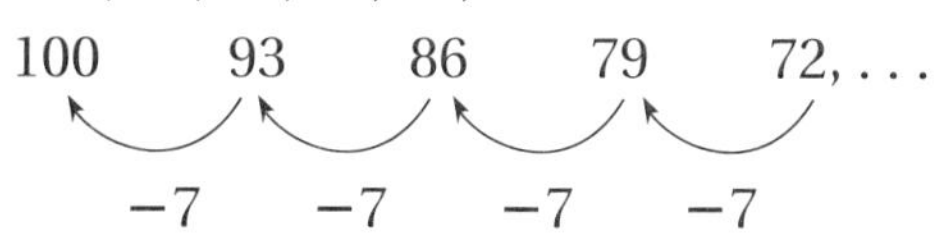

Find the difference of each term and the term before it.

The sequence could be arithmetic with a common difference of −7.

D $1, \frac{3}{2}, 2, \frac{5}{2}, 3, \frac{7}{2}, 4, \ldots$

$1 \quad \frac{3}{2} \quad 2 \quad \frac{5}{2} \quad 3 \quad \frac{7}{2} \quad 4, \ldots$

$\frac{1}{2} \quad \frac{1}{2} \quad \frac{1}{2} \quad \frac{1}{2} \quad \frac{1}{2} \quad \frac{1}{2}$

Find the difference of each term and the term before it.

The sequence could be arithmetic with a common difference of $\frac{1}{2}$.

E **5, 1, −3, −7, −11, . . .**

5 1 −3 −7 −11, . . .

−4 −4 −4 −4

Find the difference of each term and the term before it.

The sequence could be arithmetic with a common difference of −4.

Suppose you wanted to know the 100th term of the arithmetic sequence 5, 7, 9, 11, 13, ... If you do not want to find the first 99 terms, you could look for a pattern in the terms of the sequence.

Writing Math

Subscripts are used to show the positions of terms in the sequence. The first term is a_1, the second is a_2, and so on.

Term Name	a_1	a_2	a_3	a_4	a_5	a_6
Term	5	7	9	11	13	15
Pattern	5 + **0**(2)	5 + **1**(2)	5 + **2**(2)	5 + **3**(2)	5 + **4**(2)	5 + **5**(2)

The common difference d is 2. For the 2nd term, one 2 is added to a_1. For the 3rd term, two 2's are added to a_1. The pattern shows that for each term, the **number of 2's added** is one less than the **term number,** or $(n - 1)$. The 100th term is the first term, 5, plus 99 times the common difference, 2.

$$a_{100} = 5 + \mathbf{99}(2) = 5 + 198 = 203$$

FINDING THE *n*th TERM OF AN ARITHMETIC SEQUENCE

The nth term a_n of an arithmetic sequence with common difference d is

$$a_n = a_1 + (n - 1)d.$$

Determine if each sequence could be geometric. If so, give the common ratio.

D $4, -6, 9, -13.5, 20.25, \ldots$

$$4 \quad -6 \quad 9 \quad -13.5 \quad 20.25, \ldots$$

$$-1.5 \quad -1.5 \quad -1.5 \quad -1.5$$

Divide each term by the term before it.

The sequence could be geometric with a common ratio of -1.5.

Suppose you wanted to find the 15th term of the geometric sequence 2, 6, 18, 54, 162, ... If you do not want to find the first 14 terms, you could look for a pattern in the terms of the sequence.

Term Name	a_1	a_2	a_3	a_4	a_5	a_6
Term	2	6	18	54	162	486
Pattern	$2(3)^0$	$2(3)^1$	$2(3)^2$	$2(3)^3$	$2(3)^4$	$2(3)^5$

The common ratio r is 3. For the 2nd term, a_1 is multiplied by 3 once. For the 3rd term, a_1 is multiplied by 3 twice. The pattern shows that for each term, the **number of times 3 is multiplied** is one less than the **term number,** or $(n - 1)$. The 15th term is the first term, 2, times the common ratio, 3, raised to the 14th power.

$$a_{15} = 2(3)^{14} = 2(4{,}782{,}969) = 9{,}565{,}938$$

FINDING THE *n*th TERM OF A GEOMETRIC SEQUENCE

The nth term a_n of a geometric sequence with common ratio r is

$$a_n = a_1 r^{n-1}.$$

EXAMPLE 2 Finding a Given Term of a Geometric Sequence

Find the given term in each geometric sequence.

A 12th term: $6, 18, 54, 162, \ldots$

$$r = \frac{18}{6} = 3$$

$$a_{12} = 6(3)^{11} = 1{,}062{,}882$$

B 57th term: $1, -1, 1, -1, 1, \ldots$

$$r = \frac{-1}{1} = -1$$

$$a_{57} = 1(-1)^{56} = 1$$

C 10th term: $5, \frac{5}{2}, \frac{5}{4}, \frac{5}{8}, \frac{5}{16}, \ldots$

$$r = \frac{\frac{5}{2}}{5} = \frac{1}{2}$$

$$a_{10} = 5\left(\frac{1}{2}\right)^9 = \frac{5}{512}$$

D 20th term: $625, 500, 400, 320, \ldots$

$$r = \frac{500}{625} = 0.8$$

$$a_{20} = 625(0.8)^{19} \approx 9.01$$

EXAMPLE 3 Money Application

For mowing his family's yard every week, Joey has two options for payment: (1) \$10 per week or (2) 1¢ the first week, 2¢ the second week, 4¢ the third week, and so on, where he makes twice as much each week as he made the week before. If Joey will mow the yard for 15 weeks, which option should he choose?

If Joey chooses \$10 per week, he will get a total of 15(\$10) = \$150.

If Joey chooses the second option, his payment for just the 15th week will be more than the total of all the payments in option 1.

$$a_{15} = (\$0.01)(2)^{14} = (\$0.01)(16{,}384) = \$163.84$$

Option 1 gives Joey more money in the beginning, but option 2 gives him a larger total amount.

Think and Discuss

1. **Compare** arithmetic sequences with geometric sequences.
2. **Describe** how you find the common ratio in a geometric sequence.

12-2 Exercises

FOR EOG PRACTICE
see page 688

internet connect
Homework Help Online
go.hrw.com Keyword: MT4 12-2

GUIDED PRACTICE

See Example 

Determine if each sequence could be geometric. If so, give the common ratio.

1. $-4, -2, 0, 2, 4, \ldots$
2. $2, 6, 18, 54, 162, \ldots$
3. $\frac{2}{3}, -\frac{2}{3}, \frac{2}{3}, -\frac{2}{3}, \frac{2}{3}, \ldots$
4. $1, 1.5, 2.25, 3.375, \ldots$
5. $\frac{3}{16}, \frac{3}{8}, \frac{3}{4}, \frac{3}{2}, \ldots$
6. $-2, -4, -8, -16, \ldots$

See Example 

Find the given term in each geometric sequence.

7. 12th term: $3, 6, 12, 24, 48, \ldots$
8. 101st term: $\frac{1}{3}, -\frac{1}{3}, \frac{1}{3}, -\frac{1}{3}, \frac{1}{3}, \ldots$
9. 22nd term: $a_1 = 262{,}144$, $r = \frac{1}{2}$
10. 8th term: $1, 4, 16, 64, 256, \ldots$

See Example 

11. Heather makes \$6.50 per hour. Every three months, she is eligible for a 2% raise. How much will she make after 2 years if she gets a raise every time she is eligible?

INDEPENDENT PRACTICE

See Example 1 Determine if each sequence could be geometric. If so, give the common ratio.

12. 16, 8, 4, 2, 1, ... **13.** $\frac{1}{2}, \frac{1}{8}, \frac{1}{4}, \frac{1}{16}, \ldots$ **14.** 3, 6, 9, 12, ...

15. 768, 384, 192, 96, ... **16.** 1, −3, 9, −27, 81, ... **17.** $6, 2, \frac{2}{3}, \frac{2}{9}, \ldots$

See Example 2 Find the given term in each geometric sequence.

18. 6th term: $\frac{1}{2}$, 1, 2, 4, ... **19.** 5th term: $a_1 = 4096, r = \frac{7}{8}$

20. 5th term: $a_1 = 12, r = -\frac{1}{2}$ **21.** 7th term: 3, 6, 12, 24, ...

22. 22nd term: $\frac{1}{36}, \frac{1}{18}, \frac{1}{9}, \frac{2}{9}, \ldots$ **23.** 6th term: 1, 1.5, 2.25, 3.375, ...

See Example 3 **24.** A tank contains 54,000 gallons of water. One-third of the water remaining in the tank is removed each day. How much water is left in the tank on the 15th day?

PRACTICE AND PROBLEM SOLVING

Find the next three terms of each geometric sequence.

25. $a_1 = 24$, common ratio $= \frac{1}{2}$ **26.** $a_1 = 4$, common ratio = 2

27. $a_1 = \frac{1}{81}$, common ratio = −3 **28.** $a_1 = 3$, common ratio = 2.5

Find the first five terms of each geometric sequence.

29. $a_1 = 1, r = 1$ **30.** $a_1 = 5, r = -3$ **31.** $a_1 = 100, r = 1.1$

32. $a_1 = 64, r = \frac{3}{2}$ **33.** $a_1 = 10, r = 0.25$ **34.** $a_1 = 64, r = -4$

35. Find the 1st term of the geometric sequence with 6th term $\frac{64}{5}$ and common ratio 2.

36. Find the 3rd term of the geometric sequence with 7th term 256 and common ratio −4.

37. Find the 1st term of the geometric sequence with 5th term $\frac{125}{432}$ and common ratio $\frac{5}{6}$.

38. Find the 1st term of a geometric sequence with 4th term 28 and common ratio 2.

39. Find the 5th term of a geometric sequence with 3rd term 8 and 4th term 12.

40. Find the 3rd term of a geometric sequence with 4th term 5400 and 6th term 7776.

41. Find the 1st term of a geometric sequence with 3rd term 72 and 5th term 32.

42. ***ECONOMICS*** A car that was originally valued at $16,000 depreciates at 15% per year. This means that after each year, the car is worth 85% of its worth the previous year. What is the value of the car after 6 years? Round to the nearest dollar.

43. ***LIFE SCIENCE*** Under controlled conditions, a culture of bacteria doubles in size every 2 days. How many cells of the bacteria are in the culture after 2 weeks if there were originally 32 cells?

44. ***PHYSICAL SCIENCE*** A rubber ball is dropped from a height of 256 ft. After each bounce the height of the ball is recorded.

Height of Bouncing Ball					
Number of Bounces	1	2	3	4	5
Height (ft)	192	144	108	81	60.75

a. Could the heights in the table form a geometric sequence? If so, what is the common ratio?

b. Estimate the height of the ball after the 8th bounce. Round your answer to the nearest foot.

45. ***WRITE ABOUT IT*** Compare a geometric sequence with $a_1 = 2$ and $r = 3$ with a geometric sequence with $a_1 = 3$ and $r = 2$.

46. ***WHAT'S THE ERROR?*** A student is asked to find the next three terms of the geometric sequence with $a_1 = 10$ and common ratio 5. His answer is 2, $\frac{2}{5}$, $\frac{2}{25}$. What error has the student made, and what is the correct answer?

47. ***CHALLENGE*** The 5th term in a geometric sequence is 768. The 10th term is 786,432. Find the 7th term.

Spiral Review

Find the appropriate conversion factor. (Lesson 7-3)

48. meters to millimeters

49. quarts to gallons

50. gallons to pints

51. grams to centigrams

52. kilograms to grams

53. yards to inches

54. **EOG PREP** On a blueprint, a window is 2.5 inches wide. If the actual window is 85 inches wide, what scale factor was used to create the blueprint? (Lesson 7-7)

A $\frac{1}{28}$ B $\frac{1}{17}$ C $\frac{1}{24}$ D $\frac{1}{34}$

Fibonacci Sequence

Use with Lesson 12-3

internet connect
Lab Resources Online
go.hrw.com
KEYWORD: MT4 Lab12A

WHAT YOU NEED:
Square tiles

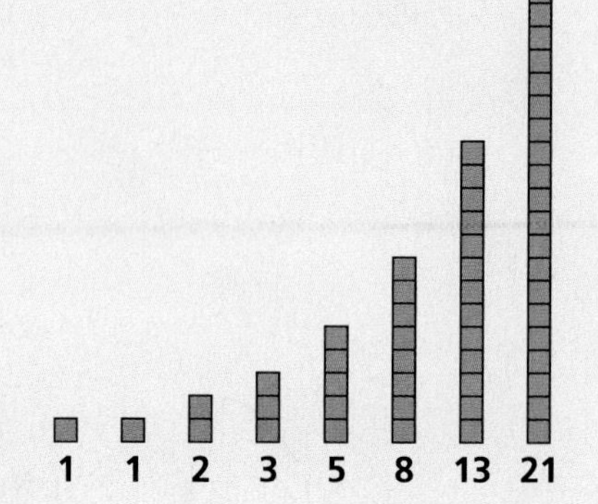

Activity

1. Use square tiles to model the following numbers:

   1 1 2 3 5 8 13 21

2. Place the first stack of tiles on top of the second stack of tiles. What do you notice?

The first two stacks added together are equal in height to the third stack.

3. Place the second stack of tiles on top of the third stack of tiles. What do you notice?

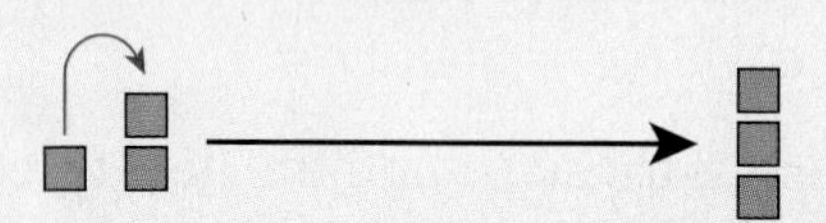

The second stack and the third stack added together are equal in height to the fourth stack.

This sequence is called the **Fibonacci sequence.** By adding two successive numbers you get the next number in the sequence. The sequence will go on forever.

Think and Discuss

1. If there were a term before the 1 in the sequence, what would it be? Explain your answer.
2. Could the numbers 144, 233, 377 be part of the Fibonacci sequence? Explain.

Try This

1. Use your square tiles to find the next two numbers in the sequence. What are they?
2. The 18th and 19th terms of the Fibonacci sequence are 2584 and 4181. What is the 20th term?

12-3 Other Sequences

Learn to find patterns in sequences.

Vocabulary

first differences

second differences

Fibonacci sequence

The first five *triangular numbers* are shown below.

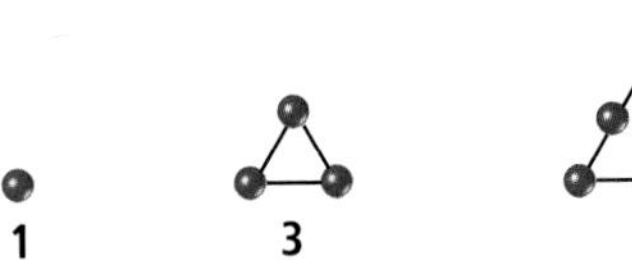

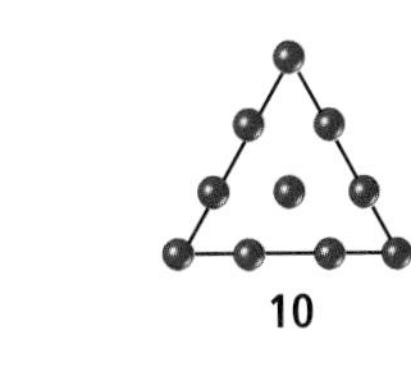

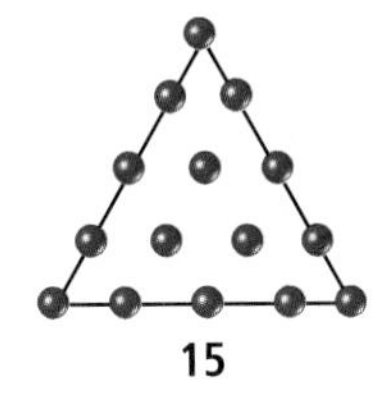

To continue the sequence, you can draw the triangles, or you can look for a pattern. If you subtract every term from the one after it, the **first differences** create a new sequence. If you do not see a pattern, you can repeat the process and find the **second differences**.

Term	1	2	3	4	5	6	7
Triangular Number	1	3	6	10	15	21	28

First differences	2	3	4	5	6	7
Second differences	1	1	1	1	1	

First and second differences can help you find terms in some sequences.

EXAMPLE 1 Using First and Second Differences

Use first and second differences to find the next three terms in each sequence.

A 1, 9, 24, 46, 75, 111, 154, . . .

Sequence	1	9	24	46	75	111	154	204	261	325
1st Differences	8	15	22	29	36	43	50	57	64	
2nd Differences	7	7	7	7	7	7	7	7		

The next three terms are 204, 261, 325.

B 5, 5, 7, 13, 25, 45, 75, . . .

Sequence	5	5	7	13	25	45	75	117	173	245
1st Differences	0	2	6	12	20	30	42	56	72	
2nd Differences	2	4	6	8	10	12	14	16		

The next three terms are 117, 173, 245.

By looking at the sequence 1, 2, 3, 4, 5, . . . , you would probably assume that the next term is 6. In fact, the next term could be any number. If no rule is given, you should use the simplest recognizable pattern in the given terms.

EXAMPLE 2 Finding a Rule, Given Terms of a Sequence

Give the next three terms in each sequence using the simplest rule you can find.

A $1, \frac{1}{2}, \frac{1}{3}, \frac{1}{4}, \frac{1}{5}, \ldots$

One possible rule is to add 1 to the denominator of the previous term. This could be written as the algebraic rule $a_n = \frac{1}{n}$.

The next three terms are $\frac{1}{6}, \frac{1}{7}, \frac{1}{8}$.

B $1, -1, 2, -2, 3, -3, \ldots$

Each positive term is followed by its opposite, and the next term is 1 more than the previous positive term.

The next three terms are 4, −4, 5.

C 2, 3, 5, 7, 11, 13, 17, . . .

The rule for the sequence could be the prime numbers from least to greatest.

The next three terms are 19, 23, 29.

D 1, 4, 9, 16, 25, 36, . . .

The rule for the sequence could be perfect squares. This could be written as the algebraic rule $a_n = n^2$.

The next three terms are 49, 64, 81.

Sometimes an algebraic rule is used to define a sequence.

EXAMPLE 3 Finding Terms of a Sequence, Given a Rule

Find the first five terms of the sequence defined by $a_n = \frac{n}{n+1}$.

$$a_1 = \frac{1}{1+1} = \frac{1}{2}$$

$$a_2 = \frac{2}{2+1} = \frac{2}{3}$$

$$a_3 = \frac{3}{3+1} = \frac{3}{4}$$

$$a_4 = \frac{4}{4+1} = \frac{4}{5}$$

$$a_5 = \frac{5}{5+1} = \frac{5}{6}$$

The first five terms are $\frac{1}{2}, \frac{2}{3}, \frac{3}{4}, \frac{4}{5}, \frac{5}{6}$.

A famous sequence called the **Fibonacci sequence** is defined by the following rule: Add the two previous terms to find the next term.

1, 1, 2, 3, 5, 8, 13, 21, . . .

1 + 1 = 2   1 + 2 = 3   2 + 3 = 5   3 + 5 = 8   5 + 8 = 13   8 + 13 = 21

EXAMPLE 4 **Using the Fibonacci Sequence**

Suppose *a, b, c,* and *d* are four consecutive numbers in the Fibonacci sequence. Complete the following table and guess the pattern.

a, b, c, d	*bc*	*ad*
1, 1, 2, 3	1(2) = 2	1(3) = 3
3, 5, 8, 13	5(8) = 40	3(13) = 39
13, 21, 34, 55	21(34) = 714	13(55) = 715
55, 89, 144, 233	89(144) = 12,816	55(233) = 12,815

The product of the two middle terms is either one more or one less than the product of the two outer terms.

Think and Discuss

1. Find the first and second differences for the sequence of pentagonal numbers: 1, 5, 12, 22, 35, 51, 70, ...

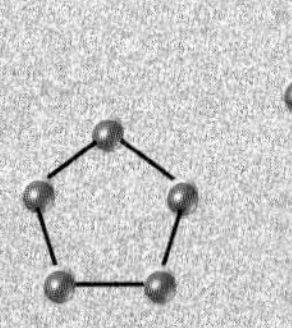
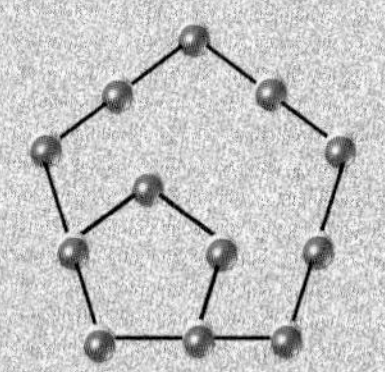

12-3 Exercises

FOR EOG PRACTICE
see page 689

GUIDED PRACTICE

See Example 1 **Use first and second differences to find the next three terms in each sequence.**

1. 1, 7, 22, 46, 79, 121, 172, . . .

2. 5, 10, 30, 65, 115, 180, . . .

3. 12, 12, 15, 24, 42, 72, 117, . . .

4. 6, 8, 19, 48, 104, 196, 333, . . .

See Example 2 **Give the next three terms in each sequence using the simplest rule you can find.**

5. $\frac{1}{2}, \frac{2}{3}, \frac{3}{4}, \frac{4}{5}, \frac{5}{6}, \frac{6}{7}, \ldots$

6. 5, −6, 7, −8, 9, −10, 11, . . .

7. 4, 5, 6, 4, 5, 6, 4, . . .

8. 1, 8, 27, 64, 125, . . .

See Example 3 Find the first five terms of each sequence defined by the given rule.

9. $a_n = \frac{4n}{n+2}$ **10.** $a_n = (n+2)(n+3)$ **11.** $a_n = \frac{2-n}{n} + 1$

See Example 4 **12.** Suppose a, b, and c are three consecutive numbers in the Fibonacci sequence. Complete the following table and guess the pattern.

a, b, c	ac	b^2
1, 1, 2	■	■
3, 5, 8	■	■
13, 21, 34	■	■
55, 89, 144	■	■

INDEPENDENT PRACTICE

See Example 1 **Use first and second differences to find the next three terms in each sequence.**

13. 11, 22, 34, 47, 61, 76, . . . **14.** −15, −11, −2, 12, 31, 55, . . .

15. 25, 26, 29, 35, 45, 60, 81, . . . **16.** 0.01, 0.02, 0.08, 0.24, 0.55, . . .

See Example 2 **Give the next three terms in each sequence using the simplest rule you can find.**

17. 1, 2, 2, 3, 3, 3, 4, 4, 4, 4, 5, . . . **18.** 3, 1, 4, 1, 5, 9, . . .

19. 1.1, 1.01, 1.001, 1.0001, . . . **20.** $1, \frac{1}{4}, \frac{1}{9}, \frac{1}{16}, \frac{1}{25}, \frac{1}{36}, \ldots$

See Example 3 **Find the first five terms of each sequence defined by the given rule.**

21. $a_n = \frac{n-1}{n+1}$ **22.** $a_n = n(n-1) - 2n$ **23.** $a_n = \frac{3n}{n+1}$

See Example 4 **24.** Suppose a, b, c, d, and e are five consecutive numbers in the Fibonacci sequence. Complete the following table and guess the pattern.

a, b, c, d, e	ae	bd	c^2
1, 1, 2, 3, 5	■	■	■
3, 5, 8, 13, 21	■	■	■
13, 21, 34, 55, 89	■	■	■
55, 89, 144, 233, 377	■	■	■

PRACTICE AND PROBLEM SOLVING

The first 14 terms of the Fibonacci sequence are 1, 1, 2, 3, 5, 8, 13, 21, 34, 55, 89, 144, 233, 377.

25. Where in this part of the sequence are the even numbers? Where do you think the next four even numbers will occur?

26. Where in this part of the sequence are the multiples of 3? Where do you think the next four multiples of 3 will occur?

Music LINK

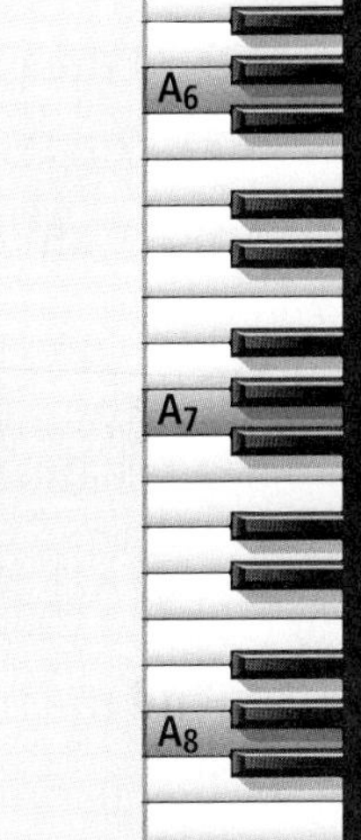

Pitch is the frequency of a musical note, measured in units called *hertz* (Hz). The lower the frequency of a pitch, the lower it sounds, and the higher the frequency of a pitch, the higher it sounds. A pitch is named by its octave. A_4 is in the 4th octave on the piano keyboard and is often called middle A.

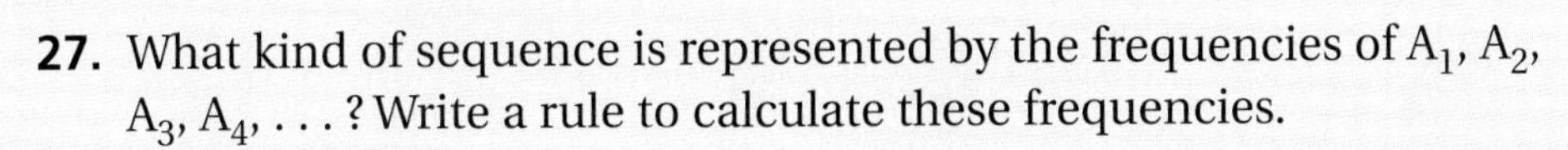

27. What kind of sequence is represented by the frequencies of A_1, A_2, A_3, A_4, . . . ? Write a rule to calculate these frequencies.

28. What is the frequency of the note A_5, which is one octave higher than A_4?

When a string of an instrument is played, its vibrations create many different frequencies at the same time. These varying frequencies are called *harmonics.*

Frequencies of Harmonics on A_1

Harmonic	Fundamental (1st)	2nd	3rd	4th	5th
Note	A_1	A_2	E_2	A_3	$C^\#_3$

29. What kind of sequence is represented by the frequencies of different harmonics? Write a rule to calculate these frequencies.

go.hrw.com
KEYWORD: MT4 Pitch
CNN Student News

30. What is the frequency of the note E_3 if it is the 6th harmonic on A_1?

31. **CHALLENGE** In music an important interval is a *fifth.* As you progress around the circle of fifths, the pitch frequencies are approximately as shown (rounded to the nearest tenth). What type of sequence do the frequencies form in clockwise order from C? Write the rule for the sequence. If the rule holds all the way around the circle, what would the frequency of the note F be?

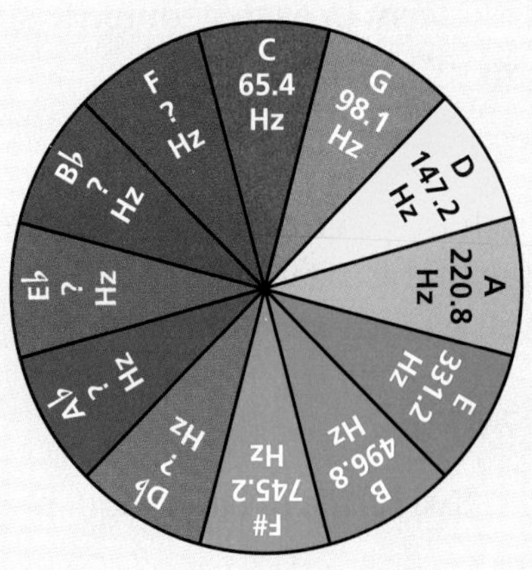

Spiral Review

Find the x-intercept and y-intercept of each line. (Lesson 11-3)

32. $3x - 8y = 48$ **33.** $5y - 15x = -45$ **34.** $13x + 2y = 26$ **35.** $9x + 27y = 81$

36. **EOG PREP** If y varies directly with x and $y = 25$ when $x = 15$, what is the equation of direct variation? (Lesson 11-5)

A $y = \frac{3}{5}x$ B $y = \frac{5}{3}x$ C $y = 15x$ D $y = 25x$

Chapter 12

Mid-Chapter Quiz

LESSON 12-1 (pp. 590–594)

Determine if each sequence could be arithmetic. If so, give the common difference.

1. 10, 11, 13, 16, ...

2. 27, 24, 21, 18, ...

3. 11, 22, 33, 44, ...

4. 17, 60, 103, 177, ...

Find the given term in each arithmetic sequence.

5. 8th term: 5, 8, 11, 14, ...

6. 11th term: 7, 6.9, 6.8, ...

7. 14th term: 9, $9\frac{1}{4}$, $9\frac{1}{2}$, ...

8. 6th term: 28, 15, 2, –11, ...

9. Frank deposited $25 in an account the first week. Each week, he deposits $5 more than the previous week. In which week will he deposit $100?

LESSON 12-2 (pp. 595–599)

Determine if each sequence could be geometric. If so, give the common ratio.

10. 1, −5, 25, −125, ...

11. 2, −5, −12, −19, ...

12. 81, 27, 9, 3, ...

13. 60, 18, 5.4, 1.62, ...

Find the given term in each geometric sequence.

14. 7th term: 12, 36, 108, ...

15. 9th term: 36, 12, 4, ...

16. 10th term: $-\frac{3}{2}$, 3, −6, ...

17. 15th term: 1000, 100, 10, ...

18. The purchase price of a machine at a factory was $500,000. Each year, the value of the machine depreciates by 5%. To the nearest dollar, what is the value of the machine after 6 years?

LESSON 12-3 (pp. 601–605)

Find the first five terms of each sequence, given its rule.

19. $a_n = 3n - 5$

20. $a_n = 2^{n-1}$

21. $a_n = (-1)^n \cdot 3n$

22. $a_n = (n + 1)^2 - 1$

Use first and second differences to find the next three terms in each sequence.

23. 9, 9, 11, 15, 21, ...

24. 3, 10, 21, 36, 55, ...

25. −6, −11, −13, −12, −8, ...

26. 0, 4, 11, 22, 38, 60, ...

Give the next three terms in each sequence using the simplest rule you can find.

27. $\frac{1}{2}, \frac{3}{4}, \frac{5}{6}, \frac{7}{8}$ ...

28. 1, 8, 27, 64, ...

1.02

Focus on Problem Solving

Solve

- **Eliminate answer choices**

When answering a multiple-choice question, you may be able to eliminate some of the choices. If the question is a word problem, check whether any answers do not make sense in the problem.

Example:

Gabrielle has a savings account with $125 in it. Each week, she deposits $5 in the account. How much will she have in 12 weeks?

A $65 **B** $185 **C** $142 **D** $190

The following sequence represents the weekly balance in dollars:

125, 130, 135, 140, 145, ...

The amount will be greater than $125, so it cannot be **A.** It will also be a multiple of 5, so it cannot be **C.**

Read each question and decide whether you can eliminate any answer choices before choosing an answer. Explain your reasoning.

1. An art gallery has 400 paintings. Each year, the curator acquires 15 new paintings. How many paintings will the gallery have in 7 years?
   A 450 **C** 505
   B 6000 **D** 295

2. There are 360 deer in a forest. The population increases each year by 10% over the previous year. How many deer will there be after 9 years?
   A 849 **C** 324
   B 450 **D** 684

3. Donna is in a book club. She has read 24 books so far, and she thinks she can read 3 books a week during the summer. How many weeks will it take for her to read a total of 60 books?
   A 20 weeks **C** 3 weeks
   B 12 weeks **D** 60 weeks

4. Oliver has $230.00 in a savings account that earns 6% interest each year. How much will he have in 12 years?
   A $230.00 **C** $395.60
   B $109.46 **D** $462.81

12-4 Functions

Input
Domain
Function
Output
Range

Learn to represent functions with tables, graphs, or equations.

Vocabulary

function
input
output
domain
range
function notation

A **function** is a rule that relates two quantities so that each **input** value corresponds to exactly one **output** value.

The **domain** is the set of all possible input values, and the **range** is the set of all possible output values.

Function
One input gives one output.

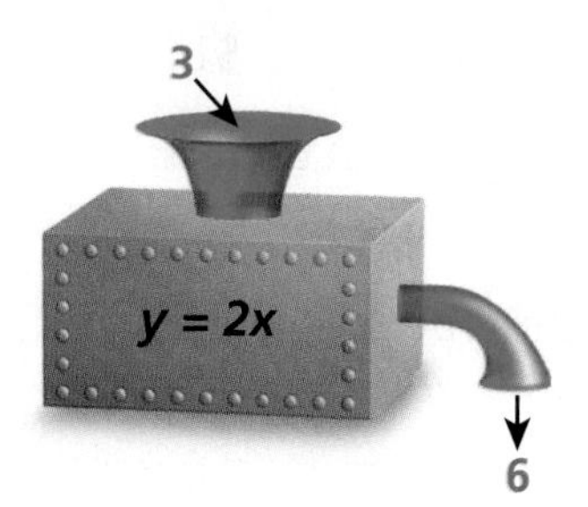

Example: The output is 2 times the input.

Not a Function
One input gives more than one output.

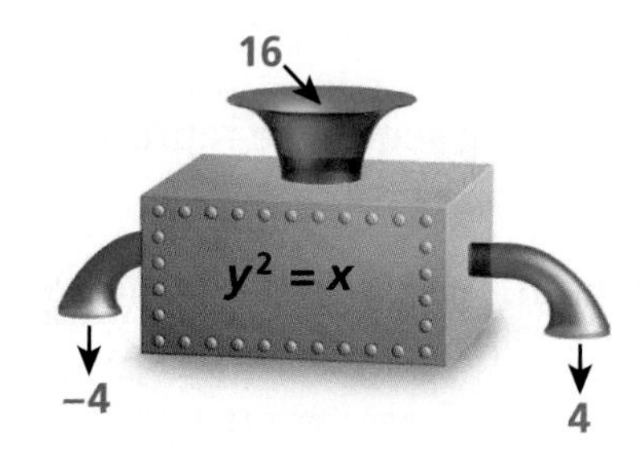

Example: The outputs are the square roots of the input.

Functions can be represented in many ways, including tables, graphs, and equations. If the domain of a function has infinitely many values, it is impossible to represent them all in a table, but a table can be used to show some of the values and to help in creating a graph.

EXAMPLE 1 Finding Different Representations of a Function

Make a table and a graph of $y = x^2 + 1$.

Make a table of inputs and outputs. Use the table to make a graph.

x	$x^2 + 1$	y
−2	$(-2)^2 + 1$	5
−1	$(-1)^2 + 1$	2
0	$(0)^2 + 1$	1
1	$(1)^2 + 1$	2
2	$(2)^2 + 1$	5

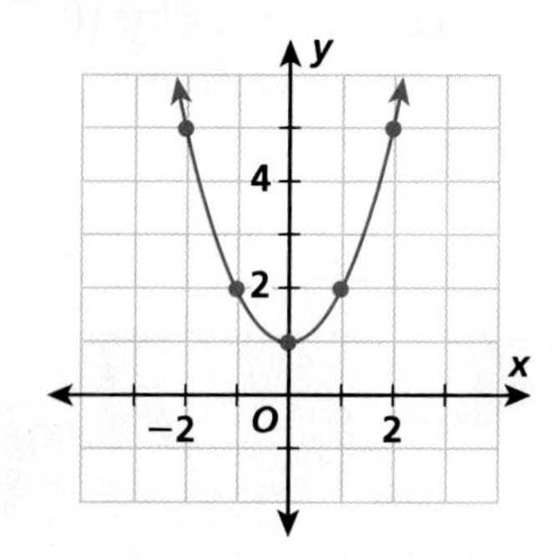

To determine if a relationship is a function, verify that each input has exactly one output.

12-5 Exercises

FOR EOG PRACTICE
see page 690

internet connect
Homework Help Online
go.hrw.com Keyword: MT4 12-5

5.01, 5.02, 5.03

GUIDED PRACTICE

See Example 1
Write the rule for each linear function.

1.

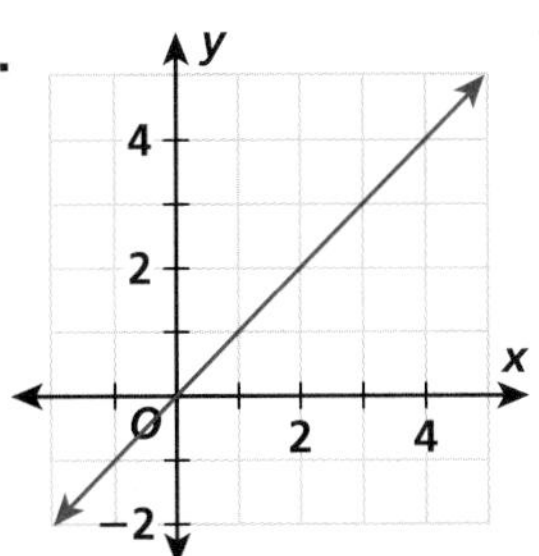

2.

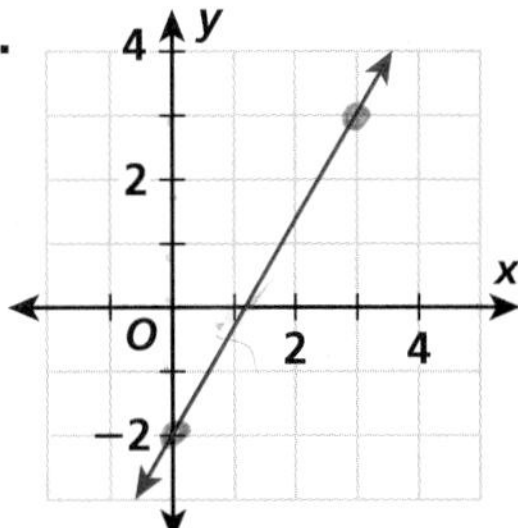

See Example 2
3.

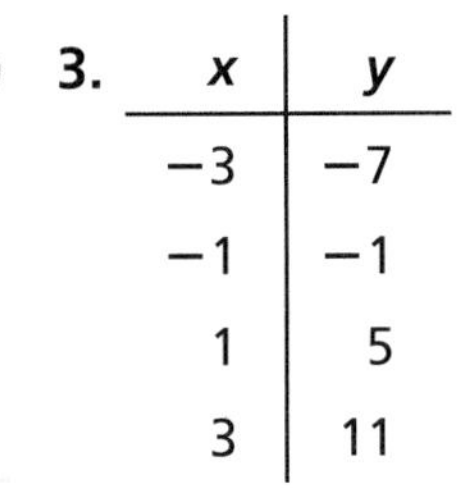

x	y
−3	−7
−1	−1
1	5
3	11

4.

x	y
−1	6
0	4
1	2
2	0

See Example 3
5. Kim earns $400 per week for 40 hours of work. If she works overtime, she makes $15 per overtime hour. Find a rule for the linear function that describes her weekly salary if she works x hours of overtime, and use it to find how much Kim earns if she works 7 hours of overtime.

INDEPENDENT PRACTICE

See Example 1
Write the rule for each linear function.

6.

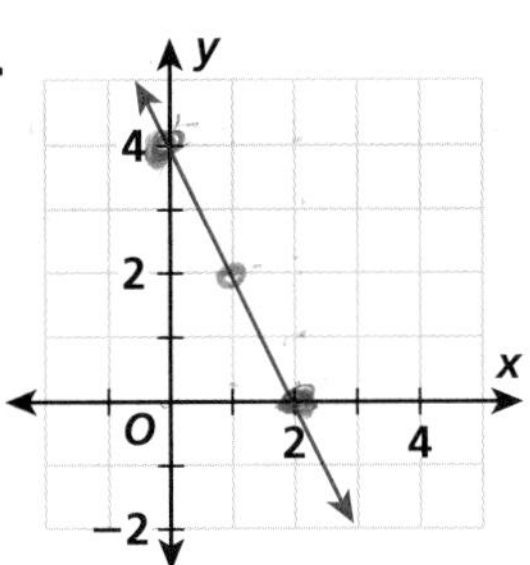

7.

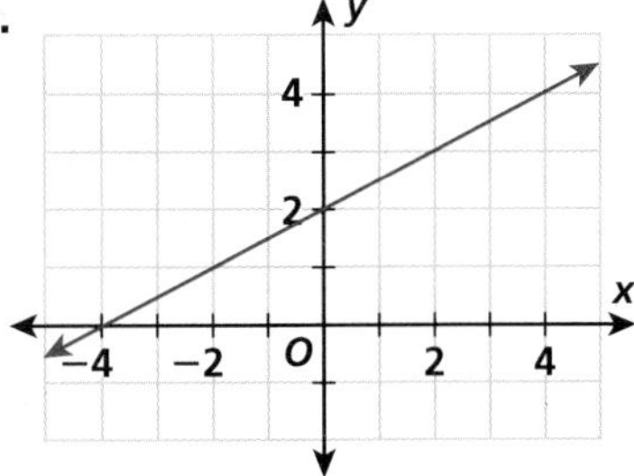

See Example 2
8.

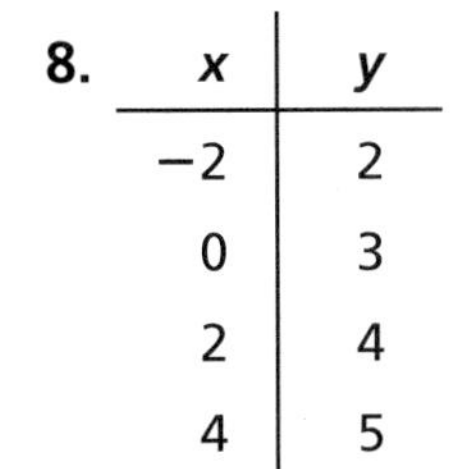

x	y
−2	2
0	3
2	4
4	5

9.

x	y
−1	−11
0	−5
1	1
2	7

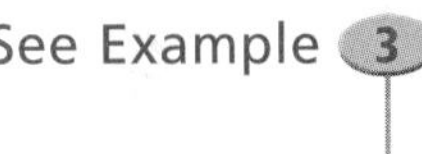
See Example 3
10. A tank contains 1200 gallons of water. The tank is being drained at a rate of 45 gallons per minute. Find a rule for the linear function that describes the amount of water in the tank, and use it to determine how much water will be in the tank after 15 minutes.

PRACTICE AND PROBLEM SOLVING

Recreation LINK

The volume of a typical hot air balloon is between 65,000 and 105,000 cubic feet. Most hot air balloons fly at altitudes of 1000 to 1500 feet.

go.hrw.com KEYWORD: MT4 Balloons

11. ***RECREATION*** A hot air balloon at a height of 1245 feet above sea level is ascending at a rate of 5 feet per second.
    a. Write a linear function that describes the balloon's height after x seconds.
    b. What will the balloon's height be in 5 minutes? How high will it have climbed from its original starting point?

12. ***ECONOMICS*** *Linear depreciation* means that the same amount is subtracted each year from the value of an item. Suppose a car valued at \$17,440 depreciates \$1375 each year for x years.
    a. Write a linear function for the car's value after x years.
    b. What will thc car's value be in 7 years?

13. ***LIFE SCIENCE*** Suppose a puppy was born weighing 4 pounds, and it gained about 3 pounds each month during the first year. Find a rule for the linear function that describes the puppy's growth and use it to find out how much the puppy would weigh after 8 months.

14. ***BUSINESS*** The table shows a retailer's cost for certain items and the price at which the retailer sells each item.

Dealer Cost	\$15	\$22	\$30.50	\$40
Selling Price	\$19.50	\$28.60	\$39.65	\$52

    a. Write a linear function for the selling price of an item that costs the retailer x dollars.
    b. What would the selling price of a television be that costs the retailer \$265?

15. ***WRITE ABOUT IT*** Explain how you can determine whether a function is linear.

16. ***WHAT'S THE QUESTION?*** Consider the function $f(x) = -3x + 9$. If the answer is -6, what is the question?

17. ***CHALLENGE*** What is the only kind of line on a coordinate plane that is not a linear function? Give an example of such a line.

Spiral Review

Find the point-slope form of each equation. (Lesson 11-4)

18. slope 5; passes through point (4, 1)

19. slope -2; passes through point $(6, -6)$

20. slope -4; passes through point (0, 12)

21. slope 1.4; passes through point $(1, -3)$

22. **EOG PREP** Which of the following ordered pairs is *not* a solution of the inequality $5x - 13y \leq 61$? (Lesson 11-6)

A (12, 6) B (0, 0) C $(-4, -3)$ D $(6, -10)$

12-6 Exponential Functions

Learn to identify and graph exponential functions.

Vocabulary

exponential function

exponential growth

exponential decay

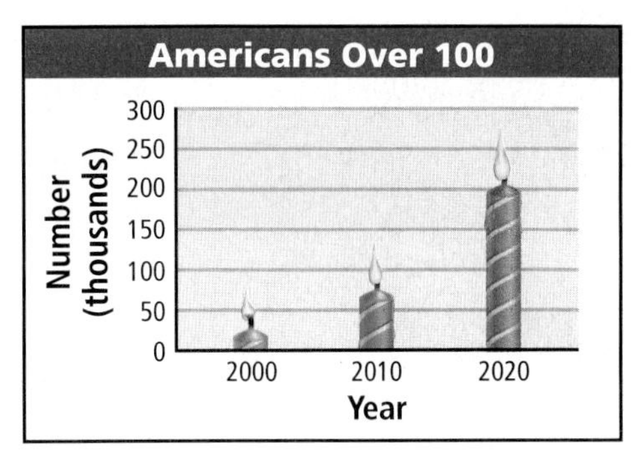

Do you think you will live to be 100? According to U.S. census data, the number of Americans over 100 nearly doubled from about 37,000 in 1990 to more than 70,000 in 2000.

Suppose the number of Americans over 100 doubles each decade. The populations would form a geometric sequence with a common ratio of 2.

Population of Americans over 100					
Year	2000	2010	2020	2030	2040
Population (thousands)	70	140	280	560	1120

An **exponential function** has the form $f(x) = p \cdot a^x$, where $a > 0$ and $a \neq 1$. If the input values are the set of whole numbers, the output values form a geometric sequence. The y-intercept is $f(0) = p$. The expression a^x is defined for all values of x, so the domain of $f(x) = p \cdot a^x$ is all real numbers.

EXAMPLE 1 Graphing an Exponential Function

Create a table for each exponential function, and use it to graph the function.

A $f(x) = \frac{1}{2} \cdot 2^x$

x	y	
−2	$\frac{1}{8}$	$\frac{1}{2} \cdot 2^{-2} = \frac{1}{2} \cdot \frac{1}{4}$
−1	$\frac{1}{4}$	$\frac{1}{2} \cdot 2^{-1} = \frac{1}{2} \cdot \frac{1}{2}$
0	$\frac{1}{2}$	$\frac{1}{2} \cdot 2^{0} = \frac{1}{2} \cdot 1$
1	1	$\frac{1}{2} \cdot 2^{1} = \frac{1}{2} \cdot 2$
2	2	$\frac{1}{2} \cdot 2^{2} = \frac{1}{2} \cdot 4$

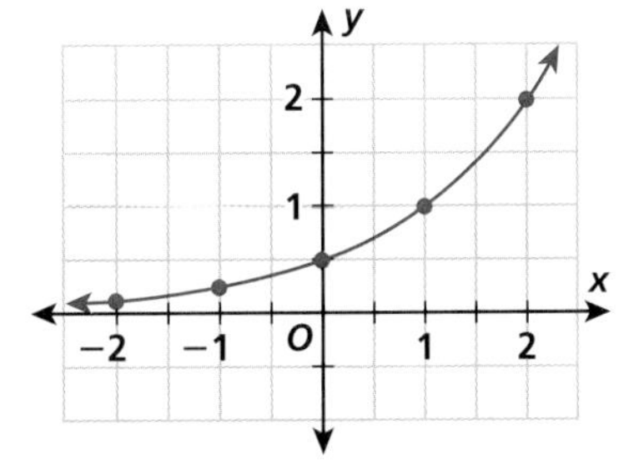

B $f(x) = 2 \cdot \left(\frac{1}{2}\right)^x$

x	y	
−2	8	$2 \cdot \left(\frac{1}{2}\right)^{-2} = 2 \cdot 4$
−1	4	$2 \cdot \left(\frac{1}{2}\right)^{-1} = 2 \cdot 2$
0	2	$2 \cdot \left(\frac{1}{2}\right)^{0} = 2 \cdot 1$
1	1	$2 \cdot \left(\frac{1}{2}\right)^{1} = 2 \cdot \frac{1}{2}$
2	$\frac{1}{2}$	$2 \cdot \left(\frac{1}{2}\right)^{2} = 2 \cdot \frac{1}{4}$

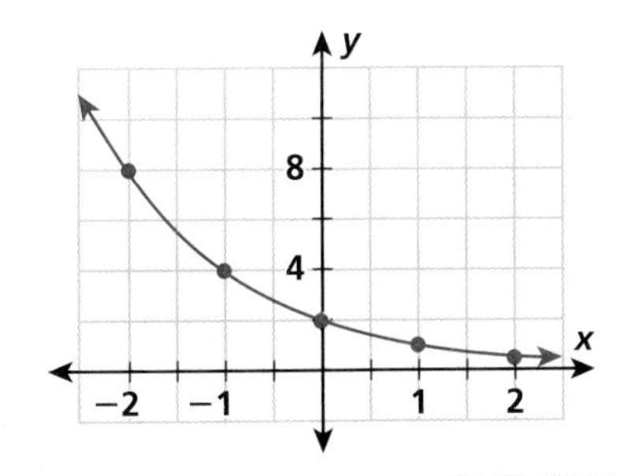

If $a > 1$, the output $f(x)$ gets larger as the input x gets larger. In this case, f is called an **exponential growth** function.

EXAMPLE 2 Using an Exponential Growth Function

The number of Americans over 100 was about 70,000 in 2000. If the population of Americans over 100 doubles each decade, estimate the population of Americans over 100 in the year 2095.

Helpful Hint

In algebra, you will learn the meaning of expressions like $2^{9.5}$. You can use a calculator to evaluate these expressions.

Year	2000	2010	2020	2030	2040
Number of Decades x	0	1	2	3	4
Population $f(x)$ (thousands)	70	140	280	560	1120

$f(x) = p \cdot a^x$

$f(x) = 70 \cdot a^x$ *$f(0) = p$*

$f(x) = 70 \cdot 2^x$ *$f(1) = 70 \cdot a^1 = 140$, so $a = 2$.*

The year 2095 is 9.5 decades after the year 2000, so let $x = 9.5$.

$f(9.5) = 70 \cdot 2^{9.5} \approx 50{,}685$ *Substitute 9.5 for x.*

If the population over 100 doubles each decade, there will be 50,685,000 Americans over 100 in 2095.

In the exponential function $f(x) = p \cdot a^x$, if $a < 1$, the output gets smaller as x gets larger. In this case, f is called an **exponential decay** function.

EXAMPLE 3 Using an Exponential Decay Function

Technetium-99m has a *half-life* of 6 hours, which means it takes 6 hours for half of the substance to decompose. Find the amount of technetium-99m remaining from a 100 mg sample after 90 hours.

Physical Science LINK

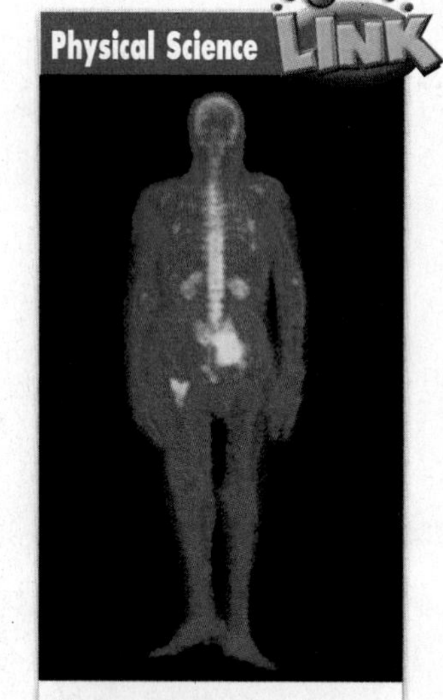

Technetium-99m is used to diagnose diseases in humans and animals.

Hours	0	6	12	18	24
Number of Half-lives x	0	1	2	3	4
Technetium-99m $f(x)$ (mg)	100	50	25	12.5	6.25

$f(x) = p \cdot a^x$

$f(x) = 100 \cdot a^x$ *$f(0) = p$*

$f(x) = 100 \cdot \left(\frac{1}{2}\right)^x$ *$f(1) = 100 \cdot a^1 = 50$, so $a = \frac{1}{2}$.*

Divide 90 hours by 6 hours to find the number of half-lives: $x = 15$.

$f(15) = 100 \cdot \left(\frac{1}{2}\right)^{15} \approx 0.003$ *Substitute 15 for x.*

There is approximately 0.003 mg left after 90 hours.

Think and Discuss

1. **Compare** the graphs of exponential growth and decay functions.

12-6 Exercises

FOR EOG PRACTICE
see page 691

internet connect
Homework Help Online
go.hrw.com Keyword: MT4 12-6

5.01a

GUIDED PRACTICE

See Example 1

Create a table for each exponential function, and use it to graph the function.

1. $f(x) = 3^x$
2. $f(x) = 50 \cdot \left(\frac{1}{3}\right)^x$
3. $f(x) = 3 \cdot 2^x$
4. $f(x) = 0.01 \cdot 5^x$

See Example 2

5. At the beginning of an experiment, a bacteria colony has a mass of 2×10^{-6} grams. If the mass of the colony doubles every 10 hours, what will the mass of the colony be after 80 hours?

See Example 3

6. Radioactive glucose is used in cancer detection. It has a half-life of 100 minutes. How much of a 100 mg sample remains after 24 hours?

INDEPENDENT PRACTICE

See Example 1

Create a table for each exponential function, and use it to graph the function.

7. $f(x) = 2 \cdot 3^x$
8. $f(x) = -2 \cdot (0.2)^x$
9. $f(x) = \left(\frac{2}{3}\right)^x$
10. $f(x) = 10 \cdot \left(\frac{1}{5}\right)^x$

See Example 2

11. Mariano invested \$500 in an account that will double his balance every 8 years. Write an exponential function to calculate his account balance. What will his balance be in 32 years?

See Example 3

12. Cesium-137 is a radioactive element with a half-life of 30 years. It is used to study upland soil erosion. How much of a 50 mg sample of cesium-137 would remain after 180 years?

PRACTICE AND PROBLEM SOLVING

For each exponential function, find $f(-5)$, $f(0)$, and $f(5)$.

13. $f(x) = 2^x$
14. $f(x) = 0.3^x$
15. $f(x) = 10^x$
16. $f(x) = 200 \cdot \left(\frac{1}{2}\right)^x$

Write the equation of the exponential function that passes through the given points. Use the form $f(x) = p \cdot a^x$.

17. (0, 3) and (1, 6)
18. (0, 4) and (1, 2)
19. (0, 1) and (2, 9)

Graph the exponential function of the form $f(x) = p \cdot a^x$.

20. $p = 6, a = 5$
21. $p = -1, a = \frac{1}{4}$
22. $p = 100, a = 0.01$
23. Carbon-14 is used by archaeologists to find the approximate age of animal and plant material. It has a half-life of 5730 years. What percent of a sample remains after 34,380 years?

Health LINK

The half-life of a substance in the body is the amount of time it takes for your body to metabolize half of the substance. An exponential decay function can be used to model the amount of the substance in the body.

Acetaminophen is the active ingredient in many pain and fever medications. Use the table for Exercises 24–26.

Acetaminophen Levels in the Body				
Elapsed Time (hr)	0	3	5	6
Substance Remaining (mg)	160	80	50.4	40

24. How much acetaminophen was present initially?

25. Find the half-life of acetaminophen. Write an exponential function that describes the level of acetaminophen in the body.

   a. How much acetaminophen will be present after 12 hours?

   b. How much acetaminophen will be present after 1 day?

26. If you take 500 mg of acetaminophen, what percent of that amount will be in your system after 9 hours?

27. The half-life of vitamin C is about 6 hours. If you take a 60 mg vitamin C tablet at 9:00 AM, how much of the vitamin will still be present in your system at 9:00 PM?

28. Caffeine has a half-life of about 5 hours in adults. Two 6 oz cups of coffee contain about 200 mg caffeine. If an adult drinks 2 cups of coffee, how much caffeine will be in his system after 12 hours?

29. ***CHALLENGE*** In children, the half-life of caffeine is about 3 hours. If a child has a 12 oz soft drink containing 40 mg caffeine at 12:00 PM and another at 6:00 PM, about how much caffeine will be present at 10:00 PM?

Vitamin deficiencies can cause serious diseases, such as scurvy, rickets, and beriberi.

Sources of caffeine include coffee, sodas, and some pain medications.

Spiral Review

Determine if each sequence could be geometric. If so, give the common ratio. (Lesson 12-2)

30. 5, 10, 15, 20, 25, ...

31. 3, 6, 12, 24, 48, ...

32. 1, −3, 9, −27, 81, ...

33. 0.1, 0.2, 0.3, 0.4, ...

34. −4, −4, −4, −4, −4, ...

35. 0.1, 0.01, 0.001, 0.0001, ...

36. **EOG PREP** The function $f(x) = 12{,}800 - 1100x$ gives the value of a car (in dollars) x years after it was purchased. What is the car's value 8 years after it was purchased? (Lesson 12-5)

A $6200   B $7300   C $4000   D $5100

12-7 Quadratic Functions

Learn to identify and graph quadratic functions.

Vocabulary

quadratic function

parabola

A **quadratic function** contains a variable that is squared. In the quadratic function

$$f(x) = ax^2 + bx + c$$

the y-intercept is c. The graphs of all quadratic functions have the same basic shape, called a **parabola**. The cross section of the large mirror in a telescope is a parabola. Because of a property of parabolas, starlight that hits the mirror is reflected toward a single point, called the *focus*.

The mirror of this telescope is made of liquid mercury that is rotated to form a parabolic shape.

EXAMPLE 1 Quadratic Functions of the Form $f(x) = ax^2 + bx + c$

Create a table for each quadratic function, and use it to make a graph.

A $f(x) = x^2 - 2$

x	$f(x) = x^2 - 2$
−3	$(-3)^2 - 2 = 7$
−2	$(-2)^2 - 2 = 2$
−1	$(-1)^2 - 2 = -1$
0	$(0)^2 - 2 = -2$
1	$(1)^2 - 2 = -1$
2	$(2)^2 - 2 = 2$
3	$(3)^2 - 2 = 7$

Plot the points and connect them with a smooth curve.

B $f(x) = x^2 + x - 2$

x	$f(x) = x^2 + x - 2$
−3	$(-3)^2 + (-3) - 2 = 4$
−2	$(-2)^2 + (-2) - 2 = 0$
−1	$(-1)^2 + (-1) - 2 = -2$
0	$(0)^2 + 0 - 2 = -2$
1	$(1)^2 + 1 - 2 = 0$
2	$(2)^2 + 2 - 2 = 4$
3	$(3)^2 + 3 - 2 = 10$

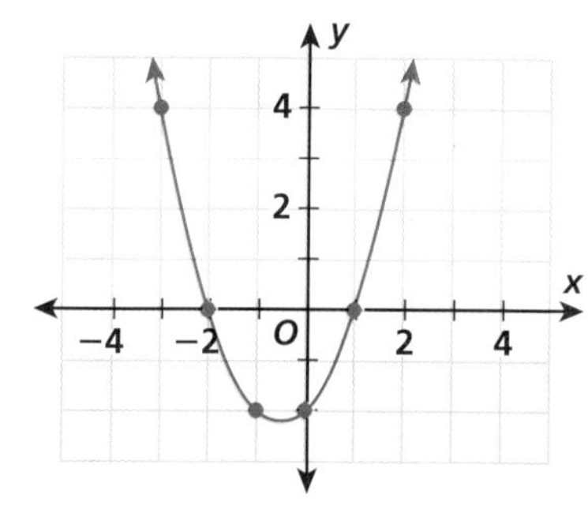

Plot the points and connect them with a smooth curve.

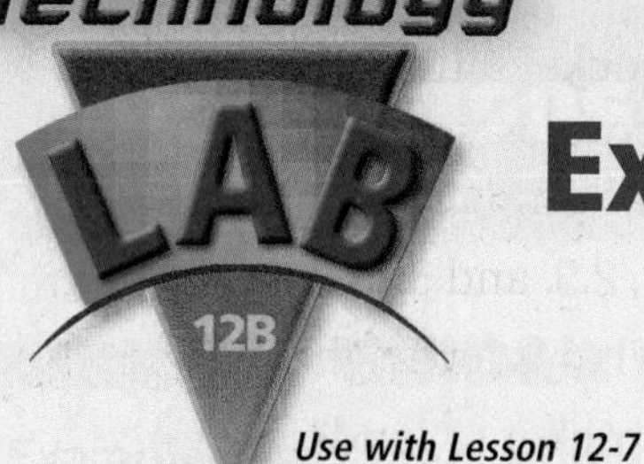

Explore Cubic Functions

Use with Lesson 12-7

1.02, 5.04

internet connect
Lab Resources Online
go.hrw.com
KEYWORD: MT4 Lab12B

You can use your graphing calculator to explore cubic functions. To graph the cubic equation $y = x^3$ in the standard graphing calculator window, press Y= ; enter the right side of the equation, X,T,θ,n ^ 3; and press ZOOM **6:ZStandard.** Notice that the graph goes from the lower left to the upper right and crosses the x-axis once, at $x = 0$.

Activity 1

1 Graph $y = -x^3$. Describe the graph.

Press Y= , and enter the right side of the equation, (−) X,T,θ,n ^ 3.

The graph goes from the upper left to the lower right and crosses the x-axis once.

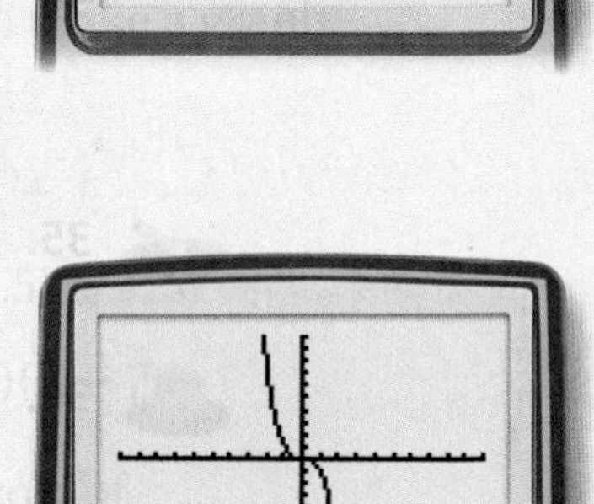

2 Graph $y = x^3 + 3x^2 - 2$. Describe the graph.

Press Y= ; enter the right side of the equation, X,T,θ,n ^ 3 + 3 X,T,θ,n x^2 − 2; and press ZOOM **6:ZStandard.** The graph goes from the lower left to the upper right and crosses the x-axis three times.

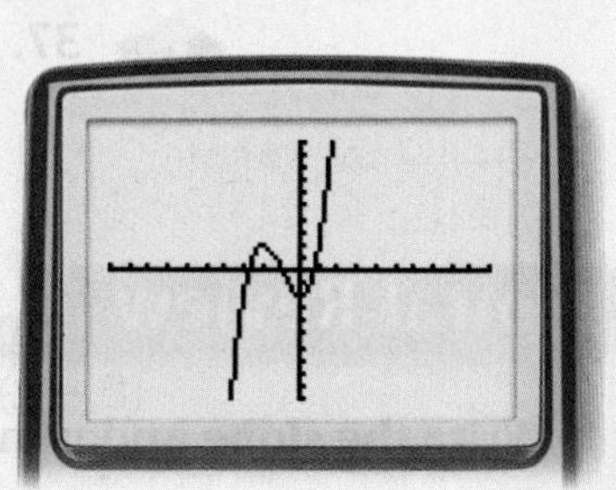

Think and Discuss

1. How does the sign of the x^3 term affect the graph of a cubic function?
2. How could you find the value of 7^3 from the graph of $y = x^3$?

Try This

Graph each function and describe the graph.

1. $y = x^3 - 2$ **2.** $y = x^3 + 3x^2 - 2$ **3.** $y = (x - 2)^3$ **4.** $y = 5 - x^3$

Activity 2

1 Compare the graphs of $y = x^3$ and $y = x^3 + 3$.

Graph **Y$_1$=X^3** and **Y$_2$=X^3+3** on the same screen, as shown. Use the TRACE button and the ◀ and ▶ buttons to trace to any integer value of x. Then use the ▲ and ▼ keys to move from one function to the other to compare the values of y for both functions for the value of x. You can also press 2nd GRAPH (TABLE) to see a table of values for both functions.

The graph of $y = x^3 + 3$ is translated up 3 units from the graph of $y = x^3$.

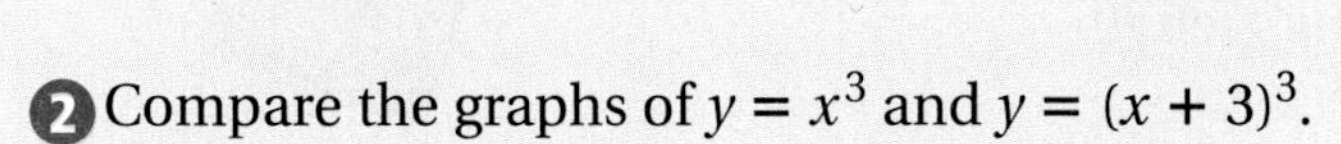

2 Compare the graphs of $y = x^3$ and $y = (x + 3)^3$.

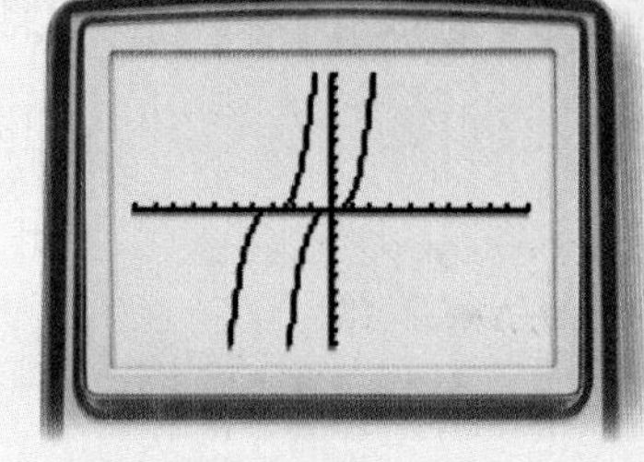

Graph **Y$_1$=X^3** and **Y$_2$=(X+3)^3** on the same screen. Notice that the graph of $y = (x + 3)^3$ is the graph of $y = x^3$ moved left 3 units. Press 2nd GRAPH (TABLE) to see a table of values. The graph of $y = (x + 3)^3$ is translated left 3 units from the graph of $y = x^3$.

3 Compare the graphs of $y = x^3$ and $y = 2x^3$.

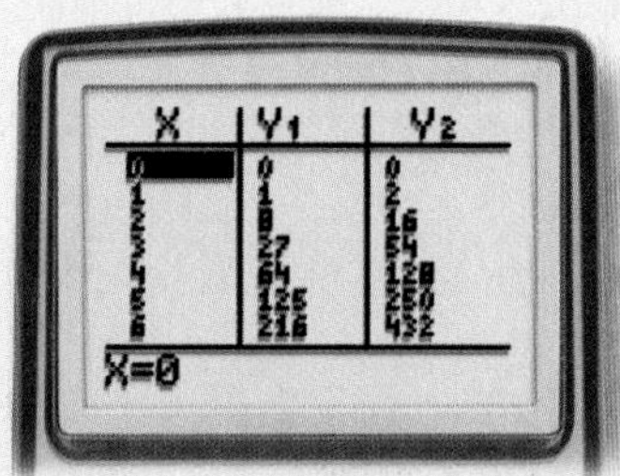

Graph **Y$_1$=X^3** and **Y$_2$=2X^3** on the same screen. Use the TRACE button and the arrow keys to see the values of y for any value of x. Press 2nd GRAPH (TABLE) to see a table of values.

The graph of $y = 2x^3$ is stretched upward from the graph of $y = x^3$. The y-value for $y = 2x^3$ increases twice as fast as it does for $y = x^3$. The table of values is shown.

Think and Discuss

1. What function would translate $y = x^3$ right 6 units?
2. Do you think that the methods shown of translating a cubic function would have the same result on a quadratic function? Explain.

Try This

Compare the graph of $y = x^3$ to the graph of each function.

1. $y = x^3 - 2$
2. $y = (x - 7)^3$
3. $y = \left(\frac{1}{2}\right)x^3$
4. $y = 5 - x^3$

12-8 Inverse Variation

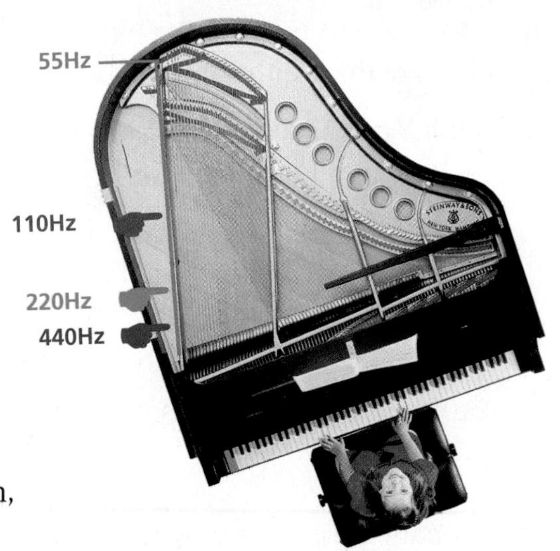

Learn to recognize inverse variation by graphing tables of data.

Vocabulary

inverse variation

The frequency of a piano string is related to its length. You can double a string's frequency by placing your finger at the halfway point of the string. The lowest note on the piano is A_1. As you place your finger at various fractions of the string's length, the frequency will *vary inversely*.

Full length: **55 Hz** $\frac{1}{2}$ the length: **110 Hz** $\frac{1}{4}$ the length: **220 Hz**

The fraction of the string length times the frequency is always 55.

INVERSE VARIATION

Words	Numbers	Algebra
An **inverse variation** is a relationship in which one variable quantity increases as another variable quantity decreases. The product of the variables is a constant.	$y = \frac{120}{x}$ $xy = 120$	$y = \frac{k}{x}$ $xy = k$

EXAMPLE 1 Identifying Inverse Variation

Tell whether each relationship is an inverse variation.

A **The table shows the number of days needed to construct a building based on the size of the work crew.**

Crew Size	2	3	5	10	20
Days of Construction	90	60	36	18	9

$20(9) = 180$; $10(18) = 180$; $5(36) = 180$; $3(60) = 180$; $2(90) = 180$

$xy = 180$ *The product is always the same.*

The relationship is an inverse variation: $y = \frac{180}{x}$.

B **The table shows the number of chips produced in a given time.**

Chips Produced	36	60	84	108	120	144
Time (min)	3	5	7	9	10	12

$36(3) = 108$; $60(5) = 300$ *The product is not always the same.*

The relationship is not an inverse variation.

Helpful Hint

To determine if a relationship is an inverse variation, check if the product of *x* and *y* is always the same number.

In the inverse variation relationship $y = \frac{k}{x}$, where $k \neq 0$, y is a function of x. The function is not defined for $x = 0$, so the domain is all real numbers except 0.

EXAMPLE 2 Graphing Inverse Variations

Graph each inverse variation function.

A $f(x) = \frac{1}{x}$

x	y
-3	$-\frac{1}{3}$
-2	$-\frac{1}{2}$
-1	-1
$-\frac{1}{2}$	-2
$\frac{1}{2}$	2
1	1
2	$\frac{1}{2}$
3	$\frac{1}{3}$

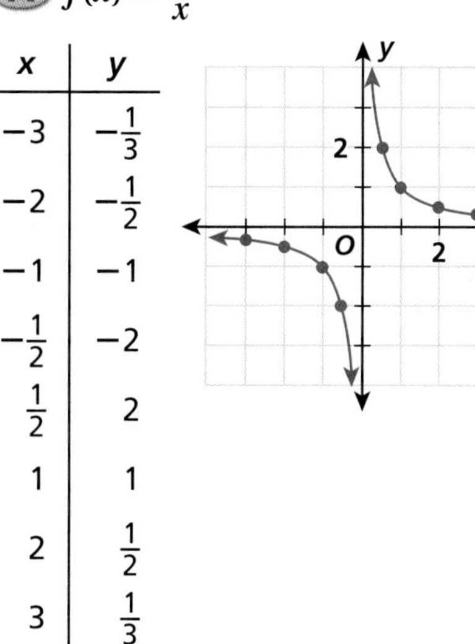

B $f(x) = \frac{-2}{x}$

x	y
-3	$\frac{2}{3}$
-2	1
-1	2
$-\frac{1}{2}$	4
$\frac{1}{2}$	-4
1	-2
2	-1
3	$-\frac{2}{3}$

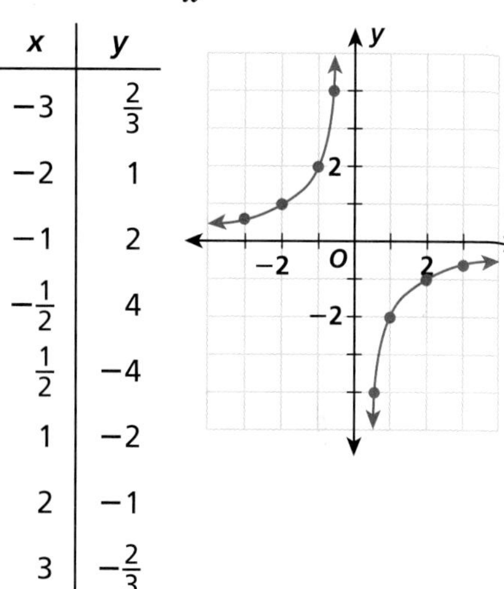

EXAMPLE 3 *Music Application*

The frequency of a piano string changes according to the fraction of its length that is allowed to vibrate. Find the inverse variation function, and use it to find the resulting frequency when $\frac{1}{16}$ of the string A_1 is allowed to vibrate.

Frequency of A_1 by Fraction of the Original String Length				
Frequency (Hz)	55	110	220	440
Fraction of the Length	1	$\frac{1}{2}$	$\frac{1}{4}$	$\frac{1}{8}$

You can see from the table that $xy = 55(1) = 55$, so $y = \frac{55}{x}$.
If the string is reduced to $\frac{1}{16}$ of its length, then its frequency will be $y = 55 \div \left(\frac{1}{16}\right) = 16 \cdot 55 = 880$ Hz.

Think and Discuss

1. **Identify** k in the inverse variation $y = \frac{3}{x}$.
2. **Describe** how you know if a relationship is an inverse variation.

12-8 Exercises

FOR EOG PRACTICE
see page 691

internet connect
Homework Help Online
go.hrw.com Keyword: MT4 12-8

5.01a

GUIDED PRACTICE

See Example 1 **Tell whether each relationship is an inverse variation.**

1. The table shows the number of CDs produced in a given time.

CDs Produced	45	120	135	165	210
Time (min)	3	8	9	11	14

2. The table shows the construction time of a wall based on the number of workers.

Construction Time (hr)	5	9	15	22.5	45
Number of Workers	9	5	3	2	1

See Example 2 **Graph each inverse variation function.**

3. $f(x) = \frac{3}{x}$
4. $f(x) = \frac{2}{x}$
5. $f(x) = \frac{1}{2x}$

See Example 3

6. Ohm's law relates the current in a circuit to the resistance. Find the inverse variation function, and use it to find the current in a 12-volt circuit with 9 ohms of resistance.

Current (amps)	0.25	0.5	1	2	4
Resistance (ohms)	48	24	12	6	3

INDEPENDENT PRACTICE

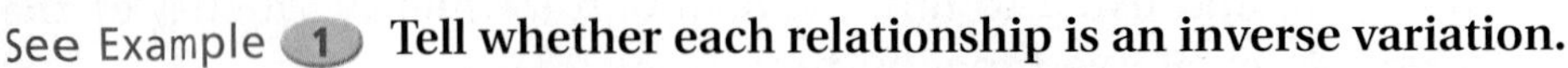

See Example 1 **Tell whether each relationship is an inverse variation.**

7. The table shows the time it takes to throw a baseball from home plate to first base depending on the speed of the throw.

Speed of Throw (ft/s)	30	36	45	60	90
Time (s)	3	2.5	2	1.5	1

8. The table shows the number of miles jogged in a given time.

Miles Jogged	1	1.5	3	4	5
Time (min)	8	12	24	32	40

See Example 2 **Graph each inverse variation function.**

9. $f(x) = -\frac{1}{x}$
10. $f(x) = \frac{1}{3x}$
11. $f(x) = -\frac{1}{2x}$

See Example 3

12. According to Boyle's law, when the volume of a gas decreases, the pressure increases. Find the inverse variation function, and use it to find the pressure of the gas if the volume is decreased to 4 liters.

Volume (L)	8	10	20	40	80
Pressure (atm)	5	4	2	1	0.5

PRACTICE AND PROBLEM SOLVING

Find the inverse variation equation, given that x and y vary inversely.

13. $y = 2$ when $x = 2$ **14.** $y = 10$ when $x = 2$ **15.** $y = 8$ when $x = 4$

16. If y varies inversely with x and $y = 27$ when $x = 3$, find the constant of variation.

17. The height of a triangle with area 50 cm^2 varies inversely with the length of its base. If $b = 25$ cm when $h = 4$ cm, find b when $h = 10$ cm.

18. ***PHYSICAL SCIENCE*** If a constant force of 30 N is applied to an object, the mass of the object varies inversely with its acceleration. The table contains data for several objects of different sizes.

Mass (kg)	3	6	30	10	5
Acceleration (m/s^2)	10	5	1	3	6

a. Use the table to write an inverse variation function.

b. What is the mass of an object if its acceleration is 15 m/s^2?

19. ***FINANCE*** Mr. Anderson wants to earn $125 in interest over a 2-year period from a savings account. The principal he must deposit varies inversely with the interest rate of the account. If the interest rate is 6.25%, he must deposit $1000. If the interest rate is 5%, how much must he deposit?

20. ***WRITE ABOUT IT*** Explain the difference between direct variation and inverse variation.

21. ***WRITE A PROBLEM*** Write a problem that can be solved using inverse variation. Use facts and formulas from your science book.

22. ***CHALLENGE*** The resistance of a 100 ft piece of wire varies inversely with the square of its diameter. If the diameter of the wire is 3 in., it has a resistance of 3 ohms. What is the resistance of a wire with a diameter of 1 in.?

Spiral Review

For each function, find $f(-1)$, $f(0)$, and $f(1)$. (Lesson 12-4)

23. $f(x) = 3x^2 - 5x + 1$ **24.** $f(x) = x^2 + 15x - 4$ **25.** $f(x) = 3(x - 9)^2$

26. $f(x) = 2x^3 - 6x - 2$ **27.** $f(x) = (x - 5)(x + 7)$ **28.** $f(x) = -144x^2 - 64x$

29. **EOG PREP** The half-life of a particular radioactive isotope of thorium is 8 minutes. If 160 grams of the isotope are initially present, how many grams will remain after 40 minutes? (Lesson 12-6)

A 10 grams B 2.5 grams C 5 grams D 1.25 grams

Problem Solving on Location

NORTH CAROLINA

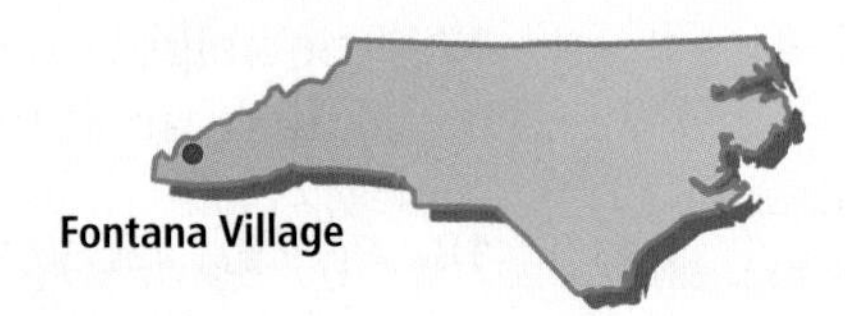

Tree Farms

The North Carolina landscape is dotted with hundreds of commercial tree farms. Choosing the type of tree to grow, buying the seeds, and growing the seedlings are the first steps to starting a tree farm. Important factors tree farmers must consider when determining how many pounds of seed to buy are the percent of the seeds in each bag that will grow into seedlings, g; the number of seeds that will be planted per square foot, s; and the number of seeds per pound, n.

Tree farmers use the formula $p = \frac{a \cdot s}{g \cdot n}$ to determine how many pounds of seed to buy in order to plant an area of a square feet. For 1–3, use the formula and the table.

1. For each type of tree in the table, write and graph the function that determines how many pounds of seed p must be bought to plant an area of a square feet. Are the functions linear or quadratic?

Seed Data by Tree Type

Tree Type	Percent of Seeds That Will Grow, g	Number of Seeds per ft^2, s	Number of Seeds per lb, n
Fraser fir	48%	40	55,000
Red cedar	35%	35	43,600
White pine	40%	25	26,000

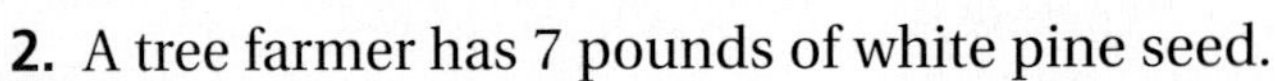

2. A tree farmer has 7 pounds of white pine seed.
   a. What function would tell the farmer how many additional pounds of seed p should be purchased to plant an area of a square feet?
   b. The farmer plans to plant 8 beds of seeds. Each bed will be 4 ft wide and 143 ft long. How many pounds of seed must the farmer buy?

3. Suppose a tree farmer is planting a square-shaped area with side length x. Rewrite your functions from problem **1** in terms of x. Graph each function. Are your new functions linear or quadratic?

Fontana Dam

In 1942, Fontana Village, North Carolina, was created to house the more than 5000 construction workers hired to build Fontana Dam. The dam stands 480 feet tall, making it the tallest dam east of the Rocky Mountains. When Fontana Lake's water level gets too low or when the water levels of the rivers that flow into the dam get too high, the dam's floodgates are opened to release water.

For 1–2, use the table.

1. What kind of sequence is formed by the total amount of water released after each second at 9 A.M. from Fonatana Dam? What is a possible rule for this sequence?

2. Assume that by exactly 10 A.M., 397,800 ft^3 of water had been released from Fontana Dam. The floodgates were then reopened. The table shows the amount of water released each second after 10 A.M. Write a possible rule for the sequence formed by the total amount of water that had been released from Fontana Dam starting at 10 A.M. If the pattern continues, what will the total amount of water released be 6 seconds after 10 A.M.?

Water Release at Fontana Dam

Number of Seconds	Total Amount of Water Released (ft^3)	
	9 A.M.	10 A.M.
1	6630	8540
2	13,260	17,080
3	19,890	25,620
4	26,520	34,160

3. Suppose water was released from Fontana Dam at 3 A.M. at a rate of 1000 cubic feet per second. At 4 A.M. water was released at a rate of 1200 cubic feet per second, at 5 A.M. water was released at a rate of 1440 cubic feet per second, and at 6 A.M. water was released at a rate of 1728 cubic feet per second.
   a. What kind of sequence do the rates of water release at each hour appear to form?
   b. Write a possible rule for the sequence.
   c. If the pattern continues, at what rate would you expect water to be released at 11 A.M.? Round your answer to the nearest whole number.

MATH-ABLES

Squared Away

How many squares can you find in the figure at right?

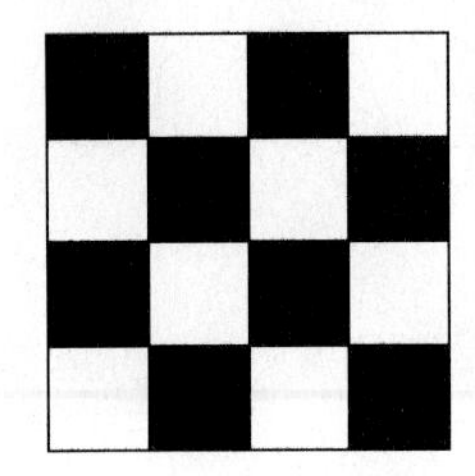

Did you find 30 squares?

There are four different-sized squares in the figure.

Size of Square	Number of Squares
4 × 4	1
3 × 3	4
2 × 2	9
1 × 1	16
	Total 30

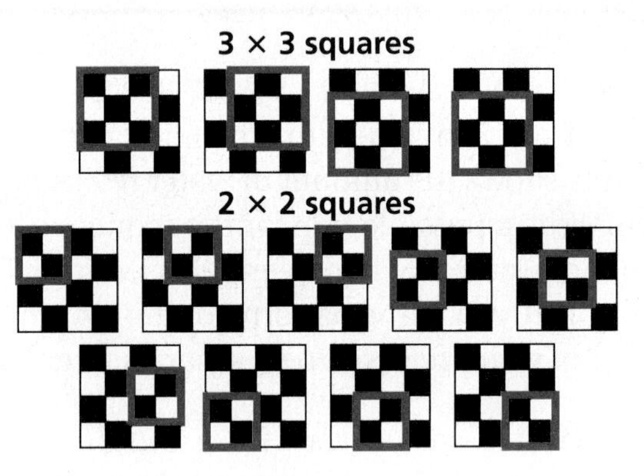

So the total number of squares is $1 + 4 + 9 + 16 = 1^2 + 2^2 + 3^2 + 4^2$.

Draw a 5 × 5 grid and count the number of squares of each size. Can you see a pattern?

What is the total number of squares on a 6 × 6 grid? a 7 × 7 grid? Can you come up with a general formula for the sum of squares on an $n \times n$ grid?

What's Your Function?

One member from the first of two teams draws a function card from the deck, and the other team tries to guess the rule of the function. The guessing team gives a function input, and the card holder must give the corresponding output. Points are awarded based on the type of function and number of inputs required. The first team to reach 20 points wins.

internet connect

Go to ***go.hrw.com*** for a complete set of rules and game cards.
KEYWORD: MT4 Game12

Generate Arithmetic and Geometric Sequences

Use with Lesson 12-2

Graphing calculators can be used to explore arithmetic and geometric sequences.

Activity

1 The command **seq(** is used to generate a sequence.

a. Press 2nd LIST STAT **OPS 5:seq.**

```
seq(
```

The **seq(** command is followed by the rule for generating the sequence, the variable used in the rule, and the positions of the first and last terms in the sequence. To find the first 20 terms of the arithmetic sequence generated by the rule **5 + (*x* − 1) · 3**, enter **seq(5 + (*x* − 1) · 3, *x*, 1, 20)**:

5 + (X,T,θ,n − 1) ×

3 , X,T,θ,n , 1 , 20 ENTER

```
seq(5+(X-1)*3,X,
1,20
{5 8 11 14 17 2…
```

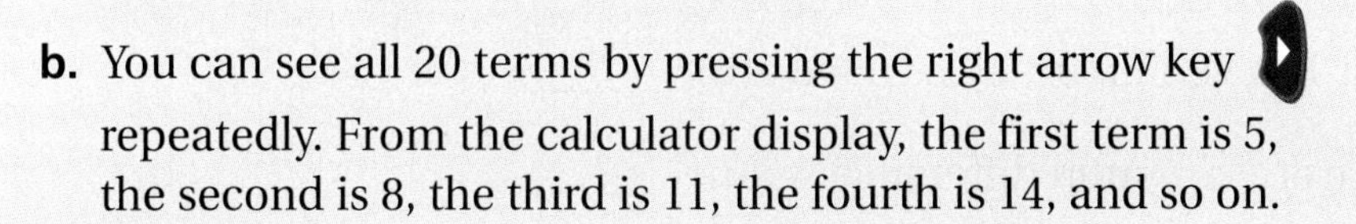

b. You can see all 20 terms by pressing the right arrow key repeatedly. From the calculator display, the first term is 5, the second is 8, the third is 11, the fourth is 14, and so on.

```
seq(5+(X-1)*3,X,
1,20
…17 20 23 26 29…
```

2 Consider the *geometric* sequence whose *n*th term is $3\left(\frac{1}{4}\right)^{n-1}$. To use a graphing calculator to find the first 15 terms in fraction form, press

2nd LIST STAT ▸ **5:seq** 3 × (1 ÷ 4) ^

(X,T,θ,n − 1) , X,T,θ,n , 1 , 15 MATH **1:Frac** ENTER .

```
seq(3*(1/4)^(X-1
),X,1,15▸Frac
{3 3/4 3/16 3/6…
```

To see all 15 terms, press the right arrow key repeatedly.

Think and Discuss

1. Why is the seventh term of the sequence in **2** *not* displayed as a fraction?

Try This

Find the first 15 terms of each sequence. Tell if the consecutive terms increase or decrease.

1. $-4 + (n - 1) \cdot 7$ **2.** $2\left(\frac{1}{5}\right)^{n-1}$ **3.** 9, 14, 19, 24, . . . **4.** $2, \frac{2}{3}, \frac{2}{9}, \frac{2}{27}, \ldots$

Chapter 12

Study Guide and Review

Vocabulary

Complete the sentences below with vocabulary words from the list above. Words may be used more than once.

1. A list of numbers or terms in a certain order is called a(n) ___?___.

2. A sequence in which there is a common difference is a(n) ___?___; a sequence in which there is a common ratio is a(n) ___?___.

3. A famous sequence in which you add the two previous terms to find the next term is the ___?___.

4. A rule that relates two quantities so that each input value corresponds to exactly one output value is a(n) ___?___. The set of all input values is the ___?___; the set of output values is the ___?___.

12-1 Arithmetic Sequences (pp. 590–594)

EXAMPLE

- **Find the 10th term of the arithmetic sequence: 12, 10, 8, 6, ...**

$d = 10 - 12 = -2$
$a_n = a_1 + (n - 1)d$
$a_{10} = 12 + (10 - 1)(-2)$
$a_{10} = 12 - 18$
$a_{10} = -6$

EXERCISES

Find the given term in each arithmetic sequence.

5. 8th term: 3, 7, 11, . . .

6. 7th term: 0.05, 0.15, 0.25, . . .

7. 9th term: $\frac{2}{3}, \frac{7}{6}, \frac{5}{3}, \ldots$

Study Guide and Review

12-2 Geometric Sequences (pp. 595–599)

EXAMPLE

- **Find the 10th term of the geometric sequence: 6, 12, 24, 48, ...**

$r = \frac{12}{6} = 2$

$a_n = a_1 r^{n-1}$

$a_{10} = 6(2)^{10-1} = 3072$

EXERCISES

Find the given term in each geometric sequence.

8. 8th term: 5, −10, 20, −40, . . .

9. 7th term: $\frac{1}{2}, \frac{1}{3}, \frac{2}{9}, \ldots$

10. 50th term: 1, −1, 1, −1, . . .

12-3 Other Sequences (pp. 601–605)

EXAMPLE

- **Find the first four terms of the sequence defined by $a_n = -2(-1)^{n-1} - 1$.**

$a_1 = -2(-1)^{1-1} - 1 = -3$
$a_2 = -2(-1)^{2-1} - 1 = 1$
$a_3 = -2(-1)^{3-1} - 1 = -3$
$a_4 = -2(-1)^{4-1} - 1 = 1$
The first four terms are −3, 1, −3, 1.

EXERCISES

Find the first four terms of the sequence defined by each rule.

11. $a_n = 3n + 1$

12. $a_n = n^2 + 1$

13. $a_n = 8(-1)^n + 2n$

14. $a_n = n! + 2$

12-4 Functions (pp. 608–612)

EXAMPLE

- **For the function $f(x) = 3x^2 + 4$, find $f(0)$, $f(3)$, and $f(-2)$.**

$f(0) = 3(0)^2 + 4 = 4$
$f(3) = 3(3)^2 + 4 = 31$
$f(-2) = 3(-2)^2 + 4 = 16$

EXERCISES

For each function, find $f(0)$, $f(2)$, and $f(-1)$.

15. $f(x) = 7x - 4$

16. $f(x) = 2x^3 + 1$

17. $f(x) = -x^2 + 3x$

18. $f(x) = -x^3 + 2x^2$

19. $f(x) = 3x^2 - x + 5$

20. $f(x) = -x^2 + x + 1$

12-5 Linear Functions (pp. 613–616)

EXAMPLE

- **Use the table to write the equation for the linear function.**

x	y
−2	−10
−1	−3
0	4
1	11

The y-intercept is $f(0) = 4$.
$f(x) = mx + 4$ $f(x) = mx + b$
Substitute and solve for m.
$11 = m(1) + 4$ $(x, y) = (1, 11)$
$m = 7$
$f(x) = 7x + 4$

EXERCISES

Write the equation for each linear function.

21.

x	y
−2	−3
−1	−2
0	−1
1	0

22.

x	y
−4	2
−2	3
0	4
2	5

Chapter 12

Performance Assessment

Show What You Know

Create a portfolio of your work from this chapter. Complete this page and include it with your four best pieces of work from Chapter 12. Choose from your homework or lab assignments, mid-chapter quiz, or any journal entries you have done. Put them together using any design you want. Make your portfolio represent what you consider your best work.

Short Response

1. Write out the next three terms of the sequence

   $\sqrt{2}, \sqrt{2+\sqrt{2}}, \sqrt{2+\sqrt{2+\sqrt{2}}}, \sqrt{2+\sqrt{2+\sqrt{2+\sqrt{2}}}}, \ldots$

   Use your calculator to evaluate each term of the sequence. Describe what seems to be happening to the terms of the sequence.

2. A basketball player throws a basketball in a path defined by the function $f(x) = -16x^2 + 20x + 7$, where x is the time in seconds and $f(x)$ is the height in feet. Graph the function, and estimate how long it would take the basketball to reach its maximum height.

3. When playing the trombone, you produce different notes by changing the effective length of the tube by moving it in and out. This movement produces a sequence of lengths that form a geometric sequence. If the length is 119.3 inches in the 2nd position and 134.0 inches in the 4th position, what is the length in the 3rd position? Write a rule that would describe this relationship.

Extended Problem Solving

4. Consider the sequence 1, 2, 6, 24, 120, 720, …

   a. Determine whether the sequence is arithmetic, geometric, or neither.

   b. Find the ratio of each pair of consecutive terms. What pattern do you notice?

   c. Write a rule for the sequence. Use your rule to find the next two terms.

Performance Assessment

internet connect
EOG Test Practice Online
go.hrw.com Keyword: MT4 TestPrep

Getting Ready for EOG

Chapter 12

Cumulative Assessment, Chapters 1–12

1. What is the next term in the sequence 1, 2, 4, 7, 11, . . . ?

   A 13
   B 14
   C 15
   D 16

2. A sequence is formed by doubling the preceding number: 2, 4, 8, 16, 32, ... What is the remainder when the 15th term of the sequence is divided by 6?

   A 0
   B 1
   C 2
   D 4

3. Which equation describes the relationship shown in the graph?

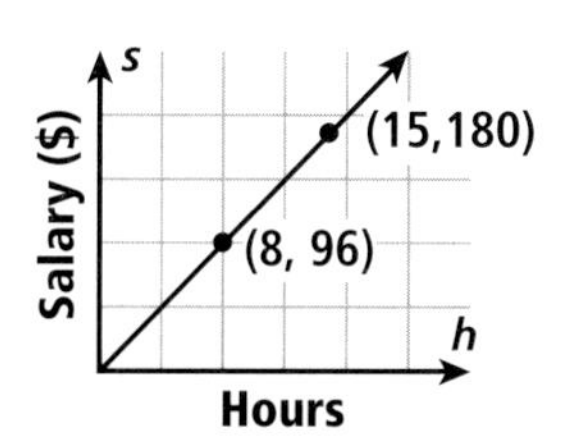

   A $h = 12s$
   B $s = 12h$
   C $h = s + 88$
   D $s = h + 88$

4. Which of the following is a solution of the system shown?

   $x > 3$
   $x + y < 2$

   A (4, −1)
   B (4, −3)
   C (5, 1)
   D (−5, 4)

5. If $r = \frac{t}{5}$ and $10r = 32$, find the value of t.

   A 64
   B 32
   C 16
   D 8

6. If $a = 3$ and $b = 4$, evaluate $b - ab^a$.

   A 64
   B 8
   C −188
   D −1724

7. In parallelogram $JKLM$, $\overline{KP}$ is perpendicular to diagonal $\overline{JL}$. Which of the following is *true?*

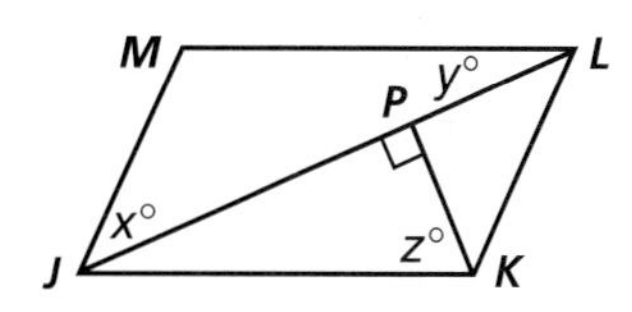

   A $x + y + z = 180$
   B $x + z = 90$
   C $y + z = 90$
   D $x + y = 90$

TIP! TEST TAKING TIP!

Reworking the given choices: It is sometimes useful to look at a choice in a form different from the given form.

8. If $2^{3x-1} = 8$, then what is the value of x?

   A $\frac{2}{3}$
   B 2
   C $1\frac{1}{3}$
   D $2\frac{1}{3}$

9. ***SHORT RESPONSE*** The length of a rectangle is 8 ft less than twice its width w. Draw a diagram of the rectangle, and label each side length. What is the perimeter of the rectangle expressed in terms of w?

10. ***SHORT RESPONSE*** If two different numbers are selected at random from the set {1, 2, 3, 4, 5, 6}, what is the probability that their product will be 12? Show your work or explain in words how you determined your answer.

Student Handbook

Preview Skills

Science Skills

Exponent

Base → 2^4

EOG Practice ■ Chapter 1

1A Equations and Inequalities

LESSON 1-1

1. What is the value of $2 + x$ when $x = 7$?

   A 2   C 9

   B 7   D 14

2. What is the value of $4m - 3$ when $m = 2$?

   A 5   C 11

   B 8   D 39

3. What is the value of $3x + y$ when $x = 2$ and $y = 4$?

   A 8   C 14

   B 10   D 16

4. What is the value of $2y + 5.7x$ when $x = 2$ and $y = 1$?

   A 7.7   C 13.4

   B 9.7   D 15.4

5. ***SHORT RESPONSE*** Evaluate $2(p + 3) - 7$ for $p = 8$. Show your work and use the order of operations to explain each step.

LESSON 1-2

6. Which expression is described by the word phrase "7 less than a number b"?

   A $b - 7$ C $b + 7$

   B $7 + b$ D $7 - b$

7. Which expression is described by the word phrase "8 more than the product of 7 and a number x"?

   A $8(7 + x)$ C $8 + 7 + x$

   B $8 + 7x$ D $7(8 + x)$

8. Which word phrase correctly describes the expression $5(c + 18)$?

   A five times c plus 18

   B five times the sum of c and 18

   C c plus 18 times 5

   D the product of 5 and c plus 18

9. Which word phrase correctly describes the expression $8 \div m$?

   A a quotient of a number m and 8

   B a number m divided by 8

   C a quotient of 8 and a number m

   D 8 divided into a number m

10. ***SHORT RESPONSE*** The formula $1.8C + 32$ will convert a temperature in degrees Celsius (°C) to degrees Fahrenheit (°F), where C = degrees Celsius. Convert 28°C to degrees Fahrenheit. Explain each step of your answer.

LESSON 1-3

11. The equation $8 + x = 13$ *cannot* be used to answer which question?

    A The sum of 8 hours and what number of hours is 13 hours?

    B How many hours longer than 8 hours is 13 hours?

    C Eight hours is less than 13 hours by what number of hours?

    D Thirteen hours is the difference between 8 hours and what number of hours?

12. ***SHORT RESPONSE*** Nathan is writing a short story that needs to be 750 words in length. He has already written 687 words. How many more words w does Nathan need to write? Show your work.

13. ***SHORT RESPONSE*** Evelyn bought 3 dozen eggs. She used e eggs to make custard for her class and had 21 eggs left. Write an equation that could be used to find the number of eggs Evelyn used. How many eggs did Evelyn use?

SHORT RESPONSE **Solve each equation. Show your work.**

14. $t - 3 = 8$ 15. $17 = m + 11$

16. $g - 16 = 23$ 17. $31 + y = 50$

EOG Practice ■ Chapter 1

1A Equations and Inequalities

LESSON 1-4

18. The equation $\frac{j}{3} = 12$ can be used to answer which question?

A A gift bag holds 3 jars of jelly. How many bags are needed to hold 12 jars?

B Three less than how many jars of jelly is 12 jars of jelly?

C How many jars of jelly can be divided into 3 equal groups of 12?

D Bo divided 12 jars of jelly into 3 equal groups. How many jars are in each group?

19. For which equation is $x = 3$ *not* the solution?

A $8x = 24$

B $\frac{x}{10} = 30$

C $x + 11 = 14$

D $2x - 3 = 3$

20. Which value of m is the solution to $4m - 7 = 17$?

A $m = 2.5$ C $m = 10$

B $m = 6$ D $m = 24$

21. ***SHORT RESPONSE*** Four friends split the cost of a pizza equally. Each friend paid \$4.17. Write an equation that could be used to determine the price of the pizza, and then find the price of the pizza.

22. ***SHORT RESPONSE*** Chaya took care of her neighbor's dog for 11 days and earned \$88. How much was Chaya paid per day? Show your work or explain in words how you determined your answer.

LESSON  1-5

23. Which comparison symbol does *not* make the statement $15 - 8$ ▒ 6 true?

A $>$ C $\leq$

B $\geq$ D $\neq$

24. Which comparison symbol makes the statement $3(7)$ ▒ 23 true?

A $>$ C $<$

B $\geq$ D $=$

25. Which inequality is graphed below?

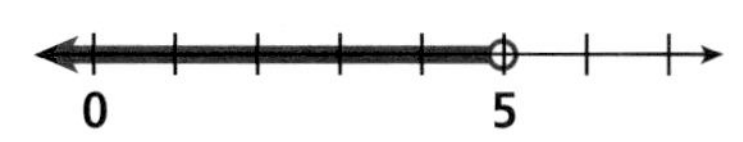

A $x > 5$ C $x < 5$

B $x \geq 5$ D $x \leq 5$

26. The solution of which inequality is graphed below?

A $t + 7 > 11$ C $7t < 28$

B $\frac{t}{4} \geq 1$ D $t - 2 \leq 2$

SHORT RESPONSE **Solve each inequality and graph your solution on a number line.**

27. $x - 3.5 \geq 7$ **28.** $5p < 40$

29. $2 \leq \frac{a}{3}$ **30.** $h - 5 \leq 13$

LESSON 1-6

31. Which algebraic expression is equivalent to $6k + 17$?

A $k + 8 + 5k + 7$

B $4k + 19 - 2 - 2k$

C $11k - 2k + 9 - 3k + 8$

D $5 + 6k - k + 13 + 2k$

32. Which algebraic expression is equivalent to $14w + 3 + z + 7 - 2w$?

A $23wz$ C $12w + z + 10$

B $13wz + 10$ D $12(w + z) + 10$

SHORT RESPONSE **Solve each equation. Show your work.**

33. $8m + 3m - 7m = 36$

34. $6b - 4b + 5b = 56$

35. $7 + 9t + 15 = 31$

36. $3x + 5 + x = 13$

EOG Practice ▪ Chapter 1

1B Graphing

LESSON 1-7

1. Which ordered pair is *not* a solution of $3x + 5y = 25$?

   A (0, 5)

   B (4, 3)

   C (5, 2)

   D (8, 0.2)

2. Which ordered pair is a solution of $y = 12x - 13$?

   A (1, 1)

   B (3, 49)

   C (5, 47)

   D (11, 2)

3. For every bag of pasta Amelia buys, she buys 3 jars of pasta sauce. The equation that can be used to find the number of jars of sauce s Amelia should purchase is $s = 3p$, where p is the number of bags of pasta she will purchase. Which ordered pair represents an accurate combination of bags of pasta and jars of sauce?

   A (0, 3) C (5, 8)

   B (2, 6) D (9, 3)

4. Nico puts $5 in his savings account each time he gets paid for babysitting his younger sister. The equation that can be used to find out how much money m Nico can spend is $m = b - 5$, where b is the amount of money he made babysitting. Which ordered pair does *not* represent an accurate combination of babysitting pay and spending money?

   A (5, 1) C (11, 6)

   B (8, 3) D (20, 15)

5. ***SHORT RESPONSE*** Create a table of ordered pair solutions to the equation $y = x - 3$ for $x = 0, 1, 2,$ and 3.

6. ***SHORT RESPONSE*** Create a table of ordered pair solutions to the equation $y = 3x - 1$ for $x = 0, 1, 2,$ and 3.

LESSON 1-8

7. Which coordinates *best* represent the point graphed on the coordinate plane below?

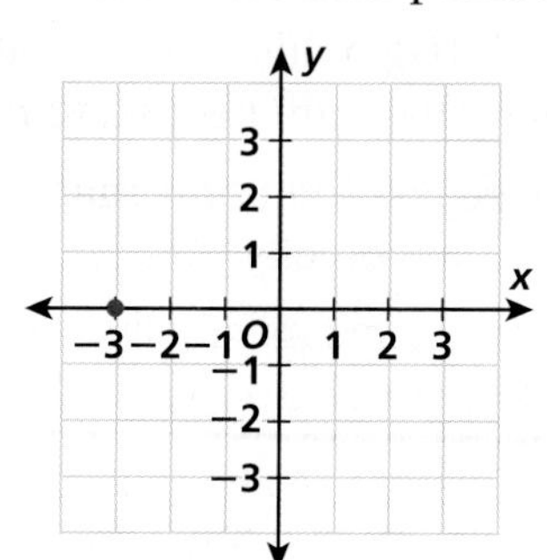

   A (−3, 0) C (0, 3)

   B (0, −3) D (3, 0)

8. For which equation does the table contain ordered pair solutions?

x	?	y	(x, y)
1	3(1) − 2	1	(1,1)
2	3(2) − 2	4	(2,4)
3	3(3) − 2	7	(3,7)
4	3(4) − 2	10	(4,10)

   A $y = 3x$ C $y = 2x - 3$

   B $y = x - 2$ D $y = 3x - 2$

SHORT RESPONSE **Graph each point on a coordinate plane.**

9. (4, 3)
10. (3, 0)
11. (−1, 3)
12. (−3, −1)

13. ***SHORT RESPONSE*** Copy and complete the table, and then graph the equation on a coordinate plane.

x	$7x$	y	(x, y)
1	■	■	■
3	■	■	■
5	■	■	■
7	■	■	■

1B Graphing

LESSON 1-9

14. A cross-country skier travels at a fairly steady rate, stops to rest for a minute, and then proceeds up a hill. Which table corresponds to the situation described?

A

Time	Speed (mi/h)
8:00	7
8:01	7.2
8:02	7
8:03	0
8:04	2
8:05	4

B

Time	Speed (mi/h)
8:00	0
8:01	2
8:02	4
8:03	6
8:04	0
8:05	2

C

Time	Speed (mi/h)
8:00	4
8:01	2
8:02	0
8:03	7
8:04	7
8:05	7

D

Time	Speed (mi/h)
8:00	4
8:01	0
8:02	0
8:03	1
8:04	2
8:05	4

15. A skier increases speed going down a ski jump. He lands and then slowly comes to a stop. Which graph matches the situation described?

A

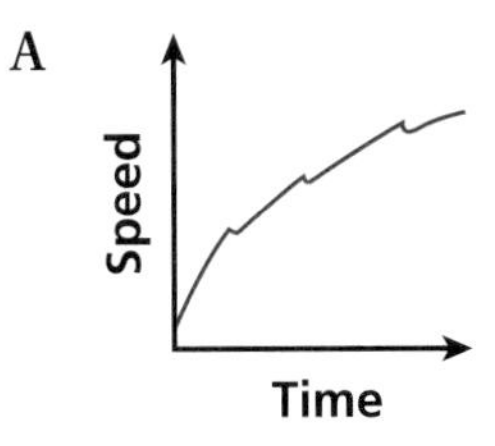

B

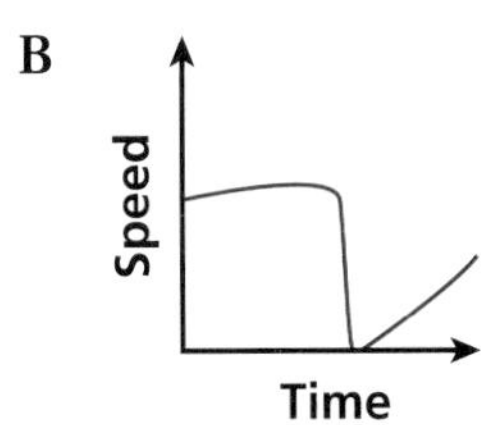

C

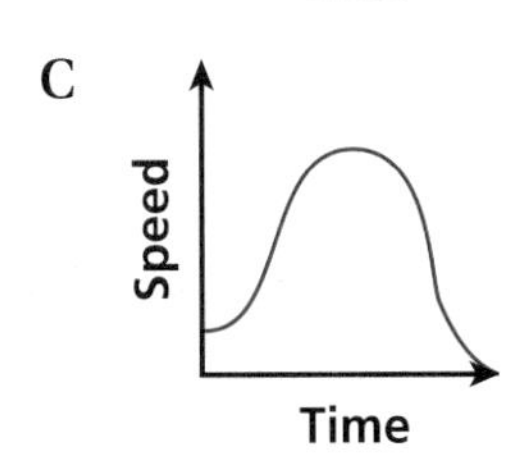

D

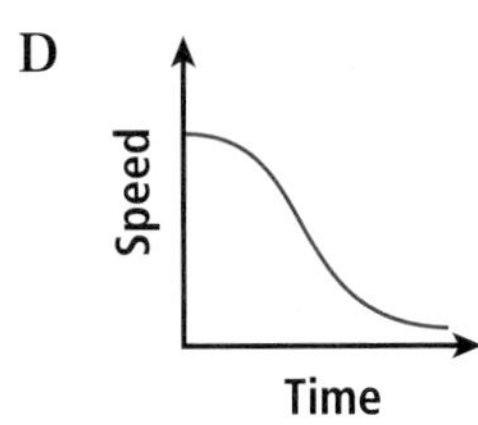

16. ***SHORT RESPONSE*** Make a graph of the information in the table. Create a story that corresponds to your graph.

Time (min)	Height (ft)
0	0
10	3000
20	6000
30	12,000
40	12,000
50	12,000
60	6000
70	3000
80	0

EOG Practice

EOG Practice ▪ Chapter 2

2A Integers

LESSON 2-1

1. Which addition expression corresponds to the number line diagram below?

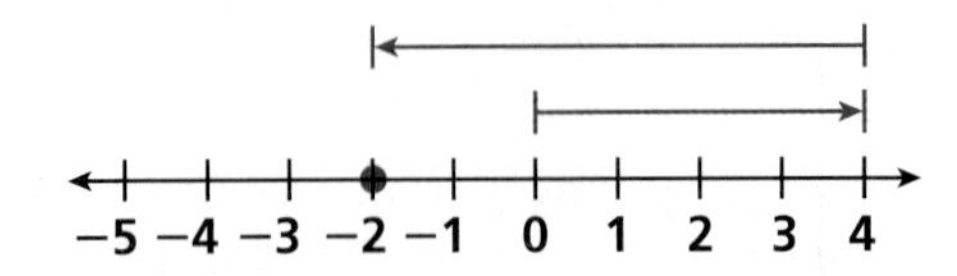

A $-4 + (-6)$ C $4 + (-6)$
B $-4 + 6$ D $4 + 6$

2. What is the value of $9 + (-5)$?

A -14 C 4
B -4 D 14

3. What is the value of $x + 2$ when $x = -3$?

A -5 C 1
B -1 D 5

4. What is the value of $-5 + x$ when $x = 2$?

A -7 C 3
B -3 D 7

5. ***SHORT RESPONSE*** Karen was adjusting the position of a painting on her wall. From its original position she moved it 7 inches to the left. Then she moved it 5 inches to the right, followed by another 3 inches to the left. Write an addition expression to determine the new position of the painting relative to its original position. Draw a number line diagram to evaluate your expression.

LESSON 2-2

6. Which subtraction expression corresponds to the number line diagram below?

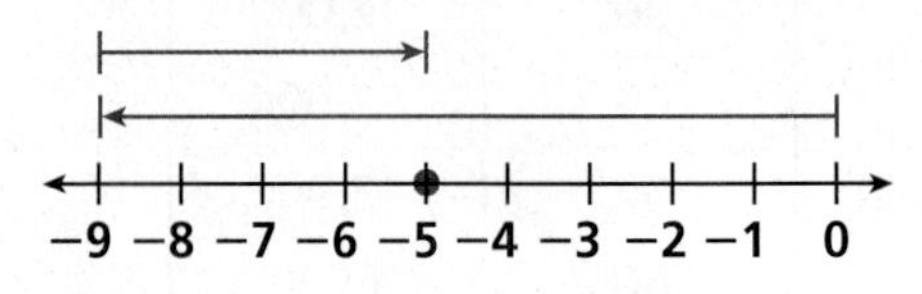

A $-5 - 9$ C $-9 - (-4)$
B $-5 - (-9)$ D $-9 - 4$

7. What is the value of $4 - (-1)$?

A -5 C 3
B -3 D 5

8. An elevator rises to a height of 281 feet above ground level. Then the elevator drops 314 feet to the basement of the building. Which subtraction expression correctly describes the position of the elevator relative to ground level?

A $-281 - 314$ C $281 - 314$
B $-281 - (-314)$ D $281 - (-314)$

SHORT RESPONSE **Evaluate each expression for the given value of the variable. Show your work.**

9. $4 - x$ for $x = -7$
10. $-7 - s$ for $s = -5$
11. $-5 - b$ for $b = 9$
12. $12 - y$ for $y = -8$

LESSON 2-3

13. What is the value of $5(-8)$?

A -40 C 3
B -3 D 40

14. What is the value of $\frac{-16}{-4}$?

A -20 C -4
B -24 D 4

15. What is the value of $-5x$ when $x = -7$?

A -35 C 2
B -12 D 35

16. What is the value of $\frac{24}{y}$ when $y = -3$?

A -21 C 8
B -8 D 21

SHORT RESPONSE **Simplify each expression. Show your work, and use the order of operations to explain each step.**

17. $(-5)(-11)(-4)$
18. $\frac{-3 - (-5)}{-4}$
19. $\frac{(-4 - 16)}{5}$
20. $2 + (-6)(-3)$

2A Integers

LESSON  2-4

21. The equation $2x = -12$ can be used to find the answer to which question?

  A The temperature fell 12°F today. This is $\frac{1}{2}$ of the total temperature change over the past 4 days. What is the total temperature change over the past 4 days?

  B The temperature is falling by 2°F every hour. How many hours will it take the temperature to fall a total of 12°F?

  C The temperature fell 12°F at a constant rate, over a period of 2 hours. What was the temperature change each hour?

  D. If the temperature is 2°F, what would the temperature change have to be for the temperature to become −12°F?

22. For which equation is $x = -4$ *not* the solution?

  A $x + 14 = 10$ C $-3x = -12$

  B $\frac{x}{-4} = 1$ D $-2x - 12 = -4$

23. If $\frac{c}{-2} = -8$, what is the value of $c - (-5)$?

  A −21 C 11

  B −11 D 21

24. If $s + 8 = 13$, what is the value of $-2s$?

  A −42 C 10

  B −10 D 42

25. ***SHORT RESPONSE*** Cyrus bought 12 pairs of socks. Each time he does the laundry, the number of socks decreases by 3. If Cyrus does the laundry once a week, how many weeks will it take for all of his socks to disappear? Explain how you found your answer.

SHORT RESPONSE **Solve each equation. Show your work.**

26. $\frac{s}{6} = -3$ 27. $-9a = 54$

28. $3k + 29 = 20$ 29. $\frac{r}{-4} + 5 = 3$

LESSON 2-5

30. Which comparison symbol makes the statement $-8(-4)$ ▪ 32 true?

  A > C <

  B = D ≠

31. Which comparison symbol does *not* make the statement −14 ▪ −11 − 4 true?

  A > C <

  B ≥ D ≠

32. The solution of which inequality is graphed below?

−14 −12 −10 −8 −6

  A $x + 9 > 3$ C $6x < 2$

  B $\frac{x}{-2} < 6$ D $x - (-7) \leq 5$

33. Which inequality does *not* have the solution graphed below?

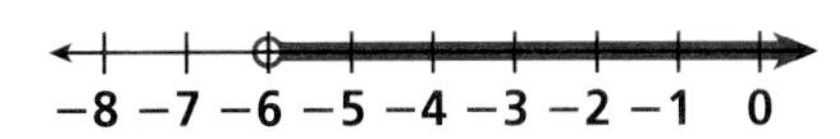

  A $m - 3 > -9$ C $4m > -24$

  B $\frac{m}{-3} > 2$ D $-1 < m + 5$

34. If $-5b \leq 70$, which of the following is a true statement?

  A $b - 9 \geq -23$ C $b + 11 \geq -25$

  B $\frac{b}{-2} \geq 7$ D $5b \leq -70$

35. If $g - 13 > -3$, which of the following is *not* a true statement?

  A $-2g < -20$ C $g + 6 > 16$

  B $\frac{g}{-5} > -2$ D $g - 23 > -13$

36. ***SHORT RESPONSE*** Ben is building a diving tank with an 11-foot-tall diving board. The distance between the top of the diving board and the bottom of the tank will be at least 35 feet. Write an inequality that could be used to determine possible depths for the tank. Solve your inequality. Graph your solution on a number line.

2B Exponents

LESSON 2-6

1. Which is another way of expressing 3^5?

   A $3 + 5$

   B 3×5

   C $5 \times 5 \times 5$

   D $3 \times 3 \times 3 \times 3 \times 3$

2. Which is *not* another way of expressing 25?

   A $(-5)^2$ C $3^2 + 4^2$

   B 2^5 D 5^2

3. What is the value of $(-2)^8$?

   A -256 C 16

   B -28 D 256

4. How would 1024 be written in exponential notation?

   A 2^8 C 6^3

   B 4^5 D 10^{24}

5. Which is equivalent to $19 + 3(2 \cdot 4^4)$?

   A 43 C 1536

   B 283 D 1555

6. ***SHORT RESPONSE*** Each morning Madison will add 3 times as much money to her money jar as she did the day before. The first day she adds 3 pennies. How many pennies will she add on the fifth day? Write the total amount she has saved by the end of the fifth day as a sum of powers expressed in dollars.

7. ***SHORT RESPONSE*** Make a table with the column headings x, x^5, and $5x$. Complete the table for $x = -3, -2, -1, 0, 1, 2,$ and 3.

SHORT RESPONSE **Simplify each expression. Show your work, and use the order of operations to explain each step.**

8. $5(-7)^3$

9. $11 + (4 - 9)^6$

10. $3(9 + 6^2)$

11. $20 + 2(2^5)$

12. $14 - 5(3^4)$

13. $17 + 2(5^3 - 7)$

LESSON 2-7

14. Which is another way of expressing $5^4 \cdot 5^3$?

    A 5^7 C 25^7

    B 5^{12} D 25^{12}

15. Which is another way of expressing $\frac{8^8}{8^4}$?

    A 1^2 C 8^2

    B 1^4 D 8^4

16. The approximate area of the city of Florence, Italy, is 10^2 square kilometers. Canada, the world's second largest country, has an approximate area of 10^7 square kilometers. How many times greater is the area of Canada than the area of Florence?

    A 10 C 10^9

    B 10^5 D 100^5

17. ***SHORT RESPONSE*** A common house spider has a mass of about 10^3 milligrams. Write an expression that could be used to determine the mass of 10^7 house spiders. Express the mass of 10^7 house spiders as one power.

18. ***SHORT RESPONSE*** Kerry measured the decibel levels of several different sounds. A soft whisper registered as 36 decibels, and a jet airplane's engines registered as 216 decibels. Express each decibel level as a power. How many times the decibel level for a soft whisper is the decibel level of a jet airplane's engines? Show your work.

SHORT RESPONSE **Write each expression as one power. Explain in words how you determined your answer.**

19. $a^8 \cdot a^5$

20. $\frac{c^6}{c^2}$

21. $\frac{9^4}{3^2}$

22. $7^5 \cdot 2^4$

23. $16\left(\frac{16^4}{16^3}\right)$

24. $\frac{(2^6 \cdot 2^3)}{2^5}$

EOG Practice ■ Chapter 2

2B Exponents

LESSON 2-8

25. Which is another way of expressing 10^{-3}?

A -1000 C $\frac{1}{1000}$

B $-\frac{1}{1000}$ D 1000

26. Which is *not* another way of expressing 5^{-4}?

A -5^4 C $\frac{1}{5^4}$

B 0.0016 D $\frac{1}{625}$

27. Which is another way of expressing $\frac{1}{243}$?

A -3^{-5} C 3^{-5}

B -3^5 D 0.243

28. Which is *not* another way of expressing $\frac{1}{256}$?

A -2^8 C $\frac{1}{2^8}$

B 2^{-8} D 4^{-4}

29. What is the value of $\frac{3^2}{3^4}$?

A 1^{-2} C $3^{\frac{1}{2}}$

B 3^{-2} D 3^6

30. What is the value of $10^5 \cdot 10^{-5}$?

A 100^{-25} C 0

B 10^{-25} D 1

31. ***SHORT RESPONSE*** The place value of each digit in base 5 is a power of 5. The number 0.0315 written in base 5 is equivalent to $0 \times 5^0 + 0 \times 5^{-1} + 1 \times 5^{-2} + 3 \times 5^{-3} + 5 \times 5^{-4} = 0.72$ in the decimal (base-ten) system. If 0.2012 is written in base 5, what is its value in base 10? Show your work.

SHORT RESPONSE **Simplify each expression. Show your work, and explain each step.**

32. $(-6)^{-4}$

33. $9^6 \cdot 9^{-8}$

34. $2^5 \cdot 2^{-2}$

35. $(-4)^{-3}$

36. $\frac{5^{-3}}{5^4}$

37. $\frac{6^4}{6^7}$

LESSON 2-9

38. How would 0.000483 be written in scientific notation?

A 0.483×10^{-3} C 48.3×10^{-5}

B 4.83×10^{-4} D 483×10^{-6}

39. How would 5,410,000,000 be written in scientific notation?

A 0.541×10^{10} C 54.1×10^8

B 5.41×10^9 D 541×10^7

40. How would 1.13×10^{-5} be written in standard notation?

A $-113{,}000$ C 0.0000113

B -0.0000113 D $113{,}000$

41. How would 3.12×10^7 be written in standard notation?

A 0.0000000312 C 31,200,000

B 0.000000312 D 3,120,000,000

42. A polar bear has an approximate mass of 322,000,000 milligrams. What is this number written in scientific notation?

A 0.322×10^9 C 32.2×10^7

B 3.22×10^8 D 322×10^6

43. ***SHORT RESPONSE*** An African elephant has an approximate mass of 6.3×10^6 grams. Write the mass of an African elephant in standard notation. Draw a diagram to illustrate how you determined the placement of the decimal point.

SHORT RESPONSE **Write each number in scientific notation. Explain in words how you determined your answer.**

44. 12,600,000

45. 0.00328

SHORT RESPONSE **Write each number in standard notation. Explain in words how you determined your answer.**

46. 8.39×10^{-7}

47. 5.62×10^5

EOG Practice ▪ Chapter 3

3A Rational Numbers

LESSON 3-1

1. What is 0.4 written as a fraction in simplest form?

   A $\frac{1}{4}$ C $\frac{2}{5}$

   B $\frac{4}{10}$ D $\frac{4}{5}$

2. Which is *not* equivalent to $\frac{1}{8}$?

   A 0.125 C $\frac{5}{40}$

   B $\frac{3}{24}$ D 0.8

3. Which is *not* equivalent to 0.75?

   A $\frac{1}{75}$ C $\frac{75}{100}$

   B $\frac{3}{4}$ D $\frac{21}{28}$

4. Which pair of fractions are *not* equivalent?

   A $\frac{9}{4}, \frac{18}{8}$ C $\frac{1}{2}, \frac{9}{18}$

   B $\frac{3}{5}, \frac{6}{8}$ D $\frac{4}{7}, \frac{12}{21}$

5. ***SHORT RESPONSE*** Write a fraction that cannot be simplified and has 28 as its denominator. Explain why your fraction cannot be simplified. Tell whether your fraction is a terminating or repeating decimal.

LESSON 3-2

6. What is the value of $\frac{2}{3} - \frac{5}{3}$?

   A −3 C $\frac{3}{3}$

   B −1 D 1

7. What is the value of $-\frac{31}{5} + \frac{24}{5}$?

   A $-\frac{7}{10}$ C $\frac{55}{10}$

   B $-\frac{7}{5}$ D $\frac{55}{5}$

8. What is the value of $\frac{17}{4} + \frac{13}{4}$?

   A $\frac{30}{8}$ C $\frac{17}{8}$

   B $\frac{15}{4}$ D $\frac{15}{2}$

SHORT RESPONSE **Evaluate each expression for the given value of the variable. Show your work. Write your answers in simplest form.**

9. $32.9 + x$ for $x = -15.8$
10. $21.3 + a$ for $a = -37.6$
11. $-\frac{8}{3} + y$ for $y = \frac{1}{3}$
12. $\frac{3}{5} + z$ for $z = 3\frac{1}{5}$

LESSON 3-3

13. What is the value of $-\frac{2}{3}\left(-\frac{5}{8}\right)$?

    A $-\frac{25}{38}$ C $\frac{10}{24}$

    B $-\frac{10}{24}$ D $\frac{25}{38}$

14. What is the value of 9(3.9)?

    A 27.81 C 278.1

    B 35.1 D 351

15. What is the value of $\frac{7}{10}x$ when $x = -\frac{3}{5}$?

    A $-\frac{21}{15}$ C $\frac{21}{50}$

    B $-\frac{21}{50}$ D $\frac{4}{5}$

16. Sunde worked on her homework for $2\frac{1}{6}$ hours. She spent $\frac{2}{3}$ of this time working on her math project. What amount of time did Sunde spend working on her math project?

    A $\frac{2}{9}$ hour

    B $\frac{1}{3}$ hour

    C $1\frac{4}{9}$ hours

    D $1\frac{2}{3}$ hours

17. ***SHORT RESPONSE*** Gunther bought a bag of 28 carrots. Gunther gave his rabbit, Priscilla, $\frac{1}{7}$ of the carrots. Then he ate $\frac{1}{4}$ of the carrots that remained. How many carrots did Gunther eat? Show your work.

EOG Practice ■ Chapter 3

3A Rational Numbers

LESSON 3-4

18. What is the value of $3\frac{2}{3} \div \frac{1}{4}$?

A $\frac{6}{12}$ C $3\frac{1}{6}$

B $\frac{11}{12}$ D $14\frac{2}{3}$

19. What is the value of $5.68 \div 0.2$?

A 0.284 C 28.4

B 2.84 D 284

20. ***SHORT RESPONSE*** Joelle saved $11.75 in quarters. How many quarters, worth $0.25 each, did she save? Show your work.

SHORT RESPONSE **Evaluate each expression for the given value of the variable. Show your work. Write your answers in simplest form.**

21. $9.45 \div m$ for $m = 0.05$

22. $5\frac{1}{5} \div s$ for $s = \frac{7}{8}$

23. $w \div \frac{2}{3}$ for $w = 6\frac{5}{8}$

LESSON 3-5

24. What is the value of $\frac{9}{10} + \frac{3}{8}$?

A $\frac{12}{80}$ C $\frac{12}{18}$

B $\frac{12}{40}$ D $\frac{102}{80}$

25. What is the value of $\frac{2}{7} - \frac{3}{4}$?

A $-\frac{13}{28}$ C $-\frac{1}{3}$

B $-\frac{1}{11}$ D $-\frac{1}{28}$

26. ***SHORT RESPONSE*** Alanna swam $3\frac{5}{6}$ miles one day. The next day she swam $1\frac{11}{15}$ miles. Write an expression that could be used to determine the total number of miles Alanna swam. Evaluate your expression. Write your answer in simplest form.

27. ***SHORT RESPONSE*** Evaluate $3\frac{9}{16} - 6\frac{3}{4}$. Show your work. Write your answer in simplest form.

LESSON 3-6

28. If $x - 3.2 = 5.1$, what is the value of $0.04x$?

A 0.076 C 33.2

B 0.332 D 76

29. If $m + \frac{3}{7} = \frac{1}{2}$, what is the value of $m \div \frac{5}{8}$?

A $\frac{5}{112}$ C $\frac{16}{25}$

B $\frac{4}{35}$ D $1\frac{17}{35}$

30. Suppose that $\frac{4}{5}w = \frac{2}{3}$ and $z + \frac{1}{4} = 1\frac{1}{12}$. Which statement is true?

A $w > z$ C $w = z$

B $z > w$ D $w \neq z$

31. ***SHORT RESPONSE*** On a combination lock numbered clockwise from 0 to 39, each number is separated by $\frac{1}{40}$ of a rotation. Randall started at 0 and turned his lock $\frac{4}{5}$ of a rotation to the right. Write an equation that could be used to determine which number Randall stopped on. Solve your equation.

LESSON 3-7

32. The solution of which inequality is graphed below?

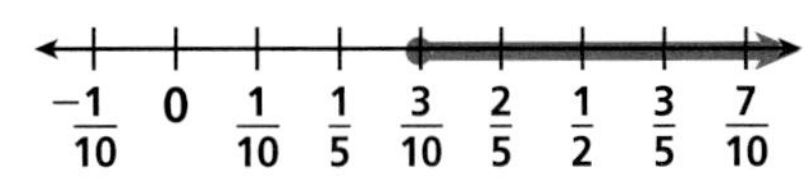

A $\frac{2}{7}t > \frac{3}{35}$ C $t - \frac{1}{6} \geq \frac{2}{15}$

B $-\frac{1}{2}t \geq \frac{15}{100}$ D $t + \frac{7}{8} < 1\frac{7}{40}$

33. Which inequality does *not* have the solution graphed below?

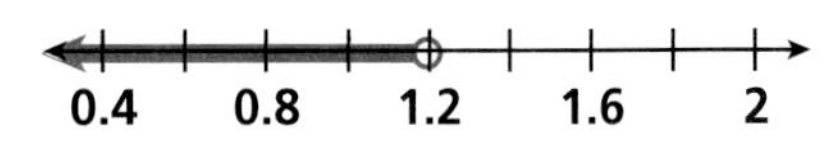

A $1.4x < 1.68$ C $x + 5.83 < 7.03$

B $\frac{x}{-0.03} < -40$ D $x - 1.2 < 0$

SHORT RESPONSE **Solve each inequality. Graph your solution on a number line.**

34. $w + \frac{2}{3} \geq \frac{2}{5}$

35. $-2\frac{1}{4}x < 9$

EOG Practice ■ Chapter 3

3B Real Numbers

LESSON 3-8

1. If $\frac{x}{9} = 16$, what are the two square roots of x?

   A −14 and 14
   B −13 and 13
   C −12 and 12
   D −11 and 11

2. If $2.5t = 722.5$, what are the two square roots of t?

   A −23 and 23
   B −17 and 17
   C −13 and 13
   D −9 and 9

3. Which ordered pair represents a number followed by one of its square roots?

   A (−25, −5)
   B (−4, 2)
   C (8, 64)
   D (16, −4)

4. Which ordered pair does *not* represent a number followed by one of its square roots?

   A (−9, −3)
   B (36, 6)
   C (49, −7)
   D (121, 11)

5. Which is equivalent to a square root of 625?

   A $-5 - 30$
   B $5\sqrt{-25}$
   C 5^2
   D $25\sqrt{25}$

6. Which is *not* equivalent to a square root of 196?

   A $\frac{-\sqrt{784}}{2}$
   B $-8 - 6$
   C $-2(-7)$
   D $4\sqrt{49}$

7. A gardener has 63 tulip bulbs and plans to plant a square tulip garden with an equal number of rows and columns. What is the greatest number of rows the gardener can plant?

   A 6
   B 7
   C 8
   D 9

8. What is the value of $3\sqrt{9}$?

   A $\sqrt{27}$
   B 9
   C 13.5
   D 27

9. Which accurately represents the value of $\sqrt{97 + 24}$?

   A $-\sqrt{97} + (-\sqrt{24})$
   B 11
   C $\sqrt{97} + \sqrt{24}$
   D 121

10. Which accurately represents the value of $\sqrt{5(45)}$?

    A $-5(-\sqrt{45})$
    B 15
    C $5(\sqrt{45})$
    D 225

11. Which accurately represents the value of $\sqrt{\frac{25}{144}}$?

    A $-\frac{25}{12}$
    B $\frac{5}{144}$
    C $\frac{5}{12}$
    D $\frac{25}{12}$

12. ***SHORT RESPONSE*** Wade counted 64 square ceiling tiles in his square classroom. The area of each ceiling tile is 4 square feet. Draw a diagram that could be used to help determine the dimensions of Wade's classroom. What are the dimensions of his classroom?

13. ***SHORT RESPONSE*** The answer to the following problem is 8 in. × 8 in. Show two ways to get this answer.

    Helen used 81 identical square tiles to tile a square floor that has an area of 5184 square inches. What are the dimensions of each square tile?

14. ***SHORT RESPONSE*** The height of square B is x. The area of square A is 361 times the area of square B. How many times the height of square B is the height of square A? Show your work and explain each step.

SHORT RESPONSE **Evaluate each expression. Show your work.**

15. $5\sqrt{36}$

16. $\sqrt{231 + 253}$

17. $\sqrt{6(54)}$

18. $\sqrt{\frac{49}{169}}$

3B Real Numbers

LESSON 3-9

19. Which two integers is $\sqrt{29}$ between?

A 3 and 4 | C 5 and 6
B 4 and 5 | D 6 and 7

20. If $k - 197 = 486$, then $\sqrt{k}$ is between which two integers?

A 25 and 26 | C 27 and 28
B 26 and 27 | D 29 and 30

21. If $y + 54 = 266$, what is the value of $\sqrt{y}$ to the nearest tenth?

A 14.5 | C 17.8
B 14.6 | D 17.9

22. The area of a square room is 700 square feet. The width of the room is between which two integers?

A 23 and 24 | C 26 and 27
B 24 and 25 | D 28 and 29

23. The area of a square room is 800 square feet. The width of the room is between which two integers?

A 23 and 24
B 24 and 25
C 26 and 27
D 28 and 29

24. ***SHORT RESPONSE*** A square game board can be folded into fourths for storage purposes. The area the top of the folded game board is 68 square inches. Write an expression that can be used to determine the side length of the unfolded game board. What is the side length of the unfolded game board, rounded to the nearest tenth of an inch?

25. ***SHORT RESPONSE*** Notice that 530 is halfway between 484 and 576. Is it true that $\sqrt{530}$ is exactly halfway between $\sqrt{484}$ and $\sqrt{576}$? Show your work, and explain in words how you determined your answer.

LESSON 3-10

26. Which name does *not* apply to $\frac{\sqrt{16}}{2}$?

A integer
B irrational number
C rational number
D real number

27. Which name applies to $\sqrt{12}$?

A integer
B irrational number
C rational number
D whole number

28. Which is a rational number and an integer but *not* a whole number?

A $-\frac{5}{4}$ | C $-(7^2)$
B $-1.\overline{03}$ | D $\sqrt{42}$

29. Which is a real number and a rational number but *not* an integer?

A -7.38 | C $\sqrt{15}$
B $-\frac{18}{3}$ | D $(-3)^3$

30. Which number is *not* between $5\frac{1}{8}$ and $5\frac{2}{8}$?

A $5\frac{5}{32}$ | C $5\frac{7}{32}$
B $5\frac{3}{16}$ | D $5\frac{5}{16}$

31. Which number is between $2\frac{1}{3}$ and $2\frac{2}{3}$?

A $\frac{10}{6}$ | C $\frac{15}{6}$
B $\frac{13}{6}$ | D $\frac{17}{6}$

32. ***SHORT RESPONSE*** Find two real numbers between $10\frac{3}{7}$ and $10\frac{4}{7}$. Show your work.

SHORT RESPONSE **State whether each number is rational, irrational, or not a real number. Explain in words how you determined your answer.**

33. $\sqrt{\frac{9}{16}}$ | **34.** $\sqrt{-4}$

35. $\frac{8}{0}$ | **36.** 12

EOG Practice • Chapter 4

4A Collecting and Describing Data

LESSON 4-1

1. In a poll, 75 math teachers were asked to identify their favorite subject. Which is the *best* reason why the sample could be biased?

   A All teachers like school.

   B The sample was too small.

   C Math teachers probably like math best.

   D Not all teachers have a favorite subject.

2. A company chooses 2000 veterinarians who belong to the same veterinary association to give their opinion about a new dog medicine. Which is the *best* reason why the sample could be biased?

   A Not all veterinarians like dogs.

   B Not all veterinarians use dog medicine.

   C Not every dog goes to a veterinarian.

   D Not all veterinarians belong to the association.

3. A questionnaire is distributed to 30 shoppers at a grocery store. Which represents a systematic sampling method?

   A Every fifth shopper is chosen.

   B Two different 2 hour blocks of time are chosen, and 15 shoppers are randomly chosen during each block of time.

   C Shoppers are chosen at varying times throughout the shopping day.

   D Three different aisles of the grocery store are chosen, and 10 shoppers are randomly chosen from each aisle.

4. ***SHORT RESPONSE*** In a nationwide survey, 7 states are chosen at random, and 150 people are chosen from each state. Identify the population and the sampling method. Give a reason why the sample could be biased.

5. ***SHORT RESPONSE*** A movie-theater manager wants to find out what people's reactions are to a new brand of popcorn. What population should be considered? Give an example of a random sampling method that could be used to conduct this survey.

LESSON 4-2

6. Which value is *not* represented in the stem-and-leaf plot?

Stem	Leaves
1	1 2 4
2	3 5 7 9
3	0 5

Key: 1|2 means 12

   A 4 **C** 35

   B 12 **D** 29

7. What are all the values represented in the back-to-back stem-and-leaf plot?

Team A		Team B
3	0	4 7
9 6 2	1	
	2	2 2

Key: |2|2 means 22
2|1| means 12

   A 21, 22, 22, 30, 40, 61, 70, 91

   B 3, 4, 7, 12, 16, 19, 22, 22

   C 0, 1, 2, 3, 22, 47, 962

   D 0, 1, 2, 2, 2, 2, 3, 4, 6, 7, 9

8. ***SHORT RESPONSE*** Use the given data to make a back-to-back stem-and-leaf plot. Which value occurs the most?

World Series Win/Loss Records of Selected Teams (through 2001)

Team	Wins	Losses
Dodgers	6	12
Giants	5	11
Pirates	5	2
Tigers	4	5
Yankees	26	15

EOG Practice

4A Collecting and Describing Data

LESSON 4-3

9. What is the median of the data set?

8, 3, 9, 10, 8, 4, 5, 7, 6, 7, 8, 5

A 5 C 7
B 6 D 8

10. What is the mode of the data set?

8, 3, 9, 10, 8, 4, 5, 7, 6, 7, 8, 5

A 5 C 7
B 6 D 8

11. What is the mean of the data set?

31, 28, 52, 40, 34, 24, 43, 21, 24

A 24 C 33
B 29.5 D 34

12. A survey questionnaire was given to 15 people. So far, 14 people have completed the questionnaire and returned it. Their ages are 18, 27, 24, 35, 43, 20, 51, 33, 21, 33, 18, 42, 23, and 60. What age will the fifteenth person have to be for the average age of the people surveyed to be 32?

A 32 C 36
B 35 D 38

13. Which value in the data set could be an outlier?

7, 6, 4, 0, 19, 5, 3, 8, 5, 2, 6, 5, 3

A 0 C 7
B 3 D 19

14. ***SHORT RESPONSE*** Compare the average number of miles per gallon of gas for each car in the table.

Number of Miles per Gallon of Gas			
	Week 1	Week 2	Week 3
Car A	28	26	29
Car B	31	34	27
Car C	19	22	19

LESSON 4-4

15. What is the range of the data set?

18, 20, 15, 13, 20, 17, 20, 20, 15, 13

A 7 C 20
B 13 D 33

16. What is the first quartile of the data set?

20, 17, 42, 26, 27, 12, 31

A 12 C 26
B 17 D 31

17. What is the third quartile of the data set?

2, 6, 6, 2, 2, 1, 1, 2, 2, 1, 0, 0, 4, 7

A 1 C 4
B 2 D 7

18. Which data set corresponds to the box-and-whisker plot?

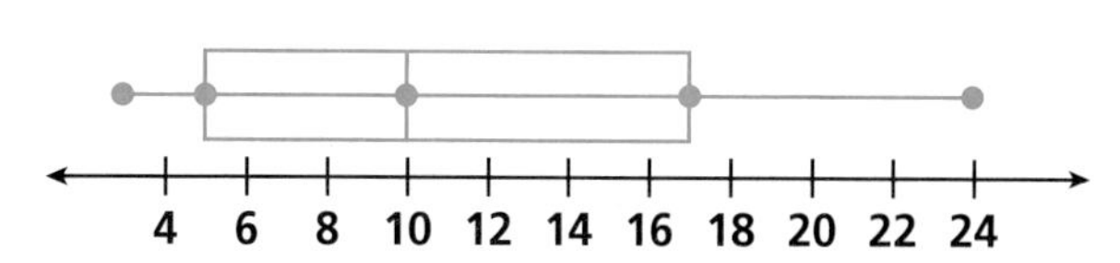

A 3, 12, 21, 9, 8, 4, 13, 24, 17, 19, 5
B 9, 13, 12, 10, 24, 3, 8, 17, 19, 5, 17
C 17, 3, 9, 4, 10, 12, 24, 8, 19, 17, 5
D 12, 5, 9, 17, 8, 4, 13, 24, 17, 3, 19

19. ***SHORT RESPONSE*** Use the data about SAT scores to make a box-and-whisker plot. Explain what the box-and-whisker plot tells you about the data set.

National Average for SAT Scores	
Year	Total Score
1997	1016
1998	1017
1999	1016
2000	1019
2001	1020
2002	1020

EOG Practice

EOG Practice ■ Chapter 4

4B Displaying Data

LESSON 4-5

1. A group of 25 students were polled to determine the average number of movies they see during a year. The data was used to create the frequency table below. Which data set corresponds to the frequency table?

Number of Movies	Frequency
0–4	2
5–9	5
10–14	11
15–19	6
20–24	1

A 2, 4, 10, 11, 20, 17, 12, 5, 13, 15, 12, 22, 16, 13, 7, 14, 6, 10, 15, 12, 11, 8, 7, 15, 10

B 2, 4, 10, 11, 18, 17, 12, 5, 13, 15, 12, 22, 16, 13, 7, 14, 6, 10, 15, 12, 11, 8, 7, 15, 10

C 2, 4, 10, 11, 18, 17, 12, 5, 13, 15, 12, 22, 16, 13, 7, 14, 6, 10, 15, 7, 11, 8, 7, 15, 10

D 2, 4, 10, 11, 18, 17, 12, 5, 13, 15, 12, 22, 16, 13, 3, 14, 6, 10, 15, 12, 11, 8, 7, 15, 10

2. A survey was conducted to determine people's favorite Italian meal. The data from the survey was recorded in the table below. Which of the following would *best* display the data?

Favorite Italian Meal	
Food	Number of People
Cannelloni	3
Lasagna	12
Manicotti	7
Spaghetti	18

A line graph

B bar graph

C histogram

D stem-and-leaf plot

3. ***SHORT RESPONSE*** *Population density* is the number of people per square mile in a geographic location. Make a line graph of the data about population density. Use your graph to estimate the population density in 1975.

Year	Population Density
1950	42.6
1960	50.6
1970	57.5
1980	64.0
1990	70.3
2000	79.6

4. ***SHORT RESPONSE*** The following are the ages at which a randomly chosen group of 20 students graduated from college: 20, 21, 23, 19, 20, 21, 21, 21, 19, 22, 21, 21, 21, 20, 21, 20, 21, 21, 22, 21. Explain why a histogram is not the best choice for displaying this data. What is the best choice? Why? Use the data to make a bar graph.

5. ***SHORT RESPONSE*** The table shows the approximate wait time for a popular amusement-park ride at each hour of the day. Organize the wait time data into a frequency table. Explain your choice of interval. Use your frequency table to make a histogram.

Hour	Wait Time (min)	Hour	Wait Time (min)
9 A.M.	15	2 P.M.	97
10 A.M.	32	3 P.M.	100
11 A.M.	64	4 P.M.	78
Noon	53	5 P.M.	45
1 P.M.	96	6 P.M.	28

4B Displaying Data

LESSON 4-6

6. A real-estate agency sold houses for $75,000, $420,000, $88,000, $80,000, and $82,000. Its ads boast an average selling price of $149,000. Why is this misleading?
   - A The average of the house prices is not $149,000.
   - B An extreme outlier distorts the average.
   - C The agency has sold only 5 houses.
   - D There is not enough data to give an average.

7. Why is the graph below misleading?

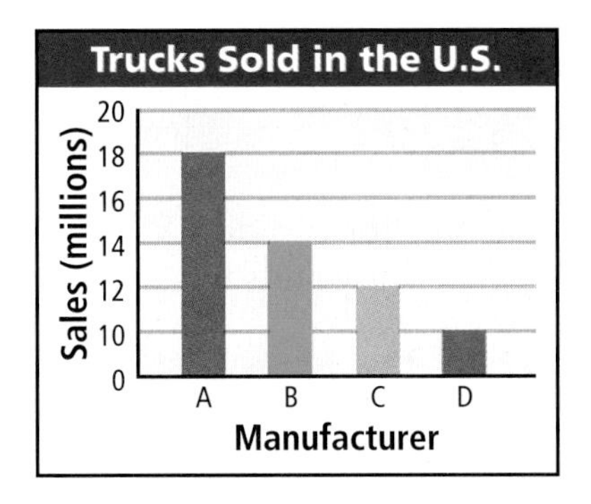

   - A The graph does not give the names of the manufacturers.
   - B The graph does not say which type of trucks are being sold.
   - C The sales amounts are in millions, but the graph is labeled from 0 to 20.
   - D The intervals on the scale are unequal.

8. ***SHORT RESPONSE*** A market researcher randomly selects 12 shoppers to sample 3 brands of sausage labeled A, B, and C. Of the shoppers, 8 selected B, 2 selected A, and 2 selected C. An ad for brand B reads, "Preferred 4 to 1 over other brands." Explain why this is misleading.

9. ***SHORT RESPONSE*** Manny's Hamburgers served 98 customers between 11 A.M. and 1 P.M. Bubba's Burgers served 14 customers between 8 P.M. and 9 P.M. Why is this statistic misleading? What change is necessary to make this statistic more accurate?

LESSON 4-7

10. Which data sets have a *negative* correlation?
    - A the temperature of an oven and the time it takes a turkey to brown
    - B the size of a turkey and its price
    - C the price of a turkey and the temperature of the oven
    - D the time it takes to brown a turkey and the price of a turkey

11. Which data sets have *no* correlation?
    - A the difficulty level of a test and the test scores
    - B the amount of time allowed for taking a test and the lengths of the pencils used on a test
    - C test scores and the amount of time allowed for taking the test
    - D the number of problems on a test and the amount of time allowed for taking the test

12. Which data sets have a *positive* correlation?
    - A the age of a tree and the size of a yard
    - B the size of a yard and the color of the house
    - C the color of a house and the age of a tree
    - D the age of a tree and the size of a tree

13. ***SHORT RESPONSE*** The table shows some relationships between the number of years of post-high-school education and salary. Use the data to make a scatter plot. Explain what you would expect the salary of a person with 7 years of post-high-school education to be.

1	$18,000	4	$51,000	6	$64,000
1	$20,500	4	$43,000	6	$58,000
3	$28,000	5	$48,000	8	$75,000
4	$35,000	5	$52,000	8	$73,500

EOG Practice ▪ Chapter 5

5A Plane Figures

LESSON 5-1

1. Which of the following is an acute angle?

A

C 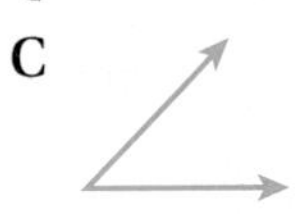

B

D 

2. Which term describes the figure below?

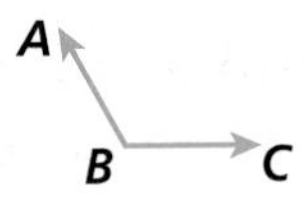

A ray C right angle

B acute angle D obtuse angle

3. ∠1 and ∠2 are vertical angles. If m∠1 = 83°, what is m∠2?

A 7° C 83°

B 45° D 97°

4. Which of the angles below are *complementary*?

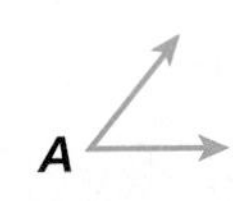

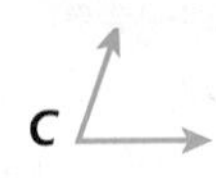

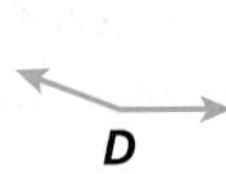

A ∠*A* and ∠*B*

B ∠*A* and ∠*C*

C ∠*B* and ∠*C*

D ∠*B* and ∠*D*

5. ***SHORT RESPONSE*** If two angles are supplementary and one angle is *not* a right angle, could the second angle be right? Draw a diagram to justify your answer.

6. ***SHORT RESPONSE*** Suppose ∠1 ≅ ∠2, ∠3 is complementary to ∠2, and m∠1 = 37°. What is m∠3? Explain in words how you determined your answer.

LESSON 5-2

7. In the figure, line *d* || line *f*. What are the measures of angles 1, 2, and 3?

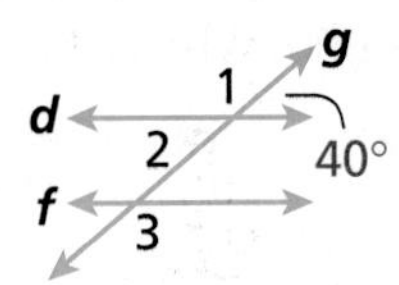

A m∠1 = 50°, m∠2 = 40°, and m∠3 = 50°

B m∠1 = 130°, m∠2 = 50°, and m∠3 = 130°

C m∠1 = 140°, m∠2 = 40°, and m∠3 = 140°

D m∠1 = 40°, m∠2 = 140°, and m∠3 = 40°

8. In the figure, line *r* || line *s*. Which angle in the figure corresponds to ∠1?

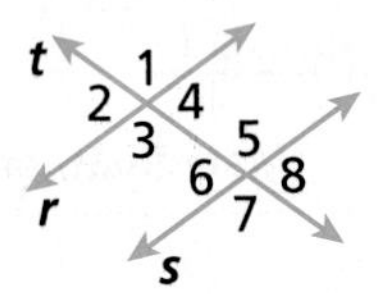

A ∠2 C ∠4

B ∠3 D ∠5

9. In the figure, ∠1 ≅ ∠2. What can be said about lines *p* and *q*?

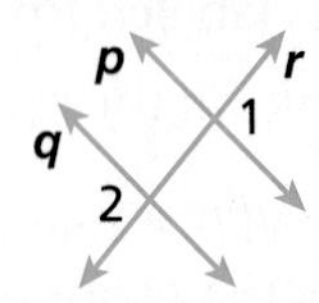

A Line *p* and line *q* are parallel.

B Line *p* and line *q* are perpendicular.

C Line *p* and line *q* intersect.

D Line *p* and line *q* are transversals.

10. ***SHORT RESPONSE*** Draw two parallel lines that are intersected by a transversal. Label the angles, and identify all congruent angles.

5A Plane Figures

LESSON 5-3

11. What is the missing measure in the triangle?

A 47°
B 63°
C 70°
D 80°

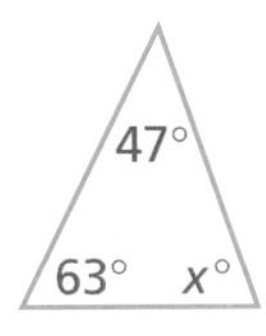

12. What is the missing measure in the triangle?

A 34°
B 56°
C 90°
D 124°

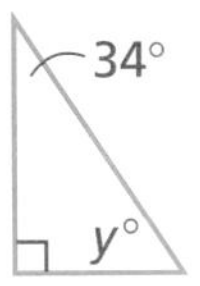

13. What is the missing measure in the triangle?

A 28°
B 35°
C 117°
D 145°

14. ***SHORT RESPONSE*** The second angle in a triangle is 6 times as large as the first. The third angle is $\frac{1}{3}$ as large as the second. Draw a diagram of the triangle, and label each angle in terms of the measure of the first angle. What are the angle measures of the triangle? Show your work.

LESSON 5-4

15. What are all the names that apply to the figure below?

A square, rectangle, and quadrilateral
B square, rhombus, and quadrilateral
C square, rectangle, parallelogram, and quadrilateral
D square, rectangle, parallelogram, rhombus, and quadrilateral

16. Which of the following would be a correct response to the question
"Is a trapezoid a parallelogram?"

A Yes, because a trapezoid has parallel sides.
B No, because a trapezoid does not have any parallel sides.
C Yes, because a trapezoid is a quadrilateral.
D No, because a trapezoid has only one set of parallel sides.

17. ***SHORT RESPONSE*** Draw a nonagon (nine sides). Divide your figure into triangles to determine the sum of the angle measures in a nonagon. What is the sum of the angle measures in a nonagon? Show your work.

LESSON 5-5

18. Which *best* describes the slope of the line graphed below?

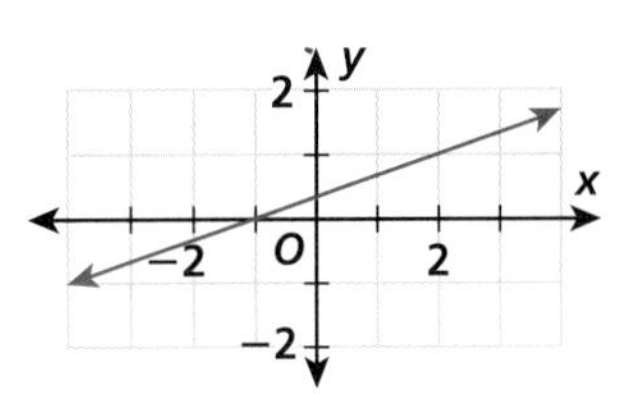

A negative slope
B positive slope
C 0 slope
D undefined slope

19. What is the slope of the line graphed below?

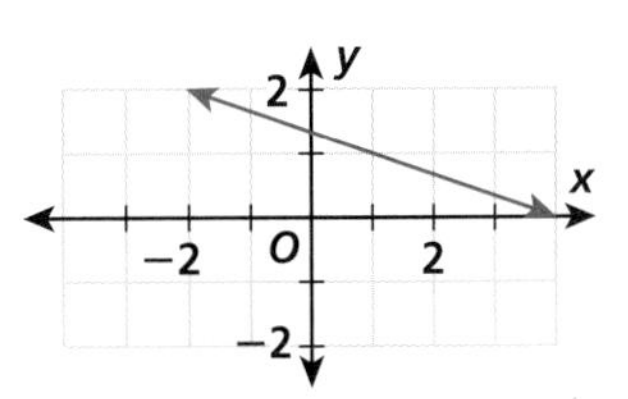

A -3
B $-\frac{1}{3}$
C $\frac{1}{3}$
D 3

20. ***SHORT RESPONSE*** Graph the quadrilateral with vertices $(-1, 0)$, $(-3, 1)$, $(-1, 3)$, and $(1, 2)$. Label the slope of each side of the figure, and give all of the names that apply to the figure. Explain in words how you determined your answer.

EOG Practice ■ Chapter 5

5B Patterns in Geometry

LESSON 5-6

1. Triangle ABC is congruent to triangle DEF. Which congruence statement is correct?

   A $\angle ACB \cong \angle DEF$ C $\angle ABC \cong \angle DFE$

   B $\angle ACB \cong \angle DFE$ D $\angle ABC \cong \angle EDF$

2. Pentagon $ABCDE$ is congruent to pentagon $GHJKL$. What is the value of t?

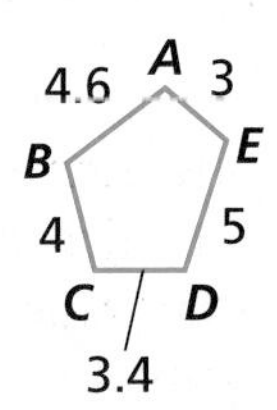

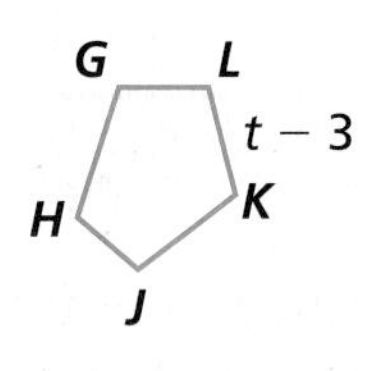

   A 6 C 7.6

   B 7 D 8

3. Quadrilateral $ABCD$ is congruent to quadrilateral $GHJK$. What is the value of s?

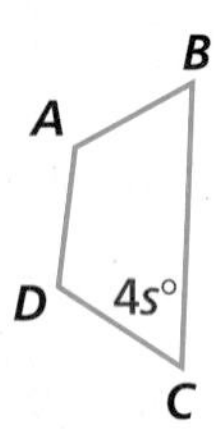

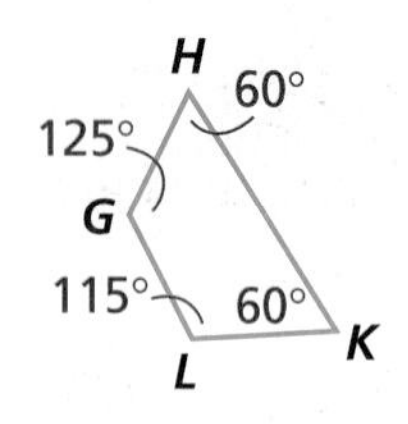

   A 15 C 31.25

   B 28.75 D 60

4. Which of the triangles below are congruent?

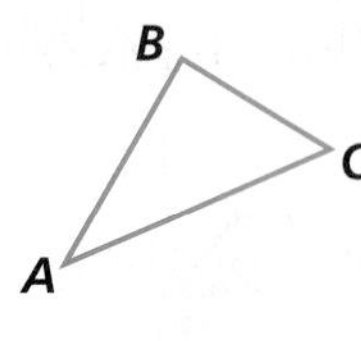

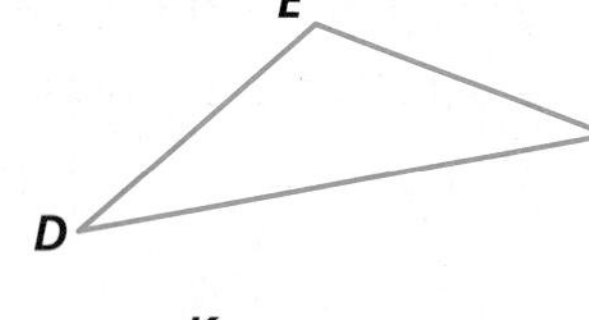

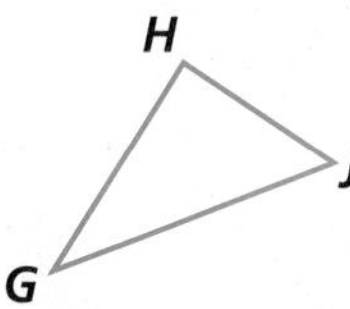

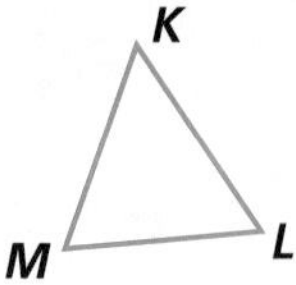

   A triangle ABC and triangle DEF

   B triangle DEF and triangle GHJ

   C triangle ABC and triangle GHJ

   D triangle GHJ and triangle KLM

5. ***SHORT RESPONSE*** Quadrilateral $ABCD \cong$ quadrilateral $KLMN$. Find the values of x, y, and z. Show your work.

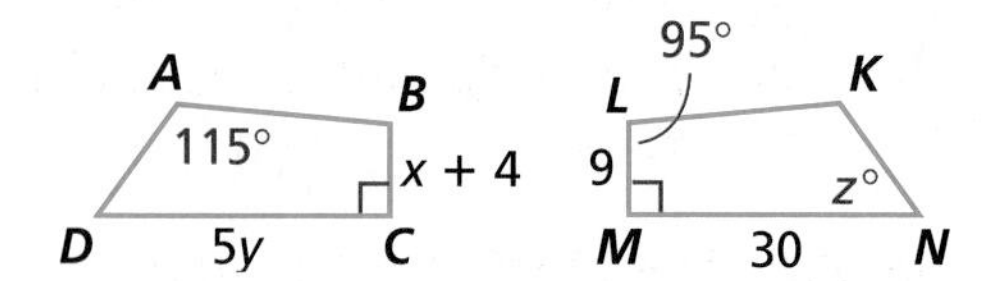

LESSON 5-7

6. Which type of transformation has been performed?

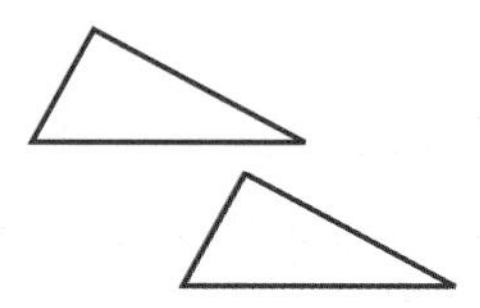

   A translation C rotation

   B reflection D none of these

7. Which type of transformation has been performed?

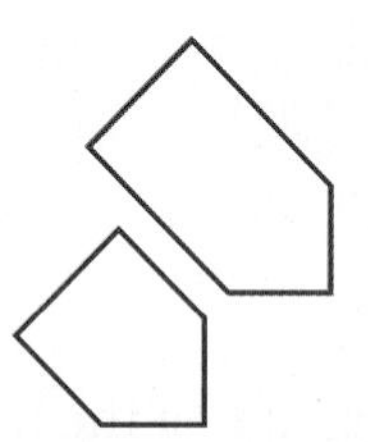

   A translation C rotation

   B reflection D none of these

8. Which type of transformation has been performed?

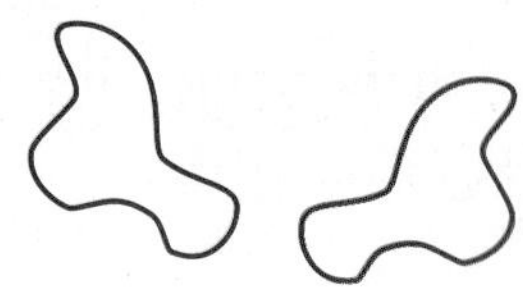

   A translation C rotation

   B reflection D none of these

EOG Practice ■ Chapter 5

5B Patterns in Geometry

9. ***SHORT RESPONSE*** On a coordinate plane, draw the image of a quadrilateral with vertices (1, 3), (4, −1), (5, 3), and (6, 1) after a 180° rotation about the point (0, 0). What are the coordinates of the image's vertices?

10. ***SHORT RESPONSE*** On a coordinate plane, draw the image of a triangle with vertices (0, 1), (2, 4), and (3, 0) after a translation of 5 units left. Explain in words how you determined the coordinates of the image's vertices.

LESSON 5-8

11. Which type of symmetry does the figure below have?

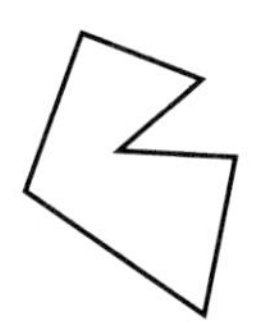

A line symmetry
B line symmetry and rotational symmetry
C no symmetry
D rotational symmetry

12. Which type of rotational symmetry does a figure have if the figure coincides with itself every 40°?

A 4-fold
C 36-fold
B 9-fold
D 40-fold

13. How many lines of symmetry does the figure below have?

A 0
C 2
B 1
D 4

14. ***SHORT RESPONSE*** The figure below has 4-fold rotational symmetry. The point is the center of rotation. Complete the figure. What is the least number of degrees you must rotate the figure for it to coincide with itself?

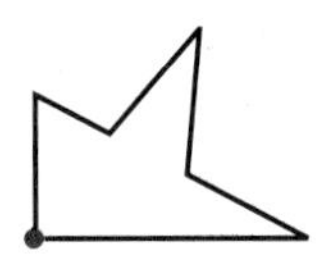

15. ***SHORT RESPONSE*** In the figure below, the dashed line is the line of symmetry. Explain in words how to complete the figure. Complete the figure.

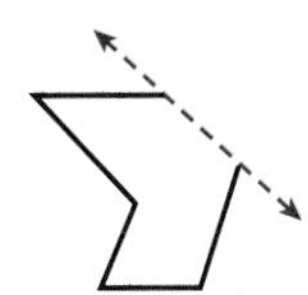

LESSON 5-9

16. Which of the figures will *not* tessellate?

A 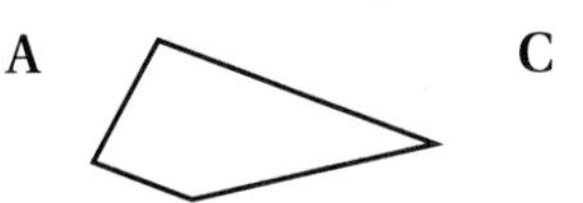

C 

B 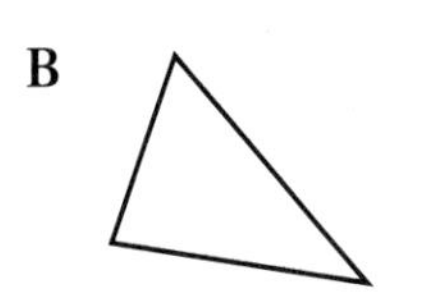

D 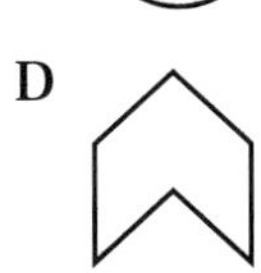

17. Which of the figures can be used to create a regular tessellation?

A 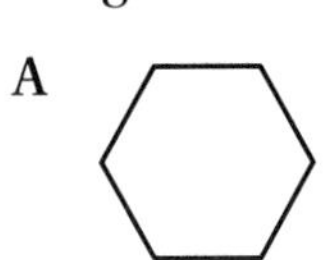

C 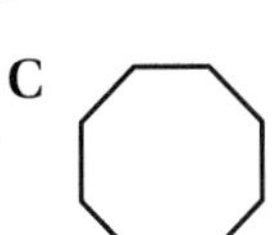

B

D 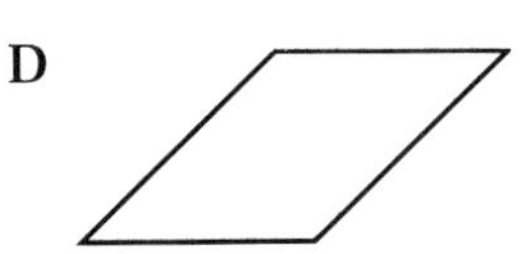

18. ***SHORT RESPONSE*** Use regular octagons and squares to create a tessellation. Explain whether your tessellation is regular or semiregular.

EOG Practice ■ Chapter 6

6A Perimeter and Area

LESSON 6-1

1. What is the perimeter of the parallelogram below?

   A 24 in.
   B 39 in.
   C 48 in.
   D 135 in.

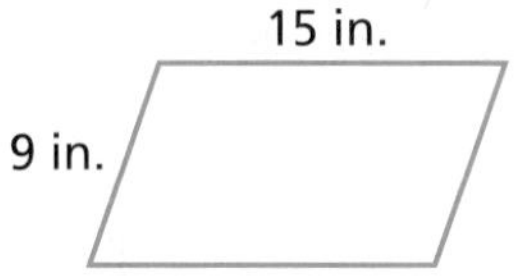

2. What is the area of the rectangle below?

   A 20 m^2
   B 32 m^2
   C 40 m^2
   D 96 m^2

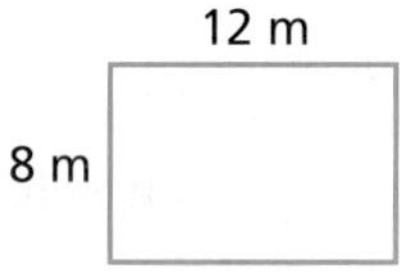

3. What is the perimeter of the figure graphed below?

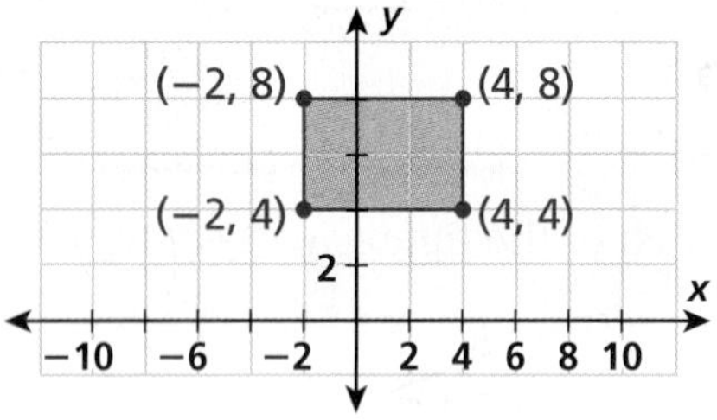

   A 10 units C 20 units
   B 16 units D 24 units

4. What is the area of the figure graphed below?

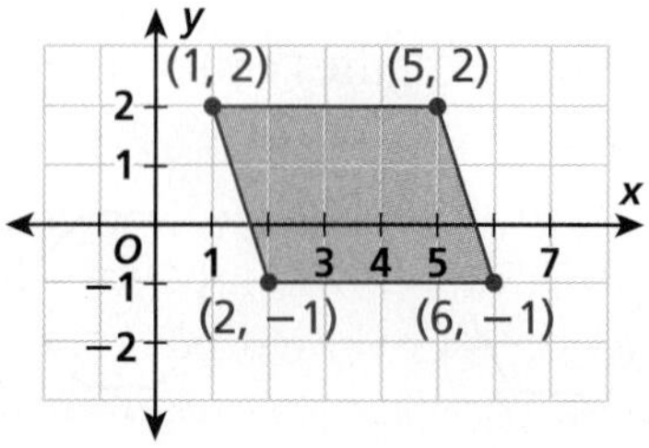

   A 7 $units^2$ C 12 $units^2$
   B 10 $units^2$ D 14 $units^2$

5. ***SHORT RESPONSE*** The area of a rectangular mirror is 60 square inches. What are three possible perimeters of the mirror? Explain in words how you determined your answer.

6. ***SHORT RESPONSE*** Find the perimeter and area of the composite figure below. Show your work.

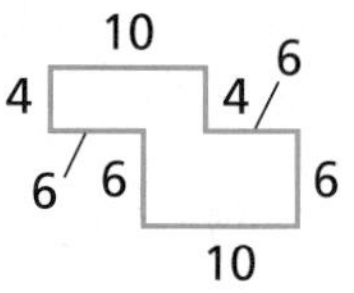

LESSON 6-2

7. Shown below are a triangle and a trapezoid with the same height. What fraction of the area of the trapezoid is the area of the triangle?

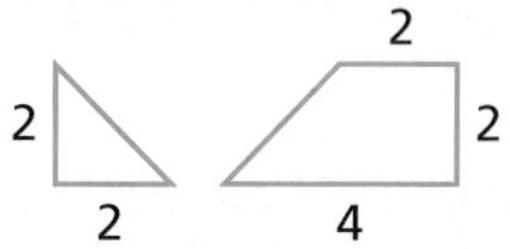

   A $\frac{1}{4}$ C $\frac{1}{2}$
   B $\frac{1}{3}$ D $\frac{2}{3}$

8. Which are the vertices of a triangle with an area of 6 square units?

   A (−1, 1), (−1, 3), and (3, 1)
   B (−1, 0), (−1, 3), and (3, 0)
   C (−1, −1), (−1, 3), and (3, −1)
   D (1, −1), (1, 3), and (3, −1)

9. The trapezoid below has an area of 24 square inches. What is the value of x?

   A 3 inches
   B 4 inches
   C 5 inches
   D 9 inches

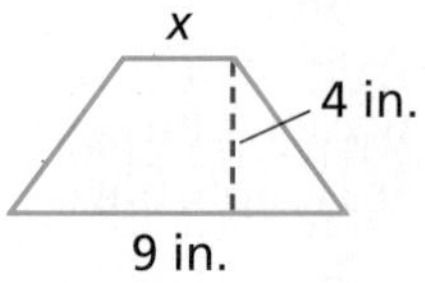

10. ***SHORT RESPONSE*** Graph the figure with vertices (2, 1), (−2, 1), (5, −3), and (−4, −3) on a coordinate plane. What is the area of the figure that you graphed?

11. ***SHORT RESPONSE*** Draw two different triangles with the same area. Label the dimensions of your triangles. What is the area of each triangle?

EOG Practice ■ Chapter 6

6A Perimeter and Area

LESSON 6-3

12. What is the missing measure in the triangle below?

A 11 mm

B 15 mm

C 108 mm

D 225 mm

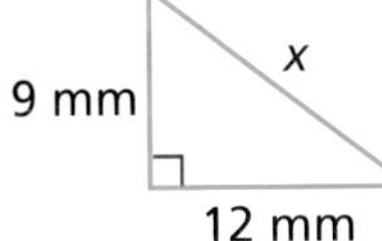

13. What is the missing measure in the triangle below?

A 4 in.

B 8 in.

C 16 in.

D 34 in.

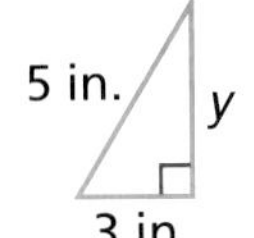

14. A right triangle has side lengths 8 meters and 10 meters. Which of the following is a possible third side length?

A 2 meters

B 6 meters

C 13 meters

D 18 meters

15. What is the length of the hypotenuse of the right triangle with vertices $(-3, -2)$, $(-3, 6)$, and $(12, -2)$?

A 8

B 15

C 17

D 19

16. Which set of numbers is a Pythagorean triple?

A 3, 4, 6

B 5, 12, 13

C 7, 8, 11

D 12, 15, 20

17. ***SHORT RESPONSE*** What is the area of the triangle below? Show your work, and explain each step.

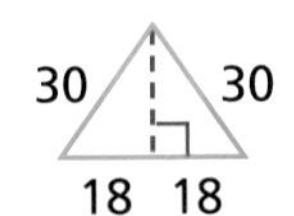

18. ***SHORT RESPONSE*** Draw a diagram of a rectangle with an area of 240 square centimeters and a perimeter of 68 centimeters. Label the dimensions of your rectangle. What is the length of the diagonal of the rectangle you drew?

LESSON 6-4

19. What is the circumference of the circle below, in terms of π?

A 2π cm

B 4π cm

C 8π cm

D 16π cm

20. What is the area of the circle below, in terms of π?

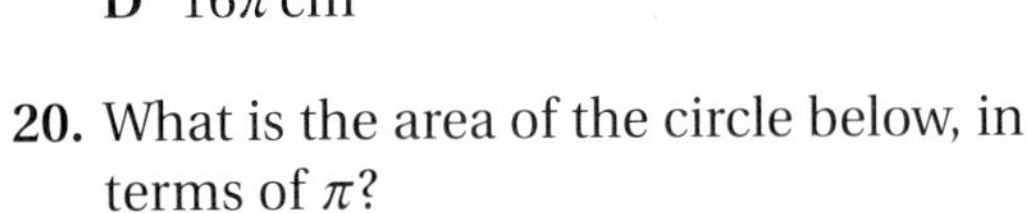

A $7.5\pi\ \text{ft}^2$

B $15\pi\ \text{ft}^2$

C $56.25\pi\ \text{ft}^2$

D $225\pi\ \text{ft}^2$

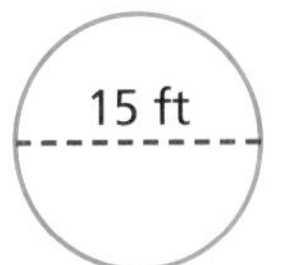

21. What is the area of the shaded region of the figure below, to the nearest tenth?

A $2.1\ \text{in}^2$

B $3.1\ \text{in}^2$

C $16.8\ \text{in}^2$

D $18.8\ \text{in}^2$

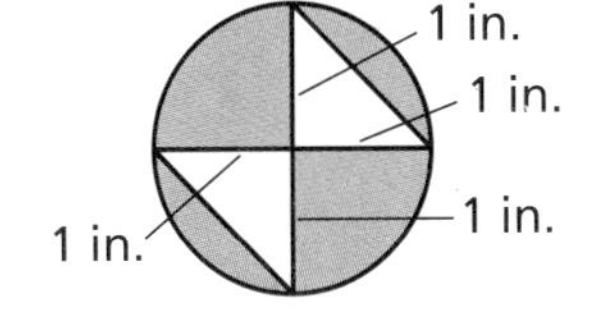

22. ***SHORT RESPONSE*** Find the perimeter and area of the composite figure below. Show your work, and round your answers to the nearest tenth.

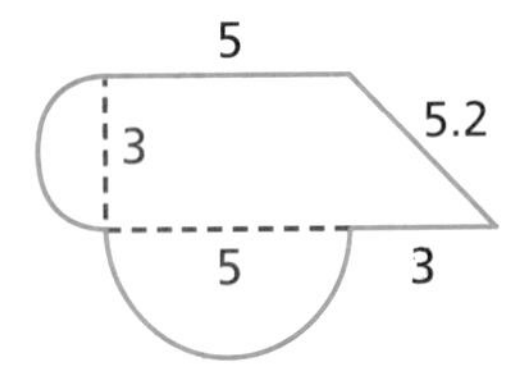

23. ***SHORT RESPONSE*** Graph the circle that has its center at (2, –1) and that passes through (6, –1) on a coordinate plane. Find the area and circumference of the circle you graphed, both in terms of π and to the nearest tenth. Use 3.14 for π.

24. ***SHORT RESPONSE*** A passenger on a 15-foot-tall Ferris wheel begins at the bottom. The Ferris wheel makes 8 complete rotations before stopping. What distance has the passenger traveled? Show your work.

EOG Practice ■ Chapter 7

7B Similarity and Scale

LESSON 7-5

1. What is the center of dilation in the dilation shown below?

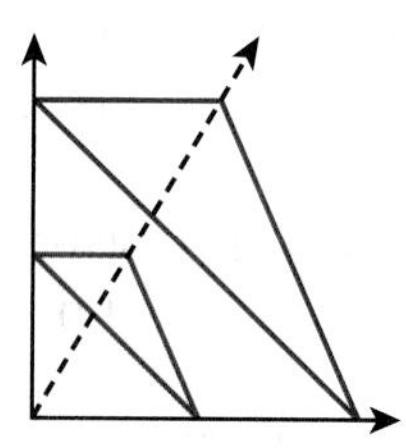

A (0, 0)

B (1, 1)

C (4, 1)

D (7, 1)

2. What scale factor was used in the dilation below?

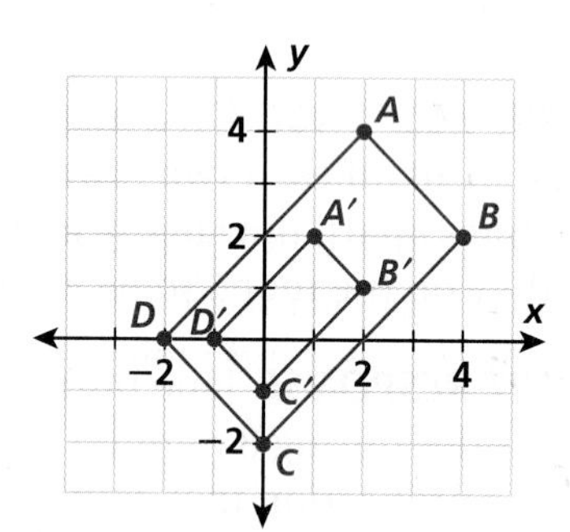

A $\frac{1}{2}$

B 2

C 1.5

D 4

3. ***SHORT RESPONSE*** Dilate the figure below by a scale factor of $\frac{3}{4}$ using P as the center of dilation. Show your work.

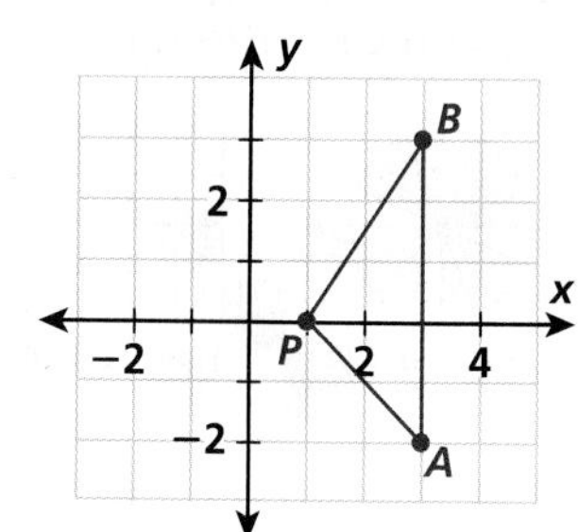

4. ***SHORT RESPONSE*** A rectangle was dilated by a scale factor of 3.2. The dimensions of the dilation are 4.8 inches by 12.8 inches. Draw a diagram of the original rectangle and label its dimensions. What is the area of the original rectangle?

5. ***SHORT RESPONSE*** A figure has vertices at (1, 2), (2, 5), (5, 6), and (6, 1). The figure is dilated by a scale factor of 2.5 with the origin as the center of dilation. What are the coordinates of the new image? How are the new coordinates related to the original coordinates?

LESSON 7-6

6. In Tiffany's backyard there are three rectangular gardens. The vegetable garden, garden A, is 12 feet long and 8 feet wide. The fruit garden, garden B, is 8 feet long and 9 feet wide. The rose garden, garden C, is 18 feet long and 12 feet wide. Which of these gardens are similar?

A gardens A and B

B gardens B and C

C gardens A and C

D gardens A, B, and C

7. ***SHORT RESPONSE*** A rectangle is 15 centimeters long and 8 centimeters tall. Another rectangle is 20 centimeters long and 12 centimeters tall. Explain whether the rectangles are similar.

8. ***SHORT RESPONSE*** $\triangle ABC \sim \triangle XYZ$. Find the missing dimensions of $\triangle XYZ$. Show your work.

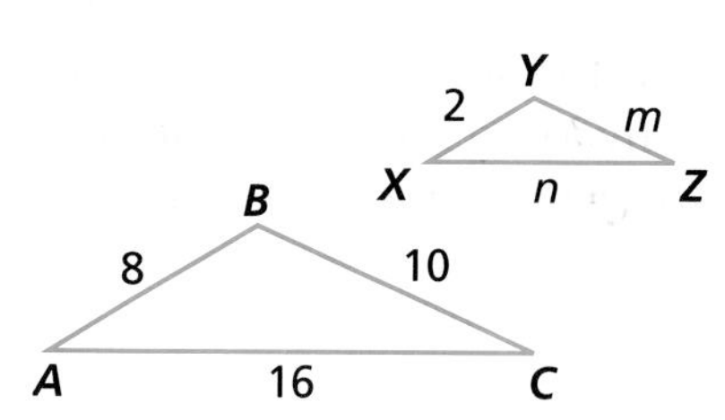

EOG Practice ■ Chapter 7

7B Similarity and Scale

LESSON 7-7

9. The length of a sidewalk on a scale drawing is 5 inches. The actual length of the sidewalk is 35 feet. What scale was used to create the drawing?

   A 1 in.:5 ft

   B 1 in.:7 ft

   C 1 in.:10 ft

   D 1 in.:30 ft

10. ***SHORT RESPONSE*** Julio uses a scale of $\frac{1}{8}$ in. = 1 foot when he paints landscapes. In one painting, a giant sequoia tree is 34.375 inches tall. How tall is the real tree? How tall would the tree in the painting be if Julio used a scale of $\frac{1}{11}$ in. = 1 foot? Show your work.

11. ***SHORT RESPONSE*** On a scale drawing of a house plan, the master bathroom is $1\frac{1}{2}$ inches wide and $2\frac{5}{8}$ inches long. If the scale of the drawing is $\frac{3}{16}$ inch = 1 foot, what are the actual dimensions of the bathroom? Show your work.

LESSON 7-8

12. Which scale does *not* preserve the size of the actual object?

    A 10 cm:1 dm

    B 5 km:25 m

    C 1 ft:12 in.

    D 1760 yd:5280 ft

13. Which scale does *not* enlarge the size of the actual object?

    A 250 cm:2 m

    B 3 yd:3 ft

    C 74 in:2 yd

    D 2 mi:10,560 ft

14. ***SHORT RESPONSE*** Give two different scales with a scale factor of 12. Show your work.

15. ***SHORT RESPONSE*** Find the scale factor that relates a 15-inch-long model train to a 20-yard-long train. If the same scale factor were used, how many inches long would the model of a 50-foot train be?

16. ***SHORT RESPONSE*** A model of a skyscraper was made with a scale of 0.5 in.:5 ft. What scale factor does the scale represent? If the actual skyscraper is 570 feet tall, how tall is the model?

LESSON 7-9

17. A 9-centimeter cube and a 2-centimeter cube are both part of a demonstration kit for architects. Which is *not* a correct comparison of the surface area of the cubes?

    A 22:1   C 162:8

    B 60.75:3   D 486:24

18. Cube A has a side length of 32 centimeters, and cube B has a side length of 4 centimeters. Which pair of numbers represents a comparison of the volume of cube A to the volume of cube B, followed by a comparison of the surface area of cube A to the surface area of cube B?

    A $\frac{1}{512}$ and $\frac{1}{64}$ C 64 and 512

    B $\frac{1}{64}$ and $\frac{1}{512}$ D 512 and 64

19. ***SHORT RESPONSE*** A popcorn machine makes enough popcorn to fill a rectangular box that measures 5 inches × 8 inches × 2 inches in 45 seconds. How long would it take the same machine to fill a rectangular box that measures 10 inches × 16 inches × 4 inches? What scale factor relates the volume of the first box to the volume of the second box?

EOG Practice ■ Chapter 8

8A Numbers and Percents

LESSON 8-1

1. Which of the following is equivalent to $\frac{5}{6}$?

   A 56% C 0.56

   B $\frac{7}{8}$ D $0.8\overline{3}$

2. Which of the following is equivalent to 11.1%?

   A 0.111 C $\frac{1}{3}$

   B $\frac{1}{9}$ D 111

3. Which of the following is *not* equivalent to $\frac{1}{2}$?

   A 0.2 C 50%

   B $\frac{13}{26}$ D 0.5

4. Which of the following is *not* equivalent to 0.085%?

   A $\frac{8.5}{100}$ C 0.00085

   B $\frac{85}{1000}$ D 8.5

5. Which of the following is equivalent to the percent of the shapes below that are *not* triangles?

   A 0.375 C $\frac{2}{3}$

   B $\frac{3}{8}$ D 0.6

6. What percent of the figure below is shaded?

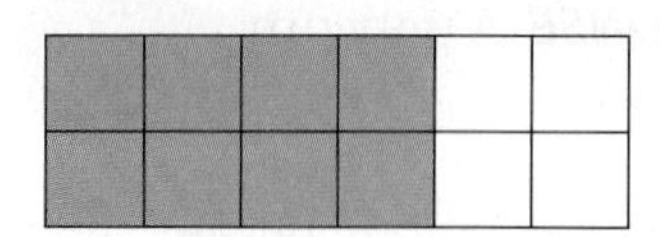

   A 8% C $66\frac{2}{3}\%$

   B 12% D 80%

7. Which figure is more than 75% shaded?

   A 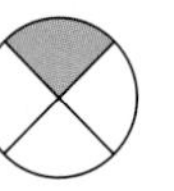 C 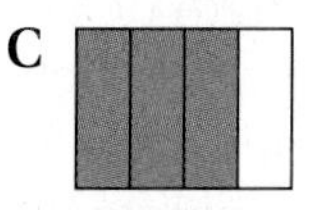

   B 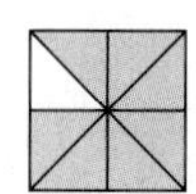 D 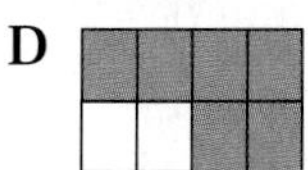

8. ***SHORT RESPONSE*** Copy and complete the table. Write each fraction in simplest form.

Fraction	Decimal	Percent
■	0.45	■
■	0.25	■
■	■	94%
■	■	85%
$\frac{14}{15}$	■	■
$\frac{19}{20}$	■	■

9. ***SHORT RESPONSE*** What percent of the squares on the checkerboard below are covered with a game piece? What are two values that are equivalent to the percent of checkerboard squares covered with a game piece? How many pieces must be removed for 6.25% of the checkerboard squares to be covered with game pieces? Explain in words how you determined your answers.

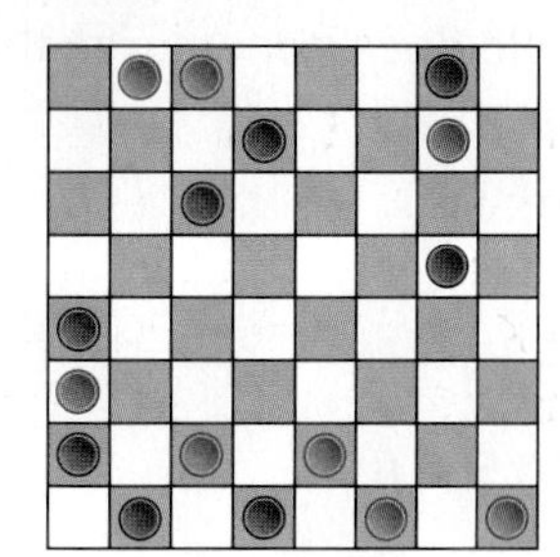

8A Numbers and Percents

10. ***SHORT RESPONSE*** Helen was shopping and found a store that had all merchandise prices dicounted by $33\frac{1}{3}$%. What fraction of the price did Helen pay for each item she bought? Show your work.

11. ***SHORT RESPONSE*** Draw a diagram of a figure that is 45% shaded. What are two equivalent ways of expressing the percent of your figure that is shaded?

LESSON 8-2

12. Approximately what percent of 264 is 93?

    A 24.5% C 35.2%
    B 28.3% D 245.5%

13. What number is to 100 as 54 is to 72?

    A 0.75 C 18
    B 5.4 D 75

14. Kayla purchased 7 oranges, 6 bananas, and 5 apples at the grocery store. To the nearest tenth, what percent of the fruit that she purchased was oranges?

    A 3.9% C 38.9%
    B 4% D 388.9%

15. Jeffrey gave 24% of his monthly allowance to his little brother as a birthday gift. If Jeffrey gets a $25 allowance each month, how many dollars did he give his little brother?

    A $0.60 C $6.00
    B $2.40 D $24.00

16. ***SHORT RESPONSE*** Mt. McKinley, in Alaska, is 20,320 feet tall. The height of Mt. Everest is about 142.9% of the height of Mt. McKinley. To the nearest foot, what is the height of Mt. Everest? Show your work.

17. ***SHORT RESPONSE*** The area of Alexander Island is 1192.9% of the area of Adelaide Island. Explain what additional information you would need to find the area of Alexander Island.

LESSON 8-3

18. To the nearest tenth, 26 is 53% of what number?

    A 20.3 C 49.1
    B 39.5 D 52

19. To the nearest tenth, 215 is 94% of what number?

    A 202.1 C 309
    B 228.7 D 437

20. Harold paid $11.99 for a new shirt. He knows that the price of the shirt was 75% of the original cost. What was the original cost of the shirt?

    A $3.00 C $15.99
    B $8.99 D $47.96

21. ***SHORT RESPONSE*** A certain rock is a compound of several elements. Tests show that the rock contains 20.2 grams of quartz. If 37.5% of the rock is quartz, find the mass in grams of the entire rock. Show your work.

22. ***SHORT RESPONSE*** The Cumberland River is 720 miles in length, or about 31% of the length of the Mississippi River. Find the approximate length of the Mississippi River to the nearest mile. Show your work.

23. ***SHORT RESPONSE*** A pair of jeans had a ticket price of $35 at store A. This is 85% of the ticket price of the same pair of jeans at store B. Write an equation that can be used to determine the ticket price of the jeans at store B. Solve your equation.

EOG Practice ■ Chapter 8

8B Applying Percents

LESSON 8-4

1. What is the percent increase or decrease from 15 to 27?

   A 29% C 80%

   B 44% D 280%

2. To the nearest percent, what is the percent increase or decrease from 85 to 22?

   A 74% C 286%

   B 126% D 486%

3. Maya is buying a shelving unit for her closet. The unit she wants to buy is on sale for $833. If the regular price is $980, what is the percent discount that Maya is receiving?

   A 15% C 82%

   B 18% D 85%

4. Philip is performing a science experiment about growing crystals. Each day he weighs his crystals to measure the amount of growth that has occurred. On Friday his crystals weighed 18 grams, and on Monday they weighed 27 grams. What was the percent increase in the weight of Philip's crystals over the weekend?

   A $33\frac{2}{3}\%$

   B 50%

   C $66\frac{2}{3}\%$

   D 150%

5. ***SHORT RESPONSE*** A computer that sells for $1295 is on sale for 30% off the regular price. What is the sale price of the computer? Show your work.

6. ***SHORT RESPONSE*** The wholesale price of a recliner is $430. The store increases the price by 40% before selling them. What is the store's selling price for each recliner? Show your work.

7. ***SHORT RESPONSE*** Mr. Escobar received a shipment of dining room tables that cost him $450 each. He marks up the price of each table by 25% before selling them to his customers. How much money does Mr. Escobar make on each table that he sells? What is the price that his customers pay for each table?

LESSON 8-5

8. Which is the *best* estimate of 26% of 37?

   A 10 C 19

   B 13 D 26

9. Which is the *best* estimate of 32% of 61?

   A 10 C 20

   B 15 D 30

10. A store is offering 31% off all coat prices. Which is the *best* estimate of the amount of money you will save if you buy a coat that is normally $59.99?

    A $15 C $24

    B $20 D $40

11. A pizza parlor is offering a 26% discount on the purchase of a slice of pepperoni pizza and a medium drink. If the combination is normally $5.99, which is the *best* estimate of how much you will pay after the discount?

    A $1.50 C $3.60

    B $3.00 D $4.50

12. ***SHORT RESPONSE*** The highest point in Australia is Mount Kosciusko. The height of Mount Kosciusko is 32% of the height of the highest point in South America, Mount Aconcagua, at 22,834 feet. Kaya estimated the height of Mount Kosciusko to be 6900 feet. Explain whether this is reasonable.

8B Applying Percents

LESSON 8-6

13. A furniture salesperson sold $8759 worth of furniture last month. If he makes 4% commission on all sales and earns a monthly salary of $1500, what was his total pay last month?

A $350.36 C $1910.36

B $1850.36 D $5003.60

14. Antwaan earns $1250 per month. Of that, $89.38 is withheld for social security and taxes. Approximately what percent of Antwaan's earnings are withheld for social security and taxes?

A 6.5% C 7.15%

B 9.3% D 14%

15. ***SHORT RESPONSE*** Simon bought a set of speakers for $279 and a new tuner for $549. Sales tax on these items was 7.5%. Write a numeric expression that could be used to determine Simon's total bill for these items. What was Simon's total bill for these items?

16. ***SHORT RESPONSE*** Ashley earns 22% on all the glassware she sells. This month she earned $2750. How much did her glassware sales total? Show your work.

17. ***SHORT RESPONSE*** Delia bought 5 pounds of potatoes and 2 pounds of carrots. The potatoes cost $0.89 per pound, and the carrots cost $0.55 per pound. The sales tax rate was 7.15%. How much tax was Delia charged? What was Delia's total bill for potatoes and carrots?

18. ***SHORT RESPONSE*** Tiare figures out that 19% of his job earnings go to taxes, insurance, and social security. If Tiare works 38 hours for $6.50 per hour, how much money will he get paid? Show your work.

LESSON 8-7

19. Nigel borrowed $7500 to add a new bathroom to his house. The bank charges $6\frac{1}{2}$% simple interest on his 3 year loan. What is the total amount Nigel will repay the bank?

A $1462.50 C $8906.25

B $7987.50 D $8962.50

20. Rich borrowed $16,000 for 12 years at an annual simple interest rate to pay for college. If he repaid a total of $31,360, at what interest rate did he borrow the money?

A 6.12% C 26.13%

B 8% D 196%

21. ***SHORT RESPONSE*** Gwen invested $10,000 in a mutual fund at a yearly rate of 7%. She earned $5600 in simple interest. How long was the money invested? Show your work.

22. ***SHORT RESPONSE*** Ray earned $5000, which he used to buy a 5-year certificate of deposit (CD). The CD paid an annual simple interest rate of 6%. What will the CD be worth at the end of 5 years? How much interest will it have earned?

23. ***SHORT RESPONSE*** The Murphy's determined that the value of their home increases at a simple yearly interest rate of 3.7%. They have owned their home for 18 years, and it has increased in value by $86,580. How much was their house worth when they bought it? How much will it be worth 5 years from now?

24. ***SHORT RESPONSE*** Marissa invested $7000 at a 7.25% yearly simple interest rate, and $5000 at a 3.5% yearly simple interest rate. Write an expression that could be used to determine the total amount of interest Marissa will earn in x years. What will the total for the two investments be after 6 years?

EOG Practice ▪ Chapter 9

9A Experimental Probability

LESSON  9-1

Use the spinner below for problems 1–4.

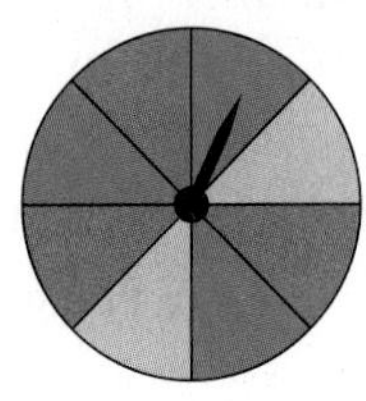

1. What is the probability that the spinner will stop on red?

   A $\frac{1}{4}$ C $\frac{3}{8}$

   B $\frac{5}{8}$ D $\frac{3}{4}$

2. What is the probability that the spinner will stop on red or blue?

   A $\frac{1}{4}$ C $\frac{3}{8}$

   B $\frac{5}{8}$ D $\frac{3}{4}$

3. Which is *not* a way to show the probability that the spinner will stop on yellow?

   A $\frac{1}{4}$ C $\frac{3}{16}$

   B $\frac{2}{8}$ D $\frac{5}{20}$

4. Which is *not* a way to show the probability that the spinner will stop on red or yellow?

   A $\frac{3}{8}$ C $\frac{10}{16}$

   B $\frac{5}{8}$ D $\frac{15}{24}$

5. What is the probability of randomly choosing one of the four answers below and getting this problem *incorrect*?

   A $\frac{1}{4}$ C $\frac{3}{4}$

   B $\frac{1}{2}$ D 1

6. A store is holding a contest. The probability of winning a car in the contest is $\frac{1}{50,000}$, the probability of winning a television is $\frac{1}{5000}$, and the probability of winning a $5 gift certificate for the store is $\frac{1}{500}$. What is the probability of winning the car or a gift certificate?

   A $\frac{99}{50,000}$

   B $\frac{101}{50,000}$

   C $\frac{11}{5000}$

   D $\frac{9}{5000}$

7. ***SHORT RESPONSE*** Marci decided to wear her new black shoes to school. She went into her closet without turning on the light, and randomly chose a pair of shoes from her closet. If Marci has 11 pairs of shoes, what is the probability that the shoes she chose are not the pair that she wanted? Show your work.

Use the table for problems 8–10.

Table of Outcomes				
Outcome	A	B	C	D
Probability	0.124	0.501	0.25	0.125

8. ***SHORT RESPONSE*** What is the probability that A, B, or D will occur? Explain in words how you determined your answer.

9. ***SHORT RESPONSE*** What is the probability that B or C will not occur? Explain in words how you determined your answer.

10. ***SHORT RESPONSE*** Suppose that a fifth possible outcome E exists for the situation described in the table. Explain what the probability of E occurring must be.

9A Experimental Probability

LESSON 9-2

A utensil is drawn at random from a drawer and then replaced. The table shows the results after 100 draws. Use the table for problems 11–13.

Outcome	Draws
Spoon	37
Knife	32
Fork	31

11. What is the experimental probability of drawing a spoon?

A 37%　　C 68%

B 63%　　D 69%

12. What is the experimental probability of *not* drawing a knife or a spoon?

A 31%　　C 68%

B 37%　　D 69%

13. After an additional 25 draws, the result for spoon draws increased by 5, the result for knife draws increased by 9, and the result for fork draws increased by 11. What is the new experimental probability of drawing a spoon?

A 5%　　C 33.6%

B 20%　　D 66.4%

14. ***SHORT RESPONSE*** A sales assistant tracks the sales of a particular sweater. The table shows the data after 1000 sales.

Outcome	Sales
Turquoise	361
Lavender	207
Pink	189
Green	243

Estimate the probability that the next sweater sold will *not* be pink. Explain in words how you determined your answer.

LESSON 9-3

Use the entire table for problems 15–17.

53736	85815	87649	31119
16635	65161	27919	86585
32848	94425	61378	41256
11632	46278	38783	87649
13325	60848	74681	54238
94228	82794	23426	46498
46278	65264	13906	24794
85976	98713	51876	25847
65972	41973	58927	16842
58147	52697	28467	21358
20650	59731	20587	20648
91845	27364	59421	18579

15. ***SHORT RESPONSE*** A golfer has an 81% chance of making a putt on the first try. Use the table of random numbers to create a simulation, and estimate the probability that she will make the putt on the first try at least 8 out of 10 times. What range of values did you choose to represent a successful first putt? Explain in words how you used the table of random numbers to determine your answer.

16. ***SHORT RESPONSE*** A field-goal kicker has a 94% chance of making successful field goals. Estimate the probability that he will make at least 9 of his next 10 field-goal attempts. Use the table of random numbers to create a simulation. Show your work.

17. ***SHORT RESPONSE*** At a local ice cream parlor, about 45% of the customers order chocolate ice cream. Estimate the probability that 3 out of 5 of the next customers will order chocolate. Use the table of random numbers to create a simulation. Show your work.

EOG Practice ■ Chapter 10

10B Solving Equations and Inequalities

LESSON 10-4

1. Which does *not* apply to the inequality $4a + 3 < 11$ and the graph of its solution?

   A $a < 2$

   B There is an open circle on the point 2.

   C Solutions are greater than 2.

   D Solutions are less than 2.

2. Which does *not* apply to the inequality $\frac{5d}{6} - \frac{1}{3} \leq \frac{15}{9}$ and the graph of its solution?

   A $d \leq 2.4$

   B There is a closed circle on the point $\frac{12}{5}$.

   C Solutions are greater than $\frac{12}{5}$.

   D $d \leq \frac{12}{5}$

3. Which is the correct graph of the solution of the inequality $4z + 8 - z \leq -1$?

   A −5 −4 −3 −2 −1 0 1

   B −5 −4 −3 −2 −1 0 1

   C −3 −2 −1 0 1 2 3

   D −3 −2 −1 0 1 2 3

4. Which is the correct graph of the solution of the inequality $-6f + 4 \leq 10$?

   A −3 −2 −1 0 1 2 3

   B −3 −2 −1 0 1 2 3

   C −3 −2 −1 0 1 2 3

   D −3 −2 −1 0 1 2 3

5. ***SHORT RESPONSE*** Shelly makes doll dresses and sells them for $12 each. The cost of the material for one dress is $4, and the fixed cost of the sewing machine is $360. Write an inequality that could be used to determine how many dresses Shelly must sell in order to make a profit. Solve your inequality.

LESSON 10-5

6. Which is *not* a correct way to solve for one of the variables in the equation $P_{\Delta} = s_1 + s_2 + s_3$?

   A $s_1 = P_{\Delta} - s_2 - s_3$

   B $s_2 = s_1 + s_3 - P_{\Delta}$

   C $s_3 = P_{\Delta} - s_1 - s_2$

   D $s_1 = -s_3 - s_2 + P_{\Delta}$

7. Which equation is represented by the table of ordered pairs shown below?

x	$\frac{2}{3}x + 2$	y	(x, y)
−2	$\frac{2}{3}(-2) + 2$	$\frac{2}{3}$	$(-2, \frac{2}{3})$
−1	$\frac{2}{3}(-1) + 2$	$\frac{4}{3}$	$(-1, \frac{4}{3})$
0	$\frac{2}{3}(0) + 2$	2	$(0, 2)$
1	$\frac{2}{3}(1) + 2$	$\frac{8}{3}$	$(1, \frac{8}{3})$
2	$\frac{2}{3}(2) + 2$	$\frac{10}{3}$	$(2, \frac{10}{3})$

   A $2x - 3y = -6$ C $2x - 3y = 6$

   B $3x - 2y = -6$ D $2x + 3y = -6$

8. ***SHORT RESPONSE*** Solve the equation $a^2 + b^2 = c^2$ for a. Show your work or explain in words how you determined your answer.

SHORT RESPONSE **Solve each equation for *y*. Graph each equation on a coordinate plane.**

9. $3x - y = 5$

10. $4y - 2x = 4$

11. $2x - 2y = 0$

10B Solving Equations and Inequalities

LESSON 10-6

12. What is the solution to the system of equations below?

$$3x + 2y = -10$$
$$3y - 2x = -2$$

A $(-2, -2)$
C $(2, 2)$
B $(-2, 1)$
D $(-2, -1)$

13. What is the solution to the system of equations below?

$$y - x = -1$$
$$y - 2x = -2$$

A $(1, 2)$
C $(0, 2)$
B $(1, 0)$
D $(0, -2)$

14. What is the solution to the system of equations below?

$$-y = 2$$
$$3y - x = 0$$

A $(7, 1)$
C $(1, 3)$
B $(3, 13)$
D $(-6, -2)$

15. The ordered pair $(-1, -2)$ is the solution to which system of equations?

A $2y - 2x = 2$
$3x + y = -5$

B $2y - 2x = -2$
$3x + y = 5$

C $2y - 2x = -2$
$3x - y = -5$

D $2y - 2x = -2$
$3x + y = -5$

16. The ordered pair $(5, 7)$ is the solution to which system of equations?

A $-3y - 4x = 1$
$y + x = 12$

B $3y - 4x = 1$
$-y + x = 12$

C $3y - 4x = 1$
$y + x = 12$

D $3y + 4x = 1$
$y + x = 12$

17. ***SHORT RESPONSE*** Julie bought supplies for a birthday party. Three rolls of streamers and 15 party horns cost $30. Later Julie went back and bought 2 more rolls of streamers and 4 more party horns for $11. Write a system of equations that can be used to determine the price of each roll of streamers and each party horn. Solve your system.

18. ***SHORT RESPONSE*** A group of friends go out for lunch. If 2 people have hamburgers and 5 have hot dogs, the bill will be $8.00. If 5 people have hamburgers and 2 people have hot dogs, the bill will be $9.50. Write a system of equations that could be used to determine the price of a hamburger. What is the price of each hamburger?

19. ***SHORT RESPONSE*** A test is made up of 40 questions for a total of 100 points. Some questions are worth 2 points, and some questions are worth 4 points. Write a system of equations that could be used to determine the number of questions of each type that are on the test. How many 4-point questions are on the test?

20. ***SHORT RESPONSE*** Luigi and Stephen volunteered to pick up trash along the highway. Luigi collected 3 more bags than Stephen did. Explain what other information you would need to find the number of bags collected by each boy.

SHORT RESPONSE **Solve each system of equations. Show your work. Graph each system of equations.**

21. $y = x - 1$
$y = -2x + 5$

22. $y = x + 1$
$y = -2x - 4$

23. $x - 2y = 11$
$3y + 5x = 3$

24. $x + y = 4$
$2y - 2x = 8$

EOG Practice ▪ Chapter 11

11A Linear Equations

LESSON 11-1

SHORT RESPONSE **Graph each equation and explain whether it is linear.**

1. $y = 3x - 4$
2. $y = -2x + 1$
3. $y = x^2 - 3$
4. $y = -x - 2$

5. ***SHORT RESPONSE*** A limousine company charges a base fee of $200, plus $50 for each hour of rental. This is represented by the equation $C = 50h + 200$, where C is the total cost based on the number of hours h. Find the total cost for 2 hours, 3 hours, 4 hours, 5 hours, and 6 hours. Is this a linear equation? Graph the relationship between the total cost and the number of hours of rental.

LESSON 11-2

6. What is the slope of the line that passes through the points (2, 4) and (−3, 1)?

   A $\frac{1}{7}$ C 1

   B $\frac{3}{5}$ D $\frac{5}{3}$

7. What is the slope of the line that passes through the points (2, 6) and (1, 4)?

   A 2 C −2

   B −1 D 1

8. What is the slope of a line perpendicular to the line that passes through the points (3, 3) and (1, −4)?

   A $-\frac{7}{2}$ C $\frac{2}{7}$

   B $-\frac{2}{7}$ D $\frac{7}{2}$

9. What is the slope of a line parallel to the line that passes through the points (2, 5) and (−7, 3)?

   A $-\frac{9}{2}$ C $\frac{2}{9}$

   B $-\frac{2}{9}$ D $\frac{9}{2}$

10. What is the slope of the line graphed below?

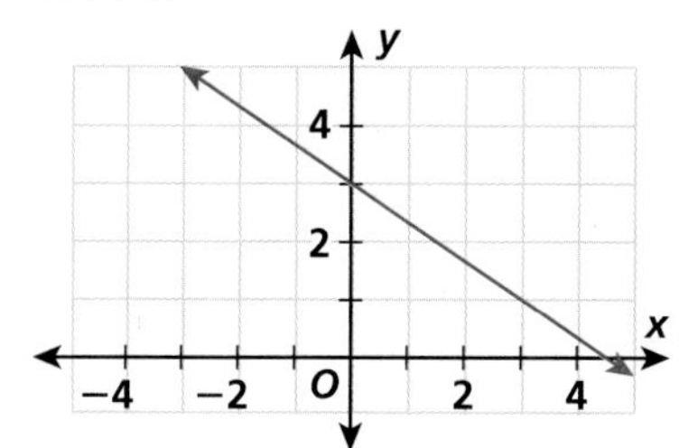

   A $-\frac{3}{2}$ C $\frac{2}{3}$

   B $-\frac{2}{3}$ D $\frac{3}{2}$

11. What is the slope of the line graphed below?

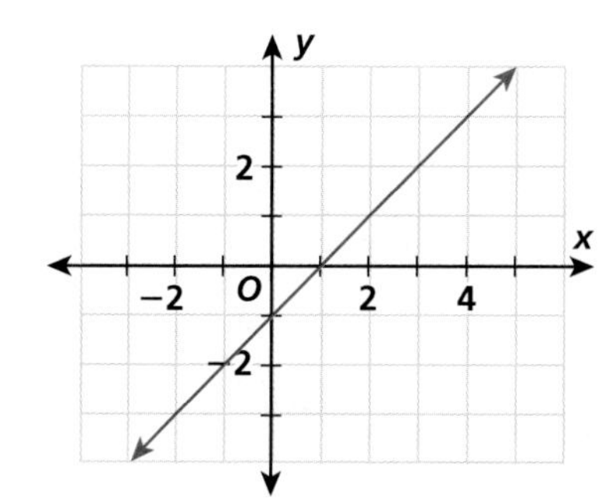

   A $-\frac{5}{4}$ C $\frac{4}{5}$

   B $-\frac{4}{5}$ D $\frac{5}{4}$

12. ***SHORT RESPONSE*** Line a passes through (−2, −6) and (2, −4). Line b passes through (−4, 1) and (4, 5). Explain whether the two lines are parallel, perpendicular, or neither.

13. ***SHORT RESPONSE*** Graph the line that passes through (4, −2) and (−2, −11). Give the slope of a line that is perpendicular to the line you graphed.

14. ***SHORT RESPONSE*** Rico and his family went on a road trip. They traveled at a constant rate, and after 2 hours on the road they had traveled 86 miles. After 5 hours on the road, they had traveled 215 miles. Give two points on the line that describes the distance they had traveled with relation to time. What is the slope of the line?

EOG Practice ■ Chapter 11

11A Linear Equations

LESSON 11-3

15. What is the slope-intercept form of the equation $4x - y = 7$?

A $y = 4x - 7$ C $y = 4x + 7$

B $y = -4x + 7$ D $-y = -4x + 7$

16. What is the slope-intercept form of the equation $3y + 5 = 2x$?

A $y = \frac{3}{2}x + 5$ C $y = \frac{2}{3}x - \frac{3}{5}$

B $y = \frac{2}{3}x - 5$ D $y = \frac{2}{3}x - \frac{5}{3}$

17. What is the equation in slope-intercept form of the line graphed below?

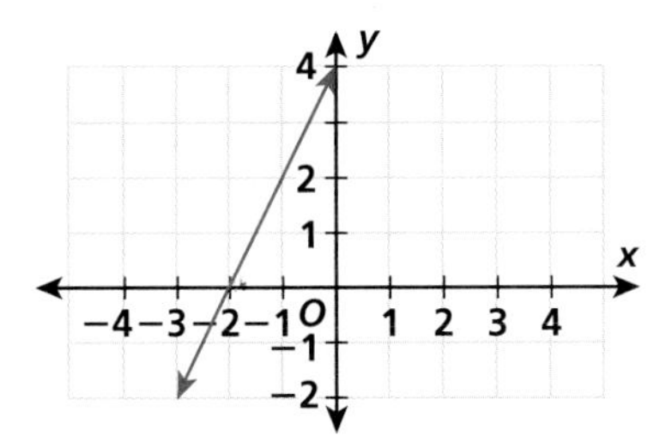

A $2x + y = -4$ C $\frac{2x}{y} + 4 = 0$

B $y = 2x + 4$ D $2x - y = -4$

18. What is the equation in slope-intercept form of a line parallel to the line graphed below?

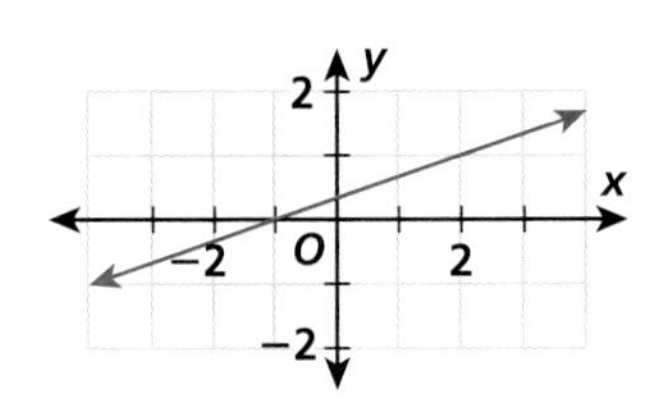

A $y = -x + 3$ C $y = \frac{x}{3}$

B $3y = -x$ D $y = x + 3$

SHORT RESPONSE **Find the x- and y-intercepts of each line. Use the intercepts to graph each line.**

19. $4x - 3y = 7$ 20. $5x + 3 = 4y$

SHORT RESPONSE **Write the equation of the line in slope-intercept form that passes through the given points. Show your work.**

21. $(2, -3)$ and $(-4, 5)$ 22. $(4, 1)$ and $(-1, -4)$

$x_1\ y_1 \quad x_2\ y_2$

LESSON 11-4

23. What is the equation in point-slope form of the line graphed below?

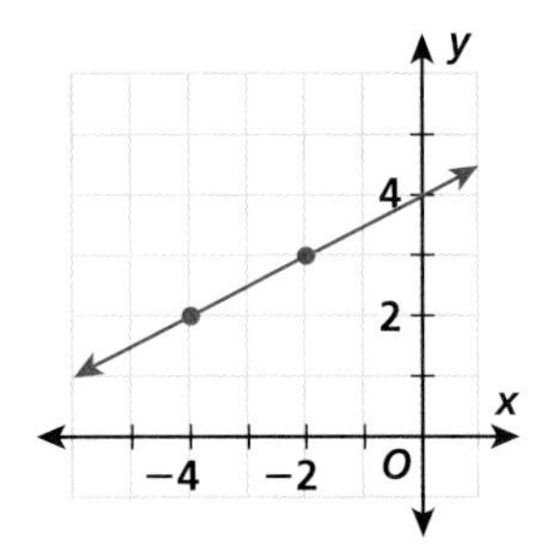

A $y - 3 = \frac{1}{2}(x - 2)$

B $y + 3 = \frac{1}{2}(x + 2)$

C $y - 3 = \frac{1}{2}(x + 2)$

D $y - 3 = -\frac{1}{2}(x + 2)$

24. Which is a point that the line $y + 5 = \frac{1}{4}(x - 3)$ passes through?

A $(-5, 3)$ C $(3, -5)$

B $(-3, 5)$ D $(5, -3)$

25. What is the slope of a line perpendicular to the line $y + 2 = 2(x - 1)$?

A -2 C $\frac{1}{2}$

B $-\frac{1}{2}$ D 2

26. ***SHORT RESPONSE*** Identify a point that the line $y - 6 = -\frac{4}{5}(x + 3)$ passes through. Give the slope of the line. Graph the line on a coordinate plane.

27. ***SHORT RESPONSE*** Robert bought a rubber-band ball made from 75 rubber bands. Each day he adds 22 rubber bands to the ball. Write an equation in point-slope form that represents the total number of rubber bands in the ball after x days. Graph your equation on a coordinate plane.

28. ***SHORT RESPONSE*** Write the point-slope form of the equation with slope 2 that passes through the point $(1, 4)$. Show your work.

EOG Practice ■ Chapter 11

11B Linear Relationships

LESSON 11-5

1. Which of the following data sets does *not* show a direct variation?

A

Serving Size (g)	Sugar Added (g)
100	25
200	50
300	75
400	100

B

Serving Size (g)	Sugar Added (g)
5	2
10	4
15	6
20	8

C

Serving Size (g)	Sugar Added (g)
10	4
20	8
30	12
40	16

D

Serving Size (g)	Sugar Added (g)
20	5
25	10
30	15
35	20

2. If y varies directly with x and y is 90 when x is 18, what is the constant of proportionality?

A $\frac{1}{5}$

B 5

C 72

D 108

SHORT RESPONSE **Given that y varies directly with x, find each equation of direct variation. Show your work.**

3. When x is 3, y is 6.
4. When x is 10, y is 15.
5. When x is 2, y is 84.
6. When x is 9, y is 36.

SHORT RESPONSE **Tell whether each data set shows a direct variation. Explain in words how you determined your answer.**

7.

Weight of Patient (lb)	Medication Prescribed (mg)
100	10
120	12
140	14
160	16
180	18

8.

Cost of Item	Sales Tax Applied
$12.50	$0.94
$34.97	$2.62
$52.10	$3.91
$64.00	$4.80
$87.56	$6.57

9. ***SHORT RESPONSE*** The distance d an object falls from a resting position varies directly with the square of the time t of the fall. This can be expressed by the formula $d = kt^2$. An object falls 90 feet in 3 seconds. What is the value of the constant of proportionality? How far will the object fall in 15 seconds?

10. ***SHORT RESPONSE*** Instructions for a cleaning-fluid concentrate state that 3 ounces of the concentrate should be added for every $2\frac{1}{2}$ gallons of water used. How much concentrate should be added to 20 gallons of water? Show your work.

EOG Practice ■ Chapter 11

11B Linear Relationships

LESSON 11-6

11. Which of the following inequalities is graphed below?

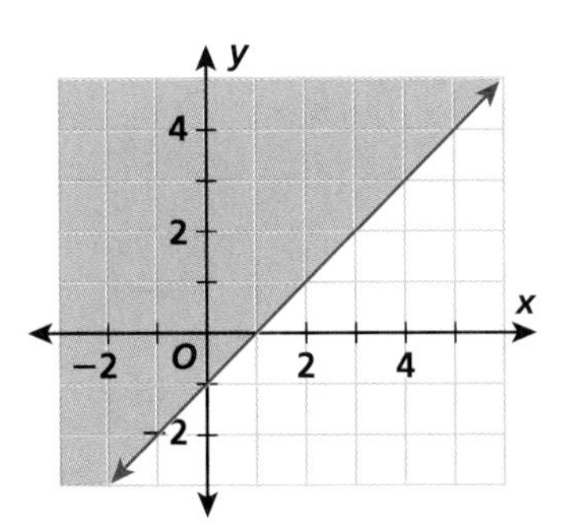

A $y \le x - 1$ C $y \ge x - 1$

B $y > x - 1$ D $y < x - 1$

12. Which of the following inequalities is graphed below?

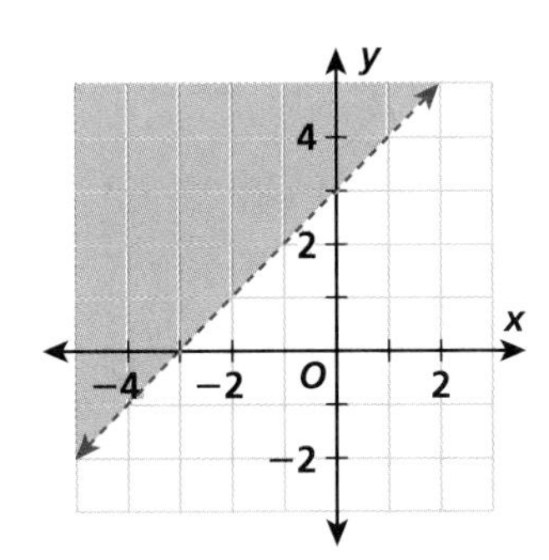

A $y \ge x + 3$ C $y \le x + 3$

B $y > x + 3$ D $y < x + 3$

13. Which of the following inequalities is graphed below?

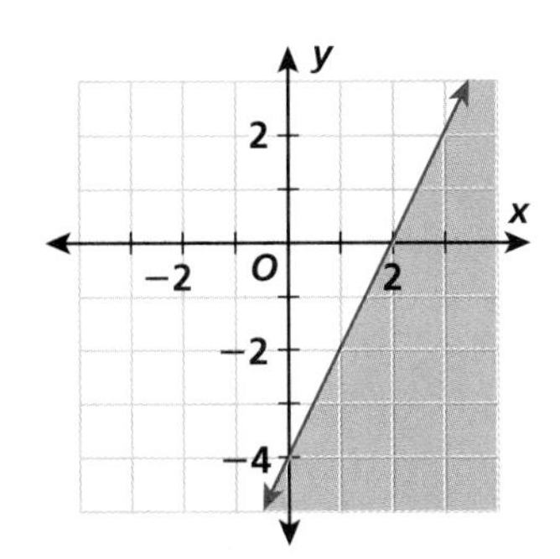

A $4x - 2y < 8$ C $4x - 2y \le 8$

B $4x - 2y > 8$ D $4x - 2y \ge 8$

14. ***SHORT RESPONSE*** A golf cart with a 2.3-gallon gas tank gets at most 3 miles per gallon of gas. On a coordinate plane, graph the relationship between the miles traveled by the golf cart and the number of gallons of gas that have been used. If one trip around the golf course is 3.8 miles, explain whether the driver will be able to make 2 full trips around the golf course without refueling.

SHORT RESPONSE **Graph each inequality, and explain whether the point (2, 1) is a solution.**

15. $6y - 12 < 3x$

16. $-2y \ge x - 3$

LESSON 11-7

17. ***SHORT RESPONSE*** Plot the data from the table below, and find the line of best fit. Write the equation of your line of best fit in slope-intercept form.

Olympic Winning Times for Selected Years in the 200 m Backstroke

Year	Winning Time (min:s)
1964	2:10
1972	2:03
1980	2:02
1988	1:59
1996	1:59
2000	1:58

18. ***SHORT RESPONSE*** Every year on her birthday, Margaret records her height on a scatter plot. Explain the relationship between the correlation of the data and the slope of the line of best fit. Would you expect a line of best fit for this data to have a positive or negative slope?

EOG Practice ■ Chapter 12

12A Sequences

LESSON 12-1

1. What is the 17th term in the arithmetic sequence below?

   7, 14, 21, 28, ...

   A 103 C 119
   B 112 D 126

2. What is the 25th term in the arithmetic sequence below?

   100, 97, 94, 91, ...

   A 25 C 31
   B 28 D 52

3. What is the common difference in the arithmetic sequence below?

   761, 748, 735, 722, ...

   A −27 C 13
   B −13 D 27

4. Look at the arithmetic sequence below.

   52, 41, 30, 19, ...

   What is the rule that describes the sequence?

   A $a_n = 52 - 11(n - 1)$
   B $a_n = 52 + 11(n - 1)$
   C $a_n = 52 - 11(n + 1)$
   D $a_n = 52 + 11(n + 1)$

5. ***SHORT RESPONSE*** Al received 200 bonus points when he signed up for a savings card at a grocery store. For every $100 he spends, he will receive 50 bonus points. Write the rule for an arithmetic sequence that could be used to determine how much money Al must spend to have 1500 bonus points. How much money must Al spend to have 1500 bonus points?

SHORT RESPONSE **Explain whether each sequence could be arithmetic. If so, give the common difference, and write the rule for the sequence.**

6. 203, 195, 187, 179, 171, 163, ...
7. 13, 24, 36, 49, 63, 78, ...

LESSON 12-2

8. What is the 13th term in the geometric sequence?

   2, 4, 8, 16, ...

   A 30 C 8192
   B 208 D 16,384

9. What is the 9th term in the geometric sequence?

   212, 106, 53, 26.5, . . .

   A 0.828125 C 3.3125
   B 0.65625 D 18

10. What is the 23rd term in the geometric sequence?

    3, 6, 12, 24, ...

    A 3,145,728 C 12,582,912
    B 6,291,456 D 25,165,824

11. To the nearest ten-thousandth, what is the 7th term in the geometric sequence?

    300, 100, 33.3, 11.1, ...

    A 0.1371 C 1.234
    B 0.4115 D 3.704

12. ***SHORT RESPONSE*** The water in a 16,000-gallon swimming pool evaporates at a rate of 2% per week during the hot summer months. Find the amount of water remaining in the pool at the end of the first week. Assuming that the pool is not refilled, write the rule for a geometric sequence that could be used to determine how many gallons of water remain in the pool at the end of each week. How much water would be left after 8 weeks?

SHORT RESPONSE **Explain whether each sequence could be geometric. If so, give the common ratio, and write the rule for the sequence.**

13. 6561, 2187, 729, 243, 81, 27, ...
14. 1, 7, 49, 343, 2401, 16,807, ...
15. 4, 8, 24, 120, 720, 5040, ...
16. 18, 54, 162, 486, 1458, 4374, ...

EOG Practice

12A Sequences

LESSON 12-3

17. The next term in the sequence below can be determined using first and second differences. What is the next term?

13, 22, 36, 55, 79, 108, ...

A 113

B 117

C 137

D 142

18. The next term in the sequence below can be determined using first and second differences. What is the next term?

17, 23, 32, 44, 59, 77, ...

A 82

B 95

C 98

D 136

19. The next term in the sequence below can be determined using first and second differences. What is the next term?

10.5, 15.25, 20.75, 27, 34, 41.75, ...

A 42.5

B 49.5

C 50.25

D 72.75

20. Look at the sequence below.

8, 15, 23, 33, 46, 63, ...

What rule describes the pattern of the second difference of the sequence?

A Add 1.

B Multiply by 2.

C It remains constant.

D Subtract 1.

21. Look at the sequence below.

214, 215, 220, 245, 370, 995, ...

What rule describes the pattern of the second difference of the sequence?

A Add 5.

B Multiply by 5.

C It remains constant.

D Subtract 5.

22. What are the first five terms of the sequence defined by $a_n = \frac{n}{n+2}$?

A $\frac{1}{2}, \frac{1}{4}, \frac{1}{8}, \frac{1}{16}, \frac{1}{32}$

B $\frac{1}{3}, \frac{1}{4}, \frac{1}{5}, \frac{1}{6}, \frac{1}{7}$

C $\frac{1}{3}, \frac{1}{2}, \frac{3}{5}, \frac{2}{3}, \frac{5}{7}$

D $\frac{1}{2}, \frac{2}{4}, \frac{3}{8}, \frac{4}{16}, \frac{5}{32}$

23. What are the first five terms of the sequence defined by $a_n = 2n\left(\frac{1}{n}\right)$?

A 2, 2, 2, 2, 2

B 2, 4, 6, 8, 10

C $2, \frac{1}{2}, \frac{1}{3}, \frac{1}{4}, \frac{1}{5}$

D $2, 1, \frac{2}{3}, \frac{3}{4}, \frac{4}{5}$

24. Look at the sequence below.

$1, \frac{1}{2}, \frac{1}{4}, \frac{1}{8}, \frac{1}{16}, \frac{1}{32}, \ldots$

What is the rule that describes the sequence?

A Add 2 to the denominator.

B Multiply the denominator by 2.

C Divide the denominator by 2.

D Subtract –2 from the denominator.

25. Look at the sequence below.

800, 200, 50, 12.5, 3.125, ...

What is the rule that describes the sequence?

A Add 2^n.

B Multiply by −4.

C Divide by 4.

D Subtract 2^n.

SHORT RESPONSE **Give the next three terms in each sequence using the simplest rule you can find. Describe your rule in words.**

26. 1, 1, 3, 3, 5, 5, 7, ...

27. 1, 3, 9, 27, 81, 243, ...

28. 801, 808, 805, 812, 809, ...

12B Functions

LESSON 12-4

1. Which function is graphed below?

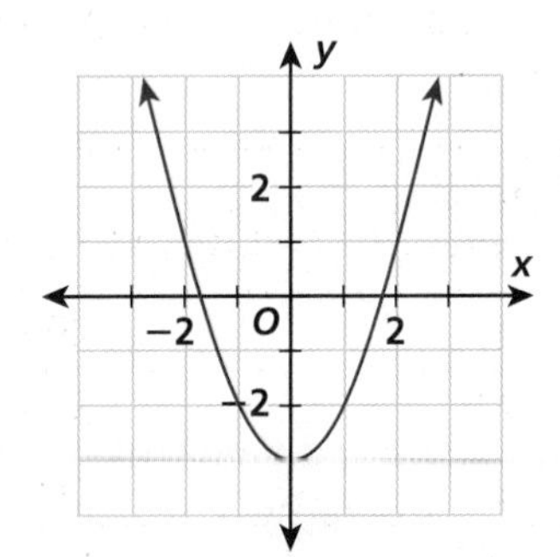

A $y = x^2 + 3$
B $y = x^2 - 3$
C $y = -x^2 - 3$
D $y = 3x^2$

2. Which function is graphed below?

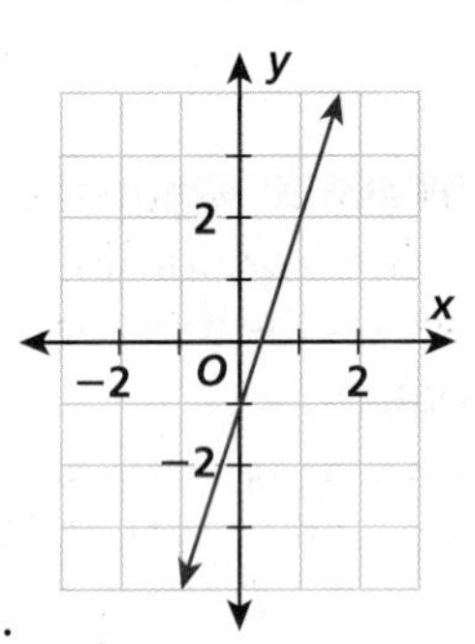

A $y = \frac{1}{3}x - 1$
B $y = -\frac{1}{3}x - 1$
C $y = 3x - 1$
D $y = -3x - 1$

3. Assuming f is the function graphed below, what is the value of $f(-1)$?

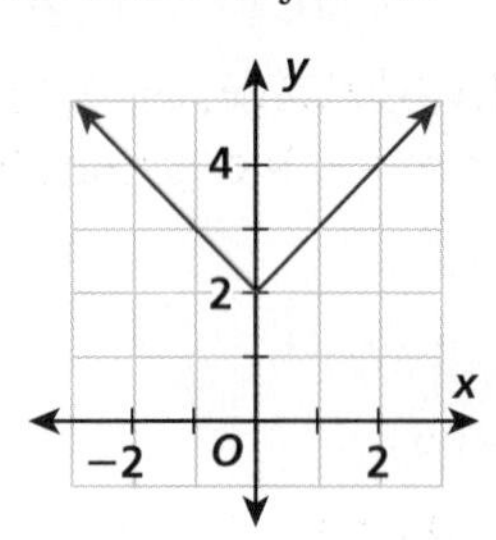

A −1
B 0
C 1
D 3

4. What is the value of $f(0)$ for the function $y = -3x + 4$?

A −3
B 0
C 4
D 7

5. ***SHORT RESPONSE*** Explain whether the given relationship represents a function.

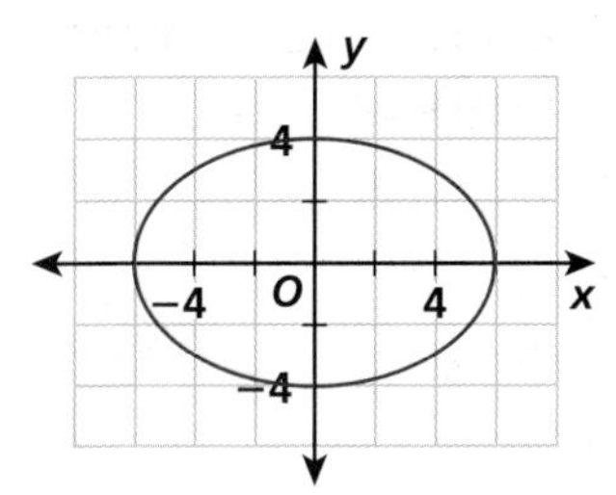

LESSON 12-5

6. What is the rule for the linear function graphed below?

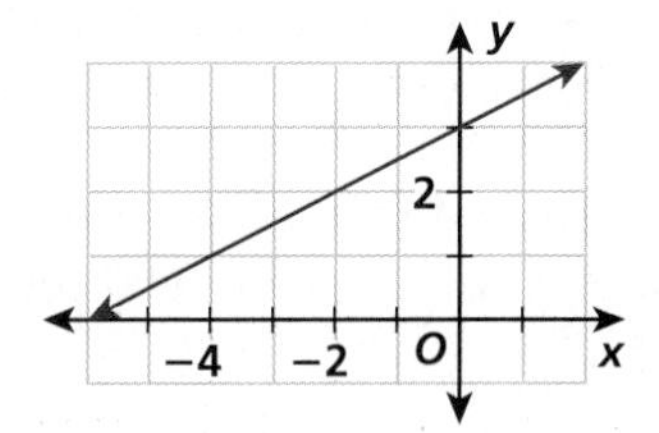

A $y = 3x + 2$
B $y = \frac{1}{2}x + 3$
C $y = 2x + 3$
D $y = -\frac{1}{2}x + 3$

7. What is the rule for the linear function represented in the table below?

x	y
−1	−8
0	−5
1	−2

A $y = \frac{1}{3}x - 5$
B $y = -\frac{1}{3}x - 5$
C $y = 3x - 5$
D $y = -3x - 5$

8. Manuel's New Year's resolution is to walk 6 miles each week. He has already walked 17 miles since New Year's Day. Which linear function could be used to determine the total number of miles he will have walked 7 weeks from now?

A $y = 6x$
B $y = 6x + 17$
C $y = 6 + 17x$
D $y = 17x$

12B Functions

LESSON 12-6

SHORT RESPONSE **Create a table for each exponential function. Graph each function on a coordinate plane.**

9. $f(x) = 3 \cdot 4^x$
10. $f(x) = \frac{1}{2} \cdot 3^x$
11. $f(x) = 0.75 \cdot 2^x$
12. $f(x) = 2 \cdot 10^x$
13. ***SHORT RESPONSE*** The isotope cobalt-60, found in radioactive waste, has a half-life of 5 years. Write a function that could be used to determine the amount of a 150-gram sample of cobalt-60 remaining after x years. Use your function to determine the amount of cobalt-60 remaining after 35 years.

LESSON 12-7

14. What is the value of $f(3)$ for the function $f(x) = x^2 + 2x - 4$?

    A 3

    B 9

    C 11

    D 15

SHORT RESPONSE **Create a table for each quadratic function. Graph each function on a coordinate plane.**

15. $f(x) = x^2 - 3$
16. $f(x) = x^2 - x + 6$
17. $f(x) = (x - 1)(x + 2)$
18. ***SHORT RESPONSE*** A penny is tossed into a wishing well that is 1386 feet deep. The initial speed of the penny is 10 feet/second. Graph the quadratic function that describes the distance the penny has fallen after t seconds. Use the formula $d = vt + 16t^2$, where d is the distance in feet, v is the initial speed in feet/second, and t is the time in seconds.

LESSON 12-8

19. Which inverse variation function describes the relationship represented in the table below?

Outdoor Temperature (°F)	Cups of Coffee Sold
40°	200
25°	320
20°	400
10°	800
5°	1600

A $y = \frac{5}{x}$

B $y = \frac{40}{x}$

C $y = \frac{200}{x}$

D $y = \frac{8000}{x}$

20. ***SHORT RESPONSE*** Graph the relationship from the table on a coordinate plane. Explain whether the relationship is an inverse variation.

Chips Produced	Time (min)
50	2
64	3
96	5
121	6
150	8

SHORT RESPONSE **Graph each inverse variation function on a coordinate plane.**

21. $f(x) = \frac{3}{x}$
22. $f(x) = \frac{-0.5}{x}$
23. ***SHORT RESPONSE*** The speed s of a gear varies inversely with the number of teeth t on the gear. One gear that has 64 teeth makes 12 revolutions per minute. Write a function that could be used to determine how many revolutions per minute would be made by a gear with t teeth. How many revolutions per minute would be made by a gear with 16 teeth?

Factors and Multiples

When two numbers are multiplied to form a third, the two numbers are said to be **factors** of the third number. **Multiples** of a number can be found by multiplying the number by 1, 2, 3, 4, and so on.

EXAMPLE

A **List all the factors of 48.**

$1 \cdot 48 = 48$, $2 \cdot 24 = 48$, $3 \cdot 16 = 48$,
$4 \cdot 12 = 48$, and $6 \cdot 8 = 48$

So the factors of 48 are
1, 2, 3, 4, 6, 8, 12, 16, 24, and 48.

B **Find the first five multiples of 3.**

$3 \cdot 1 = 3$, $3 \cdot 2 = 6$, $3 \cdot 3 = 9$,
$3 \cdot 4 = 12$, and $3 \cdot 5 = 15$

So the first five multiples of 3 are
3, 6, 9, 12, and 15.

PRACTICE

List all the factors of each number.

1. 8 **2.** 20 **3.** 9 **4.** 51 **5.** 16 **6.** 27

Write the first five multiples of each number.

7. 9 **8.** 10 **9.** 20 **10.** 15 **11.** 7 **12.** 18

Divisibility Rules

A number is divisible by another number if the division results in a remainder of 0. Some divisibility rules are shown below.

A number is divisible by . . .	Divisible	Not Divisible
2 if the last digit is an even number.	11,994	2,175
3 if the sum of the digits is divisible by 3.	216	79
4 if the last two digits form a number divisible by 4.	1,028	621
5 if the last digit is 0 or 5.	15,195	10,007
6 if the number is even and divisible by 3.	1,332	44
8 if the last three digits form a number divisible by 8.	25,016	14,100
9 if the sum of the digits is divisible by 9.	144	33
10 if the last digit is 0.	2,790	9,325

PRACTICE

Determine which of these numbers each number is divisible by: 2, 3, 4, 5, 6, 8, 9, 10

1. 56 **2.** 200 **3.** 75 **4.** 324 **5.** 42 **6.** 812

7. 784 **8.** 501 **9.** 2345 **10.** 555,555 **11.** 3009 **12.** 2001

Prime and Composite Numbers

A **prime number** has exactly two factors, 1 and the number itself.

2	Factors: 1 and 2; prime
11	Factors: 1 and 11; prime
47	Factors: 1 and 47; prime

A **composite number** has more than two factors.

4	Factors: 1, 2, and 4; composite
12	Factors: 1, 2, 3, 4, 6, and 12; composite
63	Factors: 1, 3, 7, 9, 21, and 63; composite

EXAMPLE

Determine whether each number is prime or composite.

A 17

Factors
1, 17 ⟶ prime

B 16

Factors
1, 2, 4, 8, 16 ⟶ composite

C 51

Factors
1, 3, 17, 51 ⟶ composite

PRACTICE

Determine whether each number is prime or composite.

1. 5 **2.** 14 **3.** 18 **4.** 2 **5.** 23 **6.** 27

7. 13 **8.** 39 **9.** 72 **10.** 49 **11.** 9 **12.** 89

Prime Factorization (Factor Tree)

A composite number can be expressed as a product of prime numbers. This is the **prime factorization** of the number. To find the prime factorization of a number, you can use a factor tree.

EXAMPLE

Find the prime factorization of 24 by using a factor tree.

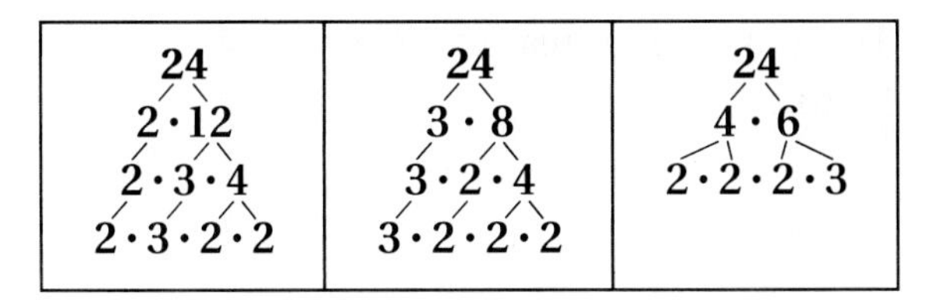

The prime factorization of 24 is $2 \cdot 2 \cdot 2 \cdot 3$, or $2^3 \cdot 3$.

PRACTICE

Find the prime factorization of each number by using a factor tree.

1. 25 **2.** 16 **3.** 56 **4.** 18 **5.** 72 **6.** 40

...eatest Common Factor (GCF)

...**greatest common factor (GCF)** of two whole numbers is the greatest ...or the numbers have in common.

EXAMPLE

Find the GCF of 24 and 32.

Method 1: List all the factors of both numbers.

Find all the common factors.

24: 1, 2, 3, 4, 6, 8, 12, 24
32: 1, 2, 4, 8, 16, 32

The common factors are 1, 2, 4, and 8.
So the GCF is 8.

Method 2: Find the prime factorizations.

Then find the common prime factors.

24: $2 \cdot 2 \cdot 2 \cdot 3$
32: $2 \cdot 2 \cdot 2 \cdot 2 \cdot 2$

The common prime factors are 2, 2, and 2.
The product of these is the GCF.
So the GCF is $2 \cdot 2 \cdot 2 = 8$.

PRACTICE

Find the GCF of each pair of numbers by either method.

1. 9, 15 **2.** 25, 75 **3.** 18, 30 **4.** 4, 10 **5.** 12, 17 **6.** 30, 96
7. 54, 72 **8.** 15, 20 **9.** 40, 60 **10.** 40, 50 **11.** 14, 21 **12.** 14, 28

Skills Bank

Least Common Multiple (LCM)

The **least common multiple (LCM)** of two numbers is the smallest common multiple the numbers share.

EXAMPLE

Find the least common multiple of 8 and 10.

Method 1: List multiples of both numbers.

8: 8, 16, 24, 32, 40, 48, 56, 64, 72, 80
10: 10, 20, 30, 40, 50, 60, 70, 80, 90

The smallest common multiple is 40.

So the LCM is 40.

Method 2: Find the prime factorizations. Then find the most occurrences of each factor.

8: $2 \cdot 2 \cdot 2$
10: $2 \cdot 5$

The LCM is the product of the factors.

$2 \cdot 2 \cdot 2 \cdot 5 = 40$ So the LCM is 40.

PRACTICE

Find the LCM of each pair of numbers by either method.

1. 2, 4 **2.** 3, 15 **3.** 10, 25 **4.** 10, 15 **5.** 3, 7 **6.** 18, 27
7. 12, 21 **8.** 9, 21 **9.** 24, 30 **10.** 9, 18 **11.** 16, 24 **12.** 8, 36

Compatible Numbers

Compatible numbers are close to the numbers in a problem and divide without a remainder. You can use compatible numbers to estimate quotients.

EXAMPLE

Use compatible numbers to estimate each quotient.

A $6134 \div 32$

$6134 \div 32$

$6000 \div 30 = 200$ ← *Estimate*

Compatible numbers (6000 and 30)

B $647 \div 7$

$647 \div 7$

$630 \div 7 = 90$ ← *Estimate*

Compatible numbers (630 and 7)

PRACTICE

Estimate the quotient by using compatible numbers.

1. $345 \div 5$
2. $5474 \div 23$
3. $46{,}170 \div 18$
4. $749 \div 7$
5. $861 \div 41$
6. $1225 \div 2$
7. $968 \div 47$
8. $3456 \div 432$
9. $5765 \div 26$
10. $25{,}012 \div 64$
11. $99{,}170 \div 105$
12. $868 \div 8$

Mixed Numbers and Fractions

Mixed numbers can be written as fractions greater than 1, and fractions greater than 1 can be written as mixed numbers.

EXAMPLE

A **Write $\frac{23}{5}$ as a mixed number.**

$\frac{23}{5}$ *Divide the numerator by the denominator.*

$$\begin{array}{r} 4 \\ 5\overline{)23} \\ \underline{20} \\ 3 \end{array} \longrightarrow 4\frac{3}{5}$$

← *Write the remainder as the numerator of a fraction.*

B **Write $6\frac{2}{7}$ as a fraction.**

Multiply the denominator by the whole number. *Add the product to the numerator.*

$6\frac{2}{7} \longrightarrow 7 \cdot 6 = 42 \longrightarrow 42 + 2 = 44$

Write the sum over the denominator. → $\frac{44}{7}$

PRACTICE

Write each mixed number as a fraction. Write each fraction as a mixed number.

1. $\frac{22}{5}$
2. $9\frac{1}{7}$
3. $\frac{41}{8}$
4. $5\frac{7}{9}$
5. $\frac{7}{3}$
6. $4\frac{9}{11}$
7. $\frac{47}{16}$
8. $3\frac{3}{8}$
9. $\frac{31}{9}$
10. $8\frac{2}{3}$
11. $\frac{33}{5}$
12. $12\frac{1}{9}$

Multiply and Divide Decimals by Powers of 10

Notice the pattern below.

$0.24 \cdot 10 = 2.4$
$0.24 \cdot 100 = 24$
$0.24 \cdot 1000 = 240$
$0.24 \cdot 10{,}000 = 2400$

$10 = 10^1$
$100 = 10^2$
$1000 = 10^3$
$10{,}000 = 10^4$

*Think: When multiplying decimals by powers of 10, move the decimal point one place to the **right** for each power of 10, or for each zero.*

Notice the pattern below.

$0.24 \div 10 = 0.024$
$0.24 \div 100 = 0.0024$
$0.24 \div 1000 = 0.00024$
$0.24 \div 10{,}000 = 0.000024$

*Think: When dividing decimals by powers of 10, move the decimal point one place to the **left** for each power of 10, or for each zero.*

PRACTICE

Find each product or quotient.

1. $10 \cdot 9.26$ **2.** $0.642 \cdot 100$ **3.** $10^3 \cdot 84.2$ **4.** $0.44 \cdot 10^4$

5. $69.7 \cdot 1000$ **6.** $11.32 \div 10$ **7.** $678 \cdot 10^8$ **8.** $1.276 \div 1000$

9. $536.5 \div 10^2$ **10.** $5.92 \div 10^3$ **11.** $25 \div 10{,}000$ **12.** $6.519 \cdot 10^2$

Multiply Decimals

When multiplying decimals, multiply as you would with whole numbers. The sum of the number of decimal places in the factors equals the number of decimal places in the product.

EXAMPLE

Find each product.

A $81.2 \cdot 6.547$

```
    6.547 ← 3 decimal places
 ×   81.2 ← 1 decimal place
   1 3094
   6 5470
 523 7600
 531.6164 ← 4 decimal places
```

B $0.376 \cdot 0.12$

```
   0.376 ← 3 decimal places
 ×  0.12 ← 2 decimal places
     752
    3760
 0.04512 ← 5 decimal places
```

PRACTICE

Find each product.

1. $6.8 \cdot 3.4$ **2.** $2.56 \cdot 4.6$ **3.** $6.787 \cdot 7.6$ **4.** $0.98 \cdot 4.6$

5. $0.97 \cdot 0.76$ **6.** $0.5 \cdot 3.761$ **7.** $42 \cdot 17.654$ **8.** $7.005 \cdot 32.1$

9. $9.76 \cdot 16.254$ **10.** $296.5 \cdot 2.4$ **11.** $7.7 \cdot 6.5$ **12.** $8.92 \cdot 2.8$

13. $3.65 \cdot 4.2$ **14.** $0.002 \cdot 8.1$ **15.** $0.03 \cdot 0.204$ **16.** $98.6 \cdot 4.9$

Divide Decimals

When dividing with decimals, set up the division as you would with whole numbers. Pay attention to the decimal places, as shown below.

EXAMPLE

Find each quotient.

A **89.6 ÷ 16**

$$\begin{array}{r} 5.6 \\ 16\overline{)89.6} \\ \underline{80} \\ 96 \\ \underline{96} \\ 0 \end{array}$$

Place decimal point.

B **3.4 ÷ 4**

$$\begin{array}{r} 0.85 \\ 4\overline{)3.40} \\ \underline{3\,2} \\ 20 \\ \underline{20} \\ 0 \end{array}$$

Place decimal point.
Insert zeros if necessary.

PRACTICE

Find each quotient.

1. 242.76 ÷ 68 **2.** 40.5 ÷ 18 **3.** 121.03 ÷ 98 **4.** 3.6 ÷ 4

5. 1.58 ÷ 5 **6.** 0.2835 ÷ 2.7 **7.** 8.1 ÷ 0.09 **8.** 0.42 ÷ 0.28

9. 480.48 ÷ 7.7 **10.** 36.9 ÷ 0.003 **11.** 0.784 ÷ 0.04 **12.** 15.12 ÷ 0.063

Terminating and Repeating Decimals

You can change a fraction to a decimal by dividing. If the resulting decimal has a finite number of digits, it is **terminating**. Otherwise, it is **repeating**.

EXAMPLE

Write $\frac{4}{5}$ and $\frac{2}{3}$ as decimals. Are the decimals terminating or repeating?

$\frac{4}{5} = 4 \div 5$

$$\begin{array}{r} 0.8 \\ 5\overline{)4.0} \\ \underline{4\,0} \\ 0 \end{array} \longrightarrow \frac{4}{5} = 0.8$$

$\frac{2}{3} = 2 \div 3$

$$\begin{array}{r} 0.6666 \\ 3\overline{)2.0000} \\ \underline{1\,8} \\ 20 \end{array} \longrightarrow \frac{2}{3} = 0.6666...$$

This pattern will repeat.

The number 0.8 is a terminating decimal. The number 0.6666 . . . is a repeating decimal.

PRACTICE

Write as a decimal. Is the decimal terminating or repeating?

1. $\frac{1}{5}$ **2.** $\frac{1}{3}$ **3.** $\frac{3}{11}$ **4.** $\frac{3}{8}$ **5.** $\frac{7}{9}$ **6.** $\frac{7}{15}$

7. $\frac{3}{4}$ **8.** $\frac{5}{6}$ **9.** $\frac{4}{11}$ **10.** $\frac{5}{10}$ **11.** $\frac{1}{9}$ **12.** $\frac{11}{12}$

13. $\frac{5}{9}$ **14.** $\frac{8}{11}$ **15.** $\frac{7}{8}$ **16.** $\frac{23}{25}$ **17.** $\frac{3}{20}$ **18.** $\frac{5}{11}$

Order of Operations

When simplifying expressions, follow the order of operations.

1. Simplify within parentheses.
2. Evaluate exponents and roots.
3. Multiply and divide from left to right.
4. Add and subtract from left to right.

EXAMPLE

A Simplify the expression $3^2 \times (11 - 4)$.

$3^2 \times (11 - 4)$

$3^2 \times 7$ *Simplify within parentheses.*

9×7 *Evaluate the exponent.*

63 *Multiply.*

B Use a calculator to simplify the expression $19 - 100 \div 5^2$.

If your calculator follows the order of operations, enter the following keystrokes:

19 − 100 ÷ 5 [x^2] [ENTER] The result is 15.

If your calculator does not follow the order of operations, insert parentheses so that the expression is simplified correctly.

19 − (100 ÷ 5 [x^2]) [ENTER] The result is 15.

PRACTICE

Simplify each expression.

1. $45 - 15 \div 3$
2. $51 + 48 \div 8$
3. $35 \div (15 - 8)$
4. $\sqrt{9} \times 5 - 15$
5. $24 \div 3 - 6 + 12$
6. $(6 \times 8) \div 2^2$
7. $20 - 3 \times 4 + 30 \div 6$
8. $3^2 - 10 \div 2 + 4 \times 2$
9. $27 \div (3 + 6) + 6^2$
10. $4 \div 2 + 8 \times 2^3 - 4$
11. $33 - \sqrt{64} \times 3 - 5$
12. $(8^2 \times 4) - 12 \times 13 + 5$

Use a calculator to simplify each expression.

13. $6 + 20 \div 4$
14. $37 - 21 \div 7$
15. $9^2 - 32 \div 8$
16. $10 \div 2 + 8 \times 2$
17. $\sqrt{25} + 4 \times 6$
18. $4 \times 12 - 4 + 8 \div 2$
19. $28 - 3^2 + 27 \div 3$
20. $9 + (50 - 16) \div 2$
21. $4^2 - (10 \times 8) \div 5$
22. $30 + 22 \div 11 - 7 - 3^2$
23. $3 + 7 \times 5 - 1$
24. $38 \div 2 + \sqrt{81} \times 4 - 31$

Properties

The following are basic properties of addition and multiplication when a, b, and c are real numbers.

Addition

Closure: $a + b$ is a real number.
Commutative: $a + b = b + a$
Associative: $(a + b) + c = a + (b + c)$
Identity Property of Zero: $a + 0 = a$ and $0 + a = a$

Multiplication

Closure: $a \cdot b$ is a real number.
Commutative: $a \cdot b = b \cdot a$
Associative: $(a \cdot b) \cdot c = a \cdot (b \cdot c)$
Identity Property of One: $a \cdot 1 = a$ and $1 \cdot a = a$
Multiplication Property of Zero: $a \cdot 0 = 0$ and $0 \cdot a = 0$

The following properties are true when a, b, and c are real numbers.

Distributive: $a \cdot (b + c) = a \cdot b + a \cdot c$

Transitive: If $a = b$ and $b = c$, then $a = c$.

EXAMPLE

Name the property shown.

A $4 \cdot (7 \cdot 2) = (4 \cdot 7) \cdot 2$
Associative Property of Multiplication

B $4 \cdot (7 + 2) = (4 \cdot 7) + (4 \cdot 2)$
Distributive Property

PRACTICE

Give an example of each of the following properties, using real numbers.

1. Associative Property of Addition
2. Commutative Property of Multiplication
3. Closure Property of Multiplication
4. Distributive Property
5. Multiplication Property of Zero
6. Identity Property of Addition
7. Transitive Property
8. Closure Property of Addition

Name the property shown.

9. $4 + 0 = 4$
10. $(6 + 3) + 1 = 6 + (3 + 1)$
11. $7 \cdot 51 = 51 \cdot 7$
12. $5 \cdot 456 = 456 \cdot 5$
13. $17 \cdot (1 + 3) = 17 \cdot 1 + 17 \cdot 3$
14. $1 \cdot 5 = 5$
15. $(8 \cdot 2) \cdot 5 = 8 \cdot (2 \cdot 5)$
16. $72 + 1234 = 1234 + 72$
17. $0 \cdot 12 = 0$
18. $15.7 \cdot 1.3 = 1.3 \cdot 15.7$
19. $8.2 + (9.3 + 7) = (8.2 + 9.3) + 7$
20. $85.98 \cdot 0 = 0$
21. If $x = 3.5$ and $3.5 = y$, then $x = y$.
22. $12a \cdot 15b = 15b \cdot 12a$
23. $(2x + 3y) + 8z = 2x + (3y + 8z)$
24. $0 \cdot 6m^2n = 0$
25. $8j + 32k = 32k + 8j$
26. If $3 + 8 = 11$ and $11 = x$, then $3 + 8 = x$.

Iteration

An **iteration** is a step in the process of repeating something over and over again. You can show the steps of the process in an **iteration diagram**.

EXAMPLE

A **Use the iteration diagram below, and complete the process three times.**

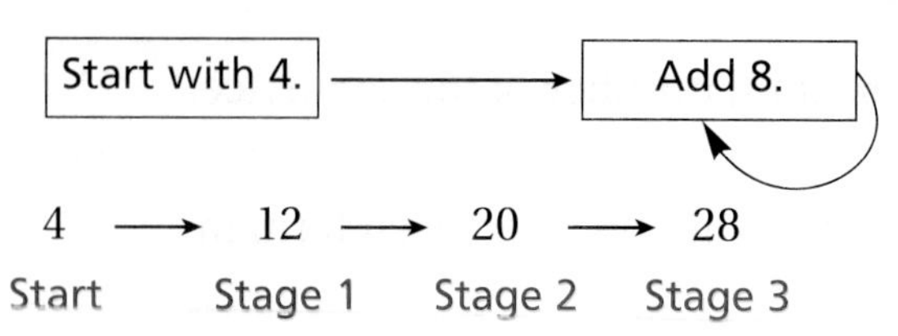

4 ⟶ 12 ⟶ 20 ⟶ 28

Start   Stage 1   Stage 2   Stage 3

B **For the pattern below, state the iteration and give the next three numbers in the pattern.**

1, 5, 25, 125, . . .

To get from one stage to the next, the iteration is to multiply by 5.

$125 \cdot 5 = 625$ $625 \cdot 5 = 3125$ $3125 \cdot 5 = 15{,}625$

The next three numbers in the pattern are 625, 3125, and 15,625.

PRACTICE

Use the diagram at right. Write the results of the first three iterations.

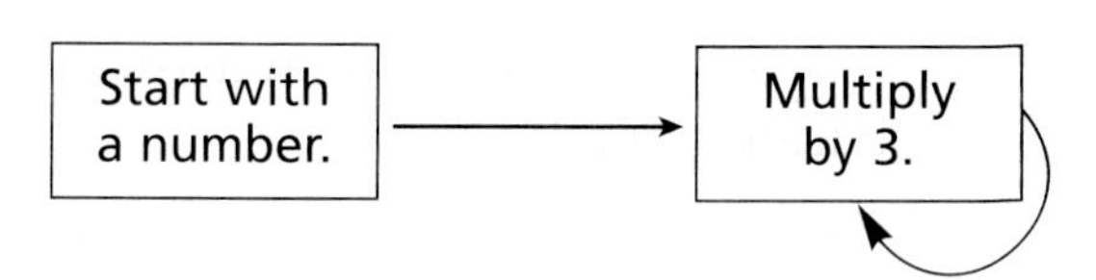

1. Start with 1.
2. Start with 8.
3. Start with 2.
4. Start with 25.
5. Start with −3.
6. Start with −7.

For each pattern, state the iteration and give the next three numbers in the pattern.

7. 11, 17, 23, 29, . . .
8. 5, 10, 20, 40, . . .
9. 345, 323, 301, 279, . . .
10. 30, 75, 120, 165, . . .
11. 15, 7, −1, −9, . . .
12. $1, 1\frac{2}{3}, 2\frac{1}{3}, 3, \ldots$

A **fractal** is a geometric pattern that is *self similar,* so each stage of the pattern is similar to a portion of another stage of the pattern. For example, the Koch snowflake is a fractal formed by beginning with a triangle and then adding an equilateral triangle to each segment of the triangle.

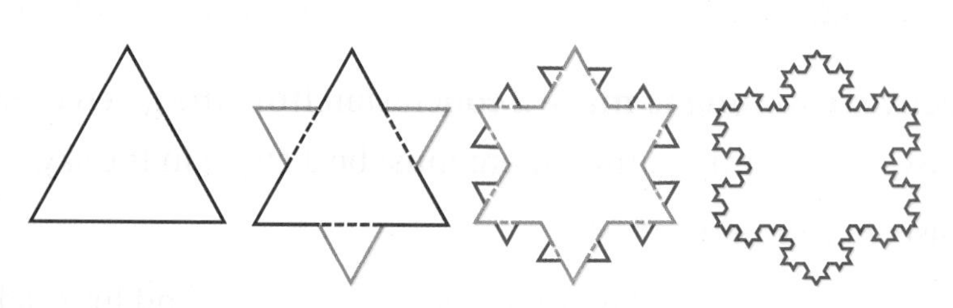

Draw the next two stages of each fractal.

13.

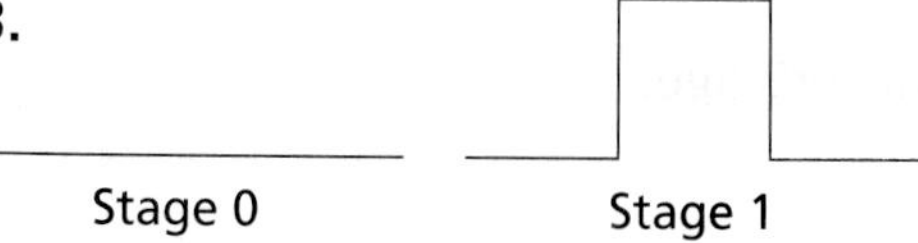

14.

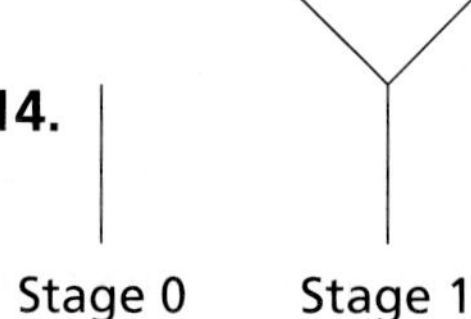

Changing Geometric Dimensions 2.01

The perimeter and area of a geometric figure are affected when the dimensions of the figure are changed.

EXAMPLE

A **Square A has side length 1 in., and square B has side length 2 in. Make a table that compares the perimeters and areas of squares A and B.**

Square	Side Length	Permeter	Area
A	1 in.	$P = 4s$ $= 4s$ $= 4$ in.	$A = s^2$ $= (1)^2$ $= 1$ in^2
B	2 in.	$P = 4s$ $= 4(2)$ $= 8$ in.	$A = s^2$ $= (2)^2$ $= 4$ in^2

The side length of square A is $\frac{1}{2}$ the side length of square B, the perimeter of square A is $\frac{1}{2}$ the perimeter of square B, and the area of square A is $\frac{1}{4}$ the area of square B.

B **Square A has side length 3 in. What is the side length of square B if its perimeter is 3 times the perimeter of square A?**

	Square A	Square B
Perimeter:	$P = 4s$ $= 4(3)$ $= 12$ in.	$3P = 3(12)$ $= 36$ in. You need to find the side length of a square with perimeter 36 in. Squares have 4 sides of equal length, so each side of square B will have a length of 9 in., which is 3 times the side length of square A.

PRACTICE

1. Circle A has radius 3 m, and circle B has radius 12 m. Make a table that compares the circumferences and areas of circles A and B. Then, complete the following statements.

   The radius of circle B is __?__ times the radius of circle A.
   The area of circle B is __?__ times the area of circle A.
   The circumference of circle B is __?__ times the circumference of circle A.

2. Circle A has radius 3 cm. How does the radius of circle B relate to the radius of circle A, if the area of circle B is $\frac{1}{9}$ the area of circle A? Explain the relationship between the circumference of circle A and the circumference of circle B.

3. Callie folded a piece of 8.5 in. × 11 in. paper such that its new perimeter is $\frac{1}{4}$ its original perimeter. How do the new dimensions of the paper relate to the original dimensions of the paper? Explain the relationship between the area of the folded paper and the area of the original paper.

Frequency Polygons

A **histogram** is a common way to represent frequency tables. A histogram is a bar graph with no space between the bars. Each bar can represent a range of values of a data set.

A **frequency polygon** is made by connecting the midpoints of the tops of all of the bars of a histogram.

EXAMPLE

A The frequency table shows the frequency of the number of push-ups done by the students in a gym class. Draw a histogram and frequency polygon of the data.

Label the horizontal axis with the number of push-ups.

Label the vertical axis with the frequency.

Push-ups Done in 1 Minute	
Number of Push-ups	Frequency
0–9	3
10–19	6
20–29	11
30–39	10
40–49	4
50–59	2

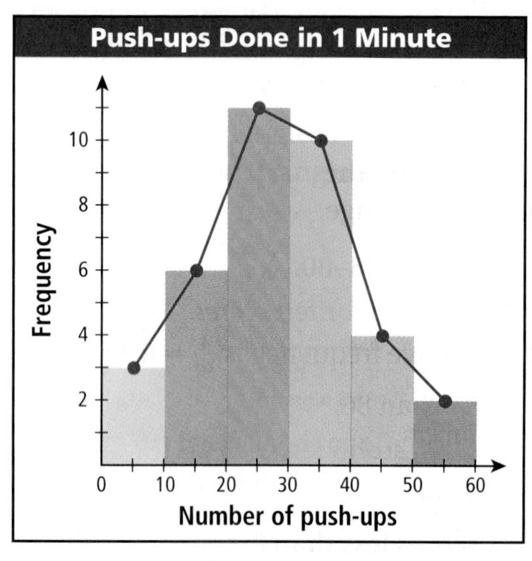

The frequency polygon is made up of the red points and red segments connecting the points.

PRACTICE

Use each frequency table to draw a histogram and frequency polygon of the data.

1.

Books Read over the Summer	
Number of Books	Frequency
0–2	5
3–5	8
6–8	12
9–11	6
12–14	4
15–17	2

2.

Miles Driven One Way to Work	
Number of Miles	Frequency
0–4	6
5–9	5
10–14	13
15–19	9
20–24	4
25–29	1

Exponential Growth and Quadratic Behavior

An **exponential growth function** is in the form $y = C(1 + r)^t$, where C is the starting amount, r is the percent increase, and t is the time.

EXAMPLE

A Patrick invested $2000 for 5 years at a 3% annual interest rate. Write an exponential growth function to represent this situation.

C = starting amount = $2000

r = percent increase = 3% = 0.03

t = time = 5 years

$y = 2000(1 + 0.03)^5$

$y = 2000(1.03)^5$

A function of the form $y = ax^2 + bx + c$ is called a **quadratic function**. The graph of a quadratic function is called a **parabola**. The most basic quadratic function is $y = x^2$. The graph of $y = x^2$ is shown at right. By examining the value of a in $y = ax^2$, you can determine the effect it will have on the graph of $y = x^2$.

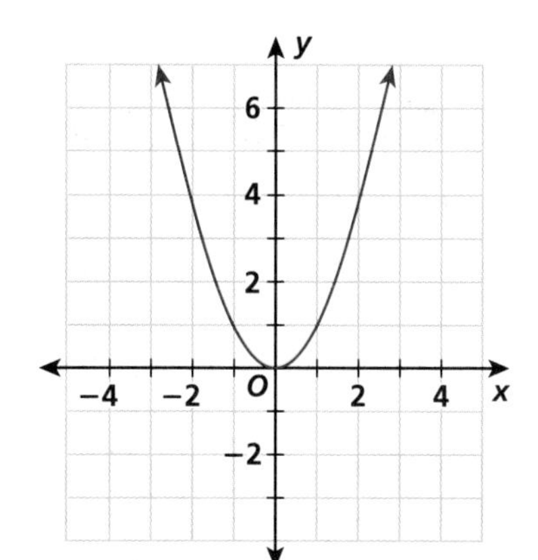

- If a is positive, the graph opens upward.
- If a is negative, the graph opens downward.
- If $|a| < 1$, the graph is wider than the graph of $y = x^2$.
- If $|a| > 1$, the graph is narrower than the graph of $y = x^2$.

EXAMPLE

B Compare the graph of $y = -2x^2$ with the graph of $y = x^2$.

Since a is negative, the graph will open downward. Since $|a| = 2 (2 > 1)$, the graph will be narrower than the graph of $y = x^2$.

PRACTICE

Write an exponential growth function to represent each situation.

1. The population of a small town in 1997 was 25,500. Over a 5-year period, the population of the town increased at a rate of 2% each year.
2. Shante invested $1800 at a 4.5% annual interest rate for 10 years.
3. Tyler took a job that paid $30,000 annually with a 4% salary increase each year. He stayed at that job for 8 years.

Compare the graph of each quadratic function with the graph of $y = x^2$.

4. $y = -x^2$
5. $y = \frac{1}{2}x^2$
6. $y = 3x^2$
7. $y = -\frac{1}{4}x^2$
8. $y = -5x^2$
9. $y = 0.2x^2$
10. $y = -\frac{3}{2}x^2$
11. $6x^2 = y$

Circles

A circle can be named by its center, using the ⊙ symbol. A circle with a center labeled C would be named $\odot C$. An unbroken part of a circle is called an **arc.** There are major arcs and minor arcs.

A **minor arc** of a circle is an arc that is shorter than half the circle and named by its endpoints. A **major arc** of a circle is an arc that is longer than half the circle and named by its endpoints and one other point on the arc.

$\widehat{AB}$ is a minor arc.

$\widehat{BAC}$ is a major arc.

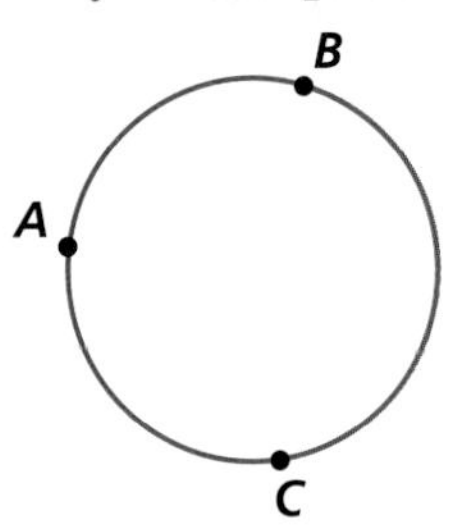

A **radius** connects the center with a point on a circle.

radius $\overline{CD}$

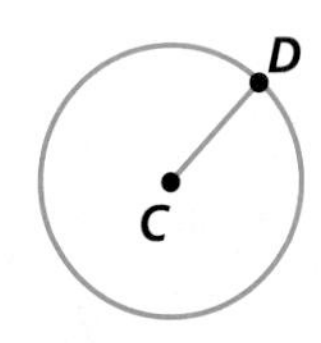

A **chord** connects two points point on a circle. A **diameter** is a chord that passes through the center of a circle.

chord $\overline{AB}$

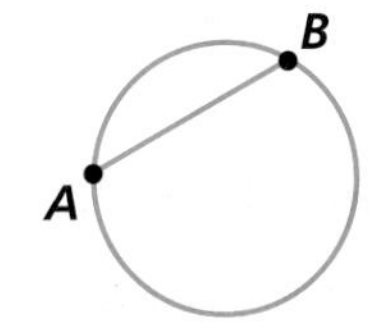

A **secant** is a line that intersects a circle at two points.

secant $\overleftrightarrow{EF}$

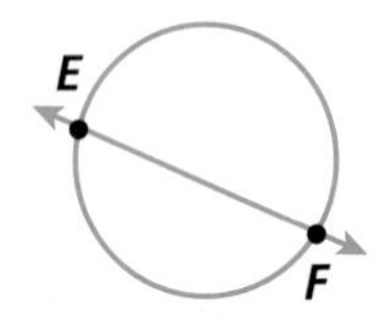

A **tangent** is a line that intersects a circle at one point.

tangent $\overleftrightarrow{GH}$

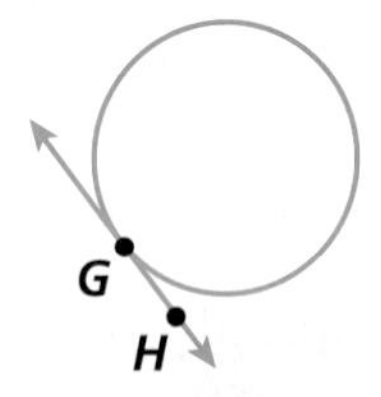

A **central angle** has its vertex at the center of the circle.

central angle
$\angle JKL$

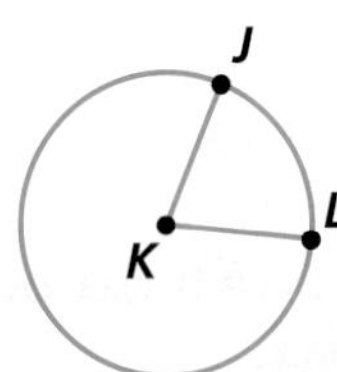

An **inscribed angle** has its vertex on the circle.

inscribed angle
$\angle MNP$

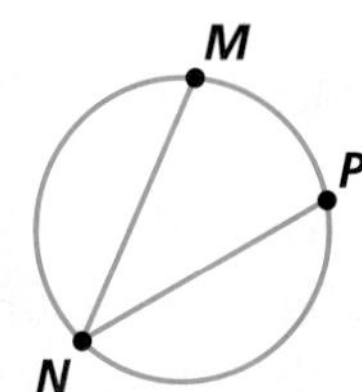

PRACTICE

Use the given diagram of $\odot A$ for exercises 1–6.

1. Name a radius.
2. What two chords make up the inscribed angle?
3. Name a secant.
4. Give the tangent line.
5. Name the central angle.
6. Name the inscribed angle.

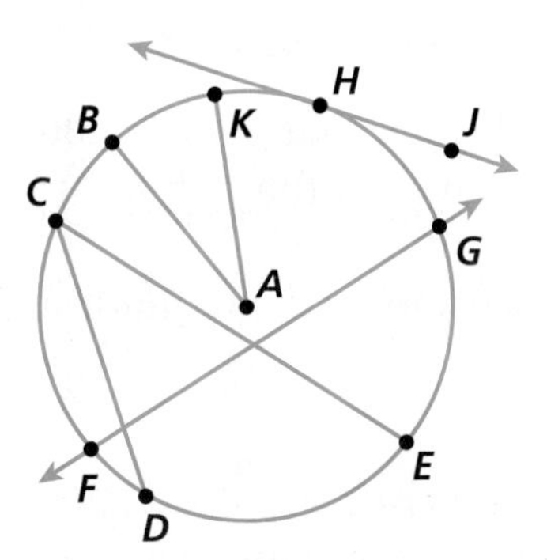

Matrices

A **matrix** is a rectangular arrangement of data enclosed in brackets. Matrices are used to list, organize, and sort data.

The **dimensions** of a matrix are given by the number of horizontal **rows** and vertical columns in the matrix. For example, Matrix A below is an example of a 3 × 2 ("3-by-2") matrix because it has 3 rows and 2 columns, for a total of 6 **elements**. The number of rows is always given first. So a 3 × 2 matrix is not the same as a 2 × 3 matrix.

$$A = \begin{bmatrix} 86 & 137 \\ 103 & 0 \\ 115 & 78 \end{bmatrix} \begin{matrix} \leftarrow \text{Row 1} \\ \leftarrow \text{Row 2} \\ \leftarrow \text{Row 3} \end{matrix}$$

↑ Column 1 ↑ Column 2

Each matrix element is identified by its row and column. The element in row 2 column 1 is 103. You can use the notation $a_{21} = 103$ to express this.

EXAMPLE

Use the data shown in the bar graph to create a matrix.

The matrix can be organized with the votes in each year as the columns:

$$\begin{bmatrix} 12 & 5 \\ 6 & 11 \\ 2 & 4 \end{bmatrix}$$

or with the votes in each year as the rows:

$$\begin{bmatrix} 12 & 6 & 2 \\ 5 & 11 & 4 \end{bmatrix}$$

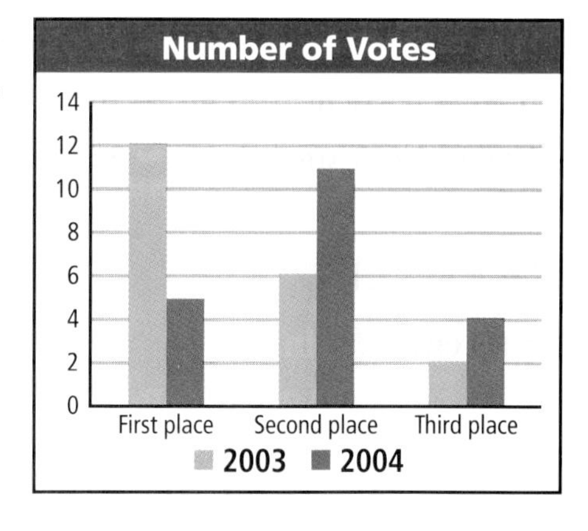

PRACTICE

Use matrix B for Exercises 1–3.

$$B = \begin{bmatrix} 1 & 0 & 7 & 4 \\ 0 & 1 & 3 & 8 \\ 6 & 5 & 2 & 9 \end{bmatrix}$$

1. B is a ▢ × ▢ matrix.

2. Name the element with a value of 5.

3. What is the value of b_{13}?

4. A football team scored 24, 13, and 35 points in three playoff games. Use this data to write a 3 × 1 matrix.

5. The greatest length and average weight of some whale species are as follows: finback whale—50 ft, 82 tons; humpback whale—33 ft, 49 tons; bowhead whale—50 ft, 59 tons; blue whale—84 ft, 98 tons; right whale—50 ft, 56 tons. Organize this data in a matrix.

6. The second matrix in the example is called the *transpose* of the first matrix. Write the transpose of matrix B above. What are its dimensions?

Customary and Metric Rulers

A metric ruler is divided into centimeter units, and each centimeter is divided into 10 millimeter units. A metric ruler that is 1 meter long is a *meter stick*.

1 m = 100 cm

1 cm = 10 mm

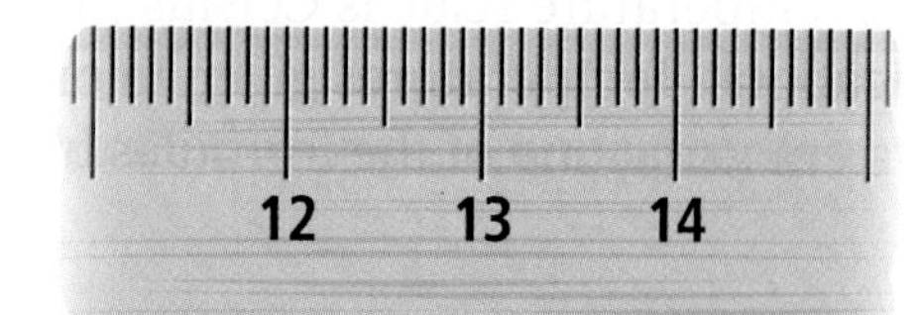

EXAMPLE

What is the length of the segment?

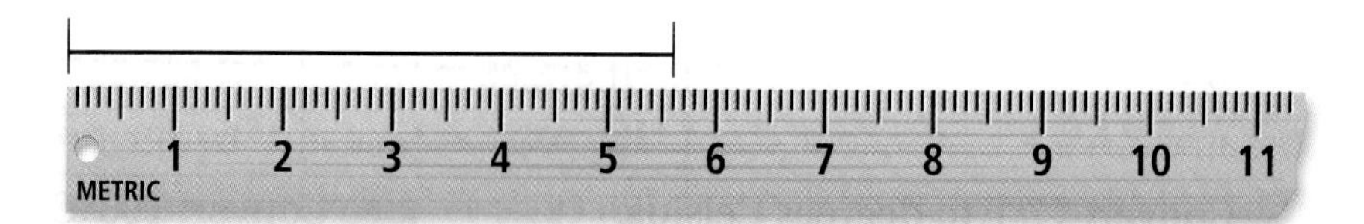

Since the segment is longer than 5 cm and shorter than 6 cm, its length is a decimal value between these measurements. The digit in the ones place is the number of centimeters and the digit in the tenths place is the number of millimeters. The length of the segment is 5.6 cm.

PRACTICE

Use a metric ruler to find the length of each segment.

1.

2.

A customary ruler is usually 12 inches long. The ruler is read in fractional units rather than in decimals. Each inch typically has a long mark at $\frac{1}{2}$ inch, shorter marks at $\frac{1}{4}$ and $\frac{3}{4}$ inch, even shorter marks at $\frac{1}{8}$, $\frac{3}{8}$, $\frac{5}{8}$, and $\frac{7}{8}$ inch, and the shortest marks at the remaining 16ths inches.

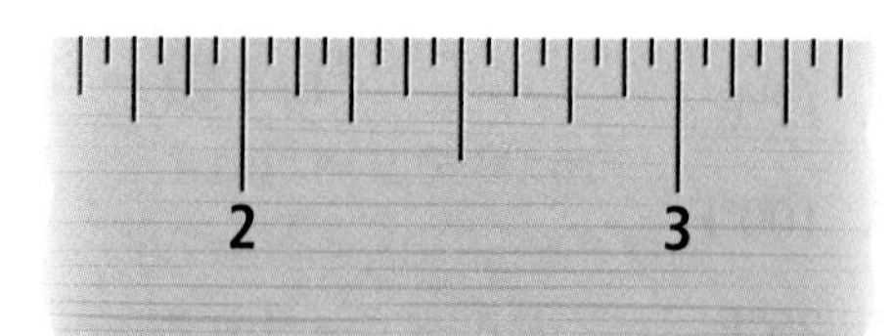

EXAMPLE

What is the length of the segment?

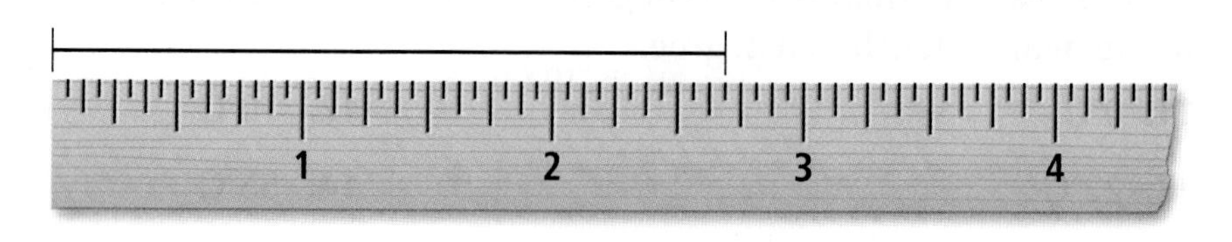

Since the segment is longer than 2 inches and shorter than 3 inches, its length is a mixed number with 2 as the whole number part. The fractional part is $\frac{11}{16}$. The length of the segment is $2\frac{11}{16}$ inches.

PRACTICE

Use a customary ruler to find the length of each segment.

3.

4.

Precision and Significant Digits

In a measurement, all digits that are known with certainty are called **significant digits** . The more precise a measurement is, the more significant digits there are in the measurement. The table shows some rules for identifying significant digits.

Rule	Example	Number of Significant Digits
All nonzero digits	15.32	All 4
Zeros beween significant digits	43,001	All 5
Zeros after the last nonzero digit that are to the right of the decimal point	0.0070	2; 0.0070

Zeros at the end of a whole number are assumed to be nonsignificant. (Example: 500)

EXAMPLE

A **Which is a more precise measurement, 14 ft or 14.2 ft?**

Because 14.2 ft has three significant digits and 14 has only two, 14.2 ft is more precise. In the measurement 14.2 ft, each 0.1 ft is measured.

B **Determine the number of significant digits in 20.04 m, 200 m, and 200.0 m.**

20.04 All 4 digits are significant.
200 There is 1 significant digit.
200.0 All 4 digits are significant.

When calculating with measurements, the answer can only be as precise as the least precise measurement.

C **Multiply 16.3 m by 2.5 m. Use the correct number of significant digits in your answer.**

When muliplying or dividing, use the least number of significant digits of the numbers.

$16.3 \text{ m} \cdot 2.5 \text{ m} = 40.75$

Round to 2 significant digits. ⟶ 41 m^2

D **Add 4500 in. and 70 in. Use the correct number of significant digits in your answer.**

When adding or subtracting, line up the numbers. Round the answer to the last significant digit that is farthest to the left.

4500 in. *5 is farthest left. Round to*
<u>+ 70 in.</u> *hundreds.*
4570 Round to the hundreds. ⟶ 4600 in.

PRACTICE

Tell which is more precise.

1. 31.8 g or 32 g
2. 496.5 mi or 496.50 mi
3. 3.0 ft or 3.001 ft

Determine the number of significant digits in each measurement.

4. 12 lb
5. 14.00 mm
6. 1.009 yd
7. 20.87 s

Perform the indicated operation. Use the correct number of significant digits in your answer.

8. 210 m + 43 m
9. $4.7 \text{ ft} \cdot 1.04 \text{ ft}$
10. 6.7 s − 0.08 s

Skills Bank

Greatest Possible Error

The smaller the units used to measure something, the greater the precision of the measurement. The **greatest possible error** of a measurement is half the smallest unit. This is written as ± 0.5 unit, which is read as "plus or minus 0.5 unit."

EXAMPLES

A **Which is a more precise measurement, 292 cm or 3 m?**

The more precise measurement is 292 cm because its unit of measurement, 1 cm, is smaller than 1 m.

B **Find the greatest possible error for a measurement of 2.4 cm.**

The smallest unit is 0.1 cm.

$0.5 \times \mathbf{0.1} = 0.05$

The greatest possible error is ± **0.05** cm.

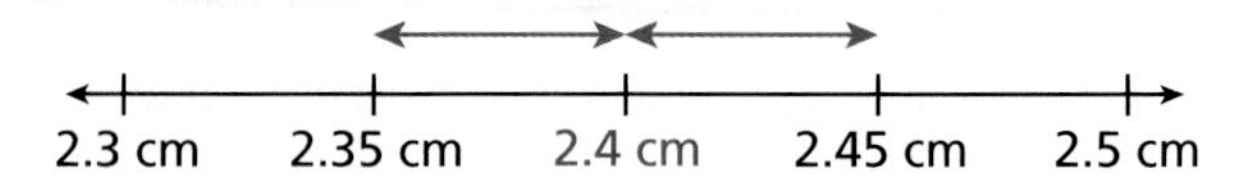

PRACTICE

Tell which is a more precise measurement.

1. 40 cm or 412 mm **2.** 3.2 ft or 1 yd **3.** 7 ft or 87 in.

4. 3116 m or 3 km **5.** 1 mi or 5281 ft **6.** 0.04 m or 4.2 cm

Find the greatest possible error of each measurement.

7. 5 ft **8.** 22 mm **9.** 12.5 mi

10. 60 km **11.** 2.06 cm **12.** 0.08 g

pH (Logarithmic Scale)

pH is a measure of the concentration of hydrogen ions in a solution. pH ranges from 0 to 14. An *acid* has a pH below 7 and a *base* has a pH above 7. A pH of 7 is *neutral* and a hydrogen ion concentration of 1×10^{-7} mol/L. The exponent is the opposite of the pH.

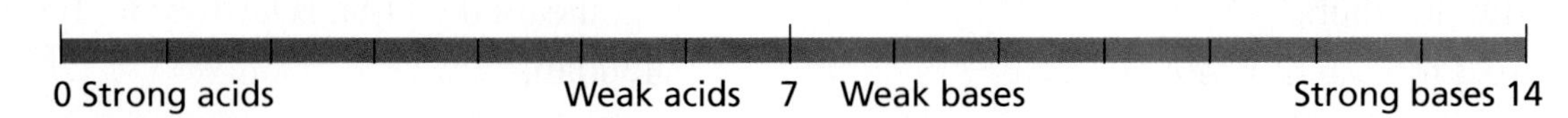

EXAMPLES

A **Write the pH of the solution, given the hydrogen ion concentration.**

coffee: 1×10^{-5} mol/L

The coffee is acidic, with a pH of 5.

B **Write the hydrogen ion concentration of the solution in mol/L.**

antacid solution: pH = 10.0

1×10^{-10} mol/L in the antacid solution

PRACTICE

Write the pH of each solution, given the hydrogen ion concentration.

1. seawater: 1×10^{-8} mol/L **2.** lye: 1×10^{-13} mol/L **3.** borax: 1×10^{-9} mol/L

Write the hydrogen ion concentration in mol/L.

4. drain cleaner: pH = 14.0 **5.** lemon juice: pH = 2.0 **6.** milk: pH = 7.0

Richter Scale

An earthquake is classified according to its magnitude. The Richter scale is a mathematical system that compares the sizes and magnitudes of earthquakes.

The magnitude is related to the height, or *amplitude,* of seismic waves as recorded by a seismograph during an earthquake. The higher the number is on the Richter scale, the greater the amplitude of the earthquake's waves.

Earthquakes per Year	Magnitude on the Richter Scale	Severity
1	8.0 and higher	Great
18	7.0–7.9	Major
120	6.0–6.9	Strong
800	5.0–5.9	Moderate
6200	4.0–4.9	Light
49,000	3.0–3.9	Minor
≈ 3,300,000	below 3.0	Very minor

The Richter scale is a *logarithmic scale,* which means that the numbers in the scale measure factors of 10. An earthquake that measures 6.0 on the Richter scale is 10 times as great as one that measures 5.0.

The largest earthquake ever measured registered 8.9 on the Richter scale.

EXAMPLE

How many times greater is an earthquake that measures 5.0 on the Richter scale than one that measures 3.0?

You can divide powers of 10, with the magnitudes as the exponents.

$$\frac{10^5}{10^3} = 10^2$$

A 5.0 quake is 100 times greater than a 3.0 quake.

PRACTICE

Describe the severity of an earthquake with each given Richter scale reading.

1. 7.6 **2.** 4.2 **3.** 5.0

4. 2.0 **5.** 3.6 **6.** 8.4

Each pair of numbers repesents two earthquake magnitudes on the Richter scale. How many times greater is the first earthquake in each pair? (Use a calculator for 10–12.)

7. 6.0 and 4.0 **8.** 8.0 and 5.0 **9.** 7.0 and 3.0

10. 7.5 and 5.5 **11.** 5.7 and 5.3 **12.** 8.6 and 7.1

Selected Answers

Chapter 1

1-1 Exercises

1. 17 **2.** 23 **3.** 3 **4.** 44 **5.** 1.8 **6.** 5 tbsp **7.** 8 tbsp **8.** 11.5 tbsp **9.** 17 tbsp **11.** 33 **13.** 67 **15.** 4 gal **17.** 2 gal **19.** 0 **21.** 22 **23.** 9 **25.** 6 **27.** 10 **29.** 16 **31.** 11 **33.** 20 **35.** 34 **37.** 12.6 **39.** 18 **41.** 105 **43.** 17 **45.** 30.5 **47.** 24 **49.** 0 **51.** Possible range: 204 to 208 beats per minute **53. b.** 165,600 frames **57.** 15, 21, 71 **59.** 49, 81 **61.** A

1-2 Exercises

1. $6 \div t$ **2.** $y - 25$ **3.** $7(m + 6)$ **4.** $7m + 6$ **5. a.** $8n$ **b.** $8(23) = \$184$ **6.** $\$15 + d$; \$17.50 **7.** $k + 34$ **9.** $5 + 5z$ **11. a.** $42 \div p$ **b.** 7 students **13.** $\$1.75n$; \$14.00 **15.** $6(4 + y)$ **17.** $\frac{1}{2}(m + 5)$ **19.** $13y - 6$ **21.** $2\left(\frac{m}{35}\right)$ **25.** $2(r - 1)$; $2(2.50 - 1) = \$3$

27.

24 + 4(2 − 2)	24
24 + 4(3 − 2)	28
24 + 4(4 − 2)	32
24 + 4(5 − 2)	36
24 + 4(6 − 2)	40

31. 202 **33.** 400 **35.** 200.2 **37.** 40 **39.** D

1-3 Exercises

1. 5 **2.** 21 **3.** $m = 32$ **4.** $t = 5$ **5.** $w = 17$ **6.** 15,635 feet **7.** 22 **9.** $w = 1$ **11.** $t = 12$ **13.** 20 **15.** 30 **17.** 7 **19.** 0 **21.** $t = 5$ **23.** $m = 24$ **25.** $h = 3$ **27.** $t = 2621$ **29.** $x = 110$ **31.** $n = 45$ **33.** $t = 0.5$ **35.** $w = 1.9$ **37. a.** $497 + m = 1696$; 1199 miles **b.** $1278 + m = 1696$; 418 miles **39. a.** $0.24 + c = 4.23$; \$3.99 **b.** $c - 3.82 = 0.53$; \$4.35 **43.** 22 **45.** 26

1-4 Exercises

1. $x = 7$ **2.** $t = 7$ **3.** $y = 14$ **4.** $w = 13$ **5.** $l = 60$ **6.** $k = 72$ **7.** $h = 57$ **8.** $m = 6$ **9.** $8n = 32$; $n = 4$ servings **10.** $\frac{1}{4}c = \$60$ or $\frac{c}{4} = \$60$; $c = \$240$ **11.** $x = 7$ **12.** $k = 40$ **13.** $y = 3$ **14.** $m = 36$ **15.** $d = 19$ **17.** $g = 10$ **19.** $n = 567$ **21.** $a = 612$ **23.** $10n = 80$; $n = 8$ mg **25.** $x = 2$ **27.** $y = 2$ **29.** $x = 7$ **31.** $y = 2$ **33.** $k = 56$ **35.** $b = 72$ **37.** $x = 17$ **39.** $y = 3$ **41.** $b = 48$ **43.** $n = 35$ **45.** $16m = 42{,}000$; $m = 2625$ miles **47.** $\frac{1}{6}m = 22$ or $\frac{m}{6} = 22$; $m = 132$ miles **49.** $x = 8$ **51.** $w = 2$ **53.** A

1-5 Exercises

1. $<$ **2.** $>$ **3.** $>$ **4.** $>$ **5.** $>$ **6.** $<$ **7.** $>$ **8.** $>$ **9.** $x < 1$ **10.** $b \geq 5$ **11.** $m \leq 32$ **12.** $15 > x$ **13.** $y \geq 17$ **14.** $f < 5$ **15.** $z > 21$ **16.** $14 \leq x$ **17.** $m > 40$; more than 40 members **19.** $<$ **21.** $>$ **23.** $<$ **25.** $<$ **27.** $x \geq 7$ **29.** $4 < t$ **31.** $x \geq 4$ **33.** $6 < a$ **35.** $x < 6$ **37.** $x > 4$ **39.** $x < 1$ **41.** $x \geq 5$ **43.** $50(50) > 2200$; $2500 > 2200$; no **45.** $x \geq 53$ **51.** 22; 19; 16; 13 **53.** 13; 21; 29; 37 **55.** 15; 13; 11; 9 **57.** C

1-6 Exercises

1. $4x$ **2.** $5z + 5$ **3.** $8f + 8$ **4.** $17g$ **5.** $4p - 8$ **6.** $4x + 12$ **7.** $3x + 5y$ **8.** $9x + y$ **9.** $5x + y$ **10.** $9p + 3z$ **11.** $7g + 5h - 12$ **12.** $10h$ **13.** $r + 6$ **14.** $10 + 8x$ **15.** $2t + 56$ **16.** $n = 42$ **17.** $y = 24$ **18.** $p = 17$ **19.** $13y$ **21.** $7a + 11$ **23.** $3x + 2$ **25.** $5p$ **27.** $9x + 3$ **29.** $5a + z$ **31.** $7x + 5q + 3$ **33.** $9a + 7c + 3$ **35.** $20y - 18$ **37.** $6y + 17$ **39.** $11x - 9$ **41.** $p = 5$ **43.** $y = 8$ **45.** $x = 14$ **47.** $8d + 1$ **49.** $x = 2$ **51.** $52g$; $41s$; $49b$ **57.** $x = 13$ **59.** $x = 8$ **61.** $x = 32$ **63.** $x = 16$ **65.** B

1-7 Exercises

1. no **2.** yes **3.** yes **4.** no

5.

x	y	(x, y)
1	2	(1, 2)
2	4	(2, 4)
3	6	(3, 6)
4	8	(4, 8)
5	10	(5, 10)
6	12	(6, 12)

6.

x	y	(x, y)
1	1	(1, 1)
2	4	(2, 4)
3	7	(3, 7)
4	10	(4, 10)
5	13	(5, 13)
6	16	(6, 16)

7. \$1.29 **9.** no **11.** no

13.

x	y	(x, y)
1	10	(1, 10)
2	12	(2, 12)
3	14	(3, 14)
4	16	(4, 16)
5	18	(5, 18)
6	20	(6, 20)

15.

x	y	(x, y)
2	2	(2, 2)
4	8	(4, 8)
6	14	(6, 14)
8	20	(8, 20)
10	26	(10, 26)

17. yes **19.** yes **21.** no **23.** yes

25.

x	y	(x, y)
1	1	(1, 1)
2	5	(2, 5)
3	9	(3, 9)
4	13	(4, 13)
5	17	(5, 17)
6	21	(6, 21)

27.

x	y	(x, y)
1	9	(1, 9)
2	10	(2, 10)
3	11	(3, 11)
4	12	(4, 12)
5	13	(5, 13)
6	14	(6, 14)

29.

x	y	(x, y)
2	8	(2, 8)
4	12	(4, 12)
6	16	(6, 16)
8	20	(8, 20)
10	24	(10, 24)

31. Possible answer: $x = y$ **33.** no; (13, 52) or (12.75, 51) **35. a.** (1980, 74) **b.** (2020, 81) **39.** 7 **41.** 4 **43.** 12 **45.** B

1-8 Exercises

1. (–2, 3) **2.** (3, 5) **3.** (2, –3) **4.** (5, –1) **5.** (5, 5) **6.** (–3, –4)

7–10.

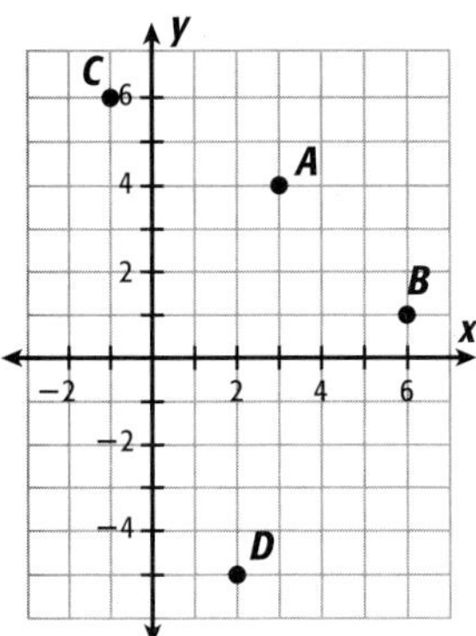

11.

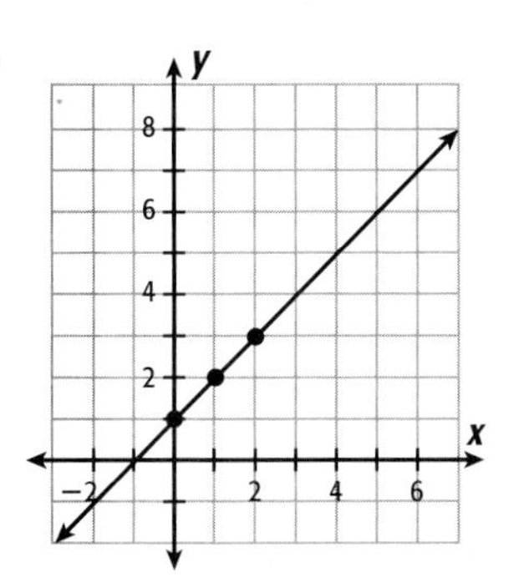

12.

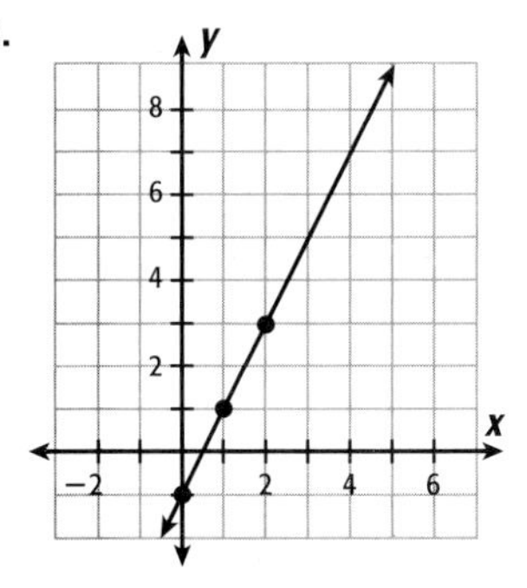

13. (0, 3) **15.** (2, −4) **17.** (−2, 5)

19–21.

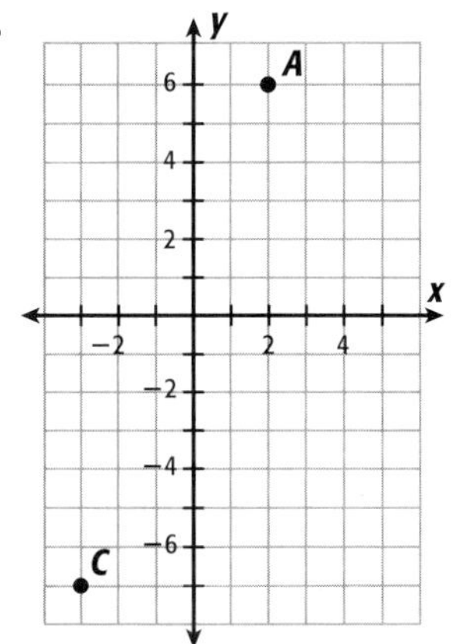

23.

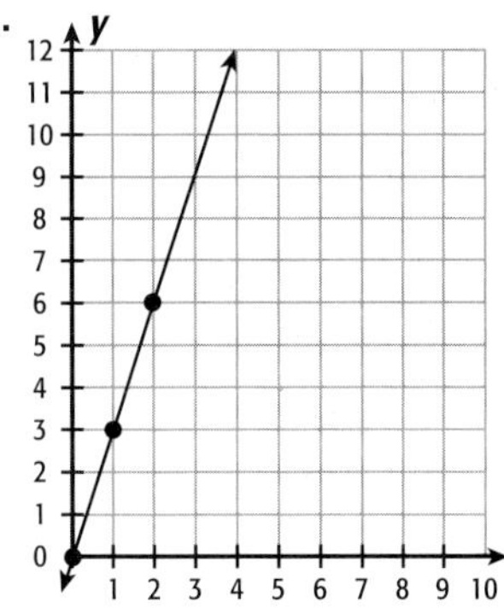

25–31. Possible answers given. **25.** (1, 0), (2, 0) **27.** (2, 7), (4, 7) **29.** (4, 3), (4, 5) **31.** (0, 4), (0, 5) **33.** 75 beats

35.

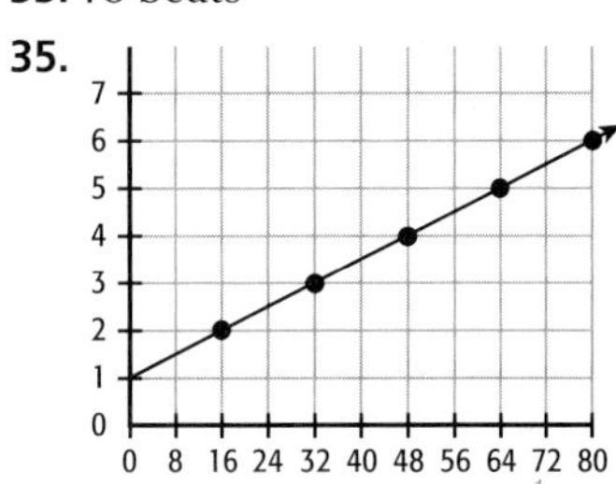

7 studs

39. $x - 13$ **41.** $x + 31$ **43.** A

1-9 Exercises

1. table 2 **2.** table 2; table 1; table 3; none

3.

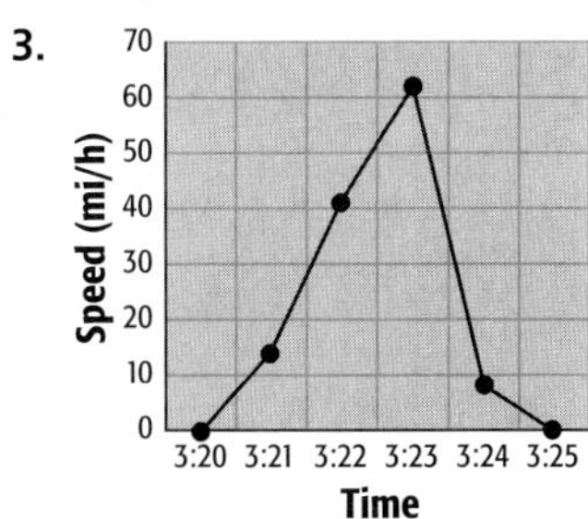

5. table 1; table 3; table 2

7.

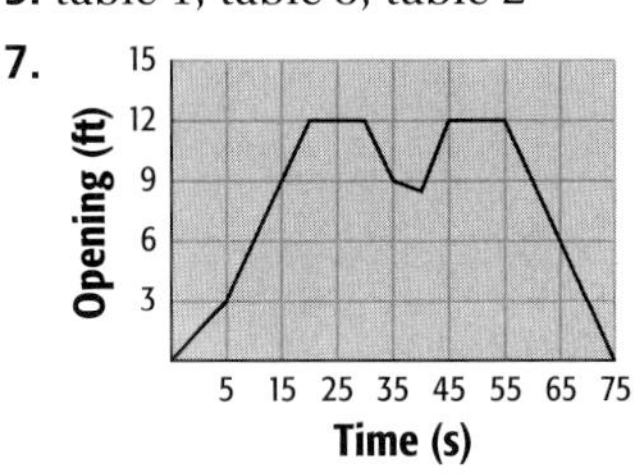

9. a. Old Faithful **b.** Riverside **11.** $x = 9$ **13.** $x = 11$ **15.** D

Chapter 1 Study Guide and Review

1. ordered pair; x-coordinate; y-coordinate **2.** solution set; inequality **3.** 147 **4.** 152 **5.** 278 **6.** $2(k + 4)$ **7.** $4t + 5$ **8.** $z = 23$ **9.** $t = 8$ **10.** $k = 15$ **11.** $x = 11$ **12.** 1300 lb. **13.** 3300 mi^2 **14.** $g = 8$ **15.** $k = 9$ **16.** $p = 80$ **17.** $w = 48$ **18.** $y = 40$ **19.** $z = 19.2$ **20.** 352.5 mi **21.** 24 months **22.** $h < 4$ **23.** $y > 7$ **24.** $x \geq 4$ **25.** $p < \frac{1}{2}$ **26.** $m > 2.3$ **27.** $q \leq 0$ **28.** $w \geq 8$ **29.** $x \leq 3$ **30.** $y > 16$ **31.** $x > 3$ **32.** $y > 6$ **33.** $x \leq 2$ **34.** $11m - 4$ **35.** $14w + 6$ **36.** $y = 5$ **37.** $z = 8$ **38.** yes **39.** no

40.

x	y	(x, y)
0	2	(0, 2)
1	5	(1, 5)
2	8	(2, 8)
3	11	(3, 11)
4	14	(4, 14)

41–46.

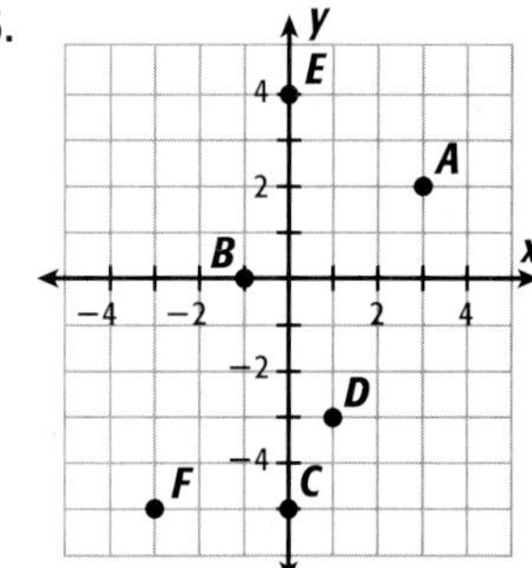

47. 5 **48.** 8 **49.** 20 **50.** Oven E

Chapter 2

2-1 Exercises

1. 5 **2.** 2 **3.** 4 **4.** −6 **5.** −8 **6.** 6 **7.** 3 **8.** −16 **9.** 11 **10.** 4 **11.** −8 **12.** \$297 **13.** −2 **15.** −3 **17.** 21 **19.** −18 **21.** 22 **23.** 9 **25.** $-6 + (-2) = -8$ **27.** −13 **29.** −18 **31.** −2 **33.** 43 **35.** 0 **37.** −19 **39.** 8 **41.** −20 **43.** −15 **45.** 5 **51.** $f = 6$ **53.** $q = 6$

2-2 Exercises

1. −15 **2.** −3 **3.** 14 **4.** −7 **5.** 13 **6.** −6 **7.** −15 **8.** 49°F **9.** −11 **11.** 17 **13.** 3 **15.** 4 **17.** 16 **19.** −17 **21.** −14 **23.** 40 m below sea level, or −40 m **25.** $5 - 8 = -3$ **27.** 51 **29.** −62 **31.** −16 **33.** 13 **35.** 2 **37.** −42 **39.** Great Pyramid to Cleopatra; about 500 years **41.** Cleopatra takes throne and Napoleon invades Egypt. **45.** no like terms **47.** C

2-3 Exercises

1. −27 **2.** −8 **3.** 30 **4.** −4 **5.** 49 **6.** −77 **7.** −24 **8.** −72

9.

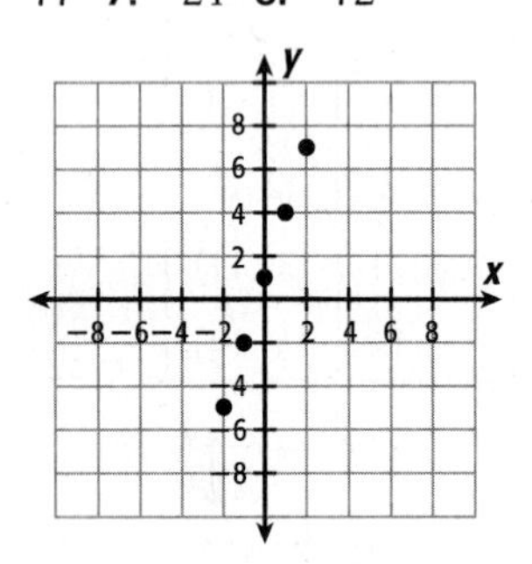

10.

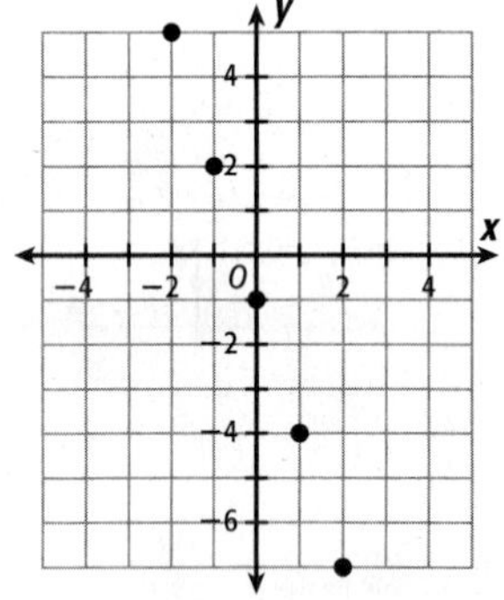

11.

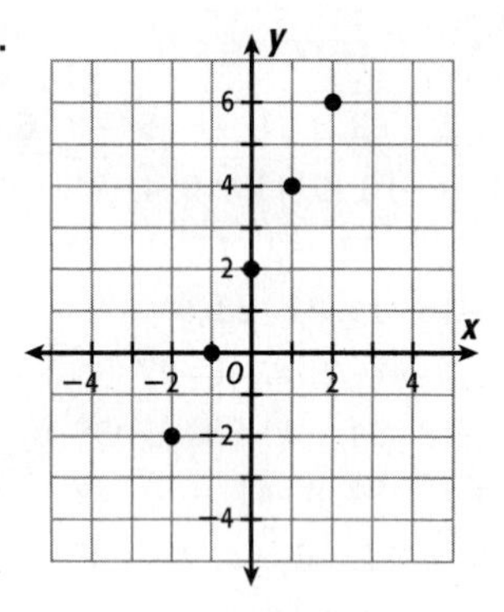

13. −11 **15.** −7 **17.** 130 **19.** −2

21.

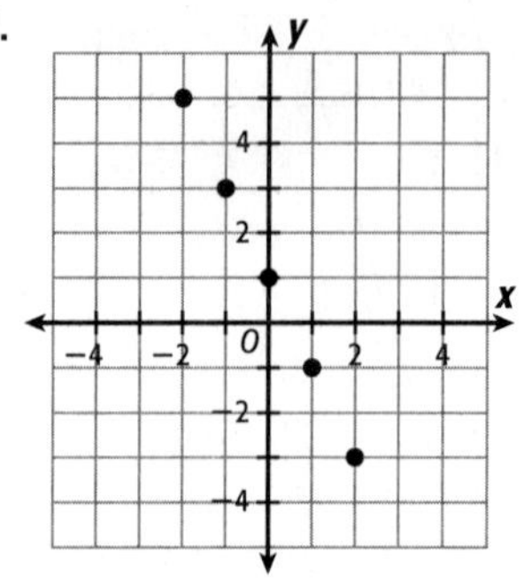

23. −45 **25.** 36 **27.** 24 **29.** −72 **31.** −80 **33.** 63 **35.** −19 **37.** 14 **39.** 3

41.

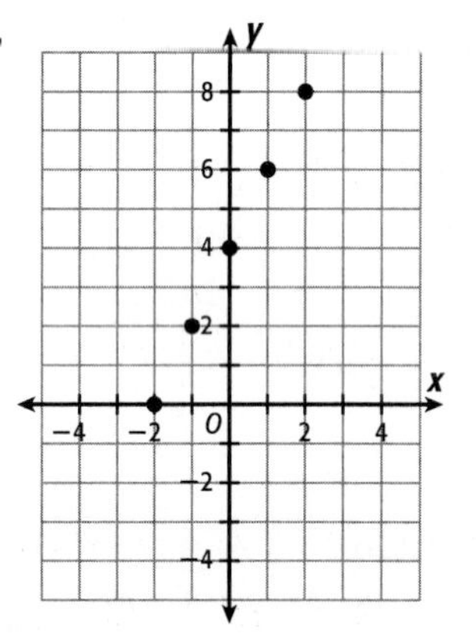

43.

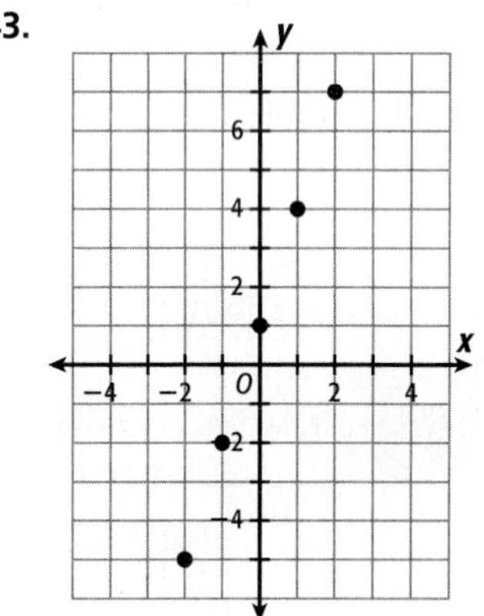

45. 32 days **51.** $w = 11$ **53.** $h = 0$ **55.** B

2-4 Exercises

1. $y = 6$ **2.** $d = 12$ **3.** $x = -11$ **4.** $b = -7$ **5.** $t = -16$ **6.** $g = -4$ **7.** $a = 12$ **8.** $f = -5$ **9.** 427°C **11.** $a = 13$ **13.** $b = -3$ **15.** $y = -37$ **17.** $h = -31$ **19.** $n = -39$ **21.** $c = 84$ **23.** $a = 45$ **25.** $r = -64$ **27.** $s = -11$ **29.** $x = 4$ **31.** $m = -27$ **33.** $z = 16$ **35.** $h = -4$ **37.** $y = -105$ **39.** $x = 24$ **41.** $p = -6$ **43. a.** $-4t = d$, t is time in minutes and d is depth. **b.** −68 m **c.** $-4t = -24$; $t = 6$ minutes **49.** $w = 2$ **51.** C

2-5 Exercises

1. $x \geq -5$ **2.** $y < 2$ **3.** $b \leq -7$ **4.** $h < 1$ **5.** $f > 4$ **6.** $k \leq 5$ **7.** $x < -3$ **8.** $y < -2$ **9.** $w \leq 3$ **10.** $x \geq -3$ **11.** $z > -8$ **12.** $n \leq 6$ **13.** $k > -3$ **15.** $x < -1$ **17.** $r \geq 2$ **19.** $n > 5$ **21.** $x \geq -4$ **23.** $x > -5$ **25.** $x > -2$ **27.** $k \geq 10$ **29.** $a \leq -12$ **31.** $r \leq -1$ **33.** $t = 2$ **35.** $b > 0$ **37.** $f = -18$ **39.** $c \leq 2$ **41.** $n < -6$ **43.** $g = 8$ **45.** $p = -9$ **47.** $3x + (-7x) > -12$; $x < 3$ **49.** $-1 + x < -7$; $x < -6$; less than 6 under par **55.** 9 **57.** −254 **59.** −16 **61.** 3 **63.** A

2-6 Exercises

1. 14^1 **2.** 15^2 **3.** b^4 **4.** $(-1)^3$ **5.** 81 **6.** 25 **7.** −243 **8.** 2401 **9.** −33 **10.** 90 **11.** −117 **12.** −47 **13.** 78 **15.** $(-7)^3$ **17.** c^5 **19.** 256 **21.** −512 **23.** 77 **25.** −360 **27.** $(-2)^3$ **29.** 4^4 **31.** 343 **33.** −1728 **35.** 729 **37.** 4 **39.** −116 **41.** −166 **43.** −4 **45.** −1 **47.** 216 **49.** 257 **51.** $2^{18} = 262{,}144$ bacteria **59.** 9 **61.** 104 **63.** C

2-7 Exercises

1. 3^{11} **2.** 12^5 **3.** m^6 **4.** cannot combine **5.** 8^2 **6.** a^8 **7.** $12^0 = 1$ **8.** 7^{12} **9.** 10^2 plants **11.** 2^6 **13.** 16^4 **15.** cannot combine **17.** $10^0 = 1$ **19.** 6^3 **21.** a **23.** x^{10} **25.** 6^6 **27.** cannot combine **29.** $y^0 = 1$ **31.** x^8 **33.** 4^6 **35.** 10^{14} **37.** n^{16} **39.** 4^4 **41.** 6^9 **43.** 26^2, or 676 more ways **45.** 12^2; 12^1 **47.** 22^3 trips **51.** 3 **53.** −12 **55.** −16 **57.** −12 **59.** D

2-8 Exercises

1. 0.0000001 **2.** 0.001 **3.** 0.000001 **4.** 0.1 **5.** $\frac{1}{16}$ **6.** $\frac{1}{9}$ **7.** $\frac{1}{8}$ **8.** $-\frac{1}{32}$ **9.** 1000 **10.** $\frac{1}{9}$ **11.** 216 **12.** $\frac{1}{27}$ **13.** 0.01 **15.** 0.00001 **17.** $-\frac{1}{64}$ **19.** 0.0001 **21.** 10,000 **23.** 1 **25.** $\frac{1}{8}$ **27.** 0.001 **29.** 128 **31.** m^7 **33.** $\frac{1}{9}$ **35.** 1024 **37.** $\frac{1}{2}$ **39.** $\frac{1}{4}$ **41.** $\frac{1}{144}$ **43.** 4 **45.** 1 kilometer **47. a.** $10^{-5} \cdot 10^3 = 10^{-2}$ g **b.** $10^{-2} \cdot 10^7 = 10^5$ g **c.** $10^5 \div 10^1 = 10^{5-1} = 10^4$; 10^4 decagrams **51.** 30 **53.** 85

2-9 Exercises

1. 3150 **2.** 0.000000125 **3.** 410,000 **4.** 0.00039 **5.** 5.7×10^{-5} **6.** 3×10^{-4} **7.** 4.89×10^6 **8.** 1.4×10^{-7} **9.** (1.485×10^6)°C **11.** 0.00067 **13.** 63,700,000 **15.** 7.8×10^6 **17.** 3×10^{-8} **19.** 13,000 **21.** 56 **23.** 0.000000053 **25.** 8,580,000 **27.** 9,112,000 **29.** 0.00029 **31.** 4.67×10^{-3} **33.** 5.6×10^7 **35.** 7.6×10^{-3} **37.** 3.5×10^3 **39.** 9×10^2 **41.** 6×10^6 **43. a.** $\approx 2.21 \times 10^7$; $\approx 1.4 \times 10^4$ mi^2 **b.** 6.35×10^{-4} mi^2/person **45.** 0.000078 **51.** −20 **53.** 21 **55.** $t = -9$ **57.** $b = -27$

Chapter 2 Study Guide and Review

1. opposite **2.** scientific notation; power **3.** exponent; base **4.** −2 **5.** −12 **6.** −3 **7.** 1 **8.** −24 **9.** 8 **10.** −8 **11.** −16 **12.** 17 **13.** 3 **14.** 15 **15.** −22 **16.** −4 **17.** 16 **18.** −5 **19.** −35 **20.** −18 **21.** 52 **22.** 25 **23.** 120 **24.** 2 **25.** $p = 9$ **26.** $t = 3$ **27.** $k = 3$ **28.** $g = -6$ **29.** $w = -80$ **30.** $b = -20$ **31.** $a = -4$ **32.** $h = -91$ **33.** $S = 38$ **34.** $b < -2$ **35.** $r > 6$ **36.** $m \geq 3$ **37.** $p < -2$ **38.** $z < -5$ **39.** $q \geq 3$ **40.** $m \geq 4$ **41.** $x > -3$ **42.** $y < 4$ **43.** $x > -3$ **44.** $b \leq 0$ **45.** $y < 6$ **46.** 7^3 **47.** $(-3)^2$ **48.** K^4 **49.** 625 **50.** −32 **51.** −1 **52.** 4^7 **53.** 9^6 **54.** p^4 **55.** 8^3 **56.** 9^2 **57.** m^5 **58.** 5^3 **59.** y^5 **60.** k^0 **61.** $\frac{1}{125}$ **62.** $-\frac{1}{64}$ **63.** $\frac{1}{11}$ **64.** 1 **65.** 1 **66.** 1 **67.** $\frac{1}{8}$ **68.** $-\frac{1}{27}$ **69.** 1620 **70.** 0.00162 **71.** 910,000 **72.** 0.000091 **73.** 8.0×10^{-9} **74.** 7.3×10^7 **75.** 9.6×10^{-6} **76.** 5.64×10^{10}

Chapter 3

3-1 Exercises

1. $\frac{4}{5}$ **2.** $\frac{3}{5}$ **3.** $-\frac{2}{3}$ **4.** $\frac{11}{27}$ **5.** $\frac{19}{23}$ **6.** $-\frac{5}{6}$ **7.** $-\frac{7}{27}$ **8.** $\frac{7}{16}$ **9.** $\frac{3}{4}$ **10.** $1\frac{1}{8}$ **11.** $\frac{431}{1000}$ **12.** $\frac{4}{5}$ **13.** $-2\frac{1}{5}$ **14.** $\frac{5}{8}$ **15.** $3\frac{21}{100}$ **16.** $-\frac{1939}{5000}$ **17.** 0.875 **18.** 0.6 **19.** $0.41\overline{6}$ **20.** 0.75 **21.** 4.0 **22.** 0.125 **23.** 2.4 **24.** 2.25 **25.** $\frac{3}{4}$ **27.** $-\frac{1}{2}$ **29.** $\frac{13}{17}$ **31.** $\frac{16}{19}$ **33.** $\frac{2}{5}$ **35.** $\frac{71}{100}$ **37.** $1\frac{377}{1000}$ **39.** $-1\frac{2}{5}$ **41.** 0.375 **43.** 1.4 **45.** 0.68 **47.** 1.16 **49.** Possible answer: $\frac{25}{36}$ **51. a.** $\frac{3}{4}$; $\frac{1}{6}$; $\frac{5}{9}$; $\frac{17}{20}$; $\frac{13}{32}$; $\frac{11}{25}$; $\frac{19}{24}$; $\frac{8}{15}$ **b.** 2×2; 2×3; 3×3; $2 \times 2 \times 5$; $2 \times 2 \times 2 \times 2 \times 2$; 5×5; $2 \times 2 \times 2 \times 3$; 3×5 **c.** 0.75 terminating; $0.1\overline{6}$ repeating; $0.\overline{5}$ repeating; 0.85 terminating; 0.40625 terminating; 0.44 terminating; $0.791\overline{6}$ repeating; $0.5\overline{3}$ repeating **53.** GCF = 4; $\frac{12}{19}$; No **59.** 28; 48 **61.** 35; 14 **63.** C

3-2 Exercises

1. 9.693 seconds **2.** 1.4 **3.** −2 **4.** −0.4 **5.** $-2\frac{1}{2}$ **6.** −1.5 **7.** $-\frac{5}{9}$ **8.** −1.9 **9.** −3 **10.** $-\frac{1}{3}$ **11.** $-1\frac{1}{3}$ **12.** $\frac{4}{5}$ **13.** $\frac{2}{5}$ **14.** $\frac{1}{2}$ **15.** $\frac{5}{17}$ **16.** $4\frac{1}{5}$ **17.** $-2\frac{5}{9}$ **18.** 4.2 **19.** $\frac{2}{5}$ **20.** 21.4 **21.** $\frac{2}{5}$ **23.** −1.6 **25.** 1.6 **27.** 1.9 **29.** −2.7 **31.** $\frac{5}{11}$ **33.** $1\frac{8}{17}$ **35.** $-\frac{1}{2}$ **37.** $1\frac{2}{21}$ **39.** 28.7 **41.** −16.34 **43. a.** $\frac{29}{32}$ in. **b.** $1\frac{7}{32}$ in. **c.** $\frac{19}{32}$ in. **45. a.** 3.63 quadrillion Btu **b.** 2.717 quadrillion Btu **49.** $7x - 5y + 18$ **51.** $16x + 22y + 11$ **53.** C

3-3 Exercises

1. $1\frac{1}{3}$ **2.** $-14\frac{2}{5}$ **3.** $1\frac{7}{8}$ **4.** $-3\frac{4}{5}$ **5.** $3\frac{1}{9}$ **6.** $-8\frac{7}{11}$ **7.** $6\frac{3}{4}$ **8.** $6\frac{3}{8}$ **9.** $\frac{4}{21}$ **10.** $-\frac{21}{80}$ **11.** $3\frac{5}{9}$ **12.** $\frac{1}{4}$ **13.** $-\frac{25}{78}$ **14.** $2\frac{1}{32}$ **15.** $\frac{7}{12}$ **16.** $-\frac{55}{192}$ **17.** 12.4 **18.** 0.144 **19.** 36.5 **20.** −0.42 **21.** 41.3 **22.** 3.65 **23.** 14.1 **24.** −0.416 **25.** $13\frac{1}{7}$ **26.** $5\frac{3}{4}$ **27.** $-6\frac{4}{7}$ **28.** $-1\frac{20}{49}$ **29.** 23 **30.** $7\frac{2}{3}$ **31.** $-9\frac{6}{7}$ **32.** $-\frac{69}{70}$ **33.** $\frac{3}{5}$ **35.** $1\frac{1}{8}$ **37.** $8\frac{2}{5}$ **39.** 4 **41.** $\frac{5}{9}$ **43.** $\frac{38}{63}$ **45.** $-\frac{3}{10}$ **47.** $\frac{3}{32}$ **49.** 8.7 **51.** 43.4 **53.** 33.6 **55.** 28.8 **57.** $16\frac{1}{2}$ **59.** −11 **61.** $8\frac{1}{4}$ **63.** $-19\frac{1}{4}$ **65.** $72\frac{1}{2}$ ounces **67. a.** $1\frac{1}{4}$ tsp **b.** $1\frac{1}{2}$ tsp **c.** 2 tsp **73.** $x = 12$ **75.** $x = 34$ **77.** $x = 44$ **79.** A

3-4 Exercises

1. $\frac{4}{5}$ **2.** $\frac{45}{68}$ **3.** $-\frac{2}{7}$ **4.** $2\frac{11}{12}$ **5.** $1\frac{3}{14}$ **6.** $-\frac{5}{54}$ **7.** $1\frac{1}{2}$ **8.** $2\frac{9}{10}$ **9.** 12.4 **10.** 68 **11.** 15.3 **12.** 8.6 **13.** $3.8\overline{4}$ **14.** 17.6 **15.** 1310 **16.** 9.2 **17.** 22.5 **18.** 21 **19.** 45 **20.** 4 **21.** 13 **22.** 270 **23.** $\frac{6}{7}$ serving **25.** $1\frac{13}{15}$ **27.** $3\frac{3}{5}$ **29.** $-\frac{8}{21}$ **31.** $2\frac{1}{28}$ **33.** $\frac{1}{4}$ **35.** $-4\frac{1}{2}$ **37.** 97 **39.** 17.1 **41.** 27.4 **43.** 25.4 **45.** 32 **47.** 5.76 **49.** 13 **51.** 11 **53.** 370 **55.** 0.7 **57.** 6 chairs **59.** $2\frac{1}{2}$ tiles **61.** Yes **65.** $x = 6.5$ **67.** $x = 8$ **69.** $x = 4.5$ **71.** C

3-5 Exercises

1. $\frac{19}{24}$ **2.** $\frac{67}{112}$ **3.** $-\frac{4}{9}$ **4.** $\frac{7}{16}$ **5.** $-3\frac{7}{15}$ **6.** $-2\frac{11}{24}$ **7.** $\frac{47}{60}$ **8.** $1\frac{29}{40}$ **9.** $1\frac{19}{40}$ **10.** $-1\frac{8}{63}$ **11.** $\frac{5}{8}$ **12.** $-\frac{37}{48}$ **13.** $6\frac{5}{8}$ ft **15.** $\frac{44}{45}$ **17.** $1\frac{1}{4}$ **19.** $-\frac{11}{112}$ **21.** $1\frac{4}{45}$ **23.** $-\frac{5}{48}$ **25.** $-\frac{7}{60}$ **27.** $660\frac{779}{800}$ in. **29.** $18\frac{21}{50}$ in.

31. $47\frac{2}{25}$ meters **35.** -27 **37.** 88 **39.** 18 **41.** B

3-6 Exercises

1. $y = -75.4$ **2.** $f = -7$ **3.** $m = -19.2$ **4.** $r = 54.7$ **5.** $s = 68.692$ **6.** $g = 6.3$ **7.** $x = -\frac{4}{7}$ **8.** $k = -\frac{1}{3}$ **9.** $w = -\frac{7}{9}$ **10.** $m = 0$ **11.** $y = -9$ **12.** $t = 0$ **13.** $17\frac{24}{25}$ mm **15.** $m = -9$ **17.** $k = -2.4$ **19.** $c = 5.16$ **21.** $d = \frac{8}{15}$ **23.** $x = \frac{1}{2}$ **25.** $c = \frac{7}{20}$ **27.** $z = \frac{2}{3}$ **29.** $j = -32.4$ **31.** $g = 9$ **33.** $v = -30.25$ **35.** $y = -5.4$ **37.** $c = -\frac{1}{24}$ **39.** $y = 64.1$ **41.** $m = -2.8$ **43. a.** 15 tiles **b.** 9 tiles **c.** 5 boxes **49.** 21 **51.** 5.24×10^{-6} **53.** 6.4×10^{10}

3-7 Exercises

1. $x \geq 2$ **2.** $k > 9.3$ **3.** $g \leq 7$ **4.** $h < 0.79$ **5.** $w \leq 0.24$ **6.** $z > 0$ **7.** $k > \frac{3}{5}$ **8.** $y \geq 0$ **9.** $q \leq -\frac{1}{169}$ **10.** $x < 1\frac{2}{3}$ **11.** $f > \frac{4}{15}$ **12.** $m \geq 4$ **13.** between 6.7 and 8.1 hours **15.** $m \leq -.07$ **17.** $g \leq -24.3$ **19.** $w \leq -1.5$ **21.** $k \geq \frac{25}{36}$ **23.** $x \geq 4\frac{3}{5}$ **25.** $m \leq -1\frac{1}{7}$ **27.** $d \leq -3$ **29.** $g \geq -2$ **31.** $t > \frac{3}{13}$ **33.** $y \geq -8$ **35.** $w \leq -\frac{1}{3}$ **37.** $c > 3.1$ **39.** $c < 3\frac{1}{3}$ **41.** $t \leq 6$ **43.** at least 12.5 in., but not more than 3600 in. **47.** 0.3 **49.** -0.26 **51.** 16.8 **53.** -0.258 **55.** C

3-8 Exercises

1. ± 5 **2.** ± 12 **3.** ± 2 **4.** ± 20 **5.** ± 1 **6.** ± 9 **7.** ± 3 **8.** ± 4 **9.** 16 ft **10.** 5 **11.** 2 **12.** -55 **13.** -1 **15.** ± 15 **17.** ± 13 **19.** ± 21 **21.** ± 19 **23.** -3 **25.** -20 **27.** ± 7 **29.** ± 17 **31.** ± 30 **33.** ± 23 **35.** $\pm\frac{1}{2}$ **37.** $\pm\frac{5}{2}$ **39.** $\pm\frac{3}{2}$ **41.** $\pm\frac{1}{10}$ **43.** 26 ft **45.** 327 **47. a.** 81; 1 **b.** 18 **51.** $t = 9$ **53.** $t = 22$ **55.** $\frac{1}{9}$ **57.** 1 **59.** C

3-9 Exercises

1. 6 and 7 **2.** -8 and -9 **3.** 14 and 15 **4.** -18 and -19 **5.** ≈ 13.27 ft **6.** 9.1 **7.** 6.5 **8.** 50 **9.** 13.8 **11.** 1 and 2 **13.** -31 and -32 **15.** 8.3 **17.** 25.5 **19.** B **21.** E **23.** F **25.** 7.14 **27.** 11.62 **29.** 42.85 **31.** -11.62 **33.** -32.83 **35.** ± 5.20 **37.** ± 317.02 **39.** 800 ft/s **43.** $y = -4.4$ **45.** $m = -25.6$ **47.** $x < 5\frac{2}{3}$ **49.** $m \geq 8$ **51.** 4 and -4 **53.** 10 and -10 **55.** D

3-10 Exercises

1. irrational, real **2.** whole, integer, rational, real **3.** rational, real **4.** rational, real **5.** rational **6.** rational **7.** irrational **8.** not real **9.** rational **10.** not real **11.** not real **12.** not real **13–15.** Possible answers given. **13.** $5\frac{1}{4}$ **14.** $\frac{2199}{700}$ **15.** $\frac{3}{16}$ **17.** rational, real **19.** integer, rational, real **21.** rational **23.** irrational **25.** irrational **27.** not real **29.** $-\frac{1}{200}$ **31.** whole, integer, rational, real **33.** irrational, real **35.** rational, real **37.** rational, real **39.** rational, real **41.** integer, rational, real **43–51.** Possible answers given. **43.** $-\sqrt{50}$ **45.** $\frac{11}{18}$ **47.** $\frac{3}{4}$ **49.** 3 **51.** -4.25 **53.** $x \geq 0$ **55.** $x \geq -3$ **57.** $x \geq -\frac{2}{5}$ **63.** 6.32 **65.** 7.75 **67.** -4.12 **69.** 3.46 **71.** 2.5×10^6 **73.** 5.68×10^{15} **75.** C

Chapter 3 Study Guide and Review

1. rational number **2.** real numbers; irrational numbers **3.** relatively prime **4.** principal square root **5.** perfect square **6.** $\frac{3}{5}$ **7.** $\frac{1}{4}$ **8.** $\frac{21}{40}$ **9.** $\frac{2}{3}$ **10.** $\frac{2}{3}$ **11.** $\frac{3}{4}$ **12.** $\frac{-6}{13}$ **13.** $1\frac{2}{5}$ **14.** $\frac{5}{9}$ **15.** $\frac{1}{6}$ **16.** $-1\frac{1}{5}$ **17.** $7\frac{3}{5}$ **18.** $\frac{8}{15}$ **19.** -4 **20.** $2\frac{1}{4}$ **21.** $3\frac{1}{4}$ **22.** 6 **23.** $\frac{3}{8}$ **24.** $\frac{2}{9}$ **25.** -16 **26.** $1\frac{1}{4}$ **27.** 2 **28.** $1\frac{1}{6}$ **29.** $\frac{5}{18}$ **30.** $11\frac{3}{10}$ **31.** $4\frac{7}{20}$ **32.** $y = -21.8$ **33.** $z = -18$ **34.** $w = -\frac{5}{8}$ **35.** $p = 2$ **36.** $m > -\frac{1}{12}$ **37.** $t \geq -12$ **38.** $y \leq -3\frac{1}{4}$ **39.** $x > -\frac{1}{2}$ **40.** ± 4 **41.** ± 30 **42.** ± 26 **43.** 5 **44.** $\frac{1}{2}$ **45.** 9 **46.** 89.4 in. **47.** 167.3 cm **48.** rational **49.** irrational **50.** not real **51.** irrational **52.** rational **53.** not real

Chapter 4

4-1 Exercises

1. Population: pet store customers; sample: 100 customers; possible bias: not all customers have dogs. **2.** systematic **3.** random **5.** systematic **7.** Population: students; sample: students who buy the entrée; possible bias: the students who buy the entrée may be the people who like the food in the cafeteria. **9.** Population: restaurant customers; sample: first four customers who order the cheese sauce; possible bias: if the customers ordered cheese sauce, then they probably like cheese. **11.** systematic **13.** stratified **15.** systematic **17 a.** Possible answer: Randomly select visitors leaving the zoo. **b.** Possible answer: Select every tenth visitor leaving the zoo. **c.** Possible answer: People visiting with children might visit the zoo only because they have children. **23.** $y = -7.2$ **25.** $c = -\frac{2}{7}$ **27.** $x > 25.6$

4-2 Exercises

1.

Nutrition in Potatoes			
	Baked Potato (100 g)	French Fries (100 g)	Potato Chips (100 g)
Fiber	2.4 g	3.2 g	4.5 g
Ca	10 mg	10 mg	24 mg
Mg	27 mg	22 mg	67 mg

2. 2, 3, 3, 7, 11, 13, 17, 17, 18, 20, 20, 27, 34, 34, 35, 35 **3.** 63, 66, 68, 73, 73, 75, 77, 80, 80, 81, 81, 90, 94, 95, 99

4.

Tens	Ones
0	1 6 7
1	8
2	0 2 6
3	5 6
4	7
5	3 6

Key: 1|8 means 18

5.

Democrats		Republicans
	3	2 6 7 8
6 6	4	1 2 3 4
8 7 6 4	5	3 4
8 4 1 1	6	

Key: |4|1 means 41; 6|4| means 46

7. 50, 51, 54, 58, 62, 66, 67, 71, 74, 75, 76, 76, 82

9.

Dollars	Cents
0.9	3 5 5
1.0	2 6
1.1	1 1 3 4 7
1.2	1 3 3 4
1.3	0 8

Key: 1.1|1 means $1.11

11.

Tens	Ones
4	3
5	7
6	5 8
7	2 2 3 5 6
8	1 2 4 8
9	1

Key: 5|7 means 57

13.

Energy Use in U.S.			
	1980	1990	2000
Fossil Fuels	89%	86%	85%
Nuclear Power	3%	7%	8%
Renewable Resources	7%	7%	7%

15.

Numbers		Time	
One	9	Night	12
Two	3	Day	4
Three	6	Supper-time	1
Ten	2	Bed-time	1
Twelve	1	Evening	1
Fourteen	1		

19. 5^{11} **21.** cannot combine **23.** population: students; sample: students on every other bus

4-3 Exercises

1. ≈ 34.43; 35; no mode **2.** 4.4; 4.4; 4.4 and 6.2 **3.** 5; 5; 5 **4.** ≈ 55.67; 56; no mode **5.** 2.39 million **6.** approximately 1.43 million **7.** 3.35 million **9.** 87.6; 88; 88 **11.** 5.85; 4.4; no mode **13.** approximately 74.33 million **15.** 25; 26; no mode; no outlier **17.** 11; 12; 10 and 13; 3 **19.** 4; 2; 2; 29 **21.** 1105 million miles; 484 million miles; no mode **29.** $14x - 45$ **31.** $x = 13$ **33.** $m = 100$ **35.** C

4-4 Exercises

1. 56; 42; 66 **2.** 6; 1.5; 4.5

3.

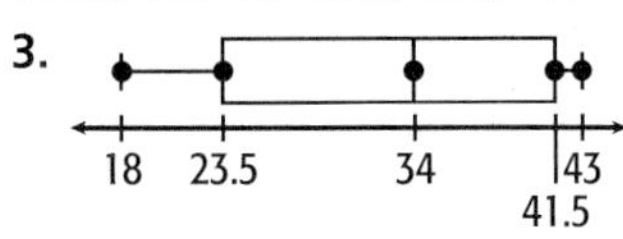

4.

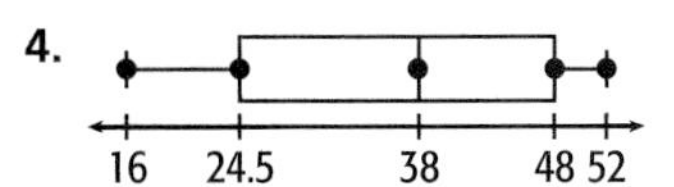

5. The medians are equal, but data set B has a much greater range. **6.** The range of the middle half of the data is greater for data set B. **7.** 30; 34.5; 46.5

9.

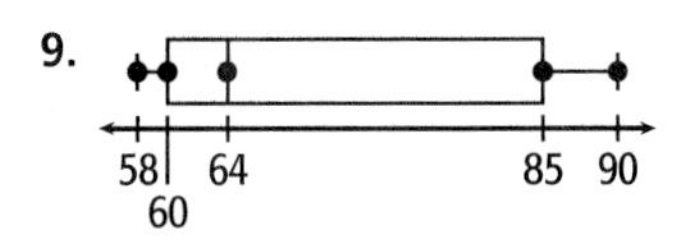

11. Data set Y has a greater median and range. **13.** 22; 78; 95 **15.** 38; 35; 57.5 **17.** 23; 9.5; 24.5

19.

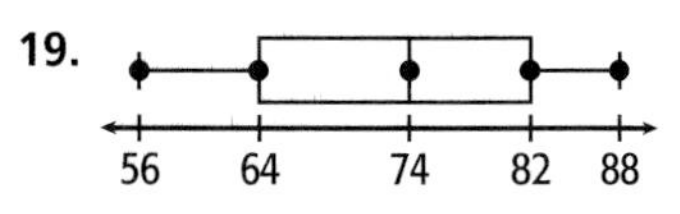

21.

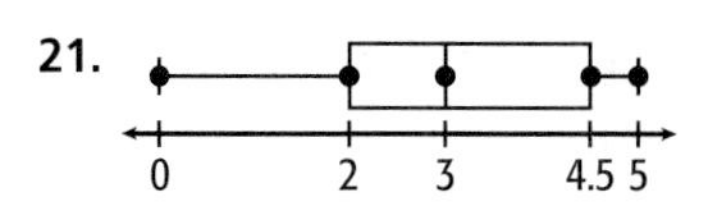

23.

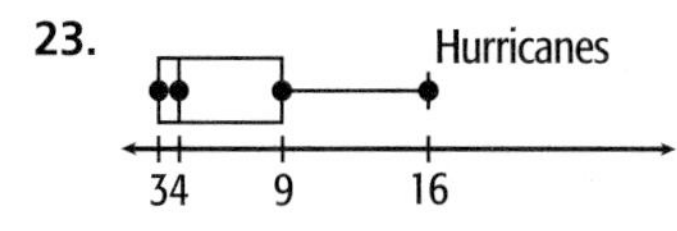

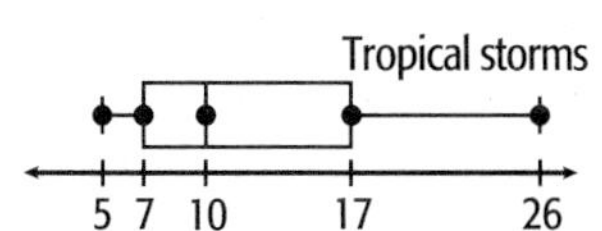

Possible answer: The median number of tropical storms is greater than the median number of hurricances.

25. a. data set C **b.** data set A **c.** data set B **29.** −2 **31.** 10 **33.** graph B **35.** graph C

4-5 Exercises

1.

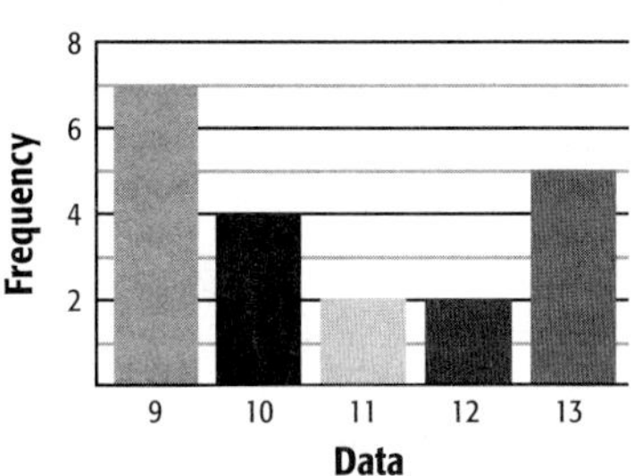

2.

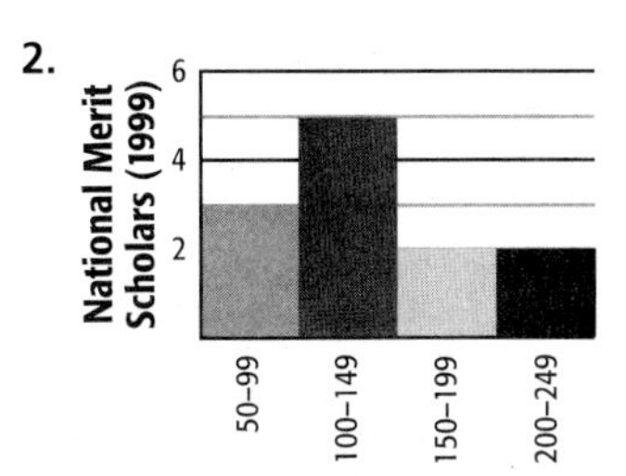

3. 74.1 years

5.

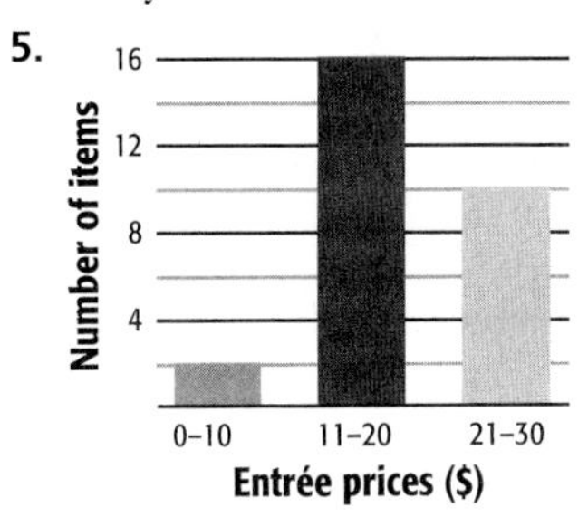

7.

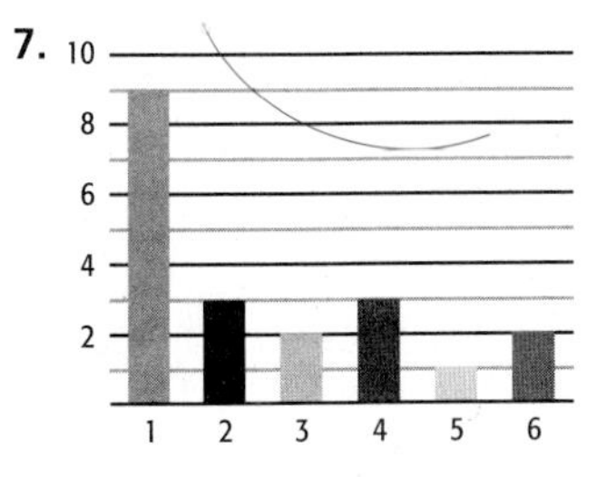

9. **a.** 34.9 hours **b.** \$11.88
13. $x < 5$ **15.** $x \leq 2$ **17.** $x > 6$
19. $6 \geq x$ **21.** B

4-6 Exercises

1–9. Possible answers given.
1. The scale does not start at zero, so changes appear exaggerated.
2. The intervals used in the histogram are not equal.
3. The fruits are all different sizes. A better comparison would be the same serving size of each fruit.
4. The sales are for different lengths of time. **5.** The graph has no scale, so it's impossible to compare the money earned.
7. The difference between the two groups' responses is only 3 people out of 1000. **9.** The areas of the sails distort the comparison. Your graph should use bars or pictures that are the same width.
15. $b = 6$ **17.** $a = 21$ **19.** $1.5 = h$
21. $f = 1.5$

4-7 Exercises

1.

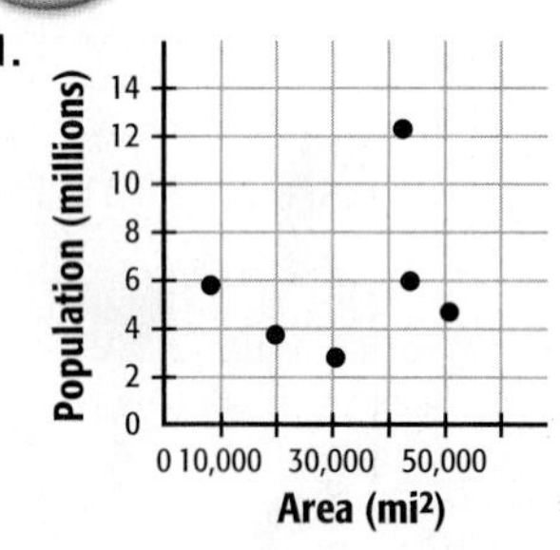

2. positive **3.** no correlation
4. 66°F
5.

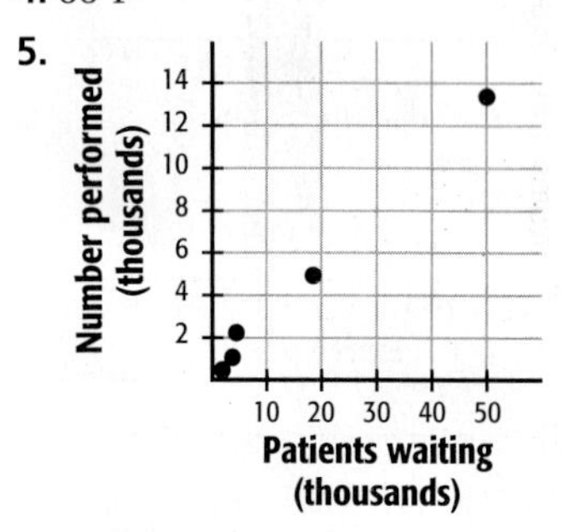

7. positive
9. There is a positive correlation between the pollen levels.
11. negative **15.** $x = 5$ **17.** $x = 6$
19. $x = 18$

Chapter 4 Extension

1. 2.4 **3.** 12.9 **5.** 2.3 **7.** 0 **9.** data set B **11. a.** week 1: 1.7; week 2: 3.1 **b.** week 2 **13.** Zero; the sum for the differences of the data values would be zero.

Chapter 4 Study Guide and Review

1. median; mode **2.** variability; variability; range **3.** line of best fit; scatter plot; correlation
4. population: moviegoers; sample: 25 people in line for a Star Wars movie; possible bias: people in line for Star Wars might have a preference for science fiction movies. **5.** population: community members; sample: 50 parents of middle-school-age children; possible bias: parents of middle-school-age children may support the field more than other community members.
6. population: constituents; sample: 75 constituents who visited the office; possible bias: constituents who visit the senator probably are strong supporters of the senator.
7.

Inaugural Age		Age at Death
3	4	6
7 7 2 1	5	6
	6	3 7
	7	
	8	3

Key: 3|4| means 43
|4|6 means 46

8. 760; 570; 500 **9.** 9.25; 9; 8, 9, and 10 **10.** 6; 6; 6 **11.** 3.1; 3.1; 3.1
12. 10; 80; 90 **13.** 32; 68; 99
14.

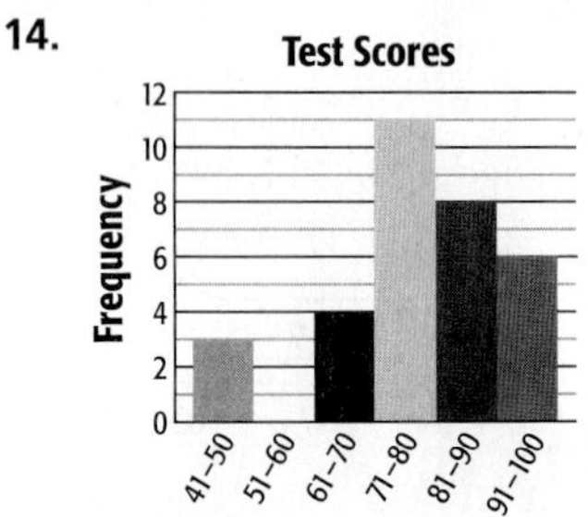

15.

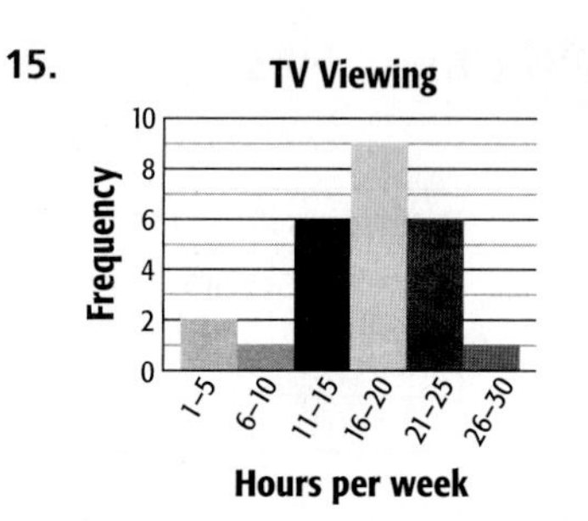

16. Possible answer: The symbols are different sizes even though they represent the same number of sightings. **17.** positive **18.** no correlation

Chapter 5

5-1 Exercises

1. points A, B, C **2.** $\overleftrightarrow{BC}$
3. plane Z or plane ABC
4. $\overline{AB}, \overline{BC}, \overline{AC}$ **5.** $\overrightarrow{BA}, \overrightarrow{BC}, \overrightarrow{CB}$
7. $\angle LJM$, $\angle MJK$ **9.** $\angle LJM$ and $\angle MJK$ **11.** 115° **13.** points V, W, X, Y **15.** plane N or plane VWX
17. $\overrightarrow{WV}, \overrightarrow{VW}, \overrightarrow{WY}, \overrightarrow{YW}, \overrightarrow{WX}$
19. $\angle DEH$, $\angle GEF$ **21.** $\angle FEG$ and $\angle HED$ **23.** 117° **25.** False
27. False **29.** False **31.** False
33. False **35. a.** 145° **b.** They are supplementary angles.
41. 18; 18; 29 **43.** A

5-2 Exercises

1. $\angle 1 \cong \angle 4 \cong \angle 5 \cong \angle 8$ (45°); $\angle 2 \cong \angle 3 \cong \angle 6 \cong \angle 7$ (135°)
2. 59° **3.** 59° **4.** 121° **5.** 59°
7. 60° **9.** 120° **11.** $\angle 4$, $\angle 5$, $\angle 8$
13. Possible answers: $\angle 1$ and $\angle 2$, $\angle 1$ and $\angle 3$, $\angle 3$ and $\angle 4$.
15. 51° **17.** 90°
19. Possible answer:

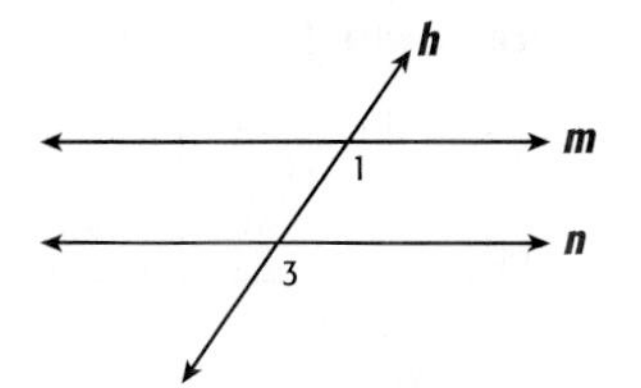

21. a. $\overline{AB}$ **b.** $m\angle 2 = m\angle 3 = m\angle 4 = 45°$ **27.** 32 **29.** 0.00000001 **31.** 128 **33.** m^{13} **35.** Population: shoppers; sample: paid shoppers at a mall; possible bias: The people may answer favorably because they are being paid.

5-3 Exercises

1. $q° = 78°$ **2.** $r° = 51°$ **3.** $s° = 120°$ **4.** $a° = 60°$ **5.** $c° = 66°$ **6.** $d° = 18°, 3d° = 54°, 6d° = 108°$ **7.** 60°, 30°, 90° **9.** $s° = 69°$ **11.** $w° = 60°$ **13.** $g° = 15°$, $4g° = 60°, 7g° = 105°$ **15.** $x° = 56°$ **17.** $w° = 40°$ **19.** $y° = 18°$ **27.** always **29.** sometimes **31.** never **33.** never **35. a.** $w° = 75°$; $y° = 75°$; two right angles **b.** $x° = 30°$; $z° = 75°$; $m° = 75°$ **c.** The two blue triangles are right scalene triangles, and the white triangle is an acute isosceles triangle. **39.** 11 **41.** C

5-4 Exercises

1. 360° **2.** 720° **3.** $t° = 90°$ **4.** $v° = 144°$ **5.** quadrilateral, trapezoid **6.** quadrilateral, parallelogram, rhombus **7.** 540° **9.** $m° = 120°$ **11.** quadrilateral, parallelogram, rhombus, rectangle, square **13.** 3240°; 162° **15.** 12,600°; 175° **17.** 2880°; 160° **19.** $x° = 110°$ **21.** $w° = 123°$ **23.** $x° = 130°$ **25.** hexagon **27.** 13-gon **29.** pentagon **35. a.** $x° = 98°$ **b.** $y° = 145°$ **39.** 6.4×10^{-7} **41.** -1.6×10^{-6} **43.** C

5-5 Exercises

1. 0 **2.** Slope is undefined. **3.** positive slope; 1 **4.** negative slope; $-\frac{1}{2}$ **5.** $\overleftrightarrow{AB} \parallel \overleftrightarrow{CD}$ **6.** $\overleftrightarrow{MN} \perp \overleftrightarrow{AB}$, $\overleftrightarrow{MN} \perp \overleftrightarrow{CD}$, and $\overleftrightarrow{AD} \perp \overleftrightarrow{BE}$ **7.** parallelogram, rhombus, rectangle, square **8.** trapezoid **9.** positive slope, 1 **11.** 0 **13.** $\overleftrightarrow{CD} \parallel \overleftrightarrow{AB}$ **15.** parallelogram, rhombus, rectangle, square **17.** 3 **19.** 0 **27.** 90° **29.** 33°

5-6 Exercises

1. triangle $ABC \cong$ triangle FED **2.** quadrilateral $LMNO \cong$ quadrilateral $STQR$ **3.** $q = 13$ **4.** $r = 4$ **5.** $s = 4$ **7.** trapezoid $PQRS \cong$ trapezoid $ZYXW$ **9.** $n = 5$ **11.** $x = 16, y = 25, z = 14.2$ **13.** $s = 120, t = 33, r = 33$ **19.** $16 = x$ **21.** $-15 = m$ **23.** $b = -6$ **25.** $a = -32$

5-7 Exercises

1. reflection **2.** rotation

3.

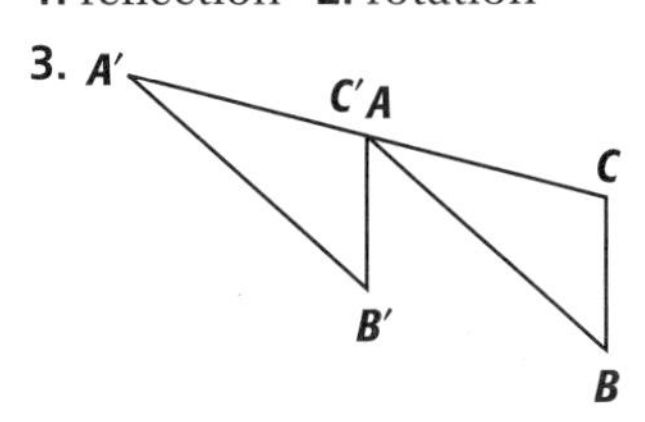

4.

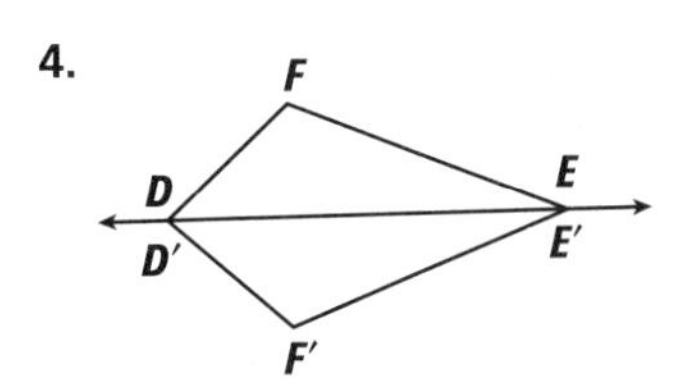

5.

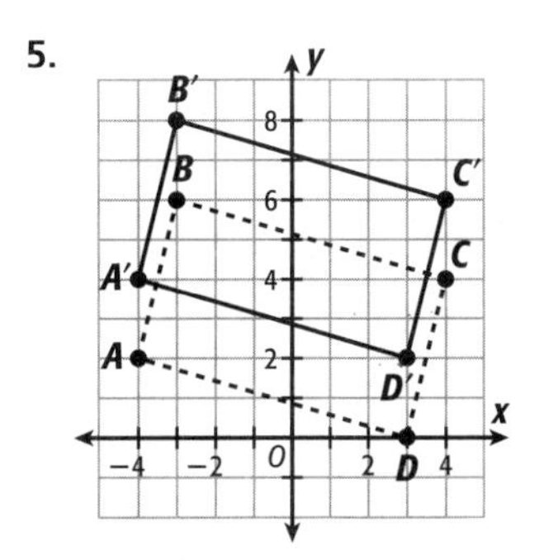

6.

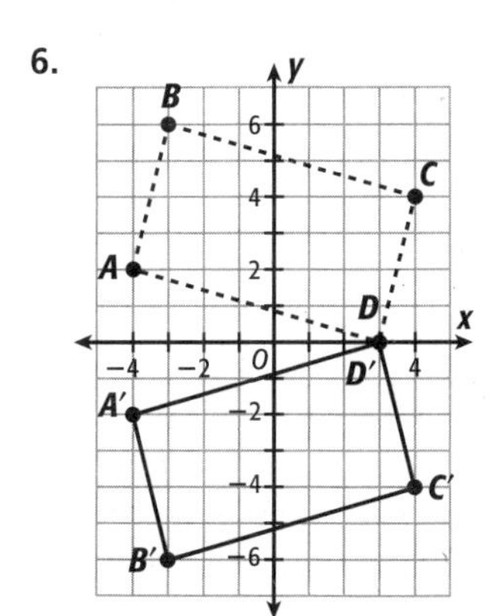

7.

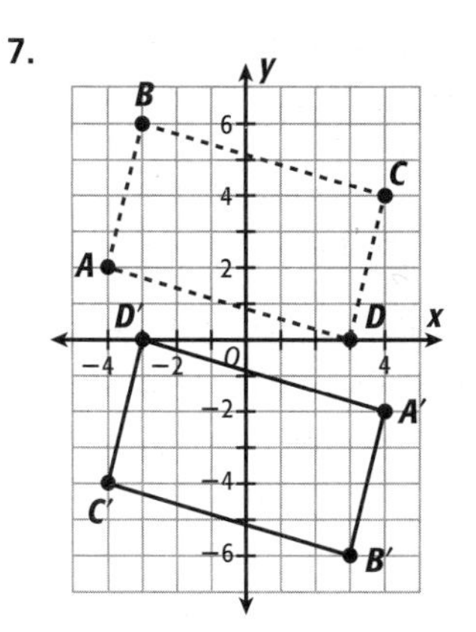

9. translation

11.

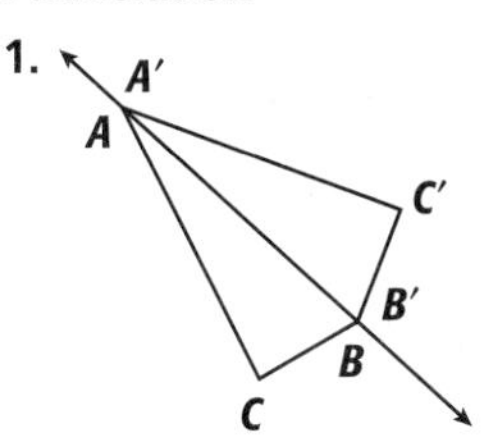

13.

15.

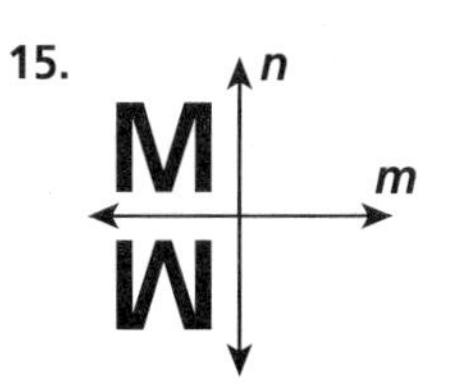

17.

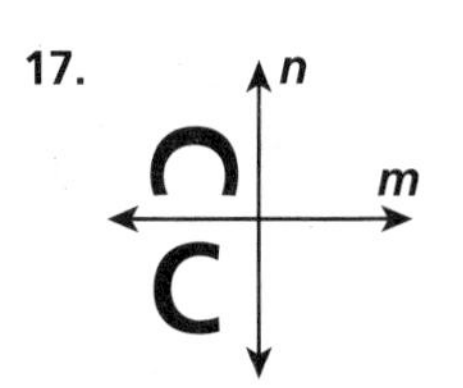

19. $(-2, -1)$ **21.** $(-4, -3)$ **23.** $(-m, n)$ **25.** $(6, -1)$

27. YLIME (mirrored)

reflection across a vertical line **31.** 32 **33.** −343 **35.** −128 **37.** 16 **39.** A

5-8 Exercises

1.

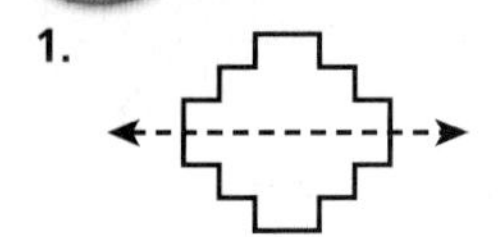

2.

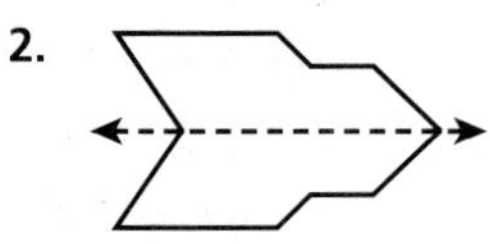

3.

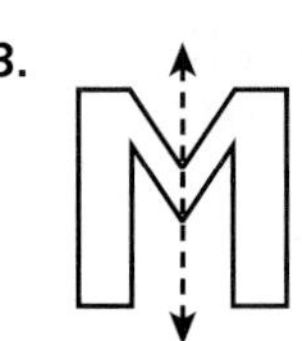

4.

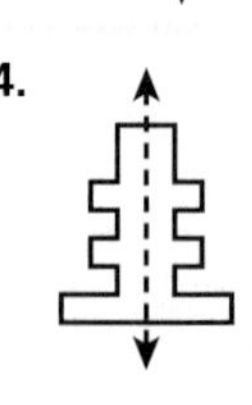

5.

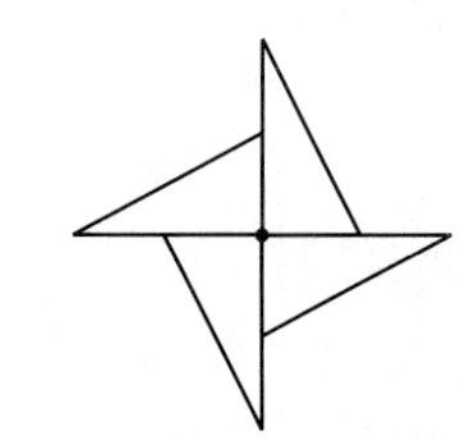

6.

7.

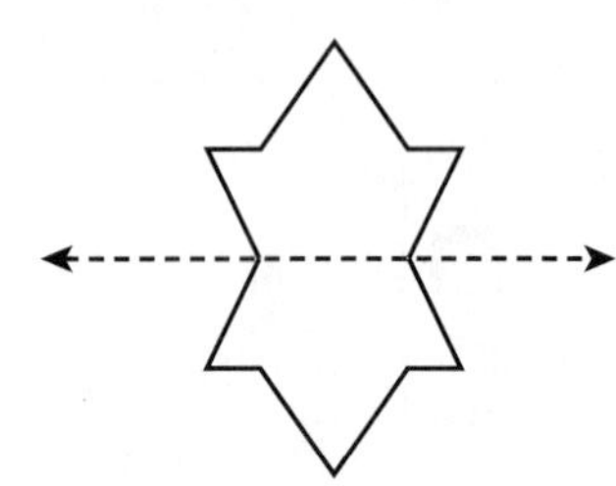

9.

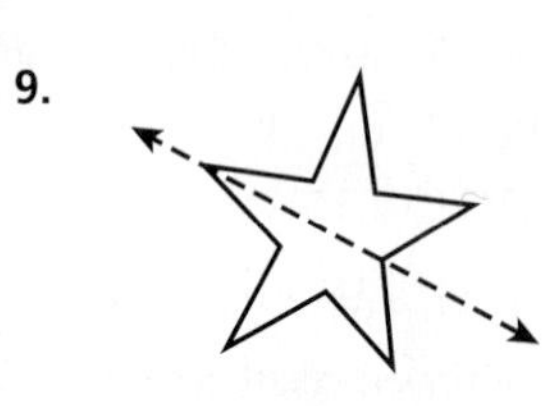

11.

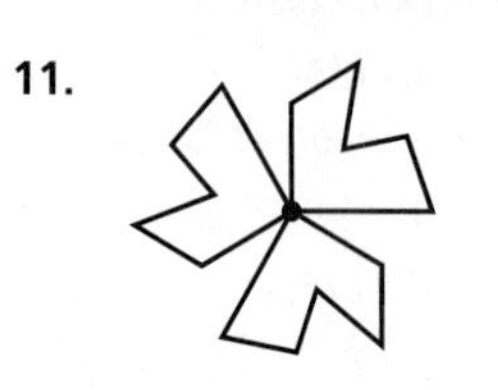

17. a.

Kage Asa no ha

There are 6 lines of symmetry and 6-fold rotational symmetry around the center.

b.

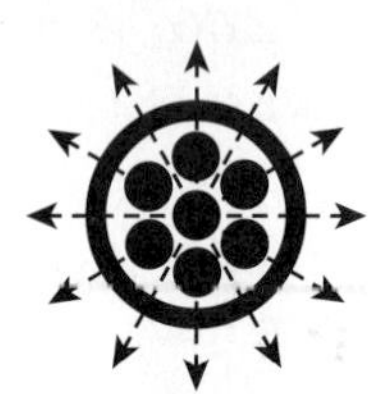

Maru ni shichiyo

There are 6 lines of symmetry and 6-fold rotational symmetry around the center.

c. There is no line symmetry and no rotational symmetry.

d.

Chukage itsutsu nenji Aoi

There is 5-fold rotational symmetry around the center.

e.

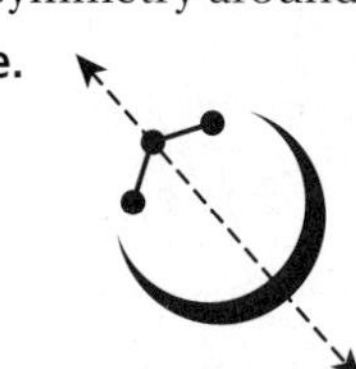

Tsuki ni sansei

There is one line of symmetry.

f.

Teuno ke

There are 16 lines of symmetry and 16-fold rotational symmetry.

21. 821,000 **23.** -1400
25. $-3.5 \cdot 10^{-5}$ **27.** B

5-9 Exercises

1. There is only one possibility: 1 square and 2 octagons

3.

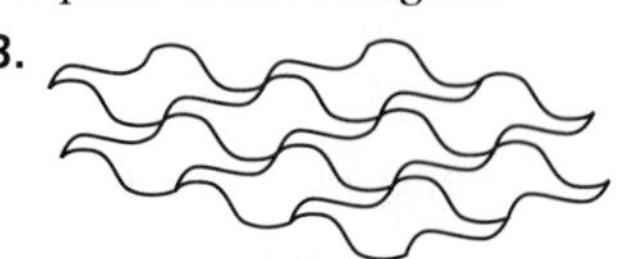

5.

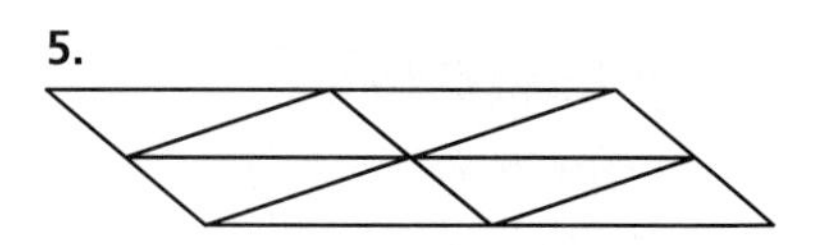

7.

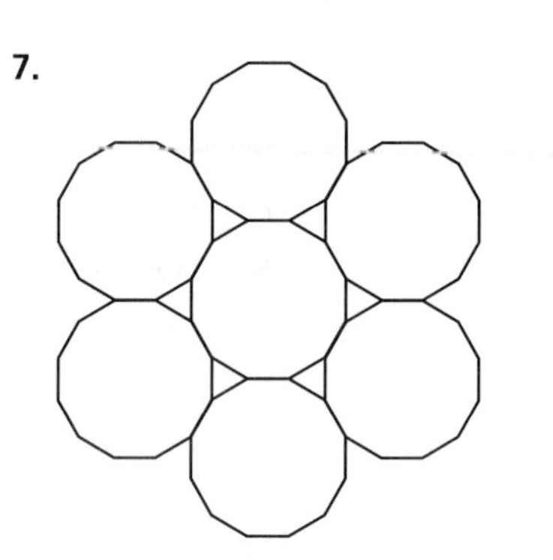

9.

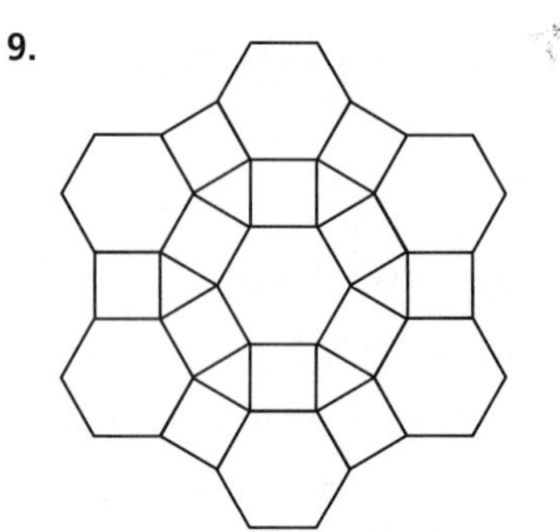

11.

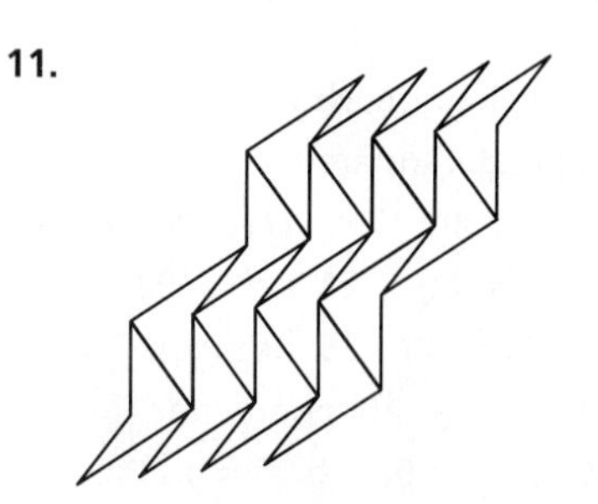

13. Yes, the shape will tessellate.
15. hexagon **19.** $p \leq 3$
21. $12 < w$ **23.** $m \leq 0$ **25.** $z < 2$

Chapter 5 Study Guide and Review

1. parallel lines; perpendicular lines **2.** rectangle; rhombus
3. 108° **4.** 72° **5.** 108° **6.** 56°
7. 124° **8.** 56° **9.** 56° **10.** 124°
11. $m° = 34°$ **12.** 120° **13.** 144°

14. trapezoid **15.** parallelogram, rhombus **16.** parallelogram
17. $x = 23$ **18.** $t = 3.2$ **19.** $q = 5$
20.

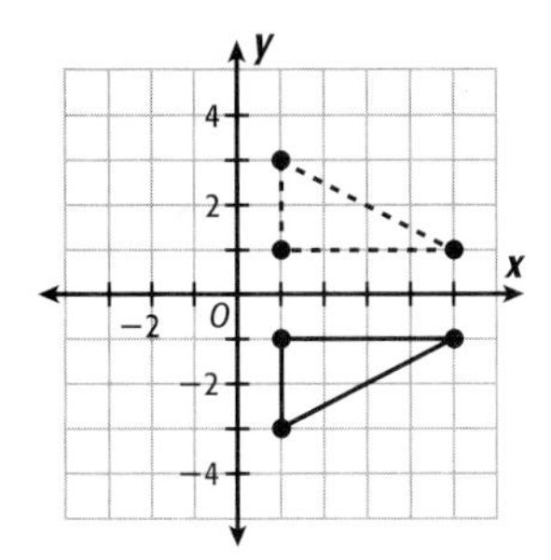

21.

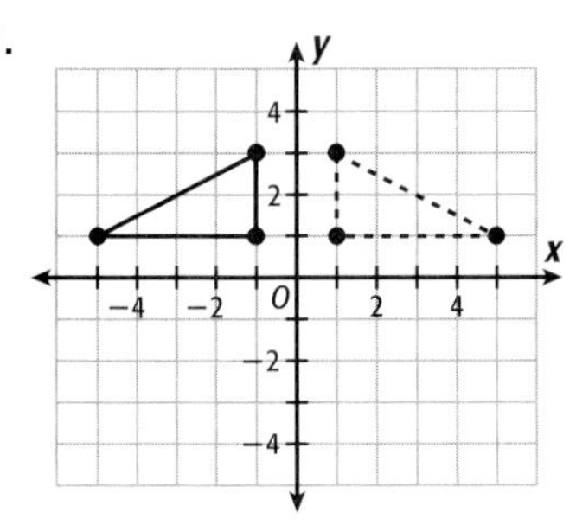

22.

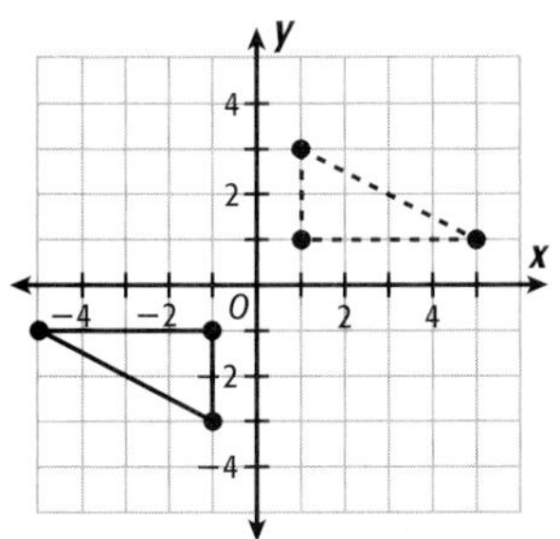

23. line symmetry: horizontal line of symmetry **24.** 2-fold rotational symmetry **25.** line symmetry: horizontal and vertical lines of symmetry; 2-fold rotational symmetry
26.

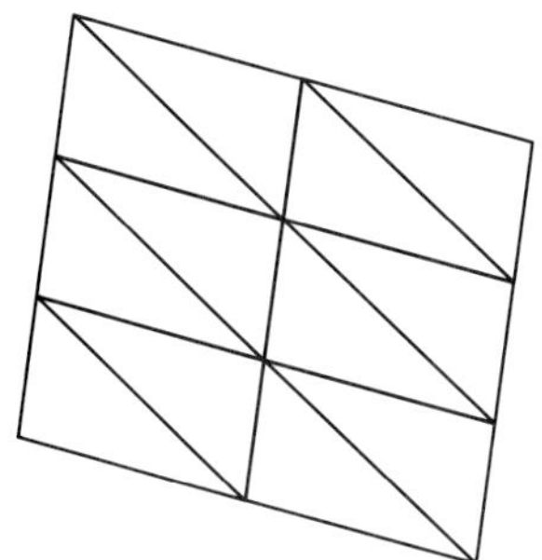

27.

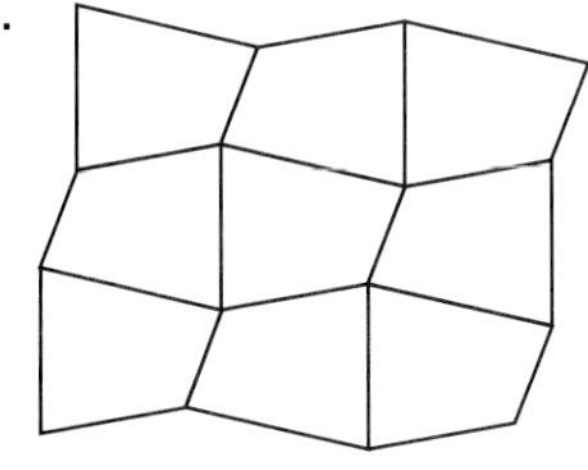

Chapter 6

6-1 Exercises

1. 20 units **2.** 36 units
3. $19.4x$ units **4.** 15 units^2
5. 28 units^2 **6.** 32 units^2
7. 14 units^2 **8.** 44 units; 53 units^2 **9.** 34 units
11. $26x$ units **13.** 24 units^2
15. 18 units^2 **17.** 64 units
19. 46 units; 72 units^2
21. 46 units; 84 units^2
23. 33 in.; 792 in^2 **25. a.** \$1125 **b.** 375 people **31.** $y < -2$
33. $w > 3$

6-2 Exercises

1. 22 units **2.** $11\frac{1}{4}$ units
3. 30 units **4.** 34.5 units
5. 84 units **6.** $(4x + 1)$ units
7. 15 units^2 **8.** 28 units^2
9. 12 units^2 **10.** 25 units^2
11. 29 units **13.** 70 units
15. $(30a + 8)$ units **17.** 20 units^2
19. 12 units^2 **21.** 49.5 units^2
23. $21x$ units^2 **25.** 9.1 ft
27. a. 1929.5 ft^2 **b.** 466.6 ft
29. 49.8 ft **31.** 874.6 ft^2; 160.4 ft
33. 0.75 **35.** 2.5 **37.** negative

6-3 Exercises

1. 5 **2.** 10.6 **3.** 7.8 **4.** 5 **5.** 5.3
6. 20 **7.** $\sqrt{24} \approx 4.9$ units; 19.6 units^2 **9.** 17 **11.** 8.9 **13.** 9.2
15. $\sqrt{80} \approx 8.9$ units; 71.2 units^2
17. 7 **19.** $\sqrt{1716} \approx 41.4$ **21.** 72
23. yes **25.** yes **27.** no **29.** yes
31. 139 km **33.** 475 mi **37.** $x = 9$
39. $y = 5$ **41.** A

6-4 Exercises

1. 8π cm; 25.1 cm **2.** 6.4π in.; 20.1 in. **3.** 2.25π ft^2; 7.1 ft^2
4. 56.25π cm^2; 176.6 cm^2
5. $A = 9\pi$ units$^2 \approx 28.3$ units^2; $C = 6\pi$ units ≈ 18.8 units
6. $\frac{175}{99} \approx 1.8$ ft **7.** 14π in.; 44.0 in.
9. 40.4π cm; 126.9 cm
11. 144π cm^2; 452.2 cm^2
13. 324π in^2; 1017.4 in^2
15. $A = 36\pi$ units$^2 \approx 113.0$ units^2; $C = 12\pi$ units ≈ 37.7 units
17. $C \approx 7.5$ m; $A \approx 4.5$ m^2
19. $C \approx 25.1$ in.; $A \approx 50.2$ in^2
21. 6.4 cm **23.** 4 cm **25.** 11.7 m
27. 297.7 m^2 **29.** $C = 12\pi \approx 37.7$ ft; $A = 36\pi \approx 113.1$ ft^2 **35.** $\frac{6}{19}$
37. $-\frac{21}{40}$

6-5 Exercises

1. Possible answer:

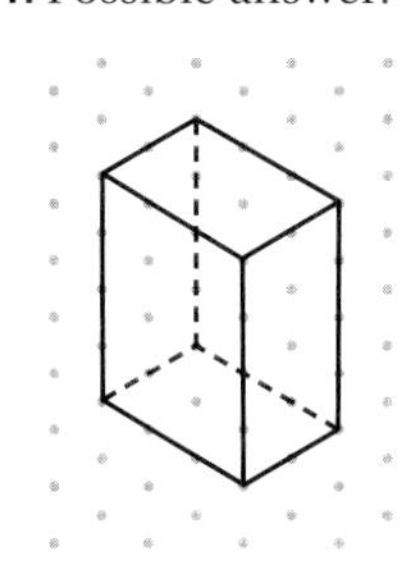

2. Possible answer:

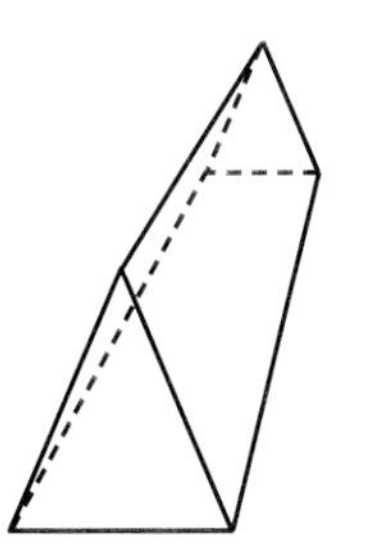

3. Possible answer:

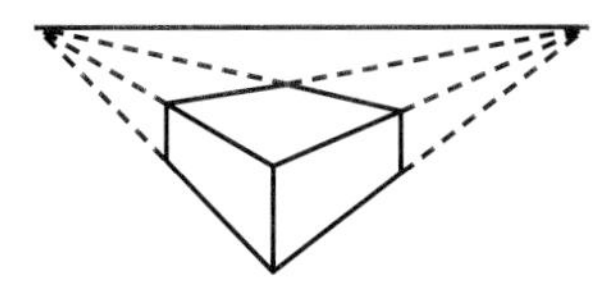

5. Possible answer:

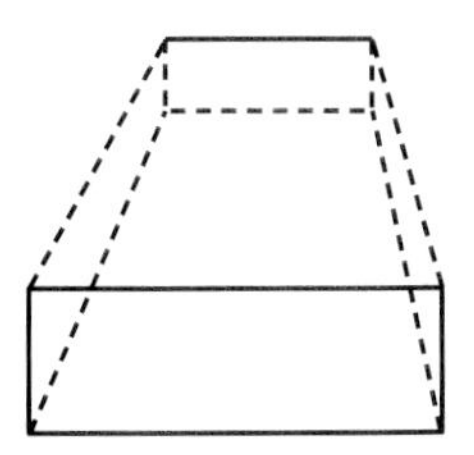

7. rectangles *JKLM*, *PQRN*, *JMNR*, *KLPQ*, *JKQR*, and *LMNP*
9. triangles *SVW* and *TUX* and rectangles *STUV*, *UVWX*, and *STXW* **15.** *A* **17.** *PQSR*
19. $\overline{RY}$, $\overline{WY}$, and $\overline{YZ}$ **21.** $\overline{UV} \parallel \overline{PQ} \parallel \overline{ST}$ **23.** $\overline{PQ}$ **25.** one-point

27. 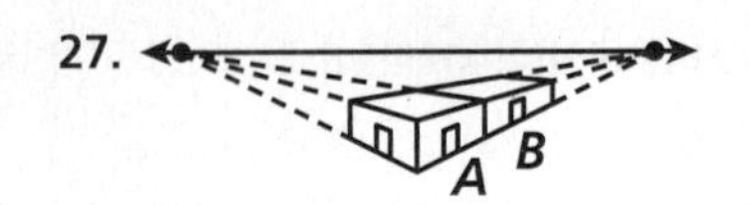

31. 2750 33. 0.00000063 35. B

6-6 Exercises

1. 210 cm^3 2. 1205.8 in^3 3. 556 in^3 4. Yes 5. No
6. 1406.25π $ft^3 \approx 4417.9$ ft^3
7. 4725 ft^3 9. 96 cm^3 11. Yes
13. 60 cm^3 15. a. 46,200,000 in^3
b. about 18.8 ft 17. about 20.5 in.
23. 15.5; 15.5; no mode 25. C

6-7 Exercises

1. 70 $units^3$ 2. 52.5 $units^3$ 3. 14.8 $units^3$ 4. 693 $units^3$ 5. 213.4 $units^3$ 6. 3159 $units^3$ 7. Yes
8. $6{,}255{,}333\frac{1}{3}$ ft^3 9. 0.2 $units^3$
11. 359.0 $units^3$ 13. 168 $units^3$
15. Yes 17. 4.0 cm 19. 9 ft
21. 301,056 ft^3
23. a. 38,520,000 ft^3 b. 27
c. 1,426,666.67 yd^3 27. 5.92
29. 7.42 31. C

6-8 Exercises

1. 351.7 in^2 2. 356 cm^2 3. No
4. 80.4 in^2 5. 768 in^2 7. No
9. 846 in^2 11. $249.6\pi \approx 783.7$ cm^2
13. 6 in. 15. 83.3 cm^2
17. 27.1 cm^2 19. $15.12
25. $\frac{1}{18}$ 27. $4\frac{2}{7}$ 29. D

6-9 Exercises

1. 144 m^2 2. 74.6 ft^2 3. No
4. ≈ 702.5 ft^2 5. 24.1 in^2 7. No
9. 765 cm^2 11. $1368\pi \approx 4295.5$ ft^2 13. $\approx 877{,}201{,}312$ mi^2
15. a. ≈ 588; ≈ 216 b. Khufu; 925,344 ft^2 c. Menkaure; $\approx 8{,}619{,}552$ ft^3 19. 0.6 21. −1.4
23. $\sqrt{709} \approx 26.63$ m

6-10 Exercises

1. 10.7π cm^3; 33.6 cm^3 2. 1333.3π ft^3; 4186.6 ft^3 3. 6.6π m^3; 20.7 m^3
4. 85.3π mi^3; 267.8 mi^3 5. 4π in^2; 12.6 in^2 6. 174.2π mm^2; 547.0 mm^2 7. 324π cm^2; 1017.4 cm^2
8. 225π yd^2; 706.5 yd^2 9. The volume of the sphere and the cube are about equal (≈ 268 in^3). The surface area of the sphere is about 201 in^2, and the surface area of the cube is about 250 in^2.
11. 147.5π cm^3; 463.2 cm^3
13. 0.17π in^3; 0.5 in^3 15. 207.4π m^2; 651.2 m^2 17. 2500π cm^2; 7850 cm^2 19. 221.83π in^3; 696.55 in^3
21. $V = 39.72\pi \approx 124.72$ yd^3; $S = 38.44\pi \approx 120.70$ yd^2
23. 30 km; $36{,}000\pi \approx 113{,}040$ km^3
25. ≈ 5392 cm^3 27. ≈ 0.0314 mm^2 31. $\frac{3}{10}$ 33. $3\frac{17}{75}$ 35. B

Chapter 6 Extension

1. rotational and bilateral
3. rotational and bilateral
5. 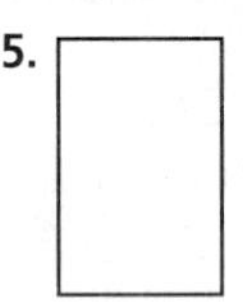

line and rotational

7. rotational 9. rotational and bilateral
11. 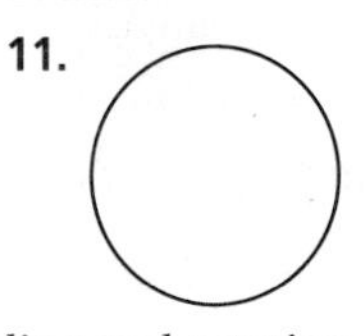

line and rotational

13. square; smaller 15. circles

Chapter 6 Study Guide and Review

1. perimeter; area 2. edge; vertex
3. great circle; hemispheres
4. $13\frac{2}{9}$ in^2; 16 in. 5. 208 m^2; 80 m
6. 16 cm^2; 20.2 cm 7. 21 in^2; 34 in. 8. $c = 10$ 9. $a = 10$
10. $A = 225\pi \approx 706.5$ in^2; $C = 30\pi \approx 94.2$ in. 11. $A = 5.8\pi \approx 18.2$ cm^2; $C = 4.8\pi \approx 15.1$ cm
12. $A = 16\pi \approx 50.2$ m^2; $C = 8\pi \approx 25.1$ m 13. $A = 0.4\pi \approx 1.3$ ft^2; $C = 1.2\pi \approx 3.8$ ft

14. 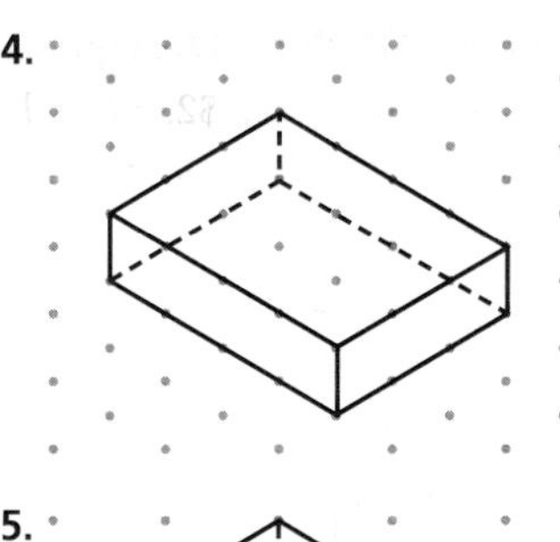

15. 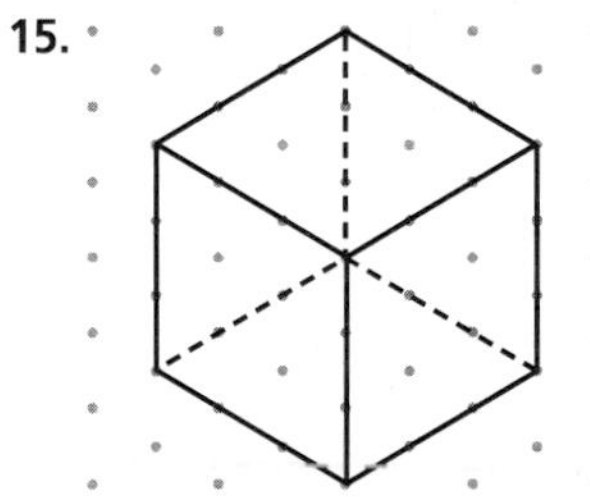

16. 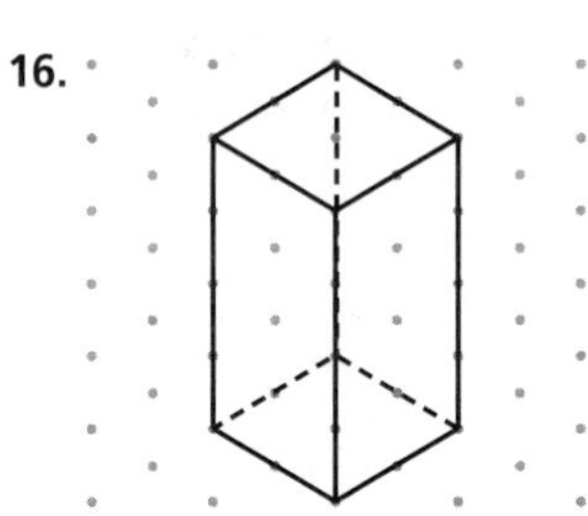

17. $216\pi \approx 678.2$ m^3 18. 1053 ft^3
19. 320 ft^3 20. $120\pi \approx 376.8$ in^3
21. 90 cm^2 22. 95 cm^2 23. 340 in^2
24. 972π $in^3 \approx 3052.1$ in^3
25. 4500π $m^3 \approx 14{,}130$ m^3

Chapter 7

7-1 Exercises

1–4. Possible answers given. 1. $\frac{2}{5}, \frac{8}{20}$ 2. $\frac{1}{3}, \frac{6}{18}$ 3. $\frac{3}{1}, \frac{42}{14}$ 4. $\frac{20}{16}, \frac{10}{8}$
5. yes 6. no 7. yes 8. No; $2\frac{1}{4}$ cups are needed. 9. Possible answers: $\frac{2}{14}, \frac{3}{21}$ 11. Possible answers: $\frac{8}{7}, \frac{32}{28}$ 13. no 15. yes
17–25. Possible answers given.
17. no; $\frac{4}{7}$ 19. no; $\frac{8}{14}$ 21. yes
23. no; $\frac{8}{42}$ 25. yes 27. no; 4 gallons 29. $\frac{39}{18}$ 35. $-1\frac{11}{36}$
37. $1\frac{5}{99}$ 39. 3 and −3 41. 13 and −13

7-2 Exercises

1. 1:5 2. 35 wpm 3. 42 wpm
4. 22 oz can 5. dozen golf balls
7. 171.6 gal/h 9. 4 boxes

11. \$26.25 per hour **13.** \$0.77 per slice **15.** \$2.49/yard; \$2.26/yard; 5 yards **17.** \$1.37/gal; \$1.42/gal; 10 gal **19. a.** Super-Cell: \$0.10/min; Easy Phone: \$0.11/min
b. Super-Cell offers a better rate.
21. a. Tom: $25\frac{3}{8}$ frames per hour; Cherise: 27 frames per hour; Tina: $28\frac{3}{8}$ frames per hour
b. Tina **c.** $1\frac{5}{8}$ **d.** 24 **25.** −4
27. −5 **29.** −4.4 **31.** D

7-3 Exercises

1. 12 in./1 ft **2.** 8 pt/1 gal
3. 1 m/100 cm **4.** 91.25 gal
5. 7.5 mi/h **6.** 0.09 m/s
7. ≈ 1.14 g **9.** 1 yd/36 in.
11. 585 ft **13.** 57,600 bricks
15. 900 radios **17.** 4 hot dogs
19. 4.98 mi **21.** A ≈ 22.88 mi/h; B ≈ 23.16 mi/h; C ≈ 21.76 mi/h
23. 200 times **29.** 14 units^2
31. 226.9 in^2 **33.** 3.8 mi^2

7-4 Exercises

1. yes **2.** yes **3.** no **4.** yes
5. no; $\frac{1}{8} \neq \frac{8}{56}$ **6.** $x = 1$ **7.** $n = 8$
8. $d = 2$ **9.** $h = 6$ **10.** $f = 9.75$
11. $t = 2$ **12.** $s = 9$ **13.** $q = 12.5$
14. ≈ 3.3 cm **15.** no **17.** no
19. yes; $\frac{18}{12} = \frac{15}{10}$ **21.** $b = 3$
23. $y = 6$ **25.** $n = 4$ **27.** $d = 0.5$
29. $\frac{6}{3}, \frac{18}{9}$ **31.** $\frac{66}{21}, \frac{22}{7}$ **33.** $\frac{0.25}{4}, \frac{1}{16}$
35. 12 molecules **37. a.** about 1.53:1 **b.** about 134.6 mm Hg
41. $-1\frac{1}{4}$ **43.** $11\frac{17}{100}$

7-5 Exercises

1. no **2.** yes
3.

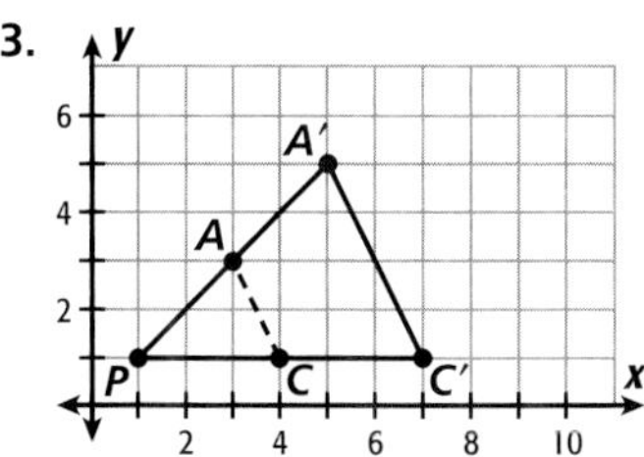

4.

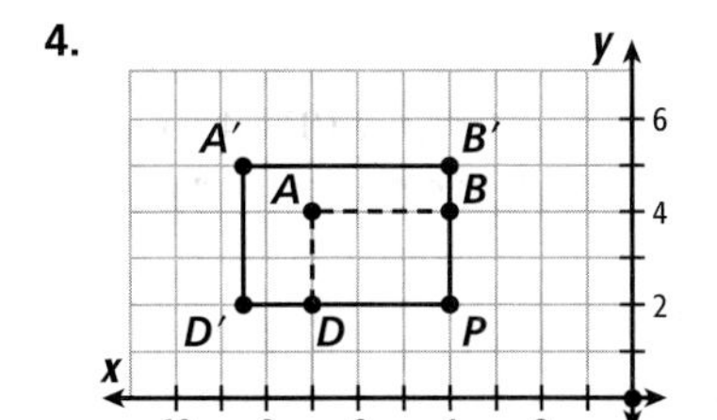

5. $A'(1.5, -1)$; $B'(1, -2.5)$; $C'(4, -3)$; $D'(5, -0.5)$ **6.** $A'(16, 4)$; $B'(28, 4)$; $C'(20, 12)$ **7.** no
9.

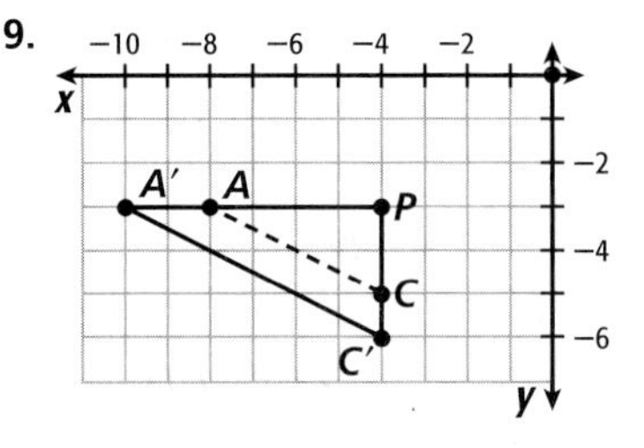

11. $A'(-9, 6)$; $B'(15, 12)$; $C'(-6, -9)$ **13.** 3 **15.** Yes **21.** 24 units^2
23. 21 units^2

7-6 Exercises

1. ≈ 5.4 in. **2.** ≈ 14.7 cm **3.** *A* and *C* are similar. **5.** ≈ 22.9 ft
7. similar **9.** similar **11.** $x = 6$ ft
13. $x = 24$ ft **15.** yes; $\frac{1}{15}$ or $\frac{4 \text{ in.}}{5 \text{ ft}}$
17. 24 ft **21.** 1256 mm^3
23. 2044.3 cm^3 **25.** D

7-7 Exercises

1. 1 in:1.25 ft **2.** 20.25 m
3. 0.0085 in. **4.** 7.5 mm **5.** 52 ft
6. 27 in. **7.** 1 cm = 1.5 m
9. 0.023 mm **11.** 20 in. **13.** 2 in.
15. 0.5 in. **17.** 18 ft **19.** 58.5 ft
21. about 580 mi **23–27.** The scale is 1.2 cm:36 in. **23.** ≈ 18 in.
25. No; each wall is only ≈ 45 in. wide. **27.** ≈ 298 ft^2 **31.** no **33.** no

7-8 Exercises

1. reduces **2.** enlarges
3. preserves **4.** preserves
5. reduces **6.** enlarges **7.** $\frac{1}{24}$
8. 14 in. **9.** 0.000028 mm
11. preserves **13.** reduces
15. reduces **17.** 7.5 ft **19.** $\frac{12}{1}$
21. $\frac{1}{45}$ **23.** $\frac{1}{12.5}$ **25.** $\frac{1}{28}$ **27.** 630 ft
33. ≈ 1869.4 ft^2 **35.** 1256 cm^2
37. \$0.17 per apple **39.** A

7-9 Exercises

1. 4:1 **2.** 16:1 **3.** 64:1 **4.** width: 30 in.; height: 10 in. **5.** 72 min
6. 7:1 **7.** 49:1 **9.** 32 cm **11.** 2 cm; 8 cubes **13.** 4 cm; 64 cubes
15. 5 cm; 125 cubes
17. 1,000,000 cm^3 **19.** 256,000
21. 14.58 oz **25.** Possible answers: $\frac{6}{10}, \frac{9}{15}$ **27.** Possible answers: $\frac{8}{22}, \frac{12}{33}$ **29.** 1.5 ft **31.** 18 ft **33.** D

Chapter 7 Extension

1. 0.777 **3.** 0.017 **5.** 45 ft
7. 16.7 m **9.** 137.7 m **11.** 11.7 yd
13. 10 ft **15.** 45°

Chapter 7 Study Guide and Review

1. ratio; proportion **2.** rate; unit rate **3.** similar; scale factor
4. dilation; enlargement; reduction **5–7.** Possible answers given. **5.** $\frac{1}{2}, \frac{2}{4}$ **6.** $\frac{3}{6}, \frac{4}{8}$ **7.** $\frac{7}{12}, \frac{14}{24}$
8. yes **9.** no **10.** yes **11.** no
12. \$0.30 per disk; \$0.29 per disk; 75 disks **13.** \$3.75 per box; \$3.75 per box; unit prices are the same.
14. \$2.89 per divider; \$4.00 per divider; 8-pack **15.** 90,000 m/h
16. 4500 ft/min **17.** $583\frac{1}{3}$ m/min
18. $80\frac{2}{3}$ ft/s **19.** 2160 m/h
20. $x = 15$ **21.** $h = 6$ **22.** $w = 21$
23. $y = 29\frac{1}{3}$
24.

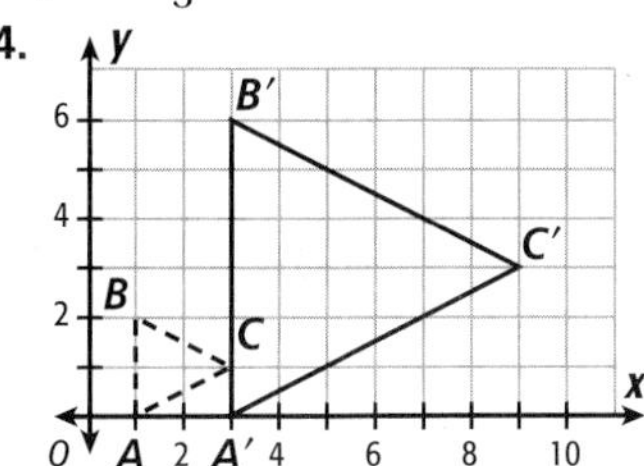

25.

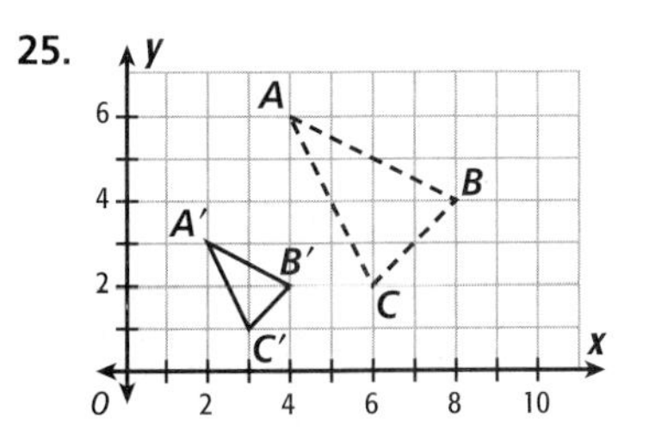

26.

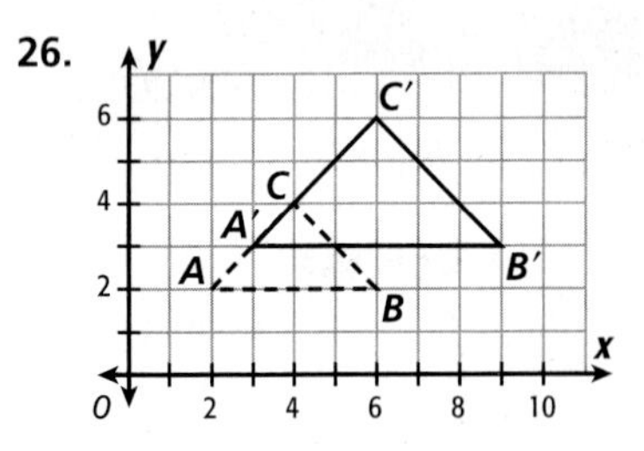

27. 12.5 in. **28.** 3.125 in. **29.** 64.8 m **30.** 6.6 in. **31.** $2.\overline{7}$:1; enlarges **32.** 2.5:1; enlarges **33.** 1:100; reduces **34.** 1:1; preserves **35.** 3:1 **36.** 9:1 **37.** 27:1

Chapter 8

8-1 Exercises

1. $\frac{3}{10}$ **2.** 46% **3.** 62.5% **4.** $\frac{17}{20}$ **5.** 40% **6.** $\frac{8}{25}$ **7.** 0.875 **8.** $33\frac{1}{3}$% **9.** 10% **11.** $\frac{3}{5}$ **13.** 0.32 **15.** $\frac{109}{200}$ **17.** 40%, 30%, 20%, 10% **19.** 40%, 30%, 25%, 5% **21.** 85% **23. a.** $\frac{4}{25}$; 0.16 **b.** 23% **27.** perpendicular **29.** parallel **31.** D

8-2 Exercises

1. 49.3% **2.** 19.9% **3.** 70.6% **4.** 31.5 pages **5.** 300% **7.** 1% **9.** 1.0% **11.** 30 **13.** 2.6 **15.** 266 **17. a.** 30 **b.** 45 **c.** 150 **19. a.** 100 **b.** 50 **c.** 25 **21.** 21.5% **23.** 16 **29.** irrational **31.** not real

8-3 Exercises

1. 34.4 **2.** 168 **3.** 166.7 **4.** 320 **5.** ≈ 1.7 oz **6.** 28 ft **7.** 315 **9.** 850 **11.** 16.7 **13.** 570 **15.** 336 **17. a.** 300 **b.** 150 **c.** 75 **19. a.** 40 **b.** 20 **c.** 10 **21.** 48.2% **23. a.** 49.5% **b.** 50.5% **c.** 49.1% **d.** 50.9% **27.** 882 **29.** 45, 65

8-4 Exercises

1. 38% increase **2.** 65% decrease **3.** 100% increase **4.** 64% increase **5.** 25% decrease **6.** 34% increase **7.** 96.6% increase **8.** $9773.60 **9.** 9% increase **11.** 44% decrease **13.** 20% decrease **15.** ≈ 8.6% **17.** 23% decrease **19.** 30% decrease **21.** 17% decrease **23.** $600 **25.** 200 **27.** 50 **29.** 40% **31. a.** $84 **b.** $156 **c.** $104 **d.** $56\frac{2}{3}$% **33.** about 8361 ft **37.** 364 m^2 **39.** 84 yd^2

8-5 Exercises

Note: All answers are estimates.

1. 100 **2.** 24 **3.** 25% **4.** 21 **5.** 50% **6.** 900 **7.** $4.50 **9.** 50% **11.** 440 **13.** 10% **15.** B **17.** A **19.** C **21.** 150 **23.** 250 **25.** 1600 **27.** 33% **29.** 400 **31.** 750 **33.** 50% **35.** 50% **39.** 120 ft^3 **41.** 132 in^3 **43.** 0.48 ft^3 **45.** D

8-6 Exercises

1. $510 **2.** $5.18 **3.** 22.5% **4.** $499 **5.** $389.50 **7.** 18% **9.** $330 **11.** $2.16 **13.** $1963.75 **15.** $2800 plus 3% of sales: $3100 to $3400 a month **17. a.** $64,208 **b.** $14,275.95 **c.** ≈ 20.0% **d.** ≈ 22.2% **21.** 40:3 **23.** 10,000:1

8-7 Exercises

1. $1794.38; $10,044.38 **2.** 5 years **3.** $1635.30 **4.** 5.5% **5.** $23,032.50 **7.** $1846.50 **9.** $33.75, $258.75 **11.** $446.25, $4696.25 **13.** $14.89, $411.89 **15.** $87.50, $787.50 **17.** $270, $1770 **19.** 6% **25.** 14.4 **27.** $16\frac{2}{3}$% **29.** 5

Chapter 8 Extension

1. $12,597.12 **3.** $14,802.44 **5.** $15,208.16 **7.** $2462.88 **9.** $6744.25

Chapter 8 Study Guide and Review

1. percent **2.** percent change **3.** commission **4.** simple interest; principal; rate of interest **5.** 0.4375 **6.** 43.75% **7.** $1\frac{1}{8}$ **8.** 112.5% **9.** $\frac{7}{10}$ **10.** 0.7 **11.** $\frac{1}{250}$ **12.** 0.4% **13.** 39% **14.** 4200 ft **15.** 3030 mi **16.** 5 lb 7 oz **17.** 20% **18.** 472,750% **19.** ≈ 12.38% **20.** ≈ 25% **21.** ≈ 25% **22.** ≈ 13 **23.** ≈ 16 **24.** ≈ 6 **25.** ≈ 4.5 **26.** $10,990 **27.** $3.04 **28.** $1796.88 **29.** $500 **30.** 7% **31.** $\frac{1}{2}$ yr, or 6 mo **32.** 2-year loan; $50

Chapter 9

9-1 Exercises

1. 0.55; 0.45 **2.** 0.262 **3.** 0.738

4.

Team	A	B	C	D
Prob.	0.25	0.3	0.15	0.3

5. $\frac{1}{3}$, $\frac{1}{3}$, $\frac{1}{6}$, $\frac{1}{6}$ **7.** 0.155 **9.** 0.319 **11.** 1 **13.** 0.465

15.

Person	Probability
Jamal	0.1
Elroy	0.2
Tina	0.2
Mel	0.2
Gina	0.3

17. 0.56 **21.** 59 in^2 **23.** ≈ 354.43 yd^2 **25.** B

9-2 Exercises

1. 0.34 **2.** 0.11 **3.** ≈ 0.186; ≈ 0.281; more likely to listen to a rock station **5.** 0.433 **7.** 0.26 **9.** 0.06 **11.** 0.36 **13.** 0.308 **17.** $x = 2$ **19.** $b = 5$ **21.** D

9-3 Exercises

1–9. Possible answers are given.

1. 90% **2.** 50% **3.** 60% **5.** 30% **7.** 30% **9.** 50% **13.** 84 ft tall **15.** 42 ft tall **17.** B

9-4 Exercises

1. $\frac{1}{2}$ **2.** $\frac{1}{3}$ **3.** $\frac{1}{6}$ **4.** $\frac{1}{36}$ **5.** $\frac{1}{4}$ **6.** $\frac{5}{18}$ **7.** $\frac{5}{18}$ **9.** $\frac{5}{6}$ **11.** $\frac{1}{36}$ **13.** 1 **15.** $\frac{1}{4}$ **17.** $\frac{1}{8}$ **19.** $\frac{3}{8}$ **21.** $\frac{7}{8}$ **23.** $\frac{1}{4}$ **25. a.** $\frac{1}{4}$ **b.** $\frac{3}{4}$ **27.** 90% **29.** 0.375 **31.** $\frac{39}{50}$

9-5 Exercises

1. 1,757,600 **2.** ≈ 0.000000569 **3.** 0.81 **4.** ≈ 0.107 **5.** 6 ways **6.** 9 combinations **7.** 17,576,000 **9.** ≈ 0.5269 **11.** 18 chairs **13.** 6 **15.** 12 **17.** 1920 **23.** 140 **25.** 20 **27.** 40

9-6 Exercises

1. 5040 **2.** 360 **3.** 20,160 **4.** 20 **5.** 3,628,800 **6.** 720 **7.** 56 **8.** 56 **9.** 6 **11.** 24 **13.** 5040 **15.** 3,838,380 **17.** 72 **19.** 39,916,800 **21.** 1 **23.** 10 **25.** n **27.** $n!$ **29.** n **31.** 1 **33.** 504 **35.** 720 **37.** 21 **39. a.** 5040 **b.** 210 **45.** $135, $885 **47.** $15.99, $425.99 **49.** $21.60, $111.60

9-7 Exercises

1. dependent **2.** independent **3.** $\frac{1}{32}$ **4.** $\frac{1}{8}$ **5.** $\frac{28}{435}$ **6.** $\frac{8}{203}$ **7.** dependent **9.** $\frac{1}{4}$ **11.** $\frac{1}{20}$ **13.** $\frac{1}{14}$ **15. a.** $\frac{9}{100} = 0.09$ **b.** $\frac{3}{275} \approx 0.01$ **c.** $\frac{19}{825} \approx 0.02$ **19.** 50% decrease **21.** 440% increase **23.** 28% decrease **25.** B

9-8 Exercises

1. 1:20 **2.** 20:1 **3.** $\frac{1}{1000}$ **4.** $\frac{1}{2250}$ **5.** 1:74 **6.** 22,749:1 **7.** 1:17 **9.** $\frac{1}{10,000}$ **11.** 1:844 **13.** 1:35, 35:1 **15.** 1:17, 17:1 **17.** 1:3, 3:1 **19.** 1:2 **21.** $\frac{1}{1275}$ **23. a.** 237:1 **b.** $\frac{237}{238}$ **27.** $\frac{1}{7}$ **29.** 0 **31.** $\frac{22}{35}$ **33.** A

Chapter 9 Study Guide and Review

1. probability; impossible; certain **2.** sample space **3.** permutation; combination **4.** 0.75; 0.25 **5.** 0.17, or 17% **6.** 0.28, or 28% **7.** Possible answer: 40% **8.** $\frac{1}{2}$ **9.** 36 **10.** 30 **11.** 210 **12.** 126 **13.** $\frac{1}{216}$ **14.** $\frac{13}{51}$ **15.** 5:21

Chapter 10

10-1 Exercises

1. 4 hr **2.** $t = 7$ **3.** $x = 3.5$ **4.** $r = 98$ **5.** $b = 2$ **6.** $q = 4$ **7.** $a = 87$ **9.** $m = -30$ **11.** $g = 3\frac{1}{3}$ **13.** $y = -5$ **15.** $w = 2.5$ **17.** $m = 15$ **19.** $q = 3$ **21.** $z = \frac{1}{3}$ **23.** $k = 5$ **25.** $n = 23$ **27.** $y = 1.3$ **29.** $b = -2$ **31.** $\frac{x+5}{7} = 12; x = 79$ **33.** 110,000 **35.** 25 in. **37.** $12x + 3$ **39.** $w - 15$ **41.** C

10-2 Exercises

1. $d = 2$ **2.** $y = 3$ **3.** $e = 6$ **4.** $c = 4$ **5.** $h = 9$ **6.** $x = -7$ **7.** $x = -1$ **8.** $y = 2$ **9.** $p = -1$ **10.** $z = \frac{2}{3}$ **11.** $24.49 **13.** $k = -10$ **15.** $w = 3$ **17.** $y = 5$ **19.** $h = 3$ **21.** $m = -2$ **23.** $x = -4$ **25.** $n = \frac{3}{4}$ **27.** $b = -13$ **29.** $x = 17$ **31.** $y = -11$ **33.** $h = 4$ **35.** $b = \frac{10}{13}$ **37.** $12.20 per hr **39.** 212°F **45.** −2.5 **47.** $-2\frac{1}{7}$ **49.** $\frac{17}{19}$

10-3 Exercises

1. $x = 1$ **2.** $a = 7$ **3.** $x = 2$ **4.** $y = -4$ **5.** $x = 1$ **6.** $n = 2$ **7.** $d = 3$ **8.** $x = 7$ **9.** 32 chairs **11.** $x = 1$ **13.** $y = 7$ **15.** no solution **17.** $x = 22.9$ **19.** $a = 11$ **21.** $y = 2$ **23.** $n = 2$ **25.** $x = 5$ **27.** $m = 3.5$ **29.** $x = 7$ **31.** 360 units **33.** 24, 25 **35. a.** 17 **b.** 11 **41.** ≈ $0.175 per oz; ≈ $0.187 per oz; 20 oz **43.** ≈ $0.199 per oz; $0.184 per oz; 20 oz **45.** $249 per monitor; $275 per monitor; 3 monitors

10-4 Exercises

1. $k > 3$ **2.** $z \le 20$ **3.** $y < -7$ **4.** $x \le -2$ **5.** $y \ge 3$ **6.** $k > 5$ **7.** $x < 3$ **8.** $b \ge -\frac{1}{4}$ **9.** $h \le 1$ **10.** $c > 2$ **11.** $d < \frac{7}{18}$ **12.** $m \le 1\frac{3}{4}$ **13.** at least 16 caps **15.** $x > 4$ **17.** $q \le 2$ **19.** $x \le -8$ **21.** $a \ge -3$ **23.** $k \ge 3$ **25.** $r < 3$ **27.** $p \le \frac{22}{3}$ **29.** $w > -1$ **31.** $a > \frac{1}{2}$ **33.** $q < 6$ **35.** $b < 2.7$ **37.** $f \le -27$ **39.** at most 26 chairs **41.** at least 38 games **47.** 650 **49.** 1.44

10-5 Exercises

1. $\ell_2 = P - \ell_1 - \ell_3$ **2.** $\ell_1 = P - \ell_2 - \ell_3$ **3.** $A = C + B - 2$ **4.** $B = A - C + 2$ **5.** $d_1 = \frac{2A}{d_2}$ **6.** $b = \sqrt{c^2 - a^2}$ **7.** $n = \frac{S}{180} + 2$ **8.** $C = \frac{5}{9}(F - 32)$ **9.** $y = -3x + 15$ **10.** $y = \frac{9}{2}x + 7$ **11.** $y = 2x - 1$ **13.** $A_3 = 180 - A_1 - A_2$ **15.** $c = p - a - 100$ **17.** $c = \sqrt{\frac{E}{m}}$ **19.** $b_1 = \frac{2A}{h} - b_2$ **21.** $y = -\frac{9}{2}x - 5$ **23.** $x = 2y + 4$ **25.** $k = \frac{10\ell^2}{3}$ **27.** $y = \frac{x}{22}$ **29.** $z = \sqrt{4y}$ **31.** $m = \frac{y - b}{x}$ **33.** $b = y - mx$ **35.** $y = -6x + 8$ **37. a.** $T = \frac{E}{P}$ **b.** 1.5 hr **39. a.** $T_1 = \frac{P_2 \cdot T_2}{P_1}$ **b.** 2.5 hr **43.** $x = 0.5$ **45.** B

10-6 Exercises

1. yes **2.** yes **3.** yes **4.** no **5.** (2, 3) **6.** (1, −1) **7.** (3, 12) **8.** (4, 13) **9.** (3, 0) **10.** (0, 7) **11.** (5, 3) **12.** (8, 12) **13.** (5, 2) **14.** (3, −7) **15.** (0, −2) **16.** (−9, 3) **17.** no **19.** no **21.** (−1, −1) **23.** (2, −1) **25.** (−6, 0) **27.** (2, 3) **29.** (1, 3) **31.** (4, 1) **33.** (2, 4) **35.** (2, −3) **37.** (0, 0) **39.** (2, 1) **41.** $\left(\frac{1}{10}, \frac{2}{15}\right)$ **43.** (12, 8) **45.** (6, −3) **47.** $\left(\frac{1}{5}, \frac{2}{3}\right)$ **49.** 14 and 4 **51. a.** $m + u = 2000$ **b.** $40m + 25u = 62,000$ **c.** 800 main-floor and 1200 upper-level **55.** 12 **57.** 9 **59.** D

Chapter 10 Study Guide and Review

1. system of equations **2.** solution of a system of equations **3.** $m = 10$ **4.** $y = -8$ **5.** $c = -16$ **6.** $r = -3$ **7.** $t = 16$ **8.** $w = 64$ **9.** $r = -6$ **10.** $h = -25$ **11.** $x = 52$ **12.** $d = -33$ **13.** $a = 67$ **14.** $c = 90$ **15.** $y = 2$ **16.** $h = 3$ **17.** $t = -1$

11.

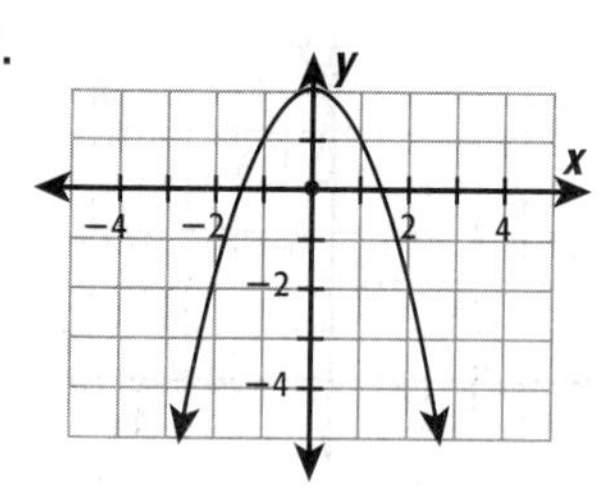

13.

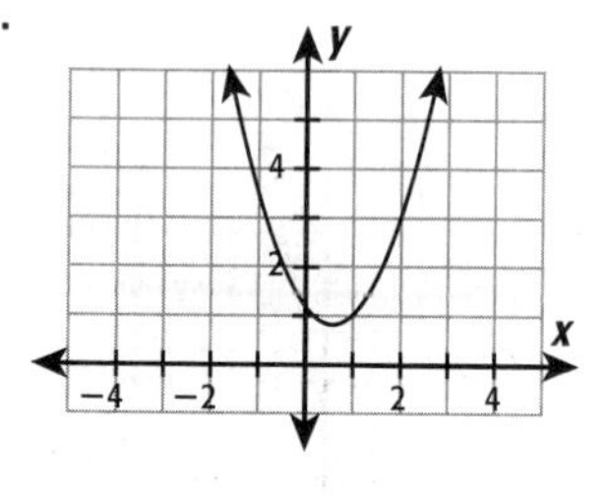

15.

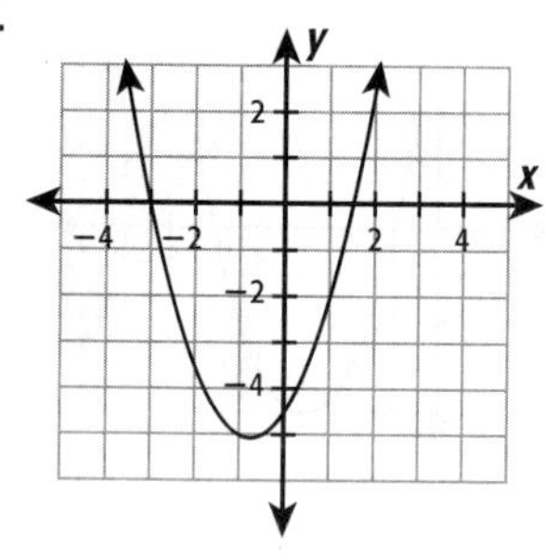

17.

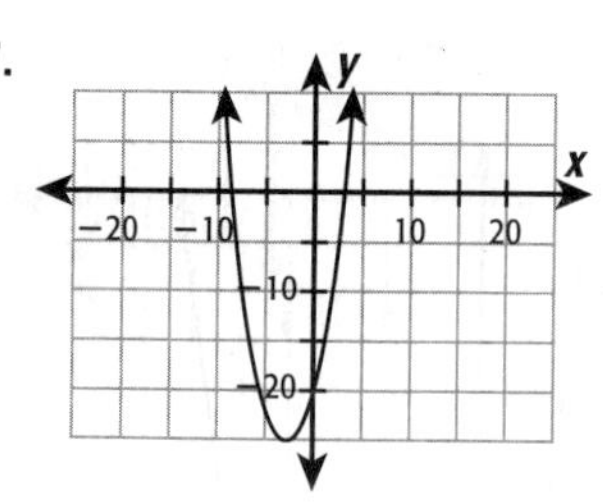

19. 14, 5, 14 **21.** 3, 0, 15 **23.** 26, 5, 20 **25.** $x = 5, x = -11$ **27.** $x = 2, x = -1$ **29.** $x = 1.8, x = -2.6$ **31.** 5 and 5; 50 **33. a.** \$4860, \$4940, \$5000, \$5040, \$5060; 7 **b.** \$115 **39.** −2, 12 **41.** −1, −4 **43.** −4, 1 **45.** −2, 4 **47.** C

12-8 Exercises

1. no **2.** yes

3.

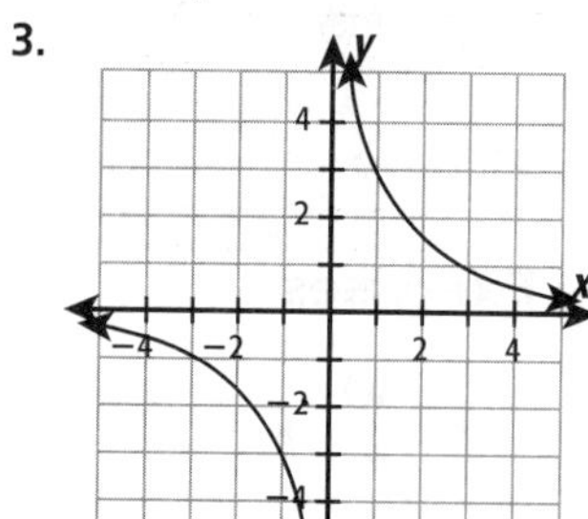

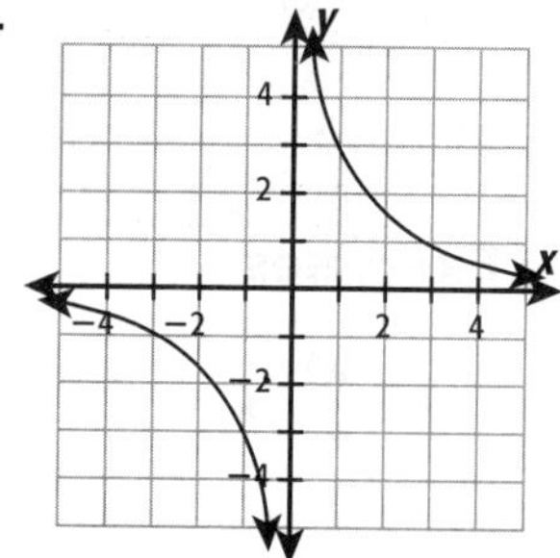

4.

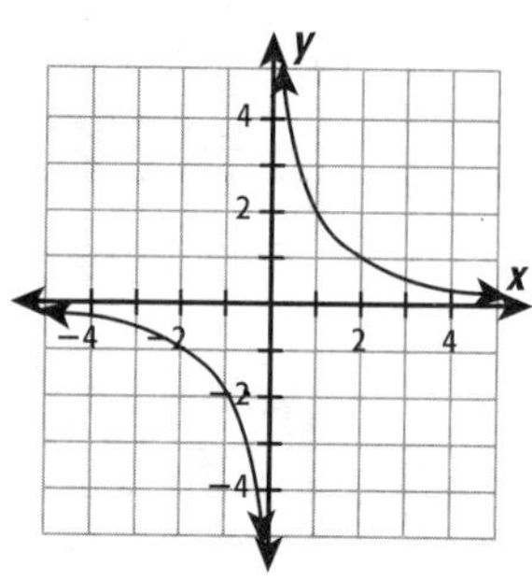

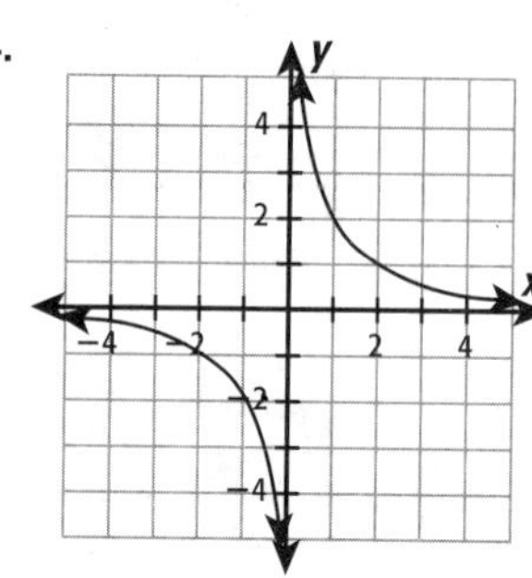

5.

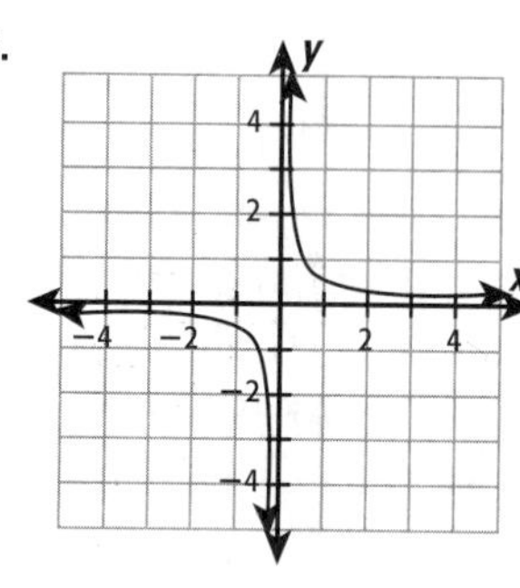

6. $y = \frac{12}{x}$; $1\frac{1}{3}$ amps **7.** yes

9.

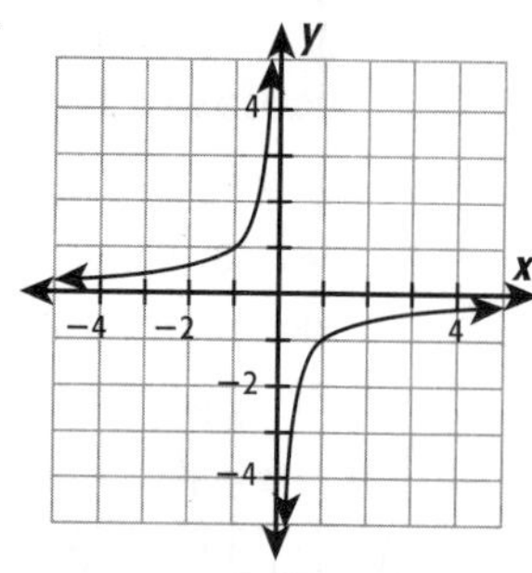

11.

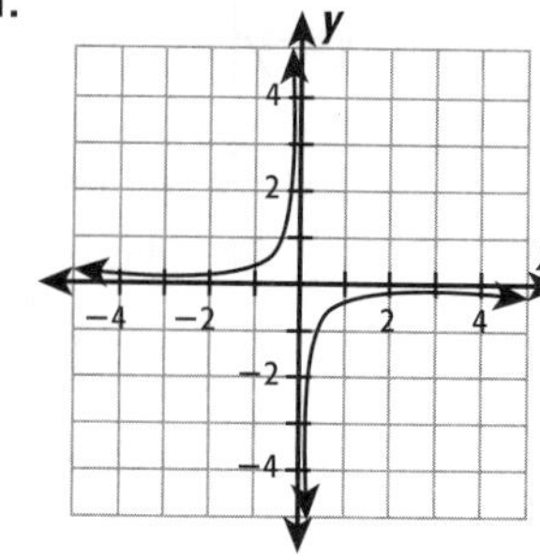

13. $y = \frac{4}{x}$ **15.** $y = \frac{32}{x}$ **17.** 10 cm **19.** \$1250 **23.** 9, 1, −1 **25.** 300, 243, 192 **27.** −36, −35, −32 **29.** C

Chapter 12 Study Guide and Review

1. sequence **2.** arithmetic sequence; geometric sequence **3.** Fibonacci sequence **4.** function; domain; range **5.** 31 **6.** 0.65 **7.** $\frac{14}{3}$ **8.** −640 **9.** $\frac{32}{729}$, or ≈ 0.0439 **10.** −1 **11.** 4, 7, 10, 13 **12.** 2, 5, 10, 17 **13.** −6, 12, −2, 16 **14.** 3, 4, 8, 26 **15.** −4, 10, −11 **16.** 1, 17, −1 **17.** 0, 2, −4 **18.** 0, 0, 3 **19.** 5, 15, 9 **20.** 1, −1, −1 **21.** $f(x) = x - 1$ **22.** $f(x) = \frac{1}{2}x + 4$

23.

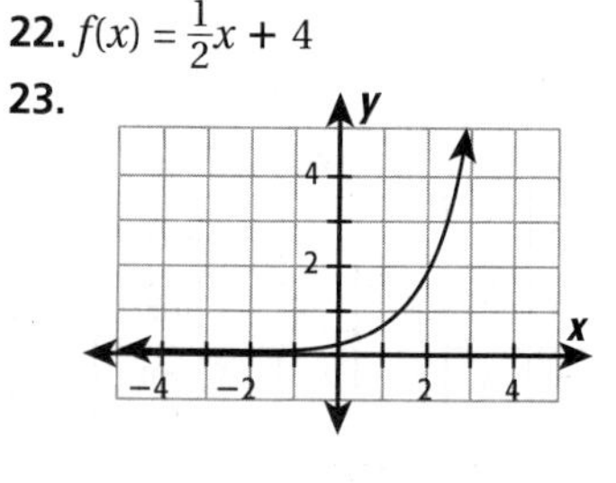

24.

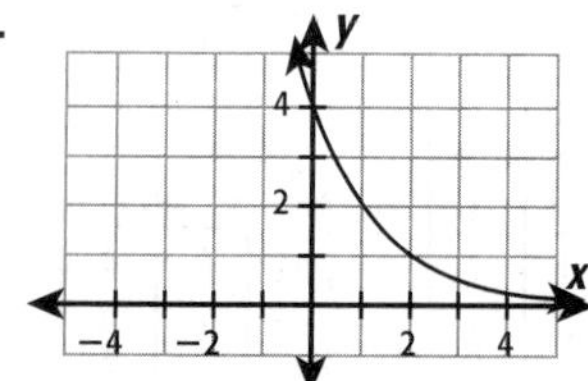

25.

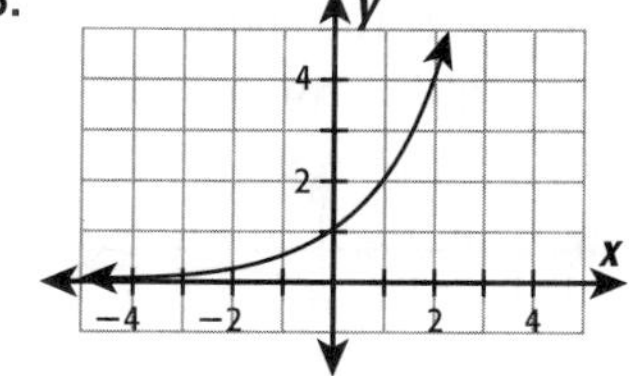

base (of a polygon or three-dimensional figure) A side of a polygon; a face of a three-dimensional figure by which the figure is measured or classified. (p. 307)

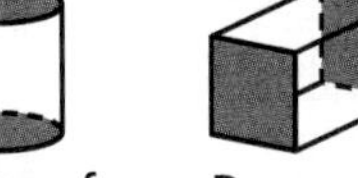

Bases of a cylinder	Bases of a prism	Base of a cone	Base of a pyramid

biased sample A sample that does not fairly represent the population. (p. 174)

binary number system A number system in which all numbers are expressed using only two digits, 0 and 1. (p. 160)

bisect To divide into two congruent parts. (p. 227)

boundary line The set of points where the two sides of a two-variable linear inequality are equal. (p. 567)

box-and-whisker plot A graph that displays the highest and lowest quarters of data as whiskers, the middle two quarters of the data as a box, and the median. (p. 189)

break (graph) A zigzag on a horizontal or vertical scale of a graph that indicates that some of the numbers on the scale have been omitted.

capacity The amount a container can hold when filled. (p. 382)

Celsius A metric scale for measuring temperature in which 0°C is the freezing point of water and 100°C is the boiling point of water; also called *centigrade.*

center (of a circle) The point inside a circle that is the same distance from all the points on the circle. (p. 294)

center (of dilation) The point of intersection of lines through each pair of corresponding vertices in a dilation. (p. 362)

center (of rotation) The point about which a figure is rotated. (p. 254)

central angle An angle formed by two radii with its vertex at the center of a circle. (p. 712)

certain (probability) Sure to happen; having a probability of 1. (p. 446)

chord A segment with its endpoints on a circle. (p. 712)

circle The set of all points in a plane that are the same distance from a given point called the center. (p. 294)

circle graph A graph that uses sectors of a circle to compare parts to the whole and parts to other parts.

circumference The distance around a circle. (p. 294)

clockwise A circular movement to the right in the direction shown.

coefficient The number that is multiplied by the variable in an algebraic expression. (p. 4)

Example: 5 is the coefficient in $5b$.

combination An arrangement of items or events in which order does not matter. (p. 472)

commission A fee paid to a person for making a sale. (p. 424)

commission rate The fee paid to a person who makes a sale expressed as a percent of the selling price. (p. 424)

common denominator A denominator that is the same in two or more fractions.

Example: The common denominator of $\frac{5}{8}$ and $\frac{2}{8}$ is 8.

common difference The difference between any two successive terms in an arithmetic sequence. (p. 590)

common factor A number that is a factor of two or more numbers. (p. 696)

Example: 8 is a common factor of 16 and 40.

common multiple A number that is a multiple of each of two or more numbers. (p. 696)

Example: 15 is a common multiple of 3 and 5.

common ratio The ratio each term is multiplied by to produce the next term in a geometric sequence. (p. 595)

Commutative Property
Addition: The property that states that two or more numbers can be added in any order without changing the sum. (p. 701)

Example: $8 + 20 = 20 + 8$; $a + b = b + a$

Multiplication: The property that states that two or more numbers can be multiplied in any order without changing the product.

Example: $6 \cdot 12 = 12 \cdot 6$; $a \cdot b = b \cdot a$ (p. 701)

compatible numbers Numbers that are close to the given numbers that make estimation or mental calculation easier. (pp. 420, 697)

complementary angles Two angles whose measures add to 90°. (p. 223)

composite number A number greater than 1 that has more than two whole-number factors. (p. 695)

compound inequality A combination of more than one inequality.

Example: $x \geq -2$ or $x < 10$, or $-2 \leq x < 10$. x is greater than or equal to -2 and less than 10.

compound interest Interest earned or paid on principal and previously earned or paid interest. (p. 432)

cone A three-dimensional figure with one vertex and one circular base. (p. 312)

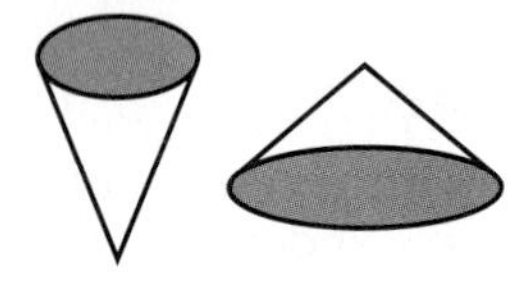

congruent Having the same size and shape. (p. 223)

congruent angles Angles that have the same measure. (p. 223)

congruent segments Segments that have the same length. (p. 223)

constant A value that does not change. (p. 4)

constant of proportionality A constant ratio of two variables related proportionally. (p. 562)

Example: $5 = k$, $10 = 2k$, and $15 = 3k$

conversion factor A fraction whose numerator and denominator represent the same quantity but use different units; the fraction is equal to 1 because the numerator and denominator are equal. (pp. 350, 714)

Example: $\frac{24 \text{ hours}}{1 \text{ day}}$ and $\frac{1 \text{ day}}{24 \text{ hours}}$

coordinate plane (coordinate grid) A plane formed by the intersection of a horizontal number line called the x-axis and a vertical number line called the y-axis. (p. 38)

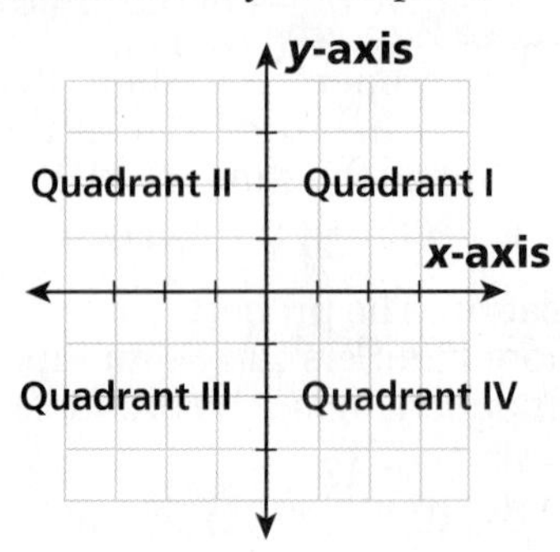

coordinate One of the numbers of an ordered pair that locate a point on a coordinate graph. (p. 38)

correlation The description of the relationship between two data sets. (p. 204)

correspondence The relationship between two or more objects that are matched. (p. 250)

corresponding angles (for lines) Angles formed by a transversal cutting two or more lines and that are in the same relative position.

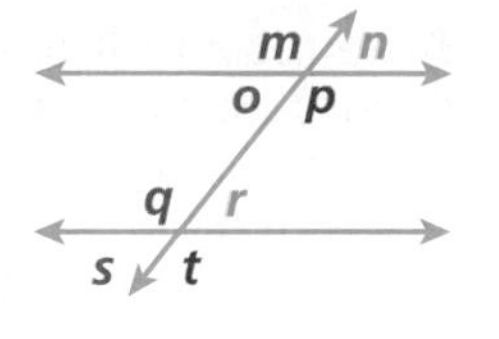

When a transversal cuts two lines as shown in the diagram, the pairs of corresponding angles are $\angle m$ and $\angle q$, $\angle n$ and $\angle r$, $\angle o$ and $\angle s$, and $\angle p$ and $\angle t$. (p. 229)

corresponding angles (in polygons) Matching angles of two or more polygons. (p. 250)

corresponding sides Matching sides of two or more polygons. (p. 250)

cosine (cos) In a right triangle, the ratio of the length of the side adjacent to an acute angle to the length of the hypotenuse. (p. 386)

counterclockwise A circular movement to the left in the direction shown.

cross product The product of numbers on the diagonal when comparing two ratios. (p. 356)

Example: 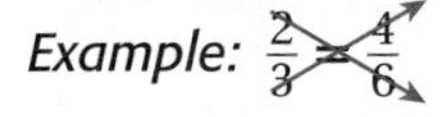 $2 \cdot 6 = 12$

$3 \cdot 4 = 12$

cube (geometric figure) A rectangular prism with six congruent square faces. (pp. 154, 300)

cube (in numeration) A number raised to the third power. (p. 154)

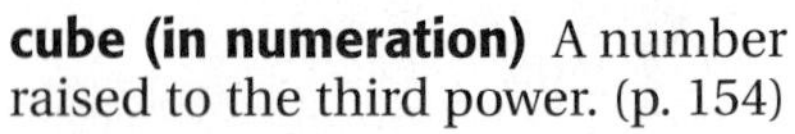

cumulative frequency The sum of successive data items. (p. 709)

customary system of measurement The measurement system often used in the United States.

Example: inches, feet, miles, ounces, pounds, tons, cups, quarts, gallons

cylinder A three-dimensional figure with two parallel, congruent circular bases connected by a curved lateral surface. (p. 307)

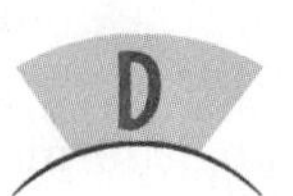

decagon A polygon with ten sides.

decimal system A base-10 place value system. (p. 160)

degree The unit of measure for angles or temperature. (p. 222)

Density Property of Real Numbers The property that states that between any two real numbers, there is always another real number. (p. 157)

denominator The bottom number of a fraction that tells how many equal parts are in the whole. (p. 112)

Example: $\frac{3}{4}$ ← denominator

dependent events Events for which the outcome of one event affects the probability of the other. (p. 477)

diagonal A line segment that connects two non-adjacent vertices of a polygon.

diameter A line segment that passes through the center of a circle and has endpoints on the circle, or the length of that segment. (p. 294)

difference The result when one number is subtracted from another.

dilation A transformation that enlarges or reduces a figure. (p. 362)

dimensions (geometry) The length, width, or height of a figure.

dimensions (of a matrix) The number of horizontal rows and vertical columns in a matrix. (p. 713)

direct variation A relationship between two variables in which the data increase or decrease together at a constant rate. (p. 562)

discount The amount by which the original price is reduced.

Distributive Property The property that states if you multiply a sum by a number, you will get the same result if you multiply each addend by that number and then add the products. (p. 701)

Example: $5 \cdot 21 = 5(20 + 1) = (5 \cdot 20) + (5 \cdot 1)$

dividend The number to be divided in a division problem.

Example: In $8 \div 4 = 2$, 8 is the dividend.

divisible Can be divided by a number without leaving a remainder. (p. 694)

Division Property of Equality The property that states that if you divide both sides of an equation by the same nonzero number, the new equation will have the same solution. (p. 18)

divisor The number you are dividing by in a division problem.

Example: In $8 \div 4 = 2$, 4 is the divisor.

dodecahedron A polyhedron with 12 faces.

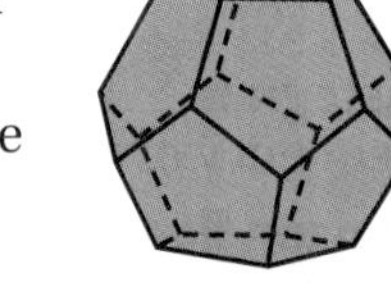

domain The set of all possible input values of a function. (p. 608)

double-bar graph A bar graph that compares two related sets of data.

double-line graph A line graph that shows how two related sets of data change over time.

edge The line segment along which two faces of a polyhedron intersect. (p. 302)

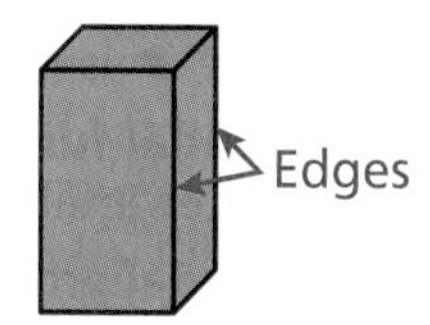

elements (of a matrix) Individual entries in a matrix. (p. 713)

elements (sets) The words, numbers, or objects in a set.

empty set A set that has no elements.

endpoint A point at the end of a line segment or ray.

enlargement An increase in size of all dimensions in the same proportions. (p. 373)

equally likely outcomes Outcomes that have the same probability. (p. 462)

equation A mathematical sentence that shows that two expressions are equivalent. (p. 13)

equilateral triangle A triangle with three congruent sides. (p. 235)

equivalent Having the same value. (p. 28)

equivalent fractions Fractions that name the same amount or part.

equivalent ratios Ratios that name the same comparison. (p. 342)

estimate (n) An answer that is close to the exact answer and is found by rounding or other method. **(v)** To find such an answer. (p. 420)

evaluate To find the value of a numerical or algebraic expression. (p. 4)

even number A whole number that is divisible by two.

event An outcome or set of outcomes of an experiment or situation. (p. 446)

expanded form A number written as the sum of the values of its digits.

Example: 236,536 written in expanded form is 200,000 + 30,000 + 6,000 + 500 + 30 + 6.

experiment (probability) In probability, any activity based on chance (such as tossing a coin). (p. 446)

experimental probability The ratio of the number of times an event occurs to the total number of trials, or times that the activity is performed. (p. 451)

exponent The number that indicates how many times the base is used as a factor. (p. 84)

Example: $2^3 = 2 \times 2 \times 2 = 8$; 3 is the exponent.

exponential decay Occurs in an exponential function when the output $f(x)$ gets smaller as the input x gets larger. (p. 618)

line symmetry A figure has line symmetry if one half is a mirror-image of the other half. (p. 259)

linear equation An equation whose solutions form a straight line on a coordinate plane. (p. 540)

linear function A function whose graph is a straight line. (p. 613)

linear inequality A mathematical sentence using $<$, $>$, $\leq$, or $\geq$ whose graph is a region with a straight-line boundary. (p. 567)

major arc An arc that is more than half of a circle. (p. 712)

matrix A rectangular arrangement of data enclosed in brackets. (p. 713)

mean The sum of a set of data divided by the number of items in the data set; also called *average.* (p. 184)

measure of central tendency A measure used to describe the middle of a data set; the mean, median, and mode are measures of central tendency. (p. 184)

median The middle number, or the mean (average) of the two middle numbers, in an ordered set of data. (p. 184)

metric system of measurement A decimal system of weights and measures that is used universally in science and commonly throughout the world.

Example: centimeters, meters, kilometers, gram, kilograms, milliliters, liters

midpoint The point that divides a line segment into two congruent line segments.

minor arc An arc that is less than half of a circle. (p. 712)

mixed number A number made up of a whole number that is not zero and a fraction. (p. 697)

mode The number or numbers that occur most frequently in a set of data; when all numbers occur with the same frequency, we say there is no mode. (p. 184)

Multiplication Property of Equality The property that states that if you multiply both sides of an equation by the same number, the new equation will have the same solution. (p. 19)

Multiplication Property of Zero The property that states that for all real numbers a, $a \cdot 0 = 0$ and $0 \cdot a = 0$. (p. 701)

multiplicative inverse A number times its multiplicative inverse is equal to 1; also called *reciprocal.* (p. 126)

Example: The multiplicative inverse of $\frac{4}{5}$ is $\frac{5}{4}$.

multiple The product of any number and a whole number is a multiple of that number. (p. 694)

mutually exclusive Two events are mutually exclusive if they cannot occur in the same trial of an experiment. (p. 464)

negative correlation Two data sets have a negative correlation if one set of data values increases while the other decreases. (p. 205)

negative integer An integer less than zero. (p. 60)

net An arrangement of two-dimensional figures that can be folded to form a polyhedron. (p. 300)

no correlation Two data sets have no correlation when there is no relationship between their data values. (p. 205)

nonlinear function A function whose graph is not a straight line.

nonterminating decimal A decimal that never ends. (p. 156)

numerator The top number of a fraction that tells how many parts of a whole are being considered. (p. 112)

Example: $\frac{4}{5}$ ← numerator

numerical expression An expression that contains only numbers and operations.

obtuse angle An angle whose measure is greater than 90° but less than 180°. (p. 223)

obtuse triangle A triangle containing one obtuse angle. (p. 234)

octagon An eight-sided polygon. (p. 239)

octahedron A polyhedron with eight faces. (p. 300)

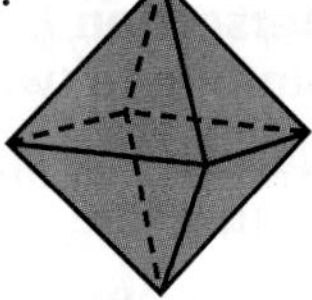

odd number A whole number that is not divisible by two.

odds A comparison of favorable outcomes and unfavorable outcomes. (p. 482)

odds against The ratio of the number of unfavorable outcomes to the number of favorable outcomes. (p. 482)

odds in favor The ratio of the number of favorable outcomes to the number of unfavorable outcomes. (p. 482)

opposites Two numbers that are an equal distance from zero on a number line; also called *additive inverse.* (p. 60)

order of operations A rule for evaluating expressions: first perform the operations in parentheses, then compute powers and roots, then perform all multiplication and division from left to right, and then perform all addition and subtraction from left to right. (p. 700)

ordered pair A pair of numbers that can be used to locate a point on a coordinate plane. (p. 34)

origin The point where the x-axis and y-axis intersect on the coordinate plane; (0, 0). (p. 38)

outcome (probability) A possible result of a probability experiment. (p. 446)

outlier A value much greater or much less than the others in a data set. (p. 185)

output The value that results from the substitution of a given input into an expression or function. (p. 608)

overestimate An estimate that is greater than the exact answer.

parabola The graph of a quadratic function. (p. 621)

parallel lines Lines in a plane that do not intersect. (p. 228)

parallelogram A quadrilateral with two pairs of parallel sides. (p. 240)

Pascal's triangle A triangular arrangement of numbers in which each row starts and ends with 1 and each other number is the sum of the two numbers above it. (p. 476)

pentagon A five-sided polygon. (p. 239)

percent A ratio comparing a number to 100. (p. 400)

Example: $45\% = \frac{45}{100}$

percent change The amount stated as a percent that a number increases or decreases. (p. 416)

percent decrease A percent change describing a decrease in a quantity. (p. 416)

percent increase A percent change describing an increase in a quantity. (p. 416)

perfect square A square of a whole number

Example: $5 \cdot 5 = 25$, and $7^2 = 49$; 25 and 49 are perfect squares. (p. 146)

perimeter The distance around a polygon. (p. 280)

permutation An arrangement of items or events in which order is important. (p. 471)

perpendicular bisector A line that intersects a segment at its midpoint and is perpendicular to the segment. (p. 227)

perpendicular lines Lines that intersect to form right angles. (p. 228)

perspective A technique used to make three-dimensional objects appear to have depth and distance on a flat surface. (p. 303)

pi (π) The ratio of the circumference of a circle to the length of its diameter; $\pi \approx 3.14$ or $\frac{22}{7}$. (p. 294)

plane A flat surface that extends forever. (p. 222)

point An exact location in space. (p. 222)

point-slope form The equation of a line in the form of $y - y_1 = m(x - x_1)$, where m is the slope and (x_1, y_1) is a specific point on the line. (p. 556)

point symmetry A figure has point symmetry if it coincides with itself after a 180° rotation.

polygon A closed plane figure formed by three or more line segments that intersect only at their endpoints (vertices). (p. 239)

polyhedron A three-dimensional figure in which all the surfaces or faces are polygons.

population The entire group of objects or individuals considered for a survey. (p. 174)

positive correlation Two data sets have a positive correlation when their data values increase or decrease together.

positive integer An integer greater than zero. (p. 60)

power A number produced by raising a base to an exponent. (p. 84)

Example: $2^3 = 8$, so 8 is the 3rd power of 2.

precision The level of detail of a measurement, determined by the unit of measure. (p. 717)

prime factorization A number written as the product of its prime factors. (p. 695)

Example: $10 = 2 \cdot 5$, $24 = 2^3 \cdot 3$

Index

A

Index

B

C

Index

D

E

Index

F

G

Index

U

Index

Formulas

Perimeter

Polygon	P = sum of the lengths of the sides
Rectangle	$P = 2(b + h)$
Square	$P = 4s$

Circumference

Circle	$C = 2\pi r$, or $C = \pi d$ $d = 2r$

Volume

Prism	$V = Bh$
Rectangular prism	$V = \ell wh$
Cube	$V = s^3$
Cylinder	$V = \pi r^2 h$
Pyramid	$V = \frac{1}{3}Bh$
Cone	$V = \frac{1}{3}\pi r^2 h$
Sphere	$V = \frac{4}{3}\pi r^3$

Area

Circle	$A = \pi r^2$
Parallelogram	$A = bh$
Rectangle	$A = bh$
Square	$A = s^2$
Triangle	$A = \frac{1}{2}bh$
Trapezoid	$A = \frac{1}{2}h(b_1 + b_2)$

Surface Area

Prism	$S = 2B + ph$
Rectangular prism	$S = 2\ell w + 2\ell h + 2wh$
Cube	$S = 6s^2$
Cylinder	$S = 2B + 2\pi rh$
Regular pyramid	$S = B + \frac{1}{2}p\ell$
Cone	$S = \pi r^2 + \pi r\ell$
Sphere	$S = 4\pi r^2$

Trigonometry

Sine	$\sin A = \frac{\text{length of side opposite } \angle A}{\text{length of hypotenuse}}$
Cosine	$\cos A = \frac{\text{length of side adjacent to } \angle A}{\text{length of hypotenuse}}$
Tangent	$\tan A = \frac{\text{length of side opposite } \angle A}{\text{length of side adjacent to } \angle A}$

Probability

Experimental	$\text{probability} \approx \frac{\text{number of times event occurs}}{\text{total number of trials}}$
Theoretical	$\text{probability} = \frac{\text{number of outcomes in the event}}{\text{number of outcomes in sample space}}$
Permutations	${}_nP_r = \frac{n!}{(n-r)!}$
Combinations	${}_nC_r = \frac{{}_nP_r}{r!} = \frac{n!}{r!(n-r)!}$
Dependent events	$P(A \text{ and } B) = P(A) \cdot P(B \text{ after } A)$
Independent events	$P(A \text{ and } B) = P(A) \cdot P(B)$